Mathematical Symbols

SYMBOLS

$=$	is equal to	$\pm$	plus or minus (for example, $\sqrt{4} = \pm 2$)
$\neq$	is not equal to	$r \to \infty$	r approaches infinity
$\equiv$	is identical to or by definition	$\Rightarrow$	implies
$a > b$	a is greater than b	Σ	the sum of
$a \gg b$	a is much greater than b	$| \ |$	the absolute value of
$a < b$	a is less than b	$|\mathbf{A}|$ or A	the magnitude of the vector $\mathbf{A}$
$a \ll b$	a is much less than b	$\oint$	a line integral around a closed path or a surface integral over a closed surface
$a \geqq b$	a is equal to or greater than b		
$a \leqq b$	a is equal to or less than b	$\cdot$	multiplication symbol
$a \sim b$	a is of the order of magnitude of b; i.e., a is within a factor of 10 or so of b	$\cdot$	(as in $\mathbf{A} \cdot \mathbf{B}$) dot product
		$\mathbf{X}$	(as in $\mathbf{A} \mathbf{X} \mathbf{B}$) cross product
$\propto$	is proportional to	$\times$	(as in 3.2×10^4) multiplication symbol in scientific notation
$\approx$	is approximately equal to		

THE GREEK ALPHABET

Alpha	A	α		Nu	N	ν
Beta	B	β		Xi	Ξ	ξ
Gamma	Γ	γ		Omicron	O	o
Delta	Δ	δ		Pi	Π	π
Epsilon	E	ϵ		Rho	P	ρ
Zeta	Z	ζ		Sigma	Σ	σ
Eta	H	η		Tau	T	τ
Theta	Θ	θ		Upsilon	Y	υ
Iota	I	ι		Phi	Φ	ϕ
Kappa	K	κ		Chi	X	χ
Lambda	Λ	λ		Psi	Ψ	ψ
Mu	M	μ		Omega	Ω	ω

SI Prefixes

MULTIPLE	PREFIX		SYMBOL	MULTIPLE	PREFIX		SYMBOL
10^{18}	exa	(ĕk′så)	E	10^{-1}	*deci	(dĕs′ĭ)	d
10^{15}	peta	(pĕt′å)	P	10^{-2}	†centi	(sĕn′tĭ)	c
10^{12}	tera	(tĕr′å)	T	10^{-3}	milli	(mĭl′ĭ)	m
10^{9}	giga	(jĭ′gå)	G	10^{-6}	micro	(mī′krō)	μ
10^{6}	mega	(mĕg′å)	M	10^{-9}	nano	(năn′ō)	n
10^{3}	kilo	(kĭl′ō)	k	10^{-12}	pico	(pē′cō)	p
10^{2}	*hecto	(hĕc′tō)	h	10^{-15}	femto	(fĕm′tō)	f
10^{1}	*deka	(dĕk′å)	da	10^{-18}	atto	(ăt′tō)	a

In each case, the accent is on the *first* syllable.
*Rarely used.
†Generally used only as *centimeter* (cm).

VOLUME TWO
UNIVERSITY PHYSICS Second Edition

Alvin Hudson
Occidental College

Rex Nelson
Occidental College

VOLUME TWO
UNIVERSITY PHYSICS
Second Edition

SAUNDERS COLLEGE PUBLISHING
Philadelphia   New York   Chicago
San Francisco   Montreal   Toronto
London   Sydney   Tokyo

Copyright © 1990, by Saunders College Publishing, a subsidiary of Holt, Rinehart and Winston, Inc.

All rights reserved. No part of this publication may be reproduced or transmitted in any form or by any means, electronic or mechanical, including photocopy, recording, or any information storage and retrieval system, without permission in writing from the publisher.

Requests for permission to make copies of any part of the work should be mailed to Permissions, Holt, Rinehart and Winston, 111 Fifth Avenue, New York, New York 10003.

The authors wish to thank *American Scientist* for permission to reprint material by Dr. Frederick Seitz from *American Scientist*, Vol. 61, May–June 1973, p. 352; and McGraw-Hill Publishing Company for permission to reprint a table from *Heat Transfer*, Fourth Edition, by J. P. Holman (McGraw-Hill, 1976).

Text Illustrations by Asterisk Group and SO CAL Graphics.

Photo and Illustration Credits appear on page I-1, which constitutes a continuation of the copyright page.

Cover: Production of a Z^0 particle. Artist's color-enhanced rendering of a computer reconstruction of particles created in a proton–antiproton collision at the Collider Detector at Fermilab (CDF). The curved lines represent trajectories of electrically charged particles that are deflected in the axial magnetic field. The energies of the two high-energy, back-to-back particles, when measured in a calorimeter, show that they are electrons with an effective mass of a Z^0 particle. The rectangular box singles out the particle with highest momentum. Its opposing particle, not shown, would appear as an almost straight diagonal line off the top right corner of this cover. See Figure 45-22. The original computer photograph was supplied courtesy of Fermilab.

Printed in the United States of America

UNIVERSITY PHYSICS, Second Edition, Volume 2

ISBN: 0-03-046983-X

Library of Congress Catalog Card Number: 89-84691

9012   041   987654321

Preface

This Second Edition of *University Physics* is designed for the calculus-based introductory course for students majoring in one of the physical sciences or engineering. The text has been extensively rewritten, in a more relaxed style that makes physics easier to study but that does not compromise the necessary rigor of the subject. Each chapter reaches the level of exact, rigorous statement that gives physics its power. Simultaneously, the occasional historical anecdotes and biographical sketches remind the reader that physics is, indeed, a human endeavor.

A number of exciting features enhance the book's accessibility to students and ensure a firm grounding in all aspects of the study of physics. Most important are the following.

Special Topics

It is important to continually remind students that "physics is all around them," and to present both modern applications and some current research activities. Though we do have a few such discussions of a page or two in length, we choose not to include long guest essays on peripheral or advanced topics, because students tend not to read them. Instead, we make liberal use of photographs showing physics in action, with extended captions that explain the concepts involved. These photographs are intended to catch the eye of the reader and pique one's curiosity, luring the student to pursue a short "aside." We believe this stratagem is more palatable and successful than the long essays.

We have made one exception to this. At the end of Chapter 45 on nuclear physics is an essay written by Thomas Ferbel, University of Rochester, on elementary particles, presenting the Standard Model of particle physics and the goal of unification. He gives due recognition to the experimental side of these quests, with inspiring insights into the future.

A Special Problem-Solving Technique

Many beginning students experience difficulty in learning to approach a problem by first seeking the general principle that applies, rather than hunting for a particular formula that works in that special circumstance. We ask our students to begin each solution by *explicitly* stating the general principle in equation form, rather than a specific relation derived from that principle. The first equation in a physics problem should be, for example, $E_0 = E$, instead of $mgh = \frac{1}{2}mv^2$ (most other textbooks simply do the latter). All of our examples illustrate this procedure. In ten years of using this method, we have found that this modest formality in solutions does help channel the student's initial thinking toward basic principles rather than specific formulas. We are convinced that this easy pedagogical tactic is effective in helping many students to "think like a physicist," a skill that becomes invaluable in future course work.

Mathematical Level

Students in an introductory course have a wide range of prior preparation and skills, so at first we use calculus gently, often in parallel with the simpler (but longer) algebraic derivation. New mathematical concepts are explained fully and introduced at the point they are needed, consistent with the progress of topics in an introductory calculus course. The Appendixes summarize all the mathematical relations used in the text, and various tables furnish the numerical data required for problems. The SI system is used throughout. However, since the transitional period to exclusive metric usage is still with us, the American Customary system is also employed at times in mechanics, disappearing in later chapters. We retain a few non-SI units, such as the *atmosphere* and the *electron-volt*, because of their great convenience and widespread usage.

Problems

Each chapter of this textbook contains an abundance of carefully worked and verified problems, arranged in three levels of difficulty. The simpler A and B problems are identified with the appropriate sections in the chapter; the more challenging C problems are not so identified. Answers for odd-numbered problems are given at the end of the book. Questions at the end of each chapter challenge the student's understanding of concepts in a way distinct from regular numerical problems. Some questions may not have precise answers and will lend themselves to class discussions.

Organization

Our topic sequence follows a traditional pattern. Volume 1 includes mechanics, wave motion, heat, and thermodynamics; Volume 2 treats electromagnetism, optics, special relativity, and quantum ideas, and briefly introduces atomic and nuclear physics. Chapter summaries highlight important concepts.

In the earlier chapters we emphasize a systematic and detailed approach to problem solving. The sophistication with which material is presented becomes greater in subsequent chapters as students become familiar with the new (for many of them) analytic procedures and linear thinking required in physics.

If desired, Chapter 41, Special Relativity, may be moved to the end of mechanics (omitting Section 41.15, Relativity and Electromagnetism). It requires only a few comments on the nature of light to set the stage for this fascinating subject. We have retained our optional Chapter 14, Accelerated Frames and Inertial Forces, believing that it is unfortunate if beginning students do not learn the physics that is taking place, for example, when they ride around a curve in an automobile. This topic does enrich and deepen one's understanding of Newton's second law.

In Chapter 42, The Quantum Nature of Radiation, and Chapter 43, The Wave Nature of Particles, we give a somewhat deeper-than-average discussion of the wave–particle duality of both matter and radiation. The atomic physics chapter (44) includes a basic presentation of the time-independent Schrödinger equation, the particle in a box, and a few hydrogen-atom wave functions. In nuclear physics (Chapter 45) we discuss the structure of the nucleus, modes of radioactive decay, nuclear reactions, and nuclear energy. We conclude with Thomas Ferbel's essay, described earlier.

The entire textbook may be covered in a three-semester course, or, if certain chapters and sections are omitted, the material can form an effective two-semester course.

Supplemental Materials

For those who adopt the book, an *Instructor's Answer Book* contains answers to all text problems. Two-color transparencies of selected figures are also provided for classroom projection. Materials for students include a *Student Study Guide*, Second Edition, by Ken Jesse,

Illinois State University, and a *Student Solutions Manual* by the authors that presents partially worked-out solutions to about 400 representative problems from the text. This should help students gain skill in the crucial first steps of analysis. (These problems are identified by an asterisk in the Instructor's Answer Book so a teacher may choose to omit or to include them in weekly assignments.)

Acknowledgments

We greatly appreciate the suggestions for improvement by many persons, including several who have had classroom experience with the First Edition. Our colleagues Professor Stuart Elliott, Professor Tim Sanders, Professor Herb Segall, and Mr. Clifford Chen have been particularly helpful, as well as our students, who continuously provided the critical responses every author seeks. The manuscript has been reviewed at various stages by many individuals. Among the reviewers and others who offered valuable suggestions are Walter Benenson, Michigan State University; Rodney Cole, University of California at Davis; Professor Alfonso Diaz-Jiminez of Bogota, Columbia; T. E. Edwards, Michigan State University; A. L. Ford, Texas A & M University; Roger Judge, University of California at San Diego; Robert L. Peterson, Shoreline Community College; S. J. Shepherd, The Pennsylvania State University; Billy S. Thomas, University of Florida; and George A. Williams, The University of Utah. Jean Nelson deserves thanks for her meticulous care in the formidable task of preparing the index. We are also grateful to the publisher's staff, who provided help and encouragement at many crucial moments. In particular, we extend heartfelt thanks to our editor, Jeff Holtmeier, our manuscript editor Cate DaPron, production editor Katherine Watson, designer Cheryl Solheid, art editor Cindy Robinson, production manager Diane Southworth, and associate editor Pamela Whiting. Their many contributions toward the finished product are praiseworthy. Every textbook contains much more than the author's contributions alone.

Al Hudson
Rex Nelson

Contents

Preface

CHAPTER 24

Coulomb's Law and the Electric Field 554

- 24.1 Introduction 554
- 24.2 Electrostatic Forces 555
- 24.3 Conductors and Insulators 557
- 24.4 Coulomb's Law 557
- 24.5 The Electric Field 563
- 24.6 The Electric Dipole 565
- 24.7 Electric Fields Due to Continuous Charge Distributions 569

CHAPTER 25

Gauss's Law 580

- 25.1 Introduction 580
- 25.2 The Electric Flux 580
- 25.3 Gauss's Law 583
- 25.4 Gauss's Law and Conductors 591

CHAPTER 26

Electric Potential 597

- 26.1 Introduction 597
- 26.2 The Electric Potential 597
- 26.3 The Gradient of V 608
- 26.4 Equipotential Surfaces 610

CHAPTER 27

Capacitance and Energy in Electric Fields 618

- 27.1 Introduction 618
- 27.2 Capacitance 618
- 27.3 Combinations of Capacitors 623
- 27.4 Dielectrics 624
- 27.5 Potential Energy of Charged Capacitors 628
- 27.6 Energy Stored in an Electric Field 630

CHAPTER 28

Electric Current and Resistance 637

- 28.1 Introduction 637
- 28.2 Electromotive Force $\mathcal{E}$ 637
- 28.3 Electric Current 638
- 28.4 Electrical Resistance 641
- 28.5 Ohm's Law 643
- 28.6 Joule's Law 645
- 28.7 Current Density and Conductivity 648

CHAPTER 29

DC Circuits 655

- **29.1** Introduction 655
- **29.2** Resistors in Series and in Parallel 655
- **29.3** Multiloop Circuits and Kirchhoff's Rules 658
- **29.4** The Superposition Principle 660
- **29.5** Applications 664
- **29.6** *RC* Circuits 670

CHAPTER 30

The Magnetic Field 684

- **30.1** Introduction 684
- **30.2** Magnetic Fields 684
- **30.3** Motion of a Charged Particle in a Magnetic Field 686
- **30.4** The Lorentz Force Law 691
- **30.5** Magnetic Force on a Current-Carrying Conductor 692
- **30.6** Magnetic Dipoles 694
- **30.7** Applications 697
- **30.8** Magnetic Flux Φ_B 703
- **30.9** Comments About Units 704

CHAPTER 31

Sources of Magnetic Field 711

- **31.1** Introduction 711
- **31.2** The Biot–Savart Law 711
- **31.3** Ampère's Law (1823) 716

CHAPTER 32

Faraday's Law and Inductance 727

- **32.1** Introduction 727
- **32.2** Faraday's Law 727
- **32.3** Motional emf 730
- **32.4** Lenz's Law 735
- **32.5** Eddy Currents 736
- **32.6** Self-Inductance 737
- **32.7** Mutual Inductance 739
- **32.8** *RL* Circuits 741
- **32.9** Energy in Inductors 744

CHAPTER 33

Magnetic Properties of Matter 752

- **33.1** Introduction 752
- **33.2** Magnetic Properties of Materials 752
- **33.3** **B** and **H** 757
- **33.4** Hysteresis 759

CHAPTER 34

AC Circuits 763

- **34.1** Introduction 763
- **34.2** Simple AC Circuits 763
- **34.3** Series *RLC* Circuits 768
- **34.4** Impedance in Series *RLC* Circuits 771
- **34.5** Impedance in Parallel *RLC* Circuits 775
- **34.6** Resonance 778
- **34.7** Power in AC Circuits 781
- **34.8** Transformers 785

CHAPTER 35

Electromagnetic Waves 794

- **35.1** Introduction 794
- **35.2** Displacement Current and Maxwell's Equations 795
- **35.3** Electromagnetic Waves 799
- **35.4** The Production of Electromagnetic Waves 807
- **35.5** Energy in Electromagnetic Waves 809
- **35.6** Momentum of Electromagnetic Waves 812

CHAPTER 36

Geometrical Optics I—Reflection 822

- **36.1** Introduction 822
- **36.2** Wavefronts and Rays 823
- **36.3** Huygens' Principle 824
- **36.4** Reflection by a Plane Mirror 825
- **36.5** Reflection by a Spherical Mirror 828
- **36.6** Ray Diagrams and Lateral Magnification 835

CHAPTER 37

Geometrical Optics II—Refraction 843

- **37.1** Introduction 843
- **37.2** Refraction at a Plane Surface 843
- **37.3** Total Internal Reflection 848
- **37.4** Refraction at a Spherical Surface 851
- **37.5** Thin Lenses 852
- **37.6** Diopter Power 856
- **37.7** Thin Lens Ray-Tracing and Image Size 857
- **37.8** Combinations of Lenses 859
- **37.9** Optical Instruments 862
- **37.10** Aberrations 870

CHAPTER 38

Physical Optics I—Interference 878

38.1 Introduction 878
38.2 Double-Slit Interference 878
38.3 Multiple-Slit Interference 887
38.4 Interference Produced by Thin Films 888
38.5 The Michelson Interferometer 892

CHAPTER 39

Physical Optics II—Diffraction 899

39.1 Introduction 899
39.2 Single-Slit Diffraction 900
39.3 Diffraction by a Circular Aperture 907
39.4 The Diffraction Grating 909
39.5 X-Ray Diffraction 916
39.6 Fresnel Diffraction—Circular Apertures and Obstacles 918
39.7 The Fresnel Zone Plate 918
39.8 Holography 921

CHAPTER 40

Polarized Light 927

40.1 Introduction 927
40.2 Polaroid 929
40.3 Polarization by Reflection and Scattering 930
40.4 Birefringence 932
40.5 Wave Plates and Circular Polarization 934
40.6 Optical Activity 937
40.7 Interference Colors and Photoelasticity 938

CHAPTER 41

Special Relativity 943

- 41.1 Introduction 943
- 41.2 The Galilean Transformation 944
- 41.3 The Fundamental Postulates of Special Relativity 948
- 41.4 Setting Clocks in Synchronism 949
- 41.5 The Lorentz Transformation 949
- 41.6 Comparison of Clock Rates 952
- 41.7 Comparison of Length Measurements Parallel to the Direction of Motion 953
- 41.8 Proper Measurements 955
- 41.9 Relativistic Momentum 955
- 41.10 A Note about Rest Mass 959
- 41.11 Relativistic Velocity Addition 959
- 41.12 Relativistic Energy 961
- 41.13 The Nonsynchronism of Moving Clocks 966
- 41.14 The Twin Paradox 969
- 41.15 Relativity and Electromagnetism 971
- 41.16 General Relativity 973

CHAPTER 42

The Quantum Nature of Radiation 981

- 42.1 Introduction 981
- 42.2 The Spectrum of Cavity Radiation 982
- 42.3 Attempts to Explain Cavity Radiation 983
- 42.4 Planck's Theory 986
- 42.5 The Photoelectric Effect 988
- 42.6 The Compton Effect and Pair Production 994
- 42.7 The Dual Nature of Electromagnetic Radiation 997

CHAPTER 43

The Wave Nature of Particles 1004

- 43.1 Introduction 1004
- 43.2 Models of an Atom 1004
- 43.3 The Correspondence Principle 1010
- 43.4 De Broglie Waves 1011
- 43.5 The Davisson–Germer Experiments 1013
- 43.6 Wave Mechanics 1016
- 43.7 Barrier Tunneling 1021
- 43.8 The Uncertainty Principle 1022
- 43.9 The Complementarity Principle 1027
- 43.10 A Brief Chronology of Quantum Theory Development 1028

CHAPTER 44

Atomic Physics 1033

- **44.1** Introduction 1033
- **44.2** The Schrödinger Wave Equation 1035
- **44.3** Electron Spin and Fine Structure 1039
- **44.4** Spin–Orbit Coupling 1039
- **44.5** Quantum States of the Hydrogen Atom 1041
- **44.6** Energy Level Diagram for Hydrogen 1042
- **44.7** The Hydrogen Atom Wave Functions 1043
- **44.8** The Pauli Exclusion Principle and the Periodic Table of the Elements 1047
- **44.9** X-Rays 1050
- **44.10** The Laser 1052

CHAPTER 45

Nuclear Physics 1059

- **45.1** Introduction 1059
- **45.2** A Description of the Nucleus 1060
- **45.3** Nuclear Mass and Binding Energy 1062
- **45.4** Radioactive Decay and Half-Life 1066
- **45.5** Modes of Radioactive Decay 1069
- **45.6** Nuclear Cross Section 1079
- **45.7** Nuclear Reactions 1081
- **45.8** Nuclear Power 1085

BRIEF HISTORY AND STATUS OF PARTICLE PHYSICS 1092

Appendixes

- **A.** SI Prefixes A-1
- **B.** Mathematical Symbols A-1
- **C.** Conversion Factors A-2
- **D.** Mathematical Formulas A-4
- **E.** Mathematical Approximations, Expansions, and Vector Relations A-6
- **F.** Fourier Analysis A-6
- **G.** Calculus Formulas A-8
- **H.** Finite Rotations A-10
- **I.** Derivation of the Lorentz Transformation A-10
- **J.** Periodic Table of the Elements A-12
- **K.** Constants and Standards A-13
- **L.** Terrestrial and Astronomical Data A-14
- **M.** SI Units A-16

Answers to Odd-Numbered Problems A-23

Index I-3

VOLUME TWO
UNIVERSITY PHYSICS Second Edition

CHAPTER 24

Coulomb's Law and the Electric Field

Electricity is of two kinds, positive and negative. The difference is, I presume, that one comes a little more expensive, but is more durable; the other is a cheaper thing, but the moths get into it.

STEPHEN LEACOCK
[*Literary Lapses* (1910)]

24.1 Introduction

We are all familiar with the fact that, after we comb our dry hair, the comb becomes "electrified" with the ability to attract bits of paper. If you stop to think about it, this phenomenon is baffling: somehow the bits of paper mysteriously sense the presence of the electrified comb without actually touching it. Magnets have similar powers of attracting iron and steel objects without touching them.

Such behavior has been observed for a long time. The ancient Greeks discovered that, when amber was rubbed by any of a variety of materials, it became capable of attracting small objects. In fact, the word *electricity* comes from the Greek word for amber: *electron*, a fossilized resin that becomes electrified when rubbed. In describing atoms, we apply the term *electron* to the negative charges surrounding the nucleus of the atom. We now trace the evolution of our understanding of electricity from the electrification of certain materials to the elegantly unified description of electric and magnetic phenomenon known as Maxwell's equations (Chapter 35).

It took many intelligent investigators a long time to unravel the story. About 200 years elapsed between the publication of Newton's *Principia* (1687) and the comparable achievement by James Clerk Maxwell in his *Treatise on Electricity and Magnetism* (1873). Despite this relatively long gap in the progress of physics, many scientists were struggling to make sense of electromagnetic phenomena during this period, and there were numerous sparks of insight that helped to illuminate the separate pieces of the puzzle. But it required the genius of Maxwell to finally fit all the pieces together in a coherent and unified theory.

Perhaps much of the delay in progress was due also to the difference between mechanical and electrical phenomena. The study of mechanical phenomena benefited from the everyday experiences of pushing and pulling objects and observing their motions. But there are no comparable sensory experiences with electromagnetism (except on the superficial level of static electricity and magnets). So the subject is inherently more abstract and more obscure from everyday observations. Furthermore, quantitative experiments in electricity and magnetism are vastly more difficult to carry out than experiments in mechanics. The electric force is so large that just a slight unknown imbalance of electrical charge easily spoils the measurements. As Richard Feynman explains it, if you were standing at arm's length from someone and each of you had just 1% more electrons than protons, the repulsive force on you would be enough to lift a "weight" equal to that of the entire earth!

Electrical forces are everywhere about us. All so-called "contact forces"— such as the forces described by Newton's third law (equal and opposite forces), which occur between adjacent links in a chain, between your pencil lead and the paper, and between a tire and the roadway—are electrical in origin. All of these originate in forces of attraction or repulsion between electric charges. We shall begin our discussion of electricity and magnetism by investigating forces between electrified objects that are *at rest* with respect to each other. This branch of electrical phenomena is known as **electrostatics**.

24.2 Electrostatic Forces

If we rub an animal fur against a hard rubber rod, the rod acquires new characteristics. For example, it readily attracts bits of paper, and it can deflect a jet of water without actually touching it. In the process of being rubbed, the rod has changed. We say it has become *electrified*, or *charged*—yet we don't really know what these terms mean.

Let us sharpen our terminology and understanding of electrical forces by carrying out some simple experiments. First, suppose we suspend a hard rubber rod by a thread as shown in Figure 24-1. If a piece of fur is brought near the rod, there is no noticeable interaction. However, when the rod is rubbed with the fur, it is then attracted to the fur even at a distance. We call the attraction an **electrostatic force** and conclude that

(a) When a fur is used to rub a hard rubber rod, that end of the rod is attracted to the fur.

Electrostatic forces (*like* gravitational forces) can be forces of attraction.

Suppose we now rub another hard rubber rod with fur. We find that the second rod repels the suspended rod that had been previously rubbed, and we conclude that

Electrostatic forces (*unlike* gravitational forces) can also be forces of repulsion.

Since the charged objects interact without touching, we further conclude that

Electrostatic forces (*like* gravitational forces) act through empty space.

We would find that the results of this experiment are the same if conducted in a vacuum.

With our knowledge of Newton's law of universal gravitation, we could estimate the force of gravity between the fur and the rod and at least qualitatively conclude that

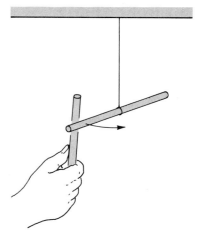

(b) When two such rods are rubbed by a fur, the rods repel each other.

FIGURE 24-1
Electrical forces may be either attractive or repulsive.

Electrostatic forces are much stronger than gravitational forces.

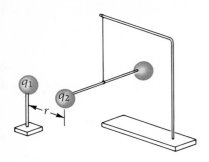

FIGURE 24-2
A *torsion balance*. The force of interaction between the charges q_1 and q_2 twists the fiber supporting the horizontal rod. (Compare with the gravitational torsion balance, Figure 16-10.)

In order to focus our attention on the nature of the interaction between two charged objects, we refine the experimental apparatus to that shown in Figure 24-2. This arrangement is a form of *torsional balance*, which the English physicist Henry Cavendish (1731–1810) used to measure gravitational forces. The charged objects are small spheres that have an *electrical charge* on them, designated by the symbols q_1 and q_2. (In this case the gravitational force between the spheres is negligible compared to the electrical force.) The numerical value of q_1 and q_2 (to be specified later) indicates the amount of charge the objects have. Since the spheres are small, they approximate *point charges*.

The force of interaction can be determined by the amount of torque required to twist the supporting fiber. The distance r between the charges is measured directly. After a series of measurements is made with differing separation of the spheres and with various amounts of charge on the spheres, we will find that, for point charges,

Electrostatic forces (*like* gravitational forces) are inverse-square forces; that is, they decrease with distance r as $1/r^2$.

Electrostatic forces are mutual forces of interaction that obey Newton's third law.

Electrostatic forces are proportional to the product of the amount of charge on each of the interacting point charges.

These results may be summarized into a single statement: for two point charges, q_1 and q_2, separated a distance r,

$$F = k \frac{q_1 q_2}{r^2} \qquad (24\text{-}1)$$

where k is a constant of proportionality. This result was first published in 1785 by the French physicist Charles Augustin de Coulomb (1736–1806), who experimented with a *torsion balance* similar to what we have described.

We have referred to *charged* objects and *charges* without really knowing what constitutes the charge. During the 1740s, Benjamin Franklin proposed that the charge was a single fluid and that all objects contained a "normal" amount of it. When he rubbed glass with a silk cloth he noted that the glass became "electrified" and attracted bits of paper. Franklin hypothesized that the rubbing did not create the charge, but merely transferred some of the "electrical fluid" from the cloth to the glass, so that the glass now had a surplus of fluid while the cloth had an equal deficiency of fluid. Franklin proposed + and − signs to signify these differences; hence the glass acquired a *positive charge* and the cloth an equal *negative charge*. Similarly, when rubbed with fur, a hard rubber or plastic rod becomes negatively charged and the fur positively charged. Indeed, all materials become more or less charged when rubbed with other substances. It would have been more fortuitous had Franklin chosen his + and − signs in the opposite sense; we now know that the positive charge on a glass rod rubbed by silk really originates because some negatively charged electrons move from the glass to the silk (instead of positive charges moving from the silk to the glass), so the actual transportation of charge is in the direction opposite to Franklin's theory. Franklin believed that the electrical fluid was *conserved*, that is, that the total amount of fluid in a closed system remains constant. Even though this single-fluid idea was later shown to be incorrect, his *conservation of charge* remains one of the fundamental principles of physics. No exception to this principle has ever been found.

In the modern view, electric charge is a basic property of matter. In addition to uncharged neutrons, atoms contain protons and electrons that are charged, respectively, positive and negative. The magnitude of the negative electron charge e is exactly equal to the magnitude of the positive proton charge (at least to within the experimental verification of 1 part in 10^{22}), though the electron and proton masses differ greatly as shown in Table 24-1.

TABLE 24-1

Particle	Symbol	Charge	Mass (kg)
Proton	p	$+e$	1.673×10^{-27}
Neutron	n	0	1.675×10^{-27}
Electron	e	$-e$	9.110×10^{-31}

24.3 Conductors and Insulators

It is convenient to classify materials in terms of their ability to conduct electrical charges. In a *conductor*, electric charges can move freely. Most metals are conductors because the outer electrons associated with each atom—the "conduction electrons"—can travel easily throughout the material, while the positively charged nuclei are held fixed. In certain conducting liquids and ionized gases, positive as well as negative charges can move. On the other hand, substances such as glass, wood, and plastics are classified as *nonconductors* or *insulators*, since electric charges are much less free to move within them. When charges are placed at a small localized region on an insulator, they remain there. While there are no perfect insulators,[1] the best of them is about 10^{25} less conducting than copper, so the range in conducting ability spans a very great scale. *Semiconductors*, such as silicon and germanium, lie between these extremes. We can alter the conducting ability of these substances dramatically by adding just a few parts per million of foreign atoms.

If you hold a hard rubber comb or a glass rod in your hand and rub them, respectively, with fur or silk, the charges on them will remain in the region where they were produced, and you can attract small pieces of paper with them, evidence that the comb or rod carries a net charge. In contrast, a piece of copper or other conducting material on which some negative charges (electrons) are placed will not attract bits of paper; electrons placed on the conductor immediately escape by readily moving through the conductor to your hand and body and then to the earth. (Had positive charges been placed on the conductor, negative electrons would have been attracted from the earth through your body to neutralize the charges electrically.) In effect, the earth acts as an infinite "sink" that can absorb or supply an almost unlimited number of electrons. To maintain a charge on a conductor, we must insulate the object from its surroundings. Figure 24-3 illustrates a process called *charging by induction* in which the charging agent (the charged rod) itself does not touch the object that acquires the charge.

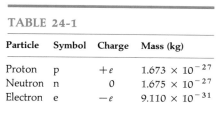

(a) A neutral, insulated metal sphere.

(b) A negatively charged rod is brought near the sphere, repelling some electrons (which move freely in the metal) to the opposite side, leaving positive charges near the rod.

(c) The sphere is grounded by a metal wire connected to the earth (symbol: ⏚). The electrons flow to the earth, repelled by the electrons on the rod.

FIGURE 24-3
Charging a metal sphere by induction.

24.4 Coulomb's Law

Equation (24-1) describes the inverse-square-law force between two point charges. To make the equation quantitative, we need to define the unit of charge and then experimentally determine the proportionality constant k. In the SI system the unit of charge is the coulomb (C). Rather than defining the coulomb through Equation (24-1), it is experimentally easier—and more precision can be attained—if we define the coulomb as the amount of charge per

---

[1] Certain materials called *superconductors* do become perfect conductors in the sense that the electrical resistance to the motion of electrons through the material becomes truly zero. Superconductivity was first discovered in metals cooled to about 4 K, but recent developments indicate that some metallic alloys and ceramic compounds become superconducting at much higher temperatures. Quantum mechanics provides an explanation of this unusual behavior. (See Figure 28-10.)

second passing through any cross-section of a wire carrying a constant current of one *ampere*. In turn, the ampere (A) is defined through the electromagnetic force between two parallel current-carrying wires, as described in Chapter 30. In this rather roundabout manner, the definition of the coulomb is connected to the SI mechanical unit for *force*. For the present, we will use the unit *coulomb*, postponing a more detailed discussion of its formal definition.

The magnitude of the fundamental charge e on a single electron is

MAGNITUDE OF THE ELECTRON CHARGE
$$e = 1.602 \times 10^{-19} \text{ C} \qquad (24\text{-}2)$$

This is the smallest electric charge that has been found; it is equal in magnitude to the positive charge on a proton.

The value of the constant k in Equation (24-1) is found experimentally to be $8.99 \times 10^9 \text{ N} \cdot \text{m}^2/\text{C}^2$. A good approximation is

$$k = 9 \times 10^9 \frac{\text{N} \cdot \text{m}^2}{\text{C}^2} \qquad (24\text{-}3)$$

However, to simplify the equations that will be developed later, it is convenient to express the constant of proportionality in another way, incorporating a factor of 4π, with the benefit that that factor will not then appear in many other equations that are used more frequently than Coulomb's law. So we express k as

$$k = \frac{1}{4\pi\varepsilon_0}$$

where ε_0, called the **permittivity of free space**, has the value

PERMITTIVITY OF FREE SPACE
$$\varepsilon_0 = 8.854 \times 10^{-12} \frac{\text{C}^2}{\text{N} \cdot \text{m}^2} \qquad (24\text{-}4)$$

Thus, Equation (24-1) becomes **Coulomb's law**,

COULOMB'S LAW
$$F = \left(\frac{1}{4\pi\varepsilon_0}\right) \frac{q_1 q_2}{r^2} \qquad (24\text{-}5)$$

or, in SI units,
$$F = \left(9 \times 10^9 \frac{\text{N} \cdot \text{m}^2}{\text{C}^2}\right) \frac{q_1 q_2}{r^2} \qquad (24\text{-}6)$$

where F is in newtons, q in coulombs, and r in meters.

The Coulomb force between two point charges is a mutual force described by Newton's third law: the force on one charge is equal and opposite to the force on the other. We express Coulomb's law in vector form as

COULOMB'S LAW (vector form)
$$\mathbf{F}_{12} = \left(\frac{1}{4\pi\varepsilon_0}\right) \frac{q_1 q_2}{r^2} \hat{\mathbf{r}}_{12} \qquad (24\text{-}7)$$

where $\mathbf{F}_{12}$ is the force charge q_1 exerts on q_2 and $\hat{\mathbf{r}}_{12}$ is the unit vector (magnitude = 1) from q_1 toward q_2, as shown in Figure 24-4. Note carefully that we always use unit vectors and subscripts that define force directions in this way: the unit vector $\hat{\mathbf{r}}$ is *always* drawn from the source of the force toward the

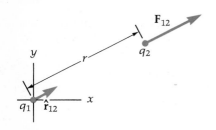

FIGURE 24-4
The force on charge q_2 is in the direction of $\hat{\mathbf{r}}_{12}$ if the product $q_1 q_2$ is positive. The situation illustrated here could be one in which q_1 and q_2 are both positive charges or both negative. The vector distance from q_1 to q_2 (not shown) is $\mathbf{r} = r\hat{\mathbf{r}}_{12}$. By Newton's third law, the force that q_2 exerts on q_1 is equal in magnitude to $\mathbf{F}$ but opposite in direction.

object upon which the force acts. (You can easily remember the order of the subscripts if you mentally insert an arrow → between them. Thus $\hat{\mathbf{r}}_{1 \to 2}$ points from 1 toward 2, and $\mathbf{F}_{1 \to 2}$ is the force that charge 1 exerts *on* charge 2.) We express the third-law character of the force by reversing the order of all subscripts to designate the equal-and-opposite force $\mathbf{F}_{21}$ that q_2 exerts on q_1. Equation (24-7) gives the correct direction of **F** if we use the following sign convention:

A *positive* charge is given the algebraic sign +.
A *negative* charge is given the algebraic sign −.

The force between "like" charges is repulsive; the force between "unlike" charges is attractive.

Coulomb's law describes the electrostatic interaction between two point charges. We now use Coulomb's law to describe the interaction of several point charges, as well as the interaction of a point charge with a distribution of charges. Such distributions may be along a line, over a surface, or throughout a volume.

As with the gravitational force between point masses, $F = Gm_1m_2/r^2$, we find experimentally that the electrostatic forces on a single charge due to the presence of many other charges may be *superposed*, or added together as vectors, a procedure called the **principle of superposition**. Since Newton's law of gravitation and Coulomb's law have the same mathematical form, similar

(a) A simple electroscope devised in the eighteenth century but still used today for indicating the presence of a net charge. Two thin metal foils are connected by a metal rod to the metal sphere. The assembly is supported by an insulating stopper in the glass bottle. When a net charge is distributed between the sphere and the foils, the foil leaves diverge because of the mutual repulsion of their "like" charges.

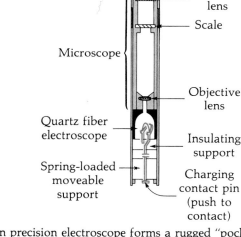

(b) A modern precision electroscope forms a rugged "pocket dosimeter" that records the presence of ionizing radiation in the vicinity. The leaves are a fixed metal electrode and a moveable quartz fiber bent into a ∪ shape and gold-plated to make it conducting. In use, the electroscope is charged, causing the quartz fiber to deflect from its uncharged position. The location of the fiber is viewed with a microscope that contains a scale. If the dosimeter is exposed to ionizing radiation, the gas in the chamber becomes slightly conductive, allowing charge to leak off in proportion to the amount of radiation, and the fiber moves across the scale. When we are viewing, illumination from below passes through the transparent supports to the fiber, lenses, and scale. (Courtesy of Dosimeter Corporation of America.)

FIGURE 24-5
The *electroscope*.

conclusions can be made for each. For example, we have shown that the gravitational attraction of two uniform, solid spheres is as though all the mass were concentrated at a point at the center of each sphere. Similarly, the electrostatic force between two uniform spheres of charge is as though the total charge of each sphere were located at its center. In those cases in which the density of electric charge within a sphere varies only with the distance from its center (that is, the object has *spherical symmetry*), the force is again the same as if each charge distribution were concentrated at its center, just as it is in the spherically symmetric gravitational case.

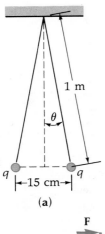

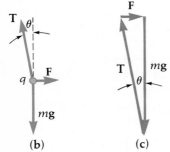

FIGURE 24-6
Example 24-1.

Two small spheres of negligible size, each of mass 2 g, are suspended from a common support by threads 1 m long. When each sphere is given an electric charge, the spheres diverge until they are 15 cm apart as shown in Figure 24-6. (a) Assuming that the charges are equal, find the charge on each sphere. (b) Is there more than one answer?

SOLUTION

First we draw a free-body diagram for the sphere on the right, as shown in Figure 24-6b. (Choosing the *right* or *left* sphere is arbitrary because of *symmetry*. That is, the spheres have identical masses and charges and are suspended by strings of the same length. Therefore, the Coulomb and gravitational forces on one sphere are the same as on the other one, except for the directions of the Coulomb force.)

(a) The net force on the sphere is zero, so the three forces add to form a closed right triangle, as indicated in Figure 24-6c. Thus:

$$\tan \theta = \frac{F}{mg} = \frac{0.075 \text{ m}}{1 \text{ m}} = 0.075$$

Substituting Coulomb's law for the force F, we have

$$\tan \theta = \left(\frac{1}{4\pi\varepsilon_0}\right) \frac{q^2}{r^2 mg}$$

We then solve for q^2 (each of the same magnitude):

$$q^2 = (4\pi\varepsilon_0)(\tan \theta) r^2 mg$$

Substituting SI values gives

$$q^2 = \left(\frac{1}{9 \times 10^9 \frac{\text{N} \cdot \text{m}^2}{\text{C}^2}}\right)(0.075)(15 \times 10^{-2} \text{ m})^2$$

$$\times (2 \times 10^{-3} \text{ kg})\left(9.80 \frac{\text{m}}{\text{s}^2}\right)$$

$$= 3.68 \times 10^{-15} \text{ C}^2$$

Since the charges are equal, we have

$$q = \sqrt{3.68 \times 10^{-15} \text{ C}^2} = \boxed{\pm 6.07 \times 10^{-8} \text{ C}}$$

(b) There are two possibilities: both could be positively charged or both could be negatively charged. Also, because the charges (electrons) have such a small mass, the two charges could have different values, as long as their product is 3.68×10^{-15} C^2, and still give the same answer to within the number of significant figures calculated.

EXAMPLE 24-2

Three different point charges are located as shown in Figure 24-7a. Charge $q_1 = 20 \ \mu C$, $q_2 = -30 \ \mu C$, and $q_3 = 40 \ \mu C$. Find the magnitude and direction of the net force on q_3.

SOLUTION

We first find the x and y components of the individual force that each charge exerts on q_3, Figure 24-7b. Here we use a double-subscript notation that will be used throughout the rest of the text. F_{13} means the force exerted *by* q_1 *on* q_3.

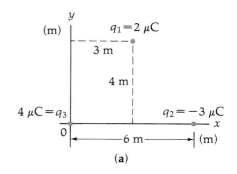

Force of q_1 on q_3

Because they are *like* charges, the force is repulsive. Noting the 3–4–5 right triangle, we find that the distance $r_{13} = 5$ m and $\theta = 53.1°$. Thus:

$$F_{13} = k \frac{q_1 q_3}{r_{13}^2}$$

$$= \left(9 \times 10^9 \ \frac{\text{N} \cdot \text{m}^2}{\text{C}^2}\right)$$

$$\times \frac{(20 \ \mu C)(40 \ \mu C)}{(5 \text{ m})^2}$$

$$F_{13} = 0.288 \text{ N}$$

The x and y components are

$$F_{13x} = F_{13} \cos(180° + 53.13°)$$
$$= \underline{\underline{-0.173 \text{ N}}}$$

$$F_{13y} = F_{13} \sin(180° + 53.13°)$$
$$F_{13y} = \underline{\underline{-0.230 \text{ N}}}$$

Force of q_2 on q_3

Because they are *unlike* charges, the force is attractive.

$$F_{23} = k \frac{q_2 q_3}{r_{23}^2}$$

$$= \left(9 \times 10^9 \ \frac{\text{N} \cdot \text{m}^2}{\text{C}^2}\right)$$

$$\times \frac{(-30 \ \mu C)(40 \ \mu C)}{(6 \text{ m})^2}$$

$$F_{23} = \underline{\underline{0.300 \text{ N}}} \qquad \text{in the } +x \text{ direction)}$$

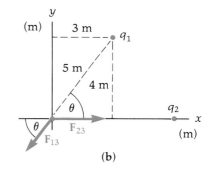

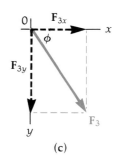

FIGURE 24-7
Example 24-2.

The x component of the net force on q_3 is $F_{3x} = (-0.173 \text{ N} + 0.300 \text{ N}) = 0.127$ N. The y component is -0.230 N. Thus the net force on q_3 is

$$F_3 = \sqrt{F_{3x}^2 + F_{3y}^2} = \sqrt{(0.127 \text{ N})^2 + (-0.230 \text{ N})^2} = \boxed{0.263 \text{ N}}$$

The direction of F_3 is given by the angle ϕ in Figure 24-7c, calculated from

$$\phi = \tan^{-1}\left(\frac{F_{3y}}{F_{3x}}\right) = \tan^{-1}\left(\frac{-0.230 \text{ N}}{0.127 \text{ N}}\right) = \boxed{-61.1°} \qquad \text{as shown}$$

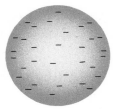

(a) When a negatively charged conducting sphere is far from other charges, the electrons distribute themselves on the surface of the sphere symmetrically.

(b) When a positive test charge q is brought nearby, the distribution of the electrons on the surface of the conducting sphere becomes asymmetric because of the attraction of unlike charges.

FIGURE 24-8
Under certain circumstances, a test charge q_0 used to determine an electric field may itself distort the very field to be determined. To sidestep this problem, we adopt the definition of Equation (24-9).

24.5 The Electric Field

Think back for a moment to the concept of a *gravitational field* (Section 16.6). The field idea is useful because it enables us to avoid the conceptual difficulties of "action-at-a-distance," which Newton's law of universal gravitation describes. For example, according to this law the earth exerts a force on a satellite in orbit even though the earth and the satellite are separated by empty space. But the idea of a force operating through empty space was repugnant to Newton and to many later scientists; "action-at-a-distance" just did not seem sensible. The concept of a field is a more modern view. This alternative way of describing the gravitational interaction is that the earth creates a *gravitational field* $\mathbf{g}$ in the surrounding space. Then, a satellite of mass m experiences a force $\mathbf{F} = m\mathbf{g}$ due to the *local* gravitational field $\mathbf{g}$ where the satellite is located. It is no longer a case of action at a distance.

The gravitational field $\mathbf{g}$ at a given location is defined as the force per unit test mass m_0 placed at that location: $\mathbf{g} = \mathbf{F}/m_0 = -(GM/r^2)\hat{\mathbf{r}}$. The electric field $\mathbf{E}$ is defined in a similar way. The force between a charge q (which produces the field) and a test positive charge q_0 is $\mathbf{F} = (1/4\pi\varepsilon_0)(qq_0/r^2)\hat{\mathbf{r}}$. Thus, the force per unit test charge q_0 is

ELECTRIC FIELD E
$$\mathbf{E} = \frac{\mathbf{F}}{q_0} = \left(\frac{1}{4\pi\varepsilon_0}\right)\left(\frac{q}{r^2}\right)\hat{\mathbf{r}} \qquad (24\text{-}8)$$

where $\mathbf{F}$ is the force on a small *positive* test charge q_0 placed in the field. In the SI system, E is in units of *newtons per coulomb* (N/C).

We need to mention a few practical concerns. We assume that the presence of the test charge q_0 does not change the original distribution of the other charges that produce the field. For example, if the charges reside on a conductor, bringing a small test charge into the vicinity will cause the charges to move around on the conductor, thus changing the field we are trying to measure,[2] Figure 24-8. To avoid this problem, we refine the definition for $\mathbf{E}$ to be the limiting value of the ratio $\mathbf{F}/q_0$ *as the charge q_0 approaches zero*:

ELECTRIC FIELD E
$$\mathbf{E} = \lim_{q_0 \to 0} \frac{\mathbf{F}}{q_0} \qquad (24\text{-}9)$$

This operational definition is logically precise and tells us to use smaller and smaller test charges q_0, with $\mathbf{E}$ being the limit as q_0 approaches zero. In this way, the influence of the test charge q_0 becomes vanishingly small.[3]

Electric Field Lines

We can visualize the concept of an electric field by introducing **field lines** (which Faraday called "lines of force"). Consider the field due to an isolated point charge q. Using a test charge q_0 and Coulomb's law in vector form,

$$\mathbf{F} = \left(\frac{1}{4\pi\varepsilon_0}\right)\frac{qq_0}{r^2}\hat{\mathbf{r}} \qquad (24\text{-}10)$$

---

[2] Such difficulties are common in measurements. For example, in measuring the temperature of a liquid, we alter the temperature by immersing a thermometer in the liquid.

[3] In practice, even this more precise definition is not often followed because of experimental difficulties. For instance, q_0 cannot be less than the electronic charge e. The field $\mathbf{E}$ is experimentally determined more easily from calculations based upon measurements of the *electric potential*, Chapter 26.

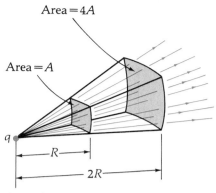

(a) A conventional two-dimensional way of depicting the electric field lines due to an isolated point charge q. The diagram is an approximate cross-section of the field pattern. For a better illustration, mentally extend the pattern to three dimensions, somewhat like the quills on a porcupine, as in (b).

(b) A three-dimensional perspective sketch of the field lines diverging from a point charge q. The lines intersect portions of the surfaces of two concentric spheres (radii R and $2R$).

FIGURE 24-9
Electric field lines associated with an isolated point charge q.

we find the field $\mathbf{E} = \mathbf{F}/q_0$ to be

ELECTRIC FIELD E
DUE TO A POINT
CHARGE q

$$\mathbf{E} = \left(\frac{1}{4\pi\varepsilon_0}\right)\frac{q}{r^2}\hat{\mathbf{r}} \qquad (24\text{-}11)$$

where the unit vector $\hat{\mathbf{r}}$ designates the radial outward direction *away* from q, the *source of the field*. If q is positive, the field is radially outward; if q is negative, the field is radially inward. The field diminishes in magnitude in accordance with the *inverse-square* dependence. These properties of the field can be visualized as equally spaced straight *electric field lines* radiating from the point charge q, Figure 24-9.

ELECTRIC
FIELD
LINES

(1) The direction of the lines is the *direction* of the electric field.

(2) The number of lines penetrating a unit area that is perpendicular to the lines is proportional to the *intensity* of the electric field.

The second statement points up a particularly useful feature regarding field lines. Where they are crowded together, the field is stronger; where they are spread apart, the field is weaker. For an isolated point charge, the inverse-square-law behavior is obvious from geometric considerations. Imagine a series of spherical surfaces concentric with the point charge, Figure 24-9b. Because the field lines extend radially (and symmetrically) from the source, the total number of lines penetrating each sphere is the same. But the area of each sphere increases with the square of the radius. Since $\mathbf{E}$ is proportional to the number of lines *per unit area*, the inverse-square relationship follows.

It is difficult to depict true three-dimensional fields in diagrams. Perhaps the best that can be done conveniently is as shown in Figure 24-9a. One must always mentally extend such two-dimensional diagrams into three dimensions to grasp the true nature of the field.

The *number* of lines we imagine to emanate from a given charge is arbitrary. For example, a 1-μC charge may be associated with 100 field lines or with 1 million field lines. We may choose any convenient "scale factor." But whatever convention we adopt, a 3-μC charge must have exactly three times as many lines emanating from it as a 1-μC charge (see Figure 24-10).

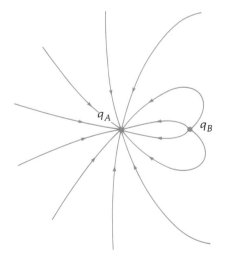

FIGURE 24-10
The electric field pattern near two isolated, unequal point charges having opposite signs. To obtain a more correct visualization of the field, mentally extend the pattern to three dimensions, preserving symmetry about the horizontal axis. From the number of lines terminating on each charge, we see that $|q_A| = 3|q_B|$ and that q_A is negative and q_B positive.

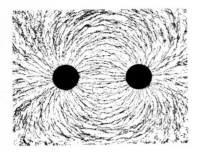

(a) The electric field near two parallel rods with opposite charges.

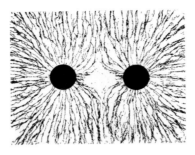

(b) The electric field near two parallel rods with the same charge.

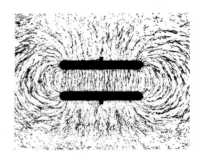

(c) The electric field near two parallel plates with opposite charges.

FIGURE 24-11
We can depict the electric field experimentally by sprinkling small, elongated, nonconducting particles on a glass plate. (Here, grass seed is used.) In the presence of a strong electric field, the particles align themselves in chains along the direction of the field.

Note that, close to each charge, the field lines are symmetrical about each point charge. At very great distances, the collection of charges appears essentially as just a single point charge (with the *net charge* of the array), so the field lines far from the array extend outward symmetrically as if they came from just a single point charge.

Electric field lines always begin at a positive charge and end at a negative charge. For isolated net charges, for which the field lines extend away from the diagram, we imagine that the lines terminate on charges "at infinity" (or, in more practical terms, on induced charges on the inner walls of the laboratory). In any case, the lines themselves should not be taken literally. Keep in mind that field <u>lines</u> do not exist in nature; they are just a convenient mental image that we use to help us think about electric <u>fields</u>. The fields themselves *do* exist in the sense that they can be operationally defined and experimentally determined.

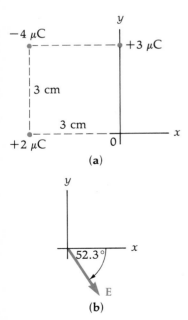

FIGURE 24-12
Example 24-3.

EXAMPLE 24-3

Three point charges are located at the corners of a square 3 cm on a side, as shown in Figure 24-12a. Find the electric field **E** at the other corner of the square.

SOLUTION

We apply the principle of superposition, noting the direction of the field produced by each point charge acting alone. The resultant field **E** is the vector sum of the individual fields: $\mathbf{E} = \mathbf{E}_2 + \mathbf{E}_3 + \mathbf{E}_4$.

$$E_2 = \frac{kq_2}{r_2^2} = \frac{(9 \times 10^9 \text{ N} \cdot \text{m}^2/\text{C}^2)(2 \times 10^{-6} \text{ C})}{(3 \times 10^{-2} \text{ m})^2}$$

$$= \left(2.00 \times 10^7 \frac{\text{N}}{\text{C}}\right) \quad \left(\begin{matrix}\text{in the } +x \\ \text{direction}\end{matrix}\right)$$

$$E_3 = \frac{kq_3}{r_3^2} = \frac{(9 \times 10^9 \text{ N} \cdot \text{m}^2/\text{C}^2)(3 \times 10^{-6} \text{ C})}{(3 \times 10^{-2} \text{ m})^2}$$

$$= \left(3.00 \times 10^7 \frac{\text{N}}{\text{C}}\right) \quad \left(\begin{matrix}\text{in the } -y \\ \text{direction}\end{matrix}\right)$$

$$E_4 = \frac{kq_4}{r_4^2} = \frac{(9 \times 10^9 \text{ N} \cdot \text{m}^2/\text{C}^2)(4 \times 10^{-6} \text{ C})}{(3\sqrt{2} \times 10^{-2} \text{ m})^2}$$

$$= \left(2.00 \times 10^7 \frac{\text{N}}{\text{C}}\right) \quad \left(\begin{matrix}\text{toward the } -4\text{-}\mu\text{C charge along} \\ \text{the diagonal of the square}\end{matrix}\right)$$

We express these fields in vector notation and add them as vectors:

$$\mathbf{E} = \mathbf{E}_2 + \mathbf{E}_3 + \mathbf{E}_4$$

$$\mathbf{E} = \left(2 \times 10^7 \frac{\text{N}}{\text{C}}\right)\hat{\mathbf{x}} - \left(3 \times 10^7 \frac{\text{N}}{\text{C}}\right)\hat{\mathbf{y}}$$

$$+ \left[\left(\frac{2 \times 10^7}{\sqrt{2}} \frac{\text{N}}{\text{C}}\right)\hat{\mathbf{x}} - \left(\frac{2 \times 10^7}{\sqrt{2}} \frac{\text{N}}{\text{C}}\right)\hat{\mathbf{y}}\right]$$

$$\mathbf{E} = \left(3.414 \times 10^7 \frac{\text{N}}{\text{C}}\right)\hat{\mathbf{x}} - \left(4.414 \times 10^7 \frac{\text{N}}{\text{C}}\right)\hat{\mathbf{y}}$$

$$\mathbf{E} = \boxed{5.58 \times 10^7 \frac{\text{N}}{\text{C}}} \text{ at } 52.3° \text{ below the } +x \text{ axis}$$

EXAMPLE 24-4

Five equal, negative point charges $-q$ are placed symmetrically around a circle of radius R. Calculate the electric field $\mathbf{E}$ at the center of the circle.

SOLUTION

We sketch the array of charges in Figure 24-13, choosing coordinate axes to match the symmetry of the distribution. Here, $\theta_1 = 360°/5$ and $\theta_2 = (2)(360°/5)$. We calculate the x and y components of the field:

$$E_x = \frac{kq}{R^2}(1 + 2\cos\theta_1 + 2\cos\theta_2)$$

$$= \frac{kq}{R^2}(1 + 0.6180 - 1.6180) = \boxed{0}$$

Similarly, $\quad E_y = \dfrac{kq}{R^2}(0 - \sin\theta_1 - \sin\theta_2 + \sin\theta_1 + \sin\theta_2) = \boxed{0}$

It can be shown that the field at the center of the circle is zero, *independent of the number of equal charges that are equally spaced around the circle*, whether there is an even or an odd number of charges.

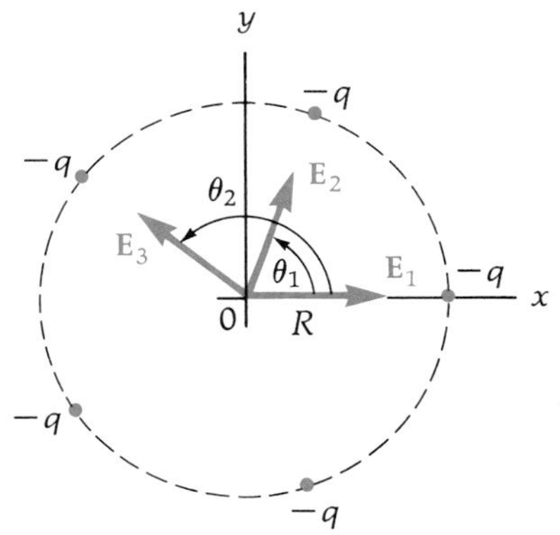

FIGURE 24-13
Example 24-4. The electric fields produced by (only) three of the charges are shown.

24.6 The Electric Dipole

One particular configuration of electric charges has application in a great number of practical cases. This configuration is the **electric dipole**: *two point charges, separated in space, with the same magnitude but with opposite signs*. Many molecules, such as water, form a permanent electric dipole. The numerous applications in atomic and molecular physics justify a rather thorough discussion of this topic. Figure 24-14a illustrates a dipole with electric field lines connecting the two charges. The *direction* of the field at any point is tangent to the field lines in the neighborhood. The *intensity* of the field is proportional to the spatial density of the lines.

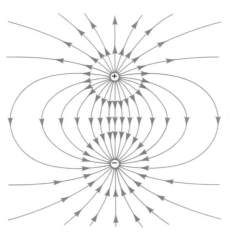

(a) The field of an **electric dipole**: point charges of equal magnitude but opposite sign.

(b) The field of point charges of equal magnitude and the same sign (positive charges illustrated).

FIGURE 24-14
Electric field patterns for two point charges. As with all diagrams representing three-dimensional fields, you should imagine the field lines filling three-dimensional space symmetrically. (In these cases, the pattern is symmetrical about the line joining the two charges.)

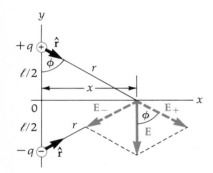

FIGURE 24-15
Example 24-5.

EXAMPLE 24-5

Consider an electric dipole aligned along the y axis as in Figure 24-15. Find the magnitude and direction of the electric field at an arbitrary distance x along the x axis.

SOLUTION

Let the separation of the charges be the distance ℓ, as shown in Figure 24-15. We will calculate the field $\mathbf{E}_+$ due to the positive charge and the field $\mathbf{E}_-$ due to the negative charge and then add them vectorially. We start with the field due to a point charge $\mathbf{E} = (kq/r^2)\hat{\mathbf{r}}$. The unit vector $\hat{\mathbf{r}}$ is from q to the point in question, and r is the distance from q to the point. Thus $r^2 = (\ell/2)^2 + x^2$, and the magnitudes of E_+ and E_- are the same.

$$E_+ = E_- = \frac{kq}{(\ell/2)^2 + x^2} \tag{24-12}$$

When we add $\mathbf{E}_+$ and $\mathbf{E}_-$ as *vectors*, the components along the x axis cancel because of symmetry, but the y components add together to yield

$$E = \left(\frac{2kq}{(\ell/2)^2 + x^2}\right) \cos\phi$$

where $\cos\phi$ may be written as

$$\cos\phi = \frac{(\ell/2)}{\sqrt{(\ell/2)^2 + x^2}}$$

When we substitute $k = 1/4\pi\varepsilon_0$, the field becomes

$$\mathbf{E} = \boxed{-\left(\frac{1}{4\pi\varepsilon_0}\right)\frac{q\ell}{[(\ell/2)^2 + x^2]^{3/2}}\hat{\mathbf{y}}} \tag{24-13}$$

The direction of $\mathbf{E}$ at this point along the x axis is in the $-y$ direction.

The Far-Field Approximation

Because most dipoles in nature are of atomic or molecular sizes, it is worthwhile to consider the limiting case of distances far from the dipole, Figure 24-16. First consider distances along the x axis. For $x \gg \ell$, Equation (24-13) reduces to[4]

$$E \approx \left(\frac{1}{4\pi\varepsilon_0}\right)\frac{q\ell}{x^3} \qquad (24\text{-}14)$$

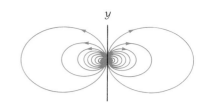

FIGURE 24-16
The electric field for the dipole *far-field approximation*. You should mentally extend the pattern to three dimensions, with field lines arrayed symmetrically about the y axis. The dipole itself is too small to be seen; the two point charges are aligned along the y axis, with the positive charge above the negative charge, so the dipole points in the $+y$ direction: ↑.

Thus, for large distances along the x axis, the field decreases with the *inverse-cube of the distance*. As demonstrated by Problem 24C-30, along the line joining the charges (the y axis), the field also falls off with the inverse-cube of the distance. In fact, it can be shown that, for *all* directions away from the dipole, an inverse-cube behavior exists at large distances. If we place the origin of the coordinate system at the center of the dipole, then distances are simply r. With this in mind, we now rewrite the previous equation in the more general notation

FAR-FIELD APPROXIMATION FOR THE ELECTRIC DIPOLE
$(r \gg \ell)$

$$E \propto \frac{q\ell}{r^3} \qquad (24\text{-}15)$$

An interesting feature about the far-field approximation is that, if q were doubled and ℓ were halved, the field would still be the same. Indeed, *any* combination of q and ℓ whose product has the same numerical value leads to the same electric field at sufficiently large distances. In other words, it is only the *product $q\ell$* that determines the field at far distances. For this reason, the product $q\ell$ is given a special name: the *electric dipole moment*.

The Electric Dipole Moment

Of special interest is the behavior of an electric dipole placed in a uniform electric field **E**, as shown in Figure 24-17. Since the field is uniform, the force $\mathbf{F}_+$ on the $+q$ charge is equal in magnitude but opposite in direction to the force $\mathbf{F}_-$ on the $-q$ charge. The net force on the dipole is zero, so the torque on the dipole may be computed from any point. Let us choose the point at the negative charge $-q$. Recall from Chapter 10 that the torque $\boldsymbol{\tau}$ about $-q$ is

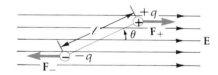

FIGURE 24-17
An electric dipole in a uniform external field **E**.

$$\boldsymbol{\tau} = \mathbf{r} \times \mathbf{F}$$

whose magnitude is $\quad \tau = F_+ \ell \sin\theta = (q\ell)E\sin\theta$

which tends to rotate the dipole toward decreasing θ. The form of this equation suggests a vector notation,

$$\boldsymbol{\tau} = (q\boldsymbol{\ell}) \times \mathbf{E}$$

---

[4] The inequality $x \gg \ell$ does not mean that x becomes infinite. Rather, when we compare the two terms in the denominator of Equation (24-13), we see that, if $x \gg \ell$, the factor $(\ell/2)^2$ is negligible compared with x^2, and thus $(\ell/2)^2$ may be dropped in the far-field approximation. Another way of stating this is that $(\ell/2)^2$ is negligible compared with x^2.

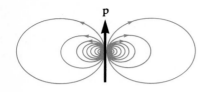

FIGURE 24-18
The *dipole moment vector* **p** points in the direction of the electric field on the axis of the dipole.

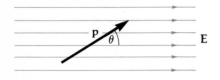

FIGURE 24-19
An electric dipole **p** in a uniform external field **E**. The angle θ is between the forward directions of **p** and **E**.

where $(q\boldsymbol{\ell})$ is the **electric dipole moment p** directed from the negative to the positive charge. The direction of $\boldsymbol{\tau}$ is specified by the cross-product.

| ELECTRIC DIPOLE MOMENT p | $\mathbf{p} = q\boldsymbol{\ell}$ | (where $\boldsymbol{\ell}$ is directed from the negative to the positive charge) | (24-16) |

The dipole moment has units of *coulomb·meters* (C·m). It is a vector whose direction is defined to be along the axis of the dipole from the negative toward the positive charge. The vector **p** thus points in the direction that the field lines come *out* of the dipole, Figure 24-18.

When the dipole is in an external electric field **E**, Figure 24-19, the torque is expressed in vector form as

| TORQUE AN ELECTRIC FIELD E EXERTS ON AN ELECTRIC DIPOLE MOMENT p | $\boldsymbol{\tau} = \mathbf{p} \times \mathbf{E}$ | (24-17) |
| | $|\boldsymbol{\tau}| = pE \sin \theta$ | (24-18) |

Note that the torque tries to align the dipole so that it points *in* the field direction. We would have to do work against this torque to rotate the dipole away from the field-parallel direction. Thus, in the presence of the external field, the dipole possesses electric potential energy when not aligned along the field direction. The electric force is *conservative*, so the change in potential energy ΔU is the *negative* of the work done by the conservative force. For linear motion (Equation 7-10), this change is

$$U_b - U_a = -\int_a^b \mathbf{F} \cdot d\mathbf{x}$$

For a torque $\boldsymbol{\tau}$ acting through an angle $d\boldsymbol{\theta}$, the relation is

$$U_\theta - U_{\theta_0} = -\int_{\theta_0}^{\theta} \boldsymbol{\tau} \cdot d\boldsymbol{\theta}$$

In Figure 24-19, by the right-hand rule the vector $d\boldsymbol{\theta}$ (representing an *increase* in θ) is <u>out of</u> the plane of the figure, while the torque vector $\boldsymbol{\tau}$ is <u>into</u> the plane of the figure. Thus the dot product $\boldsymbol{\tau} \cdot d\boldsymbol{\theta}$ introduces a minus sign: $|\boldsymbol{\tau} \cdot d\boldsymbol{\theta}| = \tau(\cos 180°) d\theta = -(\tau d\theta) = -(pE \sin\theta\, d\theta)$. (Note that here the angle θ is the angle between $\boldsymbol{\tau}$ and **E**, not the 180° angle between $\boldsymbol{\tau}$ and $d\boldsymbol{\theta}$!)

$$U_\theta - U_{\theta_0} = -\int_{\theta_0}^{\theta} (-pE \sin\theta)\, d\theta = -pE(\cos\theta - \cos\theta_0)$$

Choosing the zero reference level $U_{\theta_0} \equiv 0$ when $\theta_0 = 90°$, we have

$$U - 0 = -pE(\cos\theta - 0) = -pE\cos\theta$$

which can be written as the scalar product

| POTENTIAL ENERGY U OF AN ELECTRIC DIPOLE IN AN ELECTRIC FIELD ($U \equiv 0$ WHEN p AND E ARE AT 90°) | $U = -(\mathbf{p} \cdot \mathbf{E})$ | (24-19) |

The potential energy of the dipole is thus a *maximum* when **p** is *antiparallel* to **E** and a *minimum* when **p** is *parallel* to **E**, with the zero reference orientation midway between at 90°.

EXAMPLE 24-6

An isolated water molecule has a permanent electric dipole moment of 6.24×10^{-30} C·m. (a) Calculate the torque on this dipole when it is in an external electric field of 300 N/C, oriented with the dipole moment at 60° with respect to the field direction. (b) Find the work performed by the field in rotating the dipole from this position to an orientation parallel to the field.

SOLUTION

(a) From Equation (24-17),

$$\boldsymbol{\tau} = \mathbf{p} \times \mathbf{E}$$
$$\tau = pE \sin \theta = (6.24 \times 10^{-30} \text{ C·m})(300 \text{ N/C})(\sin 60°)$$
$$\tau = \boxed{1.62 \times 10^{-27} \text{ N·m}}$$

(b) The work done by the field is the negative of the change in electric potential energy:

$$\int_{\theta_0}^{\theta} \boldsymbol{\tau} \cdot d\boldsymbol{\theta} = -\Delta U = -[U_\theta - U_{\theta_0}] = -[-pE(\cos \theta - \cos \theta_0)]$$
$$W = [(6.24 \times 10^{-30} \text{ C·m})(300 \text{ N/C})(\cos 0° - \cos 60°)]$$
$$W = \boxed{9.36 \times 10^{-28} \text{ J}}$$

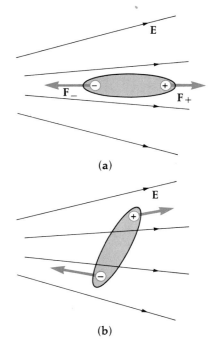

FIGURE 24-20
When a dipole is in a *nonuniform* electric field, there will be a net force on the dipole toward the region of stronger field.

A Dipole in a Nonuniform Field

When a dipole is in the presence of a *nonuniform* electric field, the force on each charge q of the dipole will not be the same if the field strength is not the same at the point where each charge is located. Thus, in addition to a possible torque, there will be a net force on the dipole *toward the region of stronger field*, Figure 24-20. Many molecules have a permanent electric dipole moment because the center of the positive charge distribution does not coincide exactly with the center of the negative charge distribution. If such *polar molecules* are free to move, they will drift toward the region of stronger field.

An electric field can create *induced dipole moments* in ordinary matter when the field causes a slight redistribution of the charges. Positive charges in the material are shifted slightly *in* the direction of the field, while the negative charges are shifted in the *opposite* direction.[5] Figure 24-21 shows an *uncharged* bit of paper or other material near an electrified rod whose diverging field lines produce a nonuniform field. The induced dipole moment in the bit of paper experiences a net force toward the region of the stronger field, because the negative charges find themselves in a stronger field than do the positive charges. Note that this effect is the same regardless of the sign of the charge on the rod.

FIGURE 24-21
The electric field near a charged rod will generate an induced dipole moment in an *uncharged* bit of paper or other material. The diverging field lines are a nonuniform field, and the paper is attracted toward the region of stronger field.

24.7 Electric Fields Due to Continuous Charge Distributions

In practice, arrays of isolated point charges are rarely encountered. Instead, charges are usually distributed closely together over a region so that we can approximate them as *smoothly continuous* charge distributions along a line, over

---

[5] We discuss the microscopic details of dipoles in Section 27.4.

TABLE 24-2

Charge Distribution	Relevant Parameter	SI Units
Along a line	λ, charge per unit length	C/m
On a surface area	σ, charge per unit area	C/m^2
Throughout a volume	ρ, charge per unit volume	C/m^3

a surface, or throughout a volume. In each case, we will pick an *element of charge* dq and calculate an element of field $d\mathbf{E}$ that it produces at a point P. The total field $\mathbf{E}$ at that point is then the *vector* sum of all the field elements at the point.

Field $d\mathbf{E}$ due to one element of charge dq

$$d\mathbf{E} = k \frac{dq}{r^2} \hat{\mathbf{r}}$$

Total field E due to all the elements of charge

$$\mathbf{E} = k \int \frac{dq}{r^2} \hat{\mathbf{r}}$$

Because of the vector nature of the integration, the mathematical procedure must be carried out with care. Fortunately, in the cases we consider, the *symmetry* of the charge distribution will usually result in a simplified calculation.

Each type of charge distribution is described by an appropriate Greek-letter parameter: λ, σ, or ρ, as shown in Table 24-2. Note the units for each. How we choose the charge element dq depends upon the particular type of charge distribution:

Charges along a line

$$dq = \lambda \, dx$$

Charges on a surface area

$$dq = \sigma \, dA$$

Charges throughout a volume

$$dq = \rho \, dV$$

In the examples that follow, note how the differential elements dx, dA, and dV are chosen so that they match the symmetry of the various charge distributions. The most difficult step in solving a problem is the initial choice of the element dq, so a good diagram that shows the element dq and the field $d\mathbf{E}$ that it produces is essential.

EXAMPLE 24-7

Five microcoulombs of charge are distributed uniformly along a thin, straight, nonconducting rod 1 m long. Find the electric field $\mathbf{E}$ at a point 0.4 m away from one end of the rod as shown in Figure 24-22.

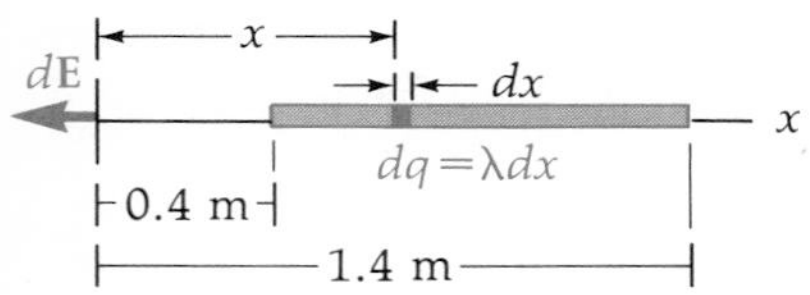

SOLUTION

The linear charge density λ along the rod is $\lambda = 5 \ \mu\text{C/m}$. We align the rod along the x axis with the origin at the point P. We next choose an element of charge[6]

FIGURE 24-22
Example 24-7.

---

[6] Note how multiplying together the units of λ and dx does result in units of *charge* for dq: [charge/length] · [length] = [charge].

$dq = \lambda\,dx$. This charge element produces the field $d\mathbf{E}$ at P in the negative x direction. As we sum over all the charge elements, we note that all the vector field elements $d\mathbf{E}$ lie in the same direction, so the $d\mathbf{E}$'s add as scalars. Thus the integral becomes a one-dimensional scalar summation with limits from $x = 0.4$ m to $x = 1.4$ m.

$$E = k \int \frac{dq}{r^2} = k \int_{0.4\text{ m}}^{1.4\text{ m}} \frac{\lambda\,dx}{x^2} = -k\lambda \left(\frac{1}{x}\right)\bigg|_{0.4\text{ m}}^{1.4\text{ m}}$$

$$= \left(9 \times 10^9 \,\frac{\text{N}}{\text{m}}\right)\left(5 \times 10^{-6} \,\frac{\text{C}}{\text{m}}\right)\left(\frac{1}{1.4\text{ m}} - \frac{1}{0.4\text{ m}}\right)$$

$$E = \boxed{-8.04 \times 10^4 \,\frac{\text{N}}{\text{C}}} \qquad \text{(in the } -x \text{ direction)}$$

EXAMPLE 24-8

A uniform line charge λ (in coulombs per meter) exists along the x axis from $x = -a$ to $x = +a$, as shown in Figure 24-23. Find the electric field E at point P a distance y along the perpendicular bisector.

SOLUTION

As in all problems involving *distributions of charge*, we first choose an element of charge dq to find the element of field $d\mathbf{E}$ it produces at the place of interest. Then we sum all such elements to find the total field $\mathbf{E}$ at that location.

Note the symmetry[7] of the situation. For each dq located at a positive value of x, there is a similar dq located at the same negative value of x. The dE_x produced by one dq is canceled by the dE_x in the opposite direction due to the other dq. Hence, *as we sum all the dq's along the line, all the dE_x components add to zero*. So we need to sum only the dE_y components, a *scalar* sum since they all point in the same direction. The element of charge is $dq = \lambda\,dx$. From Coulomb's law,

$$dE = k\,\frac{dq}{r^2} = \frac{k\lambda\,dx}{r^2}$$

and

$$dE_y = dE \cos\theta = \frac{k\lambda \cos\theta\,dx}{r^2} \qquad (24\text{-}20)$$

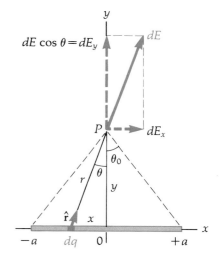

FIGURE 24-23
Example 24-8. A uniform line charge λ from $x = -a$ to $x = +a$.

We have three variables: x, r, and θ. Choosing θ as our single variable, we write the other variables in terms of θ:

$$r = \frac{y}{\cos\theta} \qquad x = y\tan\theta \qquad dx = y\sec^2\theta\,d\theta = y\,\frac{1}{\cos^2\theta}\,d\theta$$

Substituting these in Equation (24-12) gives

$$dE_y = \frac{k\lambda \cos\theta\, y}{\left(\dfrac{y}{\cos\theta}\right)^2}\left(\frac{1}{\cos^2\theta}\right)d\theta = \frac{k\lambda}{y}\cos\theta\,d\theta$$

---

[7] Symmetry arguments are very important in physics. Always look for symmetry since it usually allows a great simplification in the analysis. We will be using symmetry reasoning frequently in the next few chapters.

The parameter θ varies from $-\theta_0$ to $+\theta_0$. By symmetry, this is twice the integral from 0 to θ_0, so (from the table of integrals, Appendix G-II), the total field E_y at point P is

$$E_y = \int_{-\theta_0}^{\theta_0} dE_y = \frac{2k\lambda}{y} \int_0^{\theta_0} \cos\theta\, d\theta = \frac{2k\lambda}{y} \sin\theta \Big|_0^{\theta_0} = \frac{2k\lambda \sin\theta_0}{y}$$

We note that $\sin\theta_0 = a/\sqrt{a^2 + y^2}$ and $k = 1/4\pi\varepsilon_0$, giving

$$\boxed{E_y = \left(\frac{1}{2\pi\varepsilon_0}\right) \frac{\lambda a}{y\sqrt{a^2 + y^2}}} \quad (24\text{-}21)$$

Let us consider a limiting case. If we go very far away, so that $y \gg a$, the line of charge begins to look like just a single point charge $Q = \lambda(2a)$, and we would expect an inverse-square-law field. For $y \gg a$, Equation (24-21) does reduce to

$$E_y \Rightarrow \left(\frac{1}{4\pi\varepsilon_0}\right) \frac{Q}{r^2} \quad \text{(for } y \gg a\text{)}$$

Verifying a limiting-case situation is a useful technique for checking answers.

EXAMPLE 24-9

Find the field $\mathbf{E}$ at a distance r away from an infinitely long uniform line charge λ.

SOLUTION

By symmetry, we recognize that the field is everywhere perpendicular to the line of charge. (Reasoning: there is no asymmetry in the charge distribution to cause the field lines to bend toward either the $+x$ direction or the $-x$ direction. Nor is there any reason for the field lines to bend around the wire in any way. Thus the field can only be radially outward.) The analysis proceeds the same as in the previous example. However, we replace y by the parameter r and note that the limits of integration are from $\theta = -90°$ to $\theta = +90°$ (or twice the integral from 0 to 90°), giving

$$E_r = \frac{2k\lambda}{r} \sin\theta \Big|_0^{90°} = \frac{2k\lambda}{r}$$

FIELD DUE TO AN INFINITELY LONG UNIFORM LINE CHARGE λ

$$\boxed{E_r = \frac{\lambda}{2\pi\varepsilon_0} \frac{1}{r}} \quad \text{(radially outward)} \quad (24\text{-}22)$$

EXAMPLE 24-10

A total positive charge Q is distributed uniformly around a thin, circular, nonconducting ring of radius a. Find the electric field $\mathbf{E}$ at a point P along the axis of the ring, a distance x from the center as shown in Figure 24-24.

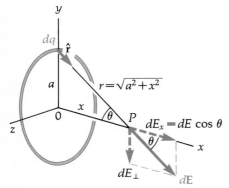

(a) A uniformly charged ring. The element of charge dq produces the element of field $d\mathbf{E}$ at point P.

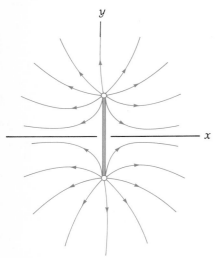

(b) The approximate electric field pattern in the xy plane.

FIGURE 24-24
Example 24-10.

SOLUTION

In problems involving *distributions of charge*, we first choose a point element of charge dq to find the element of field $d\mathbf{E}$ it produces at the place of interest. Then we sum all such elements to find the total field $\mathbf{E}$ at that location.

Note the symmetry of this situation. Every element dq can be paired with a similar element on the opposite side of the ring. Every component $dE_\perp$ perpendicular to the x axis is thus canceled by a component $dE_\perp$ in the opposite direction. Indeed, in the summation process, all the perpendicular components $dE_\perp$ add to zero. Thus we need only add the dE_x components, which all lie along the $+x$ direction, and this is a simple *scalar* integral. From Coulomb's law in vector form,

$$d\mathbf{E} = k\frac{dq}{r^2}\hat{\mathbf{r}}$$

whose magnitude is
$$dE = \frac{k\,dq}{(a^2 + x^2)}$$

The x component is
$$dE_x = \frac{k\,dq}{(a^2 + x^2)}(\cos\theta) = \frac{k\,dq}{(a^2 + x^2)}\left(\frac{x}{\sqrt{a^2 + x^2}}\right)$$

Thus:
$$E_x = \int dE_x = \int \frac{kx\,dq}{(a^2 + x^2)^{3/2}}$$

As we integrate around the ring, all the terms remain constant and $\int dq = Q$, so the total field (with k replaced by $1/4\pi\varepsilon_0$) is

$$E_x = \frac{kx}{(a^2 + x^2)^{3/2}}\int dq = \boxed{\left(\frac{1}{4\pi\varepsilon_0}\right)\frac{xQ}{(a^2 + x^2)^{3/2}}} \quad (24\text{-}23)$$

To check this result, we consider two limiting cases. When $x \to 0$ as we move toward the center of the ring, $E \to 0$. This is to be expected because, at the center, the $d\mathbf{E}$ produced by an element of charge is exactly canceled by a $d\mathbf{E}$ in the opposite direction from a similar charge element on the opposite side of the ring. By symmetry, summing all such pairs around the ring results in $E = 0$ at the center. Another limiting case is for $x \gg a$. When we go very far away along the x axis, the ring appears to be just a point charge. Equation (24-23) verifies this behavior since, for $x \gg a$, it reduces to

$$E \Rightarrow \left(\frac{1}{4\pi\varepsilon_0}\right)\frac{Q}{x^2} \quad \text{(for } x \gg a\text{)}$$

Though we will not calculate E for points off the axis, we can estimate that the field will have the general configuration shown in Figure 24-24b.

EXAMPLE 24-11

A flat, circular, nonconducting disk of radius R has a uniform charge per unit area σ on one side of the disk. Find the electric field $\mathbf{E}$ at a point P along the axis of the disk, a distance x from the center of the disk. See Figure 24-25.

SOLUTION

Making use of the answer to the previous example, we consider the disk to be made up of a set of concentric rings of radius r and width dr. From symmetry considerations, we know that the electric field $d\mathbf{E}$ at point x for each ring is

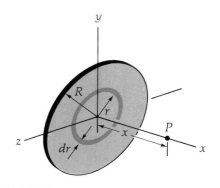

FIGURE 24-25
Example 24-11.

directed along the $+x$ axis. So the summation of $d\mathbf{E}$'s for all the rings is a simple *scalar* integral, resulting in a field E_x.

Let us write an expression for the field dE_x due to a ring of radius r and width dr. The area dA of this ring is $dA = 2\pi r\, dr$, and the charge dq on this ring is thus $dq = \sigma\, dA = 2\pi\sigma r\, dr$. From Equation (24-23), the field dE_x produced by this ring of charge dq (replacing a by r and Q by dq) is

$$dE_x = \frac{kx\, dq}{(r^2 + x^2)^{3/2}} = \frac{kx 2\pi\sigma r\, dr}{(r^2 + x^2)^{3/2}}$$

The total field E_x is the summation of the fields due to all of the rings from $r = 0$ to $r = R$. (In this integral, note that x is a constant.)

$$E_x = kx 2\pi\sigma \int_0^R \frac{r\, dr}{(r^2 + x^2)^{3/2}}$$

From Appendix G-II, the integral becomes

$$E_x = kx 2\pi\sigma \left(\frac{-1}{\sqrt{r^2 + x^2}}\right)\bigg|_0^R$$

Substituting the limits and rearranging, we obtain

$$\boxed{E_x = \frac{\sigma}{2\varepsilon_0}\left(1 - \frac{x}{\sqrt{R^2 + x^2}}\right)} \qquad (24\text{-}24)$$

We now consider two limiting cases. First, what does the field look like at very large distances along the axis from the disk—that is, when $x \gg R$? To evaluate the expression, we divide the numerator and the denominator of the last term by x, giving

$$E_x = \frac{\sigma}{2\varepsilon_0}\left(1 - \frac{1}{\sqrt{\left(\frac{R}{x}\right)^2 + 1}}\right)$$

Using the approximation $(1 + b^2)^{-1/2} \cong 1 - b^2/2$ when $b^2 \ll 1$, we obtain

$$E_x \cong \frac{\sigma}{2\varepsilon_0}\left(\frac{R^2}{2x^2}\right) = \frac{\sigma\pi R^2}{4\pi\varepsilon_0 x^2} = \left(\frac{1}{4\pi\varepsilon_0}\right)\frac{Q}{x^2} \qquad \begin{pmatrix}\text{very far from}\\ \text{the disk}\end{pmatrix}$$

which is Coulomb's law for a point charge $Q = \sigma\pi R^2$, just what we would expect.

For the second limiting case, let $x \to 0$, which is analogous to letting $R \to \infty$. That is, we approach the case of the field near the surface of an infinitely large plane sheet of charge. As $x \to 0$ in Equation (24-24), the expression becomes

$$\underline{\underline{E_x \cong \frac{\sigma}{2\varepsilon_0}}} \qquad \begin{pmatrix}\text{very close to}\\ \text{the disk}\end{pmatrix}$$

This is an interesting result since it shows that the field is *uniform* and does not depend upon the distance x from the sheet of charge. In the next chapter, we present a simple derivation of this result using Gauss's law.

TABLE 24-3

Source	Spatial Dependence of the Electric Field
Point charge	Like $\dfrac{1}{r^2}$
Uniformly charged infinite straight line	Like $\dfrac{1}{r}$
Uniformly charged infinite plane	Constant

You will find it helpful to get a "feeling" for the spatial dependence of fields produced by charge distributions that have certain simple geometries, Table 24-3.

Summary

Electric charge q, measured in units of *coulombs* (C), can be $+$ or $-$ and always occurs in multiples of the *fundamental electron charge* whose magnitude is $e = 1.602 \times 10^{-19}$ C. Charge is *conserved* so that the total charge in a closed system always remains constant.

Coulomb's law for the force between two point charges:

$$\mathbf{F}_{12} = k \frac{q_1 q_2}{r^2} \hat{\mathbf{r}}_{12} \quad \text{where } k = \left(\frac{1}{4\pi\varepsilon_0}\right) \cong 9 \times 10^9 \frac{\text{N} \cdot \text{m}^2}{\text{C}^2}$$

Here, $\mathbf{F}_{12}$ is the force that charge 1 exerts on charge 2 and $\hat{\mathbf{r}}_{12}$ is the unit vector from 1 to 2. Like charges repel; unlike charges attract.

The **electric field E** is defined as

$$\mathbf{E} = \frac{\mathbf{F}}{q_0} \quad \left(\text{or, more precisely, } \mathbf{E} = \lim_{q_0 \to 0} \frac{\mathbf{F}}{q_0}\right)$$

where q_0 is a small positive test charge. The force on a point charge q in the presence of a field $\mathbf{E}$ is $\mathbf{F} = q\mathbf{E}$. Electric forces and fields due to several charges may be added together *as vectors*—the **principle of superposition**. The concept of an electric field $\mathbf{E}$, introduced to "explain" electrostatic forces that act at a distance through empty space, is one of the most important concepts in physics.

Electric field lines (originally called "lines of force") are imaginary lines that we draw to aid our visualization of the properties of the electric field. Electric field lines always begin on a $+$ charge and end on a $-$ charge. In any region, the strength of the field is proportional to the number of lines that penetrate a unit area perpendicular to the lines.

The **electric dipole**: Two equal point charges of opposite sign, separated a distance ℓ.

Electric dipole moment: $\quad \mathbf{p} = q\boldsymbol{\ell} \quad \left(\begin{array}{l}\text{where } \boldsymbol{\ell} \text{ is the direction}\\ \text{from } -q \text{ to } +q\end{array}\right)$

Torque on an electric dipole in an external electric field E: $\quad \boldsymbol{\tau} = \mathbf{p} \times \mathbf{E}$

Potential energy of an electric dipole in an electric field E: $\quad U = -(\mathbf{p} \cdot \mathbf{E}) \quad \left(\begin{array}{l}\text{where } U \equiv 0 \text{ for}\\ \mathbf{p} \text{ and } \mathbf{E} \text{ at } 90°\end{array}\right)$

Smoothly continuous charge distributions are described by an appropriate Greek-letter parameter: λ, σ, or ρ.

Charge Distribution	Relevant Parameter	SI Units	Element of Charge
Along a line	λ, charge per unit length	C/m	$dq = \lambda\, dx$
On a surface area	σ, charge per unit area	C/m²	$dq = \sigma\, dA$
Throughout a volume	ρ, charge per unit volume	C/m³	$dq = \rho\, dV$

In calculating fields due to a distribution of charges, we first choose an *element of charge* dq to find the *element of field* $d\mathbf{E}$ that it produces at the point of interest. Then we sum all such vector elements to find the total field $\mathbf{E}$ at that location:

$$\mathbf{E} = \int d\mathbf{E} = k \int \frac{dq}{r^2} \hat{\mathbf{r}}$$

Symmetry considerations often simplify the analysis greatly.

Questions

1. An inflated toy balloon becomes slightly larger as it acquires an electrical charge. Why?
2. Why can electric field lines not cross one another or form closed loops?
3. Consider an electric field pattern composed of curved field lines. Suppose that an electron is released from rest at one point along a field line. Explain why the subsequent motion of the electron is *not* along that field line (opposite to the direction of the field).
4. An electron moves at right angles to electric field lines. Is there a force on the electron? What about motion of the electron parallel to the field lines?
5. Two nonzero point charges of unspecified signs and magnitudes are held fixed a distance D apart. Is it possible to have $\mathbf{E} = 0$ at some point off the line of length D that joins the charges (other than at ∞)? If so, explain.
6. Consider a dipole in a uniform electric field E. If you rotate the dipole through 180° so that it points opposite to its initial direction, does the work you do depend upon the dipole's initial orientation with respect to $\mathbf{E}$? Does it depend upon the plane of rotation relative to $\mathbf{E}$?
7. Read Problem 24A-1. Does it matter whether the electrons are divided equally between the earth and the moon, or divided unequally? What about putting just one electron on the moon and the rest on the earth? What about putting all of the electrons on the earth?

Problems
24.4 Coulomb's Law

24A-1 Calculate the mass of electrons that, when shared between the earth and moon, would produce a force of repulsion equal to the gravitational force between the earth and moon.

24A-2 Suppose two objects, each with a net positive charge of one coulomb, were separated by a distance equal to the distance between New York and San Francisco (about 4140 km). Calculate the mutual force of repulsion between these objects.

24A-3 Calculate the ratio of the electrostatic force to the gravitational force between the electron and the proton in a hydrogen atom.

24A-4 Two helium nuclei (each containing two protons plus two neutrons) are located 5×10^{-14} m apart. (a) Find the Coulomb force of repulsion between them. (b) Find the gravitational force of attraction. (c) If the nuclei are free to move, what is the initial acceleration of each nucleus?

24A-5 Consider three charges located in the xy plane. A charge of $+3$ μC is located at $x = 4$ cm, $y = 0$; a charge of -2 μC is located at $x = 0$, $y = 5$ cm. Find the force on a $+6$ μC charge at the origin.

24A-6 In Problem 24A-5, find the electric field (magnitude and direction) at the origin if the $+6$ μC charge were absent. Verify your answer by using the answer to Problem 24A-5.

24B-7 Two small silver spheres, each with a mass of 100 g, are separated by a distance of 1 m. Calculate the fraction of the electrons in one sphere that must be transferred to the other in order to produce an attractive force of 10^4 N (about a ton) between the spheres. (The number of electrons per atom of silver is 47, and the number of atoms per gram is Avogadro's number divided by the atomic weight of silver, 107.9.)

24B-8 Richard Feynman once said that if two persons stood at arm's length from each other and each person had 1% more electrons than protons, the force of repulsion between the two people would be enough to lift a "weight" equal to that of the entire earth. Carry out an order-of-magnitude calculation to substantiate this assertion.

24B-9 Two point charges are located as follows: a -3-μC charge at the origin and a $+2$-μC charge at $x = 0.15$ m. Find the location where a positive point charge q' may be placed so that the net force on the charge q' is zero.

24B-10 If there were a slight imbalance between the number of protons and the number of electrons in matter, the gravitational attraction between astronomical objects could be overcome by the electrostatic repulsion between these objects. Calculate the minimum fraction by which one charge would have to exceed the other for this to occur. The approximate average number of proton–electron pairs per kilogram of matter is 3×10^{26}.

24B-11 A silver dime (not the nonsilver version now in circulation) has a mass of 2.49 g. The atomic mass of silver is 107.870 and its atomic number is 47. Assume that the dime is 100% silver. For every 10^{12} electrons present, how many electrons must be removed to give the dime a net charge of 1 μC?

24.5 The Electric Field

24A-12 Express the units for an electric field in terms of the SI base units of mass (kg), length (m), time (s), and electric current (A).

24A-13 Under normal atmospheric conditions on a clear day, a downward electric field of roughly 100 N/C exists just above the surface of the earth. If a toy helium-filled balloon is barely capable of lifting a mass of 50 g, find the amount of electric charge that must be distributed over the balloon's surface so that the balloon will not rise when the mass is removed. (The amount of charge required would produce repulsive forces on the surface of the balloon that would be more than sufficient to tear the balloon apart.)

24B-14 Point charges of $+q$ and $-2q$ are located near each other. Sketch field lines to represent the approximate electric field configuration in the vicinity, making sure that two times as many field lines are associated with one charge as with the other.

24B-15 A uniform electric field is described in Cartesian coordinates by $\mathbf{E} = E_0 \hat{\mathbf{y}}$, where E_0 is a constant. A particle with a mass m and charge $+q$ is injected at the origin into the electric field with an initial velocity $\mathbf{v} = v_0 \hat{\mathbf{x}}$. Find the equation of the subsequent trajectory of the particle.

24.6 The Electric Dipole

24A-16 Many molecules possess an electric dipole moment because the center of distribution of the positive charge (protons) does not exactly coincide with that of the negative charge (electrons). The electric dipole moment of a water molecule in its gaseous state is 6.24×10^{-30} C·m. (a) If a water molecule is placed in an electric field of 10^4 N/C, calculate the maximum torque that the field can exert on the molecule. (b) Find the range of the potential energies that the molecule may have in this field.

24A-17 The electric dipole moment of a sodium fluoride molecule is 2.72×10^{-29} C·m. Assuming an (oversimplified) model of singly ionized atoms, Na^+ and F^-, for this molecule, how far apart are the centers of these atoms? (Note: the actual value is 1.93×10^{-10} m.)

24B-18 Figure 24-11 describes how grass seed can be used to visualize an electric field. Explain why small, elongated, nonconducting particles align themselves in the direction of an electric field.

24.7 Electric Fields Due to Continuous Charge Distributions

24B-19 A charge $+Q$ is distributed uniformly along a straight line of length L. Find the electric field E at a point P along the direction of the line, a distance d from one end (Figure 24-26).

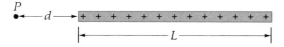

FIGURE 24-26
Problem 24B-19.

24B-20 A uniform positive charge per unit length λ exists along a thin nonconducting rod bent into the shape of a segment of a circle of radius R, subtending an angle $2\theta_0$ as shown in Figure 24-27. Find the electric field **E** at the center of curvature 0. (Hint: consider the field $d\mathbf{E}$ due to the charge dq contained within an element of length $d\ell = R\,d\theta$. Use symmetry considerations in setting up the integral between $\theta = -\theta_0$ to $\theta = +\theta_0$ to find the total field **E** at 0.)

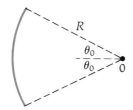

FIGURE 24-27
Problem 24B-20.

24B-21 Consider a thin, circular, nonconducting disk, radius R, that has a uniform charge per unit area σ on one side of the disk. In Example 24-11 we find the electric field E at a point along the axis a distance x from the center of the disk. Show that, as R approaches infinity (the case of a uniform infinite plane of charge), the electric field becomes $E = \sigma/2\varepsilon_0$. (Note that, for an infinite plane of uniform charge density, the field has the same constant value, independent of the distance x from the plane.)

Additional Problems

24C-22 Two protons are released from rest when they are 2×10^{-13} m apart. (a) Find the final speed of each proton. (b) If, instead, one of the protons were held fixed, what would be the speed of the other proton?

24C-23 Show that two small objects a given distance apart and sharing a given total charge will have a maximum force of repulsion when the charge is shared equally between the objects.

24C-24 Between 1909 and 1917, R. A. Millikan made the first accurate determination of the electronic charge $-e$ by observing the vertical motions of tiny charged droplets of oil in air. A vertical electric field was established between horizontal metal plates such that an upward electric force on a charged droplet just balanced the downward gravitational force. From these (and other) measurements, the highest common factor of various charges on a droplet ($-e$, $-2e$, $-3e$, ...) allowed Millikan to determine the smallest step by which the charge could increase or decrease (that is, the charge on a single electron). In a typical experiment, a droplet weighing 1.9×10^{-13} N is held stationary when 1200 V is applied to plates separated 3 mm. (a) How many surplus electrons are on the droplet? (b) If the density of the oil is 920 kg/m³, what is the radius of the droplet?

24C-25 In the Millikan Oil Drop experiment (see previous problem), the droplets are so tiny that they appear only as points of light in the microscope used to observe them. In order to find the radius (and hence the mass) of each droplet, we allow them to fall freely under gravity. The retarding force F exerted by the viscous air on a sphere of radius r moving with speed v through air is given by Stokes' law, $F = 6\pi\eta r v$, where η is the coefficient of viscosity. (a) Find the SI units for η. (b) Show that when a falling droplet achieves a constant "terminal" velocity (signifying that the viscous retarding force equals the force of gravity), the following relation is true, thus allowing the radius of the droplet to be determined:

$$v = \frac{2gr^2}{9\eta}(\rho_0 - \rho_a)$$

Here, ρ_0 and ρ_a are the respective densities of the oil and air.

24C-26 Two point charges, each of charge $+Q$, are held fixed a distance d apart. A third positive charge q is confined to move along the straight line joining the original two charges. (a) Show that if the charge q is displaced a small distance x (where $x \ll d$) from its position of equilibrium, it will execute approximately simple harmonic motion. (b) Find the "spring constant" k associated with this motion.

24C-27 Calculate the amount of work required to accumulate a charge Q on a sphere of radius R. We can build the charge by bringing infinitesimal charges dq from infinity up to the surface of the sphere until the total charge Q is reached.

24C-28 An electron with horizontal velocity $v_0 = 8 \times 10^6$ m/s enters the region midway between two horizontal deflecting plates as shown in Figure 24-28. The plates are 3 cm long and separated by 1.5 cm. A potential difference of 40 V is applied to the plates. Find the angle θ with respect to the horizontal that the electron's velocity **v** makes just as it emerges from the region between the plates. Ignore fringing field effects.

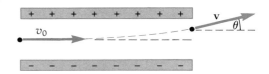

FIGURE 24-28
Problem 24C-28.

24C-29 As shown in Figure 24-29, an electron with initial speed $v_0 = 10^6$ m/s at $x_0 = 0$ moves along the $+x$ direction in a region of increasing electric field strength given by $E_x = (4 \text{ V/m})(1 + 10^3 x)$, where x is in meters. Find the distance that the electron moves before it is brought (momentarily) to rest.

24C-30 A point charge $+q$ is located at $x = \ell/2$ and a point charge $-q$ is at $x = -\ell/2$, forming an electric dipole. (a) Find an expression for the electric field $E(x)$ for all positive values of x. (b) Show that, for values of $x \gg \ell$, the electric field varies as $1/x^3$.

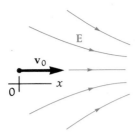

FIGURE 24-29
Problem 24C-29.

24C-31 An electric dipole is made of two point charges, $+q$ and $-q$, each of mass m, separated a distance ℓ. The dipole is placed in a uniform electric field E oriented near its lowest potential energy state. (a) Show that the dipole will undergo oscillatory rotations about its center of mass. (b) Derive an expression for the approximate period T of small-amplitude oscillations.

24C-32 *The quadrupole.* Consider three point charges in the xy plane such that a charge $-2q$ is at the origin, a charge $+q$ is at $y = +\ell/2$, and a charge $+q$ is at $y = -\ell/2$. Such an arrangement of charges is called an *electric quadrupole*. Derive expressions for the electric field along (a) the x axis as a function of x and (b) the y axis as a function of y. (c) Determine the direction of E in each case. (d) Show that $E \propto 1/r^4$ in each case for x or y much greater than ℓ.

24C-33 Positively charged particles (e,m) can be accelerated in a linear drift-tube accelerator that consists of a series of cylindrical metal tubes, of increasing lengths, inside a vacuum chamber (see Figure 24-30). Odd-numbered tubes are connected to one terminal of a high-frequency sine-wave voltage source, while even-numbered tubes are connected to the other terminal. There is no force on a particle while it travels inside a tube because the potential is constant there. However, after traveling with velocity v_1 through tube 1, the particle enters the gap at time t_1, where it is subjected to the peak portion of the time-varying force, $eE_0 \sin \omega t$, and its speed is increased. At time t_2, it enters the next tube, where it travels at the new (constant) speed through the tube (which shields it while the sine wave reverses direction), emerging at time t_3 into the next gap just in time to be accelerated again. The tube dimensions and spacings are such that the particles cross successive gaps in phase with the appropriate direction of the accelerating fields. Consider the case in which the time interval $(t_2 - t_1)$ is one-quarter of the period of the sine wave and is symmetrically situated at the peak. In terms of v_1, e, m, ω, and E_0, find expressions for the first gap distance d and the length ℓ of the second tube.

24C-34 A thin, nonconducting rod of length ℓ carries a line charge $\lambda(x)$ that varies with distance according to $\lambda(x) = Ax$ (in SI units) as shown in Figure 24-31. A point charge q is located a distance ℓ from the end of the rod as shown. (a) What are the SI units of the constant A? (b) Find the force that the line charge exerts on q.

FIGURE 24-31
Problem 24C-34.

24C-35 A positive line charge density λ exists along the x axis from $x = 0$ to $x = L$ as shown in Figure 24-32. (a) Find the value of the y component of the electric field E at the point $y = a$. (b) Find the value of the x component of the electric field at the same point.

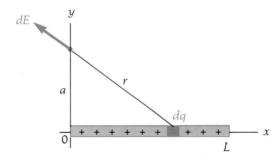

FIGURE 24-32
Problem 24C-35.

24C-36 A circular ring of radius R carries a total charge $+Q$. A negative charge $-q$ is placed at the center of the ring. When the charge $-q$ is displaced a short distance along the axis of the ring and released from rest, it will undergo oscillatory motion (if constrained so that it can move only along the axis). Derive an expression for the approximate frequency f of the oscillations. You may use the result of Example 24-10.

24C-37 A thin, nonconducting ring of radius R has a varying charge per unit length λ described by $\lambda = \lambda_0 \sin \theta$, where the angle θ is defined in Figure 24-33. (a) Sketch the charge distribution on the ring. (b) What is the direction of the electric field $\mathbf{E}$ at the center of the ring? (c) Show that the magnitude of the electric field at the center is $\lambda_0/4\varepsilon_0 R$.

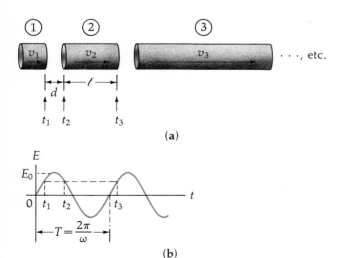

FIGURE 24-30
Problem 24C-33.

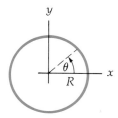

FIGURE 24-33
Problem 24C-37.

24C-38 A thin, nonconducting rod is in the shape of a semicircle of radius R. It has a varying positive charge per unit length λ described by $\lambda = \lambda_0 \sin 2\theta$, where θ is defined in Figure 24-34. (a) Sketch the charge distribution along the semicircle. (b) What is the direction of the electric field **E** at point 0, the center of the semicircle? (c) Find the magnitude of the electric field at point 0.

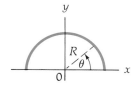

FIGURE 24-34
Problem 24C-38.

24C-39 Consider the electric field along the axis of a uniformly charged ring of radius R. Show that the maximum field $(E_x)_{max}$ on the axis is at a distance $x = R/\sqrt{2}$ from the center of the ring. Make a freehand graph of E vs. x for both positive and negative values of x. You may use the results of Example 24-10.

24C-40 A long, thin, nonconducting ribbon with a width b has a uniform surface charge density σ on both the top and bottom surfaces. Find the electric field **E** at a point P a distance a above the centerline of the ribbon, Figure 24-35. Hint: consider the ribbon as an assembly of charged "wires." Show that the charge per unit length along a wire dx wide is $2\sigma\, dx$. Each wire produces an element of field $d\mathbf{E}$ at the point P (cf. Example 24-8).

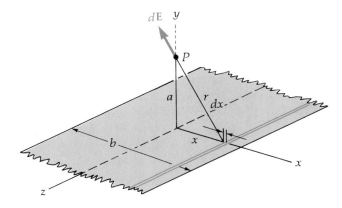

FIGURE 24-35
Problem 24C-40.

24C-41 A circular hoop of a nonconducting material with a uniform distribution of charge has zero electric field at the center of the hoop. Why? Consider such a hoop of radius R with a total charge $+Q$. A length ℓ along the circumference is now cut from the hoop. Find an expression for the field at the center of curvature of the remaining segment.

CHAPTER 25

Gauss's Law

*These immortal words of Gauss
Were posted clearly on his house:
"The outward surface field will tell
What charges in this house doth dwell."*

ANONYMOUS

25.1 Introduction

In the previous chapter we visualized an electric field as a pattern of field lines in space, which gives us a sense of the magnitude and direction of the field **E** at every point. Where the lines are closer together, the field is stronger; where they are farther apart, the field is weaker. A line is imagined to start on a positive charge and end on a negative charge, so that the direction of a line agrees with the direction of the field in that vicinity. Admittedly, field lines are a fiction—they do not exist in nature. However, they are a useful aid in our thinking about an electric field that *does* exist: at every point in space the field has a certain magnitude and direction, characteristics that (at least, in principle) we can experimentally measure by placing a small positive test charge q_0 at that location.

We now make our interpretation of field lines more quantitative. This will lead to a very useful relation known as *Gauss's law*, which provides an alternative method for calculating fields—one that for symmetric charge distributions is far easier to use than the Coulomb's-law approach used in Chapter 24.

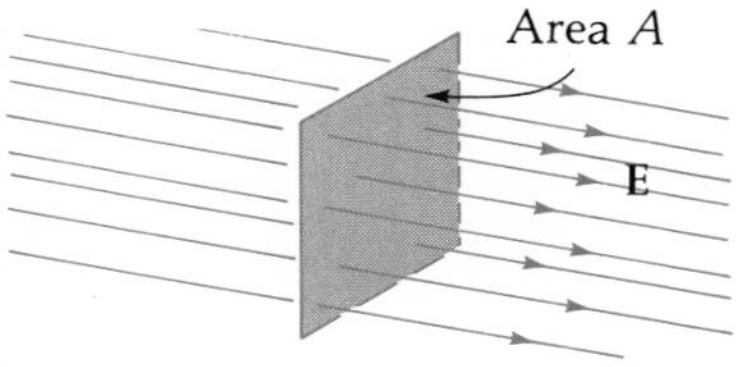

25.2 The Electric Flux

We now enlarge our interpretation of field lines so that they become quantitative, rather than just pictorial. We define the concept of *electric flux*, which is basically *a measure of the number of electric field lines that penetrate a surface*. Consider a uniform field **E** and an imaginary area A whose plane is perpendicular to the field, Figure 25-1. For this case, we define the *electric flux* Φ_E through the surface to be

$$\Phi_E = EA \qquad (25\text{-}1)$$

FIGURE 25-1
The plane of the area A is perpendicular to the uniform field **E**. The electric flux Φ_E passing through the area is $\Phi_E = EA$.

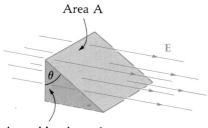
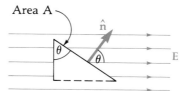

Area $A' = A \cos \theta$

(a) When the area A is tilted as shown, the projection of A to the area A' (which is perpendicular to the field lines) is $A' = A \cos \theta$.

(b) Note that the angle between A and A' is the same as the angle between the normal $\hat{n}$ to the surface and the field $\mathbf{E}$.

FIGURE 25-2
Field lines through an area A whose normal $\hat{n}$ makes an angle θ with respect to the field $\mathbf{E}$.

On occasion, we will need to deal with surfaces that are not perpendicular to field lines. If a plane area A is tilted as shown in Figure 25-2, fewer field lines penetrate the surface. Note the projection of A to the surface A', which is perpendicular to the field lines. The two areas are related according to $A' = A \cos \theta$. The same number of lines penetrates both areas, so from Equation (25-1) the electric flux Φ_E is

$$\Phi_E = EA' = EA \cos \theta \qquad (25\text{-}2)$$

We see that the flux Φ_E through the surface has a maximum value when the plane of the area is perpendicular to the field lines.

In this chapter we will need to deal with curved surfaces over which the electric field varies in both magnitude and direction. We therefore generalize our definition of electric flux by defining a vector element of area $\Delta \mathbf{A}$, defined always to be perpendicular to the surface. Furthermore, we mostly deal with *closed* surfaces. To avoid ambiguity, we always choose the direction of the vector $\Delta \mathbf{A}$ to be the *outward* normal to the surface, Figure 25-3. Making use of the vector notation for the cross product of two vectors, $\mathbf{A} \cdot \mathbf{B} \equiv AB \cos \theta$, we have for the element of flux $\Delta \Phi_E$

$$\Delta \Phi_E = E \, \Delta A \cos \theta = \mathbf{E} \cdot \Delta \mathbf{A} \qquad (25\text{-}3)$$

Finally, we let the area of each element ΔA approach zero (as the number of such elements consequently approaches infinity). In the limit, we have the *differential element of area* $d\mathbf{A}$, leading to the **differential electric flux**,

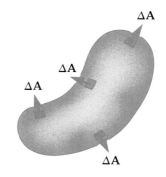

FIGURE 25-3
A closed surface with a few vector elements of area $\Delta \mathbf{A}$, where the direction of the vector $\Delta \mathbf{A}$ is always the *outward* normal to the surface.

DIFFERENTIAL ELECTRIC FLUX $d\Phi_E$

$$d\Phi_E = \mathbf{E} \cdot d\mathbf{A} \qquad (25\text{-}4)$$

where $d\mathbf{A}$ is the *outward* normal element for closed surfaces. For finite areas, we sum over all such elements to obtain[1]

ELECTRIC FLUX Φ_E
(general definition)

$$\Phi_E = \int_{\text{surface}} \mathbf{E} \cdot d\mathbf{A} \qquad (25\text{-}5)$$

The units of electric flux are $\text{N} \cdot \text{m}^2/\text{C}$.

---

[1] The integral may be over an arbitrary area, as written, or over a completely *closed* surface, in which case the symbol $\oint$ is used. (Note the similarity in notation with the integral over a *closed* path: $\oint d\ell$.)

Fortunately we need not evaluate this integral directly over such an awkward closed surface as in Figure 25-3. As discussed in the next section, an interesting property of the inverse-square field produced by an electric charge enables us to evaluate such an awkward integration *by inspection* (!) without actually carrying out the messy details. But first we show the direct calculation for some simple cases.

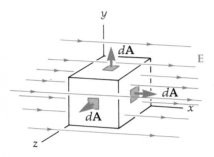

FIGURE 25-4
Example 25-1. The cube of edge length ℓ is oriented symmetrically in the uniform electric field **E**.

EXAMPLE 25-1

A uniform electric field **E** exists in the $+x$ direction. Find the net electric flux Φ_E through the surface of a cube, edge length ℓ, that is oriented with its edges along the coordinate axes as shown in Figure 25-4.

SOLUTION

Because **E** is perpendicular to $d\mathbf{A}$ on four of the faces, the flux through those four faces is zero. For each,

$$\Phi_E = \int \mathbf{E} \cdot d\mathbf{A} = \int E\underbrace{(\cos 90°)}_{=0} dA = \underline{\underline{0}}$$

For the other two faces, we note that the field lines enter the left-hand face, so the angle θ between **E** and $d\mathbf{A}$ on that face is 180°. Thus:

$$\Phi_E = \int_{\substack{\text{left}\\\text{face}}} \mathbf{E} \cdot d\mathbf{A} = \int_{\substack{\text{left}\\\text{face}}} E\underbrace{(\cos 180°)}_{=-1} dA = -E \int_{\substack{\text{left}\\\text{face}}} dA = \underline{\underline{-E\ell^2}}$$

The field lines emerge from the right-hand face, so $\theta = 0°$:

$$\Phi_E = \int_{\substack{\text{right}\\\text{face}}} \mathbf{E} \cdot d\mathbf{A} = \int_{\substack{\text{right}\\\text{face}}} E\underbrace{(\cos 0°)}_{=+1} dA = E \int_{\substack{\text{right}\\\text{face}}} dA = \underline{\underline{E\ell^2}}$$

The total flux through all of the faces of the cube is thus

$$\text{Total } \Phi_E = -E\ell^2 + E\ell^2 = \boxed{0}$$

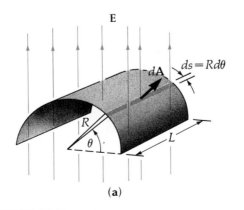
(a)

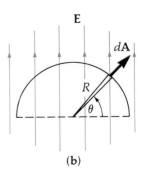
(b)

FIGURE 25-5
Example 25-2.

EXAMPLE 25-2

As shown in Figure 25-5, a uniform electric field **E** penetrates a surface in the shape of a half-cylinder. The field lines are perpendicular to the plane rectangle of length L and width $2R$. By direct integration over the curved surface, calculate the electric flux Φ_E that penetrates the curved surface.

SOLUTION

We need to calculate $\int \mathbf{E} \cdot d\mathbf{A}$ over the cylindrical surface. To simplify the integration, we seek an element of area dA whose normal makes the same angle with **E** everywhere. Noting the symmetry, we choose $d\mathbf{A}$ to be a thin strip of length L and width $ds = R\, d\theta$. Thus, $dA = LR\, d\theta$, and, as we sum all such elements, θ varies from 0 to π. The angle between **E** and $d\mathbf{A}$ is $[(\pi/2) - \theta]$. Thus:

$$\Phi_E = \int \mathbf{E} \cdot d\mathbf{A} = \int_0^\pi E\cos[(\pi/2) - \theta]LR\, d\theta = ELR \int_0^\pi \sin\theta\, d\theta$$

$$\Phi_E = ELR(-\cos\theta)\Big|_0^\pi = ELR[-(-1-1)] = \boxed{2ELR}$$

Suppose that we form a closed Gaussian surface by adding a plane surface that connects the straight edges and adding half-circle end caps. Note that the above answer is the (negative of the) flux entering the plane: $\int \mathbf{E} \cdot d\mathbf{A} = E(\cos 180°)(2LR) = -2ELR$. Thus, by formation of a *closed* surface in the region of the uniform field **E**, the total flux summed over the entire surface is zero. ($\int \mathbf{E} \cdot d\mathbf{A} = 0$ for the end caps because **E** and $d\mathbf{A}$ are at 90° there.) The next section generalizes this result to a closed surface of *any* shape, in the presence of even *non*uniform fields.

25.3 Gauss's Law

Karl Friedrich Gauss (1777–1855), one of the greatest mathematicians of the nineteenth century, gained much insight into the nature of vector fields. His mathematical conclusions are very useful in physics, and Gauss himself made many contributions in the development of electromagnetic theory. To develop Gauss's law, we start with the simplest possible case: a point charge q. Imagine a spherical surface of radius r, called a *Gaussian surface*,[2] centered on the point charge, Figure 25-6a. What is the electric flux Φ_E through this closed surface? The radially outward field lines are everywhere perpendicular to the surface, and the magnitude of **E** is the same all over the surface, so the total flux is simply

$$\Phi_E = \oint \mathbf{E} \cdot d\mathbf{A} = \oint E(\cos 0°)\, dA = E \oint dA = E(4\pi r^2)$$

From Coulomb's law, the field $E = q/4\pi\varepsilon_0 r^2$, so we have

$$\Phi_E = \left(\frac{q}{4\pi\varepsilon_0 r^2}\right)(4\pi r^2) = \frac{q}{\varepsilon_0}$$

TOTAL ELECTRIC FLUX Φ_E ASSOCIATED WITH A POINT CHARGE q

$$\Phi_E = \frac{q}{\varepsilon_0} \qquad (25\text{-}6)$$

Note that the flux Φ_E is independent of the size of the sphere.

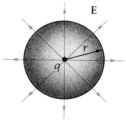

(a) A spherical Gaussian surface of radius r centered on the point charge q.

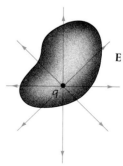

(b) An arbitrary Gaussian surface that encloses a point charge q.

FIGURE 25-6
Gaussian surfaces that enclose a charge q.

---

[2] Gaussian surfaces are hypothetical surfaces we use in calculations. They have no physical reality. They may have any convenient shape.

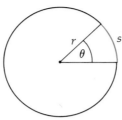

(a) The **plane angle** θ is defined as

$$\theta \equiv \frac{s}{r} \text{ (in } radians\text{)}$$

where s is the length of the arc of the circle of radius r subtended by the angle θ. For the complete circle, the whole plane angle is 2π radians.

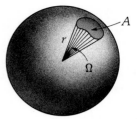

(b) The **solid angle** Ω is defined as

$$\Omega \equiv \frac{A}{r^2} \text{ (in } steradians\text{)}$$

where A is the area on the surface of the sphere of radius r subtended by the solid angle Ω. For an *element of solid angle*,

$$\Delta\Omega \equiv \frac{\Delta A}{r^2} \text{ (in } steradians\text{)}$$

The area may have any shape, but it must be everywhere perpendicular to the radius. Since the total surface area of a sphere is $4\pi r^2$, the whole solid angle surrounding the point at the center is

$$\Omega = \frac{4\pi r^2}{r^2} = 4\pi \text{ steradians}$$

In the illustration, the solid angle Ω is physically related to the more-or-less conical region extending from the origin that subtends the area ΔA.

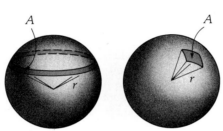

(c) The area A that defines a solid angle may have any shape.

FIGURE 25-7
The definition of a *solid angle* Ω is analogous to the definition of a plane angle. Just as the arc length s is everywhere perpendicular to the radius r, the area A must be everywhere perpendicular to the radius. Because Ω is a ratio of lengths squared, the unit *steradian* is dimensionless.

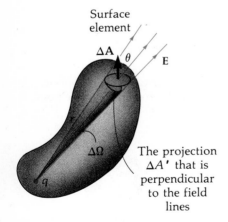

FIGURE 25-8
An arbitrary surface encloses the point charge q. The element ΔA on the surface is not perpendicular to the electric field lines from q. The projection $\Delta A' = \Delta A \cos \theta$ is perpendicular to the field lines, and it defines the element of solid angle $\Delta\Omega = (\Delta A \cos \theta)/r^2$ extending from the charge q.

What about nonspherical surfaces that enclose a charge, Figure 25-6b? We will now prove a remarkable conclusion:

> **For any arbitrary closed surface that contains a charge q anywhere inside, the integral over the entire surface $\oint \mathbf{E} \cdot d\mathbf{A}$ equals q/ε_0 for each case!**

Of course, during the integration the value of $\mathbf{E}$ will be different at various locations on the surface, and the angle between $\mathbf{E}$ and $d\mathbf{A}$ will also vary as we sum the various contributions over the surface. But, interestingly, regardless of the shape of the surface the answer is always q/ε_0.

We begin the proof by making use of the concept of the *solid angle* Ω defined in Figure 25-7. Consider a point charge q surrounded by a closed surface that has an arbitrary shape. The surface area element ΔA in Figure 25-8 is not perpendicular to the radial field lines extending outward from q. The flux $\Delta\Phi_E$ through ΔA is given by Equation (25-3):

$$\Delta\Phi_E = \mathbf{E} \cdot \Delta\mathbf{A} = E \Delta A \cos\theta = \left(\frac{kq}{r^2}\right) \Delta A \cos\theta \qquad (25\text{-}7)$$

But note that $\Delta A \cos\theta$ is the area element perpendicular to the radial field lines, so $(\Delta A \cos\theta)/r^2$ equals the solid angle element $\Delta\Omega$. Hence the total flux Φ_E through the entire closed surface is

$$\Phi_E = \oint \mathbf{E} \cdot d\mathbf{A} = kq \oint \frac{dA \cos\theta}{r^2} = kq \oint d\Omega = kq\,(4\pi)$$

$$\Phi_E = \frac{q}{4\pi\varepsilon_0}(4\pi) = \frac{q}{\varepsilon_0} \qquad (25\text{-}8)$$

This result is independent of the shape of the surface that encloses the charge q, and it also does not depend upon the particular location of q inside the surface.[3] Furthermore, we could have any number of charges inside, distributed in any arbitrary fashion, adding to a total charge inside of $q_{in} = \Sigma_i q_i$.

Here is the final step. Our reasoning has just shown that $\Phi_E = q_{in}/\varepsilon_0$ for any arbitrary surface. Since $\Phi_E = \int \mathbf{E} \cdot d\mathbf{A}$, we combine these two facts to arrive at

GAUSS'S LAW $\quad\begin{bmatrix}\text{For any closed surface}\\ \text{that contains a total}\\ \text{charge } q_{in} \text{ anywhere inside}\end{bmatrix}\quad \oint \mathbf{E} \cdot d\mathbf{A} = \dfrac{q_{in}}{\varepsilon_0} \qquad (25\text{-}9)$

Something interesting has happened here: we no longer need to deal with the electric flux Φ_E itself. (Lines of flux are just a convenient fiction, anyway!) Instead, Gauss's law connects *charges* and *fields* directly in a way that is different from Coulomb's law. For symmetrical cases, Gauss's law is a far easier approach because we can carefully choose the shape of the Gaussian surface to match the symmetry of the field, making the integral easy to calculate. In contrast to Coulomb's law (which gives us the electric field if the charge is known), Gauss's law can tell us how much charge is in a region if the electric field is known, so Gauss's law is useful "in both directions." Figure 25-9 illustrates another unusual feature of Gauss's law. Although q_{in} is the charge *inside* the Gaussian surface, the $\mathbf{E}$ that appears in the integral is due to *all* charges in the vicinity, *both inside and outside* the Gaussian surface!

Our examples deal only with symmetrical distributions of charges—the only kind that are easy to calculate with Gauss's law (though Gauss's law holds true in *all* cases). But note that, *just by reasoning from symmetry alone*, we can usually deduce the particular spatial form that the field must have before we actually calculate it. Thus we can choose a Gaussian surface that matches the field configuration, which makes Gauss's law easy to calculate.

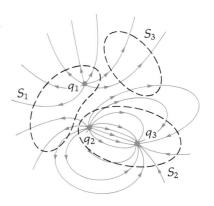

FIGURE 25-9
The total electric flux Φ_E passing through a Gaussian surface depends on only the *net* charge inside that closed surface. Thus the net flux is q_1/ε_0 for surface S_1 and $(q_2 + q_3)/\varepsilon_0$ for surface S_2. For surface S_3, the same number of lines enter the surface as leave the surface (there are no charges inside), so the net flux Φ_E is zero for S_3.

TABLE 25-1 Notation for Charge Distributions

Distribution	Symbol	Units	Element of Charge	
Point charge	q	C	q	
Line charge	λ	C/m	$dq = \lambda\,dx$,	where dx = line element
Area charge	σ	C/m^2	$dq = \sigma\,dA$,	where dA = area element*
Volume charge	ρ	C/m^3	$dq = \rho\,dV$,	where dV = volume element*

* The area and volume elements must be chosen so that all parts of the elements have the *same* charge density throughout the element.

---

[3] This simple result is a consequence of the $1/r^2$ nature of the Coulomb field.

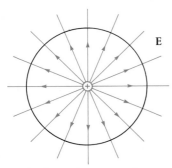

(a) The line of positive charge is perpendicular to the plane of the figure.

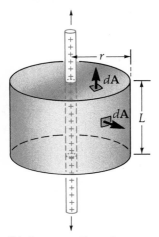

(b) A perspective view.

FIGURE 25-10
Example 25-3. A cylindrical Gaussian surface matches the symmetry of the field due to an infinitely long line of uniform positive charge. The surface area elements $d\mathbf{A}$ are either parallel or perpendicular to the field lines.

EXAMPLE 25-3

An infinitely long, straight line of uniform positive charge has a charge per unit length of λ (in units of charge per length). Find the electric field $\mathbf{E}$ at an arbitrary distance r from the line.

SOLUTION

From each incremental charge along the line, an electric field emanates equally in all directions. However, by symmetry, the superposition of the fields from all of the incremental charges results in a cancellation of fields parallel to the line of charge.[4] The result is a net field directed radially outward from the line. At all points at a given distance r from the line (in any direction), the field has the same magnitude.

Therefore, we match this symmetry with a Gaussian surface in the form of a cylinder of radius r and length L whose axis is the line of charge, Figure 25-10. At every point on the curved side of the cylinder, $\mathbf{E}$ is parallel to the area elements $d\mathbf{A}$, and it has the same magnitude everywhere. On the end caps of the cylinder, $\mathbf{E}$ is perpendicular to $d\mathbf{A}$ everywhere. The net charge q_{in} inside the cylinder is λL. Applying Gauss's law, we obtain

$$\oint \mathbf{E} \cdot d\mathbf{A} = \frac{q_{in}}{\varepsilon_0}$$

$$\underbrace{\int_{\text{curved side}} \mathbf{E} \cdot d\mathbf{A}}_{\cos 0° = 1} + \underbrace{\int_{\text{end caps}} \mathbf{E} \cdot d\mathbf{A}}_{\cos 90° = 0} = \frac{\lambda L}{\varepsilon_0}$$

$$E(2\pi r L) \quad + \quad 0 \quad = \frac{\lambda L}{\varepsilon_0}$$

Solving for E gives $\quad E = \boxed{\dfrac{\lambda}{2\pi\varepsilon_0 r}} \quad \begin{pmatrix}\text{radially}\\\text{outward}\end{pmatrix} \quad$ (25-10)

Note how much simpler this solution is than the Coulomb's-law approach used in Examples 24-4 and 24-5. The solution using Gauss's law is simple only because *we carefully choose a Gaussian surface that matches the symmetry of the electric field* (which we can determine ahead of time by symmetry reasoning), thus making the actual calculation of $\int \mathbf{E} \cdot d\mathbf{A}$ very easy. Even though Gauss's law holds true for all cases, it is only an easy calculation for fields that have obvious symmetries.

EXAMPLE 25-4

A uniform volume charge density ρ (in units of charge per volume) exists throughout the volume of an infinitely long cylinder of radius R, Figure 25-11. Find (a) the total charge Q_L in a length L of the cylinder and (b) the electric field E at a radius $r < R$.

---

[4] Another way of stating the symmetry argument is to point out that there is no asymmetry in the charge distribution that would make field lines have a component parallel to the wire in one direction instead of the opposite direction. Because the charge distribution is symmetrical along the line, the only way to make the field match this symmetry is with field lines that extend radially outward.

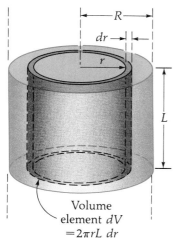
(a) The volume element dV is a thin cylindrical shell of radius r, length L, and thickness dr.

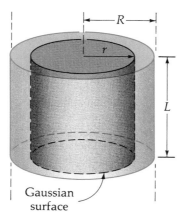
(b) The Gaussian surface is a cylinder of radius r, length L.

FIGURE 25-11
Example 25-4. A uniform charge density ρ exists throughout the volume of an infinitely long cylinder of radius R.

SOLUTION

(a) For volume charge distributions, $Q = \int \rho\, dV$. Here, we choose a volume element dV in the form of a thin cylindrical shell of radius r, thickness dr, and length L.

$$Q = \int \rho\, dV = \int_0^R \rho 2\pi r L\, dr = 2\pi\rho L \int_0^R r\, dr = \boxed{\pi\rho L R^2}$$

(b) By symmetry, we conclude that the field is radially outward and that, for a given value of r, **E** has the same magnitude everywhere. So we choose a Gaussian surface that matches this symmetry. It is a cylinder of radius r and length L that [from part (a)] encloses a charge q_{in} just within the radius r, or $q_{\text{in}} = \pi\rho L r^2$.

$$\oint \mathbf{E}\cdot d\mathbf{A} = \frac{q_{\text{in}}}{\varepsilon_0}$$

$$\underbrace{\int_0^R \mathbf{E}\cdot d\mathbf{A}}_{\cos 90° = 0} + \underbrace{\int_0^R \mathbf{E}\cdot d\mathbf{A}}_{\cos 0° = 1} = \frac{\pi\rho L r^2}{\varepsilon_0}$$

$$0 \quad + \quad E(2\pi r L) = \frac{\pi\rho L r^2}{\varepsilon_0}$$

Solving for E gives
$$E = \boxed{\left(\frac{\rho}{2\varepsilon_0}\right) r}$$

So, within the cylinder, the field is directly proportional to the distance r from the axis. Outside the cylinder, the field is the same as for Example 25-1. Figure 25-12 shows a plot of E vs. r.

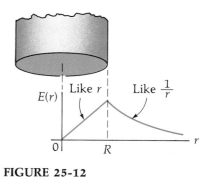

FIGURE 25-12
Examples 25-3 and 25-4. A graph of the electric field E vs. r for a uniform volume charge density throughout an infinitely long cylinder of radius R. In the graph, $E(r)$ is positive when $\mathbf{E}(r)$ is directed outward.

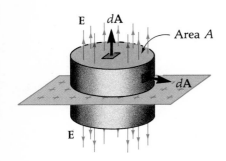

FIGURE 25-13
Example 25-5. The electric field produced by a very large plane sheet of uniform positive charge density σ. The Gaussian surface is in the form of a cylinder with flat end faces, placed symmetrically above and below the plane, matching the symmetry of the field. No field lines penetrate the curved sides of the cylinder, and the field lines are perpendicular to the end faces. (We could equally well have chosen a *rectangular* box with sides parallel to the field lines, or any other shape of box as long as the sides are parallel to the field lines and the end faces are perpendicular to the field lines.)

EXAMPLE 25-5

Find the electric field **E** produced by a very large (essentially infinite) sheet of uniform positive charge density σ (in units of charge per area).

SOLUTION

From symmetry we reason that, as long as we are not near an edge, the electric field must extend perpendicularly away from the plane on both sides. (There is no asymmetry that would cause the field lines to bend to one side or the other as they extend away from the positive charges.) We match the symmetry of this field by considering a Gaussian surface in the form of a cylinder, of cross-sectional area A, whose axis is perpendicular to the plane and whose ends are equidistant from the plane,[5] Figure 25-13. The net charge enclosed by the surface is σA. By symmetry, the field emerges uniformly and perpendicularly from each end and is tangent to the curved side of the cylinder. Applying Gauss's law, we obtain

$$\oint \mathbf{E} \cdot d\mathbf{A} = \frac{q_{in}}{\varepsilon_0}$$

$$\underbrace{\int_{\text{both ends}} \mathbf{E} \cdot d\mathbf{A}}_{\cos 0° = 1} + \underbrace{\int_{\text{curved side}} \mathbf{E} \cdot d\mathbf{A}}_{\cos 90° = 0} = \frac{\sigma A}{\varepsilon_0}$$

For the first integral, **E** is uniform and has the same magnitude over each end cap, so E may be brought out from under the integral sign $E \int dA = E(2A)$ (accounting for both end caps). The second integral is zero because $\cos 90° = 0$.

$$E(2A) + 0 = \frac{\sigma A}{\varepsilon_0}$$

Solving for E gives $\qquad E = \boxed{\dfrac{\sigma}{2\varepsilon_0}} \qquad \begin{pmatrix}\text{away from the plane,}\\ \text{above and below}\end{pmatrix}$ (25-11)

Because the distance from the surface does not appear in the expression, we conclude that the field has the same constant value for all distances on either side of the plane of charge.

EXAMPLE 25-6

A uniform positive charge density σ is in static equilibrium on the surface of a very large plane conductor. Find the electric field just above the surface.

SOLUTION

In the static case, no electric field **E** can exist within a conductor (because it would make the conduction charges *move*). So field lines extend away from the conductor, perpendicular to the surface, Figure 25-14. From symmetry considerations, we choose a cylindrical Gaussian surface as in the previous example, but

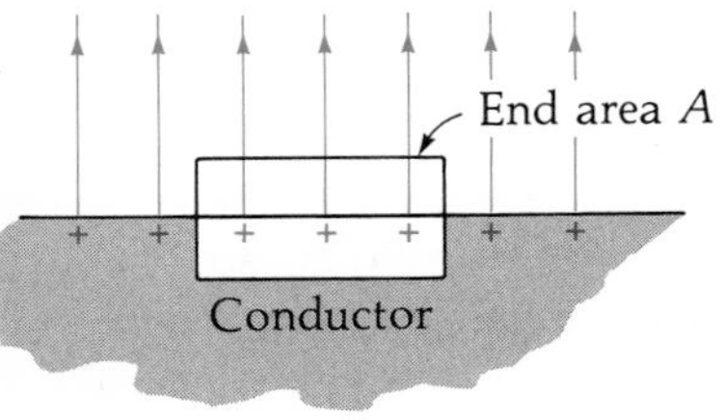

FIGURE 25-14
Example 25-6. A uniform positive surface charge density σ on a very large plane conductor.

---

[5] We cannot be certain ahead of time that the field does not vary with distance from the plane. But we do know *from symmetry* that, if such a variation is present, it must be the same above and below the plane. So we stipulate that the end faces are *equidistant* from the plane.

we note that the field lines penetrate only one end area.

$$\oint \mathbf{E} \cdot d\mathbf{A} = \frac{q_{in}}{\varepsilon_0}$$

$$\underbrace{\int_{\substack{\text{top} \\ \text{end}}} \mathbf{E} \cdot d\mathbf{A}}_{\cos 0° = 1} + \underbrace{\int_{\substack{\text{bottom} \\ \text{end}}} \mathbf{E} \cdot d\mathbf{A}}_{(E=0)} + \underbrace{\int_{\substack{\text{curved} \\ \text{side}}} \mathbf{E} \cdot d\mathbf{A}}_{\cos 90° = 0} = \frac{\sigma A}{\varepsilon_0}$$

$$EA \quad + \quad 0 \quad + \quad 0 \quad = \frac{\sigma A}{\varepsilon_0}$$

ELECTRIC FIELD JUST ABOVE A CHARGED CONDUCTOR

$$E = \frac{\sigma}{\varepsilon_0} \quad \begin{bmatrix} \text{perpendicular} \\ \text{to the surface} \\ \text{of the conductor} \end{bmatrix} \quad (25\text{-}12)$$

This field is twice the value that we found in the previous example, and it has a constant value for all distances above the infinite plane conductor.

Equation (25-12) is also valid for curved conductors in which σ may vary from point to point. *At any location, the field $\mathbf{E}$ just outside the conducting surface is σ/ε_0, normal to the surface.* Regarding this equation, it seems paradoxical that we can express the electric field just above the surface of a conductor solely in terms of the local surface charge density σ alone, even though that field is due to all charges in the vicinity: both on the conductor's surface *and anywhere else nearby!*

EXAMPLE 25-7

A uniform positive charge density ρ (in units of charge per volume) exists throughout a spherical volume of radius R, Figure 25-15. Find the electric field (a) outside the sphere and (b) inside the sphere.

SOLUTION

By symmetry, we conclude that the field $\mathbf{E}$ can only be radially outward, both inside and outside the sphere. Furthermore, for a given value of r, $\mathbf{E}$ has the same magnitude everywhere. To match this symmetry, we choose a Gaussian surface in the form of a sphere of radius r, centered on the spherical volume.
(a) For $r > R$: The total charge q_{in} inside the spherical volume (surface a) is $Q = \int \rho \, dV = \rho \int_0^R 4\pi r^2 \, dr = \rho(\frac{4}{3}\pi R^3)$. Applying Gauss's law and noting that $\oint \mathbf{E} \cdot d\mathbf{A} = E \oint dA = E(4\pi r^2)$, we find

$$\oint \mathbf{E} \cdot d\mathbf{A} = \frac{q'}{\varepsilon_0}$$

$$E(4\pi r^2) = \frac{Q}{\varepsilon_0}$$

Solving for E gives

$$E = \frac{Q}{4\pi\varepsilon_0 r^2}$$

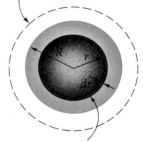

(a) To find the total charge Q in the sphere, we integrate $Q = \int_0^R \rho \, dV$, where the *volume element* dV is a thin spherical shell of radius r and thickness dr.

(b) Gauss's law involves only the charge inside the Gaussian surface.

FIGURE 25-15
Example 25-7. A uniform positive charge density throughout a spherical volume of radius R.

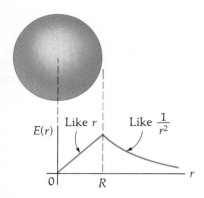

FIGURE 25-16
A graph of the electric field E vs. r for a uniform positive charge density throughout a spherical volume of radius R. Outside the sphere, the field is the same as for a point charge Q at the center. (This holds true only for charge distributions that have *spherical symmetry*.) In the graph, $E(r)$ is positive when $\mathbf{E}(r)$ is directed radially outward.

This is just the *inverse-square-law field for a point charge Q concentrated at the center of the sphere*.[6] To obtain the answer in terms of the given parameters, we substitute $Q = \rho(\tfrac{4}{3}\pi R^3)$ to obtain

$$\text{For } r > R: \quad E = \frac{(\rho \tfrac{4}{3}\pi R^3)}{4\pi\varepsilon_0 r^2} = \boxed{\left(\frac{\rho R^3}{3\varepsilon_0}\right)\frac{1}{r^2}} \quad \binom{\text{radially}}{\text{outward}} \quad (25\text{-}13)$$

(b) For $r < R$: To match the symmetry of **E**, we choose a Gaussian surface in the form of a sphere of radius $r < R$ (surface b). Gauss's law involves only the charge q_{in} inside this surface. From part (a), this is the integral $q' = \int_0^r \rho \, dV$, where the upper limit is r instead of R. Thus, $q_{\text{in}} = \rho(\tfrac{4}{3}\pi r^3)$.

$$\oint \mathbf{E} \cdot d\mathbf{A} = \frac{q_{\text{in}}}{\varepsilon_0}$$

$$E(4\pi r^2) = \frac{\rho 4\pi r^3}{3\varepsilon_0}$$

$$\text{For } r < R: \quad E = \boxed{\left(\frac{\rho}{3\varepsilon_0}\right)r} \quad \binom{\text{radially}}{\text{outward}} \quad (25\text{-}14)$$

Thus, inside the sphere of uniform charge, the field is *directly proportional to the distance r from the center*. A graph of E vs. r is shown in Figure 25-16.

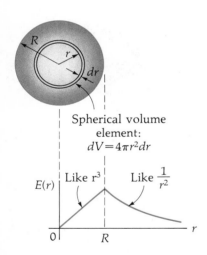

FIGURE 25-17
Example 25-8. A graph of E vs. r for a spherical charge distribution in which the charge density varies as $\rho = Br^2$. Outside the sphere, the field is the same as if the total charge Q were concentrated at the point at the center. (This holds true only for charge distributions that have *spherical symmetry*.) In the graph, $E(r)$ is positive when $\mathbf{E}(r)$ is radially outward.

EXAMPLE 25-8

A positive charge density exists throughout a spherical volume of radius R. The charge density ρ is not constant, but varies with the radius as $\rho = Br^2$, where B is a constant, Figure 25-17. Find (a) the SI units of the constant B and (b) the total charge Q in the sphere. (c) Find the electric field for $r \leq R$.

SOLUTION

(a) Since ρ is in units of charge per volume, solving for B we have

$$B = \frac{\rho}{r^2} = \frac{(\text{C/m}^3)}{\text{m}^2} = \boxed{\frac{\text{C}}{\text{m}^5}}$$

(b) Because the charge density ρ varies with the distance r from the center, we must sum elements of charge dq contained within volume elements dV, where *all of each volume element is the same distance r from the center.* We do this so that the charge density ρ has the same value throughout the volume element dV. Thus we choose elements in the form of a thin spherical shell of radius r and thickness dr (see Figure 25-17a). Its volume is $dV = 4\pi r^2 \, dr$,

---

[6] Recall an analogous result for the gravitational field due to a uniform spherical mass, Section 16.5, in which the external field is the same as if the total mass were concentrated at a point at the center. Too bad that Newton did not have Gauss's law to use. The 20-year delay in publishing his *Principia* was probably due, in part, to Newton's difficulty in trying to sum the gravitational forces on the moon due to the earth's mass distributed throughout its volume. Newton had to invent the calculus to solve this problem; his notation was very cumbersome compared with the modern version of calculus.

and the charge dq within dV is $dq = \rho 4\pi r^2 \, dr$. Substituting $\rho = Br^2$ and summing over all such shells from $r = 0$ to $r = R$, we obtain

$$Q = \int dq = \int \rho \, dV = \int_0^R Br^2(4\pi r^2) \, dr = 4\pi B \int_0^R r^4 \, dr$$

$$Q = 4\pi B \left(\frac{r^5}{5}\right)\bigg|_0^R = \boxed{\frac{4}{5}\pi BR^5}$$

(c) We choose a spherical Gaussian surface of radius $r < R$. The charge q_{in} inside this surface is

$$q_{in} = \int_0^r \rho \, dV = \int_0^r Br^2(4\pi r^2) \, dr = \tfrac{4}{5}\pi Br^5$$

Applying Gauss's law gives

$$\oint \mathbf{E} \cdot d\mathbf{A} = \frac{q_{in}}{\varepsilon_0}$$

$$E(4\pi r^2) = \frac{4\pi Br^5}{\varepsilon_0}$$

Solving for E, we get

$$E = \boxed{\frac{Br^3}{\varepsilon_0}}$$

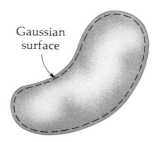

FIGURE 25-18
The dashed line is a Gaussian surface just below the surface of a solid, insulated, charged conductor of arbitrary shape. By Gauss's law, no charges can be at rest anywhere within the conductor. Charges at rest reside only at the surface.

25.4 Gauss's Law and Conductors

When charges are placed on an insulated metal conductor, they initially set up electric fields that move the charges about until *electrostatic equilibrium* is achieved, in which all charges are at rest.[7] This redistribution occurs extremely rapidly; for most circumstances the time is negligible. Gauss's law shows that *an excess charge in static equilibrium on a conductor resides entirely at the outer surface of the conductor*. To see this, consider an insulated metallic conductor of arbitrary shape, Figure 25-18. The dashed line shows a Gaussian surface that lies barely below the surface of the conductor, arbitrarily close to the surface. Under electrostatic equilibrium there can be no electric field in the conductor, so there are no field lines at *any* point on the Gaussian surface, and $\oint \mathbf{E} \cdot d\mathbf{A} = 0$. The same is true for any other Gaussian surface drawn deeper within the conductor. From Gauss's law we therefore conclude that there is no net charge anywhere within the conductor. Thus the net charge must reside only at the surface itself.[8]

In the static case, electric field lines always intersect the surface of a conductor at right angles, terminating on charges of the appropriate sign, Figure 25-19. If there were a tangential component of the electric field, it would cause conduction electrons at the surface to move, violating the static condition.

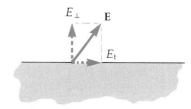

(a) There can be no tangential component E_t at the surface of a conductor because this would cause motion of free electrons in the surface.

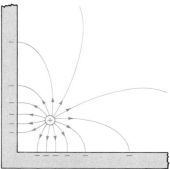

(b) A point charge near grounded conducting planes at right angles. For the static case, electric field lines always meet charges on a conducting surface at right angles to the surface.

FIGURE 25-19
The electric field near a conducting surface for *electrostatic equilibrium* conditions.

---

[7] Our discussion holds true for both positive and negative charges. Even though only negatively charged electrons are free to move in a metal, a positively charged conductor (which has a deficiency of electrons) has some atoms that lack an electron and thus are the sites of the positive charges. Since electrons are free to move in a metal, an electron from a neighboring atom will be attracted to the electron-deficient atom, leaving behind a positively charged atom. In effect, the positive charge has "moved" to the neighboring atom with the same mobility that electrons have in metals. Therefore, positive charges as well as negative charges may be considered "free" to distribute themselves within a conductor until static conditions are achieved.

[8] Are the charges *on top* of the surface or just barely *under* the surface? To answer such a question requires a definition on an atomic scale of what we mean by the word *surface*. We choose to avoid such a discussion. The phrases "on the surface" and "at the surface" both convey the right idea for our purposes.

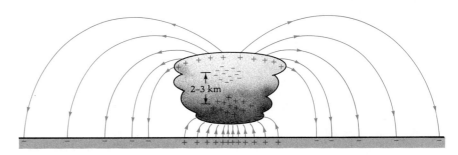

FIGURE 25-20
Cloud physicists have long sought an explanation of thunderstorm electricity. Due to the difficulty of obtaining accurate measurements, the mechanism that generates the separation of charges is controversial and not well understood. In any case, a combination of mechanical, thermodynamical, and maybe chemical energies is involved. The most common distribution of charges in a thundercloud is shown in this figure, though sometimes the polarity is reversed. (Field lines within the cloud are not shown.) The electric field above the earth's surface thus reverses direction between a cloudy day and a sunny day. Under a storm cloud, the electric field strength can be 10^4 N/C and more. When the electric field reaches the order of 10^5 to 10^6 N/C, lightning strokes can occur, both between the cloud and the ground and within the cloud itself. Typically, several tens of coulombs are neutralized in a stroke. Peak currents in a stroke are often 10–20 kiloamps. Worldwide there are probably about 100 lightning flashes occurring at any time, lasting roughly 10^{-2} s to 2 s, made up of 1 to 20 strokes per flash. Lightning occurs over continents about 10 times the frequency of lightning over oceans. (Adapted from J. V. Iribarne and H. R. Cho, *Atmospheric Physics*, Reidec Publishing Co., 1980.) Also see Figure 28-12.

What happens if we place an uncharged conducting slab in an external electric field? Originally the free electrons are distributed uniformly throughout the material, and the slab is electrically neutral everywhere. However, in response to the external field, the free electrons will quickly move in the direction *opposite* to **E** (because $\mathbf{F} = (-e)\mathbf{E}$). They accumulate on the surface to form a negative surface charge density on one face[9] and a positive charge density on the other face, Figure 25-21. These induced surface charges increase until they create an internal electric field of their own that is equal and opposite to the external field, so that *the net electric field inside the conductor is zero when all charges are at rest.*

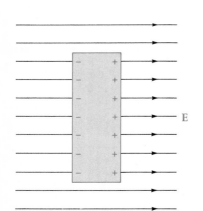

FIGURE 25-21
An uncharged metal slab placed in an external electric field acquires surface charge densities as shown. Inside the conductor, the internal field produced by these induced surface charges exactly cancels the original field **E**, resulting in a zero electric field within the conducting slab.

EXAMPLE 25-9

A hollow conducting sphere is surrounded by a larger concentric, spherical, conducting shell as shown in Figure 25-22a. The inner sphere has a net negative charge of $-Q$ and the outer sphere has a net positive charge of $+3Q$. The charges are in electrostatic equilibrium. Using Gauss's law, find the charges and the electric fields everywhere.

---

[9] Electrons are normally bound to the surface of the material, though with a sufficiently strong external field they can be pulled out of the surface in a process called *field emission*. We assume that this does not happen here.

SOLUTION

The spherical symmetry of the conductors ensures that all electric fields are spherically symmetric, either radially inward or outward.

REGION ①: A spherical Gaussian surface just inside the inner sphere encloses no net charge. By symmetry, **E** on this surface (if it existed) would have to have the same value everywhere. But $\oint \mathbf{E} \cdot d\mathbf{A} = 0$, implying that **E** must be zero everywhere in region ①.

REGION ②: Since no field lines exist in a conductor in the static case, when we apply Gauss's law to a spherical Gaussian surface barely inside the outer surface, we find $\oint \mathbf{E} \cdot d\mathbf{A} = 0$, which implies that the charge $-Q$ must reside on its outer surface.

REGION ③: Since $\oint \mathbf{E} \cdot d\mathbf{A} = -Q/\varepsilon_0$ for a spherical Gaussian surface in region ③, symmetry requires that field lines must be radially inward as if there were a point charge $-Q$ at the center. Thus:

$$\mathbf{E}_{③} = -\frac{Q}{4\pi\varepsilon_0 r^2}\hat{\mathbf{r}}$$

REGION ④: Because $\mathbf{E} = 0$ within the conductor, $\oint \mathbf{E} \cdot d\mathbf{A} = 0$, implying zero net charge within a spherical Gaussian surface. So there must be a positive charge $+Q$ on the inner wall of the outer shell to balance the $-Q$ charge on the inner sphere.

REGION ⑤: The outer shell has a net positive charge of $+3Q$. Since $-Q$ is on its inner surface, $+2Q$ must reside on its outer surface. A spherical Gaussian surface at ⑤ encloses a net charge of $-Q + 3Q = +2Q$, implying (by symmetry) a radially outward field similar to one produced by a net positive point charge of $+2Q$ at the center:

$$\mathbf{E}_{⑤} = \frac{2Q}{4\pi\varepsilon_0 r^2}\hat{\mathbf{r}}$$

The charges and fields are sketched in Figure 25-22b.

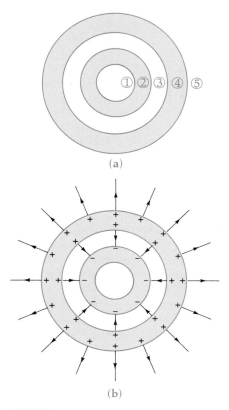

FIGURE 25-22
Example 25-9. Two concentric, conducting, spherical shells. The inner shell has a charge $-Q$ and the outer shell has a charge $+3Q$.

Summary

The *electric flux* Φ_E is a measure of *the number of electric field lines that penetrate a surface*. When the normal $\hat{\mathbf{n}}$ to a plane surface, the area element $d\mathbf{A} = \hat{\mathbf{n}}\,dA$, makes an angle θ with a uniform electric field **E**, the flux is

$$\Phi_E = EA\cos\theta$$

More generally, $\quad \Phi_E = \int_A \mathbf{E} \cdot d\mathbf{A}$

For a closed surface, the element $d\mathbf{A}$ is the *outward* normal to the surface.

Gauss's law: $\quad \oint_A \mathbf{E} \cdot d\mathbf{A} = \dfrac{q_{\text{in}}}{\varepsilon_0} \quad \begin{pmatrix}\text{where } A \text{ is any surface}\\ \text{enclosing a total charge}\\ q_{\text{in}} \text{ anywhere inside}\end{pmatrix}$

$\oint_A \mathbf{E} \cdot d\mathbf{A} = \dfrac{1}{\varepsilon_0}\int_V \rho\, dV \quad \begin{pmatrix}\text{where the charge}\\ \text{density } \rho \text{ is in the}\\ \text{volume } V \text{ enclosed}\\ \text{by the surface}\end{pmatrix}$

The notation for charge distributions:

Distribution	Symbol	Units	Element of Charge
Point charge	q	C	q
Line charge	λ	C/m	$dq = \lambda\,dx$
			(dx = line element)
Area charge	σ	C/m^2	$dq = \sigma\,dA$
			(dA = area element*)
Volume charge	ρ	C/m^3	$dq = \rho\,dV$
			(dV = volume element*)

* The area and volume elements must be chosen so that all parts of the elements have the *same* charge density throughout the element.

The *symmetry* of charge distributions often enables us to deduce the spatial form of the field before we actually calculate it. Thus a Gaussian surface can be chosen so it matches the field configuration, making Gauss's law easy to calculate. It can be perpendicular to the field everywhere and pass through points of equal field magnitude, or it can be chosen parallel to the field and thus contribute nothing to the integral.

The electric field just above a conducting surface that has a surface charge density σ is

$$E = \frac{\sigma}{\varepsilon_0}$$

The properties of charged conductors in *electrostatic equilibrium*:

(1) Excess charges in static equilibrium reside entirely at the outer surface of the conductor.
(2) The electric field everywhere inside a conductor is zero.
(3) Electric field lines just outside a conductor always intersect the surface of the conductor at right angles. The electric field has the magnitude $E = \sigma/\varepsilon_0$, where σ is the surface charge density at that location.

Questions

1. A small mass is at the center of a hollow, massive sphere. If another mass is placed external to the sphere, does the mass within the sphere experience a net gravitational force due to the presence of the external mass?
2. A charge is at the center of a hollow, uncharged metal sphere. If a charge is placed external to the sphere, does the charge within the sphere experience a net force?
3. Can the movement of a charge within a hollow conducting sphere alter the electric field outside the sphere?
4. A charge is deposited on a hollow metal sphere floating in oil. As a consequence of becoming charged, does the sphere float higher or lower or does it remain at the same level in the oil? Why?
5. How can the surface charge density on the outer surface of a hollow sphere be uniform while the surface charge density on the inner surface is not?
6. Answer Question 5 interchanging the words *outer* and *inner*.
7. Why is Gauss's law impractical for finding the electric field outside a charged metal cube?
8. Why, in general, is a charge within a hollow metal sphere attracted toward the walls of the sphere whether or not the sphere is charged?
9. Does the attraction of a small positive charge toward a large metal sphere necessarily mean that the sphere is negatively charged?

Problems

25.2 The Electric Flux
25.3 Gauss's Law

25A-1 In Figure 25-23, find the net flux Φ_E through each of the closed surfaces (a), (b), and (c).

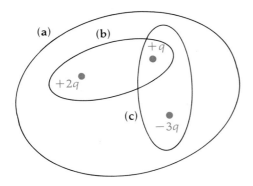

FIGURE 25-23
Problem 25A-1.

25A-2 A nonuniform electric field is in the $+x$ direction for positive x with magnitude of $20x$ N/C and the $-x$ direction for negative x with a magnitude of $20x$ N/C. A cubic box (nonconducting) of edge length 1 m is located with its center at the origin of the coordinate system and its edges parallel to the coordinate axes. (a) Make a sketch of the field. Calculate the total flux emerging from the faces of the box that are parallel to (b) the yz plane, (c) the xy plane, and (d) the zx plane. (e) Find the net charge inside the box.

25A-3 A uniform electric field $E = 30$ N/C exists parallel to the axis of a square pipe of side length $\ell = 5$ cm, Figure 25-24. Calculate the value of $\int \mathbf{E} \cdot d\mathbf{A}$ for the slanted face of the pipe to find the total electric flux Φ_E emerging from that face.

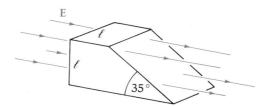

FIGURE 25-24
Problem 25A-3.

25A-4 A point charge $+Q$ is located at the center of a cubical Gaussian surface of edge length L. Suppose that 1200 electric field lines are drawn symmetrically from the charge. (a) Use a symmetry argument to find the number of field lines that emerge from one face of the cube (assuming that none coincides with edges or corners.) (b) What total number of field lines emerge from the total surface of the cube? (c) Suppose that the charge were displaced off-center, but still inside the cube. Which of the previous answers would be different? Discuss.

25A-5 Two infinite nonconducting plane sheets each have a uniform positive charge density σ. The sheets are parallel to each other. Use the superposition principle to find the electric field (a) between the sheets and (b) in the regions beyond the sheets.

25A-6 Solve the previous problem for the case in which the surface charge density on one sheet is changed to $-\sigma$.

25B-7 A uniform volume charge density ρ exists throughout a plane slab of thickness d that extends essentially to infinity in the $\pm y$ and $\pm z$ directions, Figure 25-25. The origin of the x axis is at the midplane of the slab. Find the electric field for positive values of x for (a) $0 < x < d/2$ and (b) for $x > d/2$.

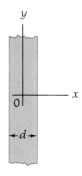

FIGURE 25-25
Problem 25B-7.

25B-8 A proton is in empty space very near the surface of the earth. (a) Find the total net charge Q that the earth would have to have (uniformly distributed over the surface) to produce an electric force of repulsion that would exactly balance the earth's gravitational force of attraction. (b) Would the same charge Q also balance forces on free protons situated at larger distances from the earth? Explain.

25.4 Gauss's Law and Conductors

25A-9 Consider an isolated conducting sphere of very large radius that possesses a uniform surface charge density σ (in coulombs per square meter). (a) Derive an expression for the electric field close to the surface of the sphere. (b) Derive an expression for the electric field close to a large, planar conducting sheet that has the same area and total charge as the sphere. Assume that the charge is distributed uniformly over the sheet, ignoring edge effects.

25A-10 Consider a hollow metallic sphere with a charge of $+10$ μC and a radius of 10 cm. The center of the sphere is at the origin of a Cartesian coordinate system. Within the sphere, at $x = 5$ cm, is a negative point charge of -3 μC. Find the electric field external to the sphere along the x axis. Make a qualitative sketch of the field lines inside and outside the sphere.

25A-11 On a clear, sunny day, there is a vertical electrical field of about 130 V/m pointing down over flat ground or water. (The field can vary considerably in magnitude and may be reversed if clouds are overhead.) What is the surface charge density on the ground for these conditions?

25B-12 A very long metal rod, radius R, has a uniform surface charge density σ. (a) Ignoring end effects, find the electric field **E** at a distance R from the surface of the cylinder. (b) Find the speed v such that an electron could travel in a circular orbit about the rod at a distance R from the rod's surface.

25B-13 The electric field near the earth's surface is due to a net surface charge density on the surface. The field may also vary with height because of free charges in the air (which terminate field lines) as shown in Figure 25-26. Suppose that at an altitude of 300 m above level ground the electric field is 100 N/C downward and that at 100 m above the ground the field is 150 N/C downward. (a) Use Gauss's law to find the average volume charge density ρ in the region between these altitudes. (b) Express this charge density in terms of a surplus or a deficiency of electrons per cubic meter.

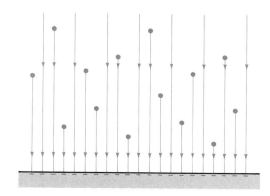

FIGURE 25-26
Problem 25B-13.

Additional Problems

25C-14 A point charge $+Q$ has 1200 electric field lines drawn symmetrically away from it in all directions. The center of a spherical Gaussian surface of radius r is located at a point $2r$ from the point charge, Figure 25-27. (a) How many field lines enter into the interior of this Gaussian surface? (Hint: see Appendix D for the definition of a solid angle Ω, measured in steradians. The whole solid angle surrounding a point is 4π steradians. The *conical solid angle* formed by a cone of half-vertex angle θ is $\Omega = 2\pi(1 - \cos \theta)$, measured in steradians.) (b) Find

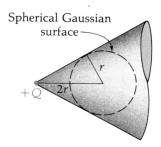

FIGURE 25-27
Problem 25C-14.

the *net* electric flux $\Phi_E = \oint \mathbf{E} \cdot d\mathbf{A}$ leaving the Gaussian surface, noting that flux lines entering are counted as negative and flux lines leaving are positive.

25C-15 As shown in Figure 25-28, a positive charge distribution exists within the volume of an infinitely long cylindrical shell between radii a and b. The charge density ρ is not uniform, but varies inversely as the radius r from the axis. That is, $\rho = \kappa/r$ for $a \leq r \leq b$, where κ is a constant in SI units. (a) Find the units of κ. (b) Find the total charge Q in a length L of the cylindrical shell. (c) Starting with Gauss's law, find the electric field E at a point r (for $a < r < b$).

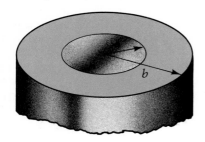

FIGURE 25-28
Problem 25C-15.

25C-16 An early (incorrect) model of the hydrogen atom, suggested by J. J. Thomson, proposed that a positive cloud of charge $+e$ was uniformly distributed throughout the volume of a sphere of radius R, with the electron an equal-magnitude negative point charge $-e$ at the center. (a) Using Gauss's law, show that the electron would be in equilibrium at the center and, if displaced from the center a distance $r < R$, would experience a restoring force of the form $F = -kr$. (b) Show that the force constant $k = e^2/4\pi\varepsilon_0 R^3$. (c) Find an expression for the frequency f of simple harmonic oscillations that an electron would undergo if displaced a short distance ($<R$) from the center and released. (d) Calculate a numerical value for R that would result in a frequency of 2.47×10^{15} Hz, the most intense line in the hydrogen spectrum.

25C-17 The Thomson model for the helium atom (see the previous problem) consists of a uniform positive charge distribution (total charge $+2e$) throughout the volume of a sphere of radius R. The two point electrons, each of charge $-e$, are symmetrically situated as shown in Figure 25-29. Show that the equilibrium separation distance d of the electrons is R.

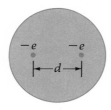

FIGURE 25-29
Problem 25C-17.

25C-18 Inside a sphere of radius R, the electric field **E** is radially outward and has a constant magnitude E_0 everywhere. Thus, $\mathbf{E} = E_0 \hat{\mathbf{r}}$, where $\hat{\mathbf{r}}$ is the unit vector in the outward radial direction. (a) Use Gauss's law to find the expression for the volume charge density $\rho(r)$ as a function of the radius r. (Hint: the fundamental theorem of calculus says that if $g(x) = \int_0^x f(t)\, dt$, then $dg/dx = f(x)$.) (b) Why does the center of the sphere present a difficulty?

25C-19 As shown in Figure 25-30, a uniform electric field **E** penetrates a closed hemisphere of radius R, entering the object perpendicular to the flat face. By direct integration over the curved surface, calculate the electric flux Φ_E that emerges through the curved surface of the hemisphere. Hint: *noting the symmetry*, choose an element of area $d\mathbf{A}$ in the form of a thin circular strip as shown so that the angle θ between **E** and $d\mathbf{A}$ is the same all over the strip. The area dA of the strip is its length, $2\pi(R \sin \theta)$, times its width, $ds = R\, d\theta$. Thus, $dA = 2\pi R^2 \sin \theta\, d\theta$. In the summation, the angle θ varies between 0 and $\pi/2$. The result is, of course, the same as the magnitude of the (negative) flux entering the flat surface, $-E(\pi R^2)$, so that the total flux over the entire closed surface is zero.

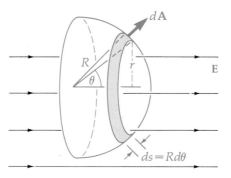

FIGURE 25-30
Problem 25C-19.

25C-20 A sphere of radius $2a$ is made of a nonconducting material that has a uniform volume charge density ρ. (Assume that the material does not affect the electric field.) A spherical cavity of radius a is now removed from the sphere as shown in Figure 25-31. Show that the electric field within the cavity is uniform and is given by $E_x = 0$ and $E_y = \rho a/3\varepsilon_0$. (Hint: the field within the cavity is the superposition of the field due to the original uncut sphere, plus the field due to a sphere the size of the cavity with a uniform negative charge density $-\rho$. This vector relation will be useful: $r\hat{\mathbf{r}} = x\hat{\mathbf{x}} + y\hat{\mathbf{y}}$.)

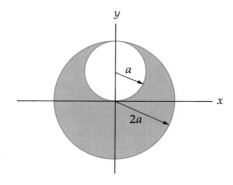

FIGURE 25-31
Problem 25C-20. The center of the spherical cavity is at the point $y = a/2$.

CHAPTER 26

Electric Potential

One electron to another:
"Now that we're together, we've achieved our greatest potential!"

26.1 Introduction

The concept of energy is so useful in physics and engineering that we now investigate the energy involved in the interaction of electric charges and fields. In mechanics, the principle of energy conservation and the work–energy relation gave us an alternative approach to solving problems: instead of working directly with the forces involved (which are *vectors*), we can use mechanical energies (which are *scalars*) in our analysis. Scalar quantities are usually easier to deal with than vectors. The analogies between gravitational and electrical forces are close. Both are inverse-square-law, *conservative* forces, and each conservative force has an associated *potential energy function* that we find useful in analyzing interactions.

26.2 The Electric Potential

Consider a test charge q_0 placed in an electric field **E**. The work dW done by the electric force $\mathbf{F} = q_0 \mathbf{E}$ in moving an incremental distance $d\boldsymbol{\ell}$ is

$$dW = \mathbf{F} \cdot d\boldsymbol{\ell} = q_0 \mathbf{E} \cdot d\boldsymbol{\ell} \tag{26-1}$$

By definition, the work done by a conservative force equals the *negative* of the change in *electric potential energy* dU, so we have

$$dU = -q_0 \mathbf{E} \cdot d\boldsymbol{\ell} \tag{26-2}$$

In moving a finite distance from position a to position b, we have

ELECTRIC POTENTIAL ENERGY U
$$U_b - U_a = -\int_a^b q_0 \mathbf{E} \cdot d\boldsymbol{\ell} \tag{26-3}$$

The SI units for energy are *joules* (J). Note that the right side of Equation (26-3) also is in energy units because the units of $q_0 E\, d\ell$ are $(C)(N/C)(m) = N \cdot m = J$. Just as we did for gravitational potential energy, we can define any convenient location as the zero reference position of q_0 and calculate the potential energy for all other positions relative to that zero reference location.

For analyzing fields and charges, however, there is a related concept that is even more useful: the *electric potential V*. Note that the work done by the electric field, $\int q_0 \mathbf{E} \cdot d\boldsymbol{\ell}$, is proportional to the magnitude of q_0. To eliminate this dependence on the property of a particle and obtain a more useful quantity that is related just to the field itself, we define the **electric potential** V as *the limit of the work per unit charge as* $q_0 \to 0$ (not to be confused with U, the electric potential *energy*). For a differential distance $d\boldsymbol{\ell}$, the change in the electric potential dV is

$$dV = \lim_{q_0 \to 0} \frac{dU}{q_0} = -\mathbf{E} \cdot d\boldsymbol{\ell} \qquad (26\text{-}4)$$

For finite distances from a to b, the potential V changes by

ELECTRIC POTENTIAL V
$$V_b - V_a = -\int_a^b \mathbf{E} \cdot d\boldsymbol{\ell} \qquad (26\text{-}5)$$

The SI units of electric potential, often called just the *potential*, are *joules per coulomb* (J/C), and are also given the name *volt* (V).[1] Because of the minus sign in $\Delta V = -\int \mathbf{E} \cdot d\boldsymbol{\ell}$, electric field lines always point in the direction of *decreasing* potential. Only potential differences ΔV are physically meaningful, and we can designate any convenient location for the zero reference potential. For fields that are due to local charge distributions, the zero reference is usually taken to be far from the charges[2] at infinity: $V \equiv 0$ at $r = \infty$.

When we analyze electric fields, the *potential* is a scalar quantity that is often more convenient to use than the electric potential *energy*. (In mechanics it is the other way around: potential energy is more useful than potential.) The two concepts differ by the factor q:

RELATION BETWEEN V **and** U
$$dV = \frac{dU}{q} \qquad (26\text{-}6)$$

When evaluating $-\int_a^b \mathbf{E} \cdot d\boldsymbol{\ell}$, we recall that the electric field is conservative. This means that the integral is *independent of the path taken between points* a *and* b. This feature will allow us to choose paths that are particularly easy to calculate. For example, consider the radially outward field $\mathbf{E}$ due to a point charge, Figure 26-1a, with the line integral $d\boldsymbol{\ell}$ along the path from a to b. In Figure 26-1b we show an increment $d\boldsymbol{\ell}$ at an arbitrary angle to the field $\mathbf{E}$. The vector $d\boldsymbol{\ell}$ has a radial component $d\mathbf{r}$ in the direction of the field ①, and two other components[3] in the directions ② and ③, each at right angles to $\mathbf{r}$. Because $\mathbf{E}$ has only a radial component E_r, the dot product $\mathbf{E} \cdot d\boldsymbol{\ell}$ is zero for components perpendicular to $\mathbf{r}$. So we can choose the easier calculation along the

TABLE 26-1 Typical Electrical Potential Differences

Nerve impulses	50 mV
Flashlight battery	1.5 V
Car battery	12 V
House wiring	120 V
Electric eel	600 V
Transmission lines	
within a city	4.4 kV
cross country	120 kV
high voltage	10^6 V
Lightning	
(cloud to ground)	10^8–10^9 V

---

[1] This unit honors Count Allesandro Volta (1745–1827), the Italian physicist who invented the *voltaic cell* (the forerunner of our modern battery), which provided the first practical method of obtaining a steady electric current. Before this time, scientists could only experiment with intermittent spark discharges or with lightning bolts.

[2] When we discuss electric circuits in a later chapter, a particular point in the circuit is assigned the zero reference $V \equiv 0$, and it is often physically connected to a metal water pipe that, in turn, is in contact with the earth. The circuit is said to be *grounded* (symbol: ⏚). Three-pronged electrical plugs achieve this grounding when one of the outlet connections is wired to a water pipe.

[3] In spherical polar coordinates, these would be the $\hat{\theta}$ and $\hat{\phi}$ directions.

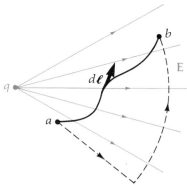
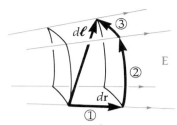

(a) Because **E** is a conservative field, the potential difference $\Delta V = -\int_a^b \mathbf{E} \cdot d\boldsymbol{\ell}$ is the same when calculated along the solid path as when calculated along the (easier-to-calculate) dashed path.

(b) The element $d\boldsymbol{\ell}$ has components in three mutually perpendicular directions: ①, ②, and ③.

FIGURE 26-1
Calculating potential differences in the Coulomb field **E** of a point charge q.

dashed path. The integral $\int \mathbf{E} \cdot d\boldsymbol{\ell}$ for the part that is perpendicular to **E** is zero ($\cos 90° = 0$), so we are left with just the radial part, which is

For radial fields: $\quad \int \mathbf{E} \cdot d\boldsymbol{\ell} \Rightarrow \int E_r \, dr \quad \begin{pmatrix}\text{where } E_r \text{ is positive} \\ \text{if } \mathbf{E} \text{ is directed} \\ \text{radially outward}\end{pmatrix} \quad$ (26-7)

We will always look for the easy paths—along **E** itself, or at right angles to **E**—to make the calculation a simple one. For fields that are linear with **E** along, say, the x axis,

For linear fields: $\quad \int \mathbf{E} \cdot d\boldsymbol{\ell} \Rightarrow \int E_x \, dx \quad$ (26-8)

From these relations, we see that electric field may be expressed in units of $E = $ (potential difference)/(distance) $=$ *volts/per meter* (V/m). These units are perhaps more commonly encountered than the equivalent units *newtons per coulomb* (N/C) used in Chapter 24.

EXAMPLE 26-1

(a) Find the electric potential V in the vicinity of a point charge q where $V \equiv 0$ at $r = \infty$. (b) Find the electric potential energy U of a system of two point charges, q_1 and q_2, a distance r apart.

SOLUTION

(a) From Equation (26-5), we have

$$V_b - V_a = -\int_a^b \mathbf{E} \cdot d\boldsymbol{\ell}$$

Here we choose the position a at infinity, where $V_a \equiv 0$, and the position b at any arbitrary distance r from the charge. The expression for the field E

due to a point charge is $E = kq/r^2$ (radially outward). Noting that the direction of $d\mathbf{r}$ is contained in the limits of integration, we substitute values in the above expression and integrate along a radially inward line to obtain

$$V - 0 = -\int_\infty^r \frac{kq}{r^2} dr = -kq \int_\infty^r \frac{1}{r^2} dr = -kq\left(-\frac{1}{r}\right)\bigg|_\infty^r = \frac{kq}{r} - 0$$

ELECTRIC POTENTIAL V NEAR A POINT CHARGE q $\qquad V = \dfrac{kq}{r} = \dfrac{q}{4\pi\varepsilon_0 r}$ $\qquad$ (26-9)
($V \equiv 0$ at $r = \infty$)

The potential V near a positive charge is positive, and it drops as $1/r$ to zero at infinity.

Using the *superposition principle*, we may generalize this result to express the electric potential V at a point due to several nearby point charges:

ELECTRIC POTENTIAL AT A POINT DUE TO SEVERAL NEARBY POINT CHARGES $\qquad V = k \sum_i \dfrac{q_i}{r_i}$ $\qquad$ (26-10)

Because potential is a scalar, this is merely an *algebraic* sum of scalars rather than a *vector* sum of electric fields that would be necessary to find the net field $\mathbf{E}$ due to several charges. Thus it is easier to calculate V than $\mathbf{E}$.

(b) To find the potential energy of two point charges a distance r apart, we use the fact that to bring a charge q from infinity (where $V \equiv 0$) to a location where the potential is V requires an amount of work qV. Thus, to bring a second charge q_2 from infinity to a distance r from a stationary charge q_1 where the potential is V, we have

ELECTRIC POTENTIAL ENERGY U OF TWO CHARGES SEPARATED A DISTANCE r $\qquad U = k\dfrac{q_1 q_2}{r} = \boxed{\dfrac{q_1 q_2}{4\pi\varepsilon_0 r}}$ $\qquad$ (26-11)

To generalize to the electric potential energy U of a system of point charges, to assemble such a system (starting with the charges infinitely far from each other), we add the potential energy associated with each *pair* of charges. For three point charges, this is

$$U = k\frac{q_1 q_2}{r_{12}} + k\frac{q_2 q_3}{r_{23}} + k\frac{q_1 q_3}{r_{13}} \qquad (26\text{-}12)$$

The total electric potential energy U of a system of point charges is the work required to bring the charges, one at a time, from an infinite separation to their final positions.

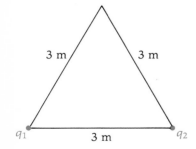

FIGURE 26-2
Example 26-2.

EXAMPLE 26-2

Two point charges, $q_1 = 2~\mu\text{C}$ and $q_2 = 3~\mu\text{C}$, are located, respectively, at two corners of an equilateral triangle of side length $\ell = 3$ m, Figure 26-2. (a) Find the potential V at the other corner of the triangle ($V \equiv 0$ at $r = \infty$). (b) Find the work required to bring a third charge $q_3 = 4~\mu\text{C}$ from infinity to the unoccupied corner of the triangle. (c) Find the total electric potential energy of the system of three charges.

SOLUTION

(a) From Equation (26-10),

$$V = k\sum_i \frac{q_i}{r_i} = k\left[\frac{q_1}{\ell_1} + \frac{q_2}{\ell_2}\right]$$

$$V = \left(9 \times 10^9 \frac{\text{N}\cdot\text{m}^2}{\text{C}^2}\right)\left[\frac{2 \times 10^{-6}\text{ C}}{3\text{ m}} + \frac{3 \times 10^{-6}\text{ C}}{3\text{ m}}\right] = \boxed{1.5 \times 10^4 \text{ V}}$$

(b) The work required to bring q_3 from infinity to this location, where the potential is V, is

$$W = q_3 V = (4 \times 10^{-6}\text{ C})(1.5 \times 10^4\text{ V}) = \boxed{6.00 \times 10^{-2}\text{ J}}$$

(c) The total potential energy U of the system of three charges is

$$U = k\frac{q_1 q_2}{r_{12}} + \frac{q_2 q_3}{r_{23}} + \frac{q_1 q_3}{r_{13}}$$

$$U = \left(9 \times 10^9 \frac{\text{N}\cdot\text{m}^2}{\text{C}^2}\right)\left[\frac{(2 \times 10^{-6}\text{ C})(3 \times 10^{-6}\text{ C})}{3\text{ m}}\right.$$

$$+ \frac{(3 \times 10^{-6}\text{ C})(4 \times 10^{-6}\text{ C})}{3\text{ m}}$$

$$\left.+ \frac{(2 \times 10^{-6}\text{ C})(4 \times 10^{-6}\text{ C})}{3\text{ m}}\right]$$

$$U = \left(\frac{9 \times 10^9}{3}\right)[(6 + 12 + 8) \times 10^{-12}]\text{ J} = \boxed{7.80 \times 10^{-2}\text{ J}}$$

A *battery* is a device that provides an electric potential difference by means of certain chemical reactions inside the battery. Consider a 12-V automobile battery with one positive and one negative terminal. The "12 V" indicates the magnitude of the potential difference between the terminals of the battery, with the positive terminal at the higher potential. If the terminals are connected to parallel metal plates separated a distance d, charges will flow from the battery to the plates until the plates also acquire a potential difference of 12 V. These charges reside at the inner surfaces of the plates, creating an electric field between them as in Figure 26-3. The field is uniform in the central region if the separation d is small compared with other dimensions. (For this preliminary discussion, we will ignore the bulging of the field, called "fringing" effects, at the edges of the plates.) The next example makes use of this arrangement to further clarify the relation between $\mathbf{E}$ and V.

EXAMPLE 26-3

In Figure 26-3, a 12-V battery is connected to two large parallel plates separated 4 mm. (a) Find the magnitude of the electric field between the plates. (b) A proton is released from rest at the top plate and is accelerated by the electric force along the dotted-line path to the negative plate. Find the change in electric potential energy of the proton during this motion. (c) Show that the change in gravitational potential energy of the proton during this motion is negligible compared with the change in electric potential energy. (d) Find the speed of the proton just as it reaches the negative plate.

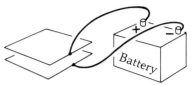

(a) A battery connected to two parallel metal plates transfers charge from one plate to the other until the potential difference between the plates equals the potential difference of the battery.

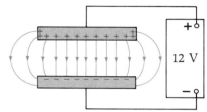

(b) A schematic diagram for the circuit of part (a). Most of the charges are at the inner surface of the plates, creating a field that is uniform except for some fringing-field effects near the edges. For plates whose edge lengths are large compared to the distance between the plates, the fringing fields are of negligible concern.

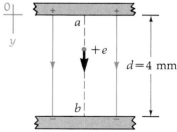

(c) The uniform electric field exerts a force $F = eE$ on the proton, accelerating it downward. (This force is very much greater than the force of gravity, which we ignore.)

FIGURE 26-3
Example 26-3. A battery connected to parallel metal plates creates an electric field between the plates.

SOLUTION

(a) The potential difference through which the proton moves is $V_b - V_a = -\int_a^b \mathbf{E} \cdot d\boldsymbol{\ell}$. Because the electric field is uniform, it can be brought out from under the integral sign:

$$V_b - V_a = -E \int_0^d dy = -Ed$$

Solving for the magnitude of E gives

$$|E| = \frac{|V_b - V_a|}{d} = \frac{12 \text{ V}}{4 \times 10^{-3} \text{ m}} = \boxed{3000 \, \frac{\text{V}}{\text{m}}}$$

(b) The change in potential energy ΔU of the proton is found from Equation (26-4):

$$\Delta U = q\,\Delta V = (1.60 \times 10^{-19} \text{ C})(-12 \text{ V}) = \boxed{-1.92 \times 10^{-18} \text{ J}}$$

(c) The change in gravitational potential energy for a proton moving vertically downward 4 mm is

$$\Delta U_g = mgh = (1.67 \times 10^{-27} \text{ kg})(9.8 \text{ m/s}^2)(-4 \times 10^{-3} \text{ m})$$
$$\Delta U_g = \boxed{-6.55 \times 10^{-29} \text{ J}}$$

This is about a factor of 30 billion smaller than the change in electric potential energy. *When analyzing the motion of fundamental charged particles in electric fields, we can almost always ignore the effects of gravity.*

(d) From the work-energy relation, the work $q\Delta V$ done by the electric field equals the change in kinetic energy:

$$q\,\Delta V = \Delta K$$
$$e\,\Delta V = \tfrac{1}{2}mv^2 - 0$$

Solving for v gives $\quad v = \sqrt{\dfrac{2e(\Delta V)}{m}} = \sqrt{\dfrac{2(1.60 \times 10^{-19} \text{ C})(12 \text{ V})}{1.67 \times 10^{-27} \text{ kg}}}$

$$v = \boxed{4.80 \times 10^4 \, \frac{\text{m}}{\text{s}}}$$

Alternate method: From Newton's second law $\Sigma F = ma$, we find the acceleration of the proton to be $a = F/m = eE/m$. Substituting this value into the kinematic equation results in

$$v^2 = v_0^2 + 2ay$$
$$v^2 = 0 + 2\left(\frac{eE}{m}\right)d$$

Solving for v, we obtain the same equation as above:

$$v = \sqrt{\frac{2eEd}{m}} = \sqrt{\frac{2e(\Delta V)}{m}} = \boxed{4.80 \times 10^4 \, \frac{\text{m}}{\text{s}}}$$

The Electron Volt

The prevalence of the electron charge in atomic and nuclear physics has led to defining a new energy unit, the *electron volt*. An **electron volt** (eV) is the amount of *energy* acquired by an object with a charge e equal in magnitude to the electronic charge when the object is accelerated through a potential difference of one volt.

$$\Delta W = e\Delta V$$
$$1 \text{ eV} = (1.602 \times 10^{-19} \text{ C})(1 \text{ J/C})$$

ELECTRON VOLT
(an energy unit)

$$1 \text{ eV} = 1.602 \times 10^{-19} \text{ J} \quad (26\text{-}13)$$

Suppose, in the previous example, that we had an alpha particle (helium nucleus) instead of a proton. Because the alpha particle has a charge of $+2e$, after moving through a potential difference of 12 V it would have a kinetic energy of $(2e)(12 \text{ V}) = 24 \text{ eV}$, *twice* that of the singly charged proton. To convert to SI units, we use a conversion ratio:

$$24 \text{ eV} \underbrace{\left(\frac{1.602 \times 10^{-19} \text{ J}}{1 \text{ eV}}\right)}_{\text{Conversion ratio}} = 3.84 \times 10^{18} \text{ J} \quad (26\text{-}14)$$

The *electron volt* as an energy unit may also be applied to nonelectrical situations. For example, an air molecule at room temperature is said to have an average kinetic energy of about $(1/40)$ eV.

EXAMPLE 26-4

Setting $V \equiv 0$ at $r = \infty$, find the potential V for regions inside and outside a uniform positive spherical charge density ρ that extends from $r = 0$ to $r = R$. Use the value of the electric fields found in Example 25-5, and express your answer in terms of the total charge $Q = \rho(\text{volume}) = \rho(\frac{4}{3}\pi R^3)$.

SOLUTION

In Example 25-5, we found the following expressions for the electric field E:

Inside ($r < R$)   Outside ($r > R$)

$$E_{in} = \frac{\rho}{3\varepsilon_0}r = \left(\frac{Q}{4\pi\varepsilon_0 R^3}\right)r \qquad E_{out} = \left(\frac{\rho R^3}{3\varepsilon_0}\right)\frac{1}{r^2} = \left(\frac{Q}{4\pi\varepsilon_0}\right)\frac{1}{r^2}$$

OUTSIDE ($r > R$). We have set the value of the potential to be zero at infinity. So we start at infinity and integrate inward to find the *change* in V as we progress inward. For this radial field, we have

$$V_b - V_a = -\int_a^b \mathbf{E} \cdot d\mathbf{r}$$

$$V_{out} - 0 = -\int_\infty^r \left(\frac{Q}{4\pi\varepsilon_0}\right)\frac{1}{r^2}\,dr = \left(\frac{Q}{4\pi\varepsilon_0}\right)\left(\frac{1}{r}\right)\bigg|_\infty^r = \boxed{\frac{Q}{4\pi\varepsilon_0 r}}$$

Outside the sphere, the field is the same as that of a point charge Q at the center.

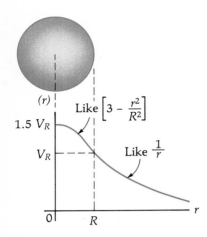

FIGURE 26-4
Example 26-4. A graph of V vs. r for a uniform charge density throughout a sphere of radius R. The maximum value of V is at the center even though $E = 0$ at the center. Outside the sphere, the potential is the same as if the total charge Q were a point charge at the center.

INSIDE ($r < R$). From the above result, we know that, at the surface of the sphere ($r = R$), the potential is $V = Q/4\pi\varepsilon_0 R$. So we start at this known value at $r = R$ and find the *change* in V as we integrate inward to an arbitrary location r inside the sphere.

$$V_b - V_a = -\int_a^b \mathbf{E} \cdot d\mathbf{r}$$

$$V_{in} - V_R = -\int_R^r \left(\frac{Q}{4\pi\varepsilon_0 R^3}\right) r\, dr = -\left(\frac{Q}{4\pi\varepsilon_0 R^3}\right)\left(\frac{r^2}{2}\right)\Big|_R^r$$

$$V_{in} = -\left(\frac{Q}{4\pi\varepsilon_0 R^3}\right)\left(\frac{r^2}{2}\right)\Big|_R^r + V_R$$

$$V_{in} = -\left(\frac{Q}{8\pi\varepsilon_0 R^3}\right)(r^2 - R^2) + \frac{Q}{4\pi\varepsilon_0 R} = \boxed{\left(\frac{Q}{8\pi\varepsilon_0 R}\right)\left[3 - \frac{r^2}{R^2}\right]}$$

Figure 26-4 is a graph of this potential $V(r)$ vs. r.

Comment: Note that, at $r = R$, the two curves join smoothly. For $r > R$, the curve is proportional to $1/r$; for $r < R$, the curve is like an inverted parabola (proportional to a constant minus r^2). Even though the electric field $E = 0$ at the center, the potential V has its maximum value at the center. This is reasonable if you think of the *change in potential* as the work/charge that you would do in bringing a positive test charge q_0 inward from infinity (where $V \equiv 0$) to the surface of the sphere. You are constantly doing positive work against the repulsive Coulomb force along this path. Even inside the sphere, the Coulomb force is still repulsive (though it falls linearly to zero toward the origin) so you must do additional positive work as you bring q_0 inward to the origin, causing V to change further in the positive direction. It is instructive to compare this result with the analogous case of the gravitational potential energy of a point mass m near the uniform spherical mass of the earth, Figure 16-14.

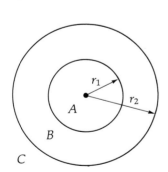

FIGURE 26-5
Example 26-5. Two concentric, thin, conducting spherical shells that carry different net charges. The charge on the inner shell is $+10$ nC, while the charge on the outer shell is -15 nC.

EXAMPLE 26-5

Consider two thin, conducting, spherical shells as in Figure 26-5. The inner shell has a radius $r_1 = 15$ cm and a charge of $+10$ nC (nC is the symbol for *nanocoulomb*, equal to 10^{-9} C). The outer shell has a radius $r_2 = 30$ cm and a charge of -15 nC. Find (a) the electric field E and (b) the electric potential V in these regions, with $V \equiv 0$ at $r = \infty$:

Region A: inside the inner shell ($r < r_1$)
Region B: between the shells ($r_1 < r < r_2$)
Region C: outside the outer shell ($r > r_2$)

SOLUTION

Because of symmetry, the charges distribute themselves symmetrically over the spheres. Also because of symmetry, it is easiest first to calculate the field using Gauss's law, then to obtain the potential from the relation between V and E (Equation 26-5). We will consider each region in turn, starting with the electric field.

(a) *Calculation of the electric field E:*

Region A: inside the inner shell. Noting the symmetry, we construct a Gaussian surface in the form of a sphere concentric with the center. We then apply Gauss's law: $\oint \mathbf{E} \cdot d\mathbf{A} = q/\varepsilon_0$. Whatever magnitude E has at one point on this surface, by symmetry it must have the same value at all points. Since there is no charge inside the Gaussian surface, we conclude that the field is zero all over the surface. Furthermore, we could construct such a surface anywhere in Region A with an arbitrary radius ($0 < r < r_1$), so we conclude that *the field $\mathbf{E}$ is zero everywhere inside the inner shell.*

$$E_A = \boxed{0} \qquad \text{(inside the inner shell)}$$

Region B: between the shells. Again, we recognize that the symmetry calls for a Gaussian surface in the form of a sphere of radius r (where $r_1 < r < r_2$) concentric with the center. We recognize this problem is similar to Example 25-7, whose result is

$$\boxed{E_B = \left(\frac{1}{4\pi\varepsilon_0}\right)\frac{q}{r^2}} \qquad \text{(radially outward for } r_1 < r < r_2)$$

Recalling that $1/(4\pi\varepsilon_0) = 9 \times 10^9$ N·m²/C², we calculate the value of E at the location just barely outside the inner shell radius r_1.

$$E_{r_1} = \left(9 \times 10^9 \frac{\text{N·m}^2}{\text{C}^2}\right)\frac{(10 \times 10^{-9}\text{ C})}{(0.15\text{ m})^2} = \boxed{4000\ \frac{\text{N}}{\text{C}}} \qquad \text{(just outside the inner shell)}$$

This field decreases as $1/r^2$ until just barely inside the outer shell, where its value is

$$E_{r_2} = \left(9 \times 10^9 \frac{\text{N·m}^2}{\text{C}^2}\right)\frac{(10 \times 10^{-9}\text{ C})}{(0.30\text{ m})^2} = \boxed{1000\ \frac{\text{N}}{\text{C}}} \qquad \text{(just inside the outer shell)}$$

Region C: outside the outer shell. Again, we construct a concentric Gaussian surface of radius r (where $r > r_2$) and apply Gauss's law, recognizing that q is the *net* charge inside the surface: $\oint \mathbf{E}\cdot d\mathbf{A} = q/\varepsilon_0$. The net charge is $q = q_1 + q_2$, or $(10\text{ nC}) + (-15\text{ nC}) = -5$ nC. As above,

$$\boxed{E_C = \left(\frac{1}{4\pi\varepsilon_0}\right)\frac{q}{r^2}} \qquad \text{(radially inward for } r > r_2 \text{ because } q \text{ is negative)}$$

The value just barely outside the outer shell (at $r = r_2$) is

$$E_{r_2} = \left(9 \times 10^9 \frac{\text{N·m}^2}{\text{C}^2}\right)\frac{(-5 \times 10^{-9}\text{ C})}{(0.30\text{ m})^2} = \boxed{-500\ \frac{\text{N}}{\text{C}}} \qquad \text{(just outside the outer shell)}$$

The minus sign indicates that the field is directed inward (in the $-\mathbf{r}$ direction). It varies as $1/r^2$, approaching zero as $r \to \infty$.

Note that the electric field is *not* a continuous function of distance. As the Gaussian surface expands across one of the shells, it suddenly encloses a new layer of charge, causing the value of $\mathbf{E}$ to change discontinuously (at least in this idealized case, where we assume that the layer of point charges has zero thickness). As we approach a discontinuity from one side or the other, we thereby learn information about the way the values change at the discontinuity itself. Figure 26-6 shows a graph of these fields.

(b) *Calculation of the electric potential V.* Since we know the field $\mathbf{E}$ everywhere, we will use

$$V_2 - V_1 = -\int_1^2 \mathbf{E}\cdot d\boldsymbol{\ell}$$

to calculate how the potential varies. First, we choose the zero reference location: $V \equiv 0$ at $r = \infty$. Then, we start at $r = \infty$ and work our way into the center of the sphere, calculating the *change* of potential as we go.

Region C: outside the outer shell. Because of the spherical symmetry of the charge distribution, both the field and the potential outside the spheres

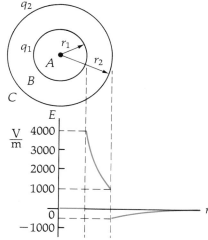

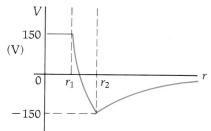

(a) The electric field E is positive if it is radially outward; negative values are radially inward fields. The curved portions of the graph vary as some function of $1/r^2$. E has discontinuities because a Gaussian surface, as it gradually expands, suddenly encloses a new layer of charge at a shell, causing E to change suddenly to a new value.

(b) The electric potential V varies as $1/r$. There are no discontinuities of V because $\int \mathbf{E}\cdot d\boldsymbol{\ell}$ may be interpreted as summing the area under the E-vs.-r graph. Integrating across a discontinuity merely changes the *rate* at which area accumulates. The area itself does not change abruptly, which implies that there is no sudden change in the work done.

FIGURE 26-6
Example 26-5. Concentric, thin, conducting spherical shells that carry different net charges.

are as though the net charge q (which equals $q_1 + q_2$) were concentrated at a point at the center. Integrating inward from ∞ to a point r (outside the spheres), we have

$$V_r - 0 = -\int_\infty^r \frac{q}{4\pi\varepsilon_0 r^2}\, dr = \frac{q}{4\pi\varepsilon_0 r}\Big|_\infty^r = \boxed{\left(\frac{1}{4\pi\varepsilon_0}\right)\frac{q}{r}} \qquad \text{(Region C: } r \geq r_2\text{)}$$

Since the net charge q is -5 nC, the numerical value at $r = r_2$ is

$$V_{r_2} = \left(9 \times 10^9 \frac{\text{N·m}^2}{\text{C}^2}\right)\left(\frac{-5 \times 10^{-9}\,\text{C}}{0.30\,\text{m}}\right) = \boxed{-150\,\text{V}} \qquad (\text{at } r = r_2)$$

Region B: between the shells. As usual, the *change* of potential between r_2 and r (where $r_1 \leq r \leq r_2$) is given by

$$V_r - V_{r_2} = -\int_{r_2}^r \mathbf{E} \cdot d\boldsymbol{\ell}$$

The magnitude of $\mathbf{E}$ is determined solely by the charge q_1 on the inner sphere (Gauss's law). For the integration from r_2 to r we have

$$V_r - V_{r_2} = -\int_{r_2}^r \frac{q_1}{4\pi\varepsilon_0 r^2}\, dr = \frac{q_1}{4\pi\varepsilon_0}\left(\frac{1}{r} - \frac{1}{r_2}\right)$$

Since $V_{r_2} = -150$ V, the value within region B is

$$V_B = \boxed{-150\,\text{V} + \frac{q_1}{4\pi\varepsilon_0}\left(\frac{1}{r} - \frac{1}{r_2}\right)} \qquad \text{(Region B: } r_1 \leq r \leq r_2\text{)}$$

The numerical value for V_{r_1} at the inner shell is

$$V_{r_1} = (-150\,\text{V}) + \left(9 \times 10^9 \frac{\text{N·m}^2}{\text{C}^2}\right)(10 \times 10^{-9}\,\text{C})\left(\frac{1}{0.15\,\text{m}} - \frac{1}{0.30\,\text{m}}\right)$$

$$V_{r_1} = -150\,\text{V} + 300\,\text{V} = \boxed{150\,\text{V}}$$

Although $\mathbf{E}$ is discontinuous at the boundaries of the shells where the charges are located, *the potential V is continuous across these boundaries.* This is plausible when you recall that integrating $\int \mathbf{E} \cdot d\boldsymbol{\ell}$ may be interpreted as summing up the area under the curve for E as a function of distance (see Figure 26-6a). Integrating across a discontinuity merely changes the *rate* at which area accumulates; the area itself does not change abruptly.

Region A: inside the inner shell. Again, we start with the same general relation:

$$V_2 - V_1 = -\int_1^2 \mathbf{E} \cdot d\boldsymbol{\ell}$$

But here $\mathbf{E}$ is zero everywhere inside the inner shell. So there is *no change* of potential as we move inward. Hence, the potential at r_1 (equal to 150 V) is the same (constant) value for all smaller values of r.

$$V_A = \boxed{150\,\text{V}} \qquad \text{(Region A: } 0 \leq r \leq r_1\text{)}$$

Figure 26-6b shows a graph of the electric potential V in all regions. Note that even though $\mathbf{E}$ is everywhere zero inside, the potential V has a finite positive value in this region.

EXAMPLE 26-6

Consider an infinitely long, straight line of uniform positive charge density λ (in units of charge per length). Find the electric potential V due to this line charge.

SOLUTION

We will calculate the potential from the electric field E that we found in Example 25-1:

For a uniform line charge λ
$$E = \frac{\lambda}{2\pi\varepsilon_0 r} \quad \text{(radially outward)} \tag{26-15}$$

We now apply
$$V_2 - V_1 = -\int_{r_1}^{r_2} E\,dr$$

Substituting for E and integrating gives

$$V_2 - V_1 = -\int_{r_1}^{r_2} \frac{\lambda}{2\pi\varepsilon_0 r}\,dr = -\frac{\lambda}{2\pi\varepsilon_0}(\ln r_2 - \ln r_1) \tag{26-16}$$

But now we have an unforeseen problem in assigning the zero reference location. We cannot set $V \equiv 0$ at $r = \infty$ or at $r = 0$, because the logarithm goes infinite at both locations. So we choose $V_1 \equiv 0$ at $r_1 = a$, a finite distance from the line charge. Thus, the potential V at a distance r from the line becomes

$$V = -\frac{\lambda}{2\pi\varepsilon_0}(\ln r - \ln a) = \boxed{-\frac{\lambda}{2\pi\varepsilon_0}\ln\left(\frac{r}{a}\right)} \quad \begin{pmatrix}\text{where } V \equiv 0 \\ \text{at } r = a\end{pmatrix} \tag{26-17}$$

Figure 26-7 graphs the result.

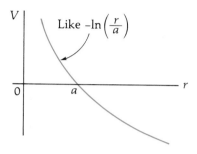

FIGURE 26-7
Example 26-6. The electric potential V near an infinitely long, straight line of uniform charge density, where $V \equiv 0$ at $r = a$.

EXAMPLE 26-7

A total positive charge Q is distributed uniformly around a thin ring of radius a. Find the electric potential V at a point P along the axis of the ring a distance x from the center, as shown in Figure 26-8. Set $V \equiv 0$ at $x = \infty$.

SOLUTION

Rather than calculating a *vector* sum $\Delta V = -\int \mathbf{E}\cdot d\boldsymbol{\ell}$ to find V, it is often easier to calculate a *scalar* sum using $dV = k\,dq/r$. This example lends itself to that approach. From the symmetry, we see that every element of charge $dq = \lambda\,ds$ around the ring is at the same distance $\sqrt{a^2 + x^2}$ from the point P. Hence, using an integral form of Equation (26-10) we have

$$V = k\int \frac{dq}{r} = k\oint_{\substack{\text{entire}\\\text{ring}}} \frac{\lambda\,ds}{\sqrt{a^2 + x^2}} = \frac{k}{\sqrt{a^2 + x^2}} \underbrace{\oint_{\substack{\text{entire}\\\text{ring}}} \lambda\,ds}_{Q} = \frac{kQ}{\sqrt{a^2 + x^2}}$$

$$V = \boxed{\frac{Q}{4\pi\varepsilon_0 \sqrt{a^2 + x^2}}}$$

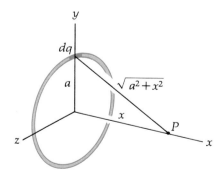

FIGURE 26-8
Example 26-7. The potential on the axis of a uniformly charged thin ring.

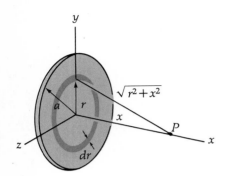

FIGURE 26-9
Example 26-9. A uniform disk of charge.

> **EXAMPLE 26-8**
>
> Find the electric potential V along the axis of a uniformly charged disk of radius a, that has a uniform surface charge σ (in units of charge per area) on one side.
>
> **SOLUTION**
>
> In Figure 26-9 we let x be the distance from the center of the disk along the axis to the point P. We divide the area of the disk into a series of charged ring elements of radius r and width dr, and we use the result of the previous example for the potential element dV produced by this ring element. Then we sum all such ring elements for the whole disk. A ring element of radius r has an area $dA = 2\pi r\, dr$ and carries a charge $dq = \sigma\, dA = \sigma 2\pi r\, dr$. Thus, the potential dV at point P due to this ring element is
>
> $$dV = \frac{k\, dq}{\sqrt{r^2 + x^2}} = \frac{k\sigma 2\pi r\, dr}{\sqrt{r^2 + x^2}}$$
>
> We now sum over all such rings elements from $r = 0$ to $r = a$. Using the result of Appendix G-II, Equation 19, we get
>
> $$V = \int_0^a \frac{k\sigma 2\pi r\, dr}{\sqrt{r^2 + x^2}} = k\sigma 2\pi \int_0^a \frac{r\, dr}{\sqrt{r^2 + x^2}} = k\sigma 2\pi [\sqrt{r^2 + x^2}]\Big|_0^a$$
>
> $$V = \boxed{k\sigma 2\pi [\sqrt{a^2 + x^2} - x]}$$

26.3 The Gradient of V

If an electric field is nonuniform—that is, if it has changing values in all three coordinates—we may still write a relation between V and $\mathbf{E}$. If a field has only one component E_x, we write

$$dV = -\mathbf{E}_x \cdot d\mathbf{x} \tag{26-18}$$

which becomes
$$E_x = -\frac{dV}{dx} \tag{26-19}$$

But if V and $\mathbf{E}$ are functions of the three variables x, y, and z, then we must find three derivatives: the derivative with respect to x (while holding the variables y and z constant), the derivative with respect to y (while holding x and z constant), and the derivative with respect to z (while holding x and y constant). There is a shorthand mathematical notation for this process. The *partial derivative* symbol, $\partial V/\partial x$, means "take the derivative of V with respect to x while holding all other variables constant." Thus, the complete form of Equation (26-19) in three dimensions is

$$E_x = -\frac{\partial V}{\partial x} \qquad E_y = -\frac{\partial V}{\partial y} \qquad E_z = -\frac{\partial V}{\partial z} \tag{26-20}$$

A specific example illustrates the process. Consider a potential of the form $V = ax^2 y$, where a is a constant. The partial derivatives are

$$E_x = -\frac{\partial V}{\partial x} = -2axy \qquad E_y = -\frac{\partial V}{\partial y} = -ax^2 \qquad E_z = \frac{\partial V}{\partial z} = 0$$

The total field **E** is written as

$$\mathbf{E} = E_x\hat{\mathbf{x}} + E_y\hat{\mathbf{y}} + E_z\hat{\mathbf{z}} = -2axy\hat{\mathbf{x}} - ax^2\hat{\mathbf{y}}$$

In general notation, the relation between **E** and V is written

$$\mathbf{E} = -\left(\frac{\partial V}{\partial x}\hat{\mathbf{x}} + \frac{\partial V}{\partial y}\hat{\mathbf{y}} + \frac{\partial V}{\partial z}\hat{\mathbf{z}}\right) \qquad (26\text{-}21)$$

The expression in parentheses is called the **gradient** of V. It is the vector that points in the direction of the greatest rate of change of potential, and thus it is always along **E**, perpendicular to the equipotential surfaces. Hence, the electric field is the negative gradient of the potential. The gradient is represented by the vector symbol $\mathbf{\nabla}$, called *del* or *grad*.

THE GRADIENT OF V
(Cartesian coordinates)
$$\mathbf{\nabla} V = \frac{\partial V}{\partial x}\hat{\mathbf{x}} + \frac{\partial V}{\partial y}\hat{\mathbf{y}} + \frac{\partial V}{\partial z}\hat{\mathbf{z}} \qquad (26\text{-}22)$$

For spherical coordinate systems (Figure (26-24)), we define the three mutually perpendicular unit vectors $\hat{\mathbf{r}}$, $\hat{\boldsymbol{\theta}}$, and $\hat{\boldsymbol{\phi}}$. Without presenting the proof, we state the gradient in spherical coordinates:

THE GRADIENT OF V
(spherical coordinates)
$$\mathbf{\nabla} V = \frac{\partial V}{\partial r}\hat{\mathbf{r}} + \frac{1}{r}\frac{\partial V}{\partial \theta}\hat{\boldsymbol{\theta}} + \frac{1}{r\sin\theta}\frac{\partial V}{\partial \phi}\hat{\boldsymbol{\phi}} \qquad (26\text{-}23)$$

Using this notation, we write the general relation:

RELATION BETWEEN
E and V
$$\mathbf{E} = -\mathbf{\nabla} V \qquad (26\text{-}24)$$

Note the convenience of the vector notation; it is easy to write and it is true for *all* coordinate systems.[4]

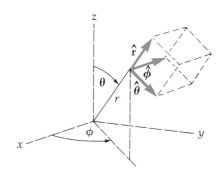

FIGURE 26-10
Spherical coordinates. The unit vectors $\hat{\mathbf{r}}$, $\hat{\boldsymbol{\theta}}$ and $\hat{\boldsymbol{\phi}}$ are mutually perpendicular. They point in the directions of the increase of the variables r, θ, and ϕ, respectively.

EXAMPLE 26-9

(a) Derive the expression for the potential V of a dipole at distances that are large compared with the separation of the charges. (b) Using Equation (26-24), find an expression for the electric field E of a dipole at large distances.

SOLUTION

(a) The potential of a dipole is the sum of the potentials for each of the two charges. For a single charge, $V = kq/r$. For both charges, the potential at

---

[4] The mathematical operator $\mathbf{\nabla}$ is a powerful and useful concept that you will use in later courses. It is a generalization of the concept "slope" to three dimensions, and it has an interesting analogy that comes from topographic contour maps. The gradient of the gravitational potential along the surface points uphill in the direction in which the rate of change of potential is the greatest. A loose rock would roll downhill in the steepest direction of $\mathbf{g} = -\mathbf{\nabla} U_g$. Similarly, the gradient of the electric potential points "uphill," while the electric field **E** points "downhill"—the direction that a free positive charge would move. Another analogy is that of a block of material that has different temperatures throughout. At any point in the block, the gradient of the (scalar) temperature is a vector that points opposite to the direction of the heat "flow."

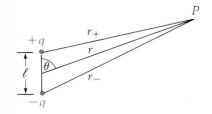

FIGURE 26-11
Example 26-9.

point P is

$$V = kq\left(\frac{1}{r_+} - \frac{1}{r_-}\right) = kq\left(\frac{r_- - r_+}{r_- r_+}\right) \quad (26\text{-}25)$$

where r_+ and r_- are defined in Figure 26-11. For $r \gg \ell$, we make the following approximations: $r_+ r_- \approx r^2$, and $(r_- - r_+) \approx \ell \cos \theta$. When we substitute these values, Equation (26-25) becomes

$$V = kq\,\frac{\ell \cos \theta}{r^2} = k\,\frac{p \cos \theta}{r^2} = \left(\frac{1}{4\pi\varepsilon_0}\right)\frac{p \cos \theta}{r^2} \quad (26\text{-}26)$$

where we have substituted the notation for the electric dipole, $q\ell = p$.

(b) To obtain the electric field E in the r, θ, and ϕ directions, we calculate the partial derivatives

$$\frac{\partial V}{\partial r} = -\left(\frac{2p \cos \theta}{4\pi\varepsilon_0 r^3}\right) \qquad \frac{\partial V}{\partial \theta} = -\left(\frac{p \sin \theta}{4\pi\varepsilon_0 r^2}\right) \quad \text{and} \quad \frac{\partial V}{\partial \phi} = 0$$

Substituting these expressions into Equation (26-24) for spherical coordinates gives

$$\mathbf{E} = -\left[\frac{\partial V}{\partial r}\hat{\mathbf{r}} + \frac{1}{r}\frac{\partial V}{\partial \theta}\hat{\boldsymbol{\theta}} + \frac{1}{r\sin\theta}\frac{\partial V}{\partial \phi}\hat{\boldsymbol{\phi}}\right]$$

$$\boxed{\mathbf{E} = \left(\frac{2p \cos\theta}{4\pi\varepsilon_0 r^3}\right)\hat{\mathbf{r}} + \left(\frac{p \sin\theta}{4\pi\varepsilon_0 r^3}\right)\hat{\boldsymbol{\theta}}} \quad \begin{pmatrix}\text{far-field approximation}\\\text{for the dipole}\end{pmatrix} \quad (26\text{-}27)$$

Note that, for distances along a direction perpendicular to the dipole axis ($\theta = \pi/2$), this result agrees with Equation (24-16). The fact that there is no component in the $\hat{\boldsymbol{\phi}}$ direction agrees with symmetry considerations and with the fact that electric field lines must terminate on charges. Equation (26-27) also reveals an interesting feature of the field. At large distances from the dipole, the field along the axis of the dipole ($\theta = 0°$) has twice the magnitude of the field at the same distance perpendicular to the dipole axis ($\theta = 90°$). At large distances in any direction, the field falls off as $1/r^3$.

26.4 Equipotential Surfaces

We have seen that diagrams of electric field lines are useful for understanding the nature of electric charges and their interactions. In a similar way, it is helpful to visualize electric potentials. Consider an imaginary surface that is everywhere perpendicular to the field lines. It would take no work to move a small test charge q_0 around on such a surface, since the force $\mathbf{F} = q_0 \mathbf{E}$ is always perpendicular to the motion. The entire surface is at the same potential: an **equipotential surface**. A family of such surfaces, spaced apart at equal intervals of potential ΔV, gives one an intuitive "feel" for the physical situation. Figure 26-12 shows several examples. For a point charge, the equipotential surfaces are spheres concentric with the charge.

Equipotentials are easier to locate experimentally than field lines. For complicated two-dimensional geometries, the field pattern is most easily found experimentally by first determining a series of equipotentials spaced at equal

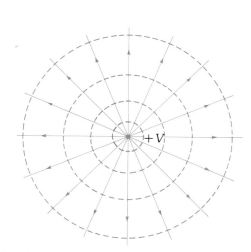

(a) A line (perpendicular to the paper) at positive potential. The field lines are imagined to extend to infinity, where they terminate on negative charges.

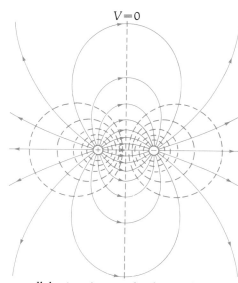

(b) Two parallel wires (perpendicular to the plane of the paper) at equal and opposite potentials. All field lines that leave the left-hand wire terminate on the right-hand wire.

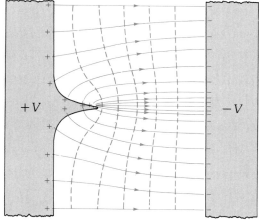

(c) Two conducting planes (perpendicular to the plane of the paper) at opposite potentials. One plane has a pointed ridge extending perpendicular to the paper.

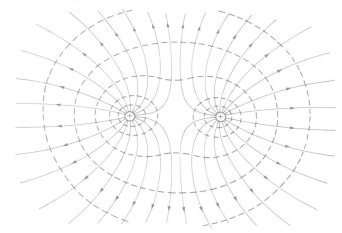

(d) Two parallel wires (perpendicular to the plane of the paper) at the same positive potential. As in (a), the field lines are imagined to extend to infinity, where they terminate on negative charges.

FIGURE 26-12
Electric field lines (solid) and cross-sections of equipotential surfaces (dashed). The field lines are everywhere perpendicular to the equipotential surfaces, a mathematical property called *orthogonality*.

intervals of potential difference. The correct field pattern can then be determined by drawing field lines perpendicular to the equipotentials.

A perfect conductor is, of course, an equipotential surface. Therefore, *electric field lines must always intersect conductors at right angles*. (If they did not, there would be a component of **E** parallel to the surface, thus requiring work to move a test charge along the surface.) Furthermore, since field lines must terminate on charges, when a field line intersects a conductor there must be a net charge at that point on the surface of the conductor. These properties make possible some interesting assertions. For example, we can place a hollow conducting sphere concentric to a point charge without altering the field outside the shell. Moreover, once the shell is in place, the charge inside may move about within the shell without changing the external field (see Figure 26-13).

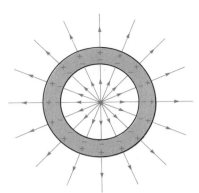

(a) The point charge induces surface charges on the inside and outside of the hollow conducting sphere.

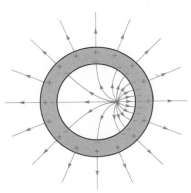

(b) Interestingly, the field outside the sphere remains symmetric even though the point charge inside is moved to various positions within the hollow interior.

FIGURE 26-13
A hollow conducting sphere, initially uncharged, with a point charge placed inside.

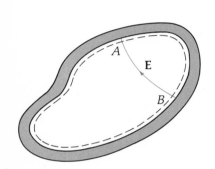

FIGURE 26-14
Contrary to this figure, no electric field can exist within an empty, closed conductor, regardless of whether the conductor is charged or not.

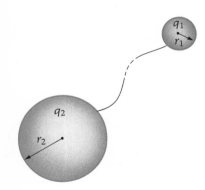

FIGURE 26-15
Two charged conducting spheres, isolated from their surroundings, are connected by a conducting wire so that they are at the *same potential V*. The wire is then removed. (The spheres should be separated much farther than shown here so that the charge distribution on each is not disturbed by the presence of the other charged sphere.) In this process, the charge will distribute itself so that the *smaller* sphere has the *larger* surface charge density σ; hence it has the larger electric field near its surface.

The concept of equipotential surfaces and associated electric fields allows us to conclude that *no electric field exists within any empty, closed conductor, whether the conductor is charged or not*. We have already shown this to be the case for a hollow conducting *sphere* (Example 26-5). Consider now an *irregular* hollow conductor, such as that in Figure 26-14. We construct a Gaussian surface just within the surface and apply Gauss's law: $\oint \mathbf{E} \cdot d\mathbf{A} = q'/\varepsilon_0$. Since there is no charge inside the Gaussian surface, $\oint \mathbf{E} \cdot d\mathbf{A} = 0$. But note that we cannot invoke symmetry arguments to assert that the field is zero. (There could be some field lines entering the surface and some leaving, so that the total integral is zero.) Let us suppose a field line enters and leaves the Gaussian surface as shown in Figure 26-14. Then an electron at A could leave the conductor, work could be done on it by the field between A and B, and it could subsequently enter the conductor at B. The electron could then be moved through the conductor *without doing work* from B to A (since the conductor is an equipotential surface). The process could be repeated, giving still more energy to the electron. The energy of the system would increase without end; it would represent a perpetual motion machine, which violates the first law of thermodynamics. *Therefore no field exists within an empty, hollow conductor.* Stated another way, a closed conductor is a perfect electrostatic shield.

Another conclusion we may draw from the use of equipotentials and field lines is that *charges tend to accumulate on the points of conductors*. Consider two charged conducting spheres, one larger than the other. A conducting wire is now connected between the spheres, causing a rearrangement of charges until both spheres are at the same potential, Figure 26-15. The wire is then removed. (We assume that the spheres are separated by a large enough distance so that the charge distribution on each sphere is not appreciably distorted by the presence of the other sphere.) The potential V of an isolated sphere with a charge q is $V = kq/r$. Because the two spheres were momentarily connected by a conductor, their potentials are equal:

$$\frac{kq_1}{r_1} = \frac{kq_2}{r_2} \qquad (26\text{-}28)$$

(a) A field ion micrograph of surface atoms in an iridium crystal needle point. Each spot corresponds to a single atom.

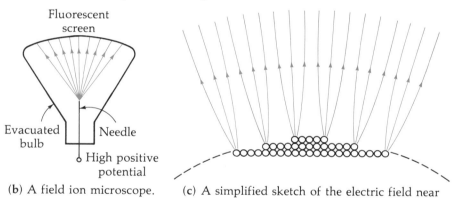

(b) A field ion microscope.

(c) A simplified sketch of the electric field near atoms on the surface of the needle point.

FIGURE 26-16
The field ion microscope developed by Erwin Müller at Pennsylvania State University gives a "picture" of individual atoms on the surface of a needle point. Its operation depends on the fact that the electric field is strongest at the sharp corners of a surface. In an evacuated glass bulb, a very sharp needle point with a tip radius of only a few tens of nanometers is held at a large positive voltage with respect to a fluorescent screen on the inside of the bulb wall. At the surface of the crystalline structure of the needle tip, atoms at the edge of a plane of atoms form sharp "corners," causing particularly strong fields just above them. Helium atoms are then introduced into the evacuated bulb. When the helium atoms encounter regions of extremely strong fields just above individual atoms, they lose an electron to the surface, become positive ions, and are accelerated along the field lines to the fluorescent screen, causing a bright spot on the screen. Each spot thus corresponds to the location of a particular atom on the needle surface. Unfortunately, the strong electric fields near the tip surface create large mechanical stresses that limit the technique to metallic substances that have very strong mechanical properties.

The surface charge density σ on a sphere of radius r is $\sigma = q/4\pi r^2$. So Equation (26-28) becomes

$$r_1\sigma_1 = r_2\sigma_2 \qquad (26\text{-}29)$$

In Example 25-4 we found that the field E just outside a conductor with a surface charge density σ is $E = \sigma/\varepsilon_0$. Substituting this relation in Equation (26-29) gives

$$E_1 r_1 = E_2 r_2 \qquad (26\text{-}30)$$

leading to the following conclusion: *for a charged conductor of irregular shape, where the radius of curvature is smallest the electric field at the surface is the largest.*[5]

---

[5] For certain geometrical shapes there can be exceptions to this general rule. See Richard H. Price and Ronald J. Crowley, "The Lightning-rod Fallacy," *American Journal of Physics* **53**, 843 (1985).

Thus, both **E** and σ can become very large near sharp points on high-voltage equipment. This becomes a problem when the small number of charged ions always present in the air (produced by cosmic-ray bombardment) are attracted toward a charged conductor of the opposite sign. Near sharp points where the electric fields are very large, these ions are accelerated to sufficiently high speeds that they collide with other air molecules, producing more ions, and an electrical breakdown of the (relatively) nonconducting air, called *corona discharge*, occurs. This discharge causes the air to glow visibly near sharp points as ions and electrons recombine. In *dry* air (STP) the electrical breakdown occurs for fields above about 3×10^6 V/m, though at low pressures (a few hundred pascal) breakdown occurs at much lower values. Until this variation was recognized, it was a source of problems in designing electric circuitry for spacecraft; the circuit functioned well at the engineer's workbench, but suffered arc discharges as the spacecraft passed through the outer limits of the earth's atmosphere. Humidity and dust also greatly lower the breakdown fields.

During an electrical storm, high potentials develop between thunderclouds and the earth. The purpose of sharp-pointed lightning rods attached to tall structures is not to "attract" lightning, but just the opposite: the strong electric fields near the points allow charges to leak off, reducing the high potential differences that might otherwise result in a lightning bolt at that location. Aircraft also have special sharp points to help reduce excess charge. *St. Elmo's fire*, named after the patron saint of sailors, refers to the glowing corona discharge from prominent points of a mast on ships at sea when a storm is brewing.

Summary

Electric *potential energy* U: $U_b - U_a = -\int_a^b q_0 \mathbf{E} \cdot d\boldsymbol{\ell}$

Electric *potential* V: $V_a - V_b = -\int_a^b \mathbf{E} \cdot d\boldsymbol{\ell}$

Only *changes* in potential, ΔV, are significant. Because the field is conservative, ΔV is the same for *any* convenient path between a and b. For localized systems of charges, the zero reference location is chosen at infinity.

The *electron volt* (eV) is the *energy* acquired by a particle with a charge equal in magnitude to the electron charge accelerated through a potential difference of one volt:

$$1 \text{ eV} = 1.602 \times 10^{-19} \text{ J}$$

For point charges ($V \equiv 0$ at $r = \infty$),

$$U = \left(\frac{1}{4\pi\varepsilon_0}\right)\frac{qq'}{r} \qquad V = \left(\frac{1}{4\pi\varepsilon_0}\right)\frac{q}{r}$$

For symmetrically distributed charges, the relation between V and $\mathbf{E}$ often permits an easier calculation of $\mathbf{E}$ than does Coulomb's law.

In integral form

$$V_b - V_a = -\int_a^b \mathbf{E} \cdot d\boldsymbol{\ell}$$

In differential form (one-dimensional)

$$E_x = -\frac{dV}{dx}$$

The vector that points in the direction of the greatest rate of change of a scalar function is called the *gradient*. Thus, the electric field **E** is the negative of the gradient of the potential V. In three dimensions, the gradient is represented by the vector symbol $\boldsymbol{\nabla}$, called "del" or "grad."

Cartesian coordinates

$$\boldsymbol{\nabla}V = \frac{\partial V}{\partial x}\hat{\mathbf{x}} + \frac{\partial V}{\partial y}\hat{\mathbf{y}} + \frac{\partial V}{\partial z}\hat{\mathbf{z}}$$

Spherical coordinates

$$\boldsymbol{\nabla}V = \frac{\partial V}{\partial r}\hat{\mathbf{r}} + \frac{1}{r}\frac{\partial V}{\partial \theta}\hat{\boldsymbol{\theta}} + \frac{1}{r\sin\theta}\frac{\partial V}{\partial \phi}\hat{\boldsymbol{\phi}}$$

The symbol $\partial/\partial x$ is the *partial derivative* with respect to x, holding all other variables constant, etc.

The electric field E just outside a surface that has a surface charge density σ is

$$E = \frac{\sigma}{\varepsilon_0}$$

For charged conductors with irregular surfaces, the field is strongest where the radius of curvature is smallest.

Questions

1. What is the distinction between electric potential energy difference and electric potential difference?
2. Do positive charges tend to seek regions of high potential or of low potential? What about electrons?
3. Consider the equations for the electric forces, fields, and potentials associated with a group of point charges. Discuss the similarities and differences with the analogous equations for the gravitational forces, fields, and potentials associated with a group of point masses.
4. Why cannot equipotential lines not cross one another?
5. Can the electric potential be zero at a point where the electric field is not zero? If so, give an example.
6. The electric field is zero at a certain point in a vacuum. Must V also equal zero at that same point? Give examples to illustrate your answer.
7. Can the electric field be zero at a point where the electric potential is not zero? If so, give an example.
8. As shown in Section 16.4, the gravitational field within a uniform, hollow, spherical shell of mass is zero. Similarly, the electric field within a hollow spherical conducting shell is zero. The electric field is also zero within a hollow conductor *of any shape* (not just one that has spherical symmetry). What about a hollow mass of any shape—say, a hollow cube? Is the gravitational field zero everywhere within a hollow mass of any shape? Can it be zero at a particular point within such a hollow mass? Can you always find a point at which the gravitational field is zero inside every hollow mass of any arbitrary shape?
9. The surface of an isolated charged conductor is an equipotential. Does this imply that the surface charge is uniform over the surface of the conductor?
10. Suppose that the electric field had the same magnitude everywhere over the surface of a conductor. What would this imply about the surface charge density? What would it imply about the physical shape of the conductor?
11. A "Faraday cage" consists of a hollow box with sides constructed of metallic wire screen. A sensitive voltmeter is connected between the screen and a probe inside the box. How does this device detect a net charge within the box?
12. Why is it impossible for the potential function of a charge distribution to have a finite discontinuity?

Problems

26.2 The Electric Potential

26A-1 A 12-V battery is connected to two large, parallel metal plates. (a) An electron released from rest at the negative plate acquires what velocity just before it strikes the positive plate? (b) Find the electron's maximum kinetic energy in electron volts and in joules. (c) If the plates are 4 mm apart, how long does the electron take to travel between the plates? (d) If the plates were a different distance apart, would this change the answers to parts (a) and (b)?

26A-2 Two parallel metal plates separated by 2 cm have a potential difference of 90 V between them (Figure 26-17). An electron passes through a small hole in the positive plate with a speed of 5×10^6 m/s. Find how close to the negative plate the electron will go.

26A-3 In the Bohr model of the hydrogen atom, the electron revolves around the proton in a circular path of radius 52.9 pm under the action of the Coulomb force between them. (Because of its much larger mass, the proton remains essentially at rest.) (a) By applying Newton's second law, find the speed of the electron. (b) Show that the magnitude of the electric potential energy is twice the electron's kinetic energy. (c) What is the total energy of the system in electron volts?

26A-4 In the Bohr model of a hydrogen atom, an electron moves in a circular path around a (stationary) proton. The radius of the path is 0.529×10^{-10} m. For the region in which the electron moves, find (a) the magnitude of the electric field E, and (b) the electric potential V (setting $V \equiv 0$ at $r = \infty$). (c) For comparison, an electrical breakdown (sparking) in air usually occurs for fields in excess of about 10^6 V/m. Why is this problem not a consideration in the Bohr model? (d) Com-

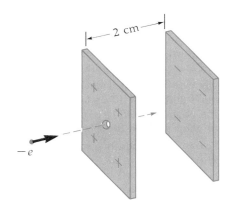

FIGURE 26-17
Problem 26A-2.

pare the electron's potential with the potential difference of a car battery.

26B-5 A point charge $+q$ is located at each vertex of an equilateral triangle with side length a. Find the potential difference ΔV between a point at the center of the triangle and a point at the center of one edge. Which point is at the higher potential?

26B-6 Four equal positive charges q form the corners of a square with a side length a. Find the potential difference between a point at the center of the square and a point midway along one side of the square. Which point is at the higher potential?

26B-7 Show that, for two positively charged, concentric conducting shells, the inner shell is always at a higher potential than the outer shell, regardless of the amount of charge on either shell.

26B-8 A point charge $q = 2\ \mu C$ is located at each vertex of the isosceles triangle shown in Figure 26-18. (a) Find the electric potential energy of this configuration of charges. (Hint: bring these charges in from infinity, one at a time. The change in potential energy when the first charge is moved in is zero.) (b) What is the electric field **E** at the origin after all three charges are in place?

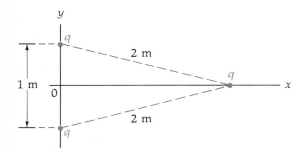

FIGURE 26-18
Problem 26B-8.

26.3 The Gradient of V
26.4 Equipotential Surfaces

26B-9 The potential V (in volts) in a region is defined by $V = (3\ \text{V/m}^2)x^2 + (0.2\ \text{V/m})y$, where x and y are expressed in meters. Find the magnitude and direction of the force on an electron placed at $x = 10$ cm, $y = 15$ cm.

26B-10 The electric potential just outside a charged conducting sphere is 200 V, and 10 cm farther from the center of the sphere the potential is 150 V. Find (a) the radius of the sphere and (b) the charge on the sphere.

26B-11 Two isolated conducting spheres, one with a radius R and the other with a radius $3R$, each carry an equal charge Q_0. The spheres are brought into contact and then separated again. Find the charge on each sphere.

26B-12 Two identical small metal spheres have net charges of q_1 and q_2, respectively. When separated a distance of 1 m, they attract each other with a force of 9×10^{-3} N. The spheres are now moved together until they touch, then again placed 1 m apart where it is found that they now repel each other with a force of 2×10^{-3} N. Find the charges q_1 and q_2.

26B-13 Consider two hollow, metallic, concentric spheres. The inner sphere has a radius of 30 cm and a charge of $-80\ \mu C$. The outer sphere has a radius of 50 cm and a charge of $40\ \mu C$. For the regions outside the spheres, between the spheres, and inside the inner sphere, find (a) the electric field and (b) the potential. (c) Sketch qualitative graphs for E and V.

26B-14 Two positive charges, each $+q$, are located on the x axis at $x = \pm a$. (a) Make a freehand sketch of the electric field pattern in the xy plane. (b) Without calculating an exact equation, sketch a qualitative graph for the electric potential $V(x)$ along the $\pm x$ axis as a function of x. ($V \equiv 0$ at $x = \pm \infty$). (c) From your graph of $V(x)$ vs. x, explain how you could obtain a qualitative graph of the electric field $E(x)$ along the x axis as a function of x. Make a freehand graph of $E(x)$ vs. x. (d) Repeat (a), (b), and (c) for equal but opposite charges, $+q$ and $-q$.

Additional Problems

26C-15 A total charge Q is spread uniformly along a thin, nonconducting rod of length ℓ. Find the electric potential V at a point P that is a distance y from the end of the rod as shown in Figure 26-19.

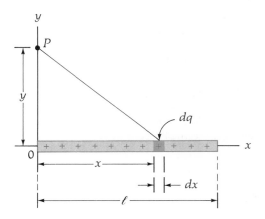

FIGURE 26-19
Problem 26C-15.

26C-16 In Figure 26-20, a positive charge distribution exists within the volume of an infinitely long cylindrical shell between radii a and b. The charge density ρ is not uniform, but varies inversely as the radius r from the axis. That is, $\rho = \kappa/r$ for $a < r < b$, where κ is a constant in SI units. Find the electric field for the regions (a) $r \leq a$, (b) $a \leq r \leq b$, and (c) $r \geq b$. (d) Find the electric potential for the same regions, setting $V = 0$ at $r = d$ (a very large distance away from the region of interest).

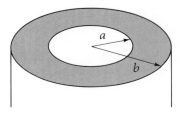

FIGURE 26-20
Problem 26C-16.

26C-17 The interior of a sphere of radius R has a volume charge density ρ that is proportional to the distance r from the center:

$$\rho = Ar \quad \text{(for } 0 < r < R\text{)}$$

where A is a constant. (a) Find the SI units for A. (b) Find the total charge Q inside the sphere in terms of A and R. (Hint: following Example 26-4, sum the charges dq in spherical shells of thickness dr.) (c) Use Gauss's law to find the electric field E inside the sphere a distance r from the center. (d) Setting $V = 0$ at $r = \infty$, find the potential V as a function of r both outside and inside the sphere.

26C-18 Repeat the previous problem for a charge distribution $\rho = Ar^2$.

26C-19 An electric field is described by $E = 2000\hat{x} + 3000\hat{y}$ (in SI units). Find the potential difference $(V_B - V_A)$ between the points A at $x = 0$, $y = 3$ m, $z = 2$ m and B at $x = 2$ m, $y = 1$ m, $z = 0$. (Hint: since E is a conservative field, $V_B - V_A$ may be calculated along any path between A and B.)

26C-20 A point charge of -20 nC is located at the origin of a coordinate system, and another point charge of $+10$ nC is located at $x = 6$ cm. An electron is released from rest at $x = 1$ cm, and it subsequently moves along the x axis toward the positive charge. Find the speed of the electron when it reaches the point $x = 5$ cm. (Hint: what is the potential difference between these points?)

26C-21 The liquid-drop model of the nucleus suggests that high-energy oscillations of certain nuclei can split the nucleus into two unequal fragments plus a few neutrons. The fragments acquire kinetic energy from their mutual Coulombic repulsion. Calculate the Coulomb potential energy (in MeV) of two spherical fragments from a uranium nucleus having the following charges and radii: $+38e$ and radius 5.5×10^{-15} m; $+54e$ and radius 6.2×10^{-15} m, respectively. Assume that the charge is distributed uniformly throughout the volume of each spherical fragment and that their surfaces are initially in contact at rest. (The electrons surrounding the nucleus can be neglected). The result agrees approximately with the observed kinetic energy associated with uranium fission.

26C-22 Two identical raindrops, each carrying surplus electrons on its surface to make a net charge $-q$ on each, collide and form a single drop of larger size. Before the collision, the characteristics of each drop are the following: (a) surface charge density σ_0, (b) electric field E_0 at the surface, (c) electric potential V_0 at the surface (where $V \equiv 0$ at $r = \infty$). For the combined drop, find these three quantities in terms of their original values.

26C-23 Two conducting parallel plates are 5 cm apart and have a potential difference of 2000 V. An electron is released from rest at the negative plate and simultaneously a proton is released from rest at the positive plate. (a) How far from the positive plate do the particles pass each other? (b) Find the speed of each particle as it strikes the other plate. (c) Find the kinetic energy (in eV and in J) of each particle as it reaches the other plate.

26C-24 A disk of radius a has a uniform surface charge σ on one side. A circular hole of radius $a/2$ is now cut in the center of the disk. (a) Using the superposition principle and the result of Example 26-8, find the electric potential V along the axis of the disk at a distance x from its center ($V \equiv 0$ at $x = \infty$). (b) What is the electric potential at the center of the hole? (c) What is the electric field at the center of the hole?

26C-25 Consider an electric quadrupole that is an assembly of three charges: $-2q$ at the origin, $+q$ at $y = \ell/2$, and $+q$ at $y = -\ell/2$. Find the potential at points (a) along the x axis and (b) along the y axis. (c) Show that, at large distances from the quadrupole (that is, x and y much larger than ℓ), the potential varies as the inverse cube of the distance. (Hint: note the approximation $1/\sqrt{1 + a^2} \approx 1 - a^2/2$.)

CHAPTER 27

Capacitance and Energy in Electric Fields

Penetrating so many secrets, we cease to believe in the unknowable. But there it sits nevertheless, calmly licking its chops.

H. L. MENCKEN
Minority Report (1956)

27.1   Introduction

In this chapter, it will become clear why we have placed so much importance on the concept of an electric field. Compact configurations of conductors can be constructed so that they contain very intense electric fields. Such devices are called *capacitors*, a name derived from their capacity for storing positive and negative charges. We will show that the external work performed in establishing the separation of charge on the capacitor appears as energy stored in the electric field that is thereby created inside the capacitor. This chapter will lead us to the important conclusion that electric fields, wherever they exist, contain energy. Capacitors are widely used in electronic circuits, and in later chapters we will illustrate some of these applications.

27.2   Capacitance

Any two conductors, separated by an insulator, form what is called a capacitor. When a potential difference[1] V (such as a battery) is applied across the two conductors, negatively charged electrons with a total charge Q are attracted from the conductor attached to the positive plate of the battery and flow to the conductor attached to the negative plate, until the potential difference V between the conductors is the same as that of the battery. The battery may then be removed and the charges remain on the conductors. The ability of a capacitor to maintain this storage of charge at a given potential difference is called **capacitance** C, defined as

CAPACITANCE $$C = \frac{Q}{V} \qquad (27\text{-}1)$$

---

[1] Only potential *difference* is important when we are dealing with capacitors, so for simplicity it is common practice to use the symbol V, rather than ΔV.

The SI units of capacitance are *coulombs per volt* (C/V), which are given the name *farad* (F).[2] The symbol for a capacitor is ─||─. In the context of its usage, there is no confusion between the letter C for capacitance, which is a *quantity*, and the *unit* coulomb (C).

When we speak of "the charge Q on a capacitor" we mean just the magnitude of the charge on *one* of the conductors. (The total net charge on the conductors is zero.) The following examples derive expressions for the capacitance of some common geometrical shapes.

EXAMPLE 27-1

The Parallel-Plate Capacitor. Two parallel plates of equal areas A, separated a small distance d, form the most common type of capacitor. One plate has a charge $+Q$ and the other a charge $-Q$ as shown in Figure 27-1. If the plate separation is very small compared with the edge lengths of the plates, the fringing field at the edges may be ignored, and we assume that the electric field between the plates is uniform everywhere. From Equation (25-10), the electric field E between the plates[3] is

$$E = \frac{\sigma}{\varepsilon_0} = \frac{Q}{\varepsilon_0 A} \tag{27-2}$$

From Equation (26-5), the magnitude of the potential difference between the plates, which we will call V (rather than ΔV), is

$$V = (-)\int \mathbf{E} \cdot d\boldsymbol{\ell} = Ed \tag{27-3}$$

Combining these equations, we obtain the capacitance C:

$$C = \frac{Q}{V} = \frac{\varepsilon_0 A E}{Ed} = \frac{\varepsilon_0 A}{d}$$

CAPACITANCE OF A PARALLEL-PLATE CAPACITOR

$$\boxed{C = \frac{\varepsilon_0 A}{d}} \tag{27-4}$$

Note that the capacitance is independent of the charge on the capacitor. *The capacitance C depends on only the physical dimensions of the capacitor* (and the constant ε_0). Here, the capacitance is directly proportional to the plate area A and inversely proportional to the plate separation d.

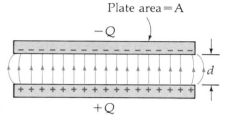

FIGURE 27-1
Example 27-1. Two parallel plates, each of area A, with a plate separation d, have equal and opposite charges. If d is very small compared with the edge lengths of the plates, the fringing field at the edges may be ignored.

---

[2] The *farad* honors the English physicist and chemist Michael Faraday (1791–1867), who investigated many electric, magnetic, optical, and chemical phenomena. Electromagnetic induction (Chapter 32) is his best known discovery. Faraday's family was very poor and he did not have the benefit of formal academic training. However, he fervently pursued his own self-education, and he had a truly outstanding knack for experimentation. At the age of 13, apprenticed to a bookseller, he became entranced with a copy of the third edition of the *Encyclopaedia Brittanica* that was brought in for repair. This edition had many articles on electricity that Faraday found specially interesting, further stimulating his interests in experimentation. Later he became an assistant to Sir Humphry Davy, the noted British chemist, who gave him rooms and an assistantship at the Royal Institution. Upon Davy's death, Faraday succeeded him at the Royal Institution, achieving fame in important research as well as giving popular lectures on scientific topics. The last nine years of his life, Faraday and his wife lived in a house in Hampton Court, provided for them by Queen Victoria.

[3] We assume a vacuum between the plates. The effects of a dielectric material between the plates is discussed in Section 27.4.

EXAMPLE 27-2

Find the capacitance of two metal plates, each 2 m² in area, separated by 1 mm. Ignore fringing effects at the edges.

SOLUTION

For parallel plates,

$$C = \frac{\varepsilon_0 A}{d} = \frac{(8.85 \times 10^{-12} \text{ F/m})(2 \text{ m}^2)}{(1 \times 10^{-3} \text{ m})} = 17.7 \times 10^{-9} \text{ F} = \boxed{17.7 \text{ nF}}$$

In spite of its physical size, this is quite a small capacitance. For a parallel-plate separation of 1 mm, a 1-F capacitor with square plates would be 10.6 km along each edge! (In Section 27.3 we will discuss methods of fabricating fairly large capacitances in small volumes.) Because the farad is a very large unit, more commonly encountered capacitances are usually expressed in units of the *microfarad* (1 μF = 10^{-6} F), the *nanofarad* (1 nF = 10^{-9} F), and the *picofarad* (1 pF = 10^{-12} F).

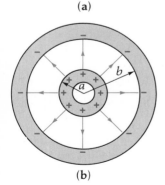

FIGURE 27-2
Example 27-3. Two long, concentric, conducting cylinders in a vacuum form a cylindrical capacitor. Equal and opposite charges per unit length, $\pm Q/L$, produce a radially outward electric field between the cylinders.

EXAMPLE 27-3

The Cylindrical Capacitor. A cylindrical capacitor consists of two concentric conducting cylinders, Figure 27-2. The outer radius of the inner conductor is a and the inner radius of the outer conductor is b. We assume that the total length of the cylinders is very great so that end effects involving fringing fields may be neglected. Consider a section of length L. In this length, the charge on the inner cylinder is $+Q$ and that on the outer cylinder is $-Q$, producing a symmetrical electric field between the cylinders that is radially outward. Applying Gauss's law to a cylindrical Gaussian surface of radius r ($a < r < b$) and length L, we find that the electric field E (see Example 25-1) is

$$\oint \mathbf{E} \cdot d\mathbf{A} = \frac{q_{\text{in}}}{\varepsilon_0}$$

$$E(2\pi r L) = \frac{Q}{\varepsilon_0}$$

$$E = \frac{Q}{2\pi \varepsilon_0 r L} \tag{27-5}$$

The potential difference $V = -\int_a^b \mathbf{E} \cdot d\boldsymbol{\ell}$ becomes

$$V = -\frac{Q}{2\pi\varepsilon_0 L} \int_a^b \frac{dr}{r} = -\left(\frac{Q}{2\pi\varepsilon_0 L}\right) \ln r \Big|_a^b = -\left(\frac{Q}{2\pi\varepsilon_0 L}\right)(\ln b - \ln a)$$

The magnitude of the potential difference V is thus

$$V = \left(\frac{Q}{2\pi\varepsilon_0 L}\right) \ln\left(\frac{b}{a}\right)$$

and the capacitance C is

$$C = \frac{Q}{V} = \frac{Q}{\left(\dfrac{Q}{2\pi\varepsilon_0 L}\right) \ln\left(\dfrac{b}{a}\right)}$$

| CAPACITANCE OF A CYLINDRICAL CAPACITOR | $C = \dfrac{2\pi\varepsilon_0 L}{\ln\left(\dfrac{b}{a}\right)}$ | (27-6) |

Again, note that only geometric factors determine the capacitance.

To transmit electrical signals between electronic equipment, a flexible *coaxial cable* is commonly used, Figure 27-3. It is basically a pair of concentric cylindrical conductors and it does have a *capacitance per unit length* that affects the electrical characteristics of the cable. Its main advantage is that, with the outer conductor grounded, the inner conductor is shielded from external electric fields that otherwise might cause undesirable voltages that would interfere with the signal.

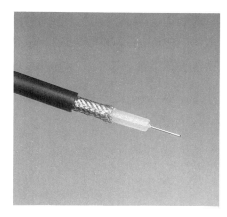

FIGURE 27-3
The components of a *coaxial cable* used to transmit electric signals between circuit components.

EXAMPLE 27-4

The Spherical Capacitor. Consider two concentric, conducting spherical shells with very thin walls, separated by a vacuum, Figure 27-4. The inner shell has a radius a, and the outer shell radius is b. Find the capacitance of this spherical capacitor.

SOLUTION

Consider a charge $+Q$ on the inner shell and an equal but opposite charge $-Q$ on the outer shell, producing a radially outward field **E** between the shells. The potential difference V between the shells is, from Equation (26-5),

$$V = -\int_a^b \mathbf{E} \cdot d\mathbf{r}$$

Noting that the field between the shells is just the Coulomb field, $E = kQ/r^2$, we have

$$V = -\int_a^b \frac{kQ}{r^2} dr = -kQ\left(-\frac{1}{r}\right)\Big|_a^b = kQ\left(\frac{1}{b} - \frac{1}{a}\right)$$

Noting that $b > a$, we make this potential difference a positive number by writing it as

$$V = kQ\left(\frac{1}{a} - \frac{1}{b}\right) = \left(\frac{1}{4\pi\varepsilon_0}\right)\left(\frac{b-a}{ab}\right)$$

The capacitance C is $\quad C = \dfrac{Q}{V} = \dfrac{Q}{\left(\dfrac{Q}{4\pi\varepsilon_0}\right)\left(\dfrac{b-a}{ab}\right)}$

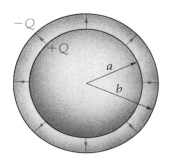

FIGURE 27-4
A *spherical capacitor* is formed of two thin, conducting concentric spheres in a vacuum. When charged as shown, the spheres have equal and opposite charges, producing a symmetrical (radially outward) electric field between them.

| CAPACITANCE OF A SPHERICAL CAPACITOR | $C = 4\pi\varepsilon_0\left(\dfrac{ab}{b-a}\right)$ | (27-7) |

As with all expressions for capacitance, the value of C depends upon only the physical dimensions of the conductors (and the constant ε_0).

EXAMPLE 27-5

Find the capacitance of a single, isolated sphere of radius R. (The second conductor may be considered as a conducting sphere at infinity where $V \equiv 0$.)

SOLUTION

In Equation (27-7), we let the outer radius b approach infinity, so that the a term in the denominator becomes insignificant. The b in the denominator then cancels the b in the numerator, resulting (in the limit) in $C = 4\pi\varepsilon_0 a$. For an isolated sphere of radius R, the capacitance is thus

CAPACITANCE OF AN ISOLATED SPHERE
$$C = 4\pi\varepsilon_0 R \qquad (27\text{-}8)$$

Note that the only significant factor is a geometric one: the radius R of the sphere.

EXAMPLE 27-6

Find the capacitance of the earth.

SOLUTION

The radius of the earth is $R_e = 6.34 \times 10^6$ m. From Equation (27-8),

$$C = 4\pi\varepsilon_0 R = \frac{6.34 \times 10^6 \text{ m}}{9 \times 10^9 \text{ N}\cdot\text{m}/\text{C}^2} = 7.04 \times 10^{-4} \text{ F}$$

This result shows that the unit *farad* (F) is an extremely large unit. For an isolated sphere to have a capacitance of 1 F, its radius would have to be more than 1400 times the radius of the earth, or about 13 times the size of the sun!

(a) The capacitance of this small capacitor is varied by moving the plates closer or farther apart. Other types have a single vane that can be rotated to achieve varying amounts of overlap with a fixed vane.

(b) One set of plates (connected together) can be rotated to vary the amount of overlap with another fixed set of plates (also connected together).

FIGURE 27-5
Two types of variable capacitors.

Many electronic circuits use capacitors whose capacitance is variable over a limited range of values. Figure 27-5 shows two common types. In practice, calculating the capacitance of arbitrary arrangements of conductors is not easy. We have illustrated three simple cases in which geometrical symmetry led to simple calculations. But for nonsymmetrical systems we find the value of C empirically by putting known charges on the conductors and measuring the potential difference between them. In electronic circuits, even this method fails because it is not possible to isolate one part of a circuit from its neighbors. Usually the *stray capacitances* between parts of a circuit are negligible, though they can sometimes be troublesome in alternating-current circuits, Chapter 34.

27.3 Combinations of Capacitors

In the construction of electronic circuits, it is often necessary to combine two or more capacitors. Combinations of capacitors consist of *parallel* and/or *series* connections, as shown in Figure 27-6. The electronic symbol ─┤├─ for a capacitor is used in the figure. (The symbol implies a parallel-plate capacitor, but it is used for any type of capacitor.)

In the *parallel* combination, *the potential difference V is the same for all capacitors*, but the charge on each may be different. The total charge on all capacitors is

$$Q = Q_1 + Q_2 + Q_3$$

Substituting gives
$$Q = C_1 V + C_2 V + C_3 V$$
$$Q = (C_1 + C_2 + C_3)V$$

Therefore, the single capacitance C_{eq} that is equivalent to this combination is

$$C_{eq} = C_1 + C_2 + C_3$$

Since the analysis could be extended to include any number of capacitors in parallel, we may write the general formula

CAPACITORS IN PARALLEL
$$C_{eq} = C_1 + C_2 + C_3 + \cdots \qquad (27\text{-}9)$$

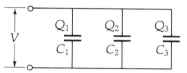

(a) In a *parallel* combination of capacitors, the voltage across each capacitor is the same.

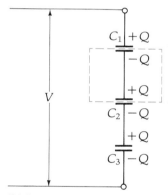

(b) In a *series* combination of capacitors, the charge on each individual capacitor is the same.

FIGURE 27-6
Combinations of capacitors.

To analyze the *series* combination, suppose that the capacitors are initially uncharged and that we connect a battery of voltage V across the ends of the series. The principle of charge conservation holds true, so the negative charge $-Q$ that flows from the battery onto one end plate must equal the negative charge that flows from the opposite end plate to the battery, leaving that plate with a charge $+Q$. Now, since the portion enclosed in the dashed box is isolated, the net charge within this region must remain zero (its initial value). However, the charged plates just outside the dashed box will cause a charge separation within the box, so that each plate of a capacitor acquires a charge equal but opposite to that on the other plate of the capacitor. Thus, *each capacitor in series acquires the same magnitude of charge Q*. The total potential V across the combination is the sum of the potentials across each capacitor:

$$V = V_1 + V_2 + V_3$$
$$V = \frac{Q}{C_1} + \frac{Q}{C_2} + \frac{Q}{C_3} = Q\left[\frac{1}{C_1} + \frac{1}{C_2} + \frac{1}{C_3}\right]$$

FIGURE 27-7
Example 27-7. The step-by-step reduction of a combination of capacitors to a single equivalent capacitance C_{eq}.

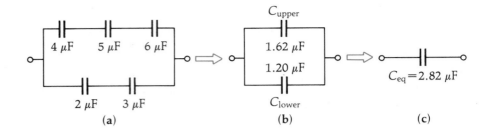

Since $V = Q/C$, the single capacitance C_{eq} that is equivalent to this series combination is

$$\frac{1}{C_{eq}} = \frac{1}{C_1} + \frac{1}{C_2} + \frac{1}{C_3}$$

Since the analysis could be extended to include any number of capacitors in series, we may write the general formula

CAPACITORS IN SERIES
$$\frac{1}{C_{eq}} = \frac{1}{C_1} + \frac{1}{C_2} + \frac{1}{C_3} + \cdots \qquad (27\text{-}10)$$

EXAMPLE 27-7

Five capacitors are connected as shown in Figure 27-7. Find the equivalent capacitance of the combination.

SOLUTION

We reduce the combination step by step, first finding the equivalent capacitances of the upper and lower branches.

Upper branch

Capacitors in series:

$$\frac{1}{C_{eq}} = \frac{1}{C_4} + \frac{1}{C_5} + \frac{1}{C_6}$$

$$\frac{1}{C_{eq}} = \frac{1}{4\ \mu F} + \frac{1}{5\ \mu F} + \frac{1}{6\ \mu F}$$

$C_{upper} = \underline{\underline{1.62\ \mu F}}$

Lower branch

Capacitors in series:

$$\frac{1}{C_{eq}} = \frac{1}{C_2} + \frac{1}{C_3}$$

$$\frac{1}{C_{eq}} = \frac{1}{2\ \mu F} + \frac{1}{3\ \mu F}$$

$C_{lower} = \underline{\underline{1.20\ \mu F}}$

We now have the equivalent circuit of (b), which is two capacitors in parallel. The equivalent capacitance of this circuit is

For *capacitors in parallel:* $\quad C_{eq} = C_{upper} + C_{lower}$

$C_{eq} = 1.62\ \mu F + 1.20\ \mu F = \boxed{2.82\ \mu F}$

27.4 Dielectrics

In our discussion of capacitors so far, we have assumed that the space between the conducting plates is a vacuum. However, it is usually impractical and even undesirable to construct vacuum capacitors. If the space between the plates is

POLAR DIELECTRIC

(a) In a polar dielectric (with zero external field), the electric dipole moments of the molecules have random orientations.

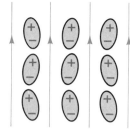

(b) When an external field E_0 is applied to a polar dielectric, the electric dipole moments tend to align themselves in the direction of the field.

FIGURE 27-8
The effect of an external electric field on the molecules of a *polar* dielectric material.

filled with certain insulating materials, the capacitance is increased and also the voltage that can be applied is increased—both desirable effects. The following discussion assumes homogeneous materials in the presence of a uniform electric field.

Suppose that we place a slab of nonconducting material called a **dielectric** between isolated charged plates of a parallel-plate capacitor. We will find that the potential difference between the plates *decreases*. To understand why, we now discuss the behavior of a dielectric material at the molecular level when it is placed in an electric field. Dielectrics may be classed as *polar* or *nonpolar*. Figure 27-8 shows a polar dielectric, so-named because its molecules have a *permanent* electric dipole moment. In the presence of the field $\mathbf{E}_0$ (produced by the charges on the plates), these dipole moments tend to align themselves in the direction of the field. In contrast, the molecules of a nonpolar dielectric, Figure 27-9, have no inherent dipole moments since the center of the positive charge distribution within a molecule coincides with the center of the negative charge distribution. However, when an external electric field $\mathbf{E}_0$ is applied, the centers of charge are drawn slightly apart to form *induced* dipole moments aligned in the direction of the field. In both types of materials, the overall effect of dipole alignments is that the surfaces of the material perpendicular to the applied field acquire *induced surface charge densities* as shown in Figure 27-10,

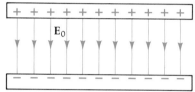

(a) The original field $\mathbf{E}_0'$ due to the isolated charged plates.

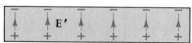

(b) The induced field $\mathbf{E}'$ due to the induced charges on the surface of the dielectric material when it is placed between the plates. (Plates are not shown.) Note the direction of the field $\mathbf{E}'$.

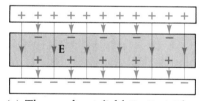

(c) The resultant field $\mathbf{E} = \mathbf{E}_0 + \mathbf{E}'$ within the dielectric is *less* than the original field $\mathbf{E}_0$.

FIGURE 27-10
A dielectric material placed between the charged (isolated) plates of a parallel-plate capacitor reduces the net field between the plates. As a result, the capacitance increases.

NONPOLAR DIELECTRIC

(a) In a nonpolar dielectric (zero external field), the center of positive charge within a molecule coincides with the center of negative charge, and the molecules have no electric dipole moments.

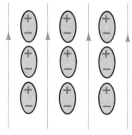

(b) When an external field E_0 is applied to a nonpolar dielectric, the centers of positive and negative charges are drawn apart, inducing a dipole moment in each molecule in the direction of the field.

FIGURE 27-9
The effect of an external electric field on the molecules of a *nonpolar* dielectric material.

TABLE 27-1 Approximate* Dielectric Constants and Dielectric Strengths

Material	Dielectric Constant κ	Dielectric Strength (10^6 V/m)
Vacuum	1	—
Air (dry)	1.000 59	3
Water	80	—
Glass	4–6	13
Castor oil	4.6	10
Polystyrene	2.5	15
Hard rubber	3	500
Mica	5	3000
Titanium dioxide	100	150

* Values are for electric fields that are constant in time. For alternating electric fields, these properties become frequency dependent. They also depend upon temperature.

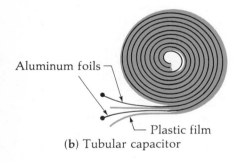

(a) High-voltage oil-filled capacitor

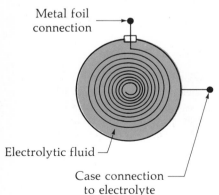

(b) Tubular capacitor

(c) Electrolytic capacitor. The metal foil has an oxide coating which forms the insulating materal between the foil and the electrolyte.

FIGURE 27-11
Some commercial capacitors.

a phenomenon called **polarization**. These induced surface charges produce an electric field $\mathbf{E}'$ of their own within the dielectric, in a direction *opposite* that of the applied field $\mathbf{E}_0$. The net electric field within the dielectric, $\mathbf{E} = \mathbf{E}_0 + \mathbf{E}'$, is thus *smaller* than the original field $\mathbf{E}_0$. By Equation (26-5),

$$V_2 - V_1 = -\int_1^2 \mathbf{E} \cdot d\boldsymbol{\ell}$$

The reduced electric field between the plates of the capacitor (for a given charge on the plates) results in a lower potential difference between them. Thus, from $C = Q/V$, the original capacitance C_0 of the capacitor is increased to a larger value C. The ratio of C to C_0 is called the **dielectric constant** κ:

DIELECTRIC CONSTANT κ $\qquad \kappa = \dfrac{C}{C_0} \quad \text{or} \quad C = \kappa C_0 \qquad$ (27-11)

where κ is the Greek letter *kappa*. The dielectric constant is larger than 1 for all materials. Table 27-1 lists some typical values. Any dielectric, if subjected to a sufficiently strong field, will become conducting. The maximum field that a dielectric can withstand without electrical breakdown is called its **dielectric strength**.

Dielectrics serve three useful functions in capacitors. (1) They provide mechanical support for very large metal sheets at very small separations. Indeed, most capacitors employ thin metal films or foils separated by paper or plastic films. (2) For a given geometry, dielectrics increase the capacitance by a factor κ. (3) Dielectrics can withstand higher fields without electrical breakdown than can air, so they increase the maximum useable voltage for the capacitor.

Electrolytic capacitors achieve a relatively large capacitance in a small volume. One conductor is a metal foil, usually aluminum or tantalum, and the other conductor is an *electrolyte*, a moist paste or liquid that conducts electricity by the motion of ions. A chemical reaction occurs on the surface of the metal foil to produce a nonconducting oxide layer, sometimes only a few atoms thick. With such an extremely thin separation between conductors the capacitance becomes enormous. Although used widely, electrolytic capacitors have certain limitations. The polarity of the metal conductor must always be positive; with a reverse polarity a chemical reaction occurs that breaks down the oxide layer.

EXAMPLE 27-8

The plates of a parallel-plate capacitor each have an area of 40 cm² and are separated by a mica sheet 0.5 mm thick. (a) Find the capacitance. Calculate (b) the maximum voltage and (c) the maximum charge that this capacitor can have without electrical breakdown.

SOLUTION

(a) From Equations (27-4) and (27-11),

$$C = C_0 = \frac{\kappa \varepsilon_0 A}{d} = \frac{(5)(8.85 \times 10^{-12} \, C^2/N \cdot m^2)(40 \times 10^{-4} \, m^2)}{(5 \times 10^{-4} \, m)}$$

$$C = 3.54 \times 10^{-10} \, F = \boxed{0.354 \text{ nF}}$$

(b) The maximum voltage is limited by the dielectric strength of the mica: 3×10^9 V/m. For a thickness $d = 5 \times 10^{-4}$ m, we have

$$V_{max} = Ed = (3 \times 10^9 \text{ V/m})(5 \times 10^{-4} \text{ m}) = \boxed{1.50 \times 10^6 \text{ V}}$$

(Because of possible irregularities in the mica, as a safety factor the maximum usable voltage would probably be set at a lower value.)

(c) The maximum charge is

$$Q_{max} = CV = (0.354 \times 10^{-9} \text{ F})(1.5 \times 10^6 \text{ V}) = \boxed{531 \, \mu C}$$

EXAMPLE 27-9

Consider the parallel-plate capacitor (plate area A) shown in Figure 27-12 where the space between the plates is filled with different thicknesses of two different dielectrics. Ignoring edge effects, find an expression for the capacitance.

SOLUTION

When the capacitor is charged, the electric field is perpendicular to the boundary between the dielectrics. Hence that boundary is an equipotential surface, and a conducting sheet could be placed at the boundary without any of the fields being altered within the capacitor. The conducting sheet could then be split as shown in Figure 27-12b, forming two capacitors in series. The capacitance C_1 of the upper capacitor, including the effect of its dielectric, is $C_1 = \kappa_1 \varepsilon_0 A / d_1$. Similarly, the capacitance C_2 of the lower capacitor is $C_2 = \kappa_2 \varepsilon_0 A / d_2$. The series combination of C_1 and C_2 becomes

$$\frac{1}{C} = \frac{1}{C_1} + \frac{1}{C_2} = \frac{1}{\varepsilon_0 A}\left[\frac{d_1}{\kappa_1} + \frac{d_2}{\kappa_2}\right]$$

Solving for C gives

$$C = \boxed{\varepsilon_0 A \left[\frac{\kappa_1 \kappa_2}{\kappa_2 d_1 + \kappa_1 d_2}\right]}$$

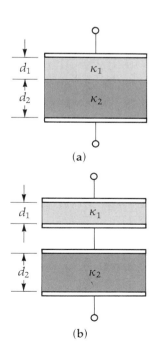

FIGURE 27-12
Example 27-9. A parallel-plate capacitor with two different dielectric materials.

27.5 Potential Energy of Charged Capacitors

As we have seen, configurations of charges have electric potential energy U. This potential energy implies the system could do work. For a charged parallel-plate capacitor, there are several ways this could be accomplished. For example, the force of attraction between the plates could do work if the plates were free to move toward each other. Or, if the charges could move, work could be done by each charge as it moves through the potential difference.

We can determine the potential energy by calculating the amount of work done by an external agent to charge the capacitor. The incremental work dW required to move a charge dq from the plate at the lower potential to the plate at the higher potential is

$$dW = V\,dq \qquad (27\text{-}12)$$

where V is the potential difference between the plates. However, the potential difference depends upon the charge q already deposited on the plates: $V = q/C$. Substituting this relation into Equation (27-12), we have

$$dW = \frac{q}{C}\,dq'$$

We obtain the total amount of work W required to charge the capacitor to a final charge Q by integrating:

$$W = \int_0^Q \frac{q}{C}\,dq = \frac{1}{C}\left(\frac{q^2}{2}\right)\bigg|_0^Q = \frac{1}{2}\left(\frac{Q^2}{C}\right)$$

Since the work done by the external agent is the gain in electric potential energy U of the capacitor; we have

$$U = \frac{1}{2}\left(\frac{Q^2}{C}\right) \qquad (27\text{-}13)$$

It is usually easier to determine the potential difference V rather than the charge Q. Since $Q = CV$, we may write Equation (27-13) in terms of V and C:

ENERGY U STORED IN A CHARGED CAPACITOR
$$U = \tfrac{1}{2}CV^2 \qquad (27\text{-}14)$$

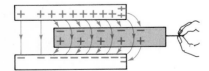

FIGURE 27-13
Example 27-10. As the dielectric is withdrawn from the capacitor, there are attractive forces between the charges on the plates and the induced charges on the surfaces of the dielectric.

EXAMPLE 27-10

A 2-nF parallel-plate capacitor is charged to an initial potential difference $V_i = 100$ V and then isolated. The dielectric material between the plates is mica ($\kappa = 5$). (a) How much work is required to withdraw the mica sheet? (b) What is the potential difference of the capacitor after the mica is withdrawn?

SOLUTION

(a) Work must be done to withdraw the mica because the charges on the plates exert forces of attraction on the induced charges of the mica (see Figure 27-13). The work required will be the difference in potential energy between the capacitor without the dielectric and the capacitor with the dielectric. Since the charge Q on the plates does not change when the dielectric is removed, we use Equation (27-13) to find the potential energy. As we will see, the potential V changes as the dielectric is withdrawn. The initial and final energies are

$$U_i = \frac{1}{2}\left(\frac{Q^2}{C_i}\right) \quad \text{and} \quad U_f = \frac{1}{2}\left(\frac{Q^2}{C_f}\right)$$

But the initial capacitance (with the dielectric) is $C_i = \kappa C_f$. Therefore:

$$U_f = \frac{1}{2}\kappa\left(\frac{Q^2}{C_i}\right)$$

Since the work done by the external force in removing the dielectric equals the change in potential energy, we have

$$W = U_f - U_i = \frac{1}{2}\kappa\left(\frac{Q^2}{C_i}\right) - \frac{1}{2}\left(\frac{Q^2}{C_i}\right) = \frac{1}{2}\left(\frac{Q^2}{C_i}\right)(\kappa - 1)$$

To express this relation in terms of the potential V_i, we substitute $Q = C_i V_i$, and evaluate:

$$W = \tfrac{1}{2}(C_i V_i^2)(\kappa - 1) = \tfrac{1}{2}(2 \times 10^{-9} \text{ F})(100 \text{ V})^2(5 - 1) = \boxed{4.00 \times 10^{-5} \text{ J}}$$

The positive result confirms that the final energy of the capacitor is greater than the initial energy. The extra energy comes from the work done *on* the system by the external force that pulled out the dielectric.
(b) The final potential difference across the capacitor is given by

$$V_f = \frac{q}{C_f}$$

Substituting $C_f = C_i/\kappa$ and $Q = C_i V_i$ gives

$$V_f = \kappa V_i = (5)(100 \text{ V}) = \boxed{500 \text{ V}}$$

Even though the capacitor is isolated and its charge remains constant, the potential difference across the plates does increase in this case.

EXAMPLE 27-11

Consider the capacitors shown in Figure 27-14a. The 4-μF and 12-μF capacitors are connected in series across a potential difference of 50 V. After becoming charged, the capacitors are disconnected from the source of potential, separated, and then rejoined in parallel, with positive plates together and negative plates together as shown in Figure 27-14b. (a) Find the initial and final potential energies. (b) Find the final voltage across the two capacitors in parallel.

SOLUTION

(a) The initial value of the series combination of two capacitors is given by Equation (27-10):

$$\frac{1}{C_i} = \frac{1}{C_1} + \frac{1}{C_2} = \frac{1}{4 \text{ } \mu\text{F}} + \frac{1}{12 \text{ } \mu\text{F}}$$

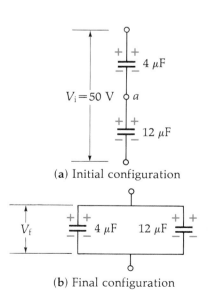

FIGURE 27-14
Example 27-11.

Solving for C_i gives $\qquad C_i = 3 \ \mu F$

The initial potential energy of the system of capacitors is given by Equation (27-14):

$$U_i = \tfrac{1}{2} C_i V_i^2$$

Evaluating, we obtain $\quad U_i = \tfrac{1}{2}(3 \times 10^{-6} \text{ F})(50 \text{ V})^2 = \boxed{3.75 \times 10^{-3} \text{ J}}$

As explained earlier, when capacitors in series are charged, *each* capacitor acquires the *same* magnitude of charge Q. This is

$$Q = C_i V_i = (3 \times 10^{-6} \text{ F})(50 \text{ V}) = 1.50 \times 10^{-4} \text{ C}$$

When connected in the new arrangement, the charge on the parallel combination of capacitors will be $2Q = 3 \times 10^{-4}$ C. The capacitance of the parallel combination is given by Equation (27-9):

$$C = C_1 + C_2$$

Evaluating, we find $\qquad C_f = 4 \ \mu F + 12 \ \mu F = 16 \ \mu F$

The final potential energy is given by Equation (27-13):

$$U = \frac{1}{2}\left(\frac{Q^2}{C}\right) = \frac{1}{2}\frac{(3 \times 10^{-4} \text{ C})^2}{(16 \times 10^{-6} \text{ F})} = \boxed{2.81 \times 10^{-3} \text{ J}}$$

Note that a *loss* in potential energy has occurred:

$$\Delta U = U_f - U_i = 2.81 \times 10^{-3} \text{ J} - 3.75 \times 10^{-3} \text{ J} = -9.4 \times 10^{-4} \text{ J}$$

(b) The final potential difference V_f is obtained from the relation $Q = CV$. In this case,

$$V_f = \frac{Q_f}{C_f} = \frac{3 \times 10^{-4} \text{ C}}{16 \times 10^{-6} \text{ F}} = \boxed{18.8 \text{ V}}$$

27.6 Energy Stored in an Electric Field

The previous example raises a few questions. Since the final energy of the system is less than the initial energy, where does the "missing" energy go? Also, where does the potential energy of a charged capacitor (or, for that matter, a single charged particle) reside? To answer the first question, we must realize that the redistribution of charge causes charges to flow through the wires connecting the capacitors. It can be shown that the resultant heating of the wires, no matter how small their electrical resistance (excluding zero resistance), exactly accounts for the energy loss of the charged capacitors. (The resistance of materials to the flow of charge is discussed in the next chapter.)

The other question, regarding *where* the potential energy resides, leads to an important new concept. Consider a charged capacitor. If an incremental charge dq is freed from the positive plate, it will be accelerated toward the negative plate by the electric field between the plates. The kinetic energy acquired by dq results in a corresponding reduction in the electric field (because the charge on the plates is now less). Therefore, it is reasonable to assume that *the potential energy of a charged capacitor resides in the electric field.*

We can derive an expression for the energy stored in an electric field by considering a parallel-plate capacitor, where the field is uniform. We have seen that, for a parallel-plate capacitor, $C = \varepsilon_0 A/d$ and $V = Ed$. Substituting these expressions for the energy U of a charged capacitor brings

$$U = \frac{1}{2}CV^2 = \frac{1}{2}\left(\frac{\varepsilon_0 A}{d}\right)(Ed)^2 = \frac{1}{2}\varepsilon_0 E^2(Ad) \qquad (27\text{-}15)$$

But (Ad) is the volume occupied by the electric field. We now define the *energy per unit volume* in the electric field as the **energy density** u_E (in joules/meter^3). Thus:

$$u_E = \frac{U}{Ad} = \frac{\frac{1}{2}\varepsilon_0 E^2 Ad}{Ad} = \frac{1}{2}\varepsilon_0 E^2$$

ENERGY DENSITY u_E IN AN ELECTRIC FIELD
(in free space)
$$u_E = \tfrac{1}{2}\varepsilon_0 E^2 \qquad (27\text{-}16)$$

Had there been a dielectric present, the capacitance C would have been increased by the factor κ. The previous analysis[4] would then lead to

ENERGY DENSITY u_E IN AN ELECTRIC FIELD
(in the presence of a dielectric)
$$u_E = \tfrac{1}{2}\kappa\varepsilon_0 E^2 \qquad (27\text{-}17)$$

Although we derived these results for the uniform electric field in a parallel-plate capacitor, Equations (27-16) and (27-17) are general expressions, valid for all field configurations.

EXAMPLE 27-12

An isolated conducting sphere of radius R has a charge Q. Show that the total energy stored in the surrounding electric field equals the energy stored in a charged capacitor, $U = \tfrac{1}{2}(Q^2/C)$, where C is the capacitance of the isolated sphere.

SOLUTION

An isolated sphere has a capacitance $C = 4\pi\varepsilon_0 R$ [Equation (27-6)]. The potential energy stored in the capacitor is thus

$$U = \frac{1}{2}\frac{Q^2}{C} = \underline{\underline{\frac{Q^2}{8\pi\varepsilon_0 R}}} \qquad (27\text{-}18)$$

The electric field outside a charged conducting sphere is the Coulomb field: $E = Q/4\pi\varepsilon_0 r^2$. At any point in this field, the energy density u_E is

$$u_E = \frac{1}{2}\varepsilon_0 E^2 = \frac{1}{2}\varepsilon_0\left(\frac{Q}{4\pi\varepsilon_0 r^2}\right)^2 = \frac{Q^2}{32\pi^2\varepsilon_0 r^4}$$

---

[4] Certain dielectric materials can be given a permanent electric dipole moment if they are melted and then allowed to solidify in the presence of an electric field. The resulting *electret* has a permanent electric field analogous to the permanent magnetic field of a magnet.

To find the total energy stored in the entire surrounding electric field, we note the spherical symmetry (E depends only on the radial distance r) and express the energy dU in the thin, spherical shell element of radius r and thickness dr. This thin shell has a volume $dV = 4\pi r^2\, dr$, and the energy dU within this shell is

$$dU = u_E\, dV = \left(\frac{Q^2}{32\pi^2 \varepsilon_0 r^4}\right)(4\pi r^2\, dr) = \frac{Q^2}{8\pi \varepsilon_0 r^2}\, dr$$

Integrating from $r = R$ to $r = \infty$, we obtain the total energy in the field:

$$U = \frac{Q^2}{8\pi \varepsilon_0} \int_R^\infty \frac{1}{r^2}\, dr = -\frac{Q^2}{8\pi \varepsilon_0}\left(\frac{1}{r}\right)\bigg|_R^\infty = \boxed{\frac{Q^2}{8\pi \varepsilon_0 R}}$$

This result is indeed the same as Equation (27-18).

Summary

A *capacitor* consists of two conductors separated by insulating material and has the ability to store charge. A capacitor with equal and opposite charges $\pm Q$ at a potential difference V has a *capacitance* C:

General definition: $\quad C = \dfrac{Q}{V}$

where Q is the magnitude of the charge on either plate and V is the potential difference. The SI unit of capacitance is the *farad* (F), or C/V. Capacitance depends solely on the geometry of the conductors. A simple capacitor formed of parallel plates of area A, separation d in a vacuum, has a capacitance of

For parallel plates: $\quad C = \dfrac{\varepsilon_0 A}{d}$

where ε_0 is the permittivity of free space. Capacitors with other geometries are discussed in the chapter.
For combinations of capacitors in circuits, the *equivalent capacitance* C_{eq} is

In parallel: $\quad C_{eq} = C_1 + C_2 + C_3 + \cdots$

In series: $\quad \dfrac{1}{C_{eq}} = \dfrac{1}{C_1} + \dfrac{1}{C_2} + \dfrac{1}{C_3} + \cdots$

The *electric potential energy* U stored in a charged capacitor is

$$U = \tfrac{1}{2} C V^2 \quad \text{(in joules)}$$

The *energy density* u_E in any electric field E is

$$u_E = \tfrac{1}{2}\varepsilon_0 E^2 \quad \text{(in joules/meter}^3\text{)}$$

When a dielectric material with a dielectric constant κ is introduced between the plates of an isolated charged capacitor, the capacitance increases by a factor of κ (always larger than 1),

$$C = \kappa C_0$$

and the potential difference V across the isolated capacitor decreases by a factor of κ,

$$V = \frac{V_0}{\kappa}$$

The potential difference decreases because the electric field produced by the charged plates aligns electric dipoles within the dielectric. The aligned dipoles produce an internal field in a direction opposite to the original field, resulting in a smaller net field.

Questions

1. The pattern of electric field lines between opposite but equal charges is undisturbed if we place a thin metal sheet halfway between the two charges so that the plane of the sheet is perpendicular to the line joining the charges. Why?
2. In terms of basic concepts, why is the capacitance of an isolated spherical conductor proportional to its radius?
3. Does the fringing effect in a parallel-plate capacitor tend to increase or decrease its actual capacitance compared with the value we calculate by ignoring the fringing effect? Why?
4. Is it possible for the plates of a capacitor to have different magnitudes of charge?

5. Why should air bubbles be avoided in oil-filled capacitors?
6. A dielectric slab is inserted between the plates of a charged parallel-plate capacitor. The capacitor is not connected to a battery. What happens to the energy of the capacitor? What happens to the potential difference across the plates of the capacitor?
7. Given three capacitors of different capacitances, how many different capacitance values can we obtain using one or more of the capacitors?
8. In view of its high dielectric constant, why is water not commonly used as a dielectric material in capacitors?
9. Capacitors are often stored with a wire connected across their terminals. Why?
10. How does the size of a given type of capacitor depend on its maximum energy storage capacity?
11. The oil in an isolated (but charged) oil-filled capacitor leaks out. What happens to the potential differences between the terminals of the capacitor?
12. The edge of a parallel-plate capacitor is placed in a pool of oil. The oil rises between the plates due to capillary action. Will the height to which the oil rises depend on the potential differences between the plates? In what way?
13. Due to the normal potential gradient in the earth's atmosphere, an electric field exists there. What are the difficulties in extracting the energy associated with this field and applying the energy for useful purposes?
14. Consider the two isolated spheres of Figure 26-15, Chapter 26. Each, alone, has a capacitance given by Equation (27-8). If we now add a fine conducting wire that connects the two spheres electrically, what is the resultant capacitance of the combination? (Are they connected in series or in parallel?)

Problems

27.2 Capacitance
27.3 Combinations of Capacitors

27A-1 A capacitor with a capacitance of 1 F, while commercially available, is difficult to visualize as a stack of plates separated by sheets of dielectric material. However, the capacitance of two parallel plates, each with an area of 1 cm^2 and separated by 1 mm of air, has a capacitance of about one 1 pF. Calculate a more exact value for the capacitance of such a capacitor.

27A-2 The *ionosphere* is a part of the earth's upper atmosphere (from about 50 km to 1000 km) that is sufficiently ionized by ultraviolet radiation from the sun so that the concentration of free electrons ($\sim 10^{11}$ m^3) affects the propagation of radio waves. The heights and intensities of ionization of these regions vary with the hour of the day, the season, sunspot activity, and other factors. Consider that the earth and a lowest ionosphere layer at 80 km altitude form a spherical capacitor. Calculate the capacitance of this earth–ionosphere system.

27A-3 Determine the equivalent capacitance for each of the networks of capacitors shown in Figure 27-15. Each capacitor has the same capacitance C.

27B-4 A collection of n identical capacitors may be connected in series or in parallel. When they are connected in parallel, the equivalent capacitance is N times larger than when the capacitors are connected in series. Express n in terms of N.

27B-5 Find the capacitance between terminals A and B of the capacitor network shown in Figure 27-16. (Hint: consider a potential difference across the terminals A and B and the way in which the charge is distributed among the capacitors.)

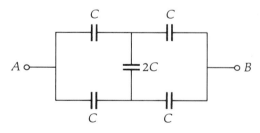

FIGURE 27-16
Problem 27B-5.

27B-6 An isolated capacitor of unknown capacitance has been charged to a potential difference of 100 V. When the charged capacitor is then connected in parallel to an uncharged 10-μF capacitor, the voltage across the combination is 30 V. Calculate the unknown capacitance.

27B-7 Consider the parallel-plate capacitor configuration shown in Figure 27-17. Derive an expression for its capacitance. Ignore fringing effects. Please explain your reasoning.

27B-8 A 2-μF capacitor and a 3-μF capacitor have the same maximum voltage rating V_{max}. Due to this voltage limitation, the maximum potential difference that can be applied to a series combination of these capacitors is 800 V. Calculate the maximum voltage rating of the individual capacitors.

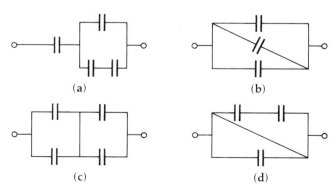

FIGURE 27-15
Problem 27A-3.

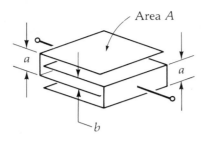

FIGURE 27-17
Problem 27B-7.

27B-9 Equation (27-7) gives the capacitance of a spherical capacitor, $C = 4\pi\varepsilon_0[ab/(b-a)]$, where a and b are the inner and outer radii, respectively. As both a and b become very large (while the difference between them remains small), over a small region the surfaces approach parallel plates. Show that this expression reduces to Equation (27-4), the capacitance of a parallel-plate capacitor.

27B-10 A potential difference of 200 V is applied to a series combination of a 2-μF capacitor and a 6-μF capacitor. (a) For each individual capacitor, find the potential difference and the charge. (b) The charged capacitors are isolated, then connected together in parallel with positive polarities joined and negative polarities joined. Find the new potential difference across the parallel combination and the charge on each capacitor. (c) If part (b) is repeated, except that the capacitors are connected in parallel with opposite polarities, what would be the final potential difference and the charge on each capacitor?

27B-11 Figure 27-18 shows a variable capacitor commonly used in the tuning circuit of radios. Alternate plates are connected together, with one group held fixed while the other group rotates together, resulting in a variable meshing of the plates. The area of each plate is A, with a spacing d between a plate of one group and the adjacent plate of the other group. The total number of plates is n. Ignoring fringing effects at the edges, show that the maximum capacitance is $C_{max} = (\varepsilon_0 A/d)(n-1)$.

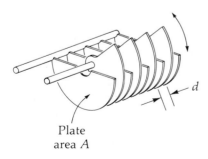

FIGURE 27-18
Problem 27B-11.

27.4 Dielectrics

27A-12 Estimate the maximum voltage to which a smooth, metallic sphere 10 cm in diameter can be charged without exceeding the dielectric strength of the dry air around the sphere.

27B-13 An isolated parallel-plate capacitor is given a charge Q. It is then filled with a dielectric material whose dielectric constant is κ. Show that the induced charge Q' that appears on the surfaces of the dielectric is $Q' = (1 - 1/\kappa)Q$.

27B-14 The plates of an isolated, charged capacitor are 1 mm apart and the potential difference across them is V_0. The plates are now separated to 4 mm (while the charge on them is preserved) and a slab of dielectric material is inserted, filling the space between the plates. The potential difference across the capacitor is now $V_0/2$. Find the dielectric constant of the material.

27B-15 A parallel-plate capacitor ($C = 5$ pF) is connected across a 20-V emf. Then the following procedure is carried out. (1) A dielectric slab ($\kappa = 4$) is inserted between the plates, filling the space completely. (2) The capacitor is disconnected from the emf. (3) The slab is withdrawn. Find (a) the final charge Q on the capacitor and (b) the final potential difference V across the capacitor.

27B-16 A detector of radiation called a Geiger tube consists of a closed, hollow, conducting cylinder with a fine wire along its axis. Suppose that the internal diameter of the cylinder is 2.5 cm and that the wire along the axis has a diameter of 0.2 mm. If the dielectric strength of the gas between the central wire and cylinder is 1.2×10^6 V/m, calculate the maximum voltage V_{max} that can be applied between the wire and the cylinder before breakdown occurs.

27B-17 A parallel-plate capacitor is constructed using a dielectric material whose dielectric constant is 3 and whose dielectric strength is 2×10^8 V/m. The desired capacitance is 0.25 μF, and the capacitor must withstand a maximum potential difference of 4000 V. Find the minimum area of the capacitor plates.

27B-18 A 1-μF, parallel-plate capacitor has a polystyrene dielectric. The maximum voltage rating of the capacitor is 1 kV. Assuming that the two identical conducting plates each occupy one-eighth of the total volume of the capacitor, find the volume of the capacitor.

27B-19 A parallel-plate capacitor with air between its plates has a capacitance C_0. A slab of dielectric material with a dielectric constant κ and a thickness equal to a fraction f of the separation of the plates is inserted between the plates in contact with one plate. Find the capacitance C in terms of f, κ, and C_0. Check your result by first letting f approach zero and then letting it approach one.

27B-20 A 0.1-μF parallel-plate capacitor has plates each with an area of 0.75 m^2 and a dielectric whose dielectric constant is 2.5. The capacitor is charged to a voltage of 600 V. (a) Find the charge on each of the plates. (b) Find the induced charge on the surfaces of the dielectric. (c) Calculate the electric field within the dielectric.

27B-21 Consider a cylindrical capacitor with two layers of dielectric material between the inner and outer cylinder, as shown in Figure 27-19. Ignoring end effects, derive an expression for the capacitance C of the capacitor in terms of the given parameters.

27B-22 The space between the plates of a parallel-plate capacitor has a volume $\mathscr{V}$ and is completely filled with a dielectric material that has a dielectric constant κ and a dielectric strength

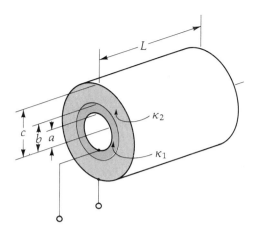

FIGURE 27-19
Problem 27B-21.

E_{max}. For a capacitor of capacitance C, derive (in terms of the given symbols) an expression for the maximum voltage V_{max} that can be applied to the capacitor.

27.5 Potential Energy of Charged Capacitors
27.6 Energy Stored in an Electric Field

27A-23 An 8-μF capacitor is placed across a potential difference of 20 V. (a) What energy is stored in the capacitor? (b) The charged capacitor is now removed from the source of potential difference and connected across the terminals of another uncharged 8-μF capacitor. After the charges redistribute themselves, what is the total energy stored in the capacitors? (c) Explain why the final stored energy is less than that initially stored on the original capacitor.

27B-24 Consider two parallel-plate capacitors that are connected in parallel as shown in Figure 27-20. The capacitors are identical except for the dielectric material in C_1. A potential difference of 150 V is applied across the terminals A and B, and then the source of potential difference is removed. (a) Find the charge on each capacitor. (b) Find the total energy stored in the capacitors. (c) If the dielectric material is now removed from C_1, determine the total energy stored in the capacitors. (d) Find the final voltage across the terminals A and B.

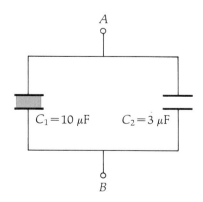

FIGURE 27-20
Problem 27B-24.

27B-25 Each plate of a parallel-plate capacitor has an area A; the plate separation is d. (a) Show that the potential energy U of a capacitor with charge Q can be written as $U = Q^2d/2\varepsilon_0 A$. (b) Using the result of Problem 27C-39, show that the force per unit area on a plate is $\frac{1}{2}\varepsilon_0 E^2$.

27B-26 A spherical capacitor is formed of two concentric metal spheres of radii 6 cm and 9 cm, respectively. The space between the spheres is filled with castor oil (see Table 27-1). Find the maximum energy that the capacitor can store without causing electrical breakdown of the dielectric.

27B-27 A parallel-plate capacitor with a polystyrene dielectric between the plates has a capacitance of 10 nF ($=10 \times 10^{-9}$ F). While the capacitor is attached to a 100-V battery, the dielectric is withdrawn. Find (a) the change in the charge on one of the plates, (b) the change in energy stored in the capacitor, and (c) the amount of work required to remove the dielectric.

27A-28 A metal sphere 50 cm in diameter is charged to a potential of 10 kV. Determine the energy density in the space just next to the outer surface of the sphere.

28B-29 Show that the energy storage capability of a parallel-plate capacitor is proportional to the volume between the plates of the capacitor.

Additional Problems

27C-30 A cylindrical capacitor is made of an inner conducting cylinder of radius a and a concentric outer conducting cylinder of radius b. The length L of the capacitor is sufficiently large that end effects may be ignored. The total charge on the inner cylinder is $+Q$, and the charge on the outer cylinder is an equal-magnitude negative charge $-Q$. (a) Starting with Gauss's law, find the electric field E between the cylinders. (b) Find the potential difference $V_b - V_a$ between the cylinders in terms of the given symbols. (c) Find the capacitance C.

27C-31 Show that the equation for the capacitance of a cylindrical capacitor of length L, $C = 2\pi\varepsilon_0 L/\ln(b/a)$ approaches the equation for the capacitance of a parallel-plate capacitor for $(b-a) \ll b$.

27C-32 A 4-μF capacitor and a 12-μF capacitor are connected in parallel across a voltage of 600 V. The voltage source is removed and the charged capacitors are isolated and then reconnected in parallel but with their polarities reversed, that is, positive-to-negative. (a) Calculate the voltage across the final parallel combination. (b) Find how much energy was lost in the reconnection.

27C-33 A 12-μF capacitor and two 2-μF capacitors, each with a maximum voltage rating of 200 V, are connected so that they produce a capacitance of 3 μF. Calculate the maximum voltage rating of the combination of capacitors.

27C-34 A parallel-plate capacitor has plate area A and separation d. A slab of copper of the same area and thickness t ($t < d$) is inserted symmetrically between the plates. (a) Find the new capacitance C of the capacitor. (b) Suppose that the copper slab is now moved closer to one plate so that the separation from that plate is half the separation between the slab and the other plate. Find the capacitance for this new geometry.

27C-35 Coaxial cable consists of a central wire surrounded by a plastic insulator, which in turn is surrounded by a woven metallic cylindrical conductor. Let κ be the dielectric constant of the insulator, a the radius of the central wire, and b the inner radius of the outer conductor. Derive an expression for the capacitance C per unit length L of the cable.

27C-36 A dielectric slab (dielectric constant κ) fills only half of the space between the plates of a parallel-plate capacitor, as shown in Figure 27-21. In terms of κ, derive an expression for the fraction f of the total energy that is stored in the dielectric.

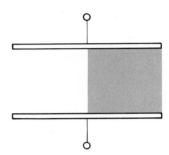

FIGURE 27-21
Problem 27C-36.

27C-37 Repeat Problem 27C-36 for the case shown in Figure 27-22.

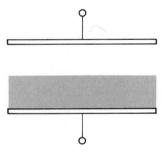

FIGURE 27-22
Problem 27C-37.

27C-38 Two capacitors, $C_1 = 2~\mu F$ and $C_2 = 6~\mu F$, originally uncharged, are connected in series and a potential difference of 200 V is applied across the combination. The capacitors are then disconnected. Then, without any charge being lost from either capacitor, they are connected together in parallel, with the positive plate of one joined to the positive plate of the other and with the negative plates joined together. (a) Find the new potential difference across the parallel combination and find the charge on each capacitor. (b) Calculate the total energy initially stored in the capacitors, and also calculate the final energy after they are connected in parallel. (If the energy changes, explain what happens to the "lost" energy.) (c) Suppose that, when the two capacitors are connected together, the positive plate of one capacitor is joined to the negative plate of the other capacitor (and the other two plates are similarly connected), again forming a parallel combination. Answer the same questions as in parts (a) and (b) for this new situation.

27C-39 Derive an expression for the force of attraction between the plates of a parallel-plate capacitor in terms of the capacitance C, the separation d of the plates, and the potential difference V between the plates. (Hint: consider the difference in stored energy dU when the plate separation is increased by an amount dx. This equals the work done $dW = F\,dx$.)

27C-40 Consider two concentric, conducting spherical shells with equal but opposite charges (a) Beginning with $u_E = \frac{1}{2}\varepsilon_0 E^2$, calculate the total energy contained in the field between the shells. (b) Show that this agrees with the energy stored in the capacitor: $\frac{1}{2}CV^2$.

27C-41 Consider two capacitors, C_1 and C_2, that are charged while connected in series. The charging voltage is removed and the capacitors isolated. The capacitors are then connected in parallel, positive-to-positive and negative-to-negative. Show that the fraction of the energy stored originally that is lost by connecting in parallel is given by $(C_1 - C_2)^2/(C_1 + C_2)^2$.

27C-42 Consider a parallel-plate capacitor that is formed of a stack of thin, square sheets of metal, edge lengths 10 cm, separated by similar slabs of dielectric of thickness d and dielectric constant 3. A total of n metal sheets is used and the stack forms a cube, 10 cm along each edge. The metal sheets are numbered consecutively. All the even-numbered sheets are connected together to form one terminal of the capacitor, and the odd-numbered sheets are connected to form the other terminal. (a) Assuming that the metal sheets have negligible thickness, find the thickness d of each slab of dielectric if the total capacitance is 1 F. (b) What is the total number of metal sheets?

27C-43 Einstein said that energy is associated with mass according to the famous relation $E = mc^2$. Estimate the radius of an electron, assuming that its charge is distributed uniformly over the surface of a sphere of radius R and that the mass energy mc^2 of the electron is equal to the total energy in the electric field between R and infinity. (Note: though this estimate is useful in some theoretical discussions, this classical model should not be taken literally. The answer one obtains depends crucially on the model chosen for calculation, and on the method of measurement used for experimental confirmation. High-energy scattering experiments suggest that the charge of the electron is concentrated in a region at least two orders of magnitude smaller than this problem assumes.)

27C-44 A charged spherical capacitor consists of two concentric spherical shells separated by a dielectric material. The inner shell has a radius a and the outer shell has a radius b. Derive an expression for the radius r (where $a < r < b$) inside of which half the energy is stored.

CHAPTER 28

Electric Current and Resistance

Don't worry—
Lightning
Never strikes twice
In the same
        BILLY BEE

28.1   Introduction

In this chapter we investigate the flow of electric charge—a *current*. Since the flow of charge occurs simultaneously throughout a conductor, there is no net accumulation of charge at any one place. The conducting path forms a *continuous closed loop* that contains an energy source to maintain the current. Networks of conductors and energy sources are called *circuits*. We will show that the current through various parts of a circuit is determined by two conservation laws: the *conservation of charge*, which means that the charge carriers are neither created nor destroyed in a circuit, and the familiar *conservation of energy*.

28.2   Electromotive Force $\mathscr{E}$

In order for a steady flow of electric charge, or *current*, to exist in a conductor, the conducting path must form a closed loop or complete circuit. The positive charges always move from a region of high potential toward a region of lower potential. Of course, after traveling around a complete loop in the direction of decreasing potential, when a charge arrives back at the starting point, it must be at the same potential as when it started. Therefore, at some location in the circuit there must be a device to do work on the charge and raise it through a potential difference. A local source of energy that performs this work on charges to raise their potential is called a **seat of electromotive force**, abbreviated *emf*. The script capital letter $\mathscr{E}$ designates the particular rise in potential.

A SEAT OF ELECTROMOTIVE FORCE (emf, $\mathscr{E}$)
    A seat of electromotive force is any device that transforms one source of energy into a source of electrical energy.

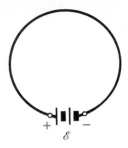

(a) The circuit diagram of a closed conducting path containing a seat of emf.

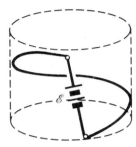

(b) A plot of the *potential V* (vertical axis) vs. *distance* for the closed circuit of (a). Positive charges entering the negative terminal of the battery are raised in potential an amount $\mathscr{E}$.

FIGURE 28-1
The source of emf $\mathscr{E}$ in a closed circuit raises the potential of charges moving through the emf.

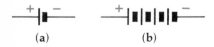

FIGURE 28-2
In the symbol for a battery, the end with the longer line designates the positive terminal at the higher potential. The + and − signs are sometimes omitted.

Examples of seats of emf are

(a) batteries, such as flashlight cells and automobile cells (a "battery" is really a series or battery of cells that transforms chemical energy into electrical energy);

(b) generators, such as a power station generator driven by a water turbine, or an automobile generator (commonly called an *alternator*, which produces an alternating current) driven by the automobile engine;

(c) solar cells, such as those that provide power for spacecraft; these transform radiant energy from the sun into electrical energy;

(d) certain biological cells that utilize chemical energy to maintain potential differences in nerves and muscle cells of living organisms.

By whatever means—chemical, mechanical, radiant, etc.—a seat of emf maintains a potential difference between its terminals. If an external circuit is connected to the terminals, electric charge will be driven around the circuit.[1] When the charge returns to the emf at the lower potential terminal, the emf does work on the charge, moving it through the seat of emf to the higher-potential terminal, ready to be driven around the external circuit again, Figure 28-1. Even though no external circuit is connected, the potential difference is maintained between the terminals.

We will restrict our discussion of electrical circuits to batteries as a source of electrical energy. However, our analysis of the circuits is valid for any type of emf. The symbol for a battery, Figure 28-2, is somewhat similar to the symbol for a capacitor, but in the context of a circuit they are seldom confused. The longer line indicates the higher-potential end.

An emf is somewhat analogous to a pump in a circulating water system that raises water vertically, increasing its gravitational potential energy. If the water pipes form a closed loop, the pump drives water around the system, Figure 28-3. A partial obstruction in the system (indicated by the portion of the pipe containing screens or gravel) will offer some mechanical resistance to the flow of water, somewhat reducing the rate of flow of the water. If the water pipe is blocked completely, so that no water can flow, the pump still exerts a pressure that will cause water to flow when the obstruction is removed. In the electrical case, a partial obstruction to the flow of charge is called an electrical *resistance* (symbolized by ⟍⋏⋏⋏⟋). If a switch (symbolized by ⟋⟋) in the electrical circuit is opened, so that no complete electrical path exists from one terminal to the other, the seat of emf still exerts an electromotive force that appears as an electrical potential difference V across the open switch terminals, ready to cause a flow of charge when the switch is closed.

28.3 Electric Current

An electric current is a flow of charge. Usually this occurs in solid conductors, such as the wire that connects your study lamp to the source of electrical energy in the wall outlet. There can be currents in liquids and gases in which both positive and negative ions move, and even in a vacuum there may be currents com-

---

[1] The first experiment to test the biological effects of electricity was perhaps made by Count Alessandro Volta (1745–1827), the Italian physicist who invented the voltaic cell. He connected 50 cells in series, put the ends of the wires in his ears, and reported that it felt like a strong blow to the head, followed by sounds of boiling soup!

posed of a beam of electrons, protons, or other charged particles that travel in the evacuated chambers of high-energy particle accelerators. For the present we will describe what is called "the classical theory of conduction" in metal conductors. In a metal there is an array of fixed positive ions and an equal number of "free" electrons that are free to move throughout the array or "lattice" of ions. Suppose we connect the ends of a long, uniform metal wire to the terminals of a battery. Because of the difference of potential between the terminals, an electric field **E** will be established throughout the wire, almost with the speed of light: $V_2 - V_1 = -\int_1^2 \mathbf{E} \cdot d\boldsymbol{\ell}$. Since the wire is uniform, the field will be constant from one end of the wire to the other.[2]

The electric field within the conductor will exert forces on the charges in the wire: $\mathbf{F} = q\mathbf{E}$. The positively charged ions are held in place in the lattice by the elastic forces between them. But the negatively charged electrons are free to move along the wire. In an oversimplified picture, an electron is accelerated by **E** (in a direction opposite to **E**) until it collides with a fixed positive ion, losing some speed in the collision. It accelerates again until the next collision, and so on. On the average, the electrons drift along the wire with an *average drift speed* v_d, ricocheting through the lattice of fixed positive ions. In these collisions, the electrons transfer some of their kinetic energies to the vibrational motion of the lattice, heating the metal. This behavior is similar to a stream of water descending a rocky rapids: *on the average*, the water does not accelerate as it falls through the gravitational field, but progresses at more or less uniform speed as it descends the rocky slope. The average distance between collisions, or *mean free path*, is about 220 ionic diameters for copper. Collisions between the electrons themselves are rare and have negligible effect on the resistivity.

The electric **current** I is defined as *the amount of charge per second that passes through a cross-section of the conductor.*

ELECTRIC CURRENT I $\qquad I = \dfrac{\Delta Q}{\Delta t} \qquad$ (28-1)

The SI unit of current is *coulombs per second* (C/s), called the *ampere* (A).[3] Milliamperes (mA = 10^{-3} A) and microamperes (μA = 10^{-6} A) are also commonly encountered. The *direction of the current*[4] is defined to be the direction that *positive* charges would move in response to the electric field. When the current is due to a flow of electrons that have negative charges, as in a metal conductor, the actual motion of the electrons is opposite to the direction we define for the current I. In certain cases, such as in an electrolyte or a semiconductor, both positive and negative charge carriers are moving simultaneously in opposite directions.

Let us examine the drift of electrons through the wire in a more quantitative way. Consider a segment of the wire shown in Figure 28-4. All of the electrons within the shaded volume will pass the plane perpendicular to the wire at P in

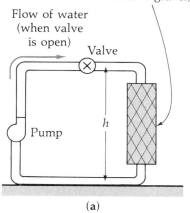

FIGURE 28-3
A water pump in a fluid-flow system is analogous to a seat of emf in an electrical circuit, and the flow of water is analogous to the electric current.

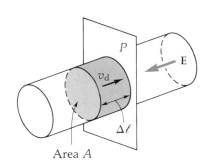

FIGURE 28-4
All the electrons in the shaded volume drift past the plane P in a time $\Delta t = \Delta\ell / v_d$.

---

[2] Note that this does not contradict the statements in previous chapters that in a perfect conductor the electric field is zero. There we dealt with the *static* case, in which charges are at rest and no battery is present to establish a potential difference between two different points on the conductor. Here, we discuss *dynamic* situations, in which charges are in motion because a battery *does* maintain two different points on a conductor at different potentials, establishing electric fields within the conductor.

[3] This unit honours André Ampère (1775–1836), a French physicist who gained considerable knowledge of the magnetic effects of currents.

[4] Although we specify a direction for the current, this does not make I a vector. The current in a wire remains the same even though we bend the wire or tie it in a knot. The arrow that designates a direction for I merely indicates the *sense* of the flow that positive charges would have.

a time Δt. The length $\Delta \ell$ of the shaded volume is

$$\Delta \ell = v_d \Delta t \qquad (28\text{-}2)$$

where v_d is the *average drift speed* of the electrons. The total charge within the volume $A \Delta \ell$ is

$$\Delta q = neA \Delta \ell \qquad (28\text{-}3)$$

where n is the number of conduction electrons per unit volume moving along the wire, e is the magnitude of the charge on the electrons, and A is the cross-sectional area of the wire. Combining Equations (28-2) and (28-3), we obtain the amount of charge ΔQ passing a given point per unit time Δt:

$$I = \frac{\Delta Q}{\Delta t} = nev_d A \qquad (28\text{-}4)$$

(Be careful not to confuse the area A with the abbreviation for ampere: A.)

EXAMPLE 28-1

Calculate the average drift speed of electrons traveling through a copper wire with a cross-sectional area of 1 mm² when carrying a current of 1 A (values similar to those for the electric wire to your study lamp). It is known that about one electron per atom of copper contributes to the current flow. The atomic weight of copper is 63.54 and its density is 8.92 g/cm³.

SOLUTION

We first calculate n, the number of current-carrying electrons per unit volume in copper. Assuming one free conduction electron per atom, $n = N_A \rho / M$, where N_A is Avogadro's number and ρ and M are the density and the atomic weight of copper, respectively.

$$n = \left(1 \frac{\text{electron}}{\text{atom}}\right) \frac{N_A \rho}{M}$$

$$n = \left(1 \frac{\text{electron}}{\text{atom}}\right)\left(6.02 \times 10^{23} \frac{\text{atoms}}{\text{mol}}\right)\left(\frac{1}{63.54 \frac{\text{g}}{\text{mol}}}\right)\left(8.92 \frac{\text{g}}{\text{cm}^3}\right)\underbrace{\left(\frac{10^6 \text{ cm}^3}{1 \text{ m}^3}\right)}_{\text{Conversion ratio}}$$

$$n = 8.45 \times 10^{28} \frac{\text{electrons}}{\text{m}^3}$$

From Equation (28-4), we obtain, for the drift speed v_d,

$$v_d = \frac{I}{neA} = \frac{1 \text{ A}}{\left(8.45 \times 10^{28} \frac{\text{electrons}}{\text{m}^3}\right)\left(1.602 \times 10^{-19} \frac{\text{C}}{\text{electron}}\right)(10^{-6} \text{ m}^2)}$$

$$v_d = \boxed{7.39 \times 10^{-5} \frac{\text{m}}{\text{s}}}$$

This is less than 0.1 millimeter per second. At this speed, it takes an electron 3.76 hours to travel just one meter!

You are probably surprised at the slow drift speed calculated in the example. If electrons typically travel through a wire at such a slow speed, why is it that, when we flip a wall switch, the light goes on almost instantaneously? The reason is that when a circuit is connected to a source of emf, the electric field is established in all parts of the circuit at nearly the speed of light. So when the final connection is made, forming a complete, closed path with the source of emf, electrons start to flow more or less simultaneously in all parts of the circuit. Even though the average drift speed of each electron is slow, all parts of the circuit feel the effects of the current almost instantaneously. Also, even though the cross-sectional area may vary along the length of the wire (thus causing the drift speed to vary), *the current has the same numerical value* I *throughout the circuit*.

The conduction electrons also take part in another motion. These electrons behave somewhat like the molecules of a gas, with random thermal velocities between collisions, whose average speed (for copper at room temperature) is about 1.6×10^6 m/s. So a typical current in a wire consists of random conduction-electron velocities of more than a million meters per second, upon which is superposed a slow drift speed of much less than a millimeter per second! See Figure 28-5.

This picturesque description of electrons moving randomly as a gas while they drift through the lattice of fixed ions, undergoing occasional collisions—the "classical" model of conduction—does lead to a quantitative theory that explains Ohm's law (discussed in the next section). However, many other phenomena disagree with this rather naive picture. A modern theory based upon quantum mechanics gives much better agreement with experimental measurements. In the more precise view we have today, the moving electrons have wavelike properties (Chapter 43). These waves interact with the array of atoms and ions such that, for a geometrically perfect lattice of identical ions, there is almost no inhibition to the electrons' motions. But if the array of ions has defects, such as a missing atom or the presence of an "impurity" atom in the array, the electron waves "scatter" from these irregularities, disrupting the motions of the electrons. Even concentrations as low as a few parts per million of impurity atoms are sufficient to make a large effect on the electrical resistance. At higher temperatures, the vibrational motions of the ions also destroy the perfect symmetry of the lattice, contributing strongly to the resistance of the material.

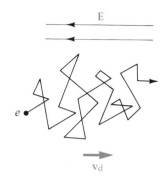

FIGURE 28-5
Free electrons in a metal have random motions similar to those of gas molecules. When an electric field is established in the metal, the electrons also experience an average drift velocity $\mathbf{v}_d$ opposite to the direction of **E**. This net drift speed of (negatively charged) electrons in one direction constitutes the (conventional) current I in the opposite direction.

28.4 Electrical Resistance

We now examine the relation $I = nev_d A$ to determine which factors are intrinsic properties of the current-carrying material itself and which are determined by the potential difference V across the material. The number of current-carrying charges per unit volume n in a metal conductor is clearly an intrinsic property of the material. The factor v_d is, in part, also an intrinsic property since it depends on the mobility of the electrons as they ricochet through the lattice of positive ions. However, this mobility also depends on the force driving the charges through the material, namely, $\mathbf{F} = e\mathbf{E}$. The electric field accelerates the electrons between collisions, but the net result of this jerky motion is an average drift velocity similar to the terminal velocity acquired by an object falling through a viscous medium. The drift velocity is such that the work done by the electric field just equals the kinetic energy "lost" in the collisions. Thus the drift velocity is proportional to the driving force $v_d \propto E$.

It is instructive to write Equation (28-4) in another form by considering a length L of conducting material that has a constant cross-sectional area A

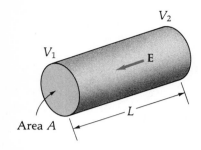

FIGURE 28-6
A uniform conductor of constant cross-sectional area A and length L. A potential difference V across the ends establishes an electric field E within the conductor, causing a current I.

across which we apply the potential difference V, Figure 28-6. Since $V_2 - V_1 = -\int_1^2 \mathbf{E} \cdot d\boldsymbol{\ell}$, the field $E = V/L$. The drift speed v_d is proportional to the driving force eE, so we have two relations, $E = V/L$ and $v_d \propto E$, which combine to give

$$v_d \propto \frac{V}{L} \qquad (28\text{-}5)$$

Substituting $I = nev_d A$, we obtain

$$I \propto \frac{VA}{L} \qquad (28\text{-}6)$$

The constant of proportionality in the above relation depends on the intrinsic properties of the particular material involved. We define this constant of proportionality to be $1/\rho$ (the Greek letter *rho*, ρ),

$$I = \left(\frac{1}{\rho}\right)\left(\frac{A}{L}\right) V \qquad (28\text{-}7)$$

where ρ is called the **resistivity** of the material. The SI units of ρ are (volt/ampere)(meter) or (V/A)(m). The unit V/A is called the *ohm* (Ω, the Greek capital letter *omega*),[5] so resistivity is usually expressed in SI units of the *ohm meter* ($\Omega \cdot$m). Occasionally the hybrid unit $\Omega \cdot$cm is also encountered. Table 28-1 gives typical resistivities at 20°C for various materials. In some contexts (see Section 28.7) it is more convenient to use the reciprocal of the resistance, defined as the **conductivity** $\sigma = 1/\rho$, in SI units of *siemens*.

We have discussed resistivity in terms of a *constant* of proportionality. In reality, the resistivity of a given material depends on a number of factors, such as moisture content, pressure, crystalline structure, and temperature. Analytically, temperature dependence is most easily handled. It is known from experiment that the fractional change in resistivity is approximately proportional to the corresponding change in temperature. That is,

$$\frac{\rho - \rho_0}{\rho_0} = \alpha(T - T_0) \qquad (28\text{-}8)$$

where α is the constant of proportionality called the **thermal coefficient of resistivity** and T_0 is the reference temperature for the handbook value of ρ_0.

Equation (28-8) is often written in the more convenient form

CHANGE OF RESISTIVITY WITH TEMPERATURE
$$\rho = \rho_0[1 + \alpha(T - T_0)] \qquad (28\text{-}9)$$

Values of ρ and α are given in Table 28-1 for a few common substances.[6]

Because R is proportional to ρ, we also have

$$R = R_0[1 + \alpha(T - T_0)] \qquad (28\text{-}10)$$

---

[5] This unit honors Georg Ohm, the German physicist who in 1827 discovered the proportionality between current and potential difference. See Ohm's law, Equation (28-11).

[6] Not listed in Table 28-1 are a number of alloys and a few elements called *superconductors* whose resistivity falls truly to *zero* at temperatures near absolute zero. See "Superconductivity," page 647.

TABLE 28-1 Resistivities and Thermal Coefficients of Resistivity

Material	Resistivity ρ at 20°C ($\Omega \cdot$m)	Thermal Coefficient of Resistivity α (1/°C)
Insulators		
Mica (clear)	2×10^{15}	-50×10^{-3}
Sulfur	1×10^{15}	-80×10^{-3}
Glass (plate)	2×10^{11}	-70×10^{-3}
Semiconductors		
Silicon	640	-75×10^{-3}
Germanium	0.46	-48×10^{-3}
Carbon (graphite)	1.4×10^{-5}	-0.5×10^{-3}
Conductors		
Aluminum	2.8×10^{-8}	3.9×10^{-3}
Bronze	18×10^{-8}	0.5×10^{-3}
Copper	1.7×10^{-8}	6.8×10^{-3}
Gold	2.4×10^{-8}	3.4×10^{-3}
Iron	10×10^{-8}	5×10^{-3}
Manganin {84% Cu, 12% Mn, 4% Ni}	44×10^{-8}	$<0.0005 \times 10^{-3}$
Mercury	96×10^{-8}	0.8×10^{-3}
Nichrome*	100×10^{-8}	0.4×10^{-3}
Platinum	10×10^{-8}	3.92×10^{-3}
Silver	1.6×10^{-8}	4.1×10^{-3}
Tungsten	5.7×10^{-8}	4.5×10^{-3}
Zinc	5.9×10^{-8}	4.2×10^{-3}

* A nickel–chromium alloy used in heating elements.

The variation of resistance with temperature forms the basis of a useful thermometer (see Problem 28A-7).

28.5 Ohm's Law

Equation (28-7) implies that, if a bar of resistive material were fitted with terminals at each end, the current through the bar would be proportional to the potential difference between the terminals. Surprisingly, this is nearly the case for a wide variety of substances. If the proportionality is exact, the substance conforms to **Ohm's law**. From Equation (28-7), we obtain

OHM'S LAW $$V = IR \tag{28-11}$$

where R is a constant called the *resistance*, which is independent of the current I through the substance. For a resistive substance of uniform cross-sectional area A and a length ℓ between the terminals, we see from Equations (28-7) and (28-11) that

$$R = \frac{\rho \ell}{A} \tag{28-12}$$

The unit of resistance is *ohms* (Ω).

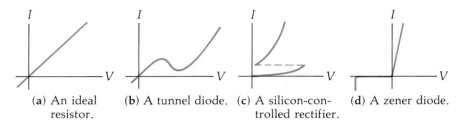

(a) An ideal resistor.   (b) A tunnel diode.   (c) A silicon-controlled rectifier.   (d) A zener diode.

FIGURE 28-7
Relationships between I and V for some common resistive devices. Only the ideal resistor in (a) obeys Ohm's law. Fortunately, many substances follow Ohm's law quite closely over a usable range of temperatures.

The functional relationship between the potential difference across the terminals of a resistive device and the current through it is not always a linear one. Figure 28-7 describes relationships between I and V for a number of devices. Of those shown, only the (ideal) resistor in (a) obeys Ohm's law, since there the value of R does not depend on the current. For all the others, R is still defined by the ratio of V to I, but it varies as the current changes. We will restrict our discussion to those resistive devices that obey Ohm's law.

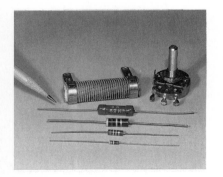

FIGURE 28-8
Typical resistors.

EXAMPLE 28-2

A resistor is constructed of a carbon rod that has a uniform cross-sectional area of 5 mm². When a potential difference of 15 V is applied across the ends of the rod, there is a current of 4×10^{-3} A in the rod. Find (a) the resistance of the rod and (b) the rod's length.

SOLUTION

(a) Applying Ohm's law (Equation 28-11), we find the resistance of the rod,

$$R = \frac{V}{I} = \frac{15 \text{ V}}{4 \times 10^{-3} \text{ A}} = 3750 \text{ }\Omega = \boxed{3.75 \text{ k}\Omega}$$

where kΩ designates the *kilohm* (1 kΩ = 10^3 Ω). Similarly, MΩ is used for the *megohm* (1 MΩ = 10^6 Ω). Note that, if R is written in units of kilohm and potential difference is written in terms of volts, then the current is in milliamperes (1 mA = 10^{-3} A), a more practical unit in modern electronics.

(b) The length of the rod is determined from Equation (28-12): $R = \rho\ell/A$. Solving for ℓ gives

$$\ell = \frac{RA}{\rho}$$

Substituting numerical values for R, A, and the values of ρ given for carbon in Table 28-1, we obtain

$$\ell = \frac{(3.75 \times 10^3 \text{ }\Omega)(5 \times 10^{-6} \text{ m}^2)}{(1.4 \times 10^{-5} \text{ }\Omega \cdot \text{m})} = \boxed{1.34 \times 10^3 \text{ m}}$$

Obviously, a resistor as large as 3.75 kΩ (a typical value) could not be constructed of pure carbon and still be part of a miniaturized electronic circuit. Resistors are constructed of a mixture of materials that is formulated not only to have the desired resistance and size, but also to contribute to its physical strength and constancy of resistance value under a variety of environmental conditions.

28.6 Joule's Law

We have seen that the potential difference across a resistor forces electrons through the resistor, with the electrons emerging with the same drift velocity they had when they entered. Recall that the potential energy lost by the electrons appears as thermal energy within the resistor. (An analogous situation is that of a boat driven at constant speed through the water by a motor; energy is dissipated in the water, heating it slightly.) For resistors, this effect is called *Joule heating*.

To calculate the Joule heating in a resistor, consider a simple circuit of a seat of emf and a resistor, as shown in Figure 28-9. The symbol ─⋀⋁⋀─ is used for resistors that obey Ohm's law, and solid lines indicate resistanceless conductors of current. Consistent with the usage introduced by Benjamin Franklin, as well as that used today, *outside a seat of emf, the current I is in the direction from a point of higher potential to one of lower potential*. This is sometimes called **conventional current**. Of course, conduction electrons in metals move in the opposite direction because of their negative charge. On the other hand, positive charges contribute to current in many substances, including liquids, gases, and certain solid-state devices. So the direction in which *positive* charges would flow (whether or not such charges are actually moving) is the direction of the conventional current.

As we have done before, we will isolate the system consisting of the seat of emf and the resistor, so that no energy enters or leaves the system. Because energy is conserved within the system, we know that the energy acquired by the charges through the work done on them as they move through the seat of emf must be equal to the thermal energy developed in the resistor. The work dW done by the emf $\mathscr{E}$ on an element of charge dq is

$$dW = \mathscr{E}\, dq$$

The time rate at which work is done by the seat of emf is

$$\frac{dW}{dt} = \mathscr{E}\frac{dq}{dt}$$

or, by the definition of current (Equation 28-1), we have

$$\frac{dW}{dt} = \mathscr{E}I$$

It is important to notice that, while $\mathscr{E}$ is the work done per unit charge by the seat of emf, the charge acquires a corresponding increase in potential energy, that is, a potential energy per unit charge (V), while moving through the seat of emf. The potential difference across the terminals of the seat of emf is therefore V, the same as that across the resistor R. Conservation of energy requires that the rate at which work is done (dW/dt) by the seat of emf must

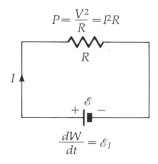

FIGURE 28-9
The work done by the seat of emf appears as thermal energy in the resistor.

equal the rate at which thermal energy is developed in the resistor. (Because this thermal energy is usually radiated away and thus "disappears" from the circuit, one often uses the phrase "dissipated in the resistor" for this thermal energy.) Using the symbol P for the power dissipated in the resistor, we have

$$P = \frac{dW}{dt} = \mathscr{E}I \tag{28-13}$$

or, since $\mathscr{E} = V$, the voltage across the resistor,

$$P = VI \tag{28-14}$$

This equation may be stated in another way. With Ohm's law, $V = IR$, Equation (28-14) becomes

$$P = I^2 R \tag{28-15}$$

The power dissipated in resistors is often called the "I squared R loss," while the total thermal energy developed is the *Joule heat*. Ohm's law also leads to $P = V^2/R$, so we have

POWER P DISSIPATED
IN RESISTIVE
CIRCUIT ELEMENTS

$$P \begin{cases} = I^2 R \\ = VI \\ = \dfrac{V^2}{R} \end{cases} \tag{28-16}$$

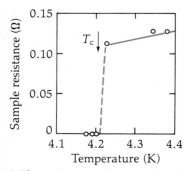

(a) The resistance of a sample of mercury versus temperature showing the sudden drop in resistance at the critical temperature T_c. Modern experiments indicate a measured resistivity for the superconducting state of no more than $10^{-25}\,\Omega\cdot\text{m}$; it may well be be truly zero.

(b) Current set up in a superconducting ring and ball produces magnetic forces that levitate the ball in space. The currents persist for many years without measurable change.

FIGURE 28-10
Superconductivity.

EXAMPLE 28-3

Referring to Figure 28-9, if $\mathscr{E} = 6$ V and $R = 12\,\Omega$, find (a) the rate at which the seat of emf does work and (b) the power dissipated in the resistor.

SOLUTION

(a) Conservation of energy requires that the answers to parts (a) and (b) be the same. Recognizing that the emf $\mathscr{E}$ is the potential difference V across the resistor, from Ohm's law we have $I = \mathscr{E}/R$. Substituting numerical values gives

$$I = \frac{\mathscr{E}}{R} = \frac{6\text{ V}}{12\,\Omega} = 0.500\text{ A}$$

The rate at which work is done by the seat of emf is given by Equation (28-13):

$$\frac{dW}{dt} = \mathscr{E}I = (6\text{ V})(0.5\text{ A}) = 3.00\,\frac{\text{J}}{\text{s}} = \boxed{3.00\text{ W}}$$

(b) The power dissipated in the resistor, although equal to dW/dt, may also be obtained from Equation (28-16):

$$P = I^2 R = (0.5\text{ A})^2 (12\,\Omega) = \boxed{3.00\text{ W}}$$

Superconductivity

In 1908, the Dutch physicist H. Kammerlingh Onnes succeeded in liquefying helium at 4.2 K and began to investigate various properties of metals at low temperatures. Three years later he discovered the phenomenon of *superconductivity*, the astonishing behavior of some materials in which the electrical resistance drops abruptly to zero below a certain *critical temperature* T_c, commonly a few degrees above absolute zero, Figure 28-10a. The consequences of zero resistance are surprising. For example, if a current is established around a superconducting ring, it continues indefinitely with no production of heat (there are no I^2R losses) and with no driving emf in the loop! Circulating currents have been set up that have persisted for years with no measurable loss.

An interesting example of the use of superconductivity is the *superconducting gravimeter*, Figure 28-11, for taking remarkably precise measurements of the earth's gravitational field. An aluminum shell, 2.54 cm in diameter, is plated with lead, which is superconducting at the temperature of liquid helium, 4.2 K. The sphere is supported in midair by currents established in two horizontal superconducting coils. As the coil currents build up, they produce an increasing magnetic field, which induces currents in the sphere. The resultant magnetic forces levitate the sphere in space without its physically touching any supports. The sphere is positioned symmetrically between six metal plates. If the value of g changes, the sphere will rise or fall vertically by a tiny amount. Its altered position changes the capacitance between the plates, which is sensed by external electrical circuits, thereby indicating a change in g. Figure 16-9 shows a graph of variations in the earth's field obtained with this instrument.

At present, superconducting electromagnets using liquid helium are employed in many scientific laboratories; in some cases the magnet windings carry many tens of thousands of amperes. An intense search is on for materials that become superconducting at higher temperatures, so that expensive liquid helium ($3 per liter) [or less costly liquid nitrogen (6¢ per liter) at 77 K] may be abolished. The discovery of a room-temperature superconductor would have far-reaching economic implications. Among the many possible benefits are cheaper electrical power transmission, efficient superconducting magnets for large particle accelerators, the magnetic levitation of trains, and powerful electric motors of compact size. If used for computer circuitry, superconductors would greatly increase the speed of computing, as well as reduce the size of computers.

The phenomenon is not just an extension of the normal conductivity of materials but is a wholly new quantum mechanical effect. Indeed, some good metallic conductors do not have this property, while certain ceramic compounds that normally are insulators do become superconducting—the latest (1988) reported at ~ 160 K. In 1972, John Bardeen, Leon Cooper, and Robert Schrieffer received the Nobel Prize for their theory of superconductivity. Its main feature is that, quantum mechanically, *pairs* of electrons can all move through the material at the same speed without giving up energy to the material itself.

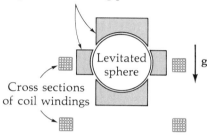

FIGURE 28-11
A superconducting gravimeter employs two horizontal current loops (shown in color) that levitate a sphere between capacitance-sensing plates. Two additional plates above and below the plane of the figure are not shown.

28.7 Current Density and Conductivity

In many cases, electric currents are not neatly confined to wires or other discrete conductors. Instead, the current is diffused over extended regions that are rather ill-defined, such as the atmosphere, the ocean, the earth, or the ionized gases of stellar atmospheres and plasmas. In these situations, we are more concerned with the **current density J** defined as *the current per unit area at a given point*:

CURRENT DENSITY J
(scalar form)
$$J = \frac{I}{A} \qquad (28\text{-}17)$$

By Equation (28-4), $I = nqv_dA$, so we define the *vector* current density **J** as

CURRENT DENSITY J
(vector form)
$$\mathbf{J} = nq\mathbf{v}_d \qquad (28\text{-}18)$$

Here, n is the number of free charge carriers per unit volume, and $\mathbf{v}_d$ is the average drift velocity. As usual, $\mathbf{v}_d$ is defined to be the velocity that *positive* charges would have. (If negative electrons are the carriers, the actual motion of the electrons is opposite to $\mathbf{v}_d$.)

By considering a conductor of length ℓ and uniform cross-sectional area A, we can derive an alternative form of Ohm's law. For such a conductor (assumed to obey Ohm's law), $I = V/R$ and $E = V/\ell$. Thus:

$$I = \frac{V}{R} = \frac{E\ell}{R} = \frac{E\ell}{\left(\rho \dfrac{\ell}{A}\right)} = \frac{EA}{\rho}$$

Dividing both sides by the area A, we have

$$J = \frac{I}{A} = \left(\frac{1}{\rho}\right) E = \sigma E \qquad (28\text{-}19)$$

OHM's LAW
(alternative form)
$$\mathbf{J} = \sigma \mathbf{E} \qquad (28\text{-}20)$$

The **conductivity** σ for the material is defined as the inverse of the resistivity, $\sigma \equiv (1/\rho)$, measured in SI units of $(\Omega \cdot \text{m})^{-1} \equiv$ *siemens* (S).

We thus have two ways of analyzing electric currents in materials:

Macroscopic approach (for a conductor of finite dimensions)	Microscopic approach (at a given point within a material)	
$I = \dfrac{V}{R}$	$\mathbf{J} = \sigma\mathbf{E}$	(28-21)

The macroscopic approach is useful for circuit elements that have finite dimensions, and we can express the electrical characteristics of V and R for the entire element. In contrast, the microscopic approach involves the *local* properties of current density and electric field *at a given point* within a material, with no reference to the physical extent of the material.

Sometimes the density of charges may vary from point to point. As a consequence, the current density **J** also varies. To obtain the total current I

through a given area A, we must integrate the current density over the total area,

$$I_{\text{total}} = \int \mathbf{J} \cdot d\mathbf{A} \tag{28-22}$$

where $d\mathbf{A}$ is the vector area element perpendicular to the area under consideration. Note that the current I is a scalar quantity, while the current density $\mathbf{J}$ is a vector.

EXAMPLE 28-4

The electron beam emerging from a certain high-energy electron accelerator has a circular cross-section of radius 1 mm. (a) If the beam current is 8 μA, find the current density in the beam, assuming that it is uniform throughout. (b) The speed of the electrons is so close to the speed of light that their speed can be taken as c with negligible error. Find the electron density in the beam. (c) How long does it take for an Avogadro's number of electrons to emerge from the accelerator?

SOLUTION

(a) $J = \dfrac{I}{A} = \dfrac{8 \times 10^{-6} \text{ A}}{\pi(1 \times 10^{-3} \text{ m})^2} = \boxed{2.55 \text{ A/m}^2}$

(b) From $I = nev_d$, we have

$$n = \dfrac{I}{ev_d} = \dfrac{2.55 \text{ A/m}^2}{(1.60 \times 10^{-19} \text{ C})(3 \times 10^8 \text{ m/s})} = \boxed{5.31 \times 10^{10} \text{ m}^{-3}}$$

(c) From $I = \Delta Q/\Delta t$, we have

$$\Delta t = \dfrac{\Delta Q}{I} = \dfrac{N_A e}{I} = \dfrac{(6.02 \times 10^{23})(1.60 \times 10^{-19} \text{ C})}{8 \times 10^{-6} \text{ A}}$$

$$= \boxed{1.20 \times 10^{10} \text{ s}} \quad \text{(or about 381 years!)}$$

EXAMPLE 28-5

The proton beam from an accelerator has a circular cross-section of radius 0.6 mm. Figure 28-12a shows how the charge density varies with radius. Find the total current in the beam.

SOLUTION

Because of the circular symmetry of the charge density, we choose an element of area in the form of a thin ring of radius r and width dr, Figure 28-12b. Thus, $dA = 2\pi r\, dr$. From the graph, the current density J (in SI units) is $(4 - r/0.6 \times 10^{-3})$ A/m^2 in the range $0 \leq r \leq 0.6$ mm. The total current is

$$I = \int_0^R J\, dA = \int_0^{0.6 \times 10^{-3} \text{ m}} (4 - r/0.6 \times 10^{-3} \text{ m})(2\pi r\, dr)$$

$$I = \left(4r - \dfrac{r^2}{1.2 \times 10^{-3} \text{ m}} \right)\Bigg|_0^{0.6 \times 10^{-3} \text{ m}}$$

$$= (2.40 - 0.30) \times 10^{-3} \text{ A} = \boxed{2.10 \text{ mA}}$$

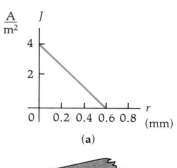

(a)

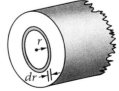

(b) A circular ring element of cross-sectional area $dA = 2\pi r\, dr$.

FIGURE 28-12
Example 28-5.

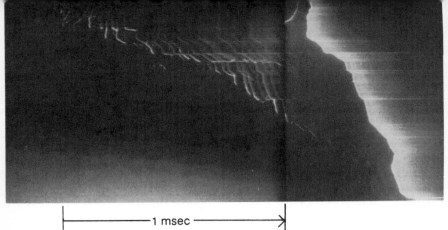
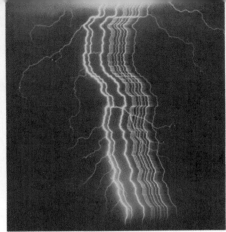

|← 1 msec →|

(a) This photograph was obtained with a moving-film camera in which a strip of film is moved at 27 m/s through the camera during the exposure. (Time increases left to right.) It shows a faint <u>stepped leader</u>: multiple discharges downward from the cloud, followed by a brilliant <u>return stroke</u> upward from the ground to the cloud. The return stroke is so rapid that it appears as a single image whose luminosity gradually dies out.

(b) The photographer resolved a multistroke lightning flash into 12 separate strokes by swinging the camera through an arc during the exposure. Usually, most of the branching occurs on the first stroke; subsequent strokes follow the same low-resistance path of ionized air, which is heated to $\sim 10^4$ K (hot enough to vaporize rocks). This tube of current has a diameter from a few millimeters to a few centimeters. The explosive expansion of the hot ionized air generates the thunder.

FIGURE 28-13

The Anatomy of a Lightning Flash. Most lightning *flashes* between a cloud and the ground are composed of about 3 to 5 *strokes* spread over a few tenths of a second, so that the eye perceives a flickering of the light intensity. Sometimes one of the strokes will persist longer than the others, producing a continuing luminosity as charges flow in the low-resistance conducting path for several tenths of a second.

There are many different types of flashes. The details can be analyzed by a moving-film camera that spreads the time sequence of events horizontally on the film. Figure 28-13a shows one type of flash that is started by a faintly glowing downward *stepped leader* from the base of the cloud toward the ground. Each successive discharge extends the conducting path of ionization about 50 m farther, with pauses of the order of 50 μs between them. Finally, as the leader nears the ground, the intense electric field between its lower tip and the ground results in a massive spark-over that initiates the large *return stroke* back to the cloud. This return stroke travels upward along the established path with speeds of one-tenth to one-half the speed of light (so it appears as a single streak on the film). While the stepped leader process may take ~ 0.02 s to travel several kilometers from the cloud base to the ground, the upward return stroke travels the same distance in only 70 μs. Peak currents of tens of thousands of amperes are typical, transferring a few ten of coulombs of charge.

The photographer took Figure 28-13b by swinging the camera through an arc while the shutter was open, thus separating the 12 strokes in this flash that lasted about 0.6 s. (These photographs are from Leon E. Salanave, *Lightning and Its Spectrum*, Univ. of Arizona Press, 1980. The book includes many fascinating photographs and offers suggestions on how you can take daytime photographs of lightning with a simple 35mm camera.)

Summary

A *seat of emf* $\mathscr{E}$ can perform work dW on a charge dQ, raising its potential by an amount $\mathscr{E}$:

$$dW = \mathscr{E}\, dq$$

An *electric current* I is the amount of charge per second that passes through a cross-section of a conductor:

$$I = \frac{dq}{dt}$$

The sense of direction of a current I is taken to be the direction that positive charges would move in response to the applied field **E**. (I is not a vector, however.) In the classical model of current in a metal conductor, the conduction electrons behave similar to molecules of a gas, undergoing random velocities in all directions. When an electric field is established within the conductor, the electrons experience forces that give them an *average drift speed* v_d (opposite to **E**), and cause them to collide with the fixed atoms and ions arrayed in the geometric pattern (or lattice) of the conductor. In a metal conductor of uniform cross-sectional area A and n conduction electrons per unit volume, each of magnitude charge e and average drift speed v_d, the current I is

$$I = nev_d A$$

OHM'S LAW $\qquad V = IR$

The *resistance* R of a rod of uniform cross-sectional area A and length ℓ is

$$R = \frac{\rho \ell}{A}$$

where ρ is the *resistivity* of the material. The *thermal coefficient of resistivity* α relates the resistivity ρ at temperature T to its value ρ_0 at a reference temperature T_0:

$$\rho = \rho_0[1 + \alpha(T - T_0)]$$

Certain elements and substances become *superconducting* at sufficiently low temperatures, where the electrical resistivity is truly zero.

On a microscopic scale, the *current density* J within a material is the *current per unit area at a given point*:

Scalar form $\qquad\qquad$ Vector form

$$J = \frac{I}{A} \qquad\qquad \mathbf{J} = nq\mathbf{v}_d$$

where n is the number of charge carriers per unit volume, q is the charge on each, and $\mathbf{v}_d$ is the average drift velocity that (positive) charges would have. The total current I through a given area A is

$$I = \int \mathbf{J} \cdot d\mathbf{A}$$

OHM's LAW
(alternative form) $\qquad \mathbf{J} = \sigma \mathbf{E}$

where $\sigma = 1/\rho$, the conductivity of the material.

Questions

1. Suppose you had a battery with unmarked terminals. How can the polarity of the terminals be determined? List as many ways as you can.
2. What are the merits, if any, in defining *conventional* current flow?
3. Why is the thermal coefficient of resistivity negative for insulators and semiconductors?
4. If the drift speed of electrons in a conductor is very slow, why does a ceiling light bulb go on so soon after the wall switch is closed?
5. What is the principal reason that resistors do not conform to Ohm's law?
6. A solid copper wire has a resistance R_1. The wire is used to form a hollow tube of the same length as the wire, so that the inside diameter is half the outside diameter. If the resistance of the tube is R_2, what is the value of the ratio R_2/R_1?
7. Early Edison light bulbs had essentially a carbon filament. Why was it necessary to operate these light bulbs with an external series resistor?
8. In Chapter 22, we were careful to point out that $\mathbf{E} = 0$ inside a conductor and that **E** is often not zero outside a conductor. Why, in this chapter, do we assert just the opposite?
9. How can the terminal voltage of a battery exceed the emf of the battery?
10. At one time automobiles utilized a 6-V electrical system. Why was a change made to the 12-V system, which is now used?
11. Of the two light bulbs designated by 25 W, 110 V and 100 W, 110 V, which has the higher filament resistance?

Problems

28.3 Electric Current

28A-1 A conductor carries a current of 5 A. How many electrons pass a given cross-section per second?

28A-2 A gas discharge tube has a metal plate (called an electrode) at each end, with a high potential difference between the two plates. An electron gun injects electrons into the gas at the negative electrode. Electrons that reach sufficiently high speeds ionize some gas atoms, producing additional electrons plus positive ions. As a result, 4×10^{17} electrons and 1×10^{17} singly charged positive ions pass a given cross-section of the tube per second, traveling in opposite directions. Find the magnitude and the sense of direction of the current in the tube.

28B-3 A silver wire 2 mm in diameter transfers a total charge of 420 C in 2 hr, 15 min. (a) Find the number of free electrons per cubic meter in the silver, assuming one conduction electron per atom. (b) What is the current in the wire? (c) Calculate the average drift speed of the electrons.

28B-4 The moving belt of a Van de Graaff generator is 30 cm wide and travels at 20 m/s. Charges are sprayed uniformly onto one side of the moving belt so that the effective current carried to the high potential sphere is 0.15 μA. Find the surface charge density σ on the belt.

28.4 Electrical Resistance
28.5 Ohm's Law

28A-5 A copper wire has a diameter of 2.60 mm. The resistivity of annealed copper is 1.77 $\mu\Omega\cdot$cm. Find the resistance of a 200-m length of this wire.

28A-6 Two solid cubes, A and B, are made of the same resistive material. Their edge lengths are, respectively, ℓ and 10ℓ. Find the ratio of their resistances R_A/R_B as measured between opposite faces of the cubes.

28A-7 One type of thermometer, a *resistance temperature detector* (RTD), utilizes the change of resistance of a platinum wire with changing temperature. A coil of platinum wire has a resistance of 100 Ω at 20°C. When the coil is immersed in liquid zinc as the zinc just begins to solidify, the resistance of the coil becomes 256 Ω. Find the melting point of zinc.

28A-8 Find the resistance of a nichrome wire 1 m long with a cross-sectional area of 0.1 mm^2 at (a) 20°C and (b) 1000°C.

28A-9 Find the temperature at which the resistance of a length of copper wire will be double its value at 20°C.

28A-10 A potential difference of 40 V exists across a 10-Ω resistor. How many electrons pass through a cross-section of the resistor in 5 min?

28B-11 We lengthen a wire with a resistance R to 1.25 times its original length by pulling it through a small hole. Find the resistance of the wire after it is stretched.

28B-12 A solid cube of silver (specific gravity = 10.50) has a mass of 90 g. (a) What is the resistance between opposite faces of the cube? (b) If there is one conduction electron for each silver atom, find the average drift speed of electrons when a potential difference of 10^{-5} V is applied to opposite faces. The atomic number of silver is 47, and its atomic mass is 107.87.

28B-13 A wire of constant diameter is composed of equal lengths of copper and iron wires joined at one end. If a potential difference of 12 V is applied across the ends of the combination, find the potential difference across the copper portion of the wire.

28.6 Joule's Law

28A-14 A 1000-Ω resistor is capable of dissipating a maximum power of 2 W. What is the maximum potential difference that should be applied to the resistor?

28A-15 In a television picture tube, electrons from the electron gun are accelerated to the screen through a potential difference of 25 kV. With an average beam current of 0.210 mA, how many watts are dissipated at the screen?

28A-16 A 12-V car battery is rated at 120 A$\cdot$hr (meaning that its initial charge is 120 ampere$\cdot$hours). While the car is parked, the two headlights, each rated 80 W, are inadvertently left on. Assuming that the terminal voltage remains constant, determine the number of hours that elapse before half the initial charge of the battery is used up. (See Problem 28C-43.)

28A-17 A 1300-W electric heater is designed to operate from 120 V. Find (a) its resistance and (b) the current it draws.

28A-18 Find the cost of electrically heating 100 L of water from 20°C to 90°C if the power company charges 8.4¢ per kW$\cdot$h.

28A-19 A generating station supplies power at 60 kV over transmission lines to a distant load. (a) If the voltage can be raised to 100 kV without damage to the power lines, how much additional power (at the same current) can be transmitted? (b) Will there be an additional transmission loss because of extra heating in the lines? Explain.

28A-20 When a light bulb at 20°C is first connected to a potential difference, the initial current through the tungsten filament is ten times the current when the lamp has heated up to its steady-state operating conditions. Find the operating temperature of the tungsten filament.

28A-21 Figure 28-14 shows a hollow cylindrical conductor of length L, with inner and outer radii a and b, respectively. The resistivity of the material is ρ. A potential difference is

FIGURE 28-14
Problems 28A-21, 28C-36, and 28C-44.

applied between the ends of the cylinder, establishing a current parallel to the axis of the cylinder. Derive an expression for the resistance R in terms of ρ, L, a, and b.

28B-22 A 500-W heating coil designed to operate from 110 V is made of nichrome wire 0.5 mm in diameter. (a) Assuming that the resistivity of the nichrome remains constant at its 20°C value, find the length of wire used. (b) Now consider the variation of resistivity with temperature. What power will the coil of part (a) actually deliver when it is heated to 1200°C?

28B-23 An electric utility company supplies a customer's house from the main power lines (120 V) with two copper wires, each 140 ft long and having a resistance of 0.108 Ω per 1000 ft. (a) Find the voltage at the customer's house for a load current of 110 A. For this load current, find (b) the power the customer is receiving and (c) the power dissipated in the copper wires.

28B-24 A certain toaster has a heating element made of nichrome resistance wire. When first connected to a 120-V voltage source (and the wire is at a room temperature of 20°C) the initial current is 1.8 A, but the current begins to decrease as the resistive element heats up. When the toaster has reached its final operating temperature, the current has dropped to 1.53 A. (a) Find the power the toaster consumes when it is at its operating temperature. (b) What is the final temperature of the heating element?

28B-25 An electric hoist operates at 240 V and uses a steady current of 9 A while lifting a 1700-lb load at the rate of 26 ft/min. Find (a) the power input to the hoist, (b) the power output (in horsepower), and (c) the efficiency of the system.

28B-26 A Van de Graaff accelerator delivers a total of 0.127 mC of charge to a target by a 4-MeV beam of alpha particles. (An α particle is the nucleus of a helium atom and contains two neutrons and two protons.) (a) Find how many α particles hit the target. (b) If the beam was on for 6 min, what was the average current in the beam? (c) Find the total energy (in joules) delivered to the target.

28B-27 A beam of high-energy alpha particles strikes an absorbing target. (An alpha particle is a helium nucleus, which has a positive charge equal in magnitude to twice the electronic charge.) If the beam current is 0.3 μA and the kinetic energy of the particles is 20 MeV, find (a) the number of particles striking the target per second and (b) the power absorbed by the target.

28.7 Current Density and Conductivity

28A-28 Beginning with $\mathbf{J} = \sigma \mathbf{E}$, derive Ohm's law $V = IR$ for a uniform cylindrical conductor.

28A-29 A current density of 6×10^{-13} A/m^2 exists in the atmosphere where the electric field (due to charged thunderclouds in the vicinity) is 100 V/m. Calculate the conductivity of the earth's atmosphere in this region.

28B-30 Find the thermal power per unit volume developed in a uniform copper wire 2.6 mm in diameter, carrying a current of 0.37 A.

28B-31 A potential difference of 5 V is applied between the ends of a nichrome wire 1.2 m long with a diameter of 0.5 mm. Find the current density J within the wire if the wire temperature is maintained at 20°C.

28B-32 The National Electrical Code for flexible copper wires used for interior electrical wiring in homes and buildings lists a maximum safe limit of 50 A for a rubber-insulated wire of diameter 0.162 in. (note the units). For a wire carrying this current, calculate (in SI units) (a) the current density J, (b) the electric field within the wire, and (c) the rate of thermal energy production for a 3-m length of wire.

28B-33 When a potential difference is applied across the ends of a conductor that has a uniform cross-section, an electric field E is established throughout the conductor (ignoring end effects). Show that the thermal power per unit volume within the conductor is σE^2, where σ is the conductivity of the conductor.

28B-34 A material of resistivity ρ has a uniform current density J throughout. Show that the power per unit volume developed in the material is ρJ^2.

Additional Problems

28C-35 A precise definition of the thermal coefficient of resistivity α is $\alpha = (1/\rho)(d\rho/dT)$, where ρ is the resistivity of the material at a temperature T. (a) If α is constant, show that $\rho = \rho_0 e^{\alpha(T - T_0)}$, where ρ_0 is the resistivity at a reference temperature T_0. (b) By making a series expansion for e^x, show that, for $\alpha(T - T_0) \ll 1$, the expression reduces to Equation (28-9).

28C-36 Refer to Problem 28A-21 and Figure 28-14. Suppose, instead, that a potential difference is applied between the inner and outer curved surfaces of the cylinder so that a current is established in the radial direction. Derive an expression (in terms of ρ, L, a, and b) for the resistance R of a length L of the cylinder when it is used in this fashion. [Hint: consider the resistance dR between the inner and outer surfaces of a thin cylindrical shell of radius r and thickness dr. The total resistance is the sum of all such elemental resistances in series.]

28C-37 A wire of length ℓ has a thermal coefficient of resistivity α. We can increase the resistance of the wire by either stretching it or increasing its temperature. Show that a fractional change in length $\Delta \ell / \ell$ corresponds to a temperature change ΔT by the relationship $\Delta \ell / \ell = \alpha \Delta T / 2$.

28C-38 A conductor of length ℓ and uniform cross-sectional area A is made from a material whose resistivity ρ varies with the distance x from one end according to $\rho = \rho_0(1 + bx)$. (a) What are the SI units of the constant b? (b) Derive a general expression for the resistance R of the conductor in terms of the given symbols.

28C-39 The current through a vacuum tube diode varies with the applied voltage as $I = (2.5 \times 10^{-4})V^{3/2}$ (in SI units). (a) Derive an expression for the resistance R as a function of the applied voltage V. (b) Derive an expression for the power P developed as a function of the applied voltage V. (c) Make qualitative graphs of these relationships and compare them with the corresponding graphs for an Ohm's-law resistance.

28C-40 A graphite (carbon) rod is attached to a nichrome rod of the same cross-sectional area. Find the ratio of the length of the graphite rod to that of the nichrome rod such that the resistance of the combination is independent of temperature over a small range of temperature.

28C-41 Figure 28-15 shows two resistors fabricated from the same resistive material. The ends are plated with a conducting substance. Assume that, in use, the current is uniform over any cross-sectional area that is perpendicular to the axes of the resistors. Show that they have the same resistance if the radius r of the cylindrical resistor is the geometrical mean $\sqrt{r_1 r_2}$ of the two radii in the truncated cone. [Hint: in (b), consider the resistance dR between opposite faces of a thin circular element, oriented perpendicular to the axis, of thickness dx and radius $y = r_1 + (r_2 - r_1)x/L$. The total resistance is the sum of all such elemental resistances in series.]

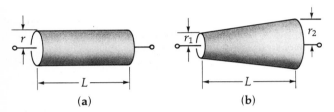

FIGURE 28-15
Problem 28C-41.

28C-42 As the applied voltage varies from 5 V to 25 V, the current through a certain electronic device remains constant at 50 mA. Make a graph of the effective resistance R of the device vs. V over the same voltage range.

28C-43 In Problem 28A-16, a more realistic assumption is that the battery voltage V drops exponentially according to $V = (12\text{ V})e^{-t/4}$, where t is in hours. Under this assumption, find the number of hours that elapse before half the initial charge of the battery is used up. (Note: the actual time is shorter than this estimate because the bulb resistances decrease as the current becomes smaller.)

28C-44 In Figure 28-14, suppose that the inner and outer curved surfaces of the object are plated with a conducting substance, forming two cylindrical conductors with a material of conductivity σ between them. A potential difference V is then applied between the two conductors, with the inner conductor at the higher potential. A current is thus established in the radially outward direction. Derive an expression for the current density J at a radius r (for $a > r > b$), in terms of a, b, L, σ, and V.

28C-45 Two thin concentric conducting spheres have radii a and b (with $a < b$). The space between the spheres is filled with a material of conductivity σ. Find the resistance R between the inner and outer spheres.

28C-46 In the previous problem, a potential difference V is established between the conducting spheres, with the inner sphere at the higher potential. Derive an expression for the current density J at a radius r (for $a < r < b$), in terms of the given symbols.

28C-47 A metal sphere of radius a is nested symmetrically inside a larger spherical metal shell of radius b. The space between the spheres is filled with a material of conductivity σ. When a potential difference V is established between the spheres, show that the current I between the spheres is $4\pi\sigma Vab/(b - a)$.

28C-48 Two rods made of iron and silver each have the same length $\ell = 80$ cm and radius $r = 2$ mm. They are joined together at one end, and a potential difference $V = 5$ V is established between the extremities of the combination. (a) Find the potential difference across each rod. (b) Determine the current density J in each and (c) the electric field E in each.

28C-49 There is a close analogy between the flow of heat because of a temperature difference (Section 19.6) and the flow of electrical charge because of a potential difference. The thermal energy dQ and the electrical charge dq are both transported by free electrons in the conducting material. Consequently, a good electrical conductor is usually also a good heat conductor. Consider a thin conducting slab of thickness dx, area A, and electrical conductivity σ, with a potential difference dV between opposite faces. Show that the current $I = dq/dt$ is given by

Charge conduction	Analogous heat conduction (Equation 19-19)
$\dfrac{dq}{dt} = -\sigma A \dfrac{dV}{dx}$	$\dfrac{dQ}{dt} = -kA \dfrac{dT}{dx}$

In the analogous heat conduction equation, the rate of heat flow dQ/dt (in SI units of joules per second) is due to a temperature gradient dT/dx, in a material of thermal conductivity k. Include a discussion of the origin of the minus sign in the charge conduction equation.

28C-50 As shown in Figure 28-16, a cylindrical conductor of conductivity σ has a cross-sectional area A_1 that tapers to a smaller cross-sectional area A_2. Because of a potential difference across the ends (not shown), an electric field exists within the conductor, causing a current. (a) Consider the closed Gaussian surface that surrounds the tapered region of the conductor. In terms of the given symbols, what is the electric flux Φ_E that enters the left-hand face? That leaves the right-hand face? What is the net flux through the entire Gaussian surface? (b) What are the current densities J_1 and J_2 in regions ① and ②? (c) Find the electric field E_2 in terms of E_1 and the areas A_1 and A_2. (d) Find the average drift speed v_2 in terms of v_1 and the areas.

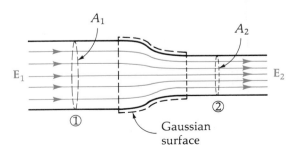

FIGURE 28-16
Problem 28C-50.

CHAPTER 29

DC Circuits

The moment man cast off his age-long belief in magic, Science bestowed upon him the blessings of the Electric Current.

JEAN GIRAUDOUX
The Enchanted (1933)

29.1 Introduction

A direct-current (DC) circuit is one in which the flow of charge is in only one direction. This chapter presents the methods of analyzing DC circuits that form networks of conducting paths containing sources of emf, resistors, and capacitors. From the conservation of energy and the conservation of electric charge, we obtain *Kirchhoff's rules*: two statements that greatly simplify circuit analysis. We then present the circuits that form a few common electrical devices, found in any laboratory, that measure currents, potential differences, and emf's. Finally, we analyze a special *RC* circuit in which the current varies with time.

29.2 Resistors in Series and in Parallel

The first step in the analysis of any circuit is to see whether we can simplify the current by combining some of its elements into simpler configurations. An array of resistors is particularly easy to reduce to an equivalent single resistor. The combination of two or more resistors connected *in series*, as in Figure 29-1a, is equivalent to a single resistance R_{eq} whose value can be found from

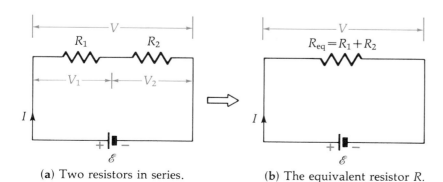

(a) Two resistors in series.   (b) The equivalent resistor R.

FIGURE 29-1
The combined resistance of two resistors in series is the sum of the two individual resistances.

the following analysis. The potential difference V across the combination is the sum of the potential differences across each resistor:

$$V = V_1 + V_2$$

Each has the same current I, so from Ohm's law ($V = IR$) we have

$$IR_{eq} = IR_1 + IR_2$$
or
$$R_{eq} = R_1 + R_2$$

If more than two resistors are connected in series, a similar reasoning shows that the equivalent single resistance R_{eq} is

RESISTORS IN SERIES
$$R_{eq} = R_1 + R_2 + R_3 + \cdots \qquad (29\text{-}1)$$

The combination of two or more resistors connected *in parallel*, as in Figure 29-2a, is equivalent to a single resistance R_{eq} whose value we can find by recognizing that, at point a, the current I splits into two parts: I_1 through R_1 and I_2 through R_2. From the **conservation of charge** we conclude that the rate $dq/dt = I$ at which charge enters point a equals the rate at which charge leaves (since no charge accumulates at point a as time goes on). Thus:

$$I = I_1 + I_2 \qquad (29\text{-}2)$$

From Ohm's law ($I = V/R$), we have

$$\frac{V}{R_{eq}} = \frac{V_1}{R_1} + \frac{V_2}{R_2} \qquad (29\text{-}3)$$

Because both resistors are connected between the same two points, a and b, the potential difference across each of the resistors is the same value: $V = V_1 = V_2$. Therefore, Equation (29-3) becomes

$$\frac{V}{R_{eq}} = \frac{V}{R_1} + \frac{V}{R_2}$$

Dividing by V, we have
$$\frac{1}{R_{eq}} = \frac{1}{R_1} + \frac{1}{R_2}$$

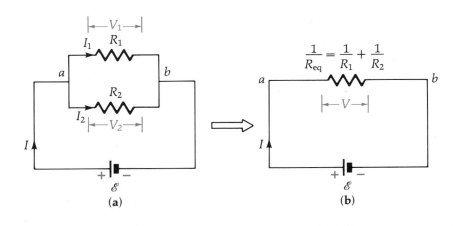

FIGURE 29-2
The equivalent resistance R of two resistors in parallel is less than the resistance of either alone.

When more than two resistors are connected in parallel between the same two junction points, a similar analysis gives

RESISTORS
IN PARALLEL
$$\frac{1}{R_{eq}} = \frac{1}{R_1} + \frac{1}{R_2} + \frac{1}{R_3} + \cdots \qquad (29\text{-}4)$$

for the equivalent resistance R_{eq}. Note that the equivalent resistance of a parallel combination is always *less* than any of the individual resistances alone. Also, it is helpful to remember that resistors add in parallel the way that capacitors add in series, and vice versa.

EXAMPLE 29-1

Find the equivalent resistance of the resistor network shown in Figure 29-3a.

SOLUTION

Usually the best procedure is to combine groups of parallel resistors to form a single equivalent resistor and groups of series resistors to form a single equivalent resistor. These combinations can then be combined further to reduce the entire network to a single equivalent resistor. In this example, we will combine R_1 and R_2 to form a single resistor R_{12}. Since they are in *parallel*, we utilize Equation (29-4):

$$\frac{1}{R_{12}} = \frac{1}{R_1} + \frac{1}{R_2} = \frac{1}{6\,\Omega} + \frac{1}{12\,\Omega}$$

Solving for R_{12} gives $\quad R_{12} = 4\,\Omega$

We next combine R_{12} and R_3 as shown in Figure 29-3b. Since these are in *series*,

$$R_{eq} = R_{12} + R_3 = 4\,\Omega + 5\,\Omega = \boxed{9.00\,\Omega}$$

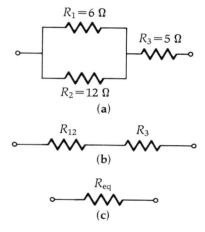

FIGURE 29-3
Example 29-1.

EXAMPLE 29-2

Three 60-W, 120-V light bulbs are connected across a 120-V power source, as shown in Figure 29-4. Find (a) the total power dissipation in the three light bulbs and (b) the voltage across each of the bulbs. Assume that the resistance of each bulb conforms to Ohm's law (even though in reality the resistance increases markedly with current).

SOLUTION

(a) The first step is to determine the resistance of each light bulb. From Equation (28-16),

$$P = \frac{V^2}{R}$$

Thus: $\quad R = \dfrac{V^2}{P} = \dfrac{(120\text{ V})^2}{60\text{ W}} = 240\,\Omega$

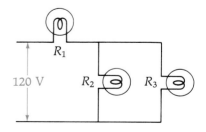

FIGURE 29-4
Example 29-2.

We obtain the equivalent resistance R_{eq} of the network of light bulbs by applying Equations (29-3) and (29-4):

$$R_{eq} = R_1 + \frac{1}{\left(\frac{1}{R_2} + \frac{1}{R_3}\right)} = 240\ \Omega + 120\ \Omega = 360\ \Omega$$

The total power dissipated in the equivalent resistance of 360 Ω is

$$P = \frac{V^2}{R_{eq}} = \frac{(120\ \text{V})^2}{360\ \Omega} = \boxed{40.0\ \text{W}}$$

(b) The current through the network is given by Equation (28-16):

$$P = I^2 R_{eq}$$

Solving for I gives $\quad I = \sqrt{\frac{P}{R_{eq}}} = \sqrt{\frac{40\ \text{W}}{360\ \Omega}} = \frac{1}{3}\ \text{A}$

The potential difference across R_1 is

$$V_1 = IR_1 = (\tfrac{1}{3}\ \text{A})(240\ \Omega) = \boxed{80.0\ \text{V}}$$

The potential difference V_{23} across the parallel combination of R_2 and R_3 is

$$V_{23} = IR_{23} = \left(\frac{1}{3}\ \text{A}\right)\left(\frac{1}{\frac{1}{240\ \Omega} + \frac{1}{240\ \Omega}}\right) = \boxed{40.0\ \text{V}}$$

29.3 Multiloop Circuits and Kirchhoff's Rules

When we analyze a network of many circuit elements, it is easiest first to reduce all parallel and series combinations of resistors to their simplest form. However, often it is not possible to reduce a circuit to just a single loop, particularly if the network has more than one emf, so we must deal with a multiloop network as shown in Figure 29-5. The procedure for obtaining currents and voltages in various parts of the circuit is greatly simplified by **Kirchhoff's Rules**. These two rules are not fundamental laws of nature, but are ways of stating the conservation of energy and the conservation of charge in an especially convenient form for analyzing networks.

We first define a few terms. A *junction*, or *branch point*, in a network is where three or more conductors are joined. In Figure 29-5, points c and f are junctions. A *branch* is one of the single paths between two junctions. A *loop* is any closed conducting path, such as *abcfa* or *abcdefa*. As a charge moves around a loop through various potential increases and decreases, the conservation of energy requires that the sum of the voltage *increases* must equal the

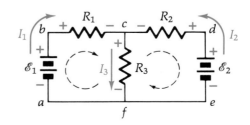

FIGURE 29-5
A network with two loops and two emf's.

sum of the voltage *decreases*. Assigning + and − signs to the voltage *increases* and *decreases*, respectively, we have the loop rule $\Sigma V = 0$. The junction rule arises from the fact that no charges can accumulate at a junction point, so if the currents entering a junction are considered as + and those leaving as −, the algebraic sum of the currents into a junction is zero: $\Sigma I = 0$.

KIRCHHOFF'S RULES

(1) **The Loop Rule: $\Sigma V = 0$.** The sum of the voltage increases and decreases around any closed loop is zero.

(2) **The Junction Rule: $\Sigma I = 0$.** The algebraic sum of all currents entering a junction is zero. (Currents entering are positive, and those leaving are negative.)

We apply these rules most easily by following a rather formal procedure. We illustrate the procedure by solving for the currents in the circuit of Figure 29-5. Here are the steps:

(1) *Label the polarity of each seat of emf with + and − signs.* Notice that, in the circuit shown, the two seats of emf oppose each other. That is, one emf may be able to force current through the other emf in the "backward" direction.

(2) *Draw an arrow showing the current direction in each branch of the circuit.* If you can guess ahead of time which direction is reasonable, choose it. If not, assign the current in some direction. (If you guess wrong, the numerical answer for that current will be a *negative* number, indicating that the actual current is in the opposite direction.)

(3) *According to the direction assumed for each current, label each resistor with a + at the end with the higher potential and a − at the other end.* Note that the current direction in a resistor is from a higher to a lower potential. Thus the end at which the current *enters* has a + label.

(4) *Establish a direction for traveling around each individual loop.* In this example, we will traverse each loop in a *clockwise* sense, as indicated by the dashed circular arrows. (The directions are arbitrary: we could have chosen a counterclockwise sense for either or both loops.) There is a third-loop path around the outer branches. But, having chosen the other two loops, we will see that this third path is redundant and need not be considered. The only criterion that must be met is that *every branch be traversed at least once by a loop path*.

(5) *Starting at any convenient point, travel around each independent loop in the direction chosen, keeping track of the potential increases and decreases.* (Increases are positive; decreases are negative.) Equate the sum to zero. For the circuit chosen, we start at point *a* in the left-hand loop and point *f* in the right-hand loop, obtaining the following equations:

$$\Sigma V = 0$$
$$\mathscr{E}_1 - I_1 R_1 - I_3 R_3 = 0 \qquad (29\text{-}5)$$
$$I_3 R_3 + I_2 R_2 - \mathscr{E}_2 = 0 \qquad (29\text{-}6)$$

Notice that, if we add these equations, we obtain the equation for the loop going clockwise around the outer branches. Therefore, traversing the outer loop adds no new information in the solution; it is redundant.

(6) *Equate the sum of the currents entering each independent junction to zero. (Currents that enter are positive and currents that leave are negative.)* For the circuit shown, at the upper junction we obtain

$$\Sigma I = 0$$
$$I_1 + I_2 - I_3 = 0 \tag{29-7}$$

The lower junction would yield the same equation except for the sign. So it is not an *independent* junction.

We now have *three* equations—(29-5), (29-6), and (29-7)—and *three* unknowns—I_1, I_2, and I_3. To organize the solution, it is helpful to rewrite these equations in a "standard" format, aligning terms for each unknown in vertical columns (and adding zeros for missing terms):

$$-R_1 I_1 + 0 - R_3 I_3 = -\mathscr{E}_1 \tag{29-8}$$
$$0 + R_2 I_2 + R_3 I_3 = \mathscr{E}_2 \tag{29-9}$$
$$I_1 + I_2 - I_3 = 0 \tag{29-10}$$

From this point on, any of the usual methods for solving simultaneous equations may be used.[1] The solutions are

$$I_1 = \frac{(R_2 + R_3)\mathscr{E}_1 - R_3 \mathscr{E}_2}{R_1 R_2 + R_1 R_3 + R_2 R_3} \tag{29-11}$$

$$I_2 = \frac{(R_1 + R_3)\mathscr{E}_2 - R_3 \mathscr{E}_1}{R_1 R_2 + R_1 R_3 + R_2 R_3} \tag{29-12}$$

$$I_3 = \frac{R_1 \mathscr{E}_2 + R_2 \mathscr{E}_1}{R_1 R_2 + R_1 R_3 + R_2 R_3} \tag{29-13}$$

If any of the currents is opposite to the direction we initially guessed, the value of I for that current will turn out to be negative when the numerical values are substituted for the network parameters in the equation.

29.4 The Superposition Principle

Provided circuit elements are *linear*—that is, resistors and emf's maintain their values regardless of the amount of current—we may solve the circuit of Figure 29-5 using the *principle of superposition*. This general procedure recognizes that the effect of just one emf working alone is independent of the effects produced by other emf's. So we may pretend that all emf's are absent except one, replacing the absent ones by conductors whose resistance is zero and then solving for the currents in the circuit due to the single remaining emf. (If the emf has an internal resistance, the emf is replaced by that resistance instead of a conducting wire.) Repeating the procedure in turn for the other emf's gives other sets of currents due to each of them working separately. The actual currents in the original network are the sum, or *superposition*, of these sets of partial currents. Figure 29-6 illustrates the procedure for the network of Figure 29-5. After replacing $\mathscr{E}_2$ by a conducting wire, we can further simplify the circuit of

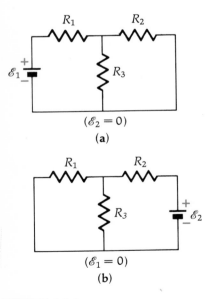

FIGURE 29-6
The superposition of the currents in circuits (a) and (b) gives the currents in the circuit shown on Figure 29-5.

---

[1] The actual calculation may be somewhat tedious. Although straightforward substitutions for unknowns will lead to the final answer, perhaps the most convenient procedure is the *determinant method* (Cramer's rule) for solving linear algebraic equations. Consult a mathematics text for details.

Figure 29-6a by finding the equivalent resistances for the parallel and series combinations. We find the currents in each resistor in the usual way. Similarly, after we replace $\mathscr{E}_1$ by a conducting wire, the circuit of Figure 29-6b can be solved. The actual current in each resistor is the sum of the currents found from solving the two simplified circuits. This procedure is often simpler than the brute-force solving of simultaneous equations.

EXAMPLE 29-3

In Figure 29-7a, find the current in each branch of the network.

SOLUTION

Method 1: Kirchhoff's rules

We choose currents in each branch as indicated in Figure 29-7a. The polarities for the assumed potential differences across each resistor are labeled with plus and minus signs (remembering that a current *enters* a resistor at the positive end). Starting at the bottom junction at *a*, we travel around the loops, equating the sum of the potential increases and decreases to zero (Kirchhoff's first rule). Omitting the units for simplicity, we obtain

$$\Sigma V = 0$$

Left loop (clockwise) $10 - 2I_1 - 4I_3 = 0$
Right loop (counterclockwise) $4 - 4I_3 = 0$

We next equate the sum of all currents entering the top junction to zero (Kirchhoff's second rule). Currents entering are positive, currents leaving are negative.

$$\Sigma I = 0$$
$$I_1 + I_2 - I_3 = 0$$

Rewriting the above equations in the standard format gives

$$-2I_1 + 0 - 4I_3 = -10$$
$$0 + 0 - 4I_3 = -4$$
$$I_1 + I_2 - I_3 = 0$$

These simultaneous equations are simple to solve by direct algebraic substitution. Substituting the value for I_3 from the second equation into the other two, we obtain

$$\boxed{I_1 = 3.00 \text{ A}} \quad \boxed{I_2 = -2.00 \text{ A}} \quad \boxed{I_3 = 1.00 \text{ A}}$$

The minus sign for I_2 signifies that the current in that branch is actually opposite to the direction assumed.

Method 2: Principle of superposition

With this method, we successively replace each of the emf's by a conductor with zero resistance and solve for the currents due to the other emf, obtaining the two simplified circuits in Figures 29-7b and c. To indicate that the currents in these circuits are only partial currents, we use single and double primes. We here assume currents in each branch that are plausible for the modified circuits. Thus, in (a), I_1' would be the current if $\mathscr{E}_2$ were the only battery; similarly, I_1'' would be the current if $\mathscr{E}_1$ were the only battery. The actual current direction in R_1 will depend on which of these currents is larger.

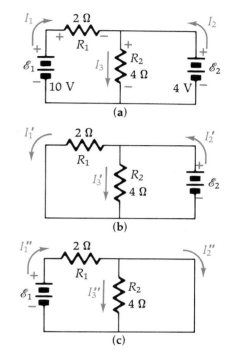

FIGURE 29-7
Example 29-3.

Circuit (b) is simply two resistors in parallel across $\mathscr{E}_2$. From Ohm's law ($I = V/R$), the current in each is

$$I'_1 = \frac{4\ V}{2\ \Omega} = 2\ A \quad \text{and} \quad I'_3 = \frac{4\ V}{4\ \Omega} = 1\ A$$

From Kirchhoff's junction rule we get

$$\Sigma I = 0$$

or

$$I'_2 = I'_1 + I'_3 = 2\ A + 1\ A = 3\ A$$

In circuit (c), the conducting wire on the right-hand side has zero resistance, so all the current flows through this parallel branch and none through R_2. (We say that R_2 has been "shorted out.") Thus:

$$I''_3 = 0 \quad \text{and} \quad I''_1 = I''_2 = \frac{10\ V}{2\ \Omega} = 5\ A$$

We now superpose these two sets of currents (noting their assumed directions) to find the actual currents in the original circuit. In resistor R_1, $I'_1 = 2\ A$ is toward the left while $I''_1 = 5\ A$ is toward the right. The actual current in that branch is therefore

$$I''_1 - I'_1 = 5\ A - 2\ A = \boxed{3.00\ A} \quad \text{(toward the right)}$$

In R_2, both I'_3 and I''_3 were assumed to be in the same direction. So

$$I'_3 + I''_3 = 1\ A + 0 = \boxed{1.00\ A} \quad \text{(down)}$$

In the right-hand branch, I'_2 and I''_2 were assumed to be in opposite directions, so the actual current is

$$I''_2 - I'_2 = 5\ A - 3\ A = \boxed{2.00\ A} \quad \text{(down)}$$

EXAMPLE 29-4

Verify that in the previous example the power exchanged between the sources of emf and the rest of the circuit does illustrate energy conservation.

SOLUTION

The current I_1 in the emf $\mathscr{E}_2$ is *in* the direction of increasing potential. Thus the seat of emf is supplying power P_1 to the rest of the circuit:

$$P_1 = \mathscr{E}_1 I_1 = (10\ V)(3.00\ A) = \underline{30.0\ W}$$

The currents I_1 and I_3 are in the resistors R_1 and R_2, respectively. Thermal energy is continuously being developed in the resistors at the rate $I^2 R$, so the Joule power P_2 in the resistors is

$$P_2 = I_1^2 R_1 + I_3^2 R_2 = (3.00\ A)^2(2\ \Omega) + (1.00\ A)^2(4\ \Omega) = \underline{22.0\ W}$$

The current I_2 in the seat of emf $\mathscr{E}_2$ is forced through this emf in the "backward" direction. (This is the process involved in charging a battery.) As we follow charges through the seat of emf they undergo a drop in potential, transferring their potential energy to the chemical energy stored within the seat of emf. Thus, the power P_3 stored in the seat of emf $\mathscr{E}_2$ is

$$P_3 = \mathscr{E}_2 I_2 = (4 \text{ V})(2.00 \text{ A}) = \underline{8.00 \text{ W}}$$

We see that the emf $\mathscr{E}_1$ supplies power to the rest of the circuit, the emf $\mathscr{E}_2$ absorbs power, and thermal power is developed in the resistors. If energy conservation holds true, then

$$\begin{bmatrix} \text{Rate of energy} \\ \text{given up by } \mathscr{E}_1 \end{bmatrix} = \begin{bmatrix} \text{Rate of thermal} \\ \text{energy developed} \end{bmatrix} + \begin{bmatrix} \text{Rate of energy} \\ \text{stored in } \mathscr{E}_2 \end{bmatrix}$$

$$\begin{array}{ccccc} P_1 & = & P_2 & + & P_3 \\ 30.0 \text{ W} & = & 22.0 \text{ W} & + & 8.00 \text{ W} \end{array}$$

This power equation balances, thus verifying the conservation of energy.

EXAMPLE 29-5

Calculate the potential difference between the points A and B for the circuit shown in Figure 29-8 and identify which point is at the higher potential.

SOLUTION

We identify the circuit as a *single*-loop circuit because points A and B are not connected. (No currents can exist in the branches containing $\mathscr{E}_2$ and R_2. Consequently, there is no potential difference across R_2.) The only current is a clockwise one in the loop at the left. The potential difference between A and B is the sum of the potential differences across $\mathscr{E}_2$, R_3, and R_2 (which is zero). We will first find the potential difference across R_3 by applying Kirchhoff's rules for the current in the loop. Assuming a clockwise direction, we sum the voltage increases and decreases around the loop:

$$\Sigma V = 0$$
$$\mathscr{E}_1 - IR_1 - IR_3 = 0$$

Solving for I and substituting numerical values gives

$$I = \frac{\mathscr{E}_1}{R_1 + R_3} = \frac{12 \text{ V}}{2 \text{ }\Omega + 4 \text{ }\Omega} = 2 \text{ A}$$

The potential difference V_3 across R_3 is thus

$$V_3 = IR_3 = (2 \text{ A})(4 \text{ }\Omega) = 8 \text{ V}$$

with the polarity indicated in Figure 29-8.
Starting at point B and moving along the network to point A, we find the potential V_{AB} of A with respect to B:

$$V_{AB} = VR_2 + IR_3 - \mathscr{E}_2 = 0 + 8 \text{ V} - 4 \text{ V} = \boxed{4.00 \text{ V}}$$

The potential at point A is thus 4 V higher than the potential at point B.

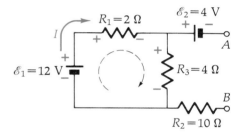

FIGURE 29-8
Example 29-5.

29.5 Applications

A number of different devices are used to measure the parameters of a circuit. They include the voltmeter, the ammeter, the Wheatstone bridge, and the potentiometer.

The Voltmeter

Potential differences across the components of a circuit are often measured with a *voltmeter*. A voltmeter usually has a sensitive current-measuring meter called a *galvanometer*, shown in Figure 29-9. The *sensitivity* of a galvanometer is the current that will cause a full-scale deflection of the needle, usually in the range of 10 μA to 1 mA. The meter movement itself has a resistance R_G. (In circuit diagrams, it is usually drawn as a separate resistor, though one should remember that this resistance is an internal part of the meter movement itself.) Usually the external voltages to be measured are much greater than that which will cause a full-scale deflection, so a resistance R is added *in series* to reduce the voltage that appears across the meter movement. Let us now calculate the resistance R for full-scale deflection when a potential difference V is applied to the terminals AB. If we let I_G be the current in the galvanometer that will produce a full-scale deflection, and R_G is the internal resistance of the galvanometer, then, from Ohm's law,

$$V = I_G(R + R_G) \qquad (29\text{-}14)$$

or

$$R = \frac{V}{I_G} - R_G \qquad (29\text{-}15)$$

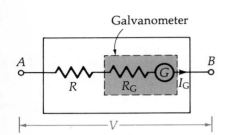

FIGURE 29-9
The voltmeter.

In order to change the range of a voltmeter, it is necessary to change the value of the series resistor. In a multirange voltmeter, this is usually accomplished by a switching arrangement. (See Problems 29B-30 and 29B-31.)

EXAMPLE 29-6

A galvanometer with a full-scale sensitivity of 1 mA requires a 900-Ω series resistor to make a voltmeter reading full scale when 1 V is across the terminals. What series resistor is required to make the same galvanometer into a 50-V (full-scale) voltmeter?

SOLUTION

We will use the values required for the 1-V voltmeter to obtain the internal resistance of the galvanometer. Applying Equation (29-14),

$$V = I_G(R + R_G)$$

we solve for R_G:

$$R_G = \frac{V}{I_G} - R = \frac{1 \text{ V}}{0.001 \text{ A}} - 900 \text{ }\Omega = 100 \text{ }\Omega$$

We then apply Equation (29-15) to obtain the series resistance required for the 50-V voltmeter:

$$R = \frac{V}{I_G} - R_G = \frac{50 \text{ V}}{0.001 \text{ A}} - 100 \text{ }\Omega = \boxed{49\,900 \text{ }\Omega}$$

Since a current I_G is required to operate a voltmeter, the introduction of a voltmeter into a circuit alters the currents in the circuit. Consequently, the voltmeter reading does not exactly represent the potential difference before the voltmeter was introduced. It is therefore desirable that a voltmeter have a very high internal resistance so that it does not draw much current from the circuit being measured. To compare various voltmeters, one calculates the *figure of merit*, or "quality," defined as the total resistance of the meter divided by the full-scale voltage reading. For the voltmeter described in Example 29-4, the quality is 1000 Ω/V. (This means that it is not a particularly good meter; a high-quality meter has a typical value of 20 000 Ω/V.) Analysis shows that a multirange meter that utilizes a given galvanometer movement will have the same figure of merit on all voltage scales. It may also be shown that the figure of merit is equal to the reciprocal of the current in the galvanometer that produces a full-scale deflection.

The Ammeter

A galvanometer measures very small currents. An *ammeter* measures larger currents by detouring, or *shunting*, some of the current around the galvanometer, as shown in Figure 29-10. Of the current I entering terminal A of the instrument, only a smaller portion I_G flows through the galvanometer movement. The voltage across R is the same as that across the galvanometer movement. Thus:

$$V_R = V_G$$
$$(I - I_G)R = I_G R_G$$

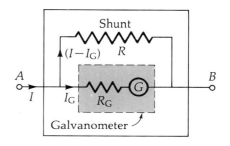

FIGURE 29-10
The ammeter.

The value of the shunt resistor R is

$$R = \frac{I_G R_G}{I - I_G} \qquad (29\text{-}16)$$

In a multirange ammeter, as in a multirange voltmeter, the value of R is usually changed by a switching arrangement.

EXAMPLE 29-7

An ammeter is constructed with a galvanometer that requires a potential difference of 50 mV across the meter movement and a current of 1 mA through the movement to cause a full-scale deflection. Find the shunt resistance R that will produce a full-scale deflection when a current of 5 A enters the ammeter.

SOLUTION

Direct application of Equation (29-16) requires a knowledge of R_G. However, R_G may be derived from the given quantities by Ohm's law:

$$R_G = \frac{V_G}{I_G}$$

Substituting this into Equation (29-16) gives

$$R = \frac{I_G R_G}{I - I_G} = \frac{I_G(V_G/I_G)}{I - I_G} = \frac{V_G}{I - I_G}$$

Note that the resulting equation is simply Ohm's law applied to the shunt resistor alone, where $I - I_G$ is the current through the shunt. Substituting the appropriate values into the equation, we have

$$R = \frac{50 \times 10^{-3} \text{ V}}{5 \text{ A} - 0.001 \text{ A}} = \boxed{0.010 \text{ } \Omega}$$

The shunt resistance is always very low for the measurement of currents that are much larger than the current requirements of the galvanometer. Just as in the construction of a voltmeter, a high-sensitivity galvanometer produces an ammeter that will introduce little change in a circuit when making a measurement.

A word of caution on the use of meters. Of course these instruments must not be connected to a circuit that will exceed their maximum range of values. But an additional hazard should be mentioned. Since an ammeter has an extremely low resistance, if it were mistakenly connected as a voltmeter *across* a source of voltage (instead of *in series* with other components), the resultant large current through the meter might easily destroy the ammeter. On the other hand, because a voltmeter has a large resistance, if it were mistakenly inserted in a circuit as an ammeter probably no damage would result. Just remember: *ammeters are connected in series in a line; voltmeters are connected in parallel across a potential difference.*

The Wheatstone Bridge

The primary use of a Wheatstone bridge is the measurement of resistance. Bridge-type circuits also have extensive application in electronics control circuits that detect small electrical imbalances.

A Wheatstone bridge circuit is shown in Figure 29-11. In the measurement of an unknown resistance R_x, the procedure is to adjust R_1 (the symbol ⇝ indicates a *variable* resistor) until no measurable current passes through the galvanometer. This is known as the *null-balance condition*. What are the conditions in the circuit at null balance? If no current passes through the galvanometer, the potential differences across R_1 and R_2 must be the same:

$$I_1 R_1 = I_2 R_2 \qquad (29\text{-}17)$$

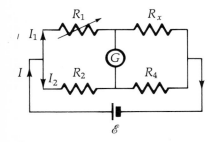

FIGURE 29-11
The Wheatstone bridge.

Moreover, the current through R_1 is the same as that through R_x. Similarly, the current through R_2 is the same as that through R_4. Therefore, the potential differences across R_x and R_4 are the same:

$$I_1 R_x = I_2 R_4 \qquad (29\text{-}18)$$

Eliminating I_1 and I_2 between Equations (29-17) and (29-18) and solving for R_x, we have

$$R_x = \left(\frac{R_4}{R_2}\right) R_1 \qquad (29\text{-}19)$$

In practice the ratio of R_4 to R_2 is known, as is the value of the adjustable resistance R_1, thereby yielding the value of the unknown resistance R_x. Note that the value of the seat of emf need not be known. (However, the magnitude of the seat of emf and the sensitivity of the galvanometer are important in the precision that the instrument can achieve, since both contribute to the galvanometer deflection when the bridge is nearly balanced.)

The Potentiometer

A potentiometer is an extremely important laboratory instrument because, in principle, it is capable of measuring potential differences in a circuit *when no current at all is being drawn from the circuit*. (This is in contrast to a voltmeter, which always requires some current for its operation.)

In Figure 29-12 the external battery causes a current in a long, uniform resistance wire called a *slide wire*. With the battery polarity shown, the potential along the slide wire drops uniformly with distance as one proceeds from the left end toward the right. The symbol $\mathscr{E}_s$ denotes a *standard cell* whose potential difference is precisely known. When the switch S introduces the standard cell in the circuit, the sliding contact (the small arrow) is moved along the slide wire until it reaches the point ℓ_s where the IR voltage decrease along the wire equals $\mathscr{E}_s$. This condition is indicated by a lack of current passing through the galvanometer G, that is, a null-balance condition. Because the potential change along the slide wire is uniform, this procedure calibrates the potential at all points along the wire in terms of the distance ℓ from the left end. Thus, the voltage V along the slide wire is proportional to the distance ℓ:

$$\frac{\mathscr{E}_s}{\ell_s} = \frac{V}{\ell} \tag{29-20}$$

After we calibrate the slide wire in this fashion, we change the switch to replace $\mathscr{E}_s$ with the voltage V_x to be measured. The sliding contact is moved again to achieve the null condition. The new setting ℓ_x then gives sufficient information to determine V_x.

$$\frac{\mathscr{E}_s}{\ell_s} = \frac{V_x}{\ell_x} \tag{29-21}$$

Solving for V_x gives

$$V_x = \left(\frac{\mathscr{E}_s}{\ell_s}\right)\ell_x \tag{29-22}$$

Standard cells are available whose emf $\mathscr{E}_s$ has been calibrated by the National Bureau of Standards or other agencies. In practice, the ratio $\mathscr{E}_s/\ell_s$ is set to a convenient value by the insertion of a variable resistance (not shown) in series with the external battery, allowing control over the amount of current in the slide wire (and thus the magnitude of the IR decrease along the wire).

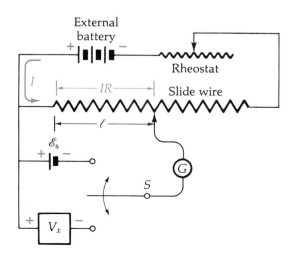

FIGURE 29-12
The basic potentiometer circuit. Note that the external battery, the standard cell $\mathscr{E}_s$, and the unknown potential V_x have the same polarity (here, positive) connected to the left end of the slide wire. The variable resistance of the *rheostat* controls the amount of current I in the slide wire to provide an appropriate range of voltage along the length of the slide wire.

Also, a protective resistance is sometimes added in series with the galvanometer to protect it from excessive currents in case the initial trial contact with the slide wire is far from the correct null-condition point. When the correct point is found (or closely approached), the protective resistance is shorted out, giving maximum sensitivity to the galvanometer reading.

The virtue of the potentiometer, *that it measures a potential difference when no current is being drawn from the circuit*, makes it a valuable instrument for measuring potential differences when no disturbance of the circuit being measured can be tolerated.

Internal Resistance and Terminal Voltage

All batteries have an *internal resistance* that we designate r. An automobile battery may have a resistance as low as 0.01 Ω, while that of an old flashlight battery may be as high as 50 Ω. This resistance is physically spread throughout the source of emf, but for circuit analysis we draw it as a separate resistance r (situated on either side of $\mathscr{E}$ between the terminals of the battery) in series with a "pure" emf $\mathscr{E}$, Figure 29-13. Of course, we can never measure the voltage at the point between $\mathscr{E}$ and r because this point exists only in the diagram. But the simplified circuit is convenient for analysis.

Drawing a current from the battery causes a potential drop across r whose polarity reduces[2] the terminal voltage V across the battery terminals. Noting the polarity of the voltages between the terminals, we have

$$V = \mathscr{E} - Ir \quad (29\text{-}23)$$

where I is the current. For this single loop circuit, the current I is

$$I = \frac{\mathscr{E}}{R + r} \quad (29\text{-}24)$$

(These two equations, while useful, are not worth memorizing since they can be written by inspection for this simple circuit.)

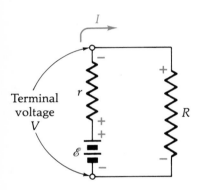

FIGURE 29-13
An external resistance R connected across the terminals of a battery that has an internal resistance r and a "pure" emf $\mathscr{E}$. The current I produces the polarities shown for the potential differences across the circuit components. The current through r reduces the terminal voltage V below that of the emf $\mathscr{E}$.

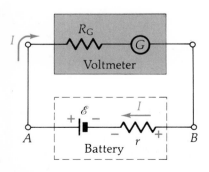

FIGURE 29-14
Example 29-8.

EXAMPLE 29-8

The terminal voltage of a particular battery is measured in two ways: first, with a potentiometer, which indicates 1.50 V, and then with a voltmeter, which indicates 1.48 V on a 2-V scale. The voltmeter is known to have a figure of merit of 1000 Ω/V. Find the internal resistance of the battery.

SOLUTION

Figure 29-14 is a circuit diagram with the internal resistance drawn as a separate resistance r between the terminals. When the potentiometer is used, at balance conditions *no current is drawn*, so there is no potential drop across r and the potentiometer measures the true emf of the battery. However, when the voltmeter is used, some current I exists, and the drop across r reduces the terminal voltage to

$$V = \mathscr{E} - Ir \quad (29\text{-}25)$$

---

[2] However, if the battery is being charged by an external seat of emf, the current is in the opposite direction and the potential difference across r reverses its polarity.

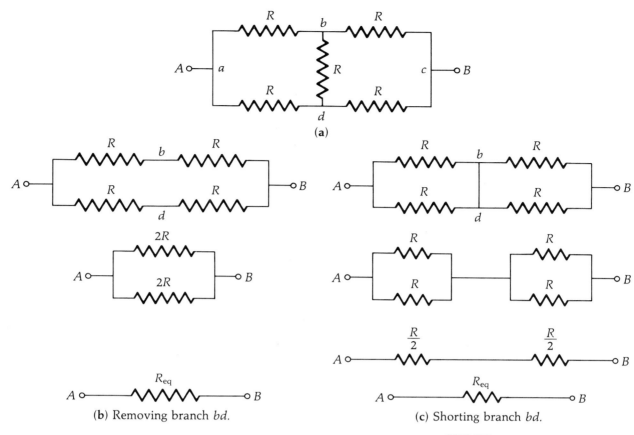

FIGURE 29-15
Example 29-9.

The figure of merit of 1000 Ω/V for the voltmeter means that its reciprocal is the current that will produce a full-scale deflection. Thus, 1 mA produces a full-scale deflection of the voltmeter. Since the meter deflects (1.48/2.00) of the full scale, the current I in the meter is

$$I = \left(\frac{1.48}{2.00}\right)(2 \text{ mA}) = 0.740 \text{ mA}$$

Solving Equation (29-25) for r gives

$$r = \frac{\mathcal{E} - V}{I} = \frac{1.50 \text{ V} - 1.48 \text{ V}}{0.74 \times 10^{-3} \text{ A}} = \boxed{27.0 \text{ Ω}}$$

EXAMPLE 29-9

In the network of Figure 29-15a, each resistor has a resistance R. Find the equivalent resistance R_{eq} between the terminals A and B.

SOLUTION

When symmetry is present, often that symmetry allows us to simplify the analysis. Here, we imagine putting a current I into terminal A (and removing the same current I from terminal B.) By symmetry, at junction a the current splits equally, so the currents in branches *ab* and *ad* are equal. Therefore, the potential drops in those branches are equal, junctions *b* and *d* are at the same potential, and no current exists in that branch. *We could thus remove that branch without disturbing the operation of the circuit*, as shown in the left-hand diagrams of Figure 29-15b. Or, as an alternative, *we could connect a resistanceless wire between b and d without disturbing the operation of the circuit*, as shown in the right-hand diagrams.

Both methods of analysis show that the equivalent resistance $R_{eq} = \boxed{R.}$

29.6 RC Circuits

Up to this point we have discussed circuits in which the currents are constant in time. We now introduce a capacitor as an additional circuit element that can cause the current to vary with time. As you will see, capacitors perform very useful functions in circuits and, indeed, are a part of nearly all practical electronic circuits.

Consider the circuit of Figure 29-16a. When the switch S is put in the left position, the seat of emf (assumed to be "ideal" with zero internal resistance) charges the capacitor with the polarities shown in (b). (The right-hand branch is isolated from the charging circuit and plays no role.)

<u>We adopt the convention of using lower-case letters for time-varying quantities and capital letters for constant quantities.</u>

Thus *i* designates the charging current. Applying Kirchhoff's loop rule to the circuit while charging, we get

$$\Sigma V = 0$$

$$\mathcal{E} - iR - \frac{q}{C} = 0 \tag{29-26}$$

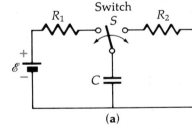

(a)

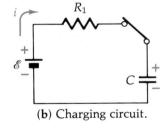

(b) Charging circuit.

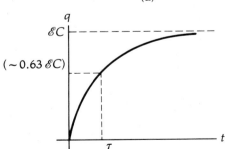

(c) While charging, the charge on the capacitor increases exponentially. In one time constant $\tau = R_1 C$, the charge q rises to $(1 - 1/e) q_0 = 0.63 q_0$.

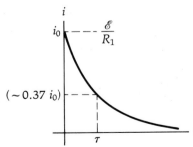

(d) While discharging, the current decreases exponentially. In one time constant $\tau = R_1 C$, the current falls to $(i_0/e) = 0.37 i_0$.

FIGURE 29-16
The switch in (a) is put in the left-hand position to charge the capacitor *C* through the resistor R_1.

where q/C is the potential difference across C. This equation indicates that, as q increases, the current i must decrease. Let us find expressions for these time-varying quantities.

Charging

Suppose that, initially, the capacitor is uncharged and the switch is closed at $t = 0$. The charge q on the capacitor will increase exponentially as shown in Figure 29-16c. We obtain the mathematical form of this variation by substituting $i = dq/dt$ in Equation (29-26) and rearranging:

$$\frac{dq}{dt} = \frac{\mathscr{E}}{R_1} - \frac{q}{R_1 C} \tag{29-27}$$

An expression for q may be found in the following way. Rearrange the equation by placing terms involving q on the left-hand side and those involving t on the right-hand side. Then integrate both sides:

$$\frac{dq}{(q - C\mathscr{E})} = -\frac{1}{R_1 C} dt$$

$$\int_0^q \frac{dq}{(q - C\mathscr{E})} = -\frac{1}{R_1 C} \int_0^t dt$$

$$\ln\left(\frac{q - C\mathscr{E}}{C\mathscr{E}}\right) = -\frac{t}{R_1 C}$$

From the definition of the natural logarithm, we can write this expression as

CHARGING A CAPACITOR THROUGH A RESISTOR

$$q = C\mathscr{E}[1 - e^{-t/R_1 C}] \tag{29-28}$$

Figure 28-16c shows this rising exponential curve for q, which changes from zero toward its final value of $C\mathscr{E}$. As time progresses, q asymptotically approaches[3] the final value $C\mathscr{E}$. The rapidity of charging the capacitor depends on the numerical values of R_1 and C. For example, if the product $R_1 C$ is made smaller, the capacitor charges more rapidly.

In this charging process, charges do not jump across the capacitor plates. Instead, the emf $\mathscr{E}$ moves charges from one plate through the resistor and battery to the other plate, producing equal and opposite charges on the plates. As the potential across the capacitor builds up to $\mathscr{E}$, the current drops to zero. To find a mathematical expression for the charging current i, we differentiate Equation (29-28) with respect to t. Finding $i = dq/dt$, we obtain

CHARGING CURRENT

$$i = \left(\frac{\mathscr{E}}{R_1}\right) e^{-t/R_1 C} \tag{29-29}$$

This falling exponential is shown in Figure 29-16d. Immediately after the switch is closed at $t = 0$, the current i has its maximum value *limited only by the*

---

[3] Mathematically, the exponential term $C\mathscr{E} e^{-t/R_1 C}$ eventually becomes smaller than the fluctuations in q due to thermal motions of electrons. So the statement that an exponential change "never" reaches its final value becomes physically unimportant.

resistance in the circuit, and all the potential drop is initially across the resistor.

At $t = 0$ $$i = I_0 = \frac{\mathscr{E}}{R_1}$$
(maximum current)

When the capacitor is fully charged to its maximum, then the current is zero and all the potential drop is across the capacitor:

At $t = \infty$ $$Q = C\mathscr{E}$$
(maximum charge)

A useful parameter associated with *RC* circuits is the characteristic time $\tau = RC$, the time at which the power of the exponential term equals -1. This value is called the **RC time constant**. It is related to the speed with which currents, voltages, and charges change. For example, in the charging of a capacitor, in *one* time constant the charge rises to $(1 - 1/e) \approx 0.63$ of its maximum final value. Similarly, in *one* time constant the charging current falls to $1/e \approx 0.37$ of its initial value. *In RC circuits, all the varying quantities have this exponential behavior, so it will be helpful to remember these 0.63 and 0.37 values.*

Discharging

After the capacitor has become fully charged, the switch is moved to the right, connecting the charged capacitor to R_2. The discharge circuit is simply the capacitor in series with R_2, Figure 29-16a. Applying Kirchhoff's loop rule during this discharge process, we have

$$\Sigma V = 0$$

$$\frac{q}{C} - iR_2 = 0 \tag{29-30}$$

where q/C is the potential difference across the capacitor as it discharges. Here, $i = -dq/dt$ (the minus sign results from the fact that q decreases as time increases), and after substituting and rearranging we have

$$\frac{dq}{q} = -\frac{dt}{R_2 C}$$

Integrating and setting $q = Q_0$ at $t = 0$ gives

$$\int_{Q_0}^{q} \frac{dq}{q} = -\frac{1}{R_2 C} \int_0^t dt$$

$$\ln\left(\frac{q}{Q_0}\right) = -\frac{t}{R_2 C}$$

DISCHARGING A CAPACITOR THROUGH A RESISTOR $$q = Q_0 e^{-t/R_2 C} \tag{29-31}$$

where $Q_0 = \mathscr{E}C$, the initial charge on the capacitor.

We obtain an expression for the current $i = dq/dt$ by differentiating the above with respect to t to obtain

DISCHARGING CURRENT
$$i = \left(\frac{\mathscr{E}}{R_2}\right)e^{-t/R_2C} \qquad (29\text{-}32)$$

where the initial (maximum) value of the current $I_0 = \mathscr{E}/R_2 = Q_0/R_2C$. Graphs of these quantities are shown in Figure 29-17. In this example, we have purposely chosen $R_2 > R_1$ to illustrate how the time constant $\tau = R_2C$ governs the rapidity of the exponential changes. The discharging process occurs more slowly than the charging process because $R_2C > R_1C$.

The voltage changes across R and C in the circuit can easily be found from

$$v_R = iR \quad \text{and} \quad v_C = \frac{q}{C} \qquad (29\text{-}33)$$

These voltages thus also change exponentially with the time constant RC. The exponential changes in q, i, and v are called **transients**; when such changes have ceased, the final conditions are called the **steady-state** values.

Two important conclusions can be drawn regarding capacitors in DC circuits:

(1) *The charge on a capacitor (and, consequently, the voltage across it) cannot change instantaneously. How fast such changes take place is governed by the RC time constant.*
(2) *After the final steady-state conditions have been reached, the DC current through a capacitor is always zero.*

Remembering these conclusions will greatly help you predict the behavior of DC circuits containing capacitors.

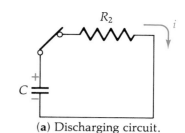

(a) Discharging circuit.

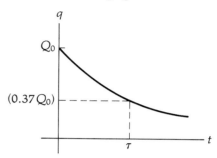

(b) The charge on the capacitor decreases exponentially. In one time constant $\tau = R_2C$, the charge q falls to $1/e = 0.37$ of its initial value $Q_0 = \mathscr{E}C$.

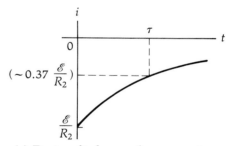

(c) During discharge, the current is negative because it is opposite to the direction of the charging current. In one time constant $\tau = R_2C$, the current falls to $1/e = 0.37$ of its initial value $I_0 = \mathscr{E}/R_2$.

FIGURE 29-17
The switch in Figure 29-16a is put in the right-hand position to discharge the charged capacitor through the resistor R_2.

EXAMPLE 29-10

The capacitance in a series RC circuit is 4 μF. (a) Find the resistance R that will result in a time constant of 2 ms for this circuit. (b) At $t = 0$, the circuit is connected to an emf of 9 V. How long will it take for the voltage on C to reach 8 V?

SOLUTION

(a) From $\tau = RC$,
$$R = \frac{\tau}{C} = \frac{2 \times 10^{-3} \text{ s}}{4 \times 10^{-6} \text{ F}} = \boxed{500 \text{ }\Omega}$$

(b) $\quad V = V_0(1 - e^{-t/RC})$

$$e^{-t/RC} = \left(1 - \frac{V}{V_0}\right)$$

$$e^{t/RC} = \left(\frac{V_0}{V_0 - V}\right) = \frac{9 \text{ V}}{1 \text{ V}} = 9$$

$$\frac{t}{RC} = \ln 9$$

$$t = RC \ln 9 = (2 \text{ ms})(\ln 9) = \boxed{4.39 \text{ ms}}$$

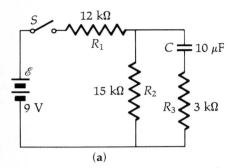

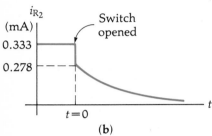

FIGURE 29-18
Example 29-11.

EXAMPLE 29-11

In Figure 29-18a, suppose that the switch has been closed sufficiently long for the capacitor to become fully charged. Find (a) the steady-state current through each resistor and (b) the charge Q on the capacitor. (c) The switch is now opened at $t = 0$. Write an equation for the current i_{R_2} through R_2 as a function of time, and (d) find the time that it takes for the charge on the capacitor to fall to $\frac{1}{5}$ of its initial value.

SOLUTION

(a) After steady-state conditions have been reached, there is no DC current through the capacitor. Thus:

For R_3: $\qquad \boxed{I_{R_3} = 0} \qquad$ (steady-state)

For the other two resistors, the steady-state current is simply determined by the 9-V emf across the 12-kΩ and 15-kΩ resistors in series:

For R_1 and R_2: $\quad I_{(R_1+R_2)} = \dfrac{\mathscr{E}}{R_1 + R_2} = \dfrac{9 \text{ V}}{(12 \text{ k}\Omega + 15 \text{ k}\Omega)}$

$\qquad\qquad\qquad\qquad = \boxed{0.333 \text{ mA}} \qquad$ (steady-state)

(b) After the transient currents have ceased, the voltage across C is the same as the voltage across R_2 ($=IR_2$) because there is no voltage drop across R_3. Therefore, the charge Q on C is

$$Q = CV_{R_2} = C(IR_2) = (10 \text{ }\mu\text{F})(0.333 \text{ mA})(15 \text{ k}\Omega) = \boxed{5.00 \text{ }\mu\text{C}}$$

(c) When the switch is opened, the branch containing R_1 is no longer part of the circuit. The capacitor discharges through $(R_2 + R_3)$ with a time constant of $(R_2 + R_3)C = (15 \text{ k}\Omega + 3 \text{ k}\Omega)(10 \text{ }\mu\text{F}) = 0.180$ s. The initial current I_0 in this discharge circuit is determined by the initial voltage across the capacitor applied to $(R_2 + R_3)$ in series:

$$I_0 = \frac{V_C}{(R_2 + R_3)} = \frac{IR_2}{(R_2 + R_3)} = \frac{(0.333 \text{ mA})(15 \text{ k}\Omega)}{(15 \text{ k}\Omega + 3 \text{ k}\Omega)} = \underline{\underline{0.278 \text{ mA}}}$$

Thus, when the switch is opened, the current through R_2 changes instantaneously from 0.333 mA (downward) to 0.278 mA (downward) as shown in Figure 29-18b. Thereafter, it decays according to

$$i_{R_2} = I_0 e^{-t/(R_2+R_3)C} = \boxed{(0.278 \text{ mA})e^{-t/(0.180 \text{ s})}} \qquad (\text{for } t > 0)$$

(d) The charge q on the capacitor decays from Q_0 to $Q_0/5$ according to

$$q = Q_0 e^{-t/(R_2+R_3)C}$$

$$\frac{Q_0}{5} = Q_0 e^{-t/(0.180 \text{ s})}$$

$$5 = e^{t/(0.180 \text{ s})}$$

$$\ln 5 = \frac{t}{0.180 \text{ s}}$$

$$t = (0.180 \text{ s})(\ln 5) = \boxed{0.290 \text{ s}}$$

EXAMPLE 29-12

We charge a capacitor C by connecting it to a seat of emf $\mathscr{E}$ with wires of total resistance R. When the capacitor is fully charged, the seat of emf will have done an amount of work W equal to

$$W = QV \qquad (29\text{-}34)$$

where Q is the total charge and V the potential difference of the seat of emf. The energy stored in the capacitor is

$$U_C = \tfrac{1}{2}CV^2 = \tfrac{1}{2}QV \qquad (29\text{-}35)$$

which is only half the work done by the seat of emf. What happened to the other half of the work done?

SOLUTION

The "missing" energy appears as I^2R heating of the resistance of the charging circuit. The current i during charging is

$$i = \left(\frac{V}{R}\right) e^{-t/RC}$$

We integrate the instantaneous power $P_{\text{inst}} = i^2R$ from $t = 0$ to $t = \infty$ to find the total Joule heating of the resistance. Letting U_{th} represent this thermal energy, we have

$$U_{\text{th}} = \int_0^\infty i^2 R \, dt = R \int_0^\infty \left(\frac{V}{R}\right)^2 e^{-2t/RC}$$

$$U_{\text{th}} = -\left(\frac{V^2}{R}\right)\left(\frac{RC}{2}\right) e^{-2t/RC}\bigg|_0^\infty = \frac{1}{2}CV^2 = \boxed{\tfrac{1}{2}QV}$$

Thus, the Joule heating of the resistance in the connecting wires accounts for the other half of the work done by the seat of emf. The Joule heating is always exactly half of the energy stored in C, *independent of the value of* R.

Summary

The *equivalent resistance* R_{eq} for combinations of resistors is

In *series*: $\quad R_{\text{eq}} = R_1 + R_2 + R_3 + \cdots$

In *parallel*: $\quad \dfrac{1}{R_{\text{eq}}} = \dfrac{1}{R_1} + \dfrac{1}{R_2} + \dfrac{1}{R_3} + \cdots$

Kirchhoff's rules for circuit analysis:

Loop rule: $\quad \Sigma V = 0 \quad$ (around any closed path)

Junction rule: $\quad \Sigma I = 0 \quad$ (currents entering a junction are $+$; those leaving are $-$)

Superposition theorem: In a linear circuit containing more than one seat of emf, the current in any branch is the superposition of all the currents contributed by each seat of emf acting individually, with all other emfs replaced by conducting wires of zero resistance.

RC circuit: In a series combination of a seat of emf, a resistor, and a capacitor:

Charge on capacitor

Charging: $\quad q = CV(1 - e^{-t/RC})$

Discharging: $\quad q = Q_0 e^{-t/RC}$
(no emf in circuit)

Current

$i = \left(\dfrac{\mathscr{E}}{R}\right) e^{-t/RC}$

$i = -\left(\dfrac{\mathscr{E}}{R}\right) e^{-t/RC}$

In one time constant, $\tau = RC$, the rising exponential rises to $1 - (1/e) \approx 0.63$ of its maximum value, and a decreasing exponential falls to $(1/e) \approx 0.37$ of its initial value.

The general behavior of a capacitor in a series *RC* circuit (with a constant emf) is as follows:

(1) *The charge on a capacitor (and, consequently, the voltage across it) cannot change instantaneously. The rapidity of the exponential changes that occur is governed by the RC time constant of the charging and discharging paths.*

(2) *After steady-state conditions have been reached, the DC current through a capacitor is always zero.*

Questions

1. A 10-W, 110-V light bulb connected to a series of batteries may produce a brighter light than a 250-W, 110-V light bulb connected to the same batteries. Why?
2. Consider a circular hoop of resistance wire with two terminals attached to different places on the hoop. How does the resistance between the terminals depend on their relative positions on the hoop?
3. Imagine a closed surface in the midst of a complicated electrical network, so that current-carrying conductors penetrate the surface and so that some of the circuit components such as resistors, batteries, and capacitors are within the surface. Is the net current through the surface zero? Does Gauss's law hold for this surface?
4. A potentiometer is often used to measure open-circuit voltages of batteries. How can the potentiometer also be used to measure current and resistance?
5. How does a somewhat run-down battery supply affect the operation of a Wheatstone bridge?
6. If the battery and the galvanometer of a Wheatstone bridge are interchanged, the circuit is still that of a Wheatstone bridge. Suppose a Wheatstone bridge is balanced. Does interchanging the battery and the galvanometer result in a *balanced* bridge?
7. A volt-ohm-meter is a single meter movement with circuits and switches that make it appropriate for use as an ammeter, a voltmeter, or an ohmmeter. When the device is not in use, why is it best to leave the switch of a volt-ohm-meter on a high-voltage scale rather than on a current scale or a resistance scale?
8. Why is it more practical to specify the meter-current sensitivity of a voltmeter in ohms per volt rather than in amperes?
9. How can a voltmeter be used to measure capacitance?
10. In the slide-wire potentiometer of Figure 29-12, a variable resistor (sometimes called a *rheostat*) is usually added in series with the external battery in order to control the amount of current through the slide wire. Suppose that this variable resistor were a combination of two variable resistors in parallel, one large and the other small, that act as "coarse" and "fine" controls of the current. Which resistor is the coarse control and which resistor is the fine control?

Problems

29.2 Resistors in Series and in Parallel

29A-1 Three resistors, *R*, 2*R*, and 3*R*, are connected in parallel, producing an equivalent resistance of 20 Ω. Find their equivalent resistance when they are connected in series.

29B-2 When *n* identical resistors are connected in series, the equivalent resistance is *N* times the equivalent resistance when they are connected in parallel. Express *n* in terms of *N*.

29B-3 Two wires, *A* and *B*, are made of the same material and have the same length, but the cross-sectional area of *A* is twice that of *B*. (a) When they are connected in parallel across a potential difference *V*, which wire will dissipate the greatest electrical power? (b) Repeat for when they are connected in series across the same potential difference. (c) Find the ratio of the total power developed in case (a) to that in case (b).

29B-4 For the circuit of Figure 29-19, find (a) the equivalent resistance between the terminals. (b) An emf $\mathscr{E} = 40$ V is now connected between the terminals. What is the potential difference across the 8-Ω resistor? (c) Find the current in the 10-Ω resistor and (d) the power developed in the 30-Ω resistor.

(e) Show how a 20-Ω resistor could be added to the circuit so that the emf would furnish a total of 4 A.

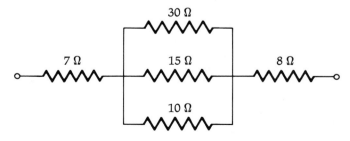

FIGURE 29-19
Problem 29B-4

29B-5 In Figure 29-20, each resistor has a resistance of 1 Ω. Suppose that a given current *I* enters at *A* and comes out at *B*. By utilizing arguments based upon the symmetry of the network, show that the equivalent resistance R_{eq} of the network from A to B is $\frac{2}{3}\,\Omega$. (Hint: what would the resistance be if the vertical resistors were absent?)

FIGURE 29-20
Problem 29B-5.

29B-6 Two resistors connected in series have an equivalent (combined) resistance of 690 Ω. When they are connected in parallel, their equivalent resistance is 150 Ω. Find the resistance of each of the resistors.

29B-7 Find the equivalent resistance between terminals A and B of the resistor network shown in Figure 29-21. (Hint: use the "delta–wye" transformations of Problems 29C-43 and 29C-45.)

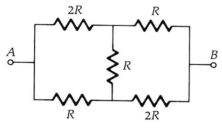

FIGURE 29-21
Problem 29B-7.

29B-8 When two resistors, R_A and R_B, are connected in series, their total resistance is R_s. When they are connected in parallel, their equivalent resistance is R_p. Find R_A and R_B in terms of R_s and R_p.

29B-9 To achieve different values of power consumption, four 40-W, 120-V light bulbs are connected in a variety of ways across a 120-V power source. Sketch nine different ways and calculate the total power consumption in each case. Assume that the resistance of the light bulb is independent of the current through it (a poor assumption).

29B-10 Using only three resistors—2 Ω, 3 Ω, and 4 Ω—find all 17 different resistance values that may be obtained by various combinations of one or more resistors. Tabulate the values in order of increasing resistance.

29.3 Multiloop Circuits and Kirchhoff's Rules

29A-11 A 12-V car battery has an internal resistance of 0.02 Ω. Find the terminal voltage while the starter motor draws 140 A from the battery. (This answer suggests a practical procedure: if your car stalls and the motor stops, it is best to turn off the headlights when restarting the engine in order to minimize the drop in terminal voltage of the battery.)

29A-12 A typical fresh AA dry cell has an emf of 1.50 V and an internal resistance of 0.311 Ω. (a) Find the terminal voltage of the battery when it supplies 58 mA to a circuit. (b) What is the resistance R of the external circuit?

29A-13 Consider a current I entering a circuit junction as shown in Figure 29-22. Show that the fraction I_1/I of I going through the branch that contains R_1 is given by $R_2/(R_1 + R_2)$.

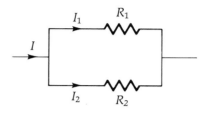

FIGURE 29-22
Problem 29A-13.

29A-14 In the circuit of Figure 29-23, find (a) the equivalent resistance in the circuit outside the battery, (b) the current through the battery, (c) the terminal voltage, and (d) the power developed in the 6-Ω resistor.

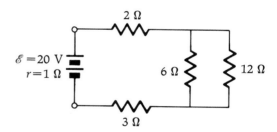

FIGURE 29-23
Problem 29A-14.

29A-15 Consider the circuit of Figure 29-24. Verify that the rate of work done by the emf $\mathscr{E}I$ equals the sum of the Joule power I^2R developed in each of the resistors.

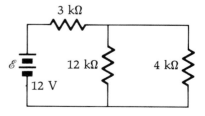

FIGURE 29-24
Problem 29A-15.

29A-16 The electrical source for the lights in a house trailer is a battery with an emf $\mathscr{E}$ and an internal resistance r. Suppose that n lights, each with a resistance R, are connected in parallel across the battery. In terms of the given symbols, find an expression for the current I that the battery supplies.

29B-17 In Figure 29-25, calculate (a) the equivalent resistance of the network outside the battery, (b) the current through the battery, and (c) the current in the 6-Ω resistor.

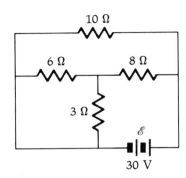

FIGURE 29-25
Problem 29B-17.

29B-18 For the circuit shown in Figure 29-26, find (a) the equivalent resistance external to the battery terminals. (b) What is the terminal voltage of the battery? (c) Find the total power that the battery supplies to the external circuit. (d) Make a table showing the power in each resistor, listing the resistors in order of increasing resistance. (Hint: for this network, you can find the currents using Ohm's law; it is not necessary to write equations from Kirchhoff's rules.)

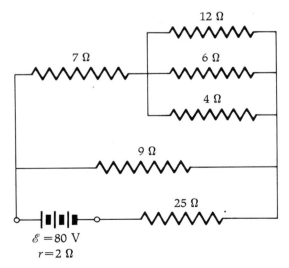

FIGURE 29-26
Problem 29B-18.

29B-19 A certain run-down battery has an open-circuit voltage across its terminals of 7.22 V. While a battery charger is charging the battery with a current of 8.60 A, the terminal voltage is 7.96 V. Find the internal resistance of the battery. (Note: a battery charger forces current into the + terminal of the battery.)

29B-20 Using Kirchhoff's rules, (a) find the current in each of the resistors in the circuit shown in Figure 29-27. (b) Find the potential difference between points c and f. Which is at the higher potential?

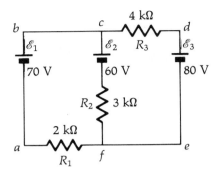

FIGURE 29-27
Problem 29B-20.

29B-21 Consider the circuit shown in Figure 29-28. Find the current in each of the resistors using Kirchhoff's rules.

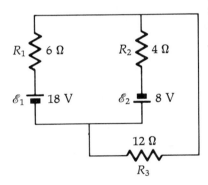

FIGURE 29-28
Problem 29B-21.

29.4 The Superposition Principle

29B-22 In Figure 29-29, use Kirchhoff's rules to find the magnitude and direction of the current in each branch.

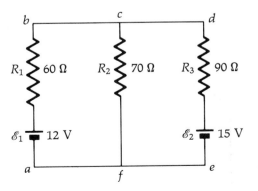

FIGURE 29-29
Problem 29B-22.

29B-23 Solve Problem 29B-22 by applying the superposition theorem.

29B-24 Solve Problem 29B-20 by applying the superposition theorem.

29B-25 Solve Problem 29B-21 by applying the superposition theorem.

29.5 Applications

29A-26 A certain meter movement has an internal resistance of 100 Ω and requires a current of 200 μA for full-scale deflection. Find the resistances that will convert the meter to (a) a 10-V voltmeter and (b) a 5-A ammeter. Include sketches showing how the resistance is connected in each case.

29B-27 The value of a resistance R may be measured with a circuit such as that shown in Figure 29-30. (a) If the ammeter, which has an equivalent resistance of 50 Ω between its terminals, reads 5 mA and the voltmeter reads 12.3 V, determine the value of R. (b) If the ammeter had zero resistance, what would the value of R be?

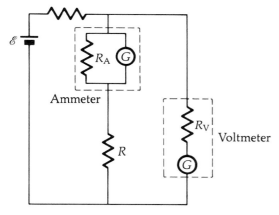

FIGURE 29-30
Problem 29B-27.

29B-28 The galvanometer G in Figure 29-31 has a resistance of 50 Ω and requires 400 μA for full-scale deflection. (a) Find the values of R_1 and R_2 that will convert the galvanometer to a two-range ammeter with full-scale currents of 1 A and 0.1 A. (b) Using the same galvanometer and two resistors R_3 and R_4, sketch the circuit that will convert the galvanometer to a two-range voltmeter whose three binding posts are marked "—," 1 V, and 10 V. Include the numerical values of R_3 and R_4.

29B-29 In the potentiometer circuit of Figure 29-32, the slide wire is 100 cm long. For an unknown emf, the null position occurs when the sliding contact is 58 cm from the left end with an uncertainty in position of 0.30 mm. (a) Find the percentage error in determining the unknown emf, assuming the instrument is accurately calibrated. (b) If the current in the slide wire is doubled, find the percentage error in the measurement, assuming the position uncertainty is still 0.30 mm.

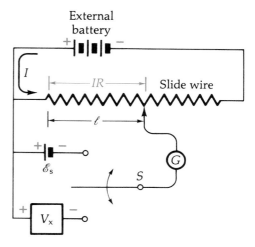

FIGURE 29-32
Problem 29B-29.

29B-30 Figure 29-33 shows the series resistances inside a multirange voltmeter. The meter movement G has an internal resistance of 500 Ω and indicates full-scale deflection when a current of 0.500 mA is present. The markings on the terminals are as indicated. Find the values of R_1, R_2, and R_3.

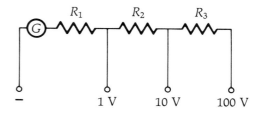

FIGURE 29-33
Problem 29B-30.

29B-31 A galvanometer is often made into a multirange ammeter through the use of an *Ayrton shunt* such as that shown in Figure 29-34. If the galvanometer has a resistance of 1000 Ω and a full-scale sensitivity of 50 μA, find the values of R_1, R_2, R_3, and R_4 such that the meter will deflect full scale for 10 mA, 100 mA, 1 A, and 10 A.

29B-32 The *figure of merit* of a voltmeter is defined as the total resistance of the meter divided by the full-scale voltage

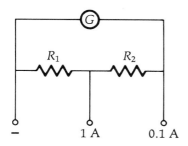

FIGURE 29-31
Problem 29B-28.

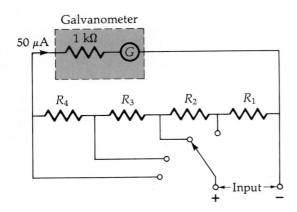

FIGURE 29-34
Problem 29B-31.

reading. Prove that, for a multirange voltmeter (see Problem 29B-30), the figure of merit is the same on all voltage scales.

29B-33 Refer to the previous problem and prove that the figure of merit is also equal to the reciprocal of the current in the galvanometer movement that produces a full-scale deflection.

29.6 RC Circuits

29A-34 A capacitance C discharges through a resistance R. How long does it take for the charge on the capacitor to reduce to $1/e^2$ of its initial value?

29A-35 Verify that the product RC has dimensions of time.

29B-36 How many time constants elapse while charging a capacitor in an RC circuit to within 2% of its maximum charge?

29B-37 A 10-μF capacitor is charged by a 10-V battery through a resistance R. The capacitor reaches a potential difference of 4 V in a period of 3 s after the charging began. Find the value of R.

29B-38 Suppose that we charge a capacitance $C = 8\ \mu$F by connecting it in series with an emf $\mathscr{E} = 20$ V (with negligible internal resistance) and a resistance $R = 500$ kΩ. (a) What is the final energy stored in the fully charged capacitor? (b) By direct integration of $\int_0^\infty i^2 R\,dt$, show that the thermal energy developed in the resistor equals the energy stored in the capacitor.

29B-39 Verify that $q = \mathscr{E}C(1 - e^{-t/RC})$ satisfies $\mathscr{E} - iR - q/C = 0$.

29B-40 A capacitor has been fully charged with a 9-V battery. A 20 000-Ω/V voltmeter set on its 10-V range is attached to the capacitor. The voltmeter reading drops from 8.00 V to 5.60 V in 5 s. Calculate the capacitance of the capacitor.

29B-41 A 20 000-Ω/V voltmeter set on a 100-V scale is connected to a charged capacitor. If the reading on the voltmeter reduces to half its initial value in 2 s, find the capacitance of the capacitor.

29B-42 A 3-μF capacitor is initially charged to a potential of 200 V, then isolated. Because of leakage through the dielectric, 5 min later the potential has dropped to 185 V. Find the leakage resistance between the plates of the capacitor.

Additional Problems

29C-43 Derive the following equations that transform the "wye" configuration of resistors shown in Figure 29-35b into the "delta" configuration shown in Figure 29-35a.

$$R_1 = (R_A R_B + R_B R_C + R_C R_A)/R_C$$
$$R_2 = (R_A R_B + R_B R_C + R_C R_A)/R_A$$
$$R_3 = (R_A R_B + R_B R_C + R_C R_A)/R_B$$

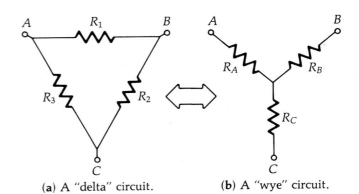

(a) A "delta" circuit.   (b) A "wye" circuit.

FIGURE 29-35
Problems 29C-43 and 29C-45.

29C-44 In Figure 29-36, a network of 12 resistors, each with a resistance R, is joined so that each resistor forms the edge of a cube. Find the resistance between diametrically opposite vertices. (Hint: apply a potential difference across these vertices and identify points of equal potential, which then may be joined by a resistanceless wire. Noting certain symmetries will be helpful.)

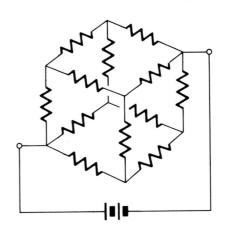

FIGURE 29-36
Problem 29C-44.

29C-45 The "delta" network of resistors, shown in Figure 29-35a may be transformed into a "wye" network, shown in Figure 29-35b, such that the resistance between corresponding terminals is equal. Derive the following values for R_A, R_B, and R_C.

$$R_A = R_1 R_3/(R_1 + R_2 + R_3)$$
$$R_B = R_1 R_2/(R_1 + R_2 + R_3)$$
$$R_C = R_2 R_3/(R_1 + R_2 + R_3)$$

29C-46 Figure 29-37 shows six terminals on an insulating circuit board. Each terminal is connected to every other terminal by a wire of resistance 2 Ω. (The wires make electrical contact only at the terminals, not where they cross one another.) Find the net resistance between any two terminals and explain why it is the same for every possible pair of terminals. (Hint: avoid a "brute-force" method of resistance calculation. Instead, consider the symmetry of the network and apply the hint given in Problem 29C-57.)

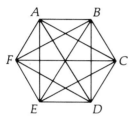

FIGURE 29-37
Problem 29C-46.

29C-47 Consider the resistor network shown in Figure 29-38. Each resistor has the same value R. Show that the equivalent resistance between the terminals A and B, as the number of the network elements becomes very large, is $R(1 + \sqrt{3})$. [Hint: if just one element is present, the equivalent resistance is $R_1 = 2R + R$. If two elements are present, the equivalent resistance is $R_2 = 2R + RR_1/(R + R_1)$. Continue the series until you can deduce R_n.]

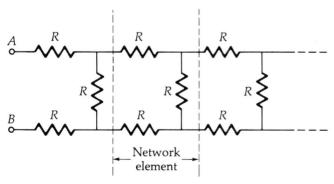

FIGURE 29-38
Problems 29C-47 and 29C-49.

29C-48 A 1000-Ω resistor is attached to the terminals of a battery. The voltage across the resistor is 45 V (measured with a 20 000-Ω/V voltmeter). When the resistor is replaced by a 3300-Ω resistor, the voltage is 47 V. Calculate the open-circuit voltage of the battery and the internal resistance of the battery.

29C-49 A long parallel pair of current-carrying wires with insulation between them may be represented by a network similar to that of Figure 29-38. However, for this problem, the horizontal resistors R_H have a very low value, while the vertical resistors R_V (representing the insulation between the wires) have a very high value. The horizontal resistors represent the resistance per unit length of the wires, $r_1/L = 2R_H/L$, and the vertical resistors represent the insulation resistance per unit length, $r_2/L = R_V/L$. Show that, if $r_2 \gg r_1$, the resistance per unit length between the terminals A and B is $\sqrt{r_1/r_2}/L$.

29C-50 A resistor R_A is in series with a resistor R_B. The equivalent resistance of the series combination is unchanged if R_A is shunted by a resistor R and R_B is increased by the resistance R. Find the value of R in terms of R_A. (The value of R is independent of R_B.)

29C-51 A power source consists of a seat of emf $\mathscr{E}$ and an internal series resistance r. The source delivers power to an external (variable) load resistance R_L. Show that, as R_L is varied, the maximum power developed in R_L occurs when $R_L = r$. This is known as the *maximum-power-transfer theorem*.

29C-52 The resistance of a resistor may be measured with a battery, a voltmeter, and an ammeter by either of the following methods: (1) the ammeter is inserted in series with the parallel combination of the resistor and voltmeter or (2) the voltmeter is in parallel with the series combination of the resistor and ammeter. Suppose that the voltmeter has a resistance of 2000 Ω, the ammeter has a resistance of 20 Ω, and the battery has negligible resistance. (a) With method (1), the voltmeter indicates a voltage of 40.0 V and the ammeter indicates 0.100 A. Determine the resistance of the resistor. (b) If method (2) is used, calculate the indications of the voltmeter and the ammeter. (c) If we determine the resistance by simply dividing the voltmeter reading by the ammeter reading, which method would provide the most accurate value of the resistance? Include clear circuit diagrams with your solution.

29C-53 A source of power with an output voltage V_1 supplies power to a load resistance R. In certain electronic applications, it is necessary to reduce the output voltage of the power source to a lower value V_2 without changing the resistance into which the source provides power. This is accomplished through the insertion of an *attenuator pad* as shown in Figure 29-39.

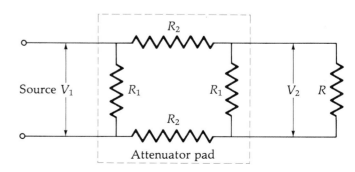

FIGURE 29-39
Problem 29C-53.

With the attenuator pad in place, the source still "sees" an equivalent resistance R. Show that $R_1 = R(V_1 + V_2)/(V_1 - V_2)$ and $R_2 = R(V_1^2 - V_2^2)/4V_1V_2$.

29C-54 Three batteries, with emf's of $\mathscr{E}_1$, $\mathscr{E}_2$, and $\mathscr{E}_3$, have internal resistances r_1, r_2, and r_3, respectively. The batteries are connected in parallel (positive terminals joined and negative terminals joined). Derive an expression for the terminal voltage of the combination of batteries when no external load resistor is present.

29C-55 We obtain the value of a resistance by measuring the current through the resistor and the voltage across it, as shown in Figure 29-40. The voltmeter indicates 30 V on a 50-V scale and the ammeter indicates 150 mA on a 500-mA scale. What is the value of the resistance R if both meters have a galvanometer with a 1-mA full-scale sensitivity?

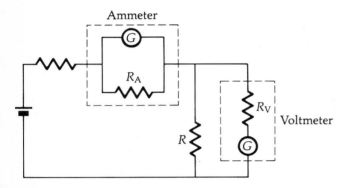

FIGURE 29-40
Problem 29C-55.

29C-56 A variable resistor is constructed from a closed circular hoop of resistance wire. (When cut and measured end to end, the resistance wire has a total resistance R.) As shown in Figure 29-41, one terminal is at a fixed point on the closed hoop and the other terminal is a sliding contact. (a) In terms of R (in ohms) and the angle θ (in radians), find an expression for the resistance r between the terminals. (b) When a potential difference is applied across the terminals, what practical difficulty might be encountered if the sliding contact is near $\theta = 0°$?

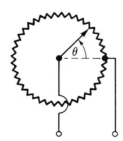

FIGURE 29-41
Problem 29C-56.

29C-57 Consider an infinite network of resistors, as shown in Figure 29-42. If each resistor has the same value R, find the resistance between points A and B. (Hint: connect a battery between point A and infinity, causing a current I into point A. Next connect another battery between point B and infinity, causing a current I out of point B. Then apply the superposition theorem.) Explain your reasoning clearly.

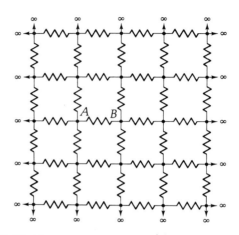

FIGURE 29-42
Problem 29C-57.

29C-58 A 4-μF capacitor, initially charged to 100 V, is in series with a 15 000-Ω resistor. The series combination is connected to an uncharged 10-μF capacitor. Calculate the current through the resistor when the voltage across the 4-μF capacitor is reduced to 50 V.

29C-59 A voltage source may be considered as a seat of emf $\mathscr{E}$ in series with an internal resistance r. When measured by a 20 000-Ω/V multirange voltmeter, the terminal voltage is 95 V on the 100-V scale and 120 V on the 200-V scale. Determine $\mathscr{E}$ and r. (The difference in the voltage readings is not a meter malfunction; the meter accurately reads the terminal voltage.)

29C-60 In Figure 29-43, the switch is put in position A and remains there until the capacitor C is fully charged. (a) What is the time constant of the charging circuit? (b) What is the initial charging current? (c) How long does the potential across C take to reach 50 V? (d) What is the energy stored in the fully charged capacitor? After the capacitor is fully charged, the switch is then moved to position B. (e) Find the time constant of the discharge circuit. (f) What is the initial discharging current? (g) What is the voltage across the capacitor 1 s after we switch to position B?

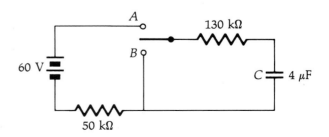

FIGURE 29-43
Problem 29C-60.

29C-61 Consider the circuit in Figure 29-44. With the capacitors initially uncharged, the switch is moved from A to B and remains there until the 10-μF capacitor is fully charged. The switch is then moved to position C. By direct calculation of $\int i^2 R\, dt$, calculate the thermal energy developed in R_1 after switching from A to B. After the switch to C, determine the energy finally developed in R_2.

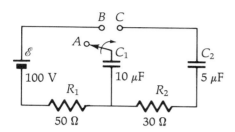

FIGURE 29-44
Problem 29C-61.

29C-62 Consider the circuit of Figure 29-45. Initially, there is no charge on the capacitor when the switch is closed at $t = 0$.

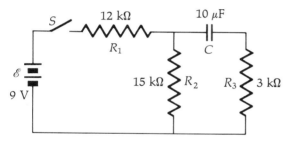

FIGURE 29-45
Problem 29C-62.

(a) Make a table showing the initial values (just after $t = 0$) of the current through each element—i_{12}, i_{15}, i_3, and i_C—and the initial voltage across each element—v_{12}, etc. (b) Repeat (a) for the final steady-state values of currents and voltages.

29C-63 The circuit for a simple *sawtooth oscillator* is shown in Figure 29-46a. The neon bulb conducts with very little resistance when the voltage across the bulb reaches 90 V, and it stops conducting when the voltage drops to 70 V. Calculate the frequency f of the oscillator.

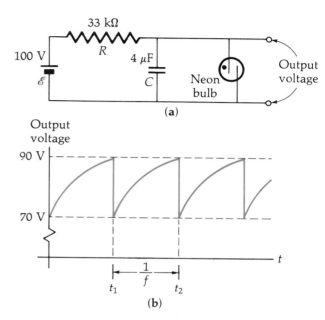

FIGURE 29-46
Problem 29C-63.

CHAPTER 30

The Magnetic Field

(Writing on his experiments and discoveries in magnetism)

We have dug them up and demonstrated them with much pain and sleepless nights and great money expense. Enjoy them, you, and, if ye can, employ them for better purposes.

WILLIAM GILBERT
On the Lodestone (published in 1600)

30.1 Introduction

In previous chapters we have discussed gravitational and Coulomb forces, both of which are inverse-square laws that do not depend on the relative motion of masses or charges. We now take up a type of force that does depend on the motion of charges. If two charges are both moving, in addition to Coulomb forces they exert a *magnetic* force on each other. The situation is a bit complicated, so we will separate the discussion into two parts. In one part we will show how a moving charge generates a magnetic field; in the other part, a second charge moves in the presence of this field and experiences a force. (We also followed this procedure in our discussion of Coulomb forces by considering one charge as the source of an electric field; the *field*, in turn, produces a force on another charge.) This chapter describes the effect that a magnetic field has on a moving charge, and the next chapter will discuss the origin of the magnetic field.

30.2 Magnetic Fields

The earliest recorded observations of magnetism were those of the Greeks about 2500 years ago. The word *magnetism* comes from the Greek *magnetis lithos*, a certain type of stone containing iron oxide (*magnetite* Fe_3O_4) found in Magnesia, a district in northern Greece. This "lodestone" could exert forces on similar stones and on pieces of iron. It would also impart this magnetic property to a piece of iron it touched. The early Chinese were perhaps the first to discover that, if a splinter of lodestone were suspended by a thread, it would align itself in a north–south direction. This suggests that the earth behaves like a large magnet. No doubt you have seen iron-filing patterns of the magnetic field surrounding a bar magnet (Figure 30-1). In the presence of

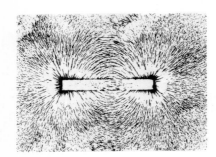

FIGURE 30-1
Iron filings sprinkled on a piece of paper covering a bar magnet arrange themselves in lines that suggest the magnetic field pattern.

a magnetic field, iron filings themselves become small magnets, aligning along the field directions and attracting each other to form chains that suggest the pattern of the field.

Since a compass needle always points in a unique direction in a magnetic field, the field has *vector* properties. How do we determine the existence of a field in a given region of space? The formal operational definition of a magnetic field is as follows. We place a test charge in the space. If there is a force exerted on the charge when it is at rest, we conclude that an *electrostatic* field is present. If still another force arises when the charge is moving, we conclude that a *magnetic* field also exists in the space. As a result of such experiments, the following facts concerning magnetic fields emerge:

> *The magnitude of the force is proportional to the magnitude of the test charge.*
> *The direction of the force is always perpendicular to the direction of motion.*
> *When the charge is moving in a given direction, the force is proportional to the speed, but for a given speed the force varies with the direction of motion. (Thus the field must be a vector.)*

The fact that the force is always perpendicular to the velocity implies a vector *cross-product* definition for the magnetic field. The following equation, based on experiment, defines the **magnetic induction**, also called the **magnetic flux density**, **B**. We will follow the current widespread (although somewhat loose) usage and call it simply the **magnetic field**.[1] The force **F** on a charge q that has a velocity **v** in the presence of a magnetic field **B** is

MAGNETIC FIELD B

$$\mathbf{F} = q\mathbf{v} \times \mathbf{B} \qquad (30\text{-}1)$$

The units for the magnetic field are newton·seconds/coulomb·meters (N·s/C·m), called[2] a *tesla* (T). Because **F** is always perpendicular to the plane containing **v** and **B**, we will often need to depict three-dimensional situations.

A magnetic field is represented graphically in the same way we represent an electric field. Lines are drawn so that their density is proportional to the magnitude of the magnetic field, and the tangent to a field line at a given point represents the direction of the field at that point. As in the representation of electric fields, the number of lines used to represent a given magnitude of magnetic field is arbitrary. For example, we may associate 10 lines/m^2 or 10^4 lines/m^2 with a given field, depending on convenience. There is no such thing in nature as a field line; we sketch the lines merely to help us visualize the properties of the magnetic field. The iron-filing patterns shown in Figure 30-1 depict the field directions fairly well, but do not give a good representation of the magnitude of the fields.

The end of a magnetized compass needle, which seeks the northerly direction, is called the *north pole* of the needle; the other end is the *south pole*. Consistent with Equation (30-1), the direction of the magnetic field created by a magnet is that the field lines *leave* the north pole and *enter* the south pole.

---

[1] Formally, the term *magnetic field strength* has been assigned to the vector $\mathbf{H} = \mathbf{B}/\mu$, where μ is the *permeability* of the space occupied by **B**.

[2] This unit honors Nikola Tesla (1856–1943), a Serbo-American engineer who devised many ingenious methods of electrical power generation and distribution. Among other accomplishments, he designed the Niagara Falls power system. One tesla is a strong field; the largest magnetic field achieved (as of Spring 1987) was a pulsed field of 68 T, lasting for 5.6 ms, at the Francis Bitter National Magnet Laboratory, Massachusetts Institute of Technology. A smaller unit (from the cgs system) is the *gauss* (G): $1\text{ G} = 10^{-4}\text{ T}$. The earth's field at the equator is roughly 0.3 G, while that of a small bar magnet may be a few hundred G. A still smaller unit, called the *gamma* (γ), is used in geophysics and space physics: $1\gamma = 10^{-5}\text{ G} = 10^{-9}\text{ T}$.

FIGURE 30-2
A way of depicting field lines perpendicular to the plane of the diagram.

(a) *Out of the paper.* (The dots suggest the points of arrows coming *toward* the reader.)

(b) *Into the paper.* (The crosses suggest the tail feathers of arrows going *away* from the reader.)

A convenient way of indicating magnetic fields *perpendicular* to the plane of a diagram is shown in Figure 30-2. In perspective sketches (refer to Figure 30-3b), idealized magnet poles are sometimes used to help establish the three-dimensionality of the diagram, with field lines emerging from the north pole and entering the south pole. The fringing fields are usually omitted in such sketches.

The spatial relationship among the force, velocity, and magnetic field vectors expressed in Equation (30-1) may be visualized by the usual right-hand rule shown in Figure 30-3a. In this convention, the fingers of the right hand curl around in the sense of rotation established when the first vector **v** is rotated (through the smallest angle) into the direction of **B**. The extended thumb then points in the direction of **F**. An alternative convention useful in dealing with fields is shown in Figure 30-3b. Here, the hand is held *flat* (with the thumb in the plane of the fingers). The fingers of the right hand point in the direction of the magnetic field. You can remember this by identifying the four fingers with field lines, which are spread through space. The thumb points in the direction of the velocity of the charged particle. (The hitchhiker putting out his thumb to ask for a ride on the moving particle!) By the definition of a vector cross-product, the force **F** is in the direction: $q\mathbf{v} \times \mathbf{B} = (qvB \sin \theta)\hat{\mathbf{n}}$ where $\hat{\mathbf{n}}$ is a unit vector perpendicular to both **v** and **B** according to the right-hand rule. The angle θ in this expression is then the angle between the thumb and the first finger. The magnetic force is outward from the palm of the hand and can be identified by the direction one would push.

When applying the right-hand rule, we will always consider q as a *positive* charge. If q is negative, we simply determine the direction of the force for a positive charge, then reverse the direction of the force. As an illustration, consider the magnetic force on a negative charge moving in an easterly direction near the equator, where the magnetic field of the earth is approximately horizontal in a northerly direction. When the right-hand rule is applied, the fingers point north and the outstretched thumb points east. The palm is upward, indicating an upward force on a positive charge. However, since the charge is negative, the force is downward.

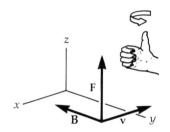

(a) The usual right-hand rule for the cross product $\mathbf{F} = q\mathbf{v} \times \mathbf{B}$. When the fingers curl around in the direction of **v** rotating into the direction of **B**, the extended thumb points in the direction of **F**.

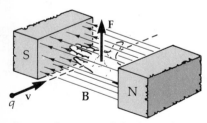

(b) Another way of thinking about the right-hand rule for cross products. When the fingers point in the direction of the field lines of **B** and the thumb points in the direction of the velocity **v**, the force **F** is in the direction your palm would push.

FIGURE 30-3
Two different ways of remembering the right-hand rule.

30.3 Motion of a Charged Particle in a Magnetic Field

An important feature of the motion of a charged particle in the presence of a magnetic field arises from the fact that the magnetic force is always at right angles to the velocity. Therefore, *the magnetic force does no work on the particle*; the particle's speed remains constant, though its direction changes in response to the sideways deflecting force of the magnetic field.

If the charged particle is given a velocity **v** at right angles to **B**, the particle will travel in a circular path at constant speed, with the magnetic force providing the centripetal force necessary to cause the centripetal acceleration: v^2/R (see Figure 30-4). Since **v** and **B** are at 90°, the magnitude of the magnetic force is

$$F = q|\mathbf{v} \times \mathbf{B}| = qvB \sin 90° = qvB$$

The radius R of the circular path may be found from Newton's second law. For the radially inward direction,

$$\Sigma \mathbf{F} = m\mathbf{a}$$

$$qvB = m\left(\frac{v^2}{R}\right) \quad (30\text{-}2)$$

$$R = \frac{mv}{qB} \quad (30\text{-}3)$$

The momentum mv of the particle is related to its kinetic energy K by[3]

$$mv = \sqrt{2mK} \quad (30\text{-}4)$$

Combining the previous two equations yields

$$R = \frac{\sqrt{2mK}}{qB} \quad (30\text{-}5)$$

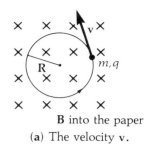

B into the paper
(a) The velocity **v**.

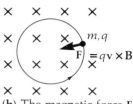

(b) The magnetic force **F**.

FIGURE 30-4
A positively charged particle traveling with velocity **v** at right angles to a uniform magnetic field **B** moves in a circle of radius R at constant speed. The frequency of revolution is called the *cyclotron frequency*.

EXAMPLE 30-1

An electron with a kinetic energy of 500 eV moves at right angles to a uniform magnetic field of 0.010 T. Find the radius of the circular motion.

SOLUTION

Making sure all numerical values are in SI units,[4] we substitute them into Equation (30-5):

$$R = \frac{\sqrt{2mK}}{qB} = \frac{\left[(2)(9.11 \times 10^{-31} \text{ kg})(500 \text{ eV})\left(\frac{1.602 \times 10^{-19} \text{ J}}{1 \text{ eV}}\right)\right]^{1/2}}{(1.602 \times 10^{-19} \text{ C})(0.010 \text{ T})}$$

$$R = \boxed{7.54 \times 10^{-3} \text{ m}}$$

Note that $1 \text{ eV} = 1.602 \times 10^{-19}$ J was used to make the conversion to SI units.

---

[3] Because electrons and even protons have such small masses, it is relatively easy to accelerate these charged particles to speeds approaching the speed of light: $c = 3 \times 10^8$ m/s, *the ultimate limiting speed for any object*. (See Chapter 41, Special Relativity.) For example, an electron accelerated from rest through a potential difference of 2500 V will be moving at about one-tenth the speed of light. Above such speeds, many of the "classical" (non-relativistic) relations, such as Equation (30-4), become noticeably incorrect. For the present, we avoid situations involving speeds near the speed of light, postponing a discussion of relativity to Chapter 41.

[4] Up to this point, including units in our numerical substitutions has allowed us to easily verify the consistency of the units by canceling the units. However, because we will now be using more *derived* units, such as the *tesla* in combination with basic units (the meter, kilogram, and second), such verification becomes difficult. For this reason, we will *ensure that all values are in SI units* and then assume with confidence that the answer is also in SI units. Though dimensional analysis is helpful in many branches of physics, in electricity and magnetism two other authors have called it "a big, buzzing, blooming confusion" and "a collection of stupidities"!

The rotational frequency of the circular motion is called the *cyclotron frequency*. The name comes from the fact that motion of this type originates in a *cyclotron*, a type of machine that accelerates charged particles (see Figure 30-8). We may obtain the cyclotron frequency from Equation (30-2):

$$qvB = m\frac{v^2}{R}$$

For circular motion, $v = 2\pi fR$. Substituting this value and solving for f leads to

CYCLOTRON FREQUENCY $\qquad f = \dfrac{B}{2\pi}\left(\dfrac{q}{m}\right) \qquad$ (30-6)

where f is the rotational frequency of circular motion in units of revolutions per unit time. This is the characteristic frequency of a particle of a given *charge-to-mass ratio* (q/m) in a uniform magnetic field. Note that *the cyclotron frequency is independent of the speed and energy of the charged particle.*

If a charged particle moves *parallel* to the field **B**, there is no force on the particle because the cross-product **v** × **B** is zero. For motion at an arbitrary angle (other than 90°) with respect to the field, its motion will be a *helix* rather than a circle (Figure 30-5). Since the velocity of the particle can be resolved into two components, *parallel* and *perpendicular* to the field, the cyclotron frequency is also the characteristic frequency for the helical motion.

The motion of charged particles in *non*uniform fields can be rather complicated. However, there is one simple example worth mentioning. Figure 30-6 depicts an axially symmetric magnetic field that is stronger at the ends than in the middle. A charged particle approaching one end as it moves in a helical path will experience a magnetic force **F** having a horizontal component that "reflects" the particle back toward the middle. This configuration is called a *magnetic bottle* because it can trap charged particles within a confined region as they oscillate in helical paths back and forth between the ends of the bottle. In recent years, magnetic bottles have been used to confine plasmas in controlled fusion experiments. Unfortunately, the bottle "leaks" somewhat, since particles traveling *along* the magnetic field lines escape out the ends. To solve this problem, the ends of the bottle are often joined together to form a toroid.

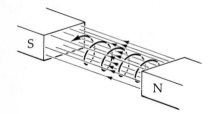

FIGURE 30-5
In a uniform magnetic field, a charged particle can travel in a helical path at constant speed. The path lies on an imaginary cylinder of constant radius.

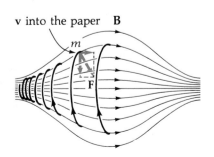

FIGURE 30-6
A *magnetic bottle* can trap charged particles by "reflecting" their helical motions at each end.

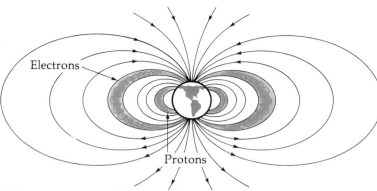

(a) A cross-section of the Van Allen belts that surround the earth. The earth's magnetic field acts as a magnetic bottle, trapping high-energy electrons and protons from the sun within two regions. The charged particles spiral between the north and south magnetic poles of the earth, with a typical round trip taking about one second. The inner belt traps mainly protons, while the other belt traps mainly electrons. [The belts are named after their discoverer, Dr. James Van Allen, who insisted that a Geiger counter to detect charged particles be carried aboard the United States' first successful earth-orbiting satellite (1958).]

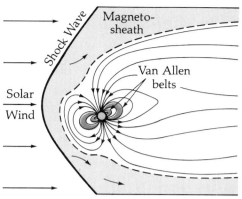

(b) The solar "wind," a stream of electrons and protons emanating from the sun, "blows away" the earth's magnetic field pattern in a direction away from the sun. The incoming particles impinge at speeds of roughly 400–800 km/s, creating a bowed shock wave and a long cometlike tail to the earth's field that extends for millions of kilometers. The magnetosheath is a region of subsonic plasma flow behind the shock wave. Some particles are trapped in the Van Allen belts and others are diverted away. The solar wind is usually fairly steady but can have occasional severe gusts due to activity on the sun, such as sunspots and solar flares.

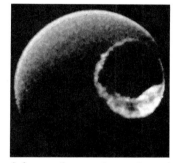

(c) A satellite photo of the southern polar region of the earth. The bright area in the upper left is the sunlit portion of the earth; the circular ring is produced by aurora. (A similar auroral ring also occurs at the north magnetic pole.) The process that produces aurora is the following. Bursts of charged particles ejected from solar flares on the sun reach the earth in a few hours or days, causing extra numbers of particles in the Van Allen belts to leak out near the magnetic poles where the magnetic-bottle effect is "leaky." Because of the configuration of the Van Allen belts, most aurora occur in circular zones about 2000 km in diameter, centered on the magnetic poles. When the charged particles collide with gases in the upper atmosphere, they cause atoms of oxygen and nitrogen to glow, producing the spectacular shimmering displays called auroras. This image was obtained in the ultraviolet (primarily atomic oxygen, 130.4 nm) and represents data received over a period of 12 min by the University of Iowa's auroral-imaging instrumentation.

(d) Auroral bands photographed in Alaska.

FIGURE 30-7
Charged particles near the earth.

The Cyclotron

Ernest O. Lawrence was awarded the Nobel Prize in 1939 for his development (with M. S. Livingston) of the cyclotron, a device that accelerates charged particles to high energies for use in nuclear experiments. Its basic components are a short cylindrical box made of copper sheet metal, divided into two sections called *dees* (see Figure 30-8). The dees are in a vacuum chamber that is evacuated so the charged particles can move without colliding with air molecules. A magnetic field is established normal to the plane of the dees. A source of alternating voltage is connected to the dees, creating an electric field across the gap between the dees that reverses its direction every half cycle. Near the center of the dees, an ion source supplies charged particles such as protons, deuterons, or alpha particles, giving them a small velocity in the plane of the dees. Within the copper dees, the metal walls shield the ions from electric fields. However, the magnetic field is not shielded, causing the ions to move in a semicircle. Consider an ion that arrives at the gap between the dees just when the electric field between them is a maximum and in a direction to accelerate the ion across the gap. Subsequently, the ion will move in a larger semicircle because of its greater speed. If the frequency of the voltage reversals is correct, the ion arrives again at the gap just as the electric field reaches its maximum value in the opposite direction, again accelerating the ion. Each time it crosses the gap, the ion thus gains kinetic energy, traveling in larger and larger radii until it approaches the circumference of the cylinder, where a negatively charged *deflecting plate* pulls the ion from its circular path and allows it to pass out of the chamber through a thin window. The key to the operation of a cyclotron is that *the travel time for each semicircular path is the same*. As Equation (30-6) shows, the cyclotron frequency is independent of the speed or of the radius of the circle.

There is an upper energy limit—about 22 million electron volts (22 MeV) for protons—because of relativity. As more work is done on a particle to increase its speed, because of relativistic effects the speed does not increase sufficiently to keep in step with voltage reversals. The difficulty is overcome in the *synchrotron*, where both the frequency and the magnetic field are varied, keeping the orbit radius essentially constant. This method has the economical advantage of requiring a magnetic field only in the region of the orbit, rather than in the entire area of the circle.

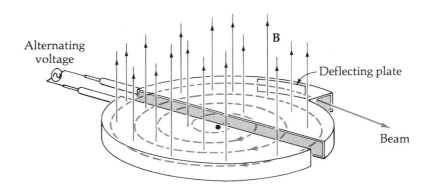

FIGURE 30-8
Under the influence of the magnetic field, the charges move in semicircular paths within the dees of the cyclotron.

EXAMPLE 30-2

Find the cyclotron frequency of an electron moving in a uniform magnetic field of 0.020 T.

SOLUTION

From Equation (30-6) we obtain

$$f = \frac{B}{2\pi}\left(\frac{q}{m}\right) = \frac{(2 \times 10^{-2}\text{ T})(1.602 \times 10^{-19}\text{ C})}{2\pi(9.11 \times 10^{-31}\text{ kg})} = \boxed{5.59 \times 10^{8}\text{ Hz}}$$

FIGURE 30-9
The proton synchrotron at the Enrico Fermi National Accelerator Laboratory, Batavia, Illinois. The main accelerator ring has a diameter of 2 km and produces protons of energy 1 TeV = 10^3 GeV = 10^6 MeV. Three experimental areas extend tangentially from the ring toward the bottom of the picture.

30.4 The Lorentz Force Law

In general, a charged particle may simultaneously experience the effects of both an electric field and a magnetic field. Since the electric and magnetic forces resulting from these fields add as vectors, the net force on a charge may be written as

LORENTZ FORCE $\mathbf{F} = q(\mathbf{E} + \mathbf{v} \times \mathbf{B})$ (30-7)

where $\mathbf{F}$ is the net force on a charge q moving with a velocity $\mathbf{v}$ in the presence of an electric field $\mathbf{E}$ and a magnetic field $\mathbf{B}$. This equation is called the **Lorentz force law**.

A useful application of the Lorentz force law is a charged-particle *velocity filter*. Consider a particle of mass m and charge q moving with speed v along a straight path defined by collimating apertures as shown in Figure 30-10. The particle will pass through the exit aperture if there is no net force on the particle while it is in the region between the collimating and exit apertures. To accomplish this, magnetic and electric fields are established in the region so that the magnetic force on the particle is equal and opposite to the electric force. The directions of these forces, for positive charges, are as shown in Figure 30-10. Both forces are reversed in direction for negative charges. We apply Equation (30-7): $\mathbf{F} = q(\mathbf{E} + \mathbf{v} \times \mathbf{B})$. For zero net force we have $\mathbf{E} + \mathbf{v} \times \mathbf{B} = 0$. In magnitude, this is

$$v = \frac{E}{B} \quad (30\text{-}8)$$

Only particles with this speed will travel in a straight line and emerge from the exit apertures, thus giving the device the name "velocity filter."

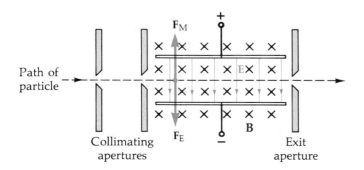

FIGURE 30-10
A *charged-particle velocity filter* is formed of magnetic and electric fields at right angles. When the magnetic force $\mathbf{F}_M$ on a moving particle just balances the electric force $\mathbf{F}_E$, the particle travels in a straight line and emerges from the exit hole. Particles traveling faster or slower than the critical velocity are bent out of the straight-line path.

EXAMPLE 30-3

A stream of electrons passes through a velocity filter when the crossed magnetic and electric fields are 2×10^{-2} T and 5×10^4 V/m, respectively. Find the kinetic energy (in electron volts) of the electrons passing through the filter.

SOLUTION

We find the speed from Equation (30-8):

$$v = \frac{E}{B} = \frac{5 \times 10^4 \text{ V/m}}{2 \times 10^{-2} \text{ T}} = 2.50 \times 10^6 \frac{\text{m}}{\text{s}}$$

Substituting numerical values in the expression for kinetic energy gives

$$K = \frac{1}{2}mv^2 = \frac{1}{2}(9.11 \times 10^{-31} \text{ kg})\left(2.50 \times 10^6 \frac{\text{m}}{\text{s}}\right)^2$$

$$K = 2.85 \times 10^{-18} \text{ J} \underbrace{\left(\frac{1 \text{ eV}}{1.602 \times 10^{-19} \text{ J}}\right)}_{\text{Conversion ratio}} = \boxed{17.8 \text{ eV}}$$

30.5 Magnetic Force on a Current-Carrying Conductor

In most applications, moving charges are confined to move through conductors. In the case of a metal wire, the charges are electrons moving with the drift velocity v_d. We shall now investigate the total force on all these moving charges when the conductor is in the presence of a magnetic field.

Equation (30-1) gives the force on one charge:

$$\mathbf{F} = q\mathbf{v} \times \mathbf{B}$$

The total number of moving charges in a wire of length ℓ is the number of conduction charges per unit volume n times the volume of the wire segment $A\ell$. Thus, the total force on a wire segment of length ℓ is

$$\mathbf{F} = q(\mathbf{v} \times \mathbf{B})nA\ell \tag{30-9}$$

In a previous chapter [Equation (28-4)], we found the current I to be

$$I = nqv_d A$$

Combining these two equations and identifying $\mathbf{v}$ with the drift velocity $\mathbf{v}_d$ of (positive) charges, we obtain

FORCE ON A CURRENT-CARRYING CONDUCTOR IN THE PRESENCE OF A MAGNETIC FIELD

$$\mathbf{F} = I\boldsymbol{\ell} \times \mathbf{B} \tag{30-10}$$

Here, we maintain the vector form of the magnetic force by defining the length of the conductor as a vector $\boldsymbol{\ell}$ *in the direction of the conventional current* (the direction *positive* charges move).

Equation (30-10) assumes the wire segment is straight and the magnetic field is uniform. If the wire segment is of arbitrary shape and if the field varies, we recognize that the force $d\mathbf{F}$ on a small element $d\boldsymbol{\ell}$ of the wire is

$$d\mathbf{F} = I\,d\boldsymbol{\ell} \times \mathbf{B} \qquad (30\text{-}11)$$

Then, to find the total force, we integrate over the entire length of the wire using the value of $\mathbf{B}$ appropriate for each element $d\boldsymbol{\ell}$.

EXAMPLE 30-4

In Figure 30-11, a straight wire carries a current of 8 A in the presence of a uniform magnetic field of 3×10^{-3} T. The field is at an angle $\theta = 48°$ with respect to the wire. Find the force per unit length that the field exerts on the current-carrying wire.

SOLUTION

From Equation (25-10),

$$F = I|\boldsymbol{\ell} \times \mathbf{B}| = I\ell B \sin\theta$$

Thus:
$$\frac{F}{\ell} = IB \sin\theta = (8\text{ A})(3 \times 10^{-3}\text{ T})(\sin 48°)$$

$$\boxed{\frac{F}{\ell} = 1.78 \times 10^{-2}\ \frac{\text{N}}{\text{m}}} \qquad \text{(out of the paper)}$$

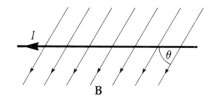

FIGURE 30-11
Example 30-4.

EXAMPLE 30-5

A current I is in a rigid semicircular loop of wire that has a radius R, as shown in Figure 30-12. A uniform magnetic field $\mathbf{B}$ is perpendicular to the loop. The current enters and leaves the loop by conductors (not shown) that are perpendicular to the plane of the paper and therefore parallel to the field. (Thus there are no forces on the wires that supply current to the semicircle.) Find the net force on the semicircular wire.

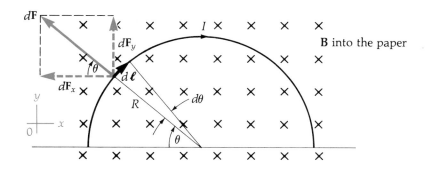

FIGURE 30-12
Example 30-5.

FIGURE 30-13
The normal to a current-carrying loop is determined by this right-hand rule: the fingers curl around in the current direction and the extended thumb points in the direction of **μ**.

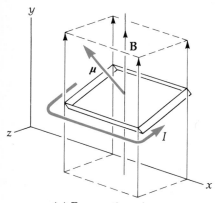

(a) Perspective view.

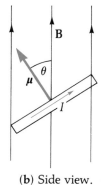

(b) Side view.

FIGURE 30-14
A rectangular current-carrying loop that has an arbitrary orientation in a uniform, vertical magnetic field **B**.

SOLUTION

Since the conductors leading into and out of the wire are parallel to the magnetic field, the cross-product in $\mathbf{F} = I\boldsymbol{\ell} \times \mathbf{B}$ involves $\sin 0° = 0$. Thus the force on these conductors is zero.

The force $d\mathbf{F}$ on an incremental length $d\boldsymbol{\ell}$ of the loop is given by Equation (30-11): $d\mathbf{F} = I\,d\boldsymbol{\ell} \times \mathbf{B}$. Since **B** is perpendicular to $d\boldsymbol{\ell}$ over the entire length of the semicircular loop, the incremental force $d\mathbf{F}$ is directed radially outward everywhere and has a magnitude

$$dF = IB\,d\ell$$

We now make use of a symmetry argument. For every incremental force $d\mathbf{F}$ on the left side of the semicircular loop, there will be a corresponding increment symmetrically located on the right-hand side. The x components of these two forces are equal but in opposite directions, so they add to zero. However, the y components are in the same direction. Therefore, as we sum up the forces for the entire semicircle, we are left with only the sum of the y components: $dF_y = IBR \sin\theta\,d\theta$. Integrating gives

$$F_y = \int dF_y = IBR \int_0^\pi \sin\theta\,d\theta = IBR(-\cos\theta)\Big|_0^\pi = \boxed{2IBR}$$

(in the $+y$ direction)

Note that this would be the force on a straight conductor *along the diameter* of the circular loop. Actually, the shape of the loop is unimportant. As shown in a problem, the net force on any arbitrarily shaped segment of wire that lies in a plane perpendicular to a field depends only on the *length of the gap* between the current input and the current output. This example leads to the conclusion that the *net force on a closed current-carrying loop (of any shape) in a uniform magnetic field is zero*. The net force is zero not because the force on each segment of the loop is zero, but because the *sum* of the forces on all segments is zero.

30.6 Magnetic Dipoles

Although a planar current-carrying loop in a uniform magnetic field experiences no net *force*, it may experience a *torque*. The behavior is analogous to that of an electric dipole in a uniform electric field. In fact, the analogy is so close that we will define a current-carrying loop as a *magnetic dipole*, in much the same way as we called a pair of charges of opposite sign an *electric dipole*.

We begin by defining a vector that is normal to a current-carrying loop. As shown in Figure 30-13, we define the direction of the normal by curling the fingers of the right hand around the loop so that the fingers circle the loop in the direction of the conventional current. The extended thumb points in the direction of the desired normal. This is the direction of the vector **μ** defined shortly.

Consider the rectangular loop in a magnetic field shown in Figure 30-14. Note how **μ** is related to the direction of the current around the loop by the angle θ between **μ** and **B**. The force on each of the sides of the rectangle is given by Equation (30-10):

$$\mathbf{F} = I\boldsymbol{\ell} \times \mathbf{B}$$

The forces $\mathbf{F}_1$ and $\mathbf{F}_2$ shown in Figure 30-15a are equal, but opposite in direction, so their contribution to the net *force* is zero. Also, since they are collinear, they produce zero net *torque*. The forces $\mathbf{F}_3$ and $\mathbf{F}_4$ are equal and opposite, so they, too, contribute zero net *force*. However, because they are not collinear,

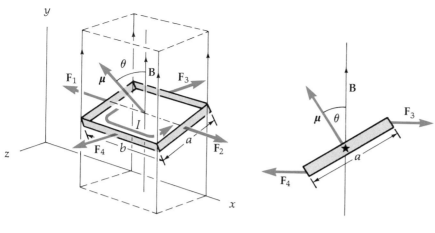

(a) Perspective view. (b) Side view.

FIGURE 30-15
The forces on each side of a rectangular current-carrying loop in a uniform magnetic field **B**. These forces exert a net torque on the loop, trying to align its magnetic moment μ in the direction of **B**.

they form a *couple* (see Figure 13-12) and produce a torque about an axis (★) normal to the view of the loop in Figure 30-15b. If θ is the angle between **B** and the normal to the loop (the vector μ), the magnitude of the torque about (★) is

$$\tau = F_3 \left(\frac{a}{2}\right) \sin\theta + F_4 \left(\frac{a}{2}\right) \sin\theta$$

Since $F_3 = F_4$, $\tau = F_3 a \sin\theta$ (or $F_4 a \sin\theta$) (30-12)

Since sides 3 and 4 are perpendicular to **B**, from $\mathbf{F} = I\boldsymbol{\ell} \times \mathbf{B}$ we have

$$F_3 = IbB$$

Substituting this value in Equation (30-12), gives

$$\tau = IabB \sin\theta \qquad (30\text{-}13)$$

Let A represent the area ab. The factor IA is called the magnitude of the **magnetic dipole moment** μ. The direction of μ is defined by the right-hand rule described previously. Using the notation **A** for the area normal *vector* (whose magnitude is the area A), we write

MAGNETIC DIPOLE MOMENT μ $\quad \boldsymbol{\mu} = I\mathbf{A} \quad$ (the direction of μ is normal to the plane of the loop of area A according to the right-hand rule) (30-14)

For N turns: $\boldsymbol{\mu} = NI\mathbf{A}$

The units of μ are ampere-meters squared (A·m²).

We write Equation (30-13) using vector notation as follows:

TORQUE ON A MAGNETIC DIPOLE IN A MAGNETIC FIELD B $\quad \boldsymbol{\tau} = \boldsymbol{\mu} \times \mathbf{B} \quad$ (30-15)

Note the close similarity to the expression for torque on an electric dipole **p** in an electric field **E**:

$$\boldsymbol{\tau} = \mathbf{p} \times \mathbf{E}$$

Although this derivation was based on a rectangular loop, it is valid for *any* planar-loop shape. That is,

$$\mu = (I)(\text{area of the loop}) \tag{30-16}$$

Figure 30-16 provides the basis for this conclusion. A current-carrying loop of arbitrary shape may be considered as a group of adjacent current-carrying rectangles. (The greater the number of rectangles, the better the approximation to the loop.) The currents in all the rectangles are clockwise. Thus, the currents of adjacent rectangles cancel out in the interior of the loop, leaving only the current around the perimeter. In this way, we generalize the derivation from that of a rectangular loop to a (planar) loop of any shape whatever.

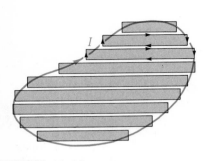

FIGURE 30-16
The clockwise currents around all the individual rectangles approximate the current I around the loop. This is because the currents in the sides of adjacent rectangles are in opposite directions, and (in the limit of infinitely thin rectangles) these currents inside the loop add to zero.

A torque on a current-carrying loop in a magnetic field implies a potential energy associated with the orientation of the loop with respect to the field direction. Following the similar development of a potential energy associated with an electric dipole in an electric field (Section 22.5), we start with the general definition of potential energy for rotation:

$$U_\theta - U_{\theta_0} = -\int_{\theta_0}^{\theta} \boldsymbol{\tau} \cdot d\boldsymbol{\theta}$$

Since θ increases counterclockwise, as indicated in Figure 30-14b, $\boldsymbol{\tau}$ and $d\boldsymbol{\theta}$ are antiparallel. Therefore, $\cos 180° = -1$, and

$$U_\theta - U_{\theta_0} = -\int_{\theta_0}^{\theta} \tau \, d\theta$$

Substituting the expression for τ given by Equations (30-13) and (30-14) and integrating, we have

$$U_\theta - U_{\theta_0} = \int_{\theta_0}^{\theta} \mu B \sin \theta \, d\theta = -\mu B (\cos \theta - \cos \theta_0)$$

Choosing the zero reference orientation for potential energy to be $U_{\theta_0} \equiv 0$ when $\theta_0 = 90°$, we have

$$U = -\mu B \cos \theta \tag{30-17}$$

This suggests the vector dot product notation:

POTENTIAL ENERGY U OF A MAGNETIC DIPOLE IN A MAGNETIC FIELD
($U \equiv 0$ when $\boldsymbol{\mu}$ and $\mathbf{B}$ are at 90°)

$$U = -(\boldsymbol{\mu} \cdot \mathbf{B}) \tag{30-18}$$

Note that the potential energy of the dipole is a maximum when $\boldsymbol{\mu}$ is antiparallel to $\mathbf{B}$ and a minimum when $\boldsymbol{\mu}$ is parallel to $\mathbf{B}$, with the zero reference orientation midway between at 90°. This is the same notation we used for the potential energy of an electric dipole in an electric field, Equation (24-20):

$$U = -\mathbf{p} \cdot \mathbf{E}$$

Since physical systems tend to move toward positions of minimum potential energy, the magnetic dipole $\boldsymbol{\mu}$ tends to align itself in the direction of the magnetic field $\mathbf{B}$.

EXAMPLE 30-6

A wire is formed into a circle with a diameter of 10 cm and placed in a uniform magnetic field of 3×10^{-3} T. A current of 5 A passes through the wire. Find (a) the maximum torque that can be experienced by the current-carrying loop and (b) the range of potential energy the loop possesses for different orientations.

SOLUTION

The magnetic dipole moment of the current-carrying loop of wire is given by Equation (30-16):

$$\mu = (I)(\text{area of the loop})$$

Substituting numerical values gives

$$\mu = (5 \text{ A})(\pi)(0.05 \text{ m})^2 = 3.93 \times 10^{-2} \text{ A} \cdot \text{m}^2$$

(a) The torque exerted on a magnetic dipole in a uniform magnetic field is given by Equation (30-15):

$$\boldsymbol{\tau} = \boldsymbol{\mu} \times \mathbf{B}$$

which has a maximum value when the field and the dipole moment are perpendicular, that is, when the plane of the wire loop is parallel to the magnetic field. Its maximum magnitude is

$$\tau = \mu B$$

Substituting yields

$$\tau = (3.93 \times 10^{-2} \text{ A} \cdot \text{m}^2)(3 \times 10^{-3} \text{ T}) = \boxed{1.18 \times 10^{-4} \text{ N} \cdot \text{m}}$$

(b) The potential energy possessed by a magnetic dipole in a magnetic field is given by Equation (30-18):

$$U = -\boldsymbol{\mu} \cdot \mathbf{B}$$

The maximum potential energy occurs when the dipole moment is antiparallel to the field, and a minimum occurs when the magnetic moment is parallel to the field. The range of potential energy is

$$\Delta U = U_{max} - U_{min} = -\mu B \cos \pi - (-\mu B \cos 0°) = 2\mu B$$

Substituting the values for μ and B, we have

$$\Delta U = 2(3.93 \times 10^{-2} \text{ A} \cdot \text{m}^2)(3 \times 10^{-3} \text{ T}) = \boxed{2.36 \times 10^{-4} \text{ J}}$$

30.7 Applications

Galvanometer

In Chapter 29 we discussed the construction of voltmeters and ammeters, both of which utilize a sensitive current-measuring device called a *galvanometer*. We shall now describe the basic principles underlying the operation of a galvanometer.

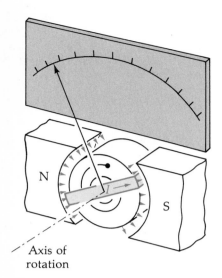

FIGURE 30-17
The basic meter movement of a galvanometer.

A galvanometer consists of a current-carrying coil in a magnetic field, as shown in Figure 30-17. Current is conducted to the coil of wire through the bearings that support the coil and allow rotation about a fixed axis. The connection from one bearing to the coil is through a spiral spring that not only conducts the current, but also exerts a restoring torque when the coil is rotated from its equilibrium position. As the loop rotates, the sides of the loop move in a region of magnetic field **B**, which is constant in magnitude and always perpendicular to the **μ** of the coil. This is achieved by specially shaped pole faces of a permanent magnet and a (fixed) iron cylinder inside the loop. Thus the torque on the coil due to the current depends only on the current and is independent of the orientation of the loop. The coil is restrained by a spiral spring that conforms to Hooke's law:

$$\tau_{\text{spring}} = -\kappa\theta$$

The torque on the coil is given by Equation (30-15): $\tau_{\text{coil}} = \boldsymbol{\mu} \times \mathbf{B}$. But since **μ** is always perpendicular to **B**,

$$\tau_{\text{coil}} = \mu B$$

When the coil is in static equilibrium,

$$\tau_{\text{spring}} = -\tau_{\text{coil}}$$
or
$$\kappa\theta = \mu B$$

Expressing the angle of rotation from equilibrium θ in terms of the current through the coil, we have

$$\theta = \left(\frac{AB}{\kappa}\right)I \qquad (30\text{-}19)$$

where A is the area of the coil. If the coil has N turns of wire, the total current I in the loop is

$$I = NI_0$$

where I_0 is the current through the wire. Then Equation (30-19) becomes

$$\theta = \left(\frac{NAB}{\kappa}\right)I_0 \qquad (30\text{-}20)$$

The angle θ is measured by a pointer attached to the coil. *The angular deflection θ is directly proportional to the current in the coil*, so the scale along which the pointer moves is linear.

EXAMPLE 30-7

A typical galvanometer has the following specifications and parameters: coil area, 1 cm^2; number of turns of wire on the coil, 100 turns; spring constant of the spiral spring, 3×10^{-7} N·m/rad; and a current sensitivity of 50 μA for a coil rotation of $\pi/2$ rad (full-scale deflection). Find the magnitude of the magnetic field through which the coil moves.

SOLUTION

From Equation (30-20), $$B = \frac{\kappa\theta}{NAI_0}$$

Substituting the appropriate values (all in SI units) yields

$$B = \frac{(3 \times 10^{-7}\ \text{N·m/rad})\left(\frac{\pi}{2}\ \text{rad}\right)}{(100\ \text{turns})(10^{-4}\ \text{m}^2)(50 \times 10^{-6}\ \text{A/turn})} = \boxed{0.942\ \text{T}}$$

Sensitive galvanometers such as this are very delicate because of the small torque exerted on a practical-size coil moving in a field that can be realistically achieved by a magnet. The bearings are often jewel bearings similar to those found in a watch, and the spiral spring consists of several turns of extremely fine spring wire.

Hall Effect

An effect used by E. H. Hall in 1879 to determine the sign of current carriers in conductors is now used extensively to measure currents and magnetic fields. The *Hall effect* describes the potential difference that develops between the sides of a current-carrying conductor when the conductor is placed in a magnetic field. In order to understand how such a potential difference develops, we will consider an idealized conductor in which the charge-carriers are free electrons.[5]

Consider an idealized conductor of rectangular cross-section placed in a magnetic field **B**, as shown in Figure 30-18. The magnetic force $\mathbf{F_M}$ on a single electron is

$$\mathbf{F_M} = (-e)\mathbf{v_d} \times \mathbf{B} \tag{30-21}$$

where $-e$ is the charge on the electron and $\mathbf{v_d}$ is the drift velocity of the electron. Initially, the magnetic force will cause the electrons to drift toward the right-hand edge of the conductor. Eventually, however, the accumulation of charge produces an electric field **E** within the conductor, thus inhibiting further lateral drift of the charge. In equilibrium, the electric force $\mathbf{F_E}$ that results from the electric field will just balance the magnetic force:

$$|\mathbf{F_E}| = |\mathbf{F_M}|$$

Applying the Lorentz force law, the forces balance if

$$eE = ev_d B$$

or $$E = v_d B \tag{30-22}$$

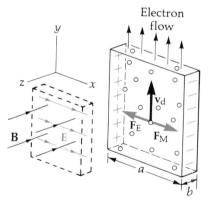

FIGURE 30-18
The Hall effect. A magnetic field in the $-z$ direction forces the moving electrons to the right edge of the conductor, creating an electric field in the $+x$ direction. When equilibrium is attained, the magnetic force $\mathbf{F_M}$ on the moving electrons is equal and opposite to the electric force $\mathbf{F_E}$.

---

[5] Monovalent metals such as copper and silver behave as nearly idealized current-carriers in the analysis of the Hall effect. The analysis of the Hall effect in magnetic current-carriers such as iron and in semiconductor current-carriers is complicated by quantum effects.

The drift speed v_d can be expressed in terms of the current and the parameters of the conductor through the definition of current,

$$I = nev_d A \qquad (30\text{-}23)$$

where n is the number of current-carriers per unit volume and A is the cross-sectional area of the conductor. In this instance, $A = ab$. Substituting and solving for v_d, we have

$$v_d = \frac{I}{neab} \qquad (30\text{-}24)$$

Further substitution into Equation (30-22) gives

$$Ea = \frac{BI}{neb} \qquad (30\text{-}25)$$

The electric field E times the width a of the conductor is the potential difference V across the width. This potential difference is referred to as the *Hall potential*, V_H.

HALL POTENTIAL
$$V_H = \frac{BI}{neb} \qquad (30\text{-}26)$$

Because the Hall potential depends upon the product BI, if we know the current, for example, we can determine the value of B by measuring the Hall potential. Hall-effect probes are commonly used to measure magnetic field strengths. The other significant feature of the Hall effect is that, if the current-carriers are positive charges (rather than negative electrons), the polarity of the Hall potential will be reversed for the same direction of magnetic field and current. So, in a known field B, the Hall effect can be used to determine the *number* and the *sign* of the current-carriers within the material.

EXAMPLE 30-8

Suppose the conductor shown in Figure 30-18 is copper and is carrying a current of 10 A in a magnetic field of 0.5 T. The width of the conductor d is 1 cm and the thickness is 1 mm. Find the Hall potential across the width of the conductor.

SOLUTION

The Hall potential is given by Equation (30-26):

$$V_H = \frac{BI}{neb}$$

Copper has a density ρ of 8.92×10^6 g/m^3, a molecular weight (mol. wt.) of 63.546 g/mol. We assume that each copper atom contributes one electron to the current, so the number n of conduction electrons per unit volume is

$$n = \frac{\rho N_A}{(\text{mol. wt.})}$$

where N_A is Avogadro's number: 6.022×10^{23} atoms/mol. Substituting the appropriate values gives

$$n = \frac{\left(8.92 \times 10^6 \, \frac{g}{m^3}\right)\left(6.022 \times 10^{23} \, \frac{electrons}{mol}\right)}{\left(63.546 \, \frac{g}{mol}\right)}$$

$$= 8.45 \times 10^{28} \, \frac{electrons}{m^3}$$

Substituting this and other values into Equation (30-26) yields

$$V_H = \frac{BI}{neb} = \frac{(0.5 \text{ T})(10 \text{ A})}{\left(8.45 \times 10^{28} \, \frac{electron}{m^3}\right)\left(1.602 \times 10^{-19} \, \frac{coulomb}{electron}\right)(1 \times 10^{-3} \text{ m})}$$

$$V_H = \boxed{3.69 \times 10^{-7} \text{ V}}$$

While this potential difference is very small for conductors, the corresponding potential difference for semiconductors is much greater. [See Equation (30-26): n is smaller for semiconductors than for ordinary conductors.] For this reason semiconductors are useful as probes in measuring magnetic fields by the Hall effect.

Analysis of the Hall effect gives us a clearer understanding of the nature of the force on a current-carrying conductor in a magnetic field. The force on the conductor is actually an electric force arising from the Hall field. Note that in Figure 30-18 the net sideways force on the moving charge is zero:

$$\mathbf{F_E} + \mathbf{F_M} = 0$$

The magnetic force $\mathbf{F_M}$ is produced by a field external to the conductor, whereas the electric force $\mathbf{F_E}$ arises within the conductor due to the Hall effect. The force on the conductor is (by Newton's third law) equal and opposite to the electric force on the charge-carriers. So we see that the magnetic force on a current-carrying conductor is actually electrical in nature.

Linear Mass Spectrometer

Charged particles may be sorted according to their charge-to-mass ratio q/m by a device illustrated in Figure 30-19. The material to be analyzed is placed in the oven and heated to a temperature high enough to produce a gas of ionized particles. The particles leave the oven with a relatively low velocity and are accelerated by a potential difference between the oven and an aperture. The particles then leave the aperture with velocities that have essentially the same component in the x direction. Since the aperture does not collimate the charged particles perfectly, the particles may also have a small component of velocity perpendicular to the x direction. After leaving the aperture, the particles enter a longitudinal magnetic field, causing them to execute a helical trajectory. Because the cyclotron frequency is the same for all particles having

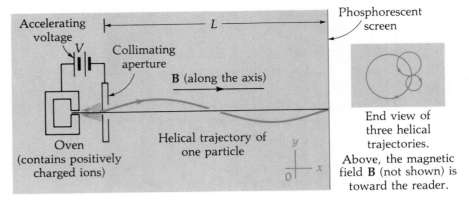

FIGURE 30-19
A *linear mass spectrometer*. All charged particles with the same q/m ratio have the same cyclotron frequency, so after one period of their motions, all converge at the same point along the axis of the spectrometer.

the same q/m ratio, after one turn all such particles will cross the axis at the same point (if their x components of velocity are the same).

Let us now solve for the charge-to-mass ratio in terms of the other parameters. The x component of the velocity v_x of a particle leaving the aperture is obtained by the energy relation

$$qV = \tfrac{1}{2}mv_x^2$$

Solving for v_x^2 gives
$$v_x^2 = 2V\left(\frac{q}{m}\right) \tag{30-27}$$

Another expression for v_x is
$$v_x = \frac{L}{T}$$

where T is the time for the particle to execute one turn of its helical trajectory. T is equal to the reciprocal of the cyclotron frequency f of the particle, given by Equation (30-6):

$$f = \frac{1}{2\pi} B\left(\frac{q}{m}\right)$$

Therefore:
$$v_x = \frac{1}{2\pi} BL\left(\frac{q}{m}\right)$$

Substituting this expression for v_x into Equation (30-27) and solving for q/m yields

$$\frac{q}{m} = \frac{8\pi^2 V}{B^2 L^2} \tag{30-28}$$

In practice, a small fixed collector of charged particles is placed on the axis of the spectrometer. The potential V is adjusted until the collection of charges is a maximum, indicating the convergence of particles. The ratio q/m can then be calculated using Equation (30-28).

EXAMPLE 30-9

An electron microscope produces a magnified image on a photographic plate, utilizing an electron beam rather than light rays. The electron beam is "focused" by a magnetic field in the same way that a linear mass spectrometer converges charged particles. (a) Find the magnitude of the minimum magnetic field that will focus 10-keV electrons at a distance of 10 cm from the source of electrons. (b) Calculate another value of the magnetic field that will also produce a focusing of the electrons.

SOLUTION

(a) Focusing an electron beam is identical to operating the linear mass spectrometer. Therefore Equation (30-2) is applicable:

$$\frac{q}{m} = \frac{8\pi^2 V}{B^2 L^2}$$

Solving for the magnetic field B, we get

$$B = \frac{\pi}{L}\left(\frac{8Vm}{q}\right)^{1/2} = \frac{\pi}{0.1 \text{ m}}\left[\frac{(8)(10^4 \text{ V})(9.11 \times 10^{-31} \text{ kg})}{(1.602 \times 10^{-19} \text{ C})}\right]^{1/2}$$

$$B = \boxed{2.12 \times 10^{-2} \text{ T}}$$

(b) This is the minimum field required to produce one turn of the helical paths of the electrons. Equation (30-27) indicates that, if the magnitude of B were doubled, *two* turns of the helical paths would be executed in the same distance L, again producing a focused spot. Therefore, focusing would also occur for

$$B = \boxed{4.24 \times 10^{-2} \text{ T}}$$

30.8 Magnetic Flux Φ_B

When we discussed electric fields, we defined in Section 25.2 the *electric flux* Φ_E as a measure of the number of electric field lines that penetrate a given surface area A:

$$\Phi_E = \int \mathbf{E} \cdot d\mathbf{A}$$

Corresponding to this definition of electric flux, the definition of **magnetic flux Φ_B** is

MAGNETIC FLUX Φ_B
$$\Phi_B = \int \mathbf{B} \cdot d\mathbf{A} \qquad (30\text{-}29)$$

Here, $d\mathbf{A}$ is the area element, and the integration is to be carried out over the entire surface area A. Magnetic flux is measured in SI units of *tesla-meters squared* (T·m²), also called a *weber* (Wb) in older texts.[6] When a plane area A

---

[6] This unit honors Wilhelm Weber (1814–1891), a German physicist who did theoretical and experimental work on magnetism. The unit is older than the *tesla*. Therefore, in many existing texts the magnetic field is referred to in units of *webers per square meter*, rather than in units of *tesla*.

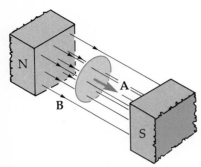

(a) The plane of the loop is perpendicular to the field lines.

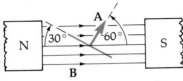

(b) The plane of the loop makes an angle of 30° with the field lines. Consequently, the normal to the area **A** makes an angle of 60° with **B**.

FIGURE 30-20
Example 30-10.

is in a uniform field **B**, the expression is simply

$$\Phi_B = \mathbf{B} \cdot \mathbf{A} = BA \cos \theta$$

where θ is the angle between **B** and the normal **A** to the plane.

EXAMPLE 30-10

A uniform magnetic field $B = 2 \times 10^{-3}$ T is perpendicular to the plane of a circular wire loop of radius 3 cm. (a) Find the magnetic flux Φ_B that the loop encloses. (b) If the loop were tilted so its plane makes an angle of 30° with respect to the field direction, find the magnetic flux that now passes through the loop.

SOLUTION

(a) As shown in Figure 30-20, the area vector **A** is parallel to the uniform field **B**. Equation (30-29) reduces to

$$\Phi_B = \mathbf{B} \cdot \mathbf{A} = BA \cos \theta$$
$$= (2 \times 10^{-3} \text{ T})(\pi)(0.03 \text{ m})^2(1) = \boxed{5.65 \times 10^{-6} \text{ Wb}}$$

(b) When the plane of the loop makes an angle of 30° with the field direction, the vector **A** (normal to the plane) makes an angle of 60° with **B**. Therefore:

$$\Phi_B = BA \cos \theta = (2 \times 10^{-3} \text{ T})(\pi)(0.03 \text{ m})^2 (\cos 60°) = \boxed{2.83 \times 10^{-6} \text{ Wb}}$$

30.9 Comments About Units

A difficulty arises in electricity and magnetism because many quantities are given special names in honor of the early investigators. This obscures the more fundamental units of meters, kilograms, seconds, and coulombs, and thus makes it difficult to check the consistency of units in a given equation. Furthermore, the same quantity may be expressed in a variety of ways, depending on the problem. For example, here is a partial list of the different units that electric and magnetic fields may have (the unit listed first is the most commonly used):

Electric Field E

$$\left[\frac{V}{m}\right] = \left[\frac{N}{C}\right] = \left[\frac{T \cdot m}{s}\right]$$

Magnetic Field B

$$[T] = \left[\frac{Wb}{m^2}\right] = \left[\frac{N}{A \cdot m}\right] = \left[\frac{N \cdot s}{C \cdot m}\right]$$
$$= \left[\frac{V \cdot s}{m^2}\right] = \left[\frac{H \cdot A}{m^2}\right] = \left[\frac{W \cdot H}{V \cdot m^2}\right]$$

(The unit H, which will be defined in the next chapter, stands for *henries*.)

Because of this variety, we again stress the importance of making certain that all numerical values are expressed in SI units before they are substituted into equations. Then one may confidently write the answer in the most appropriate SI units, even though a consistency check is not carried out for each problem.

Summary

The *magnetic induction* or *magnetic flux density* **B** (commonly called the *magnetic field*) is defined from the relation

$$\mathbf{F} = q\mathbf{v} \times \mathbf{B}$$

where **F** is the force on a charge q moving in the field with velocity **v**. The unit is the *tesla* [T].

The *Lorentz force law* expresses the forces on a charge in the presence of both **E** and **B** fields:

$$\mathbf{F} = q(\mathbf{E} + \mathbf{v} \times \mathbf{B})$$

The *force on a current-carrying conductor* of length ℓ carrying current I in the presence of a magnetic field is

$$\mathbf{F} = I\boldsymbol{\ell} \times \mathbf{B}$$

For an element of current-carrying conductor of length $d\boldsymbol{\ell}$:

$$d\mathbf{F} = I\,d\boldsymbol{\ell} \times \mathbf{B}$$

The *magnetic dipole moment* **μ** of a current loop of area A, current I, is

$$\boldsymbol{\mu} = I\mathbf{A} \qquad \text{(in A·m}^2\text{)}$$

where the direction of **μ** is given by the right-hand rule: the fingers curl around in the direction of the current and the extended thumb points in the direction of **μ**. The area vector **A** is normal to the plane of the loop.

The *torque* **τ** on a magnetic dipole in a magnetic field is

$$\boldsymbol{\tau} = \boldsymbol{\mu} \times \mathbf{B}$$

Note the similarity with the electric dipole case: $\boldsymbol{\tau} = \mathbf{p} \times \mathbf{E}$.

The *potential energy* U of a magnetic dipole in a magnetic field is

$$U = -(\boldsymbol{\mu} \cdot \mathbf{B}) \qquad \text{(where } U \equiv 0 \text{ for } \boldsymbol{\mu} \text{ and } \mathbf{B} \text{ at 90°)}$$

Note the similarity with the electric case: $U = -(\mathbf{p} \cdot \mathbf{E})$.

The *magnetic flux* Φ_B is

$$\Phi_B = \int \mathbf{B} \cdot d\mathbf{A} \qquad \text{(in T·m}^2\text{)}$$

Note the similarity with the electric flux: $\Phi_E = \int \mathbf{E} \cdot d\mathbf{A}$.

Questions

1. Which pairs of vectors in the equation $\mathbf{F} = q(\mathbf{v} \times \mathbf{B})$ are always perpendicular to each other and which are not necessarily so?

2. An oscilloscope has a cathode ray tube, which at one end produces a stream of electrons that travels the length of the tube and strikes its face, forming a light spot. By observing the spot while orienting the tube in various directions, how can you detect magnetic fields as well as electric fields? How can the fields be distinguished?

3. An electron, in passing between the poles of a magnet, experiences a change in momentum. Where is the source of the force required to produce such a change in momentum?

4. A *cloud chamber* consists of a chamber filled with supersaturated water vapor. A charged particle passing through the chamber leaves a trail of ions upon which small water droplets form, thus making the particle's path visible. A uniform magnetic field is often imposed upon the chamber, so that the *sign* of the charged particle as well as its energy can be determined. Electrons often produce spiral tracks rather than circular tracks. Why?

5. With simple equipment, is it easier to deflect an electron beam by an electric field or by a magnetic field?

6. The speed of a charged particle moving in only an electric field may or may not change, while the speed of a charged particle moving in only a magnetic field never changes. Explain.

7. An electron with a kinetic energy greater than its rest energy has a circular orbit in a magnetic field. Is the radius of the orbit larger or smaller than that predicted using nonrelativistic formulas? Explain. (See Chapter 41.)

8. A current-carrying loop lies on the top of a table. Suddenly a vertical magnetic field penetrates the table top. What changes in external forces does the loop experience?

9. Conventional current in one direction through a conductor is equivalent to electron flow in the opposite direction. Is a magnetic force on the conductor the same whether we consider the current to be electron current, conventional current, or a mixture of both?

10. A magnetic dipole is aligned with a magnetic field so that it is in stable equilibrium with the field. The work required to turn the dipole end-for-end is $2\,\mu B$. Does the work required to do this depend on the initial orientation of the dipole?

11. The magnetic moment of a magnetic dipole is antiparallel to a magnetic field. Is there a torque on the dipole? Is the dipole in stable equilibrium, in unstable equilibrium, or not in static equilibrium?

12. The precise measurement of an electric field involves the measurement of the force on a charge that is necessarily

very small. For similar reasons, does the precise measurement of a magnetic field involve the measurement of the torque on a magnetic dipole that must have a small magnetic dipole moment?

13. In *n*-type semiconductors electrons are the principal current-carriers, while in *p*-type semiconductors the current is carried by deficiencies of electrons called *holes*, which behave like positive charges. How can the Hall effect be used to determine whether a semiconductor is *n*-type or *p*-type?

14. How would you design a magnetic compass without using iron or any other magnetic material?

15. Using a galvanometer movement as a start, how would you design an electric motor?

16. Why is it usually desirable to have a large number of turns of wire in the rotating coil of a galvanometer?

17. Why is a linear mass spectrometer that is designed for the analysis of ionized atoms unsuitable for electrons?

18. Why is the Hall potential greater for semiconductors than for conductors?

19. Can we measure the drift velocity of charge carriers in a conductor using the Hall effect? If so, how?

20. How is the description of magnetic flux using no more than a number of webers incomplete? Why is magnetic flux Φ_B not a vector quantity?

21. If current were to pass through the helical turns of a stretched coil spring, would the force the spring exerts increase, decrease, or remain the same? Explain.

22. Parallel current-carrying conductors interact with each other. How do current-carrying conductors perpendicular to each other interact?

Problems
30.2 Magnetic Fields

30A-1 At a certain location, the horizontal component of the earth's magnetic field is 30 μT in a northerly direction. An electron moving westward perpendicular to this field has enough speed so that the magnetic force on the electron balances its weight. Find the speed of the electron. (The answer reveals one difficulty in "weighing" a single electron.)

30B-2 At a particular instant, a particle with a charge q moves with the velocity $\mathbf{v} = v_x\hat{\mathbf{x}} + v_y\hat{\mathbf{y}}$ under the influence of a magnetic field $\mathbf{B} = B_x\hat{\mathbf{x}}$. Derive expressions for the magnitude and direction of the force on the charge at that instant.

30B-3 An electron moves with a speed of 3×10^6 m/s outward along the x axis. Find the force on the electron if there is a magnetic field $\mathbf{B} = 0.4\hat{\mathbf{x}} + 0.7\hat{\mathbf{y}} + 0.3\hat{\mathbf{z}}$ (in tesla).

30.3 Motion of a Charged Particle in a Magnetic Field

30A-4 A 0.15-MeV beta particle (electron) emitted during the radioactive decay of ^{14}C enters a magnetic field of 0.04 T in a direction perpendicular to the magnetic field. Find the radius of curvature of the particle's trajectory.

30A-5 A proton moves in a circle perpendicular to a magnetic field. If the radius of the proton's path is 1.00 cm and the field is 0.5 T, find the kinetic energy of the proton in units of electron volts.

30A-6 A 4.2-MeV alpha particle (a helium nucleus consisting of two protons and two neutrons) emitted during the radioactive decay of ^{238}U enters a magnetic field of 0.04 T with its velocity perpendicular to the field. Find the radius of curvature of the particle's trajectory.

30A-7 One type of radar oscillator, a magnetron, utilizes the cyclotron frequency of electrons circulating in a magnetic field to determine the transmitting frequency. Find the magnitude of the magnetic field necessary to generate radar radiation with a 3-cm wavelength.

30B-8 A 1.5-keV electron moves in a circular path with a radius of 1 cm while in a uniform magnetic field $\mathbf{B}$. (a) Calculate the magnitude of $\mathbf{B}$. (b) A proton in this field also has a circular path with a radius of 1 cm. Calculate the proton energy in electron volts.

30B-9 An electron (mass = m_e and charge $-e$), a proton (mass = $1836m_e$ and charge $+e$), and an alpha particle (mass = $4 \times 1836m_e$ and charge $+2e$) all have the same kinetic energy as they move in circular orbits in a uniform magnetic field. In terms of the radius R of the electron's path, find the radii of the paths of the proton and alpha particle.

30B-10 In the mass spectrometer shown in Figure 30-21, singly charged lithium ions of mass 6 u and 7 u are accelerated by a potential difference of 900 V before they enter the uniform magnetic field $B = 0.040$ T. (One *unified mass unit* u ≡ 1.66 × 10^{-27} kg.) After traveling through a semicircle, they strike a photographic film, producing two spots on the film separated a distance x. Find x.

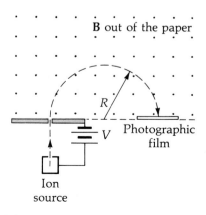

FIGURE 30-21
Problems 30B-10 and 30B-11.

30B-11 As shown in Figure 30-21, in one type of mass spectrometer charged particles (mass m and charge q) are accelerated from rest by a potential difference V. They then enter a region of uniform magnetic field B perpendicular to the plane of the diagram. Starting with Newton's second law, derive an expression for the radius R of the particles' path in the field in terms of m, q, V, and B.

30B-12 A 2-keV electron moving perpendicular to an earth's magnetic field of 50 μT has a circular trajectory. (a) Determine the radius of the trajectory. (b) Determine the time required for the electron to complete one circle. (c) Show that your answer to (b) is consistent with the cyclotron frequency of the electron.

30.4 The Lorentz Force Law

30A-13 A velocity selector for electrons employs an electric field of 1.4×10^4 V/m and a magnetic field of 18 mT. Find the speed of the electrons.

30B-14 At the equator, near the surface of the earth, the magnetic field is approximately 50 μT northward, and the electric field is about 100 N/C downward. Find the gravitational, electric, and magnetic forces on a 100-eV electron moving eastward in a straight line in this environment.

30B-15 A velocity filter consists of magnetic and electric fields described by $\mathbf{E} = E\hat{z}$ and $\mathbf{B} = B\hat{y}$. If $B = 0.015$ T, find the value of E such that a 750-eV electron moving along the $+x$ axis will be undeflected.

30.5 Magnetic Force on a Current-Carrying Conductor

30A-16 A weighing scale supports a 12-V battery, to which a rigid rectangular wire hoop is attached, as shown in Figure 30-22. The lower portion of the hoop is in a magnetic field $B = 0.10$ T. If the total mass of the battery and wire hoop is 100 g, calculate the resistance of the wire necessary for the scale to indicate zero weight. Which pole of the battery is positive?

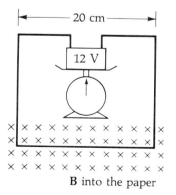

FIGURE 30-22
Problem 30A-16.

30B-17 A rectangular wire loop weighing 0.200 N is suspended halfway into a uniform horizontal magnetic field $\mathbf{B}$ as shown in Figure 30-23. When a current of 2 A exists in the loop, the tension in the supporting string is 0.370 N. (a) What is the direction of the current in the loop? (b) Find the magnitude of $\mathbf{B}$.

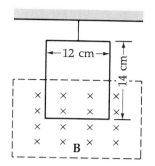

FIGURE 30-23
Problem 30B-17.

30B-18 In Figure 30-24, the cube is 40 cm on each edge. Four straight segments of wire—ab, bc, cd, and da—form a closed loop that carries a current $I = 5$ A as shown. A uniform magnetic field $\mathbf{B} = 0.02$ T is in the $+y$ direction. Make a table showing the magnitude and direction of the force on each segment, listing them in the above order.

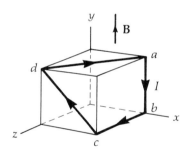

FIGURE 30-24
Problem 30B-18.

30.6 Magnetic Dipoles

30A-19 Show that the units of magnetic dipole moment, ampere-meters squared, can also be expressed as joules per tesla.

30B-20 A bar magnet is suspended from one end by a string fastened to the ceiling. A horizontal magnetic field is then established. Prove that, for the final equilibrium position of the magnet, the string is vertical.

30B-21 A rectangular loop of current-carrying wire is oriented as shown in Figure 30-25. A magnetic field $\mathbf{B} = 0.15\hat{x}$ (in tesla) exerts a torque on the loop. If $a = 8$ cm, $b = 12$ cm, $\theta = 30°$, and $I = 2$ A, calculate the torque on the loop.

30B-22 Calculate the potential energy of the current-carrying loop of Problem 30B-21.

30B-23 Calculate the magnetic dipole moment μ of the current loop shown in Figure 30-25.

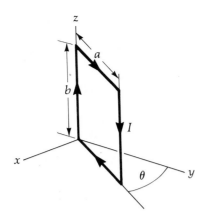

FIGURE 30-25
Problems 30A-21, 30B-22, and 30B-23.

30.7 Applications

30A-24 A silver ribbon 4 cm wide and 0.1 mm thick carries a current of 5 A. If the plane of the ribbon is perpendicular to a magnetic field of 0.15 T, calculate the Hall voltage across the ribbon. Assume that, on the average, each silver atom contributes one electron to the current flow. The density of silver is 10.5 g/cm³ and its atomic weight is 107.87 g/mol.

30A-25 A galvanometer has a full-scale sensitivity of 50 μA. By what factor must the spring constant κ of the galvanometer movement be changed in order to change the full-scale sensitivity to 10 μA?

30A-26 A Hall-effect probe is made of a semiconductor with a charge-carrier density of 10^{20} charges/m³. The dimensions of the probe are 0.8 cm wide, 0.4 mm thick, and 1 cm long. When the probe is placed appropriately in a magnetic field **B**, a current of 0.9 mA in the long direction of the probe produces a Hall voltage of 4 mV across the 0.8-cm width of the probe. Find the value of B.

30B-27 A Hall-effect probe for measuring magnetic fields is designed to operate with a 120-mA current in the probe. When the probe is placed in a uniform field of 0.08 T, it produces a voltage of 0.7 μV. (a) When it is measuring an unknown field, the voltage is 0.33 μV. What is the unknown field strength? (b) If the thickness of the probe in the direction of **B** is 2 mm, find the charge-carrier density (each of charge *e*).

30.8 Magnetic Flux Φ_B

30B-28 Consider the uniform magnetic field $\mathbf{B} = B_x\hat{\mathbf{x}} + B_y\hat{\mathbf{y}}$, where $B_x = 2B_y$. Find the magnetic flux enclosed by a circular loop in the *xz* plane of diameter *D*.

30A-29 At a certain location in Michigan, the earth's magnetic field is 5.80×10^{-5} T in a somewhat downward direction at a *dip angle* of 74°. The dip angle, or *inclination I*, is between **B** and the horizontal. Find the magnetic flux Φ_B enclosed by a flat horizontal loop of 10-cm diameter. (Note: over most of the Southern Hemisphere, **B** has an upward component, and the inclination there is considered to be negative.)

Additional Problems

30C-30 Equal positive charges *q*, each located at the corner of a cube, are each moving with an instantaneous speed *v* as shown by the arrows in Figure 30-26. There is a uniform magnetic field **B** in the $+y$ direction. (a) Make a large sketch of the figure and draw a magnetic force vector (in color) on each charge, indicating the direction of the force. Label each vector with the corresponding letter subscript. (Ignore Coulomb forces.) (b) Make a table listing these forces vertically in alphabetical order, with additional columns for the magnitude and direction of each force.

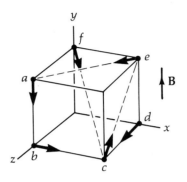

FIGURE 30-26
Problem 30C-30.

30C-31 A cyclotron at the University of California, Berkeley, has a diameter of 60 in. (1.52 m) and operates with a magnetic field of 1.6 T. It can be used to accelerate deuterons, which have the same charge as a proton, but twice its mass. (a) Find the frequency of the accelerating voltage applied to the dees when deuterons are accelerated. (b) Find the kinetic energy (in mega electron volts) of the emerging deuterons. (c) Calculate (a) and (b) for protons. (d) It is usually a major operation to change the frequency of a cyclotron, so for protons the field *B* is often reduced to a lower value with no change made to the original frequency. Find the final kinetic energy of protons following this procedure. (e) Keeping the original frequency, find the magnetic field for alpha particles (mass = $4m_p$ and charge = $2e$). (f) Again, keeping the original frequency, find the kinetic energy for alpha particles. (g) If we raised the voltage applied to the dees, which of these answers would be different?

30C-32 A uniform magnetic field $B = 27$ mT exists parallel to the $\pm x$ axis. As shown in Figure 30-27, an electron with

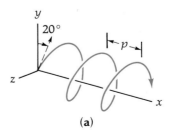
(a)

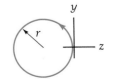
(b) The helical path as viewed along the $+x$ direction.

FIGURE 30-27
Problem 30C-32.

speed 3×10^6 m/s is injected from the origin into the region of the field. The initial velocity of the electron lies in the xy plane at an angle of 20° with the $+y$ axis. The electron subsequently travels in a helical path whose axis is along the $+x$ direction. Find (a) the direction of **B**, (b) the radius r of the helix, and (c) the pitch p of the helix.

30C-33 The color purity of a color television set requires that the electron beam strike a given location on the face of the picture tube with an error of less than one millimeter. Show that a component of the earth's magnetic field perpendicular to the electron beam of about 10 μT may well deflect a 20 keV electron beam enough to affect color purity. (Note: the deflection corresponding to a circular trajectory may be approximated using the *sagitta formula*, explained in Appendix E.) Make your own estimate of the distance from the electron gun to the screen.

30C-34 A particle with a charge-to-mass ratio q/m has a velocity $\mathbf{v} = v\hat{\mathbf{x}}$ as it passes through the origin of a rectangular coordinate system. A constant magnetic field **B** deflects the particle so that it passes through the point $\mathbf{r} = a\hat{\mathbf{x}} + b\hat{\mathbf{y}}$. (a) Determine the direction of **B**. (b) Derive an expression for b in terms of a, q, m, B, and v.

30C-35 A particle with a charge-to-mass ratio q/m moves with speed v in a circular path in the presence of a uniform magnetic field **B**. Derive an expression for the angle through which the particle's path has been deflected during a time t of the motion. Note that the angle is independent of the speed of the particle.

30C-36 An evacuated glass tube with a diameter of 8 cm has a uniform magnetic field $B' = 5 \times 10^{-5}$ T throughout its volume, parallel to the axis of the tube. Electrons are injected into the tube at a point on the axis with a speed of 2×10^6 m/s. (a) Find the largest angle θ that the electron velocity may have with respect to the axis such that the subsequent spiral motion of the electrons will not strike the tube walls. (b) How far along the tube does such an electron cross the axis again?

30C-37 A circular loop of wire with a radius R carries a current I. If the plane of the loop is perpendicular to a uniform magnetic field B, the wire experiences a tension. Derive an expression for the tension T in terms of R, I, and B. The leads that carry current to and from the loop are parallel to the magnetic field.

30C-38 A rigid rectangular loop of wire, sides a and b, is pivoted about a horizontal axis as shown in Figure 30-28. The mass of the loop is m, and a current I exists in the loop. There is a uniform magnetic field **B** in the $+y$ direction. (a) Derive an equation for B, in terms of the given symbols, that expresses the condition when the loop swings up to an equilibrium position so that its plane makes an angle θ with the yz plane. (b) What is the direction of the current in the lowest side of the loop? (c) Suppose, instead, that side b of the loop were pivoted about the horizontal z axis. Would your answer to (a) be different? Explain.

30C-39 A rigid hoop of wire (mass m and radius R) rests on a horizontal surface in a region where there is a uniform magnetic field $\mathbf{B} = B_x\hat{\mathbf{x}} + B_y\hat{\mathbf{y}}$, where $\hat{\mathbf{y}}$ is vertically upward. Find the minimum current I that will barely cause one side of the hoop to lift off the surface.

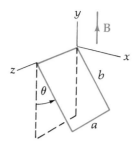

FIGURE 30-28
Problem 30C-38.

30C-40 As shown in Figure 30-29, an irregular open loop of current-carrying wire lies in the xy plane. The current input to the loop is along the z axis, and the output is parallel to the z axis at $x = h$. A uniform magnetic field is described by $\mathbf{B} = B\hat{\mathbf{z}}$. Show that the net force on the loop is independent of the shape of the loop and that the force is given by $\mathbf{F} = -BhI\hat{\mathbf{y}}$.

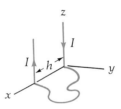

FIGURE 30-29
Problem 30C-40.

30C-41 In the Bohr model of the hydrogen atom, the electron moves in a circle about the proton, with the Coulomb force being the centripetal force necessary for circular motion. In the lowest energy state, the radius of the path is 52.9 pm (1 pm = 10^{-12} m). (a) Find the equivalent current the moving electron generates. (b) Find the magnetic dipole moment of this current loop (called the *Bohr magneton*).

30C-42 The maximum torque on a current-carrying rectangular loop of wire placed in a magnetic field depends on the shape of the loop. Show that, for a given length of wire formed in a rectangular shape, the greatest maximum torque is achieved when the loop is a square.

30C-43 A wire of length ℓ is formed into a flat, circular coil of N turns. (a) Show that, for a given current I in the coil, the greatest magnetic dipole moment is for $N = 1$. (b) Explain why a one-turn coil of any shape other than circular would have a smaller magnetic moment.

30C-44 A circular wire hoop of radius R and mass m carries a current I. The hoop hangs from its edge by a horizontal frictionless hinge in a uniform vertical magnetic field **B**. The hoop will assume an equilibrium position so that the plane of the hoop makes an angle θ with respect to the vertical. Derive an expression for the angle θ in terms of m, R, I, and B.

30C-45 A uniform disk of mass m has a total charge q distributed uniformly throughout its volume. As the disk rotates about its axis, show that its magnetic moment $\boldsymbol{\mu}$ is related to its angular momentum **L** by $\boldsymbol{\mu} = (q/2m)\mathbf{L}$. You may use the result of Problem 30C-51.

30C-46 A circular current-carrying loop of wire experiences a maximum torque τ_0 when placed in a given magnetic field. If the same loop were re-formed to a smaller circular loop containing two turns of the wire, find the maximum torque on this loop in terms of τ_0.

30C-47 A thin rod of length ℓ is made of a nonconducting material and carries a uniform charge per unit length λ. The rod is rotated with angular velocity ω about an axis through its center, perpendicular to the length of the rod. Show that the magnetic dipole moment is $\omega\lambda\ell^3/24$. (Hint: consider the charge dq located within the element dx a distance x from the axis.)

30C-48 Show that a magnetic dipole in a divergent magnetic field may experience a net force as well as a torque. Describe the condition under which the dipole moves in the direction of increasing magnetic field.

30C-49 The axis of a magnetic dipole with a dipole moment $\boldsymbol{\mu}$ and angular momentum **L** is at an angle θ with respect to a uniform magnetic field B. The vectors $\boldsymbol{\mu}$ and **L** are parallel. Show that the dipole will precess with an angular velocity $\boldsymbol{\omega}_p = -(\mu/L)\mathbf{B}$. (See Section 13.6, The Gyroscope.)

30C-50 As shown in Figure 30-30, a nonuniform magnetic field $\mathbf{B} = xB_0\hat{\mathbf{z}}$ is in the $+z$ direction (toward the reader). The field varies linearly with the distance x. A rectangular loop of dimensions a and b is oriented so that its plane is perpendicular to the field, with the left edge of the loop parallel to the y axis at a distance d from that axis. Find the total flux Φ_B through the loop. (Hint: consider the flux $d\Phi_B$ through an element of area $dA = a\,dx$. The total flux $\Phi_B = \int \mathbf{B} \cdot d\mathbf{A}$.)

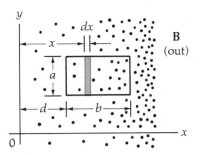

FIGURE 30-30
Problem 30C-50.

30C-51 A circular disk of nonconducting material has a radius R, and on one side there is a surface charge density σ. The disk is rotated with angular velocity ω about its axis. Show that the magnetic dipole moment is $\omega\sigma\pi R^4/4$. [Hint: consider the current loop formed by the motion of the charge within the annular ring of radius r and width dr. You may use Equation (30-14).]

30C-52 Consider a metallic conductor with a rectangular cross-section and a resistivity ρ. Show that the electric field E_H due to the Hall effect is related to the field E sustaining the current through the conductor by $E_H = (B/ne\rho)E$, where B is the magnetic field strength and n is the number of conduction electrons per unit volume, each with charge $(-)e$.

CHAPTER 31

Sources of Magnetic Field

Science walks forward on two feet, namely theory and experiment.

ROBERT A. MILLIKAN
(from his Nobel lecture, May 1924)

31.1 Introduction

The last chapter described static magnetic fields and the forces they exert on moving charges. In this chapter we will discuss the origin of static magnetic fields. One interesting fact in electromagnetism is that a steady current of electric charges produces a static magnetic field. We will also show a satisfying symmetry between electric and magnetic fields. In particular, a *changing* magnetic field produces an electric field, and a *changing* electric field produces a magnetic field. The English physicist James Clerk Maxwell (1831–1879) put the finishing touch on the elegant electromagnetic theory, which expresses this symmetry between electricity and magnetism.

31.2 The Biot–Savart Law

In 1819, the Danish scientist Hans Christian Oersted (1777–1851) was concluding a lecture on electricity and magnetism when he moved a current-carrying wire near a compass needle. The needle deflected in a new direction in response to the current.[1] This demonstration of a fundamental link between electricity and magnetism was highly significant. Other scientists, especially in France, quickly followed up on this discovery by developing new relationships that deepened our understanding of electromagnetism.

In our study of electric fields, we found that they had their origin in electrical charges. To find the field at a given point due to an arbitrary distribution of charges, we recognize that each *charge element dq* produces a field $d\mathbf{E}$ at a distance r from the charge according to Coulomb's law for electric fields. It is an *inverse-square* law:

$$d\mathbf{E} = \left(\frac{1}{4\pi\varepsilon_0}\right) \frac{dq}{r^2} \hat{\mathbf{r}}$$

---

[1] Oersted's discovery was probably accidental. Each year the American Association of Physics Teachers awards a medal to a physics teacher who has made a notable contribution to the teaching of physics. It is called the Oersted Medal, since Oersted's discovery occurred in a teaching situation.

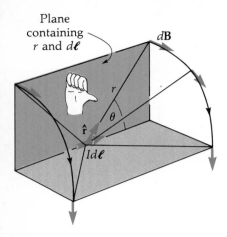

FIGURE 31-1
This figure illustrates the geometrical features of the Biot–Savart law. Here we show how to find the direction of $d\mathbf{B}$ at the top right corner of the figure by applying the right-hand rule for $I\,d\boldsymbol{\ell} \times \hat{\mathbf{r}}$. According to the right-hand rule for cross products (see Figure 10-5), the fingers of the right hand curl around in the sense of rotation established when the first vector $d\boldsymbol{\ell}$ is rotated through the angle θ into the direction of the second vector $\hat{\mathbf{r}}$ (which points toward the location where we wish to determine the field direction). The extended thumb points in the direction of $d\mathbf{B}$ at the top corner, a distance r away. Several other $d\mathbf{B}$'s are shown for other locations. (For practice, verify their directions by applying the right-hand rule.) The overall pattern of field lines, to which the $d\mathbf{B}$'s are tangent, is circles that lie in planes perpendicular to the axis of the element $I\,d\boldsymbol{\ell}$.

Here, the unit vector $\hat{\mathbf{r}}$ extends *from the source of the field* (the charge dq) *toward the point in question*. To find the total electric field $\mathbf{E}$, we sum over all the charge elements present.

We now introduce a similar equation that describes how an element of current-carrying wire $I\,d\boldsymbol{\ell}$ produces a magnetic field $d\mathbf{B}$ at a point a distance r from the element. Consider a current-carrying wire of arbitrary shape (Figure 31-1). In 1820, the French physicists Jean Baptiste Biot and Félix Savart first gave the expression for the field $d\mathbf{B}$ produced at a distance r from an element of the wire $d\boldsymbol{\ell}$ carrying a steady current I. It is an inverse-square law known as the **Biot–Savart law** (pronounced "Bee-oh–Sah-vahr"):

BIOT–SAVART LAW
$$d\mathbf{B} = \left(\frac{\mu_0}{4\pi}\right)\frac{I\,d\boldsymbol{\ell} \times \hat{\mathbf{r}}}{r^2} \qquad (31\text{-}1)$$

The direction of the vector $d\boldsymbol{\ell}$ is along the wire in the direction of the current I. The unit vector $\hat{\mathbf{r}}$ is *from the source of the field* (the current-carrying element $I\,d\boldsymbol{\ell}$) *toward the point in question*. Thus, $\mathbf{r} = r\hat{\mathbf{r}}$. To find the total magnetic field $\mathbf{B}$ at the point, we sum over all the current-carrying elements present. The constant μ_0 is called the **permeability of free space**:

PERMEABILITY OF
FREE SPACE
$$\mu_0 = 4\pi \times 10^{-7}\,\frac{\text{T}\cdot\text{m}}{\text{A}}$$

This numerical value is chosen to be consistent with the definition of the unit of current, the *ampere* (A). (The constant μ_0 should not be confused with the symbol for the magnetic dipole moment $\boldsymbol{\mu}$.) The most significant feature of Equation (31-1) is that magnetic fields, like electric fields, are *inverse-square* fields. In contrast, unlike an electric field, which is generated by an isolated electric charge, there is no isolated "magnetic charge" that generates a magnetic field.[2] Isolated current elements $I\,d\boldsymbol{\ell}$ do not exist—they are always part of a complete closed circuit. Calculations of the total field for all but very simple arrangements of conductors are quite cumbersome, so we will restrict our examples to simple, yet important, symmetrical configurations.

EXAMPLE 31-1

Calculate the magnetic field 10 cm from a very long, straight wire carrying a current of 10 A.

SOLUTION

We first develop a general expression for the field in the vicinity of a straight current-carrying conductor. In Figure 31-2b, the incremental field $d\mathbf{B}$ due to the current element $I\,d\boldsymbol{\ell}$ is directed into the plane of the figure at point P. Equation (31-1):

$$d\mathbf{B} = \left(\frac{\mu_0}{4\pi}\right)\frac{I\,d\boldsymbol{\ell} \times \hat{\mathbf{r}}}{r^2}$$

---

[2] Some advanced theories of magnetism have proposed that magnetic monopoles exist. No convincing experimental confirmation has yet been found.

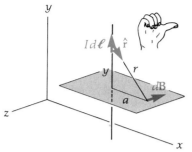
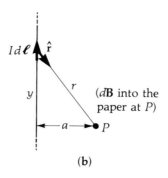

(a) The right-hand rule for the cross product $d\boldsymbol{\ell} \times \mathbf{r}$ establishes the direction of $d\mathbf{B}$.

(b)

FIGURE 31-2
Example 31-1.

The magnitude of $d\mathbf{B}$ is given by

$$dB = \left(\frac{\mu_0}{4\pi}\right) \frac{I\, d\ell}{r^2} \sin\theta$$

where θ is the angle between the forward directions of $d\boldsymbol{\ell}$ and $\hat{\mathbf{r}}$. Introducing the perpendicular distance a from the point to the wire, and letting $d\ell = dy$, we note that $r^2 = y^2 + a^2$ and that $\sin\theta = a/\sqrt{y^2 + a^2}$. Thus:

$$dB = \left(\frac{\mu_0}{4\pi}\right) \frac{Ia}{(y^2 + a^2)^{3/2}}\, dy$$

Since each element produces a field $d\mathbf{B}$ in the same direction, the total field B is merely the scalar sum $\int dB$:

$$B = \frac{\mu_0 I a}{4\pi} \int_{-\infty}^{+\infty} \frac{dy}{(y^2 + a^2)^{3/2}}$$

Using the table of integrals in Appendix G, we obtain

$$B = \frac{\mu_0 I a}{4\pi} \left[\frac{y}{a^2(y^2 + a^2)^{1/2}} \right]\bigg|_{-\infty}^{+\infty} = \frac{\mu_0 I a}{4\pi a^2}[1 - (-1)]$$

MAGNETIC FIELD DUE TO A CURRENT IN A LONG, STRAIGHT WIRE

$$B = \frac{\mu_0 I}{2\pi a} \tag{31-2}$$

Substituting numerical values in SI units gives

$$B = \frac{\left(4\pi \times 10^{-7}\, \frac{\text{T}\cdot\text{m}}{\text{A}}\right)(10\text{ A})}{2\pi(0.10\text{ m})} = \boxed{2.00 \times 10^{-5}\text{ T}}$$

The direction of $\mathbf{B}$ is found from the cross-product in the Biot–Savart law ($I\, d\boldsymbol{\ell} \times \hat{\mathbf{r}}$). From symmetry considerations, the field lines form concentric circles surrounding the wire. Their direction is easily remembered using the **right-hand rule** as defined in Figure 31-3c.

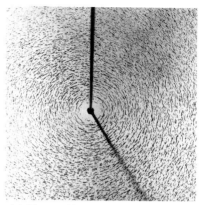

(a) If iron filings are sprinkled on a horizontal plane perpendicular to a straight, current-carrying wire, they form a pattern that suggests the magnetic field lines.

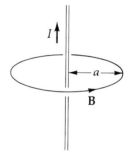

(b) One of the field lines that circle the wire symmetrically.

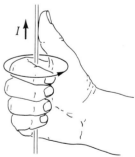

(c) The magnetic field lines circle the conductor in the direction of the fingers of the right hand when the extended thumb is in the direction of the current. This is another "right-hand rule" that describes the field lines due to a current-carrying wire.

FIGURE 31-3
The magnetic field associated with a straight, current-carrying conductor.

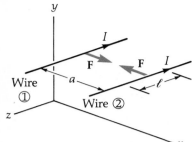

(a) Two horizontal parallel conductors one meter apart will experience a mutual force of attraction equal to exactly 2×10^{-7} N/m if each conductor carries a current of one ampere in the same direction. If the currents are antiparallel, the forces are repulsive.

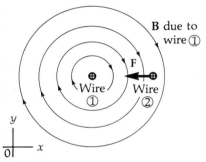

(b) The current in wire ① produces a downward magnetic field at wire ②. Consequently, a length ℓ of wire experiences a force $F = |I\boldsymbol{\ell} \times \mathbf{B}| = IB\ell$ as shown. (The situation is symmetric. The current in wire ① produces a field, not shown, that causes a force on wire ② toward the right.)

FIGURE 31-4
The definition of the ampere.

The previous example illustrates a very important characteristic of magnetic field lines: *magnetic field lines are always closed loops*. This closure of magnetic field lines is in contrast to electrostatic field lines, which always terminate on plus and minus charges.

Having developed an expression for the magnetic field around a long, straight wire enables us to define the *ampere* and thus the *coulomb*. Consider two parallel conductors, each carrying the same current I in the same direction, as shown in Figure 31-4. The field $\mathbf{B}$ produced by the current in wire ① a distance a from the wire is given by Equation (31-2): $B = \mu_0 I/2\pi a$. The direction of this field at the location of wire ② is straight down in the $-y$ direction. The magnetic force $d\mathbf{F}$ on an incremental length $d\boldsymbol{\ell}$ of wire ② is, by Equation (30-11), $d\mathbf{F} = I\, d\boldsymbol{\ell} \times \mathbf{B}$, which, since $d\boldsymbol{\ell}$ and $\mathbf{B}$ are perpendicular, equals

$$dF = IB\, d\ell$$

Substituting Equation (31-2) and rearranging, we have

$$\frac{dF}{d\ell} = \frac{\mu_0 I^2}{2\pi a} = \frac{\left(4\pi \times 10^{-7}\,\frac{\text{T}\cdot\text{m}}{\text{A}}\right) I^2}{2\pi a} = \frac{(2 \times 10^{-7}) I^2}{a}\,\frac{\text{T}\cdot\text{m}}{\text{A}}$$

If the separation of the wires is one meter and the current in each wire is one ampere, the force of attraction per unit length of wire is

$$\frac{\text{Force}}{\text{Unit length}}: \quad 2 \times 10^{-7}\,\text{T}\cdot\text{A} \quad \text{or} \quad 2 \times 10^{-7}\,\frac{\text{N}}{\text{m}}$$

DEFINITION OF THE AMPERE   If one ampere is in the same direction in each of two long, parallel conductors one meter apart, the conductors will be attracted to each other with a force of *exactly* 2×10^{-7} N per meter of length.

This basic definition of the ampere is the crucial link between electrical quantities and mechanical quantities. It extends the SI system to include electrical units by defining the *ampere* in terms of the meter, the kilogram, and the second. As mentioned in Chapter 24, it also leads to the *coulomb*, since that unit is defined as the amount of charge per second passing a cross section of a conductor carrying a steady current of one ampere. Mechanical experiments that measure forces between current-carrying wires are much easier to carry out and give greater precision than experiments that measure the Coulomb force between charges. Thus there are strong practical reasons for basing the fundamental electrical definition on the ampere rather than on the coulomb. From the above relations, we see that units of force per unit length are equivalent to $\text{T}\cdot\text{A}$, which leads to alternative units for μ_0:

$$\mu_0 = 4\pi \times 10^{-7}\,\frac{\text{N}}{\text{A}^2} \tag{31-3}$$

The two constants μ_0 and ε_0 are related. The first constant arises from forces between current-carrying elements, and the second arises from forces between charge elements. And, of course, currents and charges are intimately connected. As we will see in Chapter 35, these constants are related to the speed of light: $c = 1/\sqrt{\mu_0 \varepsilon_0}$. In 1983, the speed of light was defined to be exact. Because μ_0 is defined exactly, ε_0 also has an exact value.

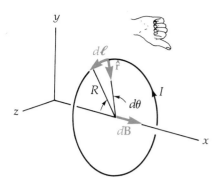
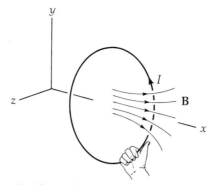

(a) The right-hand rule for cross-products $d\boldsymbol{\ell} \times \hat{\mathbf{r}}$ that identifies the direction of $d\mathbf{B}$ at the center of the circle.

(b) The right-hand rule that associates the field direction due to a current-carrying wire.

FIGURE 31-5
Example 31-2.

EXAMPLE 31-2

A wire, bent into a circle with a radius $R = 10$ cm, carries a current of 10 A, Figure 31-5. Find the magnetic field at the center of the loop.

SOLUTION

Each of the current elements $I\,d\boldsymbol{\ell}$ and the unit vector $\hat{\mathbf{r}}$ are perpendicular. Equation (31-1) therefore becomes

$$dB = \left(\frac{\mu_0}{4\pi}\right)\frac{I\,d\ell}{R^2}$$

We change to a more convenient variable of integration by expressing the element $d\ell$ as $d\ell = R\,d\theta$, where $d\theta$ is the angle subtended by $d\ell$ from the center of the loop. Then,

$$dB = \left(\frac{\mu_0}{4\pi}\right)\frac{I}{R}\,d\theta$$

At the center of the loop, each field increment $d\mathbf{B}$ is in the same direction along the $+x$ axis. So the total field B is merely the sum $\int dB$ integrated around the entire loop from $\theta = 0$ to $\theta = 2\pi$ rad. All the terms are constant except $d\theta$, so we have

$$B = \left(\frac{\mu_0}{4\pi}\right)\frac{I}{R}\int_0^{2\pi} d\theta = \left(\frac{\mu_0}{4\pi}\right)\frac{I}{R}(2\pi - 0) = \frac{\mu_0 I}{2R}$$

MAGNETIC FIELD AT THE CENTER OF A CURRENT-CARRYING LOOP

$$B = \frac{\mu_0 I}{2R} \qquad (31\text{-}4)$$

Substituting numerical values gives

$$B = \frac{\left(4\pi \times 10^{-7}\,\frac{\text{T}\cdot\text{m}}{\text{A}}\right)(10\text{ A})}{(2)(0.10\text{ m})} = \boxed{6.28 \times 10^{-5}\text{ T}}$$

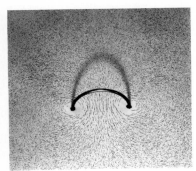

(a) A piece of paper placed horizontally is perpendicular to the plane of a current-carrying loop. If iron filings are sprinkled on the paper, they form a pattern of lines similar to the magnetic field in the plane.

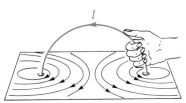

(b) The right-hand rule for determining the direction of magnetic field lines.

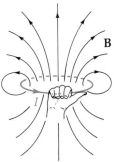

(c) A horizontal current-carrying loop with the right hand determining the direction of the magnetic field lines.

FIGURE 31-6
The magnetic field produced by a current-carrying loop.

Most practical devices used for the production of magnetic fields are constructed of loops or coils of wire. A right-hand rule determines the field direction: if the current-carrying wire is grasped with the fingers of the right hand so that the extended thumb is in the direction of the current, the curled fingers indicate the magnetic field direction. *Inside* the loop, the field at the center is along the axis of the loop. The field lines elsewhere are shown in Figure 31-6.

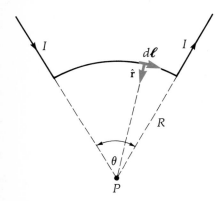

FIGURE 31-7
Example 31-3.

EXAMPLE 31-3

The wire conductor in Figure 31-7 carries a current I. The straight portions are radially outward from the point P, and the circular arc of radius R subtends an angle θ from point P. Find the magnetic field **B** at the point P.

SOLUTION

We note that the straight segments contribute nothing to the field at point P since $d\boldsymbol{\ell}$ and $\hat{\mathbf{r}}$ are parallel, so the cross-product involves $\sin 0° = 0$. For the circular arc, $d\boldsymbol{\ell}$ and $\hat{\mathbf{r}}$ are at right angles, so we have

$$d\mathbf{B} = \left(\frac{\mu_0}{4\pi}\right) \frac{I\, d\boldsymbol{\ell} \times \hat{\mathbf{r}}}{r^2}$$

The cross-product all along the arc involves $\sin 90° = 1$, and by the right-hand rule all the $d\mathbf{B}$'s are in the same direction into the plane of the paper. Here, $d\ell = ds = R\, d\theta$, so we have

$$B = \int dB = \int \frac{\mu_0 I}{4\pi R^2}\, ds = \frac{\mu_0 I}{4\pi R^2} \int_0^\theta R\, d\theta = \boxed{\frac{\mu_0 I \theta}{4\pi R}} \quad \text{(into the paper)}$$

31.3 Ampère's Law (1823)

If the configuration of a current-carrying conductor is simple, an equivalent and simpler form of the Biot–Savart law, known as Ampère's law, may be used. The basic idea involves a *closed path of integration*, sometimes called an *ampere loop.* **Ampère's law**[3] states that *the integral $\oint \mathbf{B} \cdot d\boldsymbol{\ell}$ around any closed path is $\mu_0 I$, where I is the current crossing any surface bounded by the path of integration.*

AMPÈRE'S LAW[4]
$$\oint_C \mathbf{B} \cdot d\boldsymbol{\ell} = \mu_0 I \qquad (31\text{-}5)$$

Like the Biot–Savart law, Ampère's law is true only for *steady* currents. Furthermore, just as the application of Gauss's law is feasible only for charge distributions that are highly symmetric, Ampère's law is useful only for very symmetric arrays of currents leading to symmetric fields that are known all along the path of integration. We now illustrate the use of Ampère's law for three important configurations of conductors.

I. **The field of a long, straight current-carrying conductor.** Although Ampère's law is true for any path, the calculation is feasible only when the value of $\mathbf{B} \cdot d\boldsymbol{\ell}$ is constant along the path of integration. See Figure 31-3b. From symmetry considerations (and the right-hand rule) we know that $\mathbf{B}$ is constant in magnitude on a circular path surrounding the wire and therefore may be brought out from under the integral sign. We choose a path along a field line, with $d\boldsymbol{\ell}$ defined parallel to $\mathbf{B}$, so the dot product gives $\cos 0° = 1$. Thus, the integral $\oint d\ell$ is simply around a circle of radius a: the circumference $2\pi a$. The current passing through the circular area bounded by the path is I. Therefore:

$$B \oint d\ell = \mu_0 I$$
$$B(2\pi a) = \mu_0 I$$

MAGNETIC FIELD DUE TO A CURRENT IN A LONG, STRAIGHT WIRE
$$B = \frac{\mu_0 I}{2\pi a}$$

This is the same expression obtained using the Biot–Savart law, Equation (31-2). Ampère's law and the Biot–Savart law are completely equivalent. The context of a problem determines which form is easier to use.

II. **The field of a toroid.** Our next example in the use of Ampère's law is that of a wire wound around a *toroid*: a donut-shaped coil, as illustrated in Figure 31-8. To find the field inside the windings of a toroidal coil, symmetry suggests that the appropriate path of integration is a circle of radius R in the plane of the toroid along the axis of the windings. The reason for this choice is that, because of symmetry, $\mathbf{B}$ is constant in magnitude along such a path and is parallel to $d\boldsymbol{\ell}$ everywhere on the circle. Thus B may be brought outside the integral sign. The total current through the integration

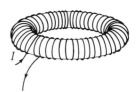

(a) We form a toroidal coil by winding a current-carrying conductor around a toroid.

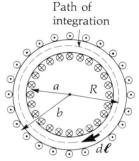

(b) In the inside edge of the coil windings, the current is into the plane of the paper; at the outside edge of the windings, the current is out of the paper.

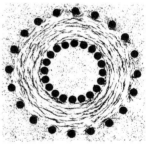

(c) Iron filings reveal the pattern of the magnetic field. Even for this loosely wound toroid, the field is confined almost wholly within the windings.

FIGURE 31-8
A toroidal coil. From the right-hand rule, the magnetic field inside the windings in (b) is clockwise.

---

[3] André Marie Ampère (1775–1836) was educated mainly by his father, a justice of the peace who opposed the French Revolution, and also by Ampère's own extensive reading, including many works in Latin. Unfortunately, at the age of 18 Ampère stood by the edge of the scaffold upon which his father was guillotined. In 1826, Ampère presented a notable paper before the French Academy in which he outlined a new theory of electrodynamics based upon his own experiments. Maxwell later remarked that this theory "seems as if it had leaped full grown and full armed from the brain of the Newton of electricity." Rare praise, indeed!

[4] If the current I is spread out over the surface S enclosed by the line integral (instead of confined to a wire), then the right-hand side of Ampère's law is calculated from

$$\mu_0 I = \mu_0 \int_S \mathbf{J} \cdot d\mathbf{A}$$

where $\mathbf{J}$ is the *current density* (see Section 28.7). The direction of the area element $d\mathbf{A}$ is given by the right-hand rule: circle the fingers of the right hand in the direction of $d\boldsymbol{\ell}$ around the closed curve; the extended thumb points in the direction of $d\mathbf{A}$.

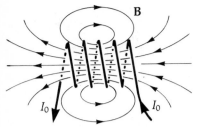

(a) A loosely wound, short solenoid.

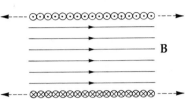

(b) An *ideal* solenoid has closely wound windings that extend to infinity in both directions, confining the field wholly within the solenoid.

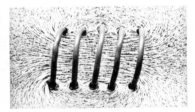

(c) Iron filings reveal the magnetic field pattern.

FIGURE 31-9
The magnetic field of a solenoid.

loop is NI, where N is the total number of turns of wire around the toroid and I is the current through the wire. Applying Ampère's law,

$$\oint \mathbf{B} \cdot d\boldsymbol{\ell} = \mu_0 I$$

we obtain
$$B(2\pi R) = \mu_0 NI$$

MAGNETIC FIELD INSIDE THE WINDINGS OF A TOROIDAL COIL
(average circumferential length of the toroid: $2\pi R$)

$$B = \frac{\mu_0 NI}{2\pi R} \qquad (31\text{-}6)$$

Since the field depends on R, the field varies slightly within the windings, being somewhat stronger near the inner radius of the toroid. (The circle of the toroid itself acts as a *single* large loop of wire of radius R carrying a current I. The external field outside the windings due to this effect is small and usually may be ignored.) For this reason, a toroid is useful in electronic circuits whenever a magnetic field must be confined.

III. **The field of a long solenoid.** A solenoid is a straight coil of wire, as shown in Figure 31-9a. Because of the relative ease of its fabrication, it is the most common configuration used to produce a magnetic field electrically. Calculation of the magnetic field is complicated for a loosely wound solenoid that is short compared with its diameter. However, the field at the center of the solenoid can be closely approximated by considering an *ideal* solenoid: one that is long compared with the diameter, with the turns of wire close together, as in Figure 31-9b. Just as we may consider a parallel-plate capacitor to be a section of large concentric spheres, we may consider a solenoid to be a short section of a toroid whose outer diameter is large compared with the cross-sectional radius of the windings. The field will then be essentially uniform within the solenoid and will be confined to the solenoid's interior. The direction of the field lines is into one end of the section and out of the other end. Rewriting Equation (31-6), we have

$$B = \left(\frac{N}{2\pi R}\right) \mu_0 I$$

For a large value of R that does not change appreciably from the inner to the outer diameter of the toroid, the quantity within the parentheses is the number of turns n per circumferential length of the toroid. Then,

MAGNETIC FIELD IN A LONG SOLENOID $B = \mu_0 n I$ (where n is the number of turns per unit length)   (31-7)

Just a short bit or reasoning will lead us to the field at *one end* of a long solenoid. Consider the point inside a long solenoid equally far from either end. (The above equation is valid for this point.) By symmetry, each half of the long coil contributes equally to the field at this midpoint. Therefore if we remove one-half of the solenoid, the field at the (newly created) open end is just half that of Equation (31-7):

MAGNETIC FIELD AT ONE END OF A LONG SOLENOID $B = \dfrac{\mu_0 n I}{2}$ (where n is the number of turns per unit length)   (31-8)

EXAMPLE 31-4

A permanent magnet similar to the one shown in Figure 31-10a has a magnetic field of 0.4 T in the air gap between its pole pieces. To produce the same magnetic field within a solenoid of comparable size, how much current would have to pass through its windings? Assume that the solenoid is 30 cm long with a small cross-section and is wound with 2000 turns of copper wire.

SOLUTION

With the assumption that the solenoid is ideal, we solve Equation (31-7), $B = \mu_0 n I$, for the current I:

$$I = \frac{B}{\mu_0 n} = \frac{0.4 \text{ T}}{\left(4\pi \times 10^{-7} \frac{\text{T} \cdot \text{m}}{\text{A}}\right)\left(\frac{2000 \text{ turns}}{0.30 \text{ m}}\right)} = \boxed{47.7 \text{ A}}$$

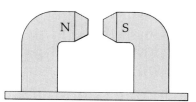

(a) A permanent magnet.

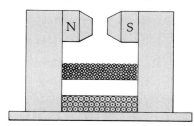

(b) An electromagnet (cross-sectional view of the windings).

FIGURE 31-10
Example 31-4. Typical laboratory magnets.

To see whether the result in Example 31-4 is consistent with a practical laboratory device, suppose that the average length of each turn of wire is 10 cm and that the cross-sectional area of the wire is 1.0 mm². The resistance of the wire is given by $R = \rho \ell / A$, where ρ is the resistivity of copper (1.8×10^{-8} ohm·m), ℓ is the length of the wire, and A is the cross-sectional area. Substituting the appropriate values in SI units, we get

$$R = (1.8 \times 10^{-8} \text{ } \Omega \cdot \text{m}) \frac{(2000 \text{ turns})(0.1 \text{ m})}{(1.0 \times 10^{-6} \text{ m}^2)} = 3.60 \text{ } \Omega$$

The rate at which heat would be generated due to Joule heating of the copper is

$$P = I^2 R = (47.7 \text{ A})^2 (3.6 \text{ } \Omega) = 8.19 \text{ kW}$$

Clearly, the cooling requirements of such a solenoid make it impractical as a source of magnetic field. However, the presence of an iron core in a solenoid greatly increases the resultant magnetic field (as will be discussed in Chapter 33). So, in practice, the solenoid would be constructed with an iron core similar to that in Figure 31-10b, greatly reducing the current requirements.

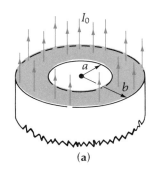

(a)

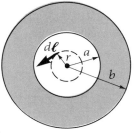

(b) The current I_0 comes toward the reader in the shaded area. Following the sign convention shown in Figure 31-3, the path of integration for $r < a$ is shown dashed.

FIGURE 31-11
Example 31-5. A long, hollow wire carries a current I_0 that is uniformly distributed over the cross-sectional area between radii a and b.

EXAMPLE 31-5

A long, hollow conducting wire carries a current I_0 that is uniformly distributed over the cross-sectional area of the wire between radii a and b, as shown in Figure 31-11. Find the magnetic field **B** for region 1, $r \leq a$; region 2, $a \leq r \leq b$; and region 3, $r \geq b$.

SOLUTION

From the symmetry of the situation, the only directions the magnetic field lines can have in Figure 31-11b are counterclockwise concentric circles about the axis of the wire (right-hand rule). Furthermore, by symmetry the magnitude B must be constant everywhere along such a line. We purposely match this symmetry by choosing paths of integration for $\oint \mathbf{B} \cdot d\boldsymbol{\ell}$ that are concentric circles about the axis, in the direction of **B**.

In region 1 $(r \leq a)$, $\quad \oint \mathbf{B} \cdot d\boldsymbol{\ell} = \mu_0 I$

The dot product gives $\cos 0° = 1$. Because B is constant along the path, it may be brought out from under the integral sign. Since $\oint d\boldsymbol{\ell} = 2\pi r$, and the value of I enclosed by the integration path is zero, we have

$$B_1(2\pi r) = \mu_0(0)$$
$$B_1 = \boxed{0}$$

In region 2 $(a \leq r \leq b)$, again, by symmetry, we choose the integration path to be a concentric circle. However, we now need to know the fraction of the total current I_0 that is enclosed by the path of integration. Since the current is distributed uniformly over the cross-sectional area, it is the fraction[5]

$$I_{\text{inside}} = \left(\frac{\text{Area inside } r}{\text{Total area}}\right) I_0 = \left[\frac{\pi(r^2 - a^2)}{\pi(b^2 - a^2)}\right] I_0$$

Therefore, $\quad \oint \mathbf{B} \cdot d\boldsymbol{\ell} = \mu_0 I$

$$B_2(2\pi r) = \mu_0 \left(\frac{r^2 - a^2}{b^2 - a^2}\right) I_0$$

$$B_2 = \boxed{\frac{\mu_0}{2\pi r}\left(\frac{r^2 - a^2}{b^2 - a^2}\right) I_0}$$

The direction of **B** is counterclockwise in the figure (right-hand rule).

In region 3 $(r \geq b)$, the path of integration encloses the entire current I.

$$\oint \mathbf{B} \cdot d\boldsymbol{\ell} = \mu_0 I$$
$$B_3(2\pi r) = \mu_0 I_0$$
$$B_3 = \boxed{\frac{\mu_0 I_0}{2\pi r}}$$

The direction of **B** is counterclockwise in the figure (right-hand rule). Note that $B_1 = B_2$ for $r = a$ and that $B_2 = B_3$ for $r = b$.

EXAMPLE 31-6

As shown in Figure 31-12a, an (essentially) infinite thin sheet lying in the xy plane carries a uniform current per unit length λ in the $+y$ direction, where "per unit length" refers to the $\pm x$ direction. Find the magnetic field B near the sheet.

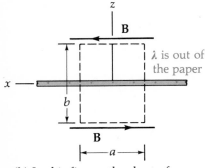

(a) A segment of an infinite sheet of current per unit x-length, λ.

(b) In this figure, the sheet of current approaches the viewer. Note that **B** has opposite directions on opposite sides of the sheet.

FIGURE 31-12
Example 31-6.

---

[5] See Footnote 4 for cases in which the current distribution is nonuniform.

SOLUTION

From symmetry arguments and the right-hand rule for determining the direction of **B** due to a line of current, we conclude that **B** is parallel to the sheet as shown in Figure 31-12b. Furthermore, from symmetry we note that, whatever magnitude B has at a given distance above the sheet, it must have the same magnitude at the same distance below the sheet. Therefore, we choose the symmetrically placed, dashed rectangular path shown for integrating $\oint \mathbf{B} \cdot d\boldsymbol{\ell}$. For the two paths that are perpendicular to the sheet, this integration is zero because **B** and $d\boldsymbol{\ell}$ are at 90°. The total current I within the rectangle is $I = \lambda a$. Applying Ampère's law, we get

$$\oint \mathbf{B} \cdot d\boldsymbol{\ell} = \mu_0 I$$

$$2Ba = \mu_0 \lambda_a$$

$$B = \boxed{\frac{\mu_0 \lambda}{2}}$$

This shows that *B is independent of the distance from the current sheet.* (The result is analogous to Example 25-5, in which we found that the electric field produced by an infinite sheet of uniform charge density σ, $E = \sigma/2\varepsilon_0$, is also independent of the distance from the infinite sheet.)

One of the aesthetically pleasing aspects of electricity and magnetism is the similarity of form among the equations describing both phenomena. As an illustration, compare the equations in Table 31-1, which describe the magnetic field of a long, straight, current-carrying conductor and the electric field of a long line of charge. In addition to the obvious symmetries, also note that, whenever ε_0 appears in the denominator of an electric field equation, μ_0 appears in the numerator of the analogous magnetic field equation.

TABLE 31-1 Similarities Between Electric and Magnetic Fields

	Magnetic Field of a Long, Straight Current-Carrying Conductor	**Electric Field of a Long Line of Charge**
1. General equations (both equations are inverse square)	$d\mathbf{B} = \left(\dfrac{\mu_0}{4\pi}\right)\dfrac{I d\boldsymbol{\ell} \times \hat{\mathbf{r}}}{r^2}$	$d\mathbf{E} = \left(\dfrac{1}{4\pi\varepsilon_0}\right)\dfrac{\lambda d\boldsymbol{\ell}}{r^2}\hat{\mathbf{r}}$ where λ is the linear charge density
2. Alternative general equations	$\oint \mathbf{B} \cdot d\boldsymbol{\ell} = \mu_0 I$ (line integral) Ampère's law	$\oint \mathbf{E} \cdot d\mathbf{A} = \dfrac{q}{\varepsilon_0}$ (surface integral) Gauss's law
3. Field equations for a long line a distance r away from the line (both equations are inverse first power)	$B = \dfrac{\mu_0 I}{2\pi r}$ where B circles the line	$E = \dfrac{\lambda}{\varepsilon_0 2\pi r}$ where E is directed away from a positively charged line

Summary

Magnetic fields are created by charges in motion. The field $d\mathbf{B}$ produced by a current-carrying element $I\,d\boldsymbol{\ell}$ is given by

Biot–Savart law: $\quad d\mathbf{B} = \left(\dfrac{\mu_0}{4\pi}\right)\dfrac{I\,d\boldsymbol{\ell} \times \hat{\mathbf{r}}}{r^2}$

where μ_0 is the *permeability of free space*:

$$\mu_0 = 4\pi \times 10^{-7}\,\dfrac{\text{T·m}}{\text{A}} \quad \left(\text{or}\ \dfrac{\text{N}}{\text{A}^2}\ \text{or}\ \dfrac{\text{Wb}}{\text{A·m}}\right)$$

Ampere: A mutual force per unit length of exactly 2×10^{-7} N/m exists between two parallel conductors one meter apart, each carrying a current of one ampere.

Ampère's law: $\quad \oint \mathbf{B} \cdot d\boldsymbol{\ell} = \mu_0 I$

where I is the total current through the area enclosed by the path of integration. Ampère's law is of practical use when symmetry indicates that $\mathbf{B}$ has a constant magnitude and a constant angle with respect to $d\boldsymbol{\ell}$ along the path of integration.

Magnetic field B produced by current-carrying conductors:

Configuration of Current-Carrying Conductors, Current I	Magnetic Field B
Long straight wire, distance r	$B = \dfrac{\mu_0 I}{2\pi r}$
Circular loop, radius R	$B_{\text{at center}} = \dfrac{\mu_0 I}{2R}$
Toroid, N turns, average circumference R	$B_{\text{inside}} = \dfrac{\mu_0 N I}{2\pi R}$
Long solenoid, n turns per unit length	$B_{\text{inside}} = \mu_0 n I$

Questions

1. Discuss the similarities and differences between Ampère's law and Gauss's law.
2. In what way is the Biot–Savart law similar to Coulomb's law? In what way are these two laws dissimilar?
3. For what kind of situation is it more appropriate to use Ampère's law rather than the Biot–Savart law for computing the magnetic field?
4. Is there a magnetic field inside hollow copper tubing that is carrying a current? If not, why not?
5. Pairs of wires carrying current in opposite directions to and from electrical devices are often twisted together to reduce stray magnetic fields. Explain how this technique works.
6. If a current is established in the helical turns of a stretched coil spring, will the force that the spring exerts increase, decrease, or remain the same? Explain.
7. Parallel current-carrying conductors interact with each other. How do current-carrying conductors perpendicular to each other interact?
8. Two concentric circular loops of wire (in the same plane) have different radii and carry currents in the same direction. Discuss the magnetic forces on each loop. If both currents are reversed, do the forces change? What happens when the currents are initially in opposite directions? What if these currents are reversed?
9. Repeat the previous question for two identical loops that are aligned coaxially near each other.
10. A plasma is a very hot, ionized gas containing equal amounts of positive ions and negative electrons. Consider a plasma contained within a cylindrical region carrying a current in the axial direction. Discuss the direction of the magnetic forces on charges moving near the outer edge of the cylinder. What are the consequences of these forces?
11. In the previous question, what happens (a) when there is a "bend" in the cylinder? (b) when the cylinder "pinches down" to a smaller diameter at a localized region? These two effects are called, respectively, the *kink instability* and the *sausage instability*.

Problems

31.2 The Biot–Savart Law

31A-1 A hiker observes a pocket compass while standing 40 m directly below a single power line that carries a steady current of 150 A. If the horizontal component of the earth's magnetic field is 3×10^{-5} T at the hiker's location, calculate the *maximum* possible error in the compass reading due to the power line.

31A-2 At the earth's magnetic poles, the magnetic field is roughly 1×10^{-4} T. If this field were produced by a current in a wire around the equator, find the current. (Assume that the current loop is symmetrically located between the poles.) You may use the result of Problem 31C-17.

31B-3 Two circular coils, each containing N turns of wire, have a radius R and are separated by a distance $2R$, as shown in Figure 31-13. Find the magnetic field at a point on the axis of the coils midway between them. Assume that the coils are in series (so that the circulation of the current I is in the same sense in both coils) and that the cross-section of the coils is small compared with R^2. You may use the result of Problem 31C-17.

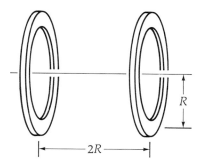

FIGURE 31-13
Problem 31B-3.

31A-4 Find the force per unit length between two long, thin, parallel wires that are separated by a distance of 5 cm. One of the wires carries a current of 10 A in one direction and the other wire carries a current of 10 A in the opposite direction. Is the force between the wires attractive or repulsive?

31A-5 Two long, thin, parallel wires carry currents different from one another. Show that the force per unit length on one wire is equal and opposite to the force per unit length on the other wire. That is, show that Newton's third law is valid.

31B-6 Find the distance x along the axis of a circular loop of radius R, carrying current I, where the magnetic field is half that at the loop's center. You may use the result of Problem 31C-17.

31B-7 In Figure 31-14, suppose that the curved segments were extended to form semicircles. (Thus, the 60° angle would become 180°.) Find the magnitude and direction of the magnetic field **B** at point P.

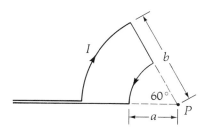

FIGURE 31-14
Problems 31B-7 and 31B-8.

31B-8 Consider the current-carrying loop shown in Figure 31-14, formed of radial lines and segments of circles whose centers are at point P. Find the magnitude and direction of the magnetic field **B** at P.

31B-9 In Figure 31-15, the rectangular wire loop and the long, straight conductor lie in the same plane. The total electrical resistance of the wire loop is 2 Ω. For a steady current I in the straight conductor, find the total magnetic flux Φ_B that passes through the loop. (Hint: choose an element of area $dA = \ell\,dr$ and find the flux $d\Phi_B$ through this area. Then integrate to find the total flux.)

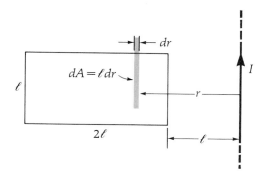

FIGURE 31-15
Problems 31B-9 and 31B-10.

31B-10 In Figure 31-15, consider a current $I_1 = 30$ A in the straight wire and a clockwise current $I_2 = 8$ A in the rectangular loop. If $\ell = 8$ cm, find the net magnetic force on the loop.

31B-11 A square loop of wire, with side length b, carries a current I. Find the magnetic field in the plane of the square at its center. (Assume that the lead-in wires for supplying the current are tightly twisted together so that their B fields cancel.) You may use the result of Problem 31C-21.

31.3 Ampère's Law

31A-12 An air-core toroid has individual windings that form loops 2 cm in diameter. The effective circumference of the toroid is 50 cm. Find the number of turns per unit length required to produce a magnetic field of 0.07 T within the windings when the current is 5 A.

31A-13 A magnetic field B of 0.07 T is required within a solenoid 50 cm long and 2 cm in diameter. (a) Calculate the total magnetic flux within the solenoid. (b) Calculate the number of turns of wire if the current is 5 A.

31B-14 Derive an expression for the magnetic field B inside a long solenoid with n turns per unit length and current I by applying Ampère's law to the rectangular path shown dashed in Figure 31-16. Assume that B is uniform inside the solenoid and negligible outside.

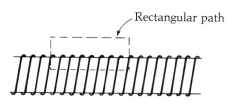

FIGURE 31-16
Problem 31B-14.

31B-15 A long, straight, solid, cylindrical conductor of radius a carries a current I. Starting with Ampère's law, derive an expression for the magnetic field B inside the wire. (Note: for steady currents, the current is spread uniformly over the cross-sectional area.) Include a graph of B vs. r for regions inside and outside the wire, specifying the mathematical behavior with respect to r.

31B-16 The uniform magnetic field between the pole pieces of a magnet cannot end abruptly at the edges of the pole pieces, as shown in Figure 31-17a. Instead, the field must fringe outward, as in Figure 31-17b. Prove this by applying Ampère's law to the region at the edge of the field, as in Figure 31-17a.

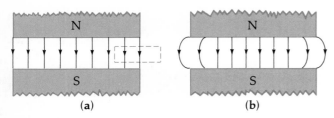

FIGURE 31-17
Problem 31B-16.

Additional Problems

31C-17 As shown in Figure 31-18, a circular loop of radius R carries a current I. Show that the magnetic field on the axis of the loop a distance x from the plane of the loop is

$$\mathbf{B} = \left(\frac{\mu_0 I}{2}\right) \frac{R^2}{(x^2 + R^2)^{3/2}} \hat{\mathbf{x}}$$

(Hint: as you sum the fields $d\mathbf{B}$ due to the current elements $I d\boldsymbol{\ell}$ around the loop, what happens to the field components $d\mathbf{B}_\perp$ perpendicular to the x direction?)

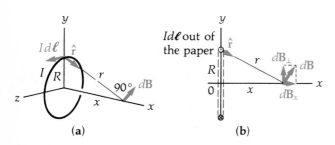

FIGURE 31-18
Problem 31C-17.

31C-18 Consider the magnetic field B at a point P near a long, straight, current-carrying wire. Starting with the Biot–Savart law, find the fraction of the field B that is due to the nearest segment of the wire that subtends an angle of $\pi/2$ rad from that point.

31C-19 A pair of *Helmholtz coils* is often used to produce a uniform magnetic field over a small region of space. The pair consists of two flat, circular coils separated by the radius of the coils, as in Figure 31-19. The current is in the same direction in both coils. Show that, for a separation equal to the radius of the coils, the magnetic field on the axis halfway between the coils is such that dB/dx and d^2B/dx^2 are both zero, where x is the distance along the axis. You may use the result of Problem 31C-17.

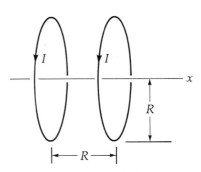

FIGURE 31-19
Problem 31C-19.

31C-20 A circular hoop with a radius of 15 cm carries a current of 10 A. A small hoop with a radius of 1 cm carrying a current of 5 A is placed at the center of the larger hoop so that their centers are coincident but the planes of the hoops are perpendicular. Calculate the torque on the smaller hoop due to the current in the larger hoop. (Assume that the field created by the larger hoop is essentially constant over the region occupied by the smaller hoop.)

31C-21 Refer to Figure 31-20. Starting with the Biot–Savart law, show that the magnetic field B at point P near the straight segment of current-carrying wire is given by $B = (\mu_0 I / 4\pi a)(\sin\theta_1 + \sin\theta_2)$.

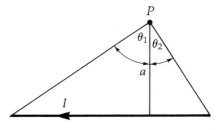

FIGURE 31-20
Problem 31C-21.

31C-22 Two long parallel wires, each having a mass per unit length of 40 g/m, are supported in a horizontal plane by strings 6 cm long as shown in Figure 31-21. Each wire carries the same current I, causing the wires to repel each other so that the angle θ between the supporting strings is 16°. (a) Are the currents in the same or opposite directions? (b) Find the magnitude of each current.

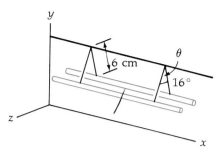

FIGURE 31-21
Problem 31C-22.

31C-23 Two long parallel wires in the xy plane carry equal currents I in opposite directions, Figure 31-22. (a) Find the direction and magnitude of the magnetic field on the z axis as a function of z. (b) Show that the field diminishes as the inverse square for z much greater than the separation of the wires.

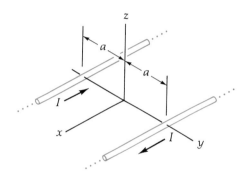

FIGURE 31-22
Problems 31C-23, 31C-24, and 31C-30.

31C-24 In Figure 31-22, assume that both currents are in the $+x$ direction. (a) Sketch the magnetic field pattern in the yz plane. (b) At what distance d along the z axis is the magnetic field a maximum?

31C-25 A horizontal magnetic compass is placed at the center of a circular coil of wire whose plane is vertical. The coil has a radius R and consists of N turns of wire. The coil (carrying no current) is oriented so that the compass needle lies in the plane of the coil. If a current is now established in the coil, the compass needle deflects through an angle θ. Derive an expression for the current I through the coil in terms of R, N, θ, and B_e, the horizontal component of the earth's magnetic field. (This device is called a *tangent galvanometer*.)

31C-26 (a) Repeat Problem 31C-23(a) for the case in which both currents are in the same direction (toward the right). (b) For very large distances $z \gg a$, how does the field vary with z? (c) Make a freehand sketch of the resultant magnetic field in the yz plane, including large distances z from the origin.

31C-27 A long, straight, hollow wire of inner radius a and outer radius b (see Figure 31-11) carries a current density J that varies directly with the radius, $J = kr$, where k is a constant. (a) What are the SI units of k? Using Ampère's law, find the magnetic field B at a distance r from the axis (b) for $r < a$, (c) for $a < r < b$, and (d) for $r > b$.

31C-28 (a) Find the magnetic field outside a very large sheet of finite thickness d that carries a uniform current density J in the $+y$ direction. (You may assume that the sheet is infinite in extent in the $\pm x$ and $\pm y$ directions.) (b) What is the magnetic field within the sheet itself? (Hint: place the origin at the center of the sheet, with the z axis perpendicular to the sheet.)

31C-29 A long, conducting cylinder, radius $2a$, has a cylindrical cavity of radius a whose axis is parallel to the axis of the cylinder but displaced a distance a from the cylinder axis. Figure 31-23 shows a cross-section of the conductor. The conductor carries a current I (out of the paper) distributed uniformly over the cross-sectional area. (a) Show that the current per unit area is $J = I/3\pi a^2$. (b) Find the magnetic field **B** along the y axis for $y \leq 2a$. (Hint: the field may be considered the superposition of the field due to a current I_1 in an uncut solid cylinder and the field of a smaller current I_2 in the opposite direction through a conductor occupying the cavity. What are the currents I_1 and I_2? You may use the result of Problem 31B-15.)

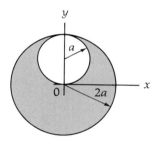

FIGURE 31-23
Problem 31C-29.

31C-30 Consider the long, parallel conductors carrying equal currents in opposite directions shown in Figure 31-22. (a) Find the magnitude and direction of the magnetic field along the $+y$ axis (in the plane of the wires) for (a) $0 < y < a$, and (b) $y > a$. (c) Show that, for $y \gg a$, the field diminishes as the inverse square.

31C-31 A uniform, thin, plastic disk of radius R has a uniform surface charge density σ over both its top and bottom surfaces. Calculate the magnetic field at the center of the disk when the disk is rotating about its axis of symmetry with an angular velocity ω. (Hint: consider the current produced by the charge contained within an annular ring of radius r and width dr.)

31C-32 Consider the long, straight coaxial cable shown in Figure 31-24. A current I is in one direction in the inner conductor and in the opposite direction in the outer conductor. The currents are uniform over the cross-sectional areas of the conductors. Find expressions for the magnetic field B in the following regions: (a) $r < a$, (b) $a < r < b$, (c) $b < r < c$, and (d) $r > c$. (e) Make a qualitative graph of the magnetic field as a function of distance r from the center of the cable.

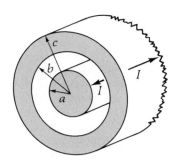

FIGURE 31-24
Problem 31C-32.

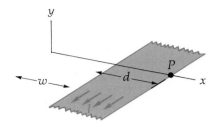

FIGURE 31-25
Problem 31C-33.

31C-33 A long, thin conducting strip of width w carries a total current I along its length, uniformly distributed over the strip as shown in Figure 31-25. Find the magnetic field B at a point P (outside the strip) in the plane of the strip at a distance d from one edge. (Hint: consider the field dB due to the current $(I/w)\,dx$ in a thin strip dx wide.)

31C-34 An electron is moving at 3×10^6 m/s parallel to and at a distance of 1.0 cm from a long, straight wire. Suddenly a steady current of 10 A passes through the wire in a direction parallel to the velocity of the electron. (a) Find the magnitude and direction of the initial acceleration of the electron. (b) Describe qualitatively the subsequent motion of the electron.

31C-35 A thin, uniform, plastic disk of mass m and radius R has a charge Q distributed uniformly over one of its surfaces. When the disk is rotating about its axis with angular velocity ω, show (a) that at the disk's center, $B = \mu_0 Q\omega/2\pi R$ and (b) that its magnetic dipole moment is $\mu = Q\omega R^2/4$. (Hint: consider the current loop due to the moving charge within the annular ring of radius r and width dr.) (c) Show that the ratio of the magnetic moment of the disk to its angular momentum (called the *gyromagnetic ratio*) is $Q/2m$.

31C-36 In Problem 31C-29, show that the magnetic field **B** within the cavity has the same constant value at all points within the cavity and is in the $-x$ direction. You may use the answer to Problem 31B-15.

31C-37 Derive the equation for the magnetic field at the center of a long solenoid by integrating the contributions of all of the individual turns of the solenoid. (Consider each turn as a current-carrying loop. You may use the answer to Problem 31C-17.)

CHAPTER 32

Faraday's Law and Inductance

Sir Robert Peel, the British Prime Minister, visited Faraday in his laboratory soon after the invention of the dynamo. Pointing to this odd machine, he inquired of what use it was. Faraday replied, "I know not, but I wager that one day your government will tax it!" Eventually they did.

32.1 Introduction

After Oersted's discovery in 1820 that a current produces a magnetic field, many investigators felt that the connection between electricity and magnetism could not be in one direction only. So they tried to find an "inverse" effect—namely, could a magnetic field produce a current? The answer is *yes*, though this did not become obvious until it was discovered that *moving* charges produce the magnetic field. Thus, perhaps a *changing* field could produce a current. The discovery was made in 1831 by the English experimenter Michael Faraday, renowned for his laboratory skills, and at the same time by Joseph Henry (1797–1878) working independently in the United States. The effect is called *electromagnetic induction*, and it is the physics behind the generators that provide the electricity used in our modern society. Previously, the only method of generating current was through chemical reactions in voltaic piles. So this discovery was of tremendous importance and began the development of electrical engineering as we know it today.

Electromagnetic induction is also the phenomenon associated with the important circuit elements known as *inductors*. Just as capacitors store energy in their electric fields, inductors store energy in their magnetic fields. In a circuit containing just a capacitor and an inductor, the stored energy can be repeatedly transferred back and forth between the electric field of the capacitor and the magnetic field of the inductor. This produces simple harmonic oscillations of the currents and voltages in the circuit—the basis of all radio transmission and other alternating-current (AC) circuits discussed in Chapter 34. In this chapter we assume that there are no magnetic materials, such as iron, anywhere in the vicinity. (See Chapter 33 for the effects of magnetic materials.)

32.2 Faraday's Law

It is easy to demonstrate that a changing magnetic field can produce a current. Consider Figure 32-1, which shows a loop of wire connected to a galvanometer. If we move a nearby magnet toward the loop as in (a), the deflection of the galvanometer needle indicates a current in the loop while the magnet is moving.

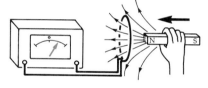

(a) Moving the magnet <u>toward</u> the loop of wire deflects the galvanometer needle as shown.

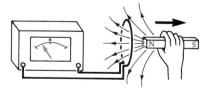

(b) Moving the magnet <u>away from</u> the loop deflects the galvanometer needle in the opposite direction from that in (a).

FIGURE 32-1
A loop of wire is connected to a galvanometer whose zero is at the center of the scale. By changing the number of magnetic field lines that thread through the loop, we induce an emf in the loop, causing an induced current as indicated by the galvanometer needle deflection.

When the magnet is moved away from the loop as in (b), the needle deflects in the opposite direction while the magnet is moving, indicating a current in the opposite direction. When the magnet is stationary, there is no deflection. We can produce similar results by holding the magnet stationary and moving the loop toward and away from the magnet, producing opposite needle deflections in the two cases. Thus it makes no difference whether we move the magnet and hold the loop fixed, or move the loop and hold the magnet fixed. Only *relative motion* between the loop and magnet is important.

The significant feature in these experiments is that *a changing magnetic field within the loop generates a current in the loop*. If the magnetic field in the loop does not change, there is no current. The currents produced in this way are called *induced currents*, and they are the result of *induced emf's* in the circuit.

We can also generate induced emf's in stationary circuits by the procedure illustrated in Figure 32-2. Here, two fixed loops are placed close together without any electrical connection between them. Closing switch S to establish a current in the right-hand loop causes the galvanometer needle momentarily to deflect and then return to zero, indicating a brief induced current in the left-hand loop. If the switch is now opened, there is a momentary current in the opposite direction, which again drops to zero. Establishing a current in the right-hand loop creates magnetic field lines, some of which thread through the left-hand loop. Only when the magnetic field is *changing* is there an induced current. With a steady current in the right-hand loop, there is no induced emf in the left-hand loop.

The common theme in these experiments is this:

An induced emf is generated whenever there is a change of the magnetic field lines that thread through the circuit.

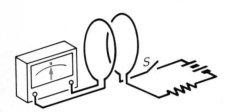

FIGURE 32-2
The two circular loops are close together but have no electrical connection between them. When switch S is closed and then opened, the galvanometer needle momentarily deflects in one direction and then in the opposite direction, indicating induced emf's in the left-hand loop as the magnetic field in that loop changes.

The important word here is *change*. The number of field lines that pass through the circuit does not matter; only the *rate of change* of these field lines determines the induced emf.

The quantity that specifies the number of magnetic field lines that thread through a closed loop is the *magnetic flux* Φ_B (Equation 30-29):

MAGNETIC FLUX $$\Phi_B = \int \mathbf{B} \cdot d\mathbf{A} \quad \text{(in units of T·m}^2\text{)} \qquad (32\text{-}1)$$

Here, $d\mathbf{A}$ is the element of surface area. The integration is carried out over the entire surface area that is defined by the circuit loop that forms its outer perimeter. The area may be a plane or an arbitrarily curved surface. The value of $\int \mathbf{B} \cdot d\mathbf{A}$ is called the **flux linkage** Φ_B through the loop. If the same flux passes through N turns in a coil, the flux linkage is $N\Phi_B$.

Faraday's law is the general statement that summarizes these experimental observations. In words,

The magnitude of the induced emf $\mathscr{E}$ in a circuit equals the time rate of change of magnetic flux through the circuit.

In equation form,

FARADAY'S LAW OF INDUCTION $$\mathscr{E} = -\frac{d\Phi_B}{dt} \quad \text{(for a single loop)} \qquad (32\text{-}2)$$

The minus sign (to be discussed in a later section) has a special meaning that indicates the polarity of the induced emf $\mathscr{E}$. If the circuit loop has N turns, this effectively puts all the individual emf's in series, increasing the induced emf by a factor N:

FARADAY'S LAW OF INDUCTION
$$\mathscr{E} = -N\frac{d\Phi_B}{dt} \quad \text{(for } N \text{ turns)} \quad (32\text{-}3)$$

Often we will deal with magnetic fields that are uniform over a plane area A (though the field may change with time). In these cases, the magnetic flux Φ_B passing through the area A is simply

$$\Phi_B = \mathbf{B} \cdot \mathbf{A} = BA \cos\theta \quad \text{(for uniform } \mathbf{B}) \quad (32\text{-}4)$$

where the angle θ is between $\mathbf{B}$ and the vector $\mathbf{A}$ normal to the plane. Therefore,

FARADAY'S LAW OF INDUCTION
$$\mathscr{E} = -\frac{d}{dt}(BA \cos\theta) \quad \text{(for uniform } \mathbf{B}) \quad (32\text{-}5)$$

There are thus several ways in which we can generate an induced emf in a circuit. We can (1) change the magnitude of $\mathbf{B}$ with time, (2) change the area A of the circuit with time, and (3) change the angle θ with time. Each method causes a change in the flux linkage $N\Phi_B$ that threads through the circuit.

EXAMPLE 32-1

Changing the magnitude of B. A flat coil of wire with 100 turns and a cross-sectional area of 40 cm² is placed with its plane perpendicular to a magnetic field $B = 0.45$ T. If the field is changing at the rate of 0.05 T/s, find the magnitude of the induced emf at the terminals of the coil.

SOLUTION

The magnitude of the field B is not relevant in determining the induced emf; only the *rate of change* of B is significant. We seek only the magnitude of $\mathscr{E}$, so we ignore the minus sign in Equation (32-3):

$$\mathscr{E} = N\frac{d\Phi_B}{dt} = NA\frac{dB}{dt} = (100)(40\text{ cm}^2)\underbrace{\left[\frac{1\text{ m}^2}{10^4\text{ cm}^2}\right]}_{\text{Conversion ratio}}\left(0.05\,\frac{\text{T}}{\text{s}}\right) = \boxed{0.0200\text{ V}}$$

EXAMPLE 32-2

Changing the orientation of the plane of the loop. A circular loop of wire, 20 cm² in area, lies on a horizontal table. At this geographical location the earth's magnetic field, $B = 50\ \mu$T, is directed downward (toward the north) at an angle of 70° with respect to the horizontal, Figure 32-3. The loop is turned completely over in 0.60 s, with its final position again horizontal. Find the average emf induced in the loop while it is being turned over.

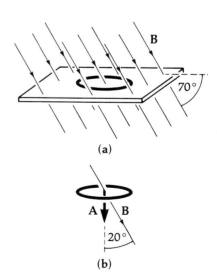

FIGURE 32-3
Example 32-2.

SOLUTION

The plane of the loop is not normal to the field lines, so the flux Φ_B threading through the loop is, from Equation (32-4),

$$\Phi_B = \mathbf{B} \cdot \mathbf{A} = BA \cos\theta = (5 \times 10^{-5} \text{ T})(20 \times 10^{-4} \text{ m}^2)(\cos 20°)$$
$$\Phi_B = 9.397 \times 10^{-8} \text{ T} \cdot \text{m}^2$$

As the loop is turned over, the flux linkages (from the loop's point of view) drop to zero, then increase to their original value in the opposite direction. So during the time $\Delta t = 0.60$ s, the *change* of flux linking the coil is twice the original value: $2(9.397 \times 10^{-8} \text{ T} \cdot \text{m}^2) = 1.879 \times 10^{-7} \text{ T} \cdot \text{m}^2$. From Equation (32-2) (again omitting the minus sign), we obtain

$$\mathscr{E} = \frac{\Delta \Phi_B}{\Delta t} = \frac{(1.879 \times 10^{-7} \text{ T} \cdot \text{m}^2)}{(0.60 \text{ s})} = \boxed{0.313 \text{ }\mu\text{V}}$$

32.3 Motional emf

As we have seen in the discussion of the experiments in the previous section, an induced emf is produced whenever there is relative motion between a magnetic field and a conductor. In this section we describe the emf induced in a conductor when it moves through a stationary magnetic field. Such an emf produced by the motion of a conductor is called a **motional emf**.

Consider a rectangular loop formed by a stationary U-shaped circuit with a sliding metal bar for one edge, Figure 32-4. As the bar slides, it maintains electrical contact with the stationary parts of the circuit. The only (appreciable) electrical resistance in the closed loop is the resistance R. A uniform field $\mathbf{B}$ exists into the plane of the figure. At any instant, the magnetic flux linking the rectangular loop is $\Phi_B = BA = B\ell x$. As the bar moves, the area A increases with time: $dA/dt = \ell\, dx/dt = \ell v$. Thus the flux linking the circuit changes with time and, by Faraday's law, the induced emf due to the motion of the bar is

$$\mathscr{E} = -\frac{d\Phi_B}{dt} = -\frac{d}{dt}(BA) = -B\frac{dA}{dt} = -B\ell v$$

MOTIONAL emf $$\mathscr{E} = -B\ell v \qquad (32\text{-}6)$$

The minus sign has a special meaning that indicates the polarity of the emf, as discussed in the next section.

It is useful to analyze the energy interchanges in this example. The emf induced in the moving bar in Figure 32-4 will cause a current to exist in the closed loop. If the velocity $\mathbf{v}$ is constant, the current I will be constant:

$$I = \frac{\mathscr{E}}{R} = \frac{B\ell v}{R} \qquad (32\text{-}7)$$

The thermal power developed in the resistance R is therefore

$$P_{\text{th}} = I^2 R = \left(\frac{B\ell v}{R}\right)^2 R = \frac{B^2 \ell^2 v^2}{R} \qquad (32\text{-}8)$$

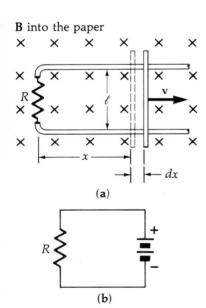

FIGURE 32-4
In (a), as the bar slides it maintains electrical contact with the stationary bars, forming a closed conducting loop that increases its area. Because of the presence of the magnetic field, an emf is induced in the moving bar, making the top end positively charged and the bottom end negatively charged. (If the bar moves in the opposite direction, the polarity is reversed.) The moving bar is a seat of emf $\mathscr{E}$, called a *motional emf*, and it causes a counterclockwise current in the loop. In (b), an analogous circuit using a battery is shown.

Where does this power come from? Recall from Section 30.5 that a current-carrying conductor in the presence of a magnetic field has a magnetic force on it of $\mathbf{F}_{mag} = I\boldsymbol{\ell} \times \mathbf{B}$. In our case, the bar length $\boldsymbol{\ell}$ and the field $\mathbf{B}$ are at right angles, so we have, for the magnetic force on the bar,

$$F_{mag} = I\ell B = \left(\frac{B\ell v}{R}\right)\ell B = \frac{B^2\ell^2 v}{R} \qquad (32\text{-}9)$$

As will be shown shortly, this magnetic force is toward the left, opposite to the (equal-magnitude) external force $\mathbf{F}$ that pulls the bar toward the right. The bar thus has *zero* net force on it, and it moves with constant velocity. The rate of doing work done by this external force is

$$P_{ext} = Fv = \left(\frac{B^2\ell^2 v}{R}\right)v = \frac{B^2\ell^2 v^2}{R} \qquad (32\text{-}10)$$

Equations (32-8) and (32-10) are equal, so we see that the power furnished to the circuit by the work done by the external force just equals the I^2R power developed in the resistor. Again, conservation of energy holds true!

It is easy to determine the direction of the magnetic force on the current-carrying bar. As the bar moves in the presence of the field, consider the Lorentz force, $\mathbf{F} = q(\mathbf{v} \times \mathbf{B})$, acting on a free (negative) conduction electron in the metal. The Lorentz force $\mathbf{F} = (-e)(\mathbf{v} \times \mathbf{B})$ is downward in Figure 32-4, moving electrons downward. The bottom of the bar becomes negatively charged, and the top end becomes positively charged. Therefore the current I circulates counter-clockwise in the loop, and the current in the moving bar is upward. Consequently, the magnetic force on that bar is $\mathbf{F}_{mag} = I\boldsymbol{\ell} \times \mathbf{B}$, or *toward the left*.

If there is no external circuit that forms a closed path, the emf is still present in the moving bar, Figure 32-5. In this case, as the bar begins to move there will be a momentary movement of conduction electrons in response to the Lorentz force, accumulating a negative charge at the bottom end and an equal positive charge at the top. Equilibrium is rapidly achieved when the Lorentz forces qvB are balanced by the electrostatic forces of attraction qE between the separated charges of opposite sign. The electric field E within the bar due to this separation of charge is related to the potential difference $V = E\ell$ between the ends of the bar. As long as the bar is in motion, the potential difference V is present across its ends. From $qE = qvB$, we choose $E = vB$. Thus:

$$V = E\ell = B\ell v \qquad (32\text{-}11)$$

This agrees with the result using Faraday's laws. Even in the case of a moving nonconductor, this same potential difference is created by the Lorentz forces, producing a slight displacement of positive and negative charges from their equilibrium positions, creating an electric field within the bar (see Section 27.4, Dielectrics).

EXAMPLE 32-3

In Figure 32-5, a metal bar 10 cm long moves through a magnetic field $B = 2$ mT as shown. (a) What speed v will produce a potential difference of 1 mV between the ends of the bar? (b) If the bar moves in the opposite direction, does the polarity change? (c) Suppose that the bar is aligned perpendicular to the field lines, but the velocity $\mathbf{v}$ is at an angle of 120° (rather than 90°) with $\mathbf{B}$. Find the potential difference if the speed is the same as in part (a).

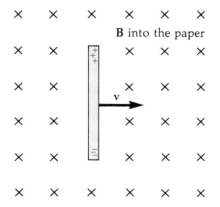

FIGURE 32-5
A conducting bar moving across a magnetic field has a motional emf $\mathscr{E}$ between the ends of the bar, whether or not an external circuit allows a current to exist.

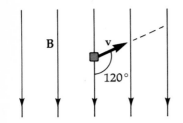

FIGURE 32-6
Example 32-3(c). The moving bar is perpendicular to the plane of the paper.

SOLUTION

(a) From Equation (32-11),

$$V = B\ell v$$
$$(1 \times 10^{-3} \text{ V}) = (2 \times 10^{-3} \text{ T})(0.1 \text{ m})(v)$$
$$v = \boxed{5.00 \text{ m/s}}$$

(b) If the bar is moved toward the left, free electrons $(-e)$ in the bar will experience a Lorentz force $\mathbf{F} = (-e)(\mathbf{v} \times \mathbf{B})$ that is upward, so

$$\boxed{\text{the polarity reverses.}}$$

(c) Figure 32-6 shows the motion from a perspective in which the bar is perpendicular to the plane of the diagram. From the Lorentz force expression, $\mathbf{F} = q(\mathbf{v} \times \mathbf{B})$, the sine of the angle between $\mathbf{v}$ and $\mathbf{B}$ is a factor that reduces the emf from its original (90°-motion) value $\mathscr{E}_0$ to

$$\mathscr{E} = \mathscr{E}_0 \sin(120°) = (1 \text{ mV})(0.866) = \boxed{0.866 \text{ mV}}$$

Note: if the length of the conductor were not at right angles to $\mathbf{B}$, a similar obliquity factor would be necessary. *For maximum emf, the conductor length ℓ, the velocity $\mathbf{v}$, and $\mathbf{B}$ must all be mutually perpendicular.*

Comments

We have illustrated two different ways of inducing an emf: by moving a conductor in the presence of a stationary magnetic field, or by using a stationary circuit in the presence of a changing field. Both are contained in Faraday's law:

$$\mathscr{E} = -\frac{d\Phi_B}{dt} = -\frac{d}{dt}(BA) = -\left[B\underbrace{\frac{dA}{dt}}_{\text{①}} + A\underbrace{\frac{dB}{dt}}_{\text{②}}\right] \qquad (32\text{-}12)$$

The moving-bar example involved term ①. But we saw that we could also obtain that result by applying the Lorentz force law. The unique contribution of Faraday really lies in term ②. The production of an emf around a stationary circuit of area A by a changing magnetic flux within that circuit was not contained in any prior physical law. Faraday demonstrated experimentally that this effect occurs, and his insight is justly honored. Furthermore, it guaranteed the equivalence of inertial frames for electromagnetism: a frame of reference with a stationary circuit and a moving field (moving a magnet toward a fixed loop) or a frame of reference with a stationary field and a moving circuit (moving the loop toward a fixed magnet). This *invariance principle* was a crucial step to Einstein's relativity. Indeed, it was the reason that Einstein's first relativity paper was titled "On the Electrodynamics of Moving Bodies."

As we have mentioned, another important conclusion of Faraday's law is that an induced electric field occurs *even when no material substance is present*. In Figure 32-7a, a magnetic field is changing at a constant rate with time. If we consider a circular conductor placed symmetrically within this field as shown,

(a) A conducting ring of radius r is placed symmetrically within the region of a uniform, increasing magnetic field. An emf $\mathscr{E} = \oint \mathbf{E} \cdot d\boldsymbol{\ell}$ exists around the ring. (If **B** were <u>decreasing</u>, the direction of **E** would be in the opposite sense.)

(b) The same field **E** exists around the path in (a) <u>even if the conductor is removed</u>.

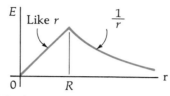

(d) The magnitude of the induced **E** field as a function of r (see Example 32-4).

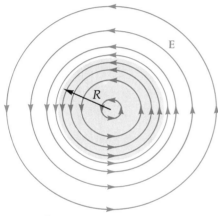

(c) When the magnetic field in (a) increases, it induces an electric field **E** both inside and outside the region of the field.

FIGURE 32-7
In a circular region of radius R, a uniform magnetic field **B** increases with time at a steady rate. That is, $dB/dt = $ constant.

a constant emf is generated in the circuit due to the changing flux that links the circuit. This emf can be written as the line integral of **E** around the closed loop, $\mathscr{E} = \oint \mathbf{E} \cdot d\boldsymbol{\ell}$, leading to the most general form of Faraday's law:

FARADAY'S LAW
$$\oint \mathbf{E} \cdot d\boldsymbol{\ell} = -\frac{d\Phi_B}{dt} \tag{32-13}$$

This expression makes no reference to any conductor, charges, or currents; it occurs in otherwise "empty" space. *A changing magnetic flux produces an electric field.* Even if we now remove the circular conductor, Figure 32-7b, the same induced electric fields still exist along the line integral path as before. Figure 32-7c shows the pattern of these induced **E** fields for this particular configuration of changing magnetic flux. Faraday's law can be applied to *any* closed path; it need not be a circular path as in our example.

> **EXAMPLE 32-4**
>
> In Figure 32-7, show that the magnitude of **E** varies with r as indicated in (d), provided that dB/dt is constant.
>
> **SOLUTION**
>
> For $r < R$, we choose a circular path of radius r for the integration (to match the symmetry of **B**). From Faraday's law,
>
> $$\oint \mathbf{E} \cdot d\boldsymbol{\ell} = -\frac{d\Phi_B}{dt} = -A\left(\frac{dB}{dt}\right)$$
>
> $$(E)(2\pi r) = -(\pi r^2)\left(\frac{dB}{dt}\right)$$
>
> For the magnitude, we drop the minus sign and rearrange:
>
> $$E = \frac{r}{2}\left(\frac{dB}{dt}\right) = \boxed{(\text{constant})(r)}$$
>
> For $r > R$, the entire flux within πR^2 is within the path of integration.
>
> $$\oint \mathbf{E} \cdot d\boldsymbol{\ell} = -\frac{d\Phi_B}{dt} = -A\left(\frac{dB}{dt}\right)$$
>
> $$(E)(2\pi r) = -(\pi R^2)\left(\frac{dB}{dt}\right)$$
>
> The magnitude is $E = \dfrac{R^2}{2r}\left(\dfrac{dB}{dt}\right) = \boxed{(\text{constant})\left(\dfrac{1}{r}\right)}$

Electric Fields and emf's

In previous chapters we investigated electric fields that arose from the presence of stationary electric charges. These were *conservative* fields because we could define a potential difference between two points that was the same for all paths between the points:

$$V_b - V_a = -\int_a^b \mathbf{E} \cdot d\boldsymbol{\ell} \tag{32-14}$$

If we choose a and b to be the same point and integrate around a closed-loop path, the integral is zero (the criterion for a conservative field):

$$\oint \mathbf{E} \cdot d\boldsymbol{\ell} = 0 \tag{32-15}$$

But now consider Faraday's law for a closed loop in the presence of a changing magnetic field, Figure 32-7b. As the field B changes, there is an emf $\mathscr{E}$ induced in the loop described as the closed line-integral $\oint \mathbf{E} \cdot d\boldsymbol{\ell}$ around the loop:

$$\mathscr{E} = \oint_C \mathbf{E} \cdot d\boldsymbol{\ell} = -\frac{d\Phi_B}{dt} = -\frac{d}{dt}\int_S \mathbf{B} \cdot d\mathbf{A} \tag{32-16}$$

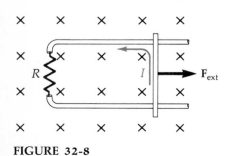

FIGURE 32-8

where $d\boldsymbol{\ell}$ is along the curve C that surrounds the surface area S. The direction of $d\mathbf{A}$ is given by the right-hand rule: circle the fingers of the right hand in the direction of $d\boldsymbol{\ell}$ around the loop; the extended thumb points in the direction of $d\mathbf{A}$. *Since this integral around the closed loop is not zero, the field E is not conservative.* That is, *an electric potential cannot be defined for induced fields.* This is an important difference between electric fields due to static charges and the induced electric fields generated by changing magnetic fields. Another distinction is that while the electric fields due to static charges always begin and end on charges, the electric fields associated with changing magnetic fields exist where no charges at all are present, and these electric field lines always form closed loops.

32.4 Lenz's Law

The information in Faraday's law was originally expressed by Faraday in a rather cumbersome form that involved several relations. Later investigators revised and reduced these relations to the succinct equation we have today. An important clarification was made by the German physicist Heinrich Lenz (1804–1865), who contributed the minus sign. This minus sign has an importance greater than it might seem at first glance, since an understanding of its meaning gives the direction of the induced emf. *Lenz's law is the interpretation we give to the minus sign.* We illustrate the law with a specific case.

Consider the movable bar in Figure 32-8 which maintains electrical contact with the stationary bars. We saw in Figure 32-4 that, as the external force $\mathbf{F}_{ext}$ moved the bar toward the right, the polarity of the induced emf was that the top end of the bar becomes positive with respect to the bottom end. This emf produces a counterclockwise current I around the circuit as shown, resulting in a magnetic force on the bar of $\mathbf{F}_{mag} = I\boldsymbol{\ell} \times \mathbf{B}$ toward the left that just balances $\mathbf{F}_{ext}$ toward the right, so the bar has zero net force on it.

Suppose, instead, that the induced current was clockwise (opposite to the true direction). The magnetic force on the bar would reverse direction so that it is toward the right. Once the bar started to move, the magnetic force would take over and accelerate the bar even faster, developing more and more Joule heating in R—a sort of perpetual motion machine that violates the conservation of energy. So we conclude that *any effects arising from induced emf's must oppose the effect that generated those emf's.* This is the insight that Lenz contributed, and the minus sign in $\mathscr{E} = -d\Phi_B/dt$ stands for this reasoning.

A convenient way to think about Lenz's law is in terms of *flux linkages.* As we pull the bar toward the right, the number of magnetic flux lines that link the circuit loop increases INTO the plane of the paper. The induced current itself produces flux lines through the loop OUT OF the plane of the paper (apply the right-hand rule for the magnetic field due to a current-carrying wire), thus opposing the <u>change</u> of flux linkages that produced the current.

> **LENZ'S LAW** The induced current in a closed loop is in a direction so as to oppose the <u>change</u> in the flux linkages that produced it.

Note carefully that the induced current produces effects that oppose the <u>change</u> of flux linkages, *not the flux itself.* Overlooking this distinction is a common error. Even if the circuit does not form a closed loop so that no induced current is actually present, we usually can imagine what would happen if it were a closed loop and thus can determine the polarity of the induced emf across the gap. Look at the examples in Figure 32-9. Before reading the analyses below, can you apply Lenz's law correctly to predict the directions of the induced currents in R for each case? Try it.

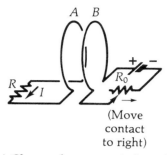

(a) Change the current in loop B by moving the sliding contact on the resistor toward the right.

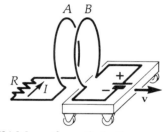

(b) Move the coils farther apart.

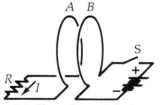

(c) Close the switch S.

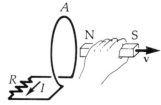

(d) Pull the magnet away from the coil or push it into the coil. (Induced current in A is shown for pulling the magnet *away*.)

FIGURE 32-9
Methods of inducing an emf in the conducting loop A by causing a change in the flux linkage through that loop. For each case, be sure that you understand the Lenz's-law reasoning that determines the <u>direction</u> of the induced current I.

Here is the application of Lenz's law to the changing-flux situations shown in Figure 32-9.

(a) Initially, the current in loop B produces magnetic field lines that thread through loop A toward the right. Moving the variable resistor contact as shown increases the current in B, causing more flux lines to thread through A toward the right. The induced current in A is such that it produces flux lines in loop A toward the left, opposing the original *change* of flux linkages.

(b) Initially, magnetic flux lines are toward the right inside both loops. Moving B toward the right results in fewer lines threading through loop A. The induced emf in A causes a current in the direction that produces flux toward the right, opposing the *change* of flux linkages in A.

(c) Closing the switch in the right-hand circuit causes magnetic flux lines to increase toward the right in loop A. Therefore, the induced current in A is such that it produces flux lines toward the left within loop A, opposing the *change* of flux linkages in A.

(d) Magnetic field lines come out of the north pole of a magnet and enter the south pole. So initially the flux line thread through loop A toward the left. As the magnet moves toward the right, these flux linkages decrease. Therefore, the induced current in loop A is as shown, itself producing flux lines toward the left in loop A, opposing the *change* of flux linkages in A.

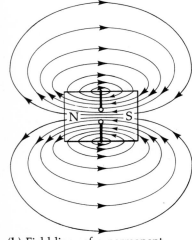

(b) Field lines of a permanent magnet form complete, closed loops. They emerge from the north end and enter the south end. Inside the magnet, the lines extend from S to N.

FIGURE 32-10
Example 32-5.

EXAMPLE 32-5

In Figure 32-10a, the wire loop with gap *ab* is held fixed while the permanent magnet is withdrawn as indicated. Find the polarity of the induced emf across the gap while the magnet is being withdrawn.

SOLUTION

Magnetic field lines always form complete, closed loops. Therefore, due to the field lines *inside* the magnet, the net flux linkages through the wire loop are initially toward the left, Figure 32-10b. If an external wire were connected across the gap, the induced current in this wire would be from *b* to *a* as the magnet is withdrawn, so that the induced current itself in the loop would create a magnetic field that tends to maintain the initial flux linkages through the loop. (It opposes the change of flux linkages.) Thus point *b* is at a higher potential than point *a*. *The points ab are, in effect, the terminals of a source of emf that would cause a current from* b *to* a *in an external wire connected between them.* The emf is present in the gap whether or not an external wire is connected across *ab*.

> Point *b* is at the higher potential.

32.5 Eddy Currents

In some instances, there is no well-defined conductor path to which the currents from induced emf's are confined. Often there will be a mass of metal moving in the presence of a magnetic field or located where magnetic fields are changing. In these cases, the induced currents circulate throughout the

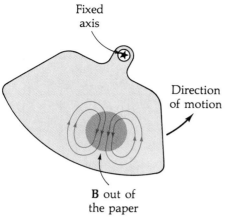

(a) A magnetic field exists <u>out</u> of the plane of the paper in the shaded region. As the metal sheet moves through the field, eddy currents are induced as shown. (The return paths for the current loops are outside the field region.) Magnetic forces on these currents impede the motion that generates them.

(b) Cutting slots in the metal interrupts the current paths, greatly reducing the eddy-current braking effect.

FIGURE 32-11
A demonstration showing the presence of eddy currents.

volume of the metal. These internal circulating currents are called *eddy currents* in analogy to the eddies sometimes occurring in fluid flows.

We can demonstrate the presence of eddy currents by allowing a sheet of nonmagnetic metal such as copper or aluminum, suspended from a horizontal axis at one end, to swing freely into a region where a magnetic field exists, Figure 32-11. As the conducting sheet moves, the field acts on the free conduction charges, causing circulating eddy currents as shown. By Lenz's law, the currents within the region of the magnetic field result in magnetic forces $\mathbf{F} = I\boldsymbol{\ell} \times \mathbf{B}$ on the metal that oppose the very motion which generated the currents. (There is no force on the return currents of the loops outside the field.) The net result is a braking action that impedes the motion of the metal. This effect has been used commercially in a type of electromagnetic brake called an *eddy-current brake*. If a series of slots is cut into the metal sheet interrupting the current paths, the eddy currents are greatly reduced, minimizing the eddy-current braking effect.

32.6 Self-Inductance

In contrast to a resistor, which restricts the amount of current in a circuit, a loop or coil of wire in a circuit restricts any <u>change</u> of current in the circuit. To understand this effect, consider the circuit of Figure 32-12 containing a resistor, a coil of wire, and a battery. When the switch S is closed, a current I is established in the direction indicated. But this current does not build up instantaneously. As I begins to increase, there is a growing magnetic field in the coil as indicated. By Faraday's and Lenz's laws, these changing flux lines in the coil windings induce an emf in the coil that opposes this change. That is, the polarity of the emf is that the bottom of the coil is *positive* with respect to the top, trying to produce a current in the opposite direction. This is called a "self-induced emf," or a "*back-emf*" $\mathscr{E}_L$, which opposes the current buildup,

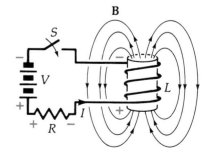

FIGURE 32-12
An inductance in series with a battery and a resistor. As the current builds up, note the polarities of the potential differences across R and L.

causing a slower increase in I than would occur without the coil. If no iron or similar magnetic materials are present, this back-emf depends only on the physical dimensions of the coil and the rate of change of current dI/dt in the coil (since these are the factors that determine the rate of change of flux linkages). We call the factor of proportionality *due to the physical dimensions alone* the **self-inductance** L of the coil. Its circuit symbol is —〰〰〰— .

BACK-emf ACROSS AN INDUCTANCE L
$$\mathcal{E}_L = -L\frac{dI}{dt} \qquad (32\text{-}17)$$

The SI unit of self-inductance[1] is the *henry* (H). Circuit elements that have *inductance* are called *inductors*. (Note the correspondence to *capacitance* and *capacitors*, and to *resistance* and *resistors*.)

A precise calculation of L for a given coil is ordinarily difficult because of end effects and "leakage" of flux lines between the windings. However, for an ideal solenoid (or a toroidal coil) with closely spaced windings, we evaluate L as follows. For each single turn, the induced emf is $\mathcal{E}_L = -d\Phi_B/dt$. Ideally the same flux Φ_B links all N turns, so

$$\mathcal{E}_L = -N\frac{d\Phi_B}{dt} \qquad (32\text{-}18)$$

Comparing this with Equation (32-17) we have

$$L\frac{dI}{dt} = N\frac{d\Phi_B}{dt}$$

Integrating both sides and noting that $\Phi_B = 0$ when $I = 0$, we obtain

$$LI = N\Phi_B$$

Solving for L gives

SELF-INDUCTANCE L
$$L \equiv \frac{N\Phi_B}{I} \qquad (32\text{-}19)$$

The product $N\Phi_B$ is called the *number of flux linkages*. Thus L is the *number of flux linkages per ampere*; this depends solely upon the physical dimensions of the coil itself. Since the SI unit for magnetic flux is the tesla·meter^2, units for L are the *henry* (H), or the *tesla*·meter^2/ampere (T·m^2/A). Usually L is called simply the *inductance* of the coil.

Earlier in this chapter we showed that the magnetic field within an ideal solenoid (ignoring end effects) is uniform and given by $B = \mu_0 nI$, where n is the number of turns per unit length and I is the current. For a cross-sectional area A, the total flux inside the coil is $\Phi_B = BA = \mu_0 nIA$. For a solenoid or toroid of length ℓ and N total turns (so $n = N/\ell$), this relation becomes

$$\Phi_B = \frac{\mu_0 ANI}{\ell}$$

---

[1] This unit honors the American physicist Joseph Henry (1797–1878), who discovered induction independently of Faraday's discoveries in England. Faraday published his results first, so he is given priority in naming the "law."

Substituting this expression for Φ_B in Equation (32-19) gives

SELF-INDUCTANCE FOR A TOROID OR IDEAL SOLENOID (ignoring end effects)

$$L = \frac{\mu_0 N^2 A}{\ell} \quad (32\text{-}20)$$

where N is the total number of turns in the length ℓ with cross-sectional area A. From Equation (32-20) we find that μ_0, the *permeability of free space*, may be expressed in units of henrys per meter (H/m) as well as other units found previously. Here is a summary of possible units for μ_0:

UNITS FOR μ_0

$$\mu_0 \equiv 4\pi \times 10^{-7} \frac{\text{H}}{\text{m}} \quad \left(\text{or } \frac{\text{N}}{\text{A}^2} \text{ or } \frac{\text{T} \cdot \text{m}}{\text{A}}\right) \quad (32\text{-}21)$$

EXAMPLE 32-6

(a) Find the self-inductance of a solenoid that has a cross-sectional area of 1 cm², a length of 10 cm, and 1000 turns of wire. (b) If the current through the inductor is increasing at the rate of 15 A/s, find the magnitude of the induced back-emf.

SOLUTION

(a) The length of the solenoid is large compared with the cross-sectional radius, and the turns of wire are closely wound. So we treat it as a "long" solenoid, ignoring end effects. Substituting the appropriate values in SI units into Equation (32-21) yields

$$L = \frac{\mu_0 N^2 A}{\ell} = \frac{(4\pi \times 10^{-7}\,\text{H/m})(1000\,\text{turns})^2(10^{-4}\,\text{m}^2)}{(0.10\,\text{m})} = \boxed{1.26\,\text{mH}}$$

(b) From Equation (32-17), we have

$$|\mathscr{E}| = (-)L\frac{dI}{dt} = (1.26 \times 10^{-3}\,\text{H})\left(15\,\frac{\text{A}}{\text{s}}\right) = \boxed{18.9\,\text{mV}}$$

32.7 Mutual Inductance

In the previous section, we defined the *self-inductance L* of a coil that involves the back-emf generated in a coil due to a changing current in the coil itself. Similar effects occur between two coils that are close enough together so that flux lines generated in one coil can link the other coil. Then, an emf will be induced in either coil due to current changes in the other coil (see Figure 32-13). This process is known as *mutual induction*, defined in the following way. The emf generated in coil 1 due to a changing current dI_2/dt in coil 2 is

$$\mathscr{E}_1 = -M_{12}\frac{dI_2}{dt}$$

where M_{12} is defined to be the **mutual inductance** of coil 1 with respect to coil 2. Mutual inductance has the same unit as self-inductance: the *henry* (H).

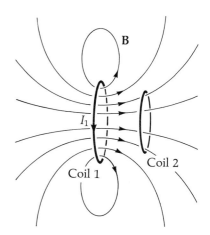

FIGURE 32-13
Mutual inductance between two coils occurs when they are close enough together so that a current in one coil causes flux linkages in the other. (Here, current in coil 1 creates flux linkages in coil 2.) It is a mutual effect: a *changing* current in either coil will induce emf's in the other coil.

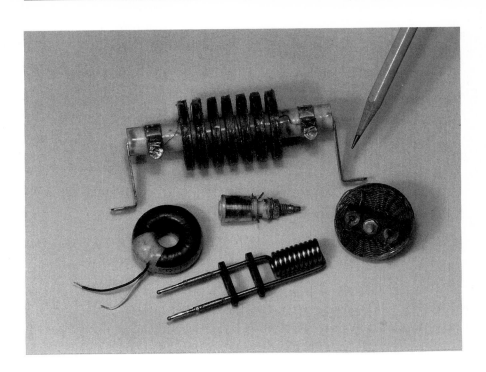

FIGURE 32-14
Small inductors used in electronic circuits.

Similarly, the emf generated in coil 2 due to a changing current dI_1/dt in coil 1 is

$$\mathscr{E}_2 = -M_{21}\frac{dI_1}{dt}$$

where M_{21} is the mutual inductance of coil 2 with respect to coil 1. We omit the proof and merely state that it can be shown that

$$M_{12} = M_{21} \tag{32-22}$$

Thus the symbol M (without subscripts) may be used for mutual inductance:

$$\mathscr{E}_1 = -M\frac{dI_2}{dt} \quad \text{and} \quad \mathscr{E}_2 = -M\frac{dI_1}{dt} \tag{32-23}$$

Since the mutual inductance depends upon the amount of flux linkages $N\Phi_B$ produced in one coil by the current I in the other coil, we may also write [by analogy with Equation (32-19)]

MUTUAL INDUCTANCE M
$$M = \frac{N_2 \Phi_{B2}}{I_1} \quad \text{and} \quad M = \frac{N_1 \Phi_{B1}}{I_2} \tag{32-24}$$

The SI unit for M is the same as for L: *henry* (H). Provided no iron or similar material is nearby, the value of M depends only on geometrical factors such as how close together the two coils are and what their orientations are. Except when the two coils are wound together so that *all* the flux from one coil links the other coil, the calculation of M may be quite complicated.

As a practical example of mutual inductance, telephone lines in a cable sometimes suffer from "cross-talk" when current changes in one line generate emf's in adjacent lines. (Capacitive effects can similarly cause trouble.) Another example is the "hum" in audio amplifiers. This occurs because alternating currents in the power supply can induce alternating emf's in nearby sensitive circuits unless these circuits are shielded from the magnetic fields or are placed sufficiently far away.

EXAMPLE 32-7

A solenoid of length $\ell_1 = 30$ cm, cross-sectional area $A_1 = 6$ cm^2, containing $N_1 = 500$ turns, has a second coil of $N_2 = 20$ turns wound tightly at its center, Figure 32-15. Calculate the mutual inductance M between the coils.

SOLUTION

Because the coils are wound tightly together, they have the same area A and so the same flux Φ_B links both coils. From Equation (32-20), the flux at the center of the solenoid when a current I_1 is present is

$$\Phi_{B1} = \Phi_{B2} = \frac{\mu_0 A N_1 I_1}{\ell_1}$$

The mutual inductance M is thus

$$M = \frac{N_2 \Phi_{B2}}{I_1} = \frac{\mu_0 A N_1 N_2 I_1}{I_1 \ell_1} = \frac{\mu_0 A N_1 N_2}{\ell_1}$$

$$M = \frac{(4\pi \times 10^{-7} \text{ H/m})(6 \times 10^{-4} \text{ m}^2)(500)(20)}{(0.30 \text{ m})} = \boxed{25.1 \text{ }\mu\text{H}}$$

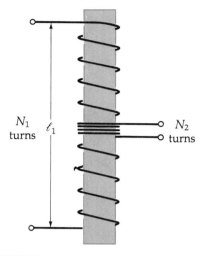

FIGURE 32-15
Example 32-7. Two coils wound together on the same air core have mutual inductance.

32.8 RL Circuits

Since the wire used to form a coil has some electrical resistance,[2] an inductor will also have some finite resistance—how much depends on the resistivity of the wire material, its cross-sectional area, and its length. We usually combine this winding resistance (if appreciable) with the other resistance in a series circuit, so we have a "pure" inductance L in series with a resistance R. Such *RL circuits* are common in electrical networks.

Consider the series RL circuit of Figure 32-16. When the switch S is closed, the battery tries to establish a current in the coil. As the current rises, however, the back-emf in the inductor acts similar to a "battery" whose polarity opposes that of the real battery. For increasing current, the potential drop across L, $\mathscr{E}_L = -L\, dI/dt$, thus has a polarity indicated by the $+$ and $-$ signs. At all times after the switch closes, Kirchhoff's loop rule must hold true. Noting the polarities across each element as the current rises, we have

$$\Sigma \mathscr{E} = 0$$

$$\mathscr{E} - IR - L\frac{dI}{dt} = 0 \qquad (32\text{-}25)$$

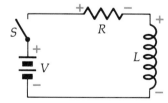

FIGURE 32-16
A series RL circuit for investigating the growth of the current in an inductor. Polarities shown are for the current buildup after S is closed.

---

[2] Superconductors are an exception. The resistivity of these materials becomes truly zero as the temperature approaches a low value.

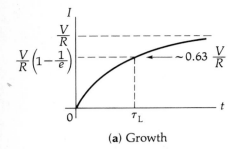

(a) Growth

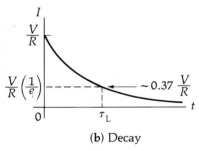

(b) Decay

FIGURE 32-17
Growth and decay of the current in an RL series circuit.

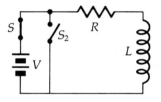

FIGURE 32-18
After a steady current is established in L, a "shorting" branch containing switch S_2 is added. (After S_2 is closed, we can remove the battery without affecting the remaining loop circuit.) The current in L then decays in the loop containing only R and L.

The solution[3] of this differential equation for the current I as a function of time is similar to that for the RC circuit, Section 29.6:

GROWTH OF CURRENT IN AN RL CIRCUIT

$$I = \frac{\mathscr{E}}{R}(1 - e^{-(R/L)t}) \qquad (32\text{-}26)$$

A graph of this equation is shown in Figure 32-17a. Because of the exponential factor, the current increases asymptotically toward the maximum value $\mathscr{E}/R$. The rate of increase depends on the ratio L/R; the larger this ratio, the more slowly the current increases. The ratio L/R is called the time constant τ_L of the RL circuit. In a time equal to *one* time constant after the switch is closed, the current will rise to $(1 - 1/e)$ of its maximum value. This is ~63% of the maximum value.

After the steady-state condition is reached (that is, the current is constant at $\mathscr{E}/R$ and there is no voltage across L), we next investigate the decay of current in an RL circuit. One way to do this is to add a "shorting" branch[4] containing a switch S_2 as shown in Figure 32-18. If we now close switch S_2, it provides a continuous path for the current in L while effectively "shorting out" the battery, and the battery can then be removed without affecting anything in the remaining loop. Because the battery is no longer present, the current in the loop immediately starts to decrease, inhibited, however, by the

---

[3] This equation is solved by the mathematical technique called *the separation of variables*. A rearrangement of the terms in Equation (32-25) produces

$$\frac{dI}{\mathscr{E} - IR} = \frac{dt}{L}$$

thus separating the variables I and t on either side of the equal sign. Both sides of the equation are integrated:

$$\int_0^I \frac{dI}{\mathscr{E} - IR} = \int_0^t \frac{dt}{L}$$

Using the table of integrals in Appendix G-II, we obtain

$$-\frac{1}{R}\ln(\mathscr{E} - IR) = \frac{t}{L} + c$$

where the constant of integration c is found from the initial conditions. Setting $t_0 = 0$ and $I_0 = 0$, we find that $c = -(1/R)\ln\mathscr{E}$. Substituting this value in the above equation, we have

$$\ln\left(\frac{\mathscr{E} - IR}{\mathscr{E}}\right) = -\frac{R}{L}t$$

If $\ln y = x$, then $y = e^x$, so

$$\left(\frac{\mathscr{E} - IR}{\mathscr{E}}\right) = e^{-(R/L)t}$$

Solving for I gives

$$I = \frac{\mathscr{E}}{R}(1 - e^{-(R/L)t})$$

[4] There are certain practical problems. If we open the switch in the original circuit, interrupting the current suddenly, the high value of dI/dt in the inductor would generate extremely high emf's that (by Kirchhoff's rule) would also appear across the gap between the switch contacts, creating a troublesome electric spark or arc. For this reason, care must always be taken to avoid arcing at switch contacts in circuits containing inductances.

 Our method of using a "shorting" switch that does not interrupt the current also has practical difficulties: some batteries would be severely damaged by a direct "short circuit," even for a second or so. However, once the switch S_2 is closed, we could immediately remove the battery without affecting the rest of the circuit. In this theoretical discussion, we ignore these annoying realities. But they must not be ignored in the laboratory!

back-emf generated in the inductor. From Kirchhoff's loop rule,

$$\Sigma \mathcal{E} = 0$$

$$IR + L\frac{dI}{dt} = 0 \qquad (32\text{-}27)$$

We solve this equation (see Problem 32C-43) by using the same method employed for the solution of Equation (32-25), leading to

DECAY OF CURRENT IN AN *RL* CIRCUIT
$$I = \frac{\mathcal{E}}{R} e^{-(R/L)t} \qquad (32\text{-}28)$$

The initial current $I_0 = \mathcal{E}/R$ drops to $1/e$ ($\sim 37\%$) of its initial value in *one* time constant: $\tau_L = L/R$. A graph of this equation is Figure 32-17b.

The exponential growth and decay of the current in *RL* circuits is similar to the exponential changes occurring in *RC* circuits. It will be helpful if you review the discussion of *RC* circuits (Section 29.6) to clarify these similarities. It is always easier to remember facts if one can relate them to similar behavior in other situations.

The general behavior of an inductor in a series *RL* circuit (with a constant-voltage source) is as follows:

(1) *The current through an inductor cannot change instantaneously. The rapidity of the exponential changes that occur are governed by the L/R time constant of the current path.*

(2) *After steady-state conditions have been reached, the voltage across a "pure" inductance is always zero.*

EXAMPLE 32-8

Consider the circuit in Figure 32-19. Find the steady-state currents in R_1, R_2, and R_3.

SOLUTION

Because of the capacitor, there can be no steady current in that branch, so $\boxed{I_2 = 0}$. (That branch is effectively an "open circuit.")

Because the steady-state current through L is constant, $\mathcal{E}_L = 0$ and the current in that branch is

$$I_3 = \frac{\mathcal{E}}{(R_1 + R_3)} = \frac{9\text{ V}}{(3\text{ k}\Omega + 2\text{ k}\Omega)} = \boxed{1.80\text{ mA}}$$

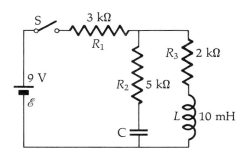

FIGURE 32-19 Example 32-8.

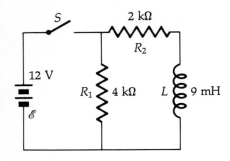

FIGURE 32-20
Example 32-9.

EXAMPLE 32-9

In Figure 32-20, the switch is closed and a steady current is established in the inductor. The switch is now opened at $t = 0$. (a) Find the initial voltage $(\mathscr{E}_L)_0$ across L just after the switch is opened. (b) How long does the current take to decrease to one-sixth of its initial value?

SOLUTION

(a) After a steady current is established with the switch closed, there is no voltage across L so the current in that branch is limited only by R_2:

$$I_2 = \frac{\mathscr{E}}{R_2} = \frac{12 \text{ V}}{2 \text{ k}\Omega} = \boxed{6.00 \text{ mA}}$$

Just after the switch is opened, the current in L must initially have the same value it had before the switch was opened: 6.00 mA. (Recall that the current in an inductor cannot change instantaneously.) To produce this current in the loop containing R_1 and R_2, the initial back-emf across the inductor must therefore be

$$(\mathscr{E}_L)_0 = IR = (6.00 \text{ mA})(6 \text{ k}\Omega) = \boxed{36.0 \text{ V}}$$

The polarity of this induced emf opposes the *change* of current, so the bottom end of the coil is positive with respect to the top end, trying to maintain the current in the same direction. (See Problem 32C-43.)

(b) The time constant for this current path is $L/(R_1 + R_2) = 9 \text{ mH}/6 \text{ k}\Omega = 1.50 \text{ } \mu\text{s}$. Thus, the time for the current to drop to one-sixth of its initial value is found from

$$I = I_0 e^{-(R/L)t}$$
$$\frac{1}{6} = e^{-(R/L)t}$$
$$\ln 6 = \left(\frac{R}{L}\right)t$$
$$t = \left(\frac{L}{R}\right)\ln 6 = (1.50 \text{ } \mu\text{s})(\ln 6) = \boxed{2.69 \text{ } \mu\text{s}}$$

32.9 Energy in Inductors

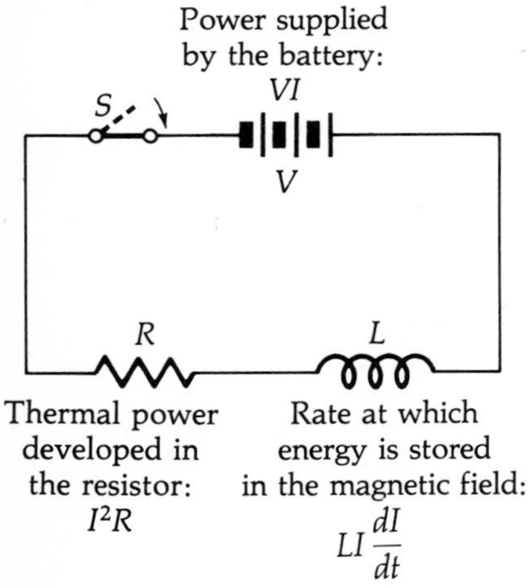

To find the energy stored within a current-carrying inductor, we apply Kirchhoff's loop rule to Figure 32-21, then multiply by I and rearrange:

$$\Sigma \mathscr{E} = 0$$
$$\mathscr{E} - IR - L\frac{dI}{dt} = 0$$
$$\mathscr{E}I = I^2 R + LI\frac{dI}{dt} \tag{32-29}$$

where
$\mathscr{E}I$ = the power supplied by the battery
$I^2 R$ = the thermal power dissipated in the resistor
$LI\dfrac{dI}{dt}$ = the rate at which energy is stored in the inductor

FIGURE 32-21
The switch is closed at $t = 0$. As the current rises toward its steady state value, part of the power supplied by the battery appears as energy stored in the inductor, part as thermal energy developed in R.

Let U_L represent the energy stored in the inductor. Then

$$\frac{dU_L}{dt} = LI\frac{dI}{dt}$$

or
$$dU_L = LI\,dI$$

Since $U_L = 0$ when $I = 0$, we integrate this equation to obtain

ENERGY STORED IN AN INDUCTOR
$$U_L = \tfrac{1}{2}LI^2 \qquad (32\text{-}30)$$

Note the similarity in form between this expression and the energy-storage equation for a capacitor:

$$U_C = \tfrac{1}{2}CV^2 \qquad (32\text{-}31)$$

An interesting difference between energy storage in a capacitor and in an inductor is that a charged capacitor may be removed from the circuit retaining its stored energy, whereas an inductor can retain its stored energy only by maintaining a current through it.

Recall that starting with the expression for the energy stored in a capacitor, $U_C = \tfrac{1}{2}CV^2$, we obtained an expression for the energy density u_E in an electric field:

$$u_E = \tfrac{1}{2}\varepsilon_0 E^2 \qquad (32\text{-}32)$$

We now follow a similar procedure for magnetic fields. The inductor that we will consider is a large toroid that we form from a long solenoid bent into a circle with the ends joined, Figure 32-22. The turns are tightly wound so that all of the magnetic field is confined inside the turns and thus we know the volume that the field occupies. If the radius of the circle is large compared with the radius of the turns [so that R in Equation (31-6) doesn't vary much across the windings], then the field B inside the windings is essentially uniform and the same as that in a long straight solenoid, Equation (31-6):

$$L = \frac{\mu_0 N^2 A}{\ell} \qquad \text{and} \qquad B = \frac{\mu_0 NI}{\ell}$$

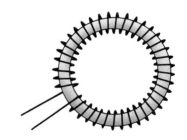

FIGURE 32-22
A tightly wound coil in the shape of a *torus* (or doughnut) forms a toroidal coil. If the radius of the torus is very large compared with the radius of the turns, the magnetic field inside the windings is essentially uniform and the same as that in a long straight solenoid.

where ℓ is the average circumferential length around the toroid. Solving for I and subsituting these values in Equation (32-30), we have

$$U_L = \frac{1}{2}LI^2 = \frac{1}{2}\left(\frac{\mu_0 N^2 A}{\ell}\right)\left(\frac{B\ell}{\mu_0 N}\right)^2 = \frac{1}{2}\left(\frac{B^2}{\mu_0}\right)(A\ell)$$

But $A\ell$ is the volume inside the windings containing the magnetic field. So the *energy per unit volume*, or **energy density** u_B of the magnetic field, is

ENERGY DENSITY u_B IN A MAGNETIC FIELD
$$u_B = \frac{1}{2\mu_0}B^2 \qquad (32\text{-}33)$$

The units of energy density are joules per cubic meter (J/m³). Though for ease of calculation we used a particular configuration for the inductor, the result is perfectly general and applies to any magnetic field B. Note the similarity[5] to the energy density in an electric field:

$$u_E = \tfrac{1}{2}\varepsilon_0 E^2$$

EXAMPLE 32-10

Find the energy stored in the gap of a permanent magnet such as the one illustrated in the previous chapter in Figure 31-10a. Assume that the field is uniform within the gap and equal to 0.5 T. The volume of the gap is 2 cm^3.

SOLUTION

From Equation (32-33), the energy density u_B is

$$u_B = \frac{1}{2\mu_0} B^2 = \frac{(0.5\ \text{T})^2}{(2)(4\pi \times 10^{-7}\ \text{H/m})} = 9.95 \times 10^4\ \text{J/m}^3$$

The total energy U_B stored in the volume V is

$$U_B = u_B V = \left(9.95 \times 10^4\ \frac{\text{J}}{\text{m}^3}\right)(2.0 \times 10^{-6}\ \text{m}^3) = \boxed{0.199\ \text{J}}$$

While it is possible to extract the energy associated with an electric field, no practical method has yet been devised to extract the considerable energy associated with the magnetic field of a permanent magnet. It is interesting to speculate about the vast reservoir of magnetic-field energy associated with the earth's magnetic field.

---

[5] Compare analogous equations for electric and magnetic fields. If ε_0 appears in the *numerator* in one equation, you will discover that μ_0 is in the *denominator* of the other equation, and vice versa.

Summary

Faraday's law:

$$\mathscr{E}_L = -N\frac{d\Phi_B}{dt}$$

or

$$\oint \mathbf{E} \cdot d\boldsymbol{\ell} = -\frac{d}{dt}\int \mathbf{B} \cdot d\mathbf{A}$$

Lenz's law: The induced current in a closed loop is in a direction so as to oppose the *change* in the flux linkages that produced it.

A *motional emf* is produced by the motion of a conductor in the presence of a (stationary) magnetic field B:

$$\mathscr{E} = -B\ell v \quad \begin{pmatrix}\text{when } \mathbf{B}, \boldsymbol{\ell}, \text{ and } \mathbf{v} \text{ are}\\ \text{mutually perpendicular}\end{pmatrix}$$

Inductance L (also called *self-inductance*):

$$\mathscr{E} = -L\frac{dI}{dt}$$

The inductance of a *long solenoid* with a length ℓ or a *large toroid* with a circumference ℓ:

$$L = \frac{\mu_0 N^2 A}{\ell} \quad (N = \text{total number of turns})$$

Mutual inductance M:

$$\mathscr{E}_1 = -M\frac{dI_2}{dt} \quad \text{and} \quad \mathscr{E}_2 = -M\frac{dI_1}{dt}$$

RL circuit: In a series combination of a seat of emf with terminal voltage V, a resistor R, and an inductor L, the current is

Growth: $\quad I = \dfrac{\mathscr{E}}{R}(1 - e^{-(R/L)t})$

Decay: $\quad I = \dfrac{\mathscr{E}}{R} e^{-(R/L)t}$

In one *time constant*, $\tau_L = L/R$, the growing exponential rises to $\sim 63\%$ of its final (maximum) value and the decaying exponential falls to $\sim 37\%$ of its original value.

Energy stored in a current-carrying inductor:

$$U_L = \tfrac{1}{2}LI^2 \quad \text{(in joules)}$$

Energy density in a magnetic field:

$$u_B = \frac{1}{2}\left(\frac{B^2}{\mu_0}\right) \quad \text{(in joules/meter}^3)$$

The general behavior of an inductor in a series RL circuit (with a constant-voltage source) is

(1) The current through an inductor cannot change instantaneously. The rapidity of the exponential changes that occur are governed by the L/R time constant of the current path.
(2) After steady-state conditions have been reached, the voltage across a "pure" inductance is always zero.

Questions

1. Where and in which direction would an airplane fly so that the earth's magnetic field would produce the greatest potential difference between the wing tips, with the right tip positive with respect to the left tip?
2. Can electric field lines form closed loops as well as originate on charges? Explain.
3. What is the connection between Lenz's law and the conservation of energy?
4. Is the net magnetic flux through a closed surface surrounding the north pole of a magnet zero? Is the net electric field flux through a surface surrounding the positive charge of an electric dipole zero?
5. A toroid inductor has essentially no external magnetic field. However, there is a small, unavoidable external field. Describe this field and its origin.
6. An airplane with a metal propeller is flying along the direction of the magnetic field lines of the earth. (a) Is there a potential difference between the tips of the propeller blades? (b) Is there a potential difference between the propeller hub and the propeller tips?
7. Two identical, rectangular loops of wire are situated so that the plane of each loop is perpendicular to a uniform magnetic field **B**. Loop A is then rotated with angular velocity ω about a central axis parallel to the longer side of the loop, while loop B is rotated with the same angular velocity about a central axis parallel to the shorter side. Is the peak value of the induced emf in loop A greater than, equal to, or less than the peak emf induced in loop B? Is the answer the same if the axes are coincident with the sides of the loops (rather than passing through their centers)?
8. Why does increased resistance increase the time constant of an RC circuit, while it decreases the time constant of an RL circuit?
9. An isolated long, straight wire does have inductance. How could its inductance be calculated? (Hint: consider the flux linkages inside the wire itself when it carries a current density J.)
10. Here is an amusing demonstration. A (nonferrous) rigid metal sheet such as aluminum is placed on a horizontal surface in a strong magnetic field. If the sheet is initially oriented with its plane almost (but not quite) vertical, and released, it falls over in "slow motion," taking several seconds to fall. Explain.
11. Compare the energy density stored in the electric field of a capacitor and in the magnetic field of a solenoid, for typical cases that are easily achieved in the laboratory.
12. When we wind a coil of resistance wire to form a resistor that has very little inductance, half the length of wire is wound in one direction and the other half in the opposite direction. Explain why such a resistor has negligible inductance.

Problems

32.2 Faraday's Law
32.4 Lenz's Law

32B-1 A circular hoop is linked through a toroidal coil as shown in Figure 32-23. The switch is closed to produce a surge of current through the coil. (a) Calculate the induced emf in the hoop when the magnetic flux within the coil is changing at the rate of 30 T·m²/s. Determine the direction of the induced current in the hoop. (b) The magnetic field produced by an ideal toroid is totally contained inside the windings of the toroid. That is, none of the field inside the toroid touches the hoop. What causes the induced current in the hoop?

32B-2 A flexible wire forms a circular loop 20 cm in diameter. It lies on a horizontal surface in the presence of a uniform magnetic field $B = 0.7$ T directed vertically upward. Opposite

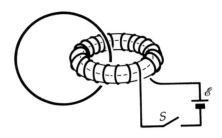

FIGURE 32-23
Problem 32B-1.

points of the loop are now rapidly pulled apart, collapsing the area of the loop to zero in 0.06 s. (a) Find the average emf induced in the loop. (b) As viewed from above, is the induced current in the collapsing loop clockwise or counterclockwise?

32B-3 A circular wire loop of radius r and resistance R lies in a plane perpendicular to a uniform magnetic field **B**. The loop is rapidly turned over (by 180°) in a time t. Find the average emf $\mathscr{E}$ induced in the loop during the time t.

32B-4 An airplane with a wingspan of 70 m is flying horizontally at 1000 km/h toward the north magnetic pole of the earth. If the vertical component of the earth's magnetic field at the airplane's position is 2×10^{-5} T, calculate the potential difference V between the wing tips. Which wing tip is at the higher potential? Explain why this potential difference cannot be used as a source of power.

32B-5 A rectangular wire loop of mass m, total resistance R, and dimensions as shown in Figure 32-24 is falling freely under gravity as it emerges from a region of uniform, horizontal magnetic field B. The plane of the loop is perpendicular to B. (a) Is the induced current in the loop clockwise or counterclockwise? (b) At a certain speed v, the loop falls without acceleration while emerging from the field. Show that this speed is $v = mgR/B^2a^2$.

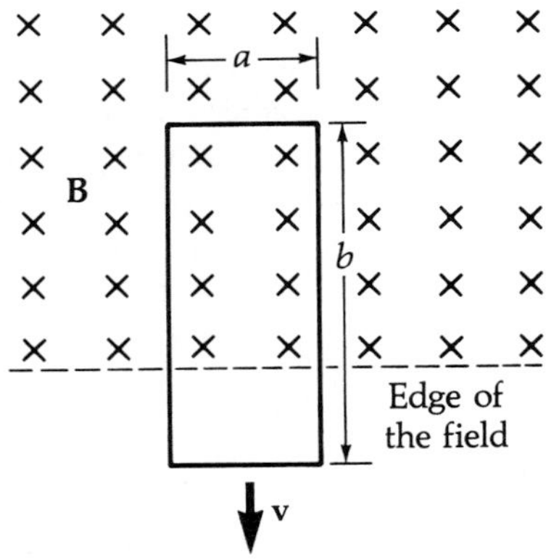

FIGURE 32-24
Problem 32B-5.

32B-6 The cube in Figure 32-25 is 50 cm along each edge and is situated in a uniform magnetic field $B = 0.3$ T directed along the $+z$ direction. One at a time, four wire segments 1, 2, 3, and 4 are moved in the directions shown with speed $v = 2$ m/s. (a) Find the motional emf generated in each wire and tabulate the values in the order that the wires are numbered. (b) Make a sketch of the figure and indicate with $+$ and $-$ signs the polarities of the induced potential differences between the ends of each wire.

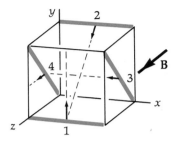

FIGURE 32-25
Problem 32B-6.

32B-7 A 30-turn flat coil of wire is placed at the end of a long solenoid wound with 4000 turns/m. The coil and solenoid have the same radius, $R = 5$ cm, and their axis are coincident. Find the rate of change of current in the solenoid if there is an induced emf of 2 mV in the coil.

32.6 Self Inductance

32A-8 A back-emf of 28 mV is produced in a 400-turn coil when the current changes at the rate of 12 A/s. Find the inductance of the coil.

32A-9 Beginning with the basic definitions of inductance L and resistance R, show that L/R has the dimensions of time.

32A-10 Ignoring end effects, find the inductance of a 1200-turn solenoid, 39 cm long, with a diameter of 3 cm.

32A-11 The field B at the center of a current-carrying circular loop of wire is [from Equation (31-4)] $B = \mu_0 I/2R$, where R is the radius of the loop. Assume that the field has this value uniformly over the plane area bounded by the loop and estimate the inductance of a flat coil of N turns, radius R.

32B-12 The current in a 12-mH inductor that has negligible resistance varies with time according to the sawtooth waveform shown in Figure 32-26. Make a graph (with numerical values) of the voltage across the inductor as a function of time.

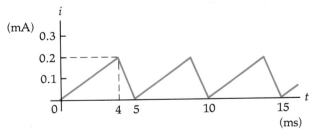

FIGURE 32-26
Problem 32B-12.

32B-13 The current in a 90-mH inductor changes with time as $I = t^2 - 6t$ (in SI units). Find the magnitude of the induced emf at (a) $t = 1$ s and (b) $t = 4$ s. (c) At what time is the emf zero?

32B-14 A time-varying current I is applied to an inductance of 5 H, as shown in Figure 32-27. Make a quantitative graph of the potential of point a relative to that at point b. The current arrow indicates the direction of conventional current.

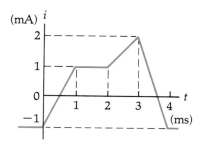
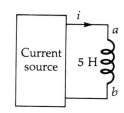

FIGURE 32-27
Problem 32B-14.

32.7 Mutual Inductance

32A-15 A toroidal solenoid has two separate sets of windings that are each spread uniformly around the toroid, with total turns N_1 and N_2, respectively. The toroid has a circumferential length ℓ and a cross-sectional area A. (a) Write expressions for the self-inductances L_1 and L_2, respectively, when each coil is used alone. (b) Derive an expression for the mutual inductance M of the two coils. (c) Show that $M^2 = L_1 L_2$. (This expression is true only when all the flux linking one coil also links the other coil.)

32A-16 Two coils, A and B, are close enough to each other to have mutual inductance. When the current in coil A is changing at the rate of 1.8 A/s, the emf induced in coil B is 24 mV. (a) Find their mutual inductance. (b) What rate of change of current in coil B will induce an emf of 30 mV in coil A?

32B-17 A long solenoid of length ℓ and cross-sectional area A contains a total of N_1 turns. A second coil of N_2 turns is closely wound around the center of the solenoid (keeping the two coils electrically insulated from each other). Find the mutual inductance M between the coil and the solenoid, ignoring end effects.

32.8 RL Circuits

32A-18 A seat of emf $\mathscr{E} = 10$ V is in a series circuit with a switch S, a resistance $R = 50\,\Omega$, and an inductance $L = 5$ H. Find the time after the switch is closed for the current to reach (a) half its final value and (b) 99% of its final value.

32B-19 Consider Equation (32-28) for the decay of current in an RL circuit. (a) Find the initial slope of the decreasing current graph. (b) Show that, if this initial rate of decrease were to continue at a constant rate (rather than to decrease exponentially), the current would reach zero in one time constant.

32B-20 A battery is in series with a switch and a 2-H inductor whose windings have a resistance R. After the switch is closed, the current rises to 80% of its final value in 0.4 s. Find the value of R.

32B-21 Verify by direct substitution that the statement $I = (\mathscr{E}/R)(1 - e^{-(R/L)t})$ is a solution of the differential equation $\mathscr{E} - IR - L\,dI/dt = 0$.

32.9 Energy in Inductors

32A-22 Find the total energy stored in a toroidal solenoid of 800 turns, circumferential length 44 cm, cross-sectional area 10 cm^2, carrying a current of 3 A.

32A-23 Calculate the energy density in the magnetic field near the center of a long solenoid that has 3800 turns/m when carrying a current of 4 A. Does the energy density depend upon the radius of the turns?

32A-24 A 60-V emf is connected across a series combination of a 40-Ω resistor and a 90-mH inductor. Find the magnetic energy stored in the inductor when the current has risen to three-fourths of its steady-state value.

32B-25 A 10-V battery, a 5-Ω resistor, and a 10-H inductor are connected in series. After the current in the circuit has reached its maximum value, calculate (a) the power supplied to the circuit by the battery, (b) the power dissipated in the resistor, (c) the power dissipated in the inductor, and (d) the energy stored in the magnetic field of the inductor.

32B-26 At $t = 0$, a source of emf, $\mathscr{E} = 500$ V, is applied to a coil that has an inductance of 0.80 H and a resistance of 30 Ω. (a) Find the energy stored in the magnetic field when the current reaches half its maximum value. (b) How long after the emf is connected does it take for the current to reach this value?

Additional Problems

32C-27 A thin metal rod of length 0.8 m falls from rest under the action of gravity. It remains horizontal with its length oriented along the magnetic east-west direction. At this location, the earth's magnetic field $\mathbf{B}$ has a magnitude of 5×10^{-5} T and a downward direction at 70° below the horizontal (the "dip" angle). (a) Find the induced emf in the rod after it falls 8 m. (b) Which end of the rod has the higher potential?

32C-28 To monitor the breathing of a hospital patient, a thin belt is girded about the patient's chest. The belt is a 200-turn coil. During inhalation, the area within the coil increases by 39 cm^2. The earth's magnetic field is 50 μT and makes an angle of 28° with the plane of the coil. If a patient takes 1.80 s to inhale, find the average induced emf in the coil while the patient is inhaling.

32C-29 An automobile has a vertical radio antenna 1.2 m long. The automobile travels at 65 km/h on a horizontal road where the earth's magnetic field is 50 μT directed downward (toward the north) at an angle of 65° below the horizontal. (a) Specify the direction that the automobile should move in order to generate the maximum motional emf in the antenna, with the top of the antenna positive relative to the bottom. (b) Calculate the magnitude of this induced emf.

32C-30 In Figure 32-28, the rolling axle, 1.5 m long, is pushed along horizontal rails at a constant speed $v = 3$ m/s. A resistor $R = 0.4\,\Omega$ is connected to the rails at points A and

B, directly opposite each other. (The wheels make good electrical contact with the rails, so the axle, rails, and R form a complete, closed-loop circuit. The only significant resistance in the circuit is R.) There is a uniform magnetic field $\mathbf{B} = 0.08$ T vertically downward. (a) Find the induced current I in the resistor. (b) What horizontal force F is required to keep the axle rolling at constant speed? (c) Which end of the resistor, A or B, is at the higher electric potential? (d) After the axle rolls past the resistor, does the current in R reverse direction?

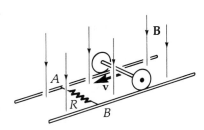

FIGURE 32-28
Problem 32C-30.

32C-31 *The search coil.* One method for determining the strength of a magnetic field B is to place a small, closely wound coil in the field, with the plane of the coil perpendicular to the field lines. When the coil is quickly withdrawn from the field, the sudden change of flux linkages induces an emf in the coil, producing a momentary current in the external circuit connected to the coil (Figure 32-29). (a) As the coil is removed, what is the direction of the current in R? (b) From Faraday's law, the average induced emf is $\mathscr{E} = (-)N\Delta\Phi/\Delta t$ and the induced current is thus $I = \mathscr{E}/R$. In terms of N, R, and Φ, find the total charge Q passing through the resistor. (Hint: recall that $I = \Delta Q/\Delta t$.) (c) Obtain an expression for the magnetic field B in terms of R, N, A (the coil area), and Q (the total charge). **Comment:** note that B is proportional to Q, the total charge passing through R. Let R be the resistance of a galvanometer movement that has a relatively large moment of inertia. As long as the total charge passes through the meter movement before it deflects appreciably, the angular impulse that the movement receives will cause a deflection proportional to the total charge Q (and thus to the field B). Importantly, the exact time it takes to remove the search coil from the field is not crucial as long as it is "sudden." A galvanometer used this way is called a *ballistic galvanometer.*

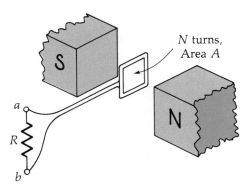

FIGURE 32-29
Problem 32C-31.

32C-32 A metal axle with conducting wheels rolls down a pair of inclined metal rails, as shown in Figure 32-30. A uniform magnetic field $\mathbf{B}$ is vertically upward between the rails. A battery $\mathscr{E}$ is connected to the rails with the polarity indicated. The axle moving in the field $\mathbf{B}$ generates a current and the system acts as a battery charger, forcing current into the $+$ terminal. A constant speed v eventually results when $|dU_g/dt|$ = the power input to the battery. Find v in terms of $\mathscr{E}$, B, ℓ, and α, the angle of the incline.

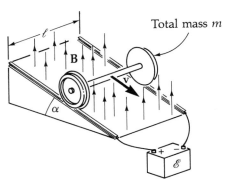

FIGURE 32-30
Problem 32C-32.

32C-33 A circular hoop of wire 30 cm in diameter has a resistance of 2 Ω. The hoop rests on a table at a location where the earth's magnetic field is 48 μT at an angle 65° below the horizontal. Calculate the net charge that passes a given point on the hoop while it is suddenly flipped over by 180°.

32C-34 A thin, horizontal metal rod 40 cm long is rotated at 6 rev/s about a vertical axis through one end. A uniform magnetic field of 0.20 T exists vertically upward. (a) What is the motional emf generated in the rod? (b) When viewed from above, the rotation is clockwise. What is the polarity of the potential difference between the ends of the rod? (c) Now suppose that the vertical axis of rotation is moved to the center of the rod, and the rod is rotated with the same angular velocity as before. For this new case, what is the motional emf generated in the rod? (d) Indicate polarities of the potential differences between the center of the rod and the ends. (e) What is the potential difference between opposite ends of the rod?

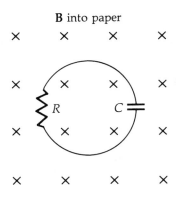

FIGURE 32-31
Problem 32C-35.

32C-35 In Figure 32-31, a uniform magnetic field decreases at a constant rate $dB/dt = -k$, where k is a positive constant. A circular loop of wire of radius a containing a resistance R and a capacitance C is placed with its plane normal to the field. (a) Find the charge Q on the capacitor when it is fully charged. (b) Which plate of the capacitor is at the higher potential? (c) Discuss the force that causes the separation of charges.

32C-36 Figure 32-32 shows a circular loop of radius r that has a resistance R spread uniformly throughout its length. The loop's plane is normal to the magnetic field B that decreases at a constant rate: $dB/dt = -k$, where k is a positive constant. (a) What is the direction of the induced current? (b) Find the value of the induced current. (c) Which point, a or b, is at the higher potential? Explain. (d) Discuss what force causes the current in the loop.

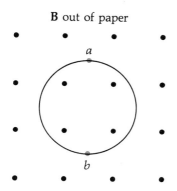

FIGURE 32-32
Problem 32C-36.

32C-37 Refer to Chapter 31, Figure 31-15 (Problem 31B-9). If the current in the straight conductor decreases uniformly from 10 A to 2 A in 2 s, find the induced current I' in the loop for the case when $\ell = 30$ cm.

32C-38 A circular loop of wire of area A and resistance R is held fixed with its plane normal to a magnetic field **B**. The field is then reduced from an initial value of B_0 so that it changes as a function of time according to $B = B_0 e^{-\alpha t}$, where α is a constant. (a) Sketch the loop, showing the magnetic field directed *into* the paper, and indicate on the diagram the direction of the induced current. (b) Do the electromagnetic forces associated with the induced current tend to make the loop expand, contract, or neither? (c) Derive an expression in terms of B_0, A, and R for the total quantity of charge Q that flows past a point in the loop during the time the field is reduced from B_0 to zero. (d) Derive an expression in terms of B_0, A, R, and α for the amount of thermal energy dissipated in the loop while the field is reduced from B_0 to zero.

32C-39 Consider two coaxial long solenoids, one inside the other. The inner solenoid has a radius R_1 and n_1 turns/m. The outer solenoid has a radius R_2 and n_2 turns/m. Show that the mutual inductance per unit length of the combination is given by $(M/\ell) = \mu_0 \pi n_1 n_2 R_1^2$.

32C-40 In Figure 32-33, the switch is closed and steady-state conditions are established in the circuit. The switch is now opened at $t = 0$. (a) Find the initial voltage $\mathscr{E}_0$ across L just after $t = 0$. Which end of the coil is at the higher potential: a or b? (b) Make freehand graphs of the currents in R_1 and in R_2 vs. t, treating the steady-state directions as positive. Show values before and after $t = 0$. (c) How long after $t = 0$ does the magnitude of the current in R_2 drop exponentially to 2 mA?

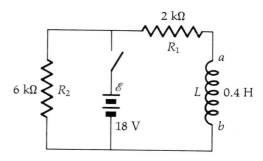

FIGURE 32-33
Problem 32C-40.

32C-41 Two inductors L_1 and L_2 are connected in series but are far enough apart so that the magnetic flux of one inductor does not link with the other inductor. (a) Show that the equivalent inductance of the combination is $L_1 + L_2$. (b) If the two inductors are close enough together so that they have a mutual inductance M, show that the combination has an equivalent inductance of $L_1 + L_2 \pm 2M$. Explain the reason for the $\pm$ sign.

32C-42 Refer to Example 32-9. By direct calculation of $\mathscr{E}_L = -L(dI/dt)$, verify that the initial back-emf induced in L just after opening the switch is 36.0 V.

32C-43 Carry out the solution of Equation (32-27) to obtain Equation (32-28). Include a circuit diagram showing polarities across R and L while the current is decreasing.

32C-44 A flat coil of wire has an inductance of 2 H and a resistance of 40 Ω. At $t = 0$, a battery of emf, $\mathscr{E} = 60$ V, is connected to the coil. Consider the state of affairs one time constant later. At this instant, find (a) the power delivered by the battery, (b) the Joule power developed in the resistance of the windings, and (c) the instantaneous rate at which energy is being stored in the magnetic field.

32C-45 A straight cylindrical conductor of radius R carries a steady current I that is distributed uniformly over a cross-sectional area of the conductor. Derive an expression for the total magnetic energy per unit length contained within the conductor. (Hint: what is the energy contained within a thin cylindrical shell of radius r ($<R$), thickness dr, and length ℓ? You may use the result of Problem 31B-13.)

32C-46 Repeat the previous problem for the case in which the current density J varies linearly with the distance r from the axis of the conductor: $J = J_0 r$. (a) Express the total current I in terms of J_0 and R. (b) Derive an expression for the total magnetic energy per unit length within the conductor.

CHAPTER 33

Magnetic Properties of Matter

The obedient [compass] steel with living instinct moves,
and veers forever to the pole it loves.

CHARLES DARWIN

33.1 Introduction

So far in our study of the magnetic fields produced by current-carrying conductors, we have assumed that the surrounding space was a vacuum. If matter is present, however, the magnetic field can be very different. Classically, we imagine electrons in atoms to undergo circulatory motions, creating microscopic magnetic-dipole fields of their own. In certain substances these dipoles can be aligned so that they contribute greatly to the resultant magnetic field.

A complete description of the magnetic effects of materials requires an understanding of quantum theory beyond the scope of this text. However, without delving too deeply into details, we will present a brief introduction to the three most familiar types of magnetic material behavior: paramagnetism, diamagnetism, and ferromagnetism.

33.2 Magnetic Properties of Materials

The origin of the magnetic properties of materials is within their atomic structures. For our purposes, we may consider an atom to be made up of a positively charged nucleus with electrons circulating in orbits about the nucleus. These microscopic current loops create magnetic dipole fields. In addition, we assume that each electron also "spins" about its own axis, similar to a spinning top, producing a "spin" magnetic dipole moment.[1] The resultant magnetic moment μ (Equation 30-14) of the atom is due partly to the orbital motions of the electrons and partly to their spins. There is a tendency for all the individual dipole moments within a single atom to combine in pairs, with opposite orientations, so that the net magnetic dipole moment for the atom as a whole can be zero.

---

[1] This model of a spinning electron is too mechanistic and should not be taken literally. The properties of spin are fully understandable only in the context of modern quantum theory. Nevertheless, this classical description of a spinning electron is useful as a first introduction to these ideas.

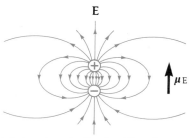
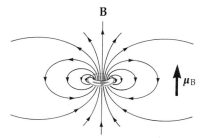

(a) An electric dipole. The electric field lines near the center point <u>opposite</u> to the direction of the electric dipole moment μ_E.

(b) A magnetic dipole formed by a current loop. The magnetic field lines near the center point <u>in</u> the direction of the magnetic dipole moment μ_B.

FIGURE 33-1
A comparison of electric and magnetic dipoles. Both dipole moments are pointing up. At distances far from the dipoles, the field patterns are identical. But near their centers, the lines of **B** and **E** are in opposite directions.

In other cases, however, the dipole moments do not exactly cancel. For example, atoms with an odd number of electrons will necessarily have an *unpaired* electron, resulting in a net magnetic moment.

Paramagnetism

For atoms that have a net dipole moment, thermal motions randomly orient their dipoles so that the bulk material has zero net dipole moment. However, as discussed in Chapter 30, in the presence of an external magnetic field the dipoles experience a torque that tends to align them parallel to the field. Depending on the field strength and the temperature (thermal agitation tends to misalign the dipoles), some materials thus exhibit a *net dipole moment* in the presence of a magnetic field. When the field is removed, thermal motions again randomize the orientation of individual atomic dipoles, and the material no longer possesses a net dipole moment. Substances that exhibit this property are called **paramagnetic**.

When magnetic dipoles are aligned, they add to the overall magnetic field, increasing its value slightly. We can see why this is so by comparing a magnetic dipole and an electric dipole, Figure 33-1. Although their far-field patterns are identical, in the regions near their centers the **B** and **E** field lines are in opposite directions. Thus, when electric dipoles are aligned in an **E** field (Section 27.3), their central field lines are *opposite* to the direction of **E**, resulting in a net reduction of the electric field within the material. But the central field lines of aligned magnetic dipoles in a **B** field are *in* the direction of **B**, resulting in a net increase of the magnetic field. The effect is small because thermal motions allow only a very small fraction of the magnetic dipoles to become aligned.

In the presence of a *non*uniform magnetic field, the dipoles experience a net force *that attracts them toward the region of the stronger field*. This is similar to the behavior of electric dipole moments. As shown in Figure 33-2a, they feel a net force toward the stronger-field region. While it is an oversimplification to imagine magnetic dipoles as tiny magnets with north and south poles as in Figure 33-2b, it does make the attractive effect understandable. The model of a dipole as a ring of current behaves similarly in a divergent field, Figure 33-3.

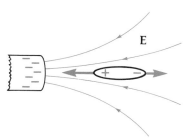

(a) In a nonuniform electric field, an electric dipole experiences a net force that pulls it into the stronger-field region.

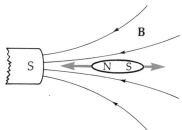

(b) In a nonuniform magnetic field, a magnetic dipole experiences a net force that pulls it into the stronger-field region.

FIGURE 33-2
Dipoles in nonuniform fields.

FIGURE 33-3
A current loop, free to move in the presence of a magnetic field **B**, will orient itself so that the magnetic dipole moment **μ** points in the direction of the field. If the field is nonuniform, the magnetic forces (shown in color—always perpendicular to **B**) on the current loop result in a net force toward the region of stronger field.

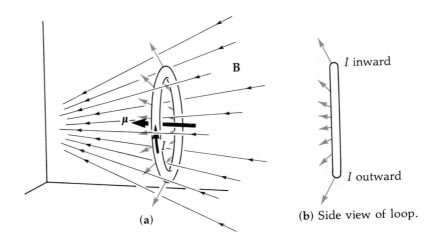

(a)

(b) Side view of loop.

EXAMPLE 33-1

In the Bohr model of an atom, an electron of charge $(-)e$ and mass m travels in a circular orbit of radius r with a speed v, thus forming a current loop. (a) Calculate the orbital magnetic moment μ_ℓ due to this orbital motion.[2] (b) Quantum theory says that the orbital angular momentum mvr is quantized such that it can only have integral multiples of $h/2\pi$, where h is Planck's constant (see Chapters 42 and 44). For an electron in the smallest orbit allowed by quantum theory—that is, one unit of $h/2\pi$—express the value of μ_ℓ in terms of h and m.

SOLUTION

(a) The orbital magnetic moment $\mu_\ell = IA$, where A is the area of the current loop. The current $I = q/T$, where T is the period of the motion: $T = 2\pi r/v$. Thus:

$$\mu_\ell = IA = \left(\frac{e}{2\pi r/v}\right)(\pi r^2) = \boxed{\frac{evr}{2}}$$

(b) The angular momentum $L = mvr = h/2\pi$. Thus:

$$\mu_\ell = \frac{evr}{2} = \left(\frac{e}{2m}\right)(mvr) = \boxed{\frac{eh}{4\pi m}}$$

This quantity is called the *Bohr magneton*, the fundamental unit of magnetic moment in atomic theory.

Diamagnetism

A few elements are *repelled* by a permanent magent. Such materials are called **diamagnetic**. Michael Faraday noticed this effect in bismuth; silver is also noticeably diamagnetic. The effect is quite weak. Elements whose atoms have zero net magnetic dipole moments are diamagnetic. Atoms that have a permanent dipole moment (and that are not *ferro*magnetic) may be either diamagnetic or paramagnetic, depending on which effect is stronger. To understand the

---

[2] The notation μ_ℓ for the *orbital* magnetic moment (and μ_s for the *spin* magnetic moment) agrees with the notation in modern quantum theory.

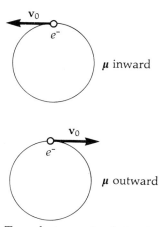
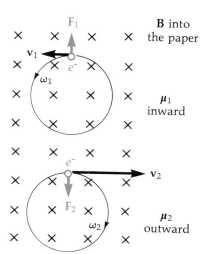

(a) Two electrons circulating in opposite directions with the same speed. The net dipole moment is zero because both dipole moments are equal in magnitude but opposite in direction.

(b) The external magnetic field does not change the size of the orbits, but it produces the additional forces F_1 and F_2 so that $v_2 > v_1$. Consequently, $\mu_2 > \mu_1$, making a net dipole moment opposite to the external field.

FIGURE 33-4
The origin of diamagnetism.

phenomenon, imagine a simple (classical) atomic model of circulating electrons held in their orbits by the electrostatic attraction of the positively charged nucleus. In diamagnetic atoms, some electrons are circulating in one direction, some in the other, with the result that their dipole moments cancel. Consider two electrons circulating in opposite directions with the same speed as shown in Figure 33-4a. (For clarity, their centers of rotation have been separated.) Because the electrons circulate in opposite directions, their combined dipole moment is zero. When an external magnetic field is applied, the circulating electrons experience an additional radial force: $\mathbf{F} = q(\mathbf{v} \times \mathbf{B})$. In one case this force *adds* to the radially inward electrostatic force, and in the other case it opposes the inward electrostatic force. It can be shown that the radius of the orbit remains the same. But the change in centripetal force ($mr\omega^2$) changes the angular velocity ω of the electron. In one case the increased speed of the circulating electron makes a larger magnetic moment *opposite* to **B**, while in the other case the slower speed makes a smaller magnetic moment *in* the direction of **B**. *Both effects result in a net induced dipole moment that is opposite in direction to the applied field* **B** (rather than *in* the field direction as in paramagnetism). Because the induced dipole moment of the material opposes the field, when placed in the nonuniform field of a nearby permanent magnet field, the material is *repelled* away from the magnet—it is diamagnetic. Since all matter contains atoms, all substances experience this *diamagnetic* effect. However, if permanent dipoles are present, this diamagnetic behavior is usually overwhelmed by the effect of the permanent dipoles and the substance is attracted toward stronger fields—the material is paramagnetic or ferromagnetic.

Ferromagnetism

The third class of magnetic materials contains five elements: iron, cobalt, nickel, gadolinium, and dysprosium as well as some alloys made from them. These are the **ferromagnetic** materials, whose magnetic effects are orders of magnitude greater than those of paramagnetic or diamagnetic substances. The basic

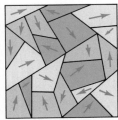

(a) Random orientation of domains in a polycrystalline specimen. Though not shown in this simplified sketch, each individual region may contain several domains oriented in different directions. The net magnetic moment for the specimen is zero.

FIGURE 33-5
Magnetization of a ferromagnetic substance.

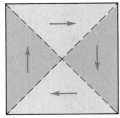

(b) An arrangement of domains within a single crystal that results in zero net magnetic moment.

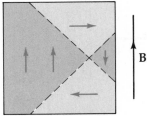

(c) When an external magnetic field **B** is applied, the volume of domains whose moments are favorably oriented in the direction of **B** will grow at the expense of adjacent regions whose moments are oriented in other directions. As a result, the domain walls shift as shown. The crystal now has a net magnetic moment in the direction of **B**.

distinction is that, because of quantum mechanical effects, the dipole moments of these ferromagnetic atoms exert forces on their neighbors, causing all the dipoles to align parallel to one another within a region called a *domain*. Magnetic domains may have volumes from about 10^{-6} cm^3 to 10^{-2} cm^3, so each may contain roughly 10^{17} to 10^{21} atoms. Boundaries between regions in which domains have different orientations are called *domain walls*. Although each domain is magnetized as strongly as it can be, neighboring domains may be aligned along different directions. In unmagnetized bulk material, the domains have sufficiently random orientations that there is no net magnetic moment, Figure 33-5a. But if an external field is applied, domains oriented in the field direction will grow in size by obtaining "converts" from adjacent, less favorably oriented domains, thus shifting the domain boundaries. The boundaries, or walls, between domains can be made visible by the technique described in Figure 33-6. If the field is strong enough, all the dipoles within a single domain may also suddenly "flip around" together to align themselves in the field direction.[3] When the external field is removed, much of the dipole moment of the bulk material remains because domains are not easily dislodged into random orientation by thermal agitation. Thus "magnetized" materials retain a permanent magnetic moment. Ferromagnetic materials are used to fabricate permanent magnets. Permalloy and Alnico are trade names of well-known examples of certain aluminum–nickel–cobalt–iron alloys that retain a high degree of permanent magnetic moment.

Each ferromagnetic material has a critical temperature, called the **Curie temperature**, above which the energies of thermal motions are great enough to upset the alignment of magnetic moments in a domain. (For iron, the Curie

---

[3] When a domain suddenly reorients its direction of magnetization, the domain as a whole does not rotate as a unit. Rather, as a result of quantum mechanical forces, almost simultaneously each *atom* within the domain reorients its magnetic moment in the new direction. If a coil of wire is wound around the material and connected to a sensitive amplifier and loudspeaker, the sudden, slight changes of flux in the coil as each domain flips are detected as tiny "ticks" in the amplifier output. This is known as the *Barkhausen effect*, after the experimenter who first discovered it in 1919.

temperature is 770°C.) Above the Curie temperature, the material becomes paramagnetic, and at still higher temperatures even paramagnetism disappears and substances become diamagnetic.

33.3 B and H

Having described three basic types of magnetic materials—*para*magnetic, *dia*magnetic, and *ferro*magnetic—we now investigate the effect of placing a paramagnetic material inside the windings of a solenoid (Figure 33-7). Consider a long solenoid whose interior magnetic field (without the paramagnetic material) is $B_0 = \mu_0 nI$, Equation (31-7). Since $n = N/\ell$, the number of turns per unit length, the magnetic field B_0 inside, due only to the current I in the windings, is

$$B_0 = \mu_0 \frac{NI}{\ell}$$

This equation is for a vacuum (and, to a close approximation, an "air core"). With a material in the core, an additional magnetic field is created due to the oriented dipoles (paramagnetic materials) or the induced dipoles (diamagnetic materials). This added field B' is proportional to the original field $B_0 = \mu_0 NI/\ell$, produced by the current in the windings:

$$B' = \chi\left(\mu_0 \frac{NI}{\ell}\right) \quad (33\text{-}1)$$

where χ, the **magnetic susceptibility**, is the factor of proportionality. It is very small, roughly 10^{-5}, and is positive for paramagnetism and negative for diamagnetism. The total field B is thus

$$B = B_0 + B'$$
$$B = \left(\mu_0 \frac{NI}{\ell}\right) + \chi\left(\mu_0 \frac{NI}{\ell}\right)$$

or, simply
$$B = \mu_0(1 + \chi)H \quad (33\text{-}2)$$

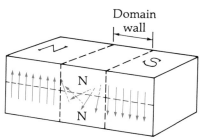

(a) As the alignment of dipoles in one domain shifts to a new direction in an adjacent domain, the change is gradual, with a transition region several hundred atoms thick in which dipoles point outward from the surface. Consequently, at these <u>walls</u> or <u>boundaries</u> between domains, a localized, intense magnetic field bulges outward from the surface. If a thin colloidal suspension of finely powdered iron oxide is spread on the surface, the walls become visible when the powder particles are attracted to the regions of intense fields protruding from the surface.

(b) Domain wall patterns for a single crystal of iron containing 3.8% silicon.

FIGURE 33-6
Magnetic domains.

(a) The current in one loop of a solenoid with an air core produces the field $\mathbf{B}_0$.

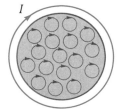

(b) A solenoid with a paramagnetic material in the core. The field $\mathbf{B}_0$ aligns magnetic moments of the paramagnetic material.

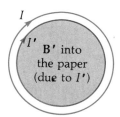

(c) The aligned moments result in an effective current I' around the outside of the paramagnetic material, producing a field B' in the same direction as $\mathbf{B}_0$.

FIGURE 33-7
The cross-section of a solenoid showing the effect of adding a paramagnetic material within its windings.

TABLE 33-1 Magnetic Susceptibilities, χ (at 20°C unless otherwise noted)

Material	χ	
Aluminum	2.1×10^{-5}	**Paramagnetic and Diamagnetic** (Negative values indicate diamagnetism.)
Air (STP)	0.036×10^{-5}	
Bismuth	$-17. \times 10^{-5}$	
Lead	-1.7×10^{-5}	
Silver	-2.6×10^{-5}	
Liquid oxygen (90 K)	$400. \times 10^{-5}$	
Cold rolled steel	2000	**Ferromagnetic** *Maximum saturation values.* (Varies widely depending on prior magnetization, the value of H, heat treatment, purity, and the history of mechanical stress.)
Iron	5000	
45 Permalloy	25 000	
Mu-Metal	100 000	
Supermalloy	800 000	

where

MAGNETIC FIELD INTENSITY H
(here expressed for a long solenoid)

$$H = \frac{NI}{\ell} \quad (33\text{-}3)$$

The symbol H is called the **magnetic field intensity**[4] in units of ampere·turns/meter (A·turns/m). The number of turns N is a dimensionless number.

We can write Equation (33-2) in a more convenient notation that recognizes B and H have directions and can thus be written as vectors:

RELATION BETWEEN B AND H

$$\mathbf{B} = \mu \mathbf{H} \quad (33\text{-}4)$$

where

$$\mu = \mu_0(1 + \chi) \quad (33\text{-}5)$$

The symbol μ is the **permeability** of the magnetic material and has the same units as μ_0. It includes the *permeability of free space*, μ_0, as well as the additional effects of paramagnetic and diamagnetic materials described by χ. The magnitude of χ is very small ($\sim 10^{-5}$), positive for paramagnetism and negative for diamagnetism, Table 33-1.

Although a relation similar to Equation (33-4) is sometimes written for ferromagnetic materials, a special problem arises in this case. For ferromagnetic substances, the value of χ is not constant but depends strongly on H as well as on the substance's prior history of magnetization. (See the discussion of *hysteresis* below.) Indeed, because of the hysteresis of a ferromagnetic substance, χ may be zero or may range to extremely large positive or negative values,

---

[4] At last our rather loose terminology catches up with us. In calling B the *magnetic field* (rather than its precise name, *magnetic induction* or *magnetic flux density*), we are following a very widespread practice. Originally, H was defined as the "magnetic field intensity." It has been proposed that these quantities be redefined to conform more to usage, but as yet the change has not occurred. In the meantime, be careful to keep B and H distinct: they *are* different. Our discussion of H is only in relation to a long solenoid. For a more fundamental definition, see a more advanced text.

up to several hundred thousand. The ferromagnetic entries in Table 33-1 show only the maximum positive values. The use of ferromagnetic materials in electromagnets, transformers, etc., immensely increases the magnetic field that can be generated by a given current in a given set of windings. As someone has remarked, "**H** is what you pay for, **B** is what you get!"

EXAMPLE 33-2

Suppose that the air core of a long solenoid is filled with iron, increasing the magnetic field inside the solenoid from its original value B_0 to B. Find the ratio B/B_0, assuming that for this value of H the susceptibility of the iron is one-quarter of its maximum value.

SOLUTION

Without the iron core, the magnetic field is $B_0 = \mu_0 H$. With the iron, it is $B = \mu_0(1 + \chi)H$. The ratio B/B_0 is thus

$$\frac{B}{B_0} = \frac{\mu_0(1 + \chi)H}{\mu_0 H} = (1 + \chi)$$

From Table 33-1, we note the maximum value of χ is 5000. One-quarter of this value is 5000/4, or 1250. Therefore,

$$\frac{B}{B_0} = (1 + \chi) = (1 + 1250) \cong \boxed{1250}$$

The use of the iron core greatly increases the magnetic flux density B inside the coil.

33.4 Hysteresis

When a ferromagnetic substance is placed in a magnetic field, a variety of interesting effects occur. The result can be quite complex since the value of χ (which describes the degree of alignment of domains in the material) depends not only on H but also on the previous history of magnetization, the prior heat treatment of the material, mechanical stresses, and other factors. Suppose we place a piece of iron with randomly oriented domains inside the windings of a solenoid. As we increase the current in the windings, we start at point a in the "B–H graph" of Figure 33-8. The parameter H is proportional to the solenoid current $H = \mu_0 NI/\ell$. As orientation of domains occurs, the net field **B** increases as shown. The curve levels off at point b, however, because of *saturation*: the majority of domains has become oriented, in the "proper" direction. If the saturation is 100%, a further increase of current would increase **B** only slightly through the $\mu_0 H$ term of Equation (33-2). If the current is now reduced to zero, the graph follows a different path to point c because some domains remain permanently oriented. The material is now a "permanent" magnet. The fact that the material does not retrace the original magnetizing curve is called **hysteresis**, from a Greek root meaning "to lag behind." Increasing the magnetizing current in the opposite direction and back again produces the characteristic curve called a *hysteresis loop*, *bcdeb*.

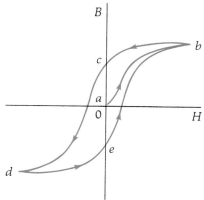

(a) Starting with an unmagnetized sample at O, the curve $a \to b$ is the *magnetization curve*. Repeated reversals of the solenoid current then trace out the outer portion (*bcdeb*), called a *hysteresis loop*.

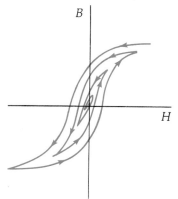

(b) Demagnetizing a ferromagnetic material involves traveling around successive hysteresis loops, gradually decreasing the magnitude of H with each cycle.

FIGURE 33-8
A graph of the magnetic field B in an iron core inside a solenoid versus the magnetic field intensity H produced by current in the coil windings.

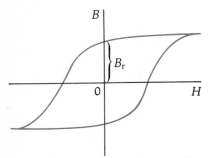

(a) Tungsten steel, a magnetically hard material.

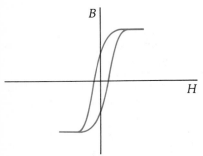

(b) Iron, a magnetically soft material.

FIGURE 33-9
Hysteresis loops for two different substances.

Ferromagnetic materials are classified as magnetically "hard" or "soft." A magnetically hard ferromagnetic material has a "fat" hysteresis loop as in Figure 33-9a, with a large *remanent* field B_r remaining after the current is cut off. A soft ferromagnetic material has a "thin" hysteresis loop, often with a small remanent field. As shown in Problem 33B-5, the area enclosed by the hysteresis loop represents *the energy loss (as heat) per unit volume per cycle*, due to the work required to overcome the resistance to domain reorientation. Since this thermal energy comes from the emf supplying the current in the coil, magnetically soft iron cores are used in transformers with alternating currents. Magnetically hard substances are used as permanent magnets in such devices as audio speakers and meter movements.

Mechanical shocks or high temperatures will cause a permanent magnet to become demagnetized. To demagnetize specimens that cannot withstand such treatment, such as watches with mechanical balance wheels and magnetic tapes, we carry the object around successive hysteresis loops by repeatedly reversing the current in the solenoid, decreasing the magnitude of the reversals with each successive cycle, Figure 33-8b.

Magnetostriction refers to the fact that, when magnetized, a rod becomes longer by about one part in 10^6. This is because the oriented magnetic moments in a domain slightly deform the crystal lattice, making it a bit longer in the direction of magnetization. It is the major source of the humming sound in transformers as the rapidly changing direction of the magnetic field causes the core to alternately stretch and contract physically, generating the sound waves. Interestingly, if we stretch an unmagnetized ferromagnetic rod, it will become somewhat magnetized because of this effect.

Finally, we should mention that there are other types of magnetic materials than the three we have discussed. Among these are the *ferrites*, whose properties of high electrical resistivity, high permeability, and low remanence make them very desirable for computer memory storage and other uses in high-frequency electronic circuits. *Antiferromagnetic* substances have adjacent magnetic dipoles that are antiparallel to each other. Consult more advanced texts for further information.

Summary

Materials may be classified magnetically in three major groups:

Diamagnetic: The atoms of the substance have no net dipole moment. When an external field is applied, induced dipole moments *opposite* to the field are created. In a nonuniform field, diamagnetic materials are repelled away from the stronger-field regions.

Paramagnetic: The atoms of the substance have net dipole moments, randomly oriented. When an external field is applied, the alignment of dipoles (or *magnetic polarization*) is *in* the field direction. In a nonuniform field, paramagnetic materials are attracted toward the stronger-field regions. Paramagnetism is temperature dependent since increased thermal motions cause disorientation of dipoles.

Ferromagnetic: The atoms of five elements—Fe, Co, Ni, Gd, and Dy—have intrinsic magnetic moments all aligned parallel within small regions called *domains*. An applied magnetic field will cause an increase in domains aligned *in* the direction of the field, increasing the resulting field as much as several hundred thousand times.

The **magnetic field intensity** H is due only to currents in wires or currents in space.

For a long solenoid: $H = \dfrac{NI}{\ell}$

When magnetic materials are present, the net *magnetic field* **B** (more precisely called the **magnetic induction** or **magnetic**

flux density) is related to **H** as follows:

$$\mathbf{B} = \mu \mathbf{H}$$

where

$$\mu = \mu_0(1 + \chi)$$

The *permeability* μ of magnetic substances includes not only the permeability of free space, μ_0, but also the effects of aligned dipole moments through the factor χ, the **magnetic susceptibility**. While χ is a constant close to zero for paramagnetism (slightly positive) and diamagnetism (slightly negative), it has extremely large values for ferromagnetic substances under *saturation*: when all the dipoles are aligned parallel to the external field. The net field inside a solenoid is $\mathbf{B} = \mathbf{B}_0 + \mathbf{B}'$, where $\mathbf{B}_0 = \mu_0 \mathbf{H}$ due to the current in the windings, plus $\mathbf{B}' = \mu_0 \chi \mathbf{H}$, due to the effects of the magnetic material.

The *hysteresis loops* of B–H graphs allow classification of ferromagnetic materials as magnetically *hard* (a "fat" loop), difficult to magnetize but retaining a large remanent field and suitable for permanent magnets; or magnetically *soft* (a "thin" loop), easy to magnetize but easy to demagnetize by mechanical or thermal shocks. The area within a hysteresis loop represents the energy "loss" (as heat) per unit volume per cycle around the loop, so soft materials are used for cores of transformers in which alternating currents repeatedly trace out a thin hysteresis loop.

Questions

1. An unmagnetized iron rod placed halfway into a solenoid will be suddenly drawn into the solenoid as soon as current begins to pass through the windings of the solenoid. What is the explanation of this phenomenon on the basis of the microscopic changes within the iron and on the basis of energy considerations?

2. Consider two iron bars that are identical except that only one of them is magnetized. If you have only the two bars, how can the magnetized bar be identified?

3. Wrenches and screwdrivers sometimes become slightly magnetized even though they are not used in electrical work. Why?

4. In order to shield a device from a magnetic field, we often enclose the device in a box made of iron or a special metal with a high permeability (called *mu-metal*). How does such a shield work?

5. The sensing element of most metal detectors is a coil of wire. How does such a metal detector work? (This effect is used for sensing automobiles at traffic-light intersections, employing a conducting loop buried in the road surface.)

6. Figure 33-10 shows coils of wires with laminated iron cores (thin sheets electrically insulated from each other). When an AC voltage is applied to the windings, the changing magnetic flux induces alternating currents in the iron, called *eddy currents*. To reduce I^2R heating in the core, should the laminations be oriented as in A or as in B?

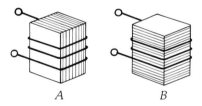

FIGURE 33-10
Question 6.

7. Must a permanent magnet have a detectable north pole and south pole? Can you devise a magnet with, say, two north poles? With *no* poles (though the material is magnetized)?

Problems

33A-1 A solenoid 20 cm long has 700 closely wound turns around an iron core of diameter 1.4 cm. Assuming that the iron is saturated, what current will produce a magnetic flux of 3×10^{-4} T·m² through the center winding?

33A-2 Find the permeability of a material that has a magnetic susceptibility of 1.8×10^{-5}.

33A-3 When a superconducting material is placed in a magnetic field, surface currents are established that make the magnetic field inside the material truly zero. (That is, the material is perfectly diamagnetic.) Suppose that a circular disk, 2 cm in diameter, is placed in a magnetic field $B = 0.02$ T with the plane of the disk perpendicular to the field lines. Find the equivalent surface current if it all lies at the circumference of the disk.

33B-4 A solenoid 25 cm long has 600 tightly wound turns that carry a current of 30 mA. Find H and B at the center (a) when there is air in the core and (b) when the core is filled with 45 Permalloy that has three-fourths of its maximum saturation susceptibility.

33B-5 Show that the product B times H has units of energy per unit volume.

33B-6 A toroidal coil with a magnetic material within its windings is called a *Rowland ring*. Consider a Rowland ring with an iron core that has a mean circumferential radius of 10 cm and carries a current of 150 mA through a winding of 250 turns. (a) Calculate the magnetic field strength H within the windings. (b) Calculate the magnetic induction B within the iron if it is 70% saturated.

33B-7 A long iron-core solenoid with 2000 turns/m carries a current of 10 mA. (a) Calculate the magnetic induction B within the solenoid, assuming that the iron is 20% saturated. (b) With the iron core removed, what current will produce the same magnetic induction as in (a)?

33B-8 A toroidal coil with an effective circumference of 50 cm has 1000 turns that carry a current of 200 mA. The core material has a magnetic susceptibility of 3000 when saturated. (a) Calculate the magnetic induction B in the core if the material is 85% saturated. (b) Find the magnetic field intensity H within the windings. (c) Calculate the fraction of B due solely to the current in the windings.

33C-9 A long solenoid with an iron core has a radius of 1.25 cm and a winding of 1200 turns/m. A secondary winding consisting of 40 turns with a total resistance of 5 Ω is wrapped tightly around the solenoid and the ends of the winding are joined. A switch is closed, and a current of 50 mA is established in the solenoid that causes the iron to become 100% saturated. Calculate the total charge that passes a given point in the secondary winding as a consequence of the change of magnetic induction within the winding.

CHAPTER 34

AC Circuits

The Buddha, the Godhead, resides quite as comfortably in the circuits of a digital computer or the gears of a cycle transmission as he does at the top of a mountain or in the petals of a flower.

ROBERT PIRSIG
(*Zen and the Art of Motorcycle Maintenance*)

34.1 Introduction

We have discussed the response of a series *RC* circuit when a battery voltage is applied. The current initially is large, limited only by the resistance, and decreases exponentially to zero. In a similar fashion, when a battery voltage is applied to an *RL* circuit the current grows exponentially from zero to a value limited only by the resistance. In both instances, the response is transient; that is, the varying part lasts only momentarily, until steady-state conditions have been achieved. The *time constant* of the circuit determines how steep the exponential curves are.

In this chapter we will investigate the response of a circuit to a constantly changing applied voltage.[1] Most electromechanical generators of electricity produce a *sinusoidally* varying voltage, resulting in *alternating current* (AC). The resultant voltages and currents are called AC voltage and AC current. (The latter is firmly entrenched in common usage, so we shall go along with it despite the redundancy.) AC circuits are used extensively for power transmission, for radio, TV, and satellite communications, in computers, and for a host of other applications in all technologically advanced societies.

34.2 Simple AC Circuits

For convenience, we use a special notation for AC circuits. Consider alternating voltages and currents of the following type[2]:

$$v = V \sin \omega t \quad \text{and} \quad i = I \sin(\omega t - \phi) \quad (34\text{-}1)$$

---

[1] Historically, formulation of the laws governing direct current was relatively simple compared with formulation of those describing alternating current. It wasn't until just before the end of the last century that the brilliant mathematician-engineer Charles Proteus Steinmetz developed the laws that describe alternating current. The initial publication of his work consisted of three volumes of detailed and complicated mathematical development of alternating-current circuit theory.

[2] The reason for the minus sign in ($\omega t - \phi$) will become evident in Section 34.3.

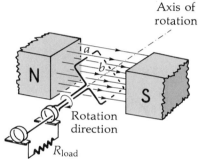

(a) A simple AC generator. An emf $\mathcal{E} = \mathcal{E}_0 \sin \omega t$ is induced in the wire loop as it rotates in the presence of a magnetic field.

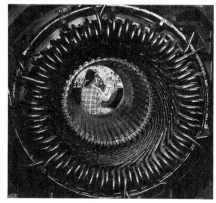

(b) The rotating armature of a modern AC generator contains many coils, which rotate in the field produced by large electromagnets (outside the photograph).

FIGURE 34-1
The AC generator.

The *amplitudes*, or *peak values*, of the voltage and current are represented by capital letters (V and I, respectively). Small letters represent voltage and current values that change in time (v and i, respectively). At any given instant, sinusoidally varying quantities have a particular *phase angle*, such as ωt and $(\omega t - \phi)$ in the above expressions. The angle ϕ is called the *phase constant* and expresses the *phase difference between two different sinusoidal variations*.

Limiting the discussion to just one frequency is justifiable even though many situations, such as the use of hi-fi amplifiers for music and speech reproduction, involve numerous frequencies simultaneously. The reason is that any complicated waveshape that is *periodic* (that is, repeats itself again and again) may be replaced by a combination of two sinusoidal variations involving a fundamental frequency (f_0) and multiples ($2f_0$, $3f_0$, $4f_0$, ...). The mathematical method is known as *Fourier analysis* (see Appendix F). Thus, more complicated (periodic) waveshapes are understandable in terms of the simple sine and cosine waves that we examine in this chapter.

We will discover that the current through a series combination of resistors, inductors, and capacitors varies sinusoidally, but the voltage across these elements will not necessarily have the same phase as the current through them. How much current is present depends not only on the value of the circuit components, but also on the frequency of the applied voltage.

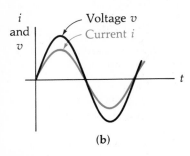

FIGURE 34-2
A purely resistive AC circuit.

Circuits with Resistance Only

Consider the circuit shown in Figure 34-2. We write the applied voltage as $v = V \sin \omega t$, implying that at $t = 0$ the voltage $v = 0$, going positive. In circuit diagrams, the symbol for an AC voltage source is —⊗—. To find the current i we use the fact that conservation-of-charge and conservation-of-energy relationships hold for AC circuits just as they do for DC circuits. *At every instant* Kirchhoff's rules apply. So we sum the voltage "rises" and "falls" around the closed-loop circuit, using minus signs for potential drops:

$$\Sigma v = 0$$
$$v - iR = 0$$

Substituting for v from Equation (34-1) and rearranging, we obtain an expression similar to Ohm's law:

$$V \sin \omega t = iR$$
$$i = \frac{V}{R} \sin \omega t \qquad (34\text{-}2)$$

Notice that the current has the *same phase* as the applied voltage, as shown in Figure 34-2b.

Circuits with Capacitance Only

Now consider the circuit shown in Figure 34-3a. As always, the sum of the potential differences around a closed loop must be zero at every instant. Thus:

$$\Sigma v = 0$$
$$V \sin \omega t - \frac{q}{C} = 0 \qquad (34\text{-}3)$$

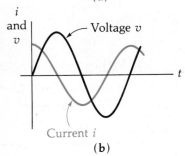

FIGURE 34-3
A purely capacitive AC circuit.

where q is the charge on the capacitor at time t. The current i through the circuit is the rate at which the charge on the plates of the capacitor is changing: $i = dq/dt$. Differentiating Equation (34-3) and solving for dq/dt, we have

$$i = \frac{dq}{dt} = V\omega C \cos \omega t \qquad (34\text{-}4)$$

We can write this expression in a form similar to Ohm's law

$$i = V\omega C \cos \omega t = \left(\frac{V}{X_C}\right) \cos \omega t$$

where we introduce the new concept

CAPACITIVE REACTANCE
$$X_C = \frac{1}{\omega C} \qquad (34\text{-}5)$$

The symbol X_C is called the **capacitive reactance**, measured in *ohms* (Ω). It limits the amplitude of the current in the way that resistance limits the current in a purely resistive circuit. (It is left as an exercise to show that capacitive reactance does have dimensions of ohms.) Notice that the current leads the applied voltage by $\pi/2$ rad (or 90°), as shown in Figure 34-3b. The phrase "leading the applied voltage" means that, as time progresses (that is, as we move along the t axis), the current reaches its positive peak value *before* the applied voltage reaches its positive peak value. The word *reactance* emphasizes the difference from *resistance*, where v_R and i are always in phase with each other.

EXAMPLE 34-1

Find the reactance of a 2-μF capacitor (a) at 60 Hz and (b) at one megahertz (1 MHz).

SOLUTION

(a) For $f = 60$ Hz, $\omega = 2\pi f = 2\pi(60 \text{ s}^{-1}) = 377$ rad/s. Thus:

$$X_C = \frac{1}{\omega C} = \frac{1}{(377 \text{ rad/s})(2 \times 10^{-6} \text{ F})} = \boxed{1326 \ \Omega}$$

Because the standard frequency for power distribution in the United States is 60 Hz, the value of ω at 60 Hz, $\omega = 377$ rad/s, is a useful number to remember.

(b) At 10^6 Hz, we have

$$X_C = \frac{1}{\omega C} = \frac{1}{(2\pi)(1 \times 10^6 \text{ Hz})(2 \times 10^{-6} \text{ F})} = \boxed{0.0796 \ \Omega}$$

The reactance of a capacitor becomes *less* as the frequency *increases*. Conversely, as the frequency decreases toward zero, the reactance becomes very large. In fact, at $\omega = 0$ (direct current), the reactance is infinite, so a capacitor completely blocks the current in that branch of a DC circuit.

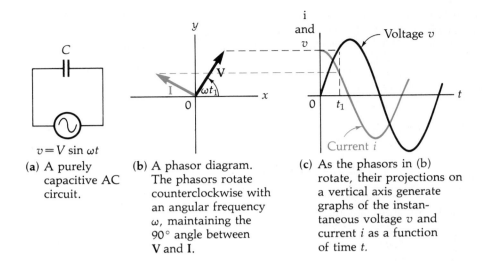

FIGURE 34-4
Voltage and current relations for a purely capacitive circuit.

(a) A purely capacitive AC circuit.

(b) A phasor diagram. The phasors rotate counterclockwise with an angular frequency ω, maintaining the 90° angle between **V** and **I**.

(c) As the phasors in (b) rotate, their projections on a vertical axis generate graphs of the instantaneous voltage v and current i as a function of time t.

Phasors

A useful way of portraying the phase relationship between the applied voltage and the resulting current is by using a **phasor diagram**. The phasor diagram for a purely capacitive circuit is shown in Figure 34-4. In this diagram, the voltage and current are represented by vectorlike arrows, **V** and **I**, called **phasors**,[3] that rotate counterclockwise with an angular frequency ω, maintaining their relative angular separations as they rotate. The lengths of the phasors are the amplitudes of the time-varying voltage and current. Their angular separation represents the **phase constant** ϕ between the voltage v and the current i. The projection of the phasors on a vertical axis is then expressed by the equations

$$v = V \sin \omega t \quad \text{and} \quad i = I \cos \omega t$$

or

$$i = I \sin\left(\omega t + \frac{\pi}{2}\right) \qquad (34\text{-}6)$$

These projections thus generate graphs of v and i versus time and are the physically "real" quantities that are actually measured. The phase constant ϕ [in the general expression $i = I \sin(\omega t - \phi)$] is $\phi = -\pi/2$ (equal to $-90°$). This means that, for a purely capacitive reactance, the current <u>leads</u> the voltage by just one-quarter cycle of the sinusoidal variation. As the phasor diagram rotates around, the current phasor **I** is ahead of the voltage phasor **V** (that is, the current leads the voltage). As time progresses, the current reaches its peak value *before* the voltage reaches its peak value. The phasor diagram helps us visualize the phase relationship between the applied voltage and the current.

Circuits with Inductance Only

Consider the circuit shown in Figure 34-5a. We assume that L represents a "pure" inductance (that is, the resistance of the windings is negligible). As always, the instantaneous sum of the potential increases and decreases around

---

[3] Although voltage and current are not vectors in the usual sense, as phasors they do follow the rules for vector addition. Their representation on a phasor diagram is a very useful mathematical technique that helps us clearly visualize the phase relationships between the applied voltage and the current.

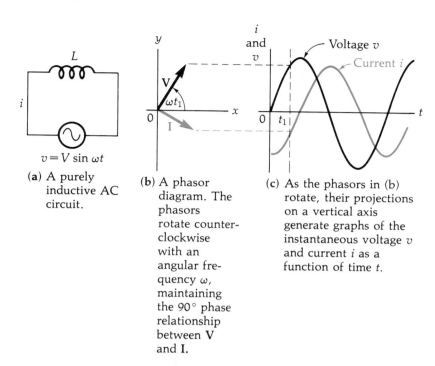

FIGURE 34-5
Voltage and current relations for a purely inductive AC circuit.

(a) A purely inductive AC circuit.

(b) A phasor diagram. The phasors rotate counter-clockwise with an angular frequency ω, maintaining the 90° phase relationship between **V** and **I**.

(c) As the phasors in (b) rotate, their projections on a vertical axis generate graphs of the instantaneous voltage v and current i as a function of time t.

the circuit loop must be zero. Recall that the voltage v_L across the inductor due to the changing current through it is $v_L = -L\,di/dt$, where the minus sign indicates opposition to the applied voltage.

$$\Sigma V = 0$$
$$v - L\frac{di}{dt} = 0 \tag{34-7}$$

When we substitute $v = V \sin \omega t$ and rearrange, Equation (34-7) becomes

$$L\frac{di}{dt} = V \sin \omega t$$

We solve this equation by separating the variables i and t, so that they appear on opposite sides of the equal sign, and integrating (see Appendix G):

$$\int di = \frac{V}{L} \int \sin \omega t\, dt$$

$$i = -\frac{V}{\omega L} \cos \omega t + c$$

Setting[4] $c = 0$ and using the fact that $-\cos \omega t = \sin(\omega t - \pi/2)$, we write this in a form similar to Ohm's law:

$$i = \frac{V}{X_L} \sin\left(\omega t - \frac{\pi}{2}\right) \tag{34-8}$$

---

[4] The constant of integration c represents a constant DC current, which could be present if a DC source of voltage were in the circuit.

where X_L is defined as the **inductive reactance**, which, like capacitive reactance, is measured in ohms (Ω).

INDUCTIVE REACTANCE X_L
$$X_L = \omega L \qquad (34\text{-}9)$$

The inductive reactance limits the amplitude of the current just as resistance limits the current in a purely resistive circuit. The reactance increases with frequency because the inductor opposes a change in current. The faster this change is made, the greater is the inductor's opposition to this change.

The phasor diagram is shown in Figure 34-5b. For a pure inductance, the phase constant ϕ is $+\pi/2$ (or $+90°$). This means that the current <u>lags</u> the applied voltage by one-quarter of the sinusoidal variation. The phrase "lags the applied voltage" means that, as time progresses, the current reaches its peak value *after* the voltage reaches its peak value. As the phasor diagram rotates, the current phasor **I** lags behind the voltage phasor **V**.

EXAMPLE 34-2

An AC voltage with an amplitude of 15 V and a frequency of 60 Hz is applied across an inductor whose inductance is 30 mH. Find the resulting AC current.

SOLUTION

From Equation (34-8), the amplitude of the current is

$$I = \frac{V}{X_L}$$

Since $X_L = \omega L = 2\pi f L$, we have

$$I = \frac{V}{2\pi f L} = \frac{(15\text{ V})}{(2\pi)(60\text{ s}^{-1})(3 \times 10^{-2}\text{ H})} = 1.33\text{ A}$$

We know that, in a pure inductance, the current lags the applied voltage by $\pi/2$ rad. The frequency of the current is the same as that of the applied voltage, 60 Hz. Thus, in SI units:

$$i = I\sin(\omega t - \phi) = \boxed{1.33\sin\left(120\pi t - \frac{\pi}{2}\right)\text{ A}} \quad \text{(where } t \text{ is in seconds)}$$

34.3 Series *RLC* Circuits

Consider the circuit shown in Figure 34-6. By Kirchhoff's loop rule, at any time t, the applied voltage $v = V\sin\omega t$ must equal the sum of the instantaneous values of the back-emf across the inductor $v_L = L\,di/dt$, the voltage drop across the resistor $v_R = iR$, and the voltage across the capacitor due to the charge on the capacitor $v_C = q/C$.

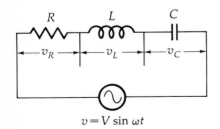

FIGURE 34-6
At any time t, the instantaneous voltages across the circuit elements add to equal the instantaneous applied voltage. That is, $v = v_R + v_L + v_C$.

$$\Sigma V = 0$$

$$V \sin \omega t - L\frac{di}{dt} - iR - \frac{q}{C} = 0$$

Rearranging gives
$$L\frac{di}{dt} + Ri + \frac{q}{C} = V \sin \omega t \qquad (34\text{-}10)$$

In order to understand the physical significance of each term in Equation (34-10), as well as to perform the initial step in the solution of the equation, we must express the current i (and its derivatives) as derivatives of the charge q:

$$L\frac{d^2q}{dt^2} + R\frac{dq}{dt} + \frac{q}{C} = V \sin \omega t \qquad (34\text{-}11)$$

This equation is identical in form to the equation that describes a forced mechanical oscillator with viscous damping [Chapter 15, Equation (15-48)]:

$$m\frac{d^2x}{dt^2} + b\frac{dx}{dt} + kx = F_0 \sin \omega t \qquad (34\text{-}12)$$

A term-by-term comparison of Equations (34-11) and (34-12) reveals the following.

(1) An inductance resists the surge of charge through an electrical circuit in a way that is analogous to mass resisting acceleration in a mechanical system.
(2) Resistance in an electrical circuit is analogous to viscosity in a mechanical system, each being responsible for energy loss in the system (in the electrical case, Joule heating).
(3) The reciprocal of capacitance provides the "resilience" to an electrical circuit in the way the spring constant in a mechanical system determines the restoring force.

These (plus other analogies) are summarized in Table 34-1.

Equation (34-11) may be solved for q as a function of time, then differentiated with respect to time to yield the current. The solution of this equation requires a mathematical technique beyond the scope of this text. The result

TABLE 34-1   Electromechanical Analogues

Mechanical System	**Electrical Circuit**
Mass M (resists change of velocity)	Inductance L (resists change of current)
Viscosity constant b (dissipates energy into thermal form)	Resistance R (dissipates energy into thermal form)
Spring constant k (determines restoring force and "elasticity" of mechanical motion)	Reciprocal of capacitance $1/C$ (provides "resilience" to an electrical current)
Displacement x	Charge q
Velocity $v = dx/dt$	Current $i = dq/dt$
Force F	Voltage V

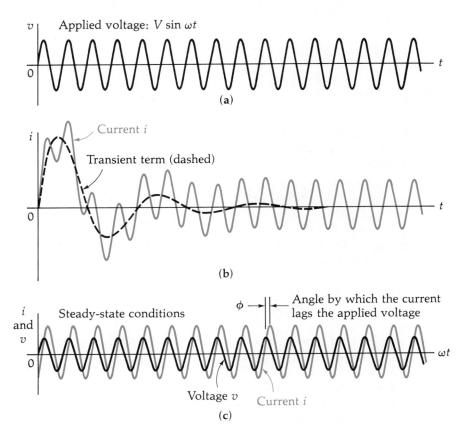

FIGURE 34-7
Voltage and current in an *RLC* circuit in which the inductive reactance is greater than the capacitive reactance, so that the circuit behaves as if it were a smaller inductive reactance in series with a resistor.

we would obtain is

$$i = \frac{V}{\sqrt{R^2 + (X_L - X_C)^2}} \sin(\omega t - \phi) + i_0(t) \tag{34-13}$$

As before, $X_L = \omega L$ and $X_C = 1/\omega C$. The term $i_0(t)$ is called the *transient term*. It describes the current variations that occur immediately after the voltage is first applied. In most circuits, it becomes essentially zero soon after the voltage is applied.[5] A typical example is shown in Figure 34-7, wherein the transient effects die out rapidly as the AC current settles down to its steady-state condition. For our purposes, we will not analyze these transient effects, but instead will concentrate on *steady-state conditions*:

$$i = I \sin(\omega t - \phi) \tag{34-14}$$

where the **phase constant** ϕ is given by

PHASE CONSTANT ϕ
$$\phi = \tan^{-1}\left(\frac{X_L - X_C}{R}\right) \tag{34-15}$$

---

[5] The size of the transient effect depends on the initial conditions. For example, what is the phase of the AC voltage at the instant it is applied? Are capacitors initially charged or uncharged? With suitable adjustment of the initial conditions, the transient can be eliminated entirely. Unfortunately, under certain adverse circumstances, the transient can cause extreme surges of current that could damage circuit components.

It may have any value between $-\pi/2$ rad and $+\pi/2$ rad, depending on the relative magnitudes of X_L and X_C. Recall that ϕ is the phase constant between the voltage v applied to the circuit and the current i in the circuit.

There is an easy way to keep track of the various phase relationships in a series RLC circuit. Because at any instant the current is the *same* in all components, **we use the current phasor I as a reference**, measuring all other phase angles with respect to the current. In Figure 34-8(a) we develop a *voltage phasor diagram* that depicts voltages and current in their correct phase relationships. We represent each as a vectorlike *phasor*: **V** or **I**. By custom, the phasor for the reference current **I** is drawn horizontally toward the right. Because the voltage across the resistor is *in phase* with the current, both $\mathbf{V}_R$ and **I** are in the same direction. In an inductor, the current *lags* the voltage across the inductor by $\pi/2$ rad, so $\mathbf{V}_L$ is shown as a phasor 90° ahead of the current phasor **I**. In a capacitor, the current *leads* the voltage across the capacitor by $\pi/2$ rad, so V_C is shown 90° behind the current phasor **I**. From Kirchhoff's loop rule, the sum of the voltages across the circuit elements equals the applied voltage phasor **V**. *To take their various phases into account, we must add the phasors as vectors* to obtain the phasor for the applied voltage **V**:

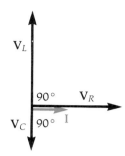

(a) A voltage phasor diagram with the current **I** as a reference.

PHASORS ADD AS VECTORS
$$\mathbf{V} = \mathbf{V}_R + \mathbf{V}_L + \mathbf{V}_C \quad (34\text{-}16)$$

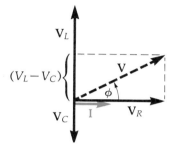

(b) The applied voltage phasor **V** is the vector sum of the voltage phasors for individual circuit elements: $\mathbf{V} = \mathbf{V}_R + \mathbf{V}_L + \mathbf{V}_C$.

The voltage phasor diagram portrays the various phase relationships in a series AC circuit. As the phasor diagram rotates counterclockwise with angular frequency ω, *the projections of the phasors on the vertical axis give the instantaneous values of all voltages and currents as functions of time.*

34.4 Impedance in Series *RLC* Circuits

The alternating current through a series RLC circuit is impeded by an amount dependent upon the value of the components as well as the frequency. The amplitude I of the current is, from Equation (34-13),

$$I = \frac{V}{\sqrt{R^2 + (X_L - X_C)^2}} \quad (34\text{-}17)$$

where
V = amplitude of the applied voltage
R = resistance
$X_L = \omega L$, the inductive reactance
$X_C = 1/\omega C$, the capacitive reactance

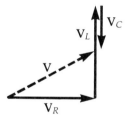

(c) Another way of sketching the vector sum of the individual voltage phasors to obtain the applied voltage phasor $\mathbf{V} = \mathbf{V}_R + \mathbf{V}_L + \mathbf{V}_C$.

FIGURE 34-8
Phase relationships between voltages and current in a series RLC circuit. (In this illustration, the net reactance for the circuit as a whole is *inductive*, so the current **I** *lags* the applied voltage **V** by the phase-constant angle ϕ.

The combination of resistance and reactances is defined as the **impedance** Z measured in ohms (Ω).

IMPEDANCE Z IN A SERIES RLC CIRCUIT
$$Z = \sqrt{R^2 + (X_L - X_C)^2} \quad (34\text{-}18)$$

Thus, the amplitude of the current is related to the amplitude of the applied voltage by the simple relation $I = V/Z$, or

OHM'S LAW FOR AC
$$V = IZ \quad (34\text{-}19)$$

For AC, the impedance Z plays a role similar to resistance in DC circuits.

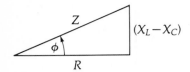

(a) A right triangle formed by R, $X_L - X_C$, and Z.

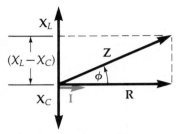

(b) An impedance diagram.

FIGURE 34-9
An *impedance diagram* for a series *RLC* circuit. The angle ϕ is the phase constant between the applied voltage and the current in the circuit. For this example, the net reactance is *inductive*; that is, $X_L > X_C$. This means that the phase angle ϕ is positive and that the current, $i = I \sin(\omega t - \phi)$, lags the applied voltage. In (b), the current phasor **I** is in the same direction as **R**, while the applied voltage phase **V** is in the direction of **Z**.

The mathematical form of Equation (34-18) suggests the Pythagorean theorem, in which R and $X_L - X_C$ are lengths of the legs of a right triangle, with Z forming the hypotenuse, as illustrated in Figure 34-9a. The angle between Z and R is defined by Equation (35-15):

PHASE CONSTANT
$$\phi = \tan^{-1}\left(\frac{X_L - X_C}{R}\right) \quad (34\text{-}20)$$

which is the phase constant ϕ between the applied voltage v and the current i in the series circuit.[6] The triangle is related to the *impedance diagram* of Figure 34-9b. To sketch an impedance diagram, we draw the resistance as a vectorlike arrow **R** along the $+x$ axis, we draw $\mathbf{X_L}$ as a vectorlike arrow along the $+y$ axis, and we draw $\mathbf{X_C}$ along the $-y$ axis. The vector sum of these three arrows is the total impedance **Z**.

The impedance diagram is closely related to the voltage phasor diagram of Figure 34-8, since the voltages in that diagram are merely the scalar I times the corresponding resistance, reactance, and impedance of Figure 34-9b. Thus, the two representations differ only by the scale factor I. (The impedance diagram does not rotate, however.)

EXAMPLE 34-3

Consider the series *RLC* circuit of Figure 34-10 with the following circuit parameters: $R = 200\ \Omega$, $L = 663$ mH, and $C = 26.5\ \mu$F. The applied voltage has an amplitude of 50 V and a frequency of 60 Hz. Find the following amplitudes:

(a) The current i, including its phase constant ϕ relative to the applied voltage v.
(b) The voltage V_R across the resistor and its phase relative to the current.
(c) The voltage V_C across the capacitor and its phase relative to the current.
(d) The voltage V_L across the inductor and its phase relative to the current.

SOLUTION

In general, the initial step is to calculate reactances, and impedances and then apply Ohm's law for AC circuits.

$$X_C = \frac{1}{\omega C} = \frac{1}{(2\pi f)(C)} = \frac{1}{(2\pi)(60\ \text{s}^{-1})(26.5 \times 10^{-6}\ \text{F})} = 100\ \Omega$$

$$X_L = \omega L = (2\pi f)(L) = (2\pi)(60\ \text{s}^{-1})(663 \times 10^{-3}\ \text{H}) = 250\ \Omega$$

$$Z = \sqrt{R^2 + (X_L - X_C)^2} = [(200)^2 + (250 - 100)^2]^{1/2}\ \Omega = 250\ \Omega$$

Figure 34-10b is an impedance diagram for this circuit. Because $X_L > X_C$, it has a net inductive reactance, so the current will *lag* the applied voltage.

(a) Applying Ohm's law for AC circuits, we obtain the magnitude of the current I:

$$I = \frac{V}{Z} = \frac{50\ \text{V}}{250\ \Omega} = \underline{\underline{0.200\ \text{A}}}$$

---

[6] The current is always *in phase* with $\mathbf{V_R}$. A *positive* phase constant ϕ means that the current lags the applied voltage **V** and vice versa. This is consistent with the minus sign in our defining relation for current, $i = I \sin(\omega t - \phi)$ in Equation (34-1). Fortunately, we can easily determine the leading or lagging relationship by inspecting a voltage phasor diagram. Just remember that, *with the current as a reference plotted horizontally to the right* (and, consequently, $\mathbf{V_R}$ plotted horizontally to the right), the *inductive* quantities plot vertically *upward* and the *capacitive* quantities plot vertically *downward*.

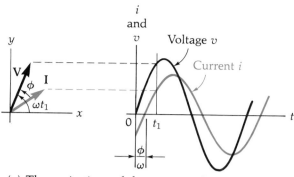
(a) The projections of the rotating phasors on a vertical axis generate graphs of the instantaneous voltage v and current i vs. the time t.

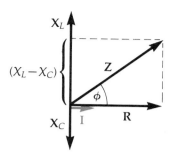
(b) An impedance diagram.

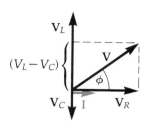
(c) A voltage phasor diagram. The current $\mathbf{I}$ is in phase with $\mathbf{V}_R$.

FIGURE 34-10
Example 34-3.

It has the same frequency, $f = 60$ Hz, as the applied voltage. The phase constant ϕ between the current and the applied voltage is found from Equation (34-20):

$$\phi = \tan^{-1}\left(\frac{X_L - X_C}{R}\right)$$
$$= \tan^{-1}\left(\frac{250\ \Omega - 100\ \Omega}{200\ \Omega}\right) = \underline{\underline{36.9°}}$$

(The net reactance is inductive, so the current *lags* the applied voltage.)

Incorporating these values in the general expression [Equation (34-14)], we get

$$i = I \sin(\omega t - \phi) = 0.200\ \sin[(2\pi)(60\ \text{s}^{-1}) - 36.9°]\ \text{A}$$
$$i = \boxed{0.200\ \sin(120\pi t - 36.9°)\ \text{A}} \qquad \text{(where } t \text{ is in seconds)}$$

The current is expressed relative to the applied voltage, $v = 50 \sin(120\pi t)$ V, and it *lags* the voltage by 36.9°.

(b) The voltage V_R across the resistor is

$$V_R = IR = (0.200\ \text{A})(200\ \Omega) = \boxed{40.0\ \text{V}}$$

The instantaneous voltage across a resistor is *in phase* with the current through it, so $\phi = 0°$.

(c) The voltage V_C across the capacitor is

$$V_C = IX_C = (0.200\ \text{A})(100\ \Omega) = \boxed{20.0\ \text{V}}$$

The instantaneous current through a pure capacitor always *leads* the voltage across it by $\pi/2$ rad.

(d) The voltage V_L across the inductor is

$$V_L = IX_L = (0.200\ \text{A})(250\ \Omega) = \boxed{50.0\ \text{V}}$$

The instantaneous current through a pure inductor always *lags* the voltage across it by $\pi/2$ rad.

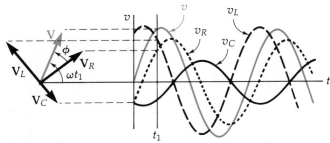
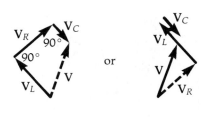

(a) A voltage phasor diagram. The projections of the rotating phasors on a vertical axis generate graphs of the instantaneous voltages vs. time for the voltage across each circuit element.

(b) The voltage phasors for each circuit element add together as vectors to give the applied voltage phasor $\mathbf{V} = \mathbf{V}_R + \mathbf{V}_L + \mathbf{V}_C$. (Two different ways of drawing the vector addition are shown.)

FIGURE 34-11
Voltage phasor diagrams for Example 34-3.

Figure 34-11 is a phasor diagram showing all voltages. Note the way in which the AC voltages combine. In particular, the *algebraic* sum of their magnitudes is *not* the applied voltage: $V \neq V_R + V_L + V_C$. (This sum is 40 V + 50 V + 20 V = 110 V, instead of the correct value of 50 V.) On the other hand, the algebraic sum of the *instantaneous* voltages across the circuit elements always equals the instantaneous applied voltage:

$$v = v_R + v_L + v_C$$

These instantaneous voltages are the *projections* on the vertical axis of the voltage phasors in Figure 34-11. The fact that the projections of the phasors add algebraically implies that the phasors themselves add *as vectors*:

$$\mathbf{V} = \mathbf{V}_R + \mathbf{V}_L + \mathbf{V}_C$$

From the vector diagram, we have

$$V^2 = V_R^2 + (V_L - V_C)^2 = (40\text{ V})^2 + (50\text{ V} - 20\text{ V})^2 = 50\text{ V}$$

which is the correct value of the applied voltage amplitude.

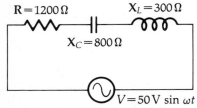

(a) A series *RLC* circuit with an applied AC voltage of 50-V amplitude.

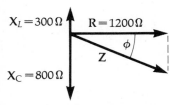

(b) The impedance diagram for the circuit shown in (a).

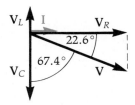

(c) The voltage phasor diagram.

FIGURE 34-12
Example 34-4.

EXAMPLE 34-4

Consider the circuit shown in Figure 34-12. Find (a) the impedance, (b) the amplitude of the current in the circuit, and (c) the phase constant between the applied voltage and current.

SOLUTION

(a) The impedance is

$$Z = \sqrt{R^2 + (X_L - X_C)^2} = [(1200\text{ }\Omega)^2 + (300\text{ }\Omega - 800\text{ }\Omega)^2]^{1/2}$$
$$= \boxed{1300\text{ }\Omega}$$

Figure 34-12b depicts the impedance diagram.

(b) From Ohm's law for AC,

$$I = \frac{V}{Z} = \frac{50 \text{ V}}{1300 \text{ }\Omega} = \boxed{0.0385 \text{ A}}$$

(c) The phase constant is

$$\phi = \tan^{-1}\left(\frac{X_L - X_C}{R}\right)$$

$$\phi = \tan^{-1}\left(\frac{300 \text{ }\Omega - 800 \text{ }\Omega}{1200 \text{ }\Omega}\right) = \boxed{-22.6°} \quad \text{(The current } leads \text{ the applied voltage.)}$$

The negative phase angle implies that the current leads the applied voltage. This agrees with the fact that since $X_C > X_L$, the net reactance is capacitive.

EXAMPLE 34-5

In Example 34-4, find the voltage across the capacitor and its phase relative to the applied voltage.

SOLUTION

The amplitude of the voltage across the capacitor is

$$V_C = IX_C = (0.0385 \text{ A})(800 \text{ }\Omega) = \boxed{30.8 \text{ V}}$$

Multiplying the impedance diagram by I, we obtain the *voltage phasor diagram*, Figure 34-12c. The voltage across the capacitor V_C lags the applied voltage by $90° - \phi = 90° - 22.6° = \boxed{67.4°}$.

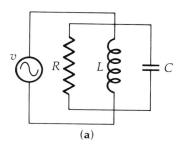

(a)

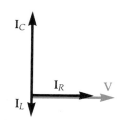

(b) The phasor diagram for currents (with the phasor **V** as a reference).

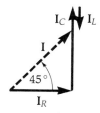

(c) The current phasors add as vectors to give the total current phasor $\mathbf{I} = \mathbf{I}_R + \mathbf{I}_C + \mathbf{I}_L$ that the source supplies.

FIGURE 34-13
Example 34-6.

34.5 Impedance in Parallel *RLC* Circuits

Consider the circuit shown in Figure 34-13a, with impedances in parallel across a voltage v. The analysis of a parallel circuit differs in one important feature from the way we analyzed a series circuit. In a series circuit, the current is common to all components, so we used the current phasor **I** as a reference for the phases of various voltages. In a parallel combination, the *voltage* across the combination is common to both branches, so **we use the voltage phasor V as a reference for phase relations**. The method is first to find the currents in each branch, then to add the currents together *vectorially* (to preserve their phase relations) to obtain the current i. This reasoning is based on the Kirchhoff junction rule: $\Sigma i = 0$. That is, the instantaneous current i entering a junction must equal the sum of the instantaneous currents leaving the junction. Just as we used a voltage phasor diagram to add voltages vectorially, we construct a current phasor diagram to add currents vectorially. Figure 34-13b shows the phase of the current through each component (with the applied voltage **V** as a reference). In Figure 34-13c, we add the current phasors *as vectors* to obtain the total current $\mathbf{I} = \mathbf{I}_R + \mathbf{I}_C + \mathbf{I}_L$ that the source supplies.

EXAMPLE 34-6

Three circuit elements R, L, and C are connected in parallel as shown in Figure 34-13a. (a) Sketch a phasor diagram showing the relative sizes of the current phasors for each branch when $R = X_L = 2X_C$. (b) Find the phase angle ϕ of the total current phasor $\mathbf{I}$ relative to the applied voltage phasor $\mathbf{V}$.

SOLUTION

(a) The same voltage V is applied to all three branches, so we use the voltage phasor $\mathbf{V}$ as a reference. The current phasors in the reactive branches are 90° out of phase with the current phasor I_R as shown. From $I = V/Z$, their relative magnitudes are $I_R = I_L$ and $I_C = 2I_L$ (because $X_L = R = 2X_C$). Thus, to an arbitrary scale, we sketch the phasor diagram of Figure 34-13b.

(b) Because $I_C = 2I_L = 2I_R$, the y component of the phasor for the total current from the source has the magnitude

$$I_y = (I_C - I_L) = (2I_L - I_L) = I_L = \underline{\underline{I_R}}$$

Similarly, the x component has the magnitude

$$I_x = \underline{\underline{I_R}}$$

Thus the phase angle ϕ between $\mathbf{I}$ and $\mathbf{V}$ is $\tan^{-1}(I_y/I_x) = \tan^{-1}(I_R/I_R) = \boxed{45°}$ as shown in Figure 34-13c.

Note that the circuit as a whole behaves as an *RC* circuit with a net *capacitive* reactance even though the individual reactances compare as $X_L > X_C$. This is similar to resistances in parallel: the branch with the *smallest* resistance dominates the circuit—it carries the most current and dissipates the most power.

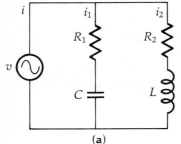

(a)

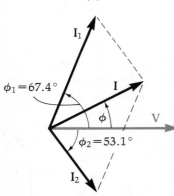

(b) The current phasor diagram using the voltage phasor V (which is common to both branches) as a reference.

FIGURE 34-14
Example 34-7.

EXAMPLE 34-7

Consider the circuit in Figure 34-14a. The applied voltage is $v = 260 \sin \omega t$, $R_1 = 5\ \Omega$, $R_2 = 12\ \Omega$, $X_C = 12\ \Omega$, and $X_L = 16\ \Omega$. Write an expression for the current i from the source, including the phase angle ϕ relative to the applied voltage v.

SOLUTION

The solution involves the following steps:

(1) Calculate the impedance of each branch.
(2) For each branch, find the current amplitude and its phase relative to the applied voltage.
(3) Construct a current phasor diagram and add the branch currents *vectorially* to find the total current i.

Branch 1	Branch 2
Step 1:	
$Z_1 = \sqrt{R_1^2 + (X_{L_1} - X_{C_1})^2}$	$Z_2 = \sqrt{R_2^2 + (X_{L_2} - X_{C_2})^2}$
$Z_1 = \sqrt{(5\ \Omega)^2 + (-12\ \Omega)^2} = 13\ \Omega$	$Z_2 = \sqrt{(12\ \Omega)^2 + (16\ \Omega)^2} = 20\ \Omega$
Step 2:	
$I_1 = \dfrac{V}{Z_1} = \dfrac{260\ V}{13\ \Omega} = 20\ A$	$I_2 = \dfrac{V}{Z_2} = \dfrac{260\ V}{20\ \Omega} = 13\ A$
$\phi_1 = \tan^{-1}\left(\dfrac{X_{L_1} - X_{C_1}}{R_1}\right)$	$\phi_2 = \tan^{-1}\left(\dfrac{X_{L_2} - X_{C_2}}{R_2}\right)$
$\phi_1 = \tan^{-1}\left(\dfrac{-12\ \Omega}{5\ \Omega}\right) = -67.4°$	$\phi_2 = \tan^{-1}\left(\dfrac{16\ \Omega}{12\ \Omega}\right) = 53.1°$
The current $\mathbf{I}_1$ *leads* the voltage $\mathbf{V}$ by 67.4°.	The current $\mathbf{I}_2$ *lags* the voltage $\mathbf{V}$ by 53.1°.

Step 3: We plot the currents as phasors in a current phasor diagram with the applied voltage $\mathbf{V}$ as a reference. We then calculate the vector addition $\mathbf{I} = \mathbf{I}_1 + \mathbf{I}_2$, using the method of component addition. Indicating the x and y axes as shown, we have:

x component	y component
$I_1:\ I_{1x} = I_1 \cos\phi_1$	$I_{1y} = I_1 \sin\phi_1$
$\quad\quad = (20\ A)(\tfrac{5}{13}) = 7.69\ A$	$\quad\quad = (20\ A)(\tfrac{12}{13}) = 18.5\ A$
$I_2:\ I_{2x} = I_2 \cos\phi_2$	$I_{2y} = I_2 \sin\phi_2$
$\quad\quad = (13\ A)(\tfrac{3}{5}) = 7.80\ A$	$\quad\quad = (13\ A)(-\tfrac{4}{5}) = -10.4\ A$
$I:\ \ I_x = I_{1x} + I_{2x}$	$I_y = I_{1y} + I_{2y}$
$\quad\quad = 7.69\ A + 7.80\ A = 15.5\ A$	$\quad\quad = 18.5\ A - 10.4\ A = 8.10\ A$

Combining I_x and I_y, we obtain

$$I = \sqrt{I_x^2 + I_y^2} = \sqrt{(15.5\ A)^2 + (8.10\ A)^2} = \underline{17.5\ A}$$

The magnitude of the phase angle ϕ is found from

$$\phi = \tan^{-1}\left(\dfrac{I_y}{I_x}\right) = \tan^{-1}\left(\dfrac{8.10\ A}{15.5\ A}\right) = \underline{27.6°}$$

From the phasor diagram we note that the current *leads* the voltage v by this angle. Therefore, in the general expression $i = I \sin(\omega t - \phi)$ we add a minus sign for $\phi = -27.6°$ to obtain

$$\boxed{i = 17.5\ \sin(\omega t + 27.6°)} \qquad \text{(current leads)}$$

Note that the parallel network as a whole behaves as a *series RC* combination. Yet the branch containing the capacitance has the lower impedance. (This is similar to the situation in DC parallel resistive circuits: the smallest-resistance branch dominates in determining the equivalent resistance of a parallel circuit, in contrast to a series combination in which the *largest* resistance dominates in determining the equivalent resistance.

FIGURE 34-15
The AC electrical field near the ground beneath a 765-kV transmission line is strong enough to light two fluorescent bulbs held in the hands.

34.6 Resonance

Even though most of us are aware of natural resonances in mechanical systems such as springboards, tuning forks, and springs, we do not usually view them as *frequency-selection mechanisms*. Yet this is exactly how they behave. If the driving frequency coincides with one of the natural frequencies of the system, large-amplitude oscillations occur. Electrical circuits composed of inductors, capacitors, and resistors behave in a similar way. If an AC voltage is applied to a resonant circuit, at the resonant frequency the current will have either a maximum or a minimum value, depending upon the design of the circuit. We will discover that the **resonant frequency** of a circuit is the frequency at which *the current through the circuit is in phase with the driving voltage*. The most practical way to view electrical resonance is as a *frequency-selection* phenomenon. Whenever we tune to a particular radio or television broadcast, we utilize this selection capability of resonant circuits.

Series Resonance

Consider a series RLC combination, as shown in Figure 34-16a. In order to examine the behavior of the circuit as the angular frequency ω changes, we will construct a series of impedance diagrams. We begin by constructing the diagram shown in Figure 34-16b, for which $X_L = X_C$; thus the impedance Z is just the resistance R. The angular frequency ω_0 corresponding to this condition is found as follows:

$$X_L = X_C$$

Substituting for X_L and X_C and solving for ω_0 gives

$$\omega_0 L = \frac{1}{\omega_0 C}$$

RESONANT ANGULAR FREQUENCY ω_0 FOR A SERIES RLC CIRCUIT

$$\omega_0 = \frac{1}{\sqrt{LC}} \qquad (34\text{-}21)$$

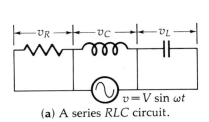

(a) A series RLC circuit.

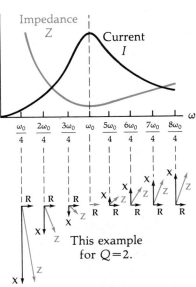

(b) Current and impedance as a function of frequency.

FIGURE 34-16
Series resonance.

We then construct impedance diagrams for $\frac{1}{4}\omega_0$, $\frac{2}{4}\omega_0$, $\frac{3}{4}\omega_0$, ... all the way to $\frac{8}{4}\omega_0$. The arrow representing **Z** has two reactive components: $\mathbf{X}_L$, pointing upward, and $\mathbf{X}_C$, pointing downward. As the frequency increases, $\mathbf{X}_L$ increases linearly and $\mathbf{X}_C$ decreases hyperbolically. The resistive component **R** remains constant. The magnitudes of the impedance from each of the impedance diagrams are plotted as a function of frequency, generating the impedance curve shown in Figure 34-16b. The corresponding values of the current I (equal to V/Z) are plotted and represented by the dashed curve. The current-vs.-frequency curve is called the *resonance curve* and reveals the following important features of the series resonant circuit:

(1) The *sharpness* of the resonance curve increases as the value of the resistance decreases relative to the inductive or capacitive reactance. The sharpness is described by the Q, or *quality factor*, of the circuit. By definition,

SHARPNESS Q
OF A RESONANT
CIRCUIT
$$Q \equiv \frac{\omega_0 L}{R} \qquad (34\text{-}22)$$

Since Q is a ratio of ohms over ohms, it is dimensionless. Typical low-frequency resonant circuits may have a Q of less than 10, while a very high-frequency resonant circuit may have a Q of several thousand (see Figure 34-17).

(2) At the resonant frequency, the current becomes very large, limited only by the value of R. At resonance,

$$I = \frac{V}{R} \qquad (34\text{-}23)$$

where V is the magnitude of the applied voltage.

(3) At resonance, the magnitude of the voltage across the inductor equals that across the capacitor. However, the voltage across one is 180° out of phase with the voltage across the other, so they add vectorially to zero. That is, $\mathbf{V}_L + \mathbf{V}_C = 0$ at every instant. *But each, by itself, may be a very large value;* in high-Q circuits, the voltage across a reactance may be thousands of times larger than the applied voltage!

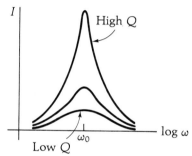

(a) Resonance curves for series RLC circuits having different sharpness Q.

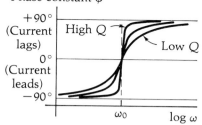

(b) The phase constant by which the current leads or lags the applied voltage in a series RLC circuit.

FIGURE 34-17
Series RLC resonance.

EXAMPLE 34-8

A series RLC circuit has the following values: $L = 20$ mH, $C = 100$ nF, $R = 20\ \Omega$, and $V = 100$ V, with $v = V \sin \omega t$. Find (a) the resonant frequency, (b) the magnitude of the current at the resonant frequency, (c) the Q of the circuit, and (d) the magnitude of the voltage across the inductor at resonance.

SOLUTION

(a) The resonant frequency is obtained from Equation (34-21):

$$\omega_0 = \frac{1}{\sqrt{LC}}$$

Substituting numerical values, we obtain

$$\omega_0 = [(20 \times 10^{-3}\text{ H})(100 \times 10^{-9}\text{ F})]^{-1/2} = 2.24 \times 10^4 \frac{\text{rad}}{\text{s}}$$

$$f_0 = \left(2.24 \times 10^4 \frac{\text{rad}}{\text{s}}\right)\underbrace{\left(\frac{1\text{ cycle}}{2\pi\text{ rad}}\right)}_{\text{Conversion ratio}} = \boxed{3.56\text{ kHz}}$$

(b) At resonance, the magnitude of the current is simply the magnitude of the applied voltage divided by the resistance:

$$I = \frac{V}{R} = \frac{100\text{ V}}{20\text{ }\Omega} = \boxed{5.00\text{ A}}$$

(c) The Q of the circuit is obtained from Equation (34-22):

$$Q = \frac{\omega_0 L}{R} = \frac{\left(2.24 \times 10^4 \frac{\text{rad}}{\text{s}}\right)(20 \times 10^{-3}\text{ H})}{20\text{ }\Omega} = \boxed{22.4}$$

Note that Q is dimensionless.

(d) The magnitude of the voltage V_L across the inductor is given by $V_L = X_L I$, where X_L is the inductive reactance at the resonant frequency and I is the magnitude of the current at resonance:

$$V_L = (\omega_0 L)(I) = \left(2.24 \times 10^4 \frac{\text{rad}}{\text{s}}\right)(20 \times 10^{-3}\text{ H})(5\text{ A}) = \boxed{2240\text{ V}}$$

Note that this voltage is considerably higher than the applied voltage of 100 V.

Parallel Resonance

One of the most common forms of a resonant circuit is a parallel combination of a capacitor and an inductor, such as that illustrated in Figure 34-18a. A resistor is shown in the branch containing the inductor to represent the resistance of the windings inherent to all inductors.[7] To analyze this circuit, we draw a phasor diagram for currents in a parallel circuit, with the applied voltage phasor **V** as a reference. At resonance, the current phasor **I** ($= \mathbf{I}_1 + \mathbf{I}_2$) is *in phase* with the applied voltage **V**. Furthermore, at resonance the current **I** is a minimum. Figure 34-18b is drawn for this resonance condition. The current phasor **I** is the vector sum of $\mathbf{I}_1$ (the current through the capacitor) and $\mathbf{I}_2$ (the current through the series RL branch). The current $\mathbf{I}_1$ leads **V** by $\pi/2$ rad, while the current $\mathbf{I}_2$ lags **V** by the angle ϕ, where

$$\phi = \tan^{-1}\left(\frac{\omega_0 L}{R}\right) \qquad (34\text{-}24)$$

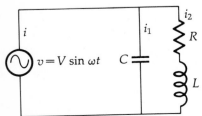

(a) A parallel resonant circuit.

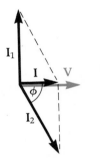

(b) The current phasor diagram at resonance.

FIGURE 34-18
Resonance in a parallel circuit.

---

[7] The reason we do not show a resistance in the capacitive branch is the following. If the dielectric material of a capacitor "leaks," allowing some current under DC conditions, this electrical resistance would be represented as a resistance *in parallel* with the capacitor. (Thus, for DC, some current would flow.) Since we usually try to design high-Q circuits, the DC resistance of capacitors can be made so high that the current through it is essentially zero and therefore the leakage resistance can be neglected in circuit analyses.

The magnitudes of $\mathbf{I}_1$ and $\mathbf{I}_2$ are

$$I_1 = \frac{V}{X_C} \quad \text{and} \quad I_2 = \frac{V}{\sqrt{X_L^2 + R^2}} \quad (34\text{-}25)$$

At resonance, the vertical components of $\mathbf{I}_1$ and $\mathbf{I}_2$ in the current phasor diagram must be equal. So the circuit behaves essentially as just a resistor. Thus:

$$\frac{V}{X_C} = \frac{V}{\sqrt{X_L^2 + R^2}} \sin \phi \quad (34\text{-}26)$$

where X_L, X_C, and ϕ are the values at resonance. Using Equation (34-24), $\tan \phi = \omega_0 L/R$. Therefore:

$$\sin \phi = \frac{\omega_0 L}{\sqrt{(\omega_0 L)^2 + R^2}}$$

Substituting the appropriate quantities into Equation (34-26) and solving for ω_0 yields

$$\omega_0 C = \left(\frac{1}{\sqrt{(\omega_0 L)^2 + R^2}}\right)\left(\frac{\omega_0 L}{\sqrt{(\omega_0 L)^2 + R^2}}\right)$$

RESONANT ANGULAR FREQUENCY ω_0 FOR THE PARALLEL RLC CIRCUIT OF FIGURE 34-18

$$\omega_0 = \sqrt{\frac{1}{LC} - \frac{R^2}{L^2}} \quad (34\text{-}27)$$

If R is small compared with L (corresponding to a high Q), the condition for the parallel resonance frequency ω_0 is the same as that for series resonance. See Figure 34-19.

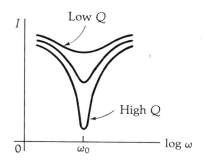

FIGURE 34-19
Resonance curves for parallel resonance. The current I into a parallel circuit is a minimum at resonance. Such a circuit may be used as a "band-stop" frequency filter, reducing currents with frequencies near ω_0 from existing in parts of the circuit that follow.

34.7 Power in AC Circuits

When we are considering DC circuits, the energy balance is quite simple. The average rate at which the seats of emf supply energy to the circuit equals the rate at which energy is lost through the Joule heating of the resistors. At any instant, the rate at which energy is supplied by the AC source must be balanced not only by the rate of Joule heating of resistors but also by the rate at which energy associated with magnetic and electric fields is stored or released in the inductors and capacitors. At any instant, the incremental work dW done by a source of varying voltage v in changing the potential of an incremental charge dq is $dW = v\,dq$. The rate at which work is being done at that instant is the *instantaneous* power p supplied by the source of the circuit: $p = dW/dt$. Combining these two equations gives

$$p = v\frac{dq}{dt} = vi \quad (34\text{-}28)$$

Since $v = V \sin \omega t$ and $i = I \sin(\omega t - \phi)$, the instantaneous power $p = vi$ is

$$p = VI \sin \omega t \sin(\omega t - \phi) \quad (34\text{-}29)$$

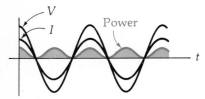

(a) A pure resistive load. Current I and voltage V are in phase: $\phi = 0°$. The power is always positive.

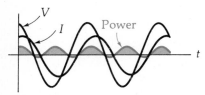

(b) An inductive reactance with resistance. The current I lags the applied voltage V: $\phi = 45°$. When the power is negative, energy is being returned from the inductance to the source.

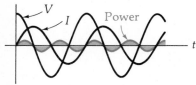

(c) A pure inductive load (no resistance). The current I lags the applied voltage V: $\phi = 90°$. The power varies equally between positive and negative values, so the average power is zero.

FIGURE 34-20
Voltage, current, and power vs. time in AC circuits. The instantaneous power P is the product of the instantaneous values of V and I and varies sinusoidally with a frequency $2f$. The average power P_{av} depends upon the phase angle ϕ: $P_{av} = VI \cos \phi$.

The power supplied to the circuit thus varies in time (Figure 34-20). However, we are most often concerned about the *average power* P_{av} supplied to the circuit. From the mathematical definition for the average over time (that is, the *time-weighted* average):

$$P_{av} = \frac{1}{T} \int_0^T p \, dt$$

where T is one period of power variation. [Note the similarity to the *mass-weighted* average used in the determination of center of mass, Equation (9-12)]. Substituting from Equation (34-29), we have

$$P_{av} = \frac{1}{T} \int_0^T VI \sin \omega t \sin(\omega t - \phi) \, dt \qquad (34\text{-}30)$$

From Appendix D, $\sin(\omega t - \phi) = (\sin \omega t \cos \phi - \cos \omega t \sin \phi)$. Therefore, Equation (34-30) becomes

$$P_{av} = \frac{VI \cos \phi}{T} \int_0^T \sin^2 \omega t \, dt - \frac{VI \sin \phi}{T} \int_0^T \sin \omega t \cos \omega t \, dt$$

Using Appendix G-II to evaluate the integrals, we have

$$P_{av} = \frac{VI \cos \phi}{T} \left(\frac{t}{2} - \frac{\sin 2\omega t}{4} \right) \Big|_0^T - \frac{VI \sin \phi}{T} \left(\frac{\sin^2 \omega t}{2} \right) \Big|_0^T$$

Substituting the limits and using the relation $T = 2\pi/\omega$, we have

$$P_{av} = \frac{VI}{2} \cos \phi \qquad (34\text{-}31)$$

where ϕ is the phase angle between the voltage v and current i. (Note that the integral of either $\sin^2 \omega t$ or $\cos^2 \omega t$ over a period T is equal to $\frac{1}{2}$.) The cosine term is called the **power factor**. From Figure 32-12b it equals

POWER FACTOR $$\cos \phi = \frac{R}{Z} \qquad (34\text{-}32)$$

The fact that the average power supplied to the circuit depends on the *cosine of the phase angle* has important implications concerning how the power is dissipated in the components of the circuit. In a purely inductive circuit, the phase angle $\phi = \pi/2$. Since the cosine of $\pi/2$ is zero, the *average* power dissipated in the inductor is zero. We may interpret this physically by realizing that the work done by the source of current in building the magnetic field of the inductor is returned to the source when the field collapses. Similarly, since for a purely capacitive circuit $\phi = -\pi/2$, the *average* power dissipated in a capacitor is also zero. The work done in creating the electric field in the capacitor is returned to the source when the field collapses. If you follow the buildup and reduction of these fields, you will discover that the processes occur exactly 180° out of phase; while the electric field is building up, the magnetic field is collapsing and vice versa. In effect, the inductance and capacitance merely exchange energy back and forth between themselves. If the reactances are pure (if there is no resistance associated with them), there is no average energy loss in the reactances. *The only energy dissipation occurs in the resistance R by Joule heating.* Since the voltage v_R across the resistor is in phase with the current i through it, $\cos \phi = \cos 0° = 1$, and Equation (34-31) becomes

$$P_{av} = \frac{V_R I}{2} \qquad (34\text{-}33)$$

We may also express the average power in terms of the *root-mean-square*[8] (rms) values of V and I:

ROOT-MEAN-SQUARE VALUES
$$V_{rms} = \frac{V}{\sqrt{2}} \quad \text{and} \quad I_{rms} = \frac{I}{\sqrt{2}} \qquad (34\text{-}34)$$

leading to
$$P_{av} = V_{rms} I_{rms} \cos \phi \qquad (34\text{-}35)$$

Using $V_R = V \cos \theta$, we may express P_{av} in still another form:

$$P_{av} = (V_R)_{rms} I_{rms} \qquad (34\text{-}36)$$

The several forms for P_{av} are listed together for easy reference:

AVERAGE POWER DISSIPATED IN AN RLC CIRCUIT

$$P_{av} \begin{cases} = \dfrac{VI}{2} \cos \phi & (34\text{-}37) \\ = V_{rms} I_{rms} \cos \phi & (34\text{-}38) \\ = \dfrac{V_{rms}^2}{Z} \cos \phi & (34\text{-}39) \\ = (V_R)_{rms} I_{rms} & (34\text{-}40) \\ = I_{rms}^2 R & (34\text{-}41) \end{cases}$$

The last two equations are similar to the expression for DC circuits, in which the power dissipated in a resistor is

(For DC circuits) $\quad P = V_R I \quad$ and $\quad P = I^2 R \qquad (34\text{-}42)$

where P is the constant power dissipated in a resistor that has a constant potential difference V_R across its terminals, resulting in a constant current I through the resistor. The similarity of Equations (34-40) and (34-41) to Equations (34-42) is the basis for describing rms values as **effective** values. *The rms values of current and voltage produce the same Joule heating in a resistor as DC current and voltage of the same magnitudes*; they are just as "effective" in producing I^2R losses in a resistor.

Remember that the above rms values are "effective" values for *sinusoidal* currents and voltages only. (Other waveshapes have different effective values.) Power-line currents and voltages are always quoted in rms values, even though the subscript is commonly omitted. For example, an electrical outlet supplying

---

[8] The root-mean-square value of any quantity is the square *root* of the average (or *mean*) value of the *square* of the quantity. Thus, in the case of a sinusoidally varying voltage,

$$V_{rms} = \left(\frac{1}{T} \int_0^T V^2 \sin^2 \omega t \, dt \right)$$

where T is the period of the variation. Since $1/T \int_0^T \sin^2 \omega t \, dt = \frac{1}{2}$, this equals $V_{rms} = V/\sqrt{2}$.

an AC voltage of 110 V, 60 Hz, has a *peak* value of $(\sqrt{2})(110\ V)$, or 156 V. Such a line voltage would be expressed analytically in SI units as 110 V, 60 Hz $\Rightarrow$ 156 sin(120πt) V.

EXAMPLE 34-9

An AC voltage of the form (in SI units)

$$v = 100 \sin(1000t) \quad \text{(in volts if } t \text{ is in seconds)}$$

is applied to a series *RLC* circuit. If $R = 400\ \Omega$, $C = 5.0\ \mu F$, and $L = 0.50\ H$, find the average power dissipated in the circuit.

SOLUTION

All three expressions for the average power involve the current, so we first solve for the current in the circuit. From Equation (34-18), the impedance Z is

$$Z = \sqrt{R^2 + (X_L - X_C)^2}$$

where
$$X_L = \omega L = \left(1000\ \frac{\text{rad}}{\text{s}}\right)(0.50\ H) = 500\ \Omega$$

and
$$X_C = \frac{1}{\omega C} = \frac{1}{\left(1000\ \frac{\text{rad}}{\text{s}}\right)(5.0 \times 10^{-6}\ F)} = 200\ \Omega$$

Substituting gives
$$Z = \sqrt{(400\ \Omega)^2 + (500\ \Omega - 200\ \Omega)^2} = 500\ \Omega$$

The amplitude of the current is
$$I = \frac{V}{Z} = \frac{100\ V}{500\ \Omega} = 0.200\ A$$

Knowing the current, we may use any one of the expressions for the average power. We will illustrate the use of all five.

Using Equation (34-37):
$$P_{av} = \frac{VI}{2} \cos \phi$$

where
$$\cos \phi = \frac{R}{Z} = \frac{400\ \Omega}{500\ \Omega} = 0.800$$

Substituting gives
$$P_{av} = \frac{(100\ V)(0.200\ A)}{2}(0.800) = \boxed{8.00\ W}$$

Using Equation (34-38):

$$P_{av} = V_{rms} I_{rms} \cos \phi = \left(\frac{100\ V}{\sqrt{2}}\right)\left(\frac{0.200\ A}{\sqrt{2}}\right)(0.800) = \boxed{8.00\ W}$$

Using Equation (32-39):

$$P_{av} = \frac{V_{rms}^2}{Z} \cos \phi = \left(\frac{100\ V}{\sqrt{2}}\right)^2 \left(\frac{1}{500\ \Omega}\right)(0.800) = \boxed{8.00\ W}$$

Using Equation (32-40):
$$P_{av} = (V_R)_{rms} I_{rms}$$

where
$$(V_R)_{rms} = (IR)_{rms} = \frac{(0.200\ A)(400\ \Omega)}{\sqrt{2}} = \frac{80}{\sqrt{2}}\ V$$

Substituting gives $P_{av} = \left(\dfrac{80}{\sqrt{2}} \text{ V}\right)\dfrac{(0.200 \text{ A})}{\sqrt{2}} = \boxed{8.00 \text{ W}}$

Using Equation (34-41):

$$P_{av} = I_{rms}^2 R = \left(\dfrac{0.200 \text{ A}}{\sqrt{2}}\right)^2 (400 \text{ }\Omega) = \boxed{8.00 \text{ W}}$$

34.8 Transformers

One of the most universally useful electrical devices is a *transformer*. It is capable of raising or lowering the amplitude of an AC voltage without appreciable loss of power. To transmit power over great distances, the sinusoidally varying voltage at the source is usually raised by a transformer to a very high value. Since the total power ($V_{rms}I_{rms}$) remains the same, raising the voltage means that the current is lower. Consequently, the I^2R losses in the transmission lines are reduced. At the consumer end, another transformer lowers the voltage to a safe and practical value for use in household appliances.

Figure 34-21 is the conventional way of indicating a transformer. An AC generator supplies the *input* voltage V_1 to the primary winding. The other side is the secondary winding, which has the *output* voltage V_2 across its terminals. The soft iron core greatly increases the magnetic flux and, because flux lines are almost entirely confined within the iron, also ensures that essentially all the flux that links the primary coil also links the secondary coil. That is, there is very little "leakage." In an *ideal* transformer, we assume no leakage and no thermal losses in the core or windings.[9] Consider first that the secondary switch is open, so that there is no secondary current and no power is transmitted through the transformer. The changing magnetic flux in the secondary winding induces an emf across the output terminals. Since both windings surround essentially the same varying magnetic flux, the emf $\mathscr{E}_1$ per turn N_1 in the primary is the same as the emf $\mathscr{E}_2$ per turn N_2 in the secondary. Mathematically, this is

$$\dfrac{\mathscr{E}_1}{N_1} = \dfrac{\mathscr{E}_2}{N_2} \qquad (34\text{-}43)$$

If the resistance of the primary windings is negligible compared with its reactance, the transformer is essentially a pure inductor connected to an AC generator. Current and voltage are 90° out of phase, so the average power that the AC generator delivers to the transformer is zero.

If we now add a resistive load R_{load} across the secondary terminals, there is a current in the secondary windings and power I^2R_{load} is developed in the load resistor. By Lenz's law, this secondary current produces a magnetic flux that opposes the flux produced by the primary current, tending to reduce the primary voltage. But the primary voltage V_1 is fixed by the AC generator, so the primary circuit draws extra current from the generator to maintain the

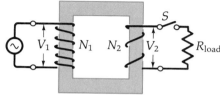

Primary   Secondary

(a) An AC voltage source supplies the input of a step-down, iron-core transformer with N_1 turns in the primary and N_2 turns in the secondary.

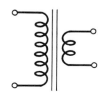

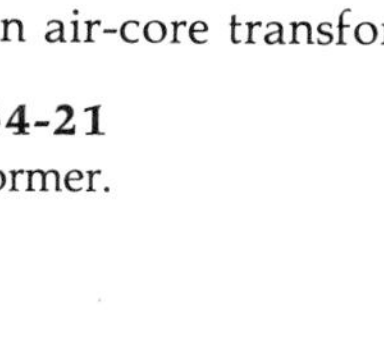

(b) The circuit symbol for a step-down, iron-core transformer. (The vertical lines are omitted for an air-core transformer.)

FIGURE 34-21
The transformer.

---

[9] There are well-designed, high-capacity transformers that approach 99% efficiency, so our assumption of an "ideal" transformer is reasonable. The Joule-heating losses in the windings are reduced by the use of low-resistance wires, the eddy-current losses are reduced by the laminations of the core, and a soft iron core with a very thin hysteresis loop reduces the hysteresis losses (Section 33.3).

original flux. Replacing the emf symbols by their respective voltage symbols, we can write the above equation as

$$\frac{N_2}{N_1} = \frac{V_2}{V_1} \tag{34-44}$$

showing that the *turns ratio* (N_2/N_1) for an ideal transformer is the same as the voltage ratio (V_2/V_1). If the output voltage is larger than the input voltage, the transformer is called a **step-up** transformer; if the output voltage is lower, it is a **step-down** transformer. Because the power input $V_1 I_1$ equals the power output $V_2 I_2$ (if ideal), the turns ratio is the inverse of the current ratio:

$$\frac{N_2}{N_1} = \frac{I_1}{I_2} \tag{34-45}$$

Thus a transformer may be used to transform a varying current as well as a varying voltage. (Note that a step-up transformer steps down the current.)

In the United States, at the generator the voltage is commonly stepped up to 350 kV for long-distance transmission, then for safety it is stepped down to 20 kV for local distribution, and finally to 110 V or 220 V for household use by the transformers on neighborhood utility poles. Three-wire systems are used, with one wire grounded. Smaller appliances utilize the 110-V voltage between one "hot" wire and the ground, while 220 V between the two hot wires serves larger appliances such as clothes driers and electric stoves.

EXAMPLE 34-10

Consider the ideal transformer shown in Figure 34-21. What is the ratio V_1/I_1 in terms of N_1, N_2, and R_2? (This is an important ratio because it is the *equivalent input resistance* that the source "sees" when a resistive load R_{load} is placed across the secondary.)

SOLUTION

The secondary voltage is

$$V_2 = R_2 I_2 \tag{34-46}$$

From Equation (34-44) we have $V_2 = (N_2/N_1)V_1$, and from Equation (34-45) we have $I_2 = (N_1 N_2)I_1$. Substituting these into Equation (34-46) and solving for $R_{eff} = V_1/I_1$ gives

$$\left(\frac{N_2}{N_1}\right) V_1 = R_2 \left(\frac{N_1}{N_2}\right) I_1$$

$$\boxed{R_{eff} = \frac{V_1}{I_1} = \left(\frac{N_1}{N_2}\right)^2 R_2} \tag{34-47}$$

R_{eff} is the *effective resistance* (for the load R_2) that the generator sees.

The technique of changing the effective resistance using a transformer is very important in power transfer from one part of an electronic circuit to another. Just as was shown for DC power sources (Problem 29C-51), the *maximum power transfer* occurs when the load that a generator "sees" has the same magnitude impedance as the internal impedance of the generator itself.[10] This is why, for example, sinusoidal voltage generators have impedance-matching transformers just before the load terminals. If mismatched, more power is dissipated in the generator than in the load.

EXAMPLE 34-11

An AC source has an internal resistance of 3200 Ω. In order for the maximum power to be transferred to an 8-Ω resistive load R_2, a transformer is used between the source and the load. Assuming an ideal transformer, (a) find the appropriate turns ratio of the transformer. If the output voltage of the source is 80 V (rms), determine (b) the rms voltage across the load resistor and (c) the rms current in the load resistor. (d) Calculate the power dissipated in the load. (e) Verify that the ratio of currents is inversely proportional to the turns ratio.

SOLUTION

(a) For maximum power transfer, the effective resistance of the 8-Ω load (as viewed from the primary side) should be 3200 Ω. From Equation (34-47),

$$R_{eff} = \left(\frac{N_1}{N_2}\right)^2 R_2$$

Thus:
$$\frac{N_1}{N_2} = \left(\frac{R_{eff}}{R_2}\right)^{1/2} = \left(\frac{3200\ \Omega}{8\ \Omega}\right)^{1/2} = \boxed{20}$$

The primary should have twenty times as many turns as the secondary.

(b) Using Equation (34-44) and substituting rms numerical values, we obtain

$$V_2 = V_1\left(\frac{N_2}{N_1}\right) = (80\text{ V rms})\frac{1}{20} = \boxed{4.00\text{ V rms}}$$

(c) The load current is
$$I_2 = \left(\frac{V_2}{R_2}\right) = \frac{4\text{ V rms}}{8} = \boxed{0.500\text{ A rms}}$$

(d) Since the load is a pure resistance, the power is

$$P_2 = (I_{rms})^2 R_2 = (0.500\text{ A})^2(8\ \Omega) = \boxed{2.00\text{ W}}$$

(If the impedance-matching transformer were omitted and the load resistor connected directly to the AC source, the power in the load would be only 7.77×10^{-5} W).

(e) The rms current in the primary is

$$I_1 = \frac{V_1}{R_1} = \frac{80\text{ V rms}}{3200} = \underline{\underline{25\text{ mA rms}}}$$

So the current ratio is

$$\frac{I_1}{I_2} = \frac{25 \times 10^{-3}\text{ A rms}}{0.500\text{ A rms}} = \boxed{\frac{1}{20}}$$

which is the inverse of the turns ratio.

---

[10] More accurately, it can be shown that, if the load has a reactive component that is capacitive, the source should have an equal-magnitude inductive component and vice versa. We do not take up such cases.

Summary

AC voltages and currents are described mathematically by

$$v = V \sin \omega t \quad \text{and} \quad i = I \sin(\omega t - \phi)$$

where i and v are the sinusoidally varying values, I and V are peak values, and ϕ is the *phase angle* between v and i. The phase relations in "pure" circuit elements are

- Resistor: i and v are *in phase*.
- Capacitor: i leads v by $\pi/2$ rad.
- Inductor: i lags v by $\pi/2$ rad.

The *reactances* of circuit elements are

- Capacitive: $X_C = 1/\omega C$
- Inductive: $X_L = \omega L$
- Total reactance: $X = X_L - X_C$

Series RLC circuit:

Impedance Z: $\quad Z = \sqrt{R^2 + (X_L - X_C)^2}$

Current i: $\quad i = \dfrac{V}{Z} \sin(\omega t - \phi)$

where $\quad \phi = \tan^{-1}\left(\dfrac{X_L - X_C}{R}\right)$

In *phasor diagrams*, the amplitudes of voltages and currents are represented by vectorlike arrows called *phasors*, drawn to depict their phase relationships. The phasor diagram rotates counterclockwise, with angular frequency ω. The projections of the phasors on a vertical axis give the instantaneous values of v and i.

An *impedance diagram* depicts R along the $+x$ axis, X_L along the $+y$ axis, and X_C along the $-y$ axis. The arrows add vectorially to give the impedance Z, with the phase angle ϕ between R and Z.

RESONANCE.

Series: $\quad X_L = X_C \;\Rightarrow\; \omega_0 = \dfrac{1}{\sqrt{LC}}$

Parallel: For a capacitor in parallel with an inductor–resistor series combination:

$$\omega_0 = \sqrt{\dfrac{1}{LC} - \dfrac{R^2}{L^2}}$$

Sharpness of resonance Q: $\quad Q = \dfrac{\omega_0 L}{R}$

The *effective* (or *rms*) value of a sinusoidally varying current or voltage is *that DC current or voltage that produces the same heating effect in a resistor*. It is related to peak values as

$$I_{\text{eff}} = I_{\text{rms}} = \dfrac{I}{\sqrt{2}} \quad \text{and} \quad V_{\text{eff}} = V_{\text{rms}} = \dfrac{V}{\sqrt{2}}$$

The *average power* in AC circuits: All the average power dissipated in AC circuits is in the resistive components. If a power supply delivers to a circuit a current I at a voltage V (peak values), then

$$P_{\text{av}} = \dfrac{VI}{2} \cos \phi = V_{\text{rms}} I_{\text{rms}} \cos \phi = \dfrac{V_{\text{rms}}^2}{Z} \cos \phi$$

$$= (V_R)_{\text{rms}} I_{\text{rms}} = I_{\text{rms}}^2 R$$

(These last two relations involve the resistive element alone.)

Transformers: Letting the subscript 1 refer to the *primary* and the subscript 2 refer to the *secondary*, in an *ideal* transformer (no $I^2 R$ losses and no flux leakage) the input power equals the output power:

$$V_1 I_1 = V_2 I_2$$

The *turns ratio* is

For the voltage

$$\dfrac{N_2}{N_1} = \dfrac{V_2}{V_1}$$

For the current

$$\dfrac{N_2}{N_1} = \dfrac{I_1}{I_2}$$

In a *step-up* transformer, $V_2 > V_1$ (with $I_2 < I_1$); in a *step-down* transformer, $V_2 < V_1$ (with $I_2 > I_1$).

The *effective resistance* R_{eff} of the load resistor R_2 viewed from the primary is

$$R_{\text{eff}} = \left(\dfrac{N_1}{N_2}\right)^2 R_2$$

Questions

1. Using nonmathematical reasoning, can you explain why the current through a capacitor leads the voltage across the capacitor and why the current through an inductor lags the voltage across the inductor?
2. A square-wave voltage is applied to a series combination of a resistor and an inductor. If the resistance is large compared with the inductive reactance (corresponding to the lowest Fourier component of the square-wave), what is the voltage waveform across the inductor?
3. If the secondary winding of a transformer is open circuit, why does a small current still pass through the primary winding?

4. An AC voltage is applied to a series *RLC* circuit. How does the phase constant change as the frequency of the applied voltage changes from zero to a very high value?

5. In what ways do Kirchhoff's junction and loop rules for DC circuits have to be modified to apply to AC circuits?

6. As the frequency of an AC power source varies from zero to a very high value, how does the behavior of a series combination of a capacitor and an inductor compare with that of a parallel combination of a capacitor and an inductor?

7. In a parallel circuit, one branch has a capacitive reactance X_C, while the other branch has an inductive reactance X_L, where $X_L > X_C$. Is the parallel combination capacitive or inductive?

8. Why is it often "hazardous to your health" to experiment with high-*Q* resonant circuits (unless you take careful precautions)?

9. In a series *RLC* circuit, how should the frequency be adjusted so as to dissipate the maximum amount of power in the resistor?

10. Is the rms current through a series *RLC* circuit at $1/N$ times the resonant frequency equal to the rms current at N times the resonant frequency, where N is any number?

11. Is it possible to have resonance in a power transmission line? If so, and if such resonance presents a serious problem to power transmission, how could the problem be avoided?

12. An *RLC* circuit is analogous to a driven mechanical oscillator. What are the analogies between the two systems?

13. An AC voltage is applied to a series *RLC* circuit. In what ways could you determine whether the circuit is above or below resonance? Repeat for a parallel *RLC* circuit (capacitance in one branch and inductance in the other).

14. The resonant power circuit of a radio transmitter has an inductor made of very heavy wire mounted on large insulators. Why?

15. Why is it inadvisable to interchange the input and output terminals of a step-down transformer in order to make it a step-up transformer? (Hint: what limits the primary current with an open-circuit secondary?)

16. Is the power dissipation in an *RLC* circuit continuous or pulsating?

17. In order to reduce household electrical power consumption, why not decrease the power factor rather than decrease the rms current?

18. A resistor is connected to the secondary winding of a transformer while a square-wave voltage is applied to the primary winding. What is the voltage waveform across the secondary?

19. The average power dissipated in an ideal inductor or capacitor is zero. How does the instantaneous power input to these devices vary with time? How does this variation lead to the conclusion that the average power input is zero?

20. Edison proposed that power distribution systems should be direct current. What are the advantages and disadvantages of such a system?

21. Why is the engineer in a commercial power station concerned about the power factor of the load the station supplies? (Hint: consider power losses in transmission lines.)

22. An AC voltage source whose frequency can be varied is applied to a series *RLC* circuit. As the frequency is raised from ω_1 to ω_2, the current gradually decreases. Suppose a capacitor is now added in series with the circuit. Will this increase or decrease the original impedance in this range of frequencies?

23. How could a transformer be used as a variable inductance? Your answer should also explain why the technique is not used.

Problems

34.2 Simple AC Circuits

34A-1 Beginning with the definitions of capacitance and inductance, show that (a) capacitive reactance and inductive reactance have the dimensions of ohms and (b) that $(LC)^{1/2}$ has the dimensions of time.

34A-2 (a) Find the reactance of an 8-μF capacitor at 60 Hz and at 6000 Hz. (b) Repeat part (a) for an 8-mH inductor. (c) At what frequency is the reactance of the capacitor equal to the reactance of the inductor?

34A-3 Show that $i = (V/\omega L) \sin(\omega t - 90°)$ is a solution to the differential equation $V \sin \omega t - L\, di/dt = 0$.

34B-4 Refer to the AC voltage generator shown in Figure 34-22. (a) Show that the torque required to turn the generator is given by $\tau = [\omega(abB)^2/R] \sin^2 \omega t$. (b) Describe the orientation of the loop relative to **B** at the instant when the torque is a maximum. The rectangular loop has side lengths a and b.

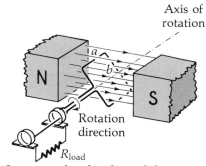

FIGURE 34-22
Problems 34B-4 and 34B-5.

34B-5 Figure 34-22 shows a simple AC generator. As the wire loop rotates in the presence of a uniform magnetic field, the induced emf in the loop is of the form $v = V \sin \omega t$. Consider a rectangular loop of sides $a = 0.2$ m and $b = 0.4$ m, rotating at 3600 rpm in the presence of a uniform field $B = 0.8$ T. (a) Write an equation for the induced emf, including numerical values for V and ω in SI units. (b) Describe the orientation of the loop relative to **B** at the instant $t = 0$.

34.4 Impedance in Series *RLC* Circuits

34A-6 A voltage $v = 100 \sin 2500t$ (in SI units) is applied to a series combination of a 30-Ω resistor and a 10-μF capacitor. (a) Make impedance and phasor diagrams for the circuit. (b) Calculate the maximum energy stored in the electric field of the capacitor.

34A-7 A voltage $v = 100 \sin 2500t$ (in SI units) is applied to a series combination of a 30-Ω resistor and a 15-mH inductor. (a) Make impedance and phasor diagrams for the circuit. (b) Calculate the maximum energy stored in the magnetic field of the inductor.

34A-8 A voltage $v = 100 \sin 2500t$ (in SI units) is applied to a series combination of a 30-Ω resistor, a 15-mH inductor, and a 10-μF capacitor. (a) Make impedance and phasor diagram for the circuit. (b) Calculate the maximum energy stored in the magnetic field of the inductor.

34B-9 A voltage $v = 10 \sin 1000t$ (in SI units) is applied across a 1.0-μF capacitor in series with a 1.5-kΩ resistor. (a) Draw a phasor diagram showing the input voltage, the voltages across the resistor and the capacitor, and the current. (b) Describe the voltage across the resistor in a functional form similar to that describing the input voltage. Include the phase constant.

34B-10 The input to a phase-shifting circuit (see Problem 34B-11) is $15 \sin 1000t$ (in SI units). The desired output is $V_0 \sin(1000t + \pi/3)$. (a) Devise the phase-shifter using a 10^4-Ω resistor and a capacitor. (b) Repeat, using the same resistor and an inductor. (c) Determine the value of V_0 in each case.

34B-11 Consider the phase-shifter circuit shown in Figure 34-23. The input voltage is described by $v = 10 \sin 200t$ (in SI units). If $L = 500$ mH, (a) find the value of R such that the output voltage v_0 lags the input voltage by 30° and (b) find the amplitude of the output voltage.

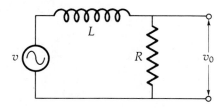

FIGURE 34-23
Problem 34B-11.

34B-12 The circuit shown in Figure 34-24 is called a "low-pass" filter. The impedance of the capacitor becomes less at higher frequencies, so the output voltage for higher frequencies is reduced. The *half-power frequency* is defined as the frequency above which the amplitude of the output voltage is smaller than $1/\sqrt{2}$ times the input voltage. (a) Derive the expression for the half-power frequency, ω, in terms of R and C. (b) Find the phase of the output voltage relative to the input voltage at this frequency.

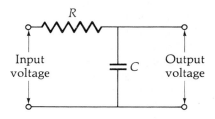

FIGURE 34-24
Problem 34B-12.

34.5 Impedance in Parallel *RLC* Circuits

34B-13 Show that the impedance Z of a resistor R, an inductor L, and a capacitor C all in parallel with one another is given by $Z^{-2} = R^{-2} + (1/\omega L - \omega C)^{-2}$.

34B-14 The voltage $v = 240 \sin 500t$ (in SI units) is applied across a parallel combination of a 600-Ω resistor and a 2.5-μF capacitor. Express the current from the voltage source in the form $i = I \sin(500t - \phi)$, including numerical values for I and ϕ in SI units.

34B-15 A voltage $v = 40 \sin 10^5 t$ (in SI units) is applied to a parallel combination of a 60-Ω resistor and a 0.2-mH inductor. Write an equation for the current i from the source in the form $i = I \sin(10^5 t - \phi)$, including numerical values for I and ϕ in SI units.

34A-16 A sinusoidally varying voltage with an amplitude of 100 V is connected across a series combination of a 10-Ω resistor, a 100-mH inductor, and a 0.1-μF capacitor. Calculate the amplitude of the voltage across the capacitor at (a) the resonant frequency and (b) $\frac{1}{10}$ the resonant frequency. (c) At each of these frequencies, is the circuit classed as mainly inductive, capacitive, or resistive?

34.6 Resonance

34A-17 Calculate the Q of the circuit of Problem 34A-16.

34A-18 A series *RLC* circuit resonates at 1070 kilocycles per second. (a) If $C = 0.2$ μF, find the value of L. (b) What is R if $Q = 70$?

34A-19 The tuning circuit of an AM radio is a parallel *LC* combination that has negligible resistance. The inductance is 0.2 mH and the capacitor is variable, so that the circuit can resonate at frequencies between 550 kHz and 1650 kHz. Find the range of values for C.

34B-20 For a parallel resonant circuit, Figure 34-18, sketch a freehand graph of the phase constant ϕ vs. ω where ϕ is the angle by which the phase of the current i differs from the applied voltage v.

34B-21 Show that the phase constant ϕ in a series *RLC* circuit may be expressed in terms of Q and the resonant frequency ω_0 by the equation $\tan \phi = Q(\omega^2 - \omega_0^2)/\omega \omega_0$.

34.7 Power in AC Circuits

34A-22 A 4.7-kW clothes drier operates from 220 V (rms), 60 Hz. Find (a) the rms current and (b) the peak current. (c) What would these values be for a 110-V (rms) source?

34A-23 The voltage at a household electrical outlet is often stated as "120 volt, 60 cycle." The "120 volt" is the rms value of the voltage and the "60 cycle" represents a frequency of 60 Hz. Describe this voltage in the form $v = V \sin \omega t$, including numerical values.

34A-24 For the circuit of Problem 34A-16, find the average power dissipated in the circuit (a) at resonance and (b) at one-tenth the resonant frequency.

34A-25 A sinusoidal voltage with an amplitude of 156 V is connected to a heater with a resistance of 100 Ω. Calculate the power dissipated in the heater.

34A-26 A voltage $v = 100 \sin 5000t$ (in SI units) is applied across a series combination of a 700-Ω resistor and a 100-mH inductor. (a) Sketch impedance and phasor diagrams for the circuit. (b) Calculate the rms current in the circuit. (c) Find the power dissipated in the resistor. (d) Calculate the power supplied to the circuit by the voltage source.

34B-27 A sinusoidal voltage with an rms amplitude V_{rms} is applied to a series combination of a resistor R, an inductor L, and a capacitor C. Show that the average power P_{av} dissipated in the circuit may be expressed as $P_{av} = RV_{rms}^2/Z^2$.

34B-28 An AC current of 0.5 A (rms) exists in an inductor that has a reactance of 39 Ω. The I^2R loss in the inductor is 8 W. Find the impedance of the inductor.

34B-29 The circuit shown in Figure 34-25 can be used as a "high-pass" filter. For a given input rms voltage V_{rms}, the power delivered to the resistor is essentially V_{rms}^2/R at high frequencies. Derive an expression for the frequency at which the power delivered to the resistor is $V_{rms}^2/2R$.

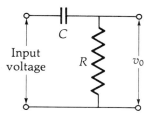

FIGURE 34-25
Problem 34B-29.

34B-30 An AC voltage with an amplitude of 100 V is applied to a series combination of a 200-μF capacitor, a 100-mH inductor, and a 20-Ω resistor. Calculate the power dissipation and the power factor for a frequency of (a) 60 Hz and (b) 50 Hz.

34B-31 A voltage $v = 200 \sin 2000t$ is applied across a series combination of a 2500-Ω resistor and a 1.5-H inductor. (a) Sketch an impedance diagram and a phasor diagram for this circuit. Calculate the rms values of (b) the applied voltage, (c) the current, (d) the voltage across the inductor, and (e) the voltage across the resistor.

34B-32 A 60-Hz, sinusoidally varying voltage with an amplitude of 156 V is applied to a 0.15-H inductor that has a resistance of 50 Ω. Calculate the rate at which heat is produced in the inductor when (a) the resistance of the inductor is considered to be a resistance in series with a resistanceless inductance and (b) when the resistance of the inductor is considered to be a resistance in parallel with a resistanceless inductance.

34B-33 The power delivered by a 110-V (rms), 60-Hz source is 480 W. The power factor is 0.70 and the current lags the voltage. (a) Find the value of the capacitor C added in series that will change the power factor to unity. (b) Find the power delivered by the source under these new conditions.

34B-34 The windings of a 150-mH inductor have 30-Ω resistance. A 20-V (rms), 60-Hz voltage is applied to the inductor. Assuming that the equivalent circuit is a resistance in series with a pure inductance, find (a) the power factor and (b) the power developed in the windings. (c) Suppose that the frequency of the applied voltage were changed to 50 Hz (with the same rms value). Find the power developed in the windings. (This problem is of practical importance when American electronic equipment designed for 60 Hz is taken to a foreign country where 50 Hz is the standard.)

34B-35 Consider a series combination of a 10-mH inductor, a 100-μF capacitor, and a 10-Ω resistor. A 50-V (rms) sinusoidal voltage is applied to the combination. Calculate the rms current for (a) the resonant frequency, (b) half the resonant frequency, and (c) double the resonant frequency.

34B-36 An AC voltage of amplitude 200 V, frequency 60 Hz, is applied to a series combination of a 900-Ω resistor and a 4-μF capacitor. (a) Sketch a phasor diagram showing V, V_R, V_C, and I, with their (peak) numerical values. (b) Find the rms value of the current in the circuit. (c) What is the phase angle between the applied voltage and the current? Does the current lead or lag the applied voltage? (d) Find the power developed in the circuit.

34.8 Transformers

34A-37 An "ideal" model-train transformer operates from 120 V (rms), 60 Hz. There are 600 turns in the primary and 100 in the secondary. When there is an rms current of 0.11 A in the primary, find the rms values of (a) the output voltage and (b) the output current in the secondary.

34B-38 A step-up transformer operating from 120 V (rms) furnishes 20 kV to a neon sign. For protection, a fuse inserted in the primary circuit is designed to blow when the secondary current exceeds 8 mA. (a) Find the turns ratio of the transformer. (b) At maximum current, what power is supplied to the transformer? (c) What is the current rating of the fuse?

34A-39 A power plant generates 400 mW at 22 kV, 60 Hz. For economy of transmission, an (ideal) transformer steps up the voltage to 440 kV. (a) Find the rms current on the generator side. (b) Find the rms current in the transmission line.

34B-40 A transformer operating from 120 V (rms) supplies a 12-V lighting system for a garden. Eight lights, each rated 40 W, are installed in parallel. (a) Find the equivalent resistance of the total lighting system. (b) What current is in the secondary circuit? (c) What single resistance, connected across the

120 V supply, would consume the same power as when the transformer is used? Show that this equals the answer to part (a) times the square of the turns ratio.

Additional Problems

34C-41 A voltage $v = 100 \sin 1000t$ (in SI units) is applied across a series combination of a 1000-Ω resistor, a 0.5-μF capacitor, and a 1.5-H inductor. (a) Sketch an impedance diagram for the circuit. At $t = 0.7$ ms, calculate the instantaneous voltage across (b) the resistor, (c) the capacitor, and (d) the inductor. (e) Calculate the algebraic sum of these voltages and compare with the applied voltage at that instant. (f) On a sketch of the circuit, indicate the instantaneous polarities of these voltages.

34C-42 Phase-shifters with only a resistor and a capacitor or an inductor can only produce phase shifts of less than 90°. Greater phase shifts can be achieved by using a series combination of a resistor, a capacitor, and an inductor. Consider such a circuit containing an 80-mH inductor, a 10-μF capacitor, and resistance R. (a) Determine the value of R and the location of the output terminals to produce an output voltage $v_0 = V_0 \sin(1000t + 120°)$, where the input voltage $v_i = 10 \sin 1000t$. (b) Find the value of V_0.

34C-43 Consider the circuit shown in Figure 34-26. The input voltage is a time-varying voltage (not necessarily sinusoidal). Show that the output voltage v_0 is approximately proportional to the integral of the input voltage v if the resistance R is much less than the inductive reactance at all frequencies present in the input voltage.

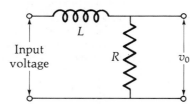

FIGURE 34-26
Problem 34C-43.

34C-44 A voltage $v = 100 \sin 2000t$ (in SI units) is applied across a series combination of a 2500-Ω resistor and a 1.5-H inductor. (a) Sketch an impedance diagram for this circuit. For the time $t = 1$ ms, calculate the instantaneous values of (b) the applied voltage, (c) the current, (d) the voltage across the resistor, and (e) the voltage across the inductor. The sum of your answers to (d) and (e) should equal the answer to (b).

34C-45 In the circuit of Figure 34-24, suppose that the inductor L is removed and a capacitor C is inserted in its place. Show that the output voltage v_0 is approximately the derivative of the input voltage v if the resistance is much less than the capacitive reactance at all frequencies present in the input voltage.

34C-46 A nonideal inductor whose windings have appreciable resistance is connected in series with a 4-μF capacitor across a 120-V (rms), 60-Hz power source. The rms voltage across the capacitor is 180 V and the rms voltage across the inductor is 75 V. If the nonideal inductor is assumed to be equivalent to a resistor in series with an ideal inductor, find (a) the inductance of such an ideal inductor and (b) the resistance of the series resistor.

34C-47 A 30-mH inductor and a 40-kΩ resistor are connected across a voltage source described by $v = 100 \sin 10^6 t$ (in SI units). Find the maximum rate at which the current is changing in the circuit.

34C-48 A sinusoidal voltage is applied to a series circuit of a 50-mH inductor, a 40-μF capacitor, and a 500-Ω resistor. Determine the frequency of the applied voltage that will create a current through the circuit that leads the applied voltage by 30°.

34C-49 Consider the circuit shown in Figure 34-27. The input voltage is time-varying (but not necessarily sinusoidal). Show that the output voltage v_0 is approximately proportional to the integral of the input voltage v if the capacitive reactance is much less than the resistance at all frequencies present in the input voltage.

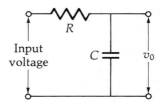

FIGURE 34-27
Problem 34C-49.

34C-50 Sketch a qualitative phasor diagram of the circuit shown in Figure 34-28 for the case in which the current in the source leads the applied voltage v.

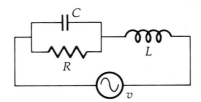

FIGURE 34-28
Problems 34C-50 and 34C-56.

34C-51 Show by direct substitution that $i = (V/Z) \sin(\omega t - \phi)$, where $\phi = \tan^{-1}[(X_L - X_C)/R]$, is a solution of Kirchhoff's loop rule for a series RLC circuit with an AC voltage source: $L(di/dt) + Ri + q/C = V \sin \omega t$.

34C-52 In the circuit of Figure 34-14a, $v = 100 \sin \omega t$, $R_1 = 0$, $X_C = 80\ \Omega$, and $i = 2 \sin(\omega t - 32.0°)$. (a) Find the total impedance Z (including phase angle) that the source "sees." (b) Find the impedance Z_2 of the inductive branch, including its phase angle ϕ_2 with respect to the resistance R_2. (c) Find the resistance R_2 and the reactance X_L of the inductance L.

34C-53 The circuit of Figure 34-14a has the following numerical values: $v = 200 \sin \omega t$, $R_1 = 4\ \Omega$, $R_2 = 15\ \Omega$, $X_C = 3\ \Omega$, and $X_L = 20\ \Omega$. Find an expression for the current i from the source, including the phase angle ϕ relative to the applied voltage.

34C-54 Using a method similar to that used to demonstrate resonance in a series inductance–capacitance–resistance circuit, plot a resonance curve for a capacitor in parallel with a series combination of a resistance and an inductance.

34C-55 An inductor is in series with an 80-Ω resistor and the combination is placed across a 110-V (rms), 60-Hz power source. If the resistor dissipates 50 W of power, find the inductance of the inductor.

34C-56 A series resonant circuit consists of an ideal inductor and a capacitor that "leaks," as indicated in Figure 34-28. Sketch a qualitative phasor diagram at the resonant frequency. Indicate the phasor representing the current through each component and the voltage across each component.

34C-57 A series RLC circuit has the following values: $R = 20\ \Omega$ and $X_L = 10\ \Omega$. The applied voltage is 50 V (rms) at $\omega = 400$ rad/s, and the value of the capacitance is unknown. The power factor is 0.800 and the current of 2 A (rms) leads the applied voltage. (a) Find the value of the capacitor. (b) There are several ways to bring the circuit into resonance. To what value should the angular frequency be changed to make resonance occur? (c) At this new resonant frequency, what power is developed in the circuit? (d) At this new resonance, what is the rms voltage across the inductor? (e) Suppose that we kept the original frequency of 400 rad/s and instead changed the value of C to achieve resonance. Find the value of a single capacitor that could be *added* to the circuit to bring it into resonance. Would it be added in series or in parallel with the original capacitor?

34C-58 A voltage $v = 100 \sin \omega t$ (in SI units) is applied across a series combination of a 2-H inductor, a 10-μF capacitor, and a 10-Ω resistor. (a) Determine the angular frequency ω_0 at which the power dissipated in the resistor is a maximum. (b) Calculate the power dissipated at that frequency. (c) Determine the two angular frequencies ω_1 and ω_2 at which the power dissipated is one-half the maximum value. [The Q of the circuit is approximately $\omega_0/(\omega_2 - \omega_1)$.]

34C-59 A certain source of AC power has an internal resistance r_s. (a) Prove that the maximum power that will be developed in a variable, external load resistor R_{load} occurs when $r_s = R_{\text{load}}$. (The matching of source and load resistances for maximum power transfer is called *impedance-matching*.)

34C-60 A 5-Ω resistor, a 2-μF capacitor, and an inductor are connected in series. An AC voltage of 20 mV (rms) at the resonant frequency of 5000 Hz is applied to the circuit. (a) What is the inductance L in the circuit? (b) Find the rms voltage across each circuit element. (c) The frequency of the applied voltage is now changed to 7500 Hz at the same rms value. Sketch an impedance diagram for the circuit at 7500 Hz. (d) Find the current in the circuit at this new frequency. Does the current lead or lag the applied voltage? (e) Find the power dissipated in the circuit at this new frequency.

# CHAPTER 35

Electromagnetic Waves

I have also a paper afloat, with an electromagnetic theory of light, which, till I am convinced to the contrary, I hold to be great guns.

J. C. MAXWELL, in a letter to C. H. Cay, 5 January 1865
[*American Journal of Physics*, 44, 676 (1976)]

One cannot escape the feeling that these [Maxwell's] mathematical formulae have an independent existence and an intelligence of their own, that they are wiser than we are, wiser even than their discoverers, that we get more out of them than was originally put into them.

HEINRICH HERTZ

35.1 Introduction

The waves that we discussed in Chapter 18 are mechanical waves, which must have a medium for their transmission from one location to another. For example, such phenomena as sound waves, water waves, and waves on a string all involve some physical medium undergoing mechanical motions as the wave disturbance passes by. We now describe electromagnetic waves, which can travel through the perfect vacuum of empty space.

In 1864, James Clerk Maxwell drew together the laws of electricity and magnetism into a single theory of *electromagnetism*.[1] It was surely a great stride forward in physics—indeed, one of the momentous intellectual achievements of humankind. Maxwell's complete unification of electricity and magnetism easily ranks with Newton's mechanics and Einstein's relativity. Maxwell's work also had a profound effect on the philosophical foundation of physics. The laws of physics began to assume a unity that was not previously apparent; this search for unification in other areas of physics continues today. Maxwell's work

---

[1] Maxwell's theory of electromagnetism rivals Newton's laws of mechanics for its elegance and wide applicability. In spite of their brevity, Maxwell's four equations include all that is known concerning macroscopic effects of electricity, magnetism, and electromagnetic waves (light, radio waves, and so on). True, on an atomic scale, quantum mechanics and relativity must be introduced. But these modern theories were purposely developed so as to reduce to the classical expressions of Maxwell and Newton in the limit of low velocities and macroscopic dimensions.

   Commenting on Maxwell's famous work, *Treatise on Electricity and Magnetism*, R. A. Millikan (1921 Nobel Prize winner) ranked it with Newton's *Principia*, "the one," he said, "creating our modern mechanical world and the other our modern electrical world."

led to the concept of the electromagnetic spectrum and to Heinrich Hertz's experimental verification of radio waves in 1890 (later exploited commercially by Marconi). His theory also made optics a branch of electromagnetism and established the basis for Einstein's work in relativity.

35.2 Displacement Current and Maxwell's Equations

We begin our discussion of Maxwell's equations by summarizing the laws of electricity and magnetism as they were known in 1870, Table 35-1. We will show how Maxwell made a crucial addition to one of them to create a unified theory that brought together the great discoveries of Coulomb, Faraday, Oersted, Ampère, and others into a single theory of electromagnetism.

By considering the following example, Maxwell recognized that Ampère's law, in the form $\oint \mathbf{B} \cdot d\boldsymbol{\ell} = \mu_0 I$, was incomplete. Suppose that a parallel-plate capacitor is being charged by a current I through the wires leading to the capacitor, as shown in Figure 35-1. We apply Ampère's law by constructing the curve C encircling the wire leading to the capacitor. If we choose S_1 as the flat surface enclosed by the curve C, the current I passes through this surface, producing a magnetic field in accordance with Ampère's law. If, however, we choose a curved surface S_2 that passes between the plates of the capacitor and is not pierced by a current-carrying conductor, Ampère's law predicts that a magnetic field does *not* exist along the curve C. To resolve this contradiction, Maxwell restated Ampère's law to cover such cases,

$$\oint_C \mathbf{B} \cdot d\boldsymbol{\ell} = \mu_0 \left(I + \varepsilon_0 \frac{d\Phi_E}{dt} \right) \tag{35-1}$$

where Φ_E is the electric field flux through the surface enclosed by the curve C.

We obtain the term $\varepsilon_0 (d\Phi_E/dt)$ by considering the circuit shown in Figure 35-2. After the switch is closed, charge flows in the conductor, charging the plates of the capacitor and creating an electric field E between the plates. If

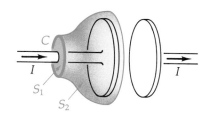

FIGURE 35-1
The integral $\oint \mathbf{B} \cdot d\boldsymbol{\ell}$ is calculated for the closed loop C that circles the wire.

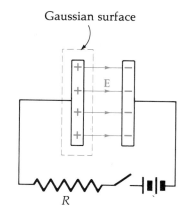

FIGURE 35-2
After the switch is closed, charges flow to the plates of the parallel-plate capacitor, creating the electric field $\mathbf{E}$ between the plates. The Gaussian surface encloses the charge q on one plate.

TABLE 35-1 The Laws of Electricity and Magnetism

Law	Phenomenon	Equation
Coulomb's law	The electrostatic force between charges	$F = \left(\dfrac{1}{4\pi\varepsilon_0}\right) \dfrac{q_1 q_2}{r^2} \hat{\mathbf{r}}$
Gauss's law	A mathematical consequence of the inverse-square form of Coulomb's law	$\oint \mathbf{E} \cdot d\mathbf{A} = \dfrac{q}{\varepsilon_0}$
Lorentz force law	The magnetic force on a moving charge (the definition of a magnetic field)	$\mathbf{F} = q\mathbf{v} \times \mathbf{B}$
	The electric force on a stationary charge (the definition of an electric field)	$\mathbf{F} = q\mathbf{E}$
Biot-Savart law	The magnetic field of a current-carrying conductor	$d\mathbf{B} = \left(\dfrac{\mu_0}{4\pi}\right) \dfrac{I\, d\boldsymbol{\ell} \times \hat{\mathbf{r}}}{r^2}$
Ampère's law (original form)	A mathematical consequence of the Biot-Savart law	$\oint \mathbf{B} \cdot d\boldsymbol{\ell} = \mu_0 I$
Faraday's law	An electric field produced by a changing magnetic flux	$\oint \mathbf{E} \cdot d\boldsymbol{\ell} = -\dfrac{d\Phi_B}{dt}$

we enclose one plate with a Gaussian surface, the charge q on the plates at any time is related to the electric flux Φ_E through the surface according to Gauss's law:

$$\frac{q}{\varepsilon_0} = \oint \mathbf{E} \cdot d\mathbf{A} = \Phi_E$$

We solve for q and find the current $I = dq/dt$ in the wire leading to the capacitor plate:

$$I = \frac{dq}{dt} = \varepsilon_0 \frac{d\Phi_E}{dt} \qquad (35\text{-}2)$$

We may interpret this result by considering Figure 35-1. Since $\oint \mathbf{B} \cdot d\boldsymbol{\ell}$ must have the same value whether the Gaussian surface encloses either of the surfaces, S_1 or S_2, then

NEW FORM OF AMPÈRE'S LAW (extended by Maxwell)

$$\oint \mathbf{B} \cdot d\boldsymbol{\ell} = \mu_0 \left(I + \varepsilon_0 \frac{d\Phi_E}{dt} \right) \qquad (35\text{-}3)$$

The term I equals zero if the surface S_2 is chosen, and the term $\varepsilon_0(d\Phi_E/dt)$ equals zero if S_1 is chosen. The term $\varepsilon_0(d\Phi_E/dt)$ has units of current and is called the *displacement current*[2]:

DISPLACEMENT CURRENT I_d

$$I_d = \varepsilon_0 \frac{d\Phi_E}{dt} \qquad (35\text{-}4)$$

We may obtain the magnetic field that exists between the plates of a charging capacitor *solely* by applying the Biot–Savart law to the conduction current in the wires leading to the plates.[3] Alternatively, we can evaluate the magnetic field using the extended form of Ampère's law, Equation (35-3). Suppose that we choose the curve C to enclose a plane between the plates of the capacitor. In this case, no current I pierces the plane, so that

$$\oint_C \mathbf{B} \cdot d\boldsymbol{\ell} = \mu_0 \varepsilon_0 \frac{d\Phi_E}{dt} \qquad (35\text{-}5)$$

In Figure 35-3, the curve C is a circle of radius $r < R$, concentric with the symmetry axis of the capacitor to take advantage of the symmetry. Integrating, we obtain

$$B(2\pi r) = \mu_0 \varepsilon_0 \frac{d\Phi_E}{dt} \left(\frac{r}{R} \right)^2$$

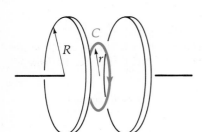

(a) The path of integration is the closed loop C.

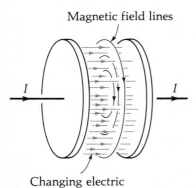

(b) The current causes the electric flux Φ_E to increase between the plates.

FIGURE 35-3
A changing electric field generates a magnetic field. Such a magnetic field between the plates of a capacitor has been experimentally verified.

---

[2] The word *displacement* comes from the fact that $\varepsilon_0 E$ is sometimes called the *electric displacement*. This term is a historical remnant from early proposals that a vacuum contained a polarizable ether analogous to dielectric materials that become polarized by the displacement of electric charges. The idea was eventually discarded, but the term remained. Interestingly, the concept of a displacement current was not favorably received by Maxwell's most distinguished contemporaries.

[3] This fact is often overlooked in discussions of the displacement current. See A. P. French and Jack R. Tessman, "Displacement Currents and Magnetic Fields," *American Journal of Physics* **31**, 201 (1963).

The factor $(r/R)^2$ is the fractional part of $d\Phi_E/dt$ that lies within the circle of radius r. Solving for B gives

$$B = \frac{\mu_0 \varepsilon_0}{2\pi} \frac{d\Phi_E}{dt} \left(\frac{r}{R^2}\right) \quad \text{(for } r \leq R\text{)} \quad (35\text{-}6)$$

We see that B increases linearly with the radius r until $r = R$. For $r \geq R$,

$$B = \frac{\mu_0 \varepsilon_0}{2\pi} \frac{d\Phi_E}{dt} \left(\frac{1}{r}\right) \quad \text{(for } r \geq R\text{)} \quad (35\text{-}7)$$

From Equation (35-2), $\varepsilon_0(d\Phi_E/dt)$ equals the current I leading to the capacitor, so the expression for B for $r \geq R$ becomes

$$B = \frac{\mu_0}{2\pi r} I \quad \text{(for } r \geq R\text{)} \quad (35\text{-}8)$$

This is the value of B around the current-carrying wire leading to the capacitor [also see Equation (31-2), Chapter 31]. Thus, the magnetic field outside the capacitor $(r > R)$ is the same as the field an equal distance from the wire.

The extended form of Ampère's law enables us to calculate the magnetic field where only a changing electric field exists, as shown in the following example.

EXAMPLE 35-1

Consider the situation illustrated in Figure 35-4. An electric field of 300 V/m is confined to a circular area 10 cm in diameter and directed outward from the plane of the figure. If the field is increasing at a rate of 20 V/m·s, what is the direction and magnitude of the magnetic field at the point P, 15 cm from the center of the circle?

SOLUTION

We use the extended form of Ampère's law, Equation (35-3). Since no moving charges are present, $I = 0$ and we have

$$\oint \mathbf{B} \cdot d\boldsymbol{\ell} = \mu_0 \varepsilon_0 \frac{d\Phi_E}{dt} \quad (35\text{-}9)$$

In order to evaluate the integral, we make use of the symmetry of the situation. Symmetry requires that no particular direction from the center can be any different from any other direction. Therefore, there must be *circular symmetry* about the central axis. From the experiment of Figure 35-3, we know the magnetic field lines are circles about the axis. Therefore, as we travel around such a magnetic field circle, the magnetic field remains constant in magnitude. Setting aside until later the determination of the *direction* of $\mathbf{B}$, we integrate $\oint \mathbf{B} \cdot d\boldsymbol{\ell}$ around the circle at $R = 0.15$ m to obtain $2\pi R B$. Differentiating the expression $\Phi_E = AE$, we have

$$\frac{d\Phi_E}{dt} = \left(\frac{\pi d^2}{4}\right) \frac{dE}{dt}$$

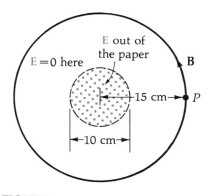

FIGURE 35-4
Example 35-1.

Equation (35-9) thus becomes
$$2\pi RB = \mu_0 \varepsilon_0 \left(\frac{\pi d^2}{4}\right)\frac{dE}{dt}$$

Solving for B gives
$$B = \frac{\mu_0 \varepsilon_0}{2\pi R}\left(\frac{\pi d^2}{4}\right)\frac{dE}{dt}$$

Substituting the numerical values yields

$$B = \frac{(4\pi \times 10^{-7}\ \text{H/m})(8.85 \times 10^{-12}\ \text{F/m})(\pi)(0.10\ \text{m})^2(20\ \text{V/m·s})}{(2\pi)(0.15\ \text{m})(4)}$$

$$= \boxed{1.85 \times 10^{-18}\ \text{T}}$$

Equation (35-2) determines the field direction, because it states that an *increasing* electric flux produces a magnetic field in the same manner as a current *I*. In Figure 35-3, the direction of the *increase* of electric field is out of the plane of the paper. By the right-hand rule, this implies that the direction of **B** is *counterclockwise*. Note that the magnitude of the electric field is irrelevant; only the *rate of change* of the electric flux determines the magnetic field.

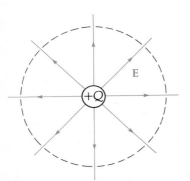

(a) An isolated electric charge.

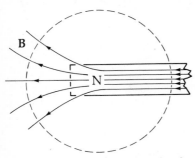

(b) One end of a magnetic dipole.

FIGURE 35-5
In (a), the electric flux through a closed Gaussian surface will not be zero if the surface encloses a net charge. In (b), the magnetic flux through the closed Gaussian surface is believed to always equal zero because monopoles apparently do not exist.

Table 35-1 has a notable omission. It does not include the fact that to our knowledge magnetic monopoles do not exist. The concept of a magnetic *monopole* has its origin in the comparison of a magnetic dipole with an electric dipole. Since electric dipoles are made up of two distinct electric charges, $+q$ and $-q$, it is tempting to visualize a magnetic dipole similarly as a pair of magnetic "charges," or *monopoles*, $+p$ and $-p$. The north pole of a magnet would contain a $+p$ monopole (from which field lines would emanate) and the south pole a $-p$ monopole (toward which field lines would converge). Of course, this way of thinking about it is contrary to the model of a magnetic dipole as a *current loop*. For a current loop, it seems inconceivable that a monopole could exist by itself: the current loop inherently generates both "poles" together, so that a magnetic *di*pole is the most fundamental magnetic structure. Breaking a long bar magnet in half produces two separate *dipoles* (not monopoles). Presumably the fragmenting process could be continued until just a single atom was left, with its inherent "loop current" and electron spin, also creating a dipole. It is interesting that some recent theories do predict that monopoles should exist. Many experiments have attempted to detect them, but to date they have not been found in nature. If monopoles were experimentally detected, it would require a change in certain electromagnetic equations.

Despite this disclaimer, it will be helpful to use the concept of a monopole for a short discussion. Figure 35-5 shows a comparison between charges and magnetic poles. Figure 35-5a demonstrates Gauss's law for electric fields:

GAUSS'S LAW FOR ELECTRIC FIELDS
$$\oint \mathbf{E} \cdot d\mathbf{A} = \frac{q}{\varepsilon_0} \tag{35-10}$$

where the total electric flux emanating from the Gaussian surface is not zero. Figure 35-5b shows the north-seeking pole of a long magnet surrounded by a Gaussian surface.

Since isolated magnetic monopoles apparently do not exist, the north pole of the magnet is always paired with a south pole. Thus, the magnetic flux emanating from any closed surface must equal that entering from the paired south pole. This fact may be formulated as

| GAUSS'S LAW FOR MAGNETIC FIELDS | $\oint \mathbf{B} \cdot d\mathbf{A} = 0$ | (35-11) |

Table 35-2 is a revised version of Table 28-1. Coulomb's law and the Biot–Savart law have been deleted because they are represented, respectively, by Gauss's law for electric fields and Ampère's law. The Lorentz force law has been deleted because it is essentially a statement of forces in terms of electric and magnetic fields. Ampère's law has been extended to include magnetic fields arising from changing electric field flux, and Gauss's law for magnetic fields has been added. The resulting Table 35-2 is a collection of four basic equations known as **Maxwell's electromagnetic field equations**,[4] in honor of Maxwell's great contribution.

To physicists, Maxwell's equations have great mathematical elegance and power. In spite of their compactness, they describe all phenomena in electricity and magnetism. Their far-reaching scope covers everything from electric motors and generators, radio, television, and high-energy particle accelerators to modern communication by fiber optics and the electromagnetic levitation of high-speed transportation vehicles. Maxwell's equations are regarded as the same kind of gigantic achievement as Newton's laws of motion.[5] An unexpected bonus (which no doubt would have pleased Maxwell greatly had he lived to see it) was the fact that Maxwell's equations survived the impact of Einstein's relativity unchanged, while Newton's laws had to be drastically altered for relative speed approaching the speed of light.

35.3 Electromagnetic Waves

As we have shown, Maxwell unified the theories of electricity and magnetism by extending Ampère's law. But another startling result of his accomplishment was the fact that his equations had *wavelike* solutions, which predicted that

TABLE 35-2 Maxwell's Equations (in vacuum)

Gauss's law for electric fields	$\oint \mathbf{E} \cdot d\mathbf{A} = \dfrac{q}{\varepsilon_0}$	(35-12)
Gauss's law for magnetic fields	$\oint \mathbf{B} \cdot d\mathbf{A} = 0$	(35-13)
Ampère's law (extended by Maxwell)	$\oint \mathbf{B} \cdot d\boldsymbol{\ell} = \mu_0 \left(I + \varepsilon_0 \dfrac{d\Phi_E}{dt} \right)$	(35-14)
Faraday's law	$\oint \mathbf{E} \cdot d\boldsymbol{\ell} = -\dfrac{d\Phi_B}{dt}$	(35-15)

---

[4] Maxwell's equations are often expressed as *differentials*. But the use of the differential form leads to mathematical procedures best postponed to a more advanced course in electromagnetism. In these equations, we can incorporate the presence of a dielectric material by simply replacing ε_0, the permittivity of free space (a vacuum), with ε, the permittivity of the dielectric material. Similarly, for magnetic materials, μ_0, the permeability of free space, is replaced by μ, the permeability of the magnetic material.

[5] The thermodynamicist Ludwig Boltzmann used a line from Goethe in commenting on them: "Was it a god who wrote these lines . . . ?". In his 1964 book *Electrons and Waves*, John R. Pierce gives a chapter the title "Maxwell's Wonderful Equations" and says, "To anyone who is motivated by anything beyond the most narrowly practical, it is worthwhile to understand Maxwell's equations simply for the good of his soul."

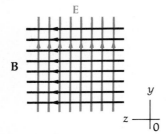

(a) The **E** and **B** fields are at right angles. (They extend like this, uniformly to infinity, all over the yz plane.)

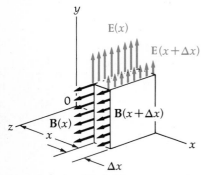

(b) The **E** and **B** field vectors at two different yz planes, spaced a distance Δx apart. The fields at x differ from the fields at $x+\Delta x$.

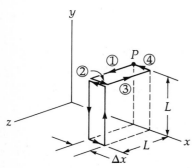

(c) The paths of integration along the edges of the top of the slab depicted in (b).

FIGURE 35-7
A plane electromagnetic wave traveling in the $+x$ direction has this pattern of "crossed" **E** and **B** fields.

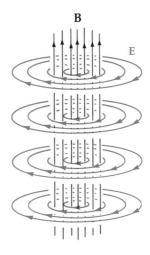

Faraday's law

$$\oint \mathbf{E} \cdot d\boldsymbol{\ell} = -\frac{d\Phi_B}{dt}$$

(a) If Φ_B increases uniformly, a constant **E** field is generated.

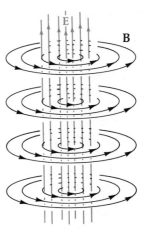

Ampere's law
(as extended by Maxwell)

$$\oint \mathbf{B} \cdot d\boldsymbol{\ell} = \mu_0 \epsilon_0 \frac{d\Phi_E}{dt}$$

(for current-free regions)

(b) If Φ_E increases uniformly, a constant **B** field is generated.

FIGURE 35-6
The symmetry of **E** and **B** in the absence of moving charges. Note that the symmetry is not quite exact, however, since the circular fields are in opposite senses. (One equation has a minus sign.)

electromagnetic waves could exist, even in a perfect vacuum. Furthermore, Maxwell showed that these waves had a numerical speed equal to c, the speed of light. This was the first definite indication that light was an electromagnetic wave phenomenon.

While you learn about electromagnetic waves, in addition to developing skill in manipulating the mathematical expressions it will also be helpful to gain a "feel" for what the wave patterns are like in space. As a first step in acquiring a pictorial acquaintance with electromagnetic waves, consult Figure 35-6. This illustrates the symmetry between **E** and **B** fields; in particular, that a changing **E** field generates a **B** field and vice versa.

It takes some mathematical manipulation to start with Maxwell's equations and derive electromagnetic waves. So we will not present the complete, step-by-step story. However, if we start with a simple combination of **E** and **B** fields, we will show that electromagnetic *waves* follow and that the equations for these waves do agree with Maxwell's equations. The starting point is a combination of "crossed" **E** and **B** fields[6] in a vacuum, Figure 35-7a. In this simplified example we assume that, in the yz plane, the fields are *uniform*, extending without change to plus and minus infinity. (We have sketched only a small segment of the field pattern.) As we will demonstrate in a later chapter, it is easy to verify experimentally that a traveling wave has **E** and **B** fields that

---

[6] Maxwell's equations do require that **E** and **B** be at right angles, though they need not be uniform. Our arrangement of crossed fields is a simpler version of the fields associated with displacement current: if you examine the space between the capacitor plates of Figure 35-3, you will see that **E** and **B** are everywhere at right angles.

are each perpendicular to the direction of propagation. Therefore, we suggest that this particular configuration is applicable to *plane wave propagation along the x axis*.

For a wave traveling along the x axis, we could expect the magnitudes of **E** and **B** to be different at different points along x, as well as to vary with time. To ferret out the ways these fields vary in both space and time, we examine the fields on either side of a thin slab of space Δx thick and parallel to the yz plane, as shown in Figure 35-7b. Both E_y and B_z on the plane at x differ from the corresponding fields on the plane at $x + \Delta x$.

We now apply Ampère's law, Equation (35-11), to the top face of the slab (Figure 35-7c). There are no actual charges in a vacuum, so there can be no current I. Thus we have only the displacement-current term

$$\oint \mathbf{B} \cdot d\boldsymbol{\ell} = \mu_0 \varepsilon_0 \frac{d\Phi_E}{dt} \qquad (35\text{-}16)$$

The dimensions of the slab are L along two edges and Δx along the other two edges. We now calculate $\oint \mathbf{B} \cdot d\boldsymbol{\ell}$ around the perimeter of this slab. Beginning at the corner marked P, the four segments of the closed-path integration give

$$\oint \mathbf{B} \cdot d\boldsymbol{\ell} = \underbrace{B_z(x)L}_{①} + \underbrace{0}_{②} - \underbrace{B_z(x+\Delta x)L}_{③} + \underbrace{0}_{④}$$

For paths ② and ④, **B** is perpendicular to $d\boldsymbol{\ell}$, so the dot product is zero for these segments.

The right-hand side of Equation (35-16) may be written in terms of **E** using the fact that $\Phi_E = AE_y$, and therefore $d\Phi_E/dt = A\, dE_y/dt$. The area A enclosed by the path is $L\,\Delta x$. Since E_y varies in both space and time, we use *partial* derivative symbols[7] to indicate that all other variables are to be held constant as we take the derivative indicated. Thus the right-hand side is

$$\mu_0 \varepsilon_0 \frac{\partial \Phi_E}{\partial t} = \mu_0 \varepsilon_0 L\,\Delta x\, \frac{\partial E_y}{\partial t}$$

Combining the previous two equations, we have

$$B_z(x)L - B_z(x+\Delta x)L = \mu_0 \varepsilon_0 L\,\Delta x\, \frac{\partial E_y}{\partial t}$$

Canceling L from both sides and allowing the thickness of the slab to become infinitesimally small, we obtain

$$\lim_{\Delta x \to 0} \frac{B_z(x+\Delta x) - B_z(x)}{\Delta x} = -\mu_0 \varepsilon_0 \frac{\partial E_y}{\partial t} \qquad (35\text{-}17)$$

The left-hand side is just the definition of the derivative dB_z/dx, so

$$\frac{\partial B_z}{\partial x} = -\mu_0 \varepsilon_0 \frac{\partial E_y}{\partial t} \qquad (35\text{-}18)$$

---

[7] For a comment on partial derivatives, see Appendix G-V.

(Again, the *partial* derivative $\partial B_z/\partial x$ acknowledges that B_z may also vary in time.) In a similar fashion, we apply Faraday's law

$$\oint \mathbf{E} \cdot d\boldsymbol{\ell} = -\frac{d\Phi_B}{dt} \tag{35-19}$$

to the face perpendicular to the z direction and obtain

$$\frac{\partial E_y}{\partial x} = -\frac{\partial B_z}{\partial t} \tag{35-20}$$

Equations (35-18) and (35-20) are now solved simultaneously to obtain two equations: one involving only the electric field E_y and the other involving only the magnetic field B_z. The procedure is not difficult. We first differentiate both sides of Equation (35-18) with respect to x and obtain

$$\frac{\partial^2 B_z}{\partial x^2} = -\mu_0 \varepsilon_0 \frac{\partial^2 E_y}{\partial t\, \partial x} \tag{35-21}$$

We next differentiate Equation (35-20) with respect to t and obtain

$$\frac{\partial^2 E_y}{\partial x\, \partial t} = -\frac{\partial^2 B_z}{\partial t^2} \tag{35-22}$$

Substituting this value for the mixed derivative $\partial^2 B_z/\partial x\, \partial t$ into Equation (35-21) we obtain an expression involving B alone:

WAVE EQUATION FOR B_z
$$\frac{\partial^2 B_z}{\partial x^2} = \mu_0 \varepsilon_0 \frac{\partial^2 B_z}{\partial t^2} \tag{35-23}$$

By a similar process, Equations (35-18) and (35-20) may be combined to obtain an expression involving E alone:

WAVE EQUATION FOR E_y
$$\frac{\partial^2 E_y}{\partial x^2} = \mu_0 \varepsilon_0 \frac{\partial^2 E_y}{\partial t^2} \tag{35-24}$$

The previous two equations have the same form as the *wave equation* we developed in Chapter 18 [Equation (18-8)] for the propagation of transverse waves on a rope. A solution to the wave equation is

$$A = A_0 \sin(kx - \omega t) \quad \begin{pmatrix} \text{for a wave moving} \\ \text{in the } +x \text{ direction} \end{pmatrix} \tag{35-25}$$

where $A_0 =$ *amplitude* of the wave

$k = \dfrac{2\pi}{\lambda}$, the *wave number* for a wave of *wavelength* λ

$\omega = \dfrac{2\pi}{T}$, the *angular frequency*

$T = \dfrac{1}{f}$, the *period* for a wave of *frequency* f

$\dfrac{\omega}{k} = v$, the *speed of propagation* of the wave

$v = \lambda f$ (in free space, $v = c$)

The electric field E_y thus varies in space and time according to

**ELECTRIC FIELD E_y
FOR PLANE WAVES**
(traveling in $+x$ direction)
$$E_y = E_{y0} \sin(kx - \omega t) \quad (35\text{-}26)$$

For a wave traveling in the $-x$ direction, the argument of the sine is $(kx + \omega t)$. A graph of E_y is shown in Figure 35-8.

Evaluating derivatives $\partial^2 E_y/\partial x^2$ and $\partial^2 E_y/\partial t^2$, and substituting into Equation (35-24), we obtain

$$\frac{\omega}{k} = \frac{1}{\sqrt{\mu_0 \varepsilon_0}} = \text{Speed of propagation} \quad (35\text{-}27)$$

This relation says that the electric field pattern propagates with a speed $(\mu_0 \varepsilon_0)^{-1/2}$ in a direction perpendicular to **E**.

Because the magnetic field B_z satisfies an equation identical to that of E_y, we also have

**MAGNETIC FIELD B_z
FOR PLANE WAVES**
(traveling in $+x$ direction)
$$B_z = B_{z0} \sin(kx - \omega t) \quad (35\text{-}28)$$

where, again,

$$\frac{\omega}{k} = \frac{1}{\sqrt{\mu_0 \varepsilon_0}} = \text{Speed of propagation} \quad (35\text{-}29)$$

The combination of E_y and B_z is called a *plane* electromagnetic wave because the *wavefronts*, which are surfaces of constant phase, are planes perpendicular to the direction of propagation. It is important to become familiar with the characteristics of plane waves since they occur in many different contexts in physics. (It will be helpful to review Section 18.6, which describes waves in three dimensions). Because the electromagnetic field equations are *linear*, if two sets of waves satisfy Maxwell's equations, so does their sum (the Principle of Superposition).

Plane Waves

(1) **The wavefronts are planes perpendicular to the direction of propagation.** (Wavefronts are surfaces of constant phase.)
(2) **The E and B fields are perpendicular to each other.** This was assumed initially but proved to be entirely consistent with Maxwell's equations (and, indeed, can be derived from Maxwell's equations).
(3) **The E and B fields are transverse waves (perpendicular to the direction of propagation) and are in phase with each other.** This is ensured by the identical form of the wave equations for **E** and **B** (see Figure 35-9a). The sine-wave curves in (a) and (b) represent the magnitudes of the **E** and **B** fields. The diagram should not be interpreted as vibrations of something like a string or water waves. Instead, the sine curves are the *envelope* of the tips of the field vectors, where the *length of the vector represents the strength of the field*. Another representation (c) of the spatial distribution makes use of the convention that the *density* of field lines corresponds to the field strength. Although the sketch in (c) is more cumbersome to draw, perhaps it gives the best impression of the actual field distribution in a plane wave. Careful study of Figure 35-9 will help you avoid misconceptions regarding the nature of plane waves.

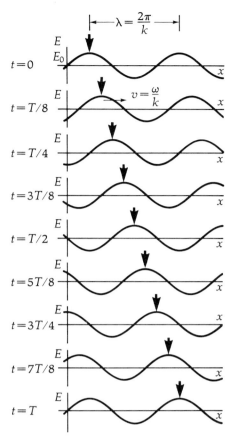

FIGURE 35-8
A series of "snapshots," taken at intervals at $T/8$ s (where $T = 2\pi/\omega$), showing the electric field variation moving in the $+x$ direction with a speed v. A *point of constant phase* (the peak positive value of E_y) is shown as it moves along in the $+x$ direction.

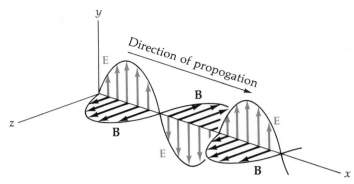

(a) A "snapshot" of the spatial variation of a plane electromagnetic wave moving in the $+x$ direction. The *length of the vectors* corresponds to the field strength. The pattern moves along the $+x$ direction with a speed c.

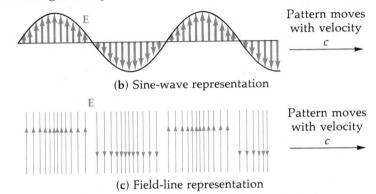

(b) and (c) In the sine-wave representation (b) for the electric field, the vectors themselves are often omitted. This curve implies that the field lines are crowded together where the field is stronger and are farther apart where the field is weaker, as shown in (c). Since it is a *plane wave* (that is, uniform over the yz plane), *the field lines should be mentally extended to infinity in the $\pm y$ direction and the pattern should be duplicated in and out of the paper to fill all space in the $\pm z$ direction.* The *wavefronts* are planes in the yz direction; they move in the $+x$ direction with the speed c. (A wave has the *same phase* at every point on a wavefront.)

FIGURE 35-9
Representations of a plane electromagnetic wave.

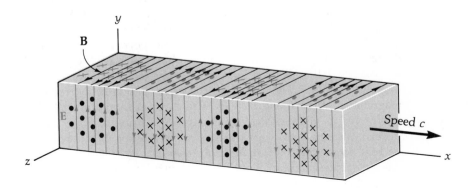

FIGURE 35-10
A "snapshot" of a portion of a plane electromagnetic wave traveling in the $+x$ direction. (Compare with Figure 35-9.)

(4) **The speed of propagation of the wave is $(\mu_0\varepsilon_0)^{-1/2}$**, a term whose numerical value constitutes one of the most remarkable aspects of the electromagnetic wave. Before evaluating this combination of μ_0 and ε_0, let us review the origin of these constants.

The constant ε_0 appears in Coulomb's law:

$$F = \left(\frac{1}{4\pi\varepsilon_0}\right)\frac{q_1 q_2}{r^2} \qquad (35\text{-}30)$$

In the modern definition, the quantity of charge q is determined in terms of the ampere. The current I of one ampere (one coulomb per second) is defined in terms of the force per unit length F/ℓ between parallel current-carrying conductors a distance d apart:

$$\frac{F}{\ell} = \frac{\mu_0 I^2}{2\pi d} \qquad (35\text{-}31)$$

On the other hand, μ_0 is an *assigned* number that fixes the value of F/ℓ in Equation (35-31) to be *exactly* $2\pi \times 10^{-7}$ N/m for $d = 1$ m. Thus:

$$\varepsilon_0 = 8.8542 \times 10^{-12} \frac{C^2}{N \cdot m^2} \qquad \left(\begin{array}{l}\text{Formerly, this was experimentally}\\ \text{determined. See Footnote 8.}\end{array}\right)$$

$$\mu_0 = 4\pi \times 10^{-7} \frac{N \cdot s^2}{C^2} \qquad \text{(defined exact)}$$

Substituting these values into the expression for the speed of propagation, we obtain

$$(\mu_0\varepsilon_0)^{-1/2} = \left[\left(4\pi \times 10^{-7}\frac{N \cdot s^2}{C^2}\right)\left(8.8542 \times 10^{-12}\frac{C^2}{N \cdot m^2}\right)\right]^{-1/2}$$

$$= 2.9979 \times 10^8 \frac{m}{s}$$

which is *the speed of light in a vacuum*. The fact that the speed of propagation of an electromagnetic wave appeared to be the speed of light led to the realization that *light is electromagnetic in nature*. Therefore:

SPEED OF LIGHT c
$$c = \frac{1}{\sqrt{\mu_0\varepsilon_0}} \qquad (35\text{-}32)$$

Thus it became accepted that light is an electromagnetic wave, making up just a small portion of the electromagnetic spectrum, which includes radio waves, microwaves, thermal radiation, x-rays, and gamma rays—all described by Maxwell's four equations.

From Equation (35-32), the speed of light was determined by purely electrical methods (rather than by a direct velocity measurement). For example, consider a parallel-plate capacitor charged by a battery. By measuring both the force between the wires leading to the capacitor and the force between the plates of the charged capacitor, we can experimentally determine the value

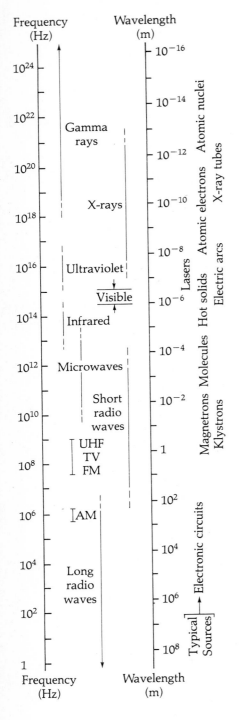

FIGURE 35-11
The *electromagnetic spectrum* presented on logarithmic scales extends without limits in frequency and wavelength. Note that the range of visible light (~440 nm to ~670 nm) extends less than a factor of 2. The ranges associated with the various names are not definite, though the Federal Communications Commission has allocated many specific bands (not shown) at different regions of the spectrum for special communications purposes. Essentially all that we know about the universe outside the earth and moon comes to us via electromagnetic waves.

of c. In 1906, E. B. Rosa and N. E. Dorsey of the National Bureau of Standards performed a beautifully precise experiment that determined the speed of light by electrical methods. It was the most accurate determination of c at that time. The value they obtained was $c = 299\,784 \pm 15$ km/s.

The last step in this story of units was taken on October 20, 1983, when the speed of light was officially adopted[8] as a **defined** SI standard equal to its "best" value at that time:

SPEED OF LIGHT c
(defined exact)
$$c \equiv (2.997\,924\,58) \times 10^8 \frac{\text{m}}{\text{s}} \qquad (35\text{-}33)$$

The value 3.00×10^8 is sufficiently accurate for most use.

(5) **The magnitudes of E and B are related.** Solutions of the wave equation are Equations (35-26) and (35-28):

$$E_y = E_{y0} \sin(kx - \omega t) \qquad \text{and} \qquad B_z = B_{z0} \sin(kx - \omega t)$$

These solutions must satisfy Equation (35-20):

$$\frac{\partial E_y}{\partial x} = -\frac{\partial B_z}{\partial t}$$

obtained in the first part of the wave-equation development. Evaluating these derivatives, we obtain

$$kE_{y0} \sin(kx - \omega t) = \omega B_{z0} \sin(kx - \omega t)$$
$$\text{or} \qquad kE_y = \omega B_z$$

Rearranging, and recalling that $\omega/k = c$, we have

$$\frac{E_y}{B_z} = \frac{\omega}{k} = c$$

RELATION BETWEEN E_y AND B_z IN ELECTROMAGNETIC WAVES
$$\frac{E_y}{B_z} = c \qquad (35\text{-}34)$$

---

[8] As a consequence, the value of ε_0 is now *defined* to be $\varepsilon_0 \equiv 1/\mu_0 c^2$, where μ_0 is chosen as $4\pi \times 10^{-7}$ N/A^2 and c is the exact value defined in Equation (35-33).

EXAMPLE 35-2

The electric field in an electromagnetic wave is described by the equation

$$E_y = 100 \sin(10^7 x - \omega t) \quad \text{(in SI units)}$$

Find (a) the amplitude of the corresponding magnetic wave, (b) the wavelength λ, and (c) the frequency f.

SOLUTION

(a) From Equation (35-34) we obtain

$$B_z = \frac{E_y}{c} = \frac{\left(100 \frac{\text{V}}{\text{m}}\right)}{\left(3 \times 10^8 \frac{\text{m}}{\text{s}}\right)} = \boxed{3.33 \times 10^{-7} \text{ T}}$$

(b) To find the wavelength λ and the frequency f, we note that the given equation is of the form $E_y = E_{y0} \sin(kx - \omega t)$. From the relations following Equation (35-25) we have

$$\lambda = \frac{2\pi}{k} = \frac{2\pi}{(10^{-7} \text{ m}^{-1})} = 6.28 \times 10^{-7} \text{ m} \underbrace{\left(\frac{10^9 \text{ nm}}{1 \text{ m}}\right)}_{\text{Conversion ratio}} = \boxed{628 \text{ nm}}$$

This is a red-orange wavelength of visible light.

(c) To find the frequency, we make the calculation

$$f = \frac{c}{\lambda} = \frac{\left(3 \times 10^8 \frac{\text{m}}{\text{s}}\right)}{(628 \times 10^{-9} \text{ m})} = \boxed{4.78 \times 10^{14} \text{ Hz}}$$

35.4 The Production of Electromagnetic Waves

There are many ways of generating electromagnetic waves. All of them rely on the phenomenon that *accelerated charges radiate electromagnetic waves*. Figure 35-12 explains the origin of this radiation. In Figure 35-12c, the charge is originally at rest at O. At $t = 0$, it accelerates for a very short time Δt to acquire a speed $v = 0.2c$ at O'. From there, it travels at constant speed to reach point P at time t. Now the "kink" in the field lines introduced by the acceleration travels outward with the speed c. It is just this kink that carries the information that the charge has accelerated. For distances away from O larger than t/c, news of the sudden acceleration has not yet arrived, so the field lines farther away point to the original location O. For distances from O' smaller than t/c, the field lines center on the charge at its present location P (similar to the field pattern in Figure 35-12b). Note an important feature of the kink: it contains a *transverse* component of the electric field. *This is the origin of the transverse* **E** *field in the traveling wave.*

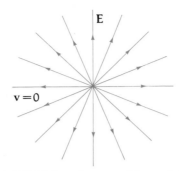

(a) Field lines for a positive point charge at rest.

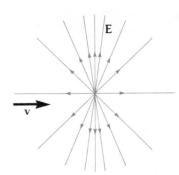

(b) Field lines for a point charge moving with constant velocity. Because of relativity, the pattern of field lines is "squashed together" along the direction of motion. As a result, the field lines are not quite so close together along the direction of motion as in (a) and are closer together perpendicular to that direction.

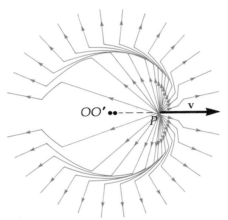

(c) Field lines for a positive point charge that has undergone a very brief acceleration from rest at O to O' and then traveled at constant speed to the point P (where it continues to move). The "kink" in the field lines produced by the acceleration travels outward with speed c from the region OO'.

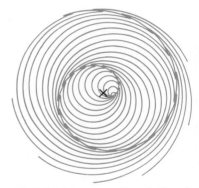

(d) Field lines around an isolated point charge moving clockwise at constant speed $v = 0.9c$ in a circle centered on the ×. The kink in the spiral pattern travels outward with speed c.

FIGURE 35-12
When a point charge accelerates, it generates a "kink" in the pattern of field lines. In (c) and (d), the kink has a component of **E** that is perpendicular to the direction of motion as the kink moves outward (at speed c) from the region where the acceleration occurred. This outward-moving component is the electromagnetic radiation from the accelerated charge.

A common example of radiation is the *dipole antenna* illustrated in Figures 35-13 and 35-14, composed of two wires that are connected to an AC voltage source. Electrons are accelerated first in one direction and then in the other, making one wire positive and the other negative and then vice versa. These oscillations produce a growing electric field pattern, as shown in Figure 35-13a. At the instant when the potential reverses, there is no net charge on the dipole, so no field lines can terminate there. Consequently, the loops of electric field are "pinched off" and propagate away from the antenna with the speed c. At

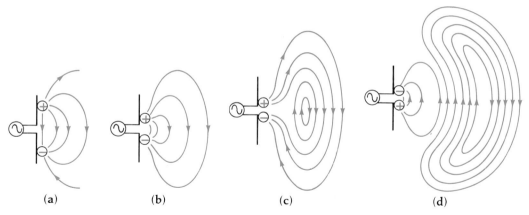

FIGURE 35-13
The generation of an electromagnetic wave by the accelerating charges in a dipole antenna. (Only the electric field is shown; the associated magnetic field is omitted for clarity.) The complete field pattern forms a figure-of-revolution about the axis of the dipole wires. (See Figure 35-14.)

any given point in space, the electric field changes in time, so according to Maxwell's equations there is also a changing magnetic field (not shown). At very large distances from the antenna, the waves become approximately *plane waves*, as described in Figure 35-10 (p. 804).

35.5 Energy in Electromagnetic Waves

Electromagnetic waves from the sun bring to the earth about 174 trillion kilowatts of power striking the top of the earth's atmosphere. This inflow of energy undoubtedly was essential to the origin of life and to the storage of immense reserves of fossil fuels. It continues to be important in driving the earth's winds and ocean currents, in the evaporation of water to produce rain which replenishes fresh-water supplies, and in other energy-transfer processes that are so important in sustaining living systems. The flow of energy to the earth appears to be in balance with enough energy radiated from the earth to maintain thermal equilibrium. Although living matter relies directly on only a few hundredths of one percent of this incoming radiant energy, life could not continue very long without this constant flow of energy from the sun.

In this section, we will explain how electromagnetic waves transport energy along the direction of propagation. As shown in previous chapters [Equations (27-16) and (32-33)], the energy per unit volume, *energy density u*, of electric and magnetic fields is

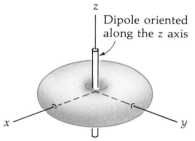

(a) Doughnut-shaped radiation pattern.

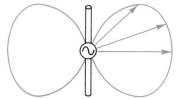

(b) Cross-section of the radiation pattern.

ENERGY DENSITY IN FIELDS
(instantaneous)

Electric: $\quad u_E = \dfrac{1}{2} \varepsilon_0 E^2 \quad$ (35-35)

Magnetic: $\quad u_B = \dfrac{1}{2\mu_0} B^2 \quad$ (35-36)

To see how this energy is carried along by the wave, we apply these equations to a thin volume of space, as illustrated in Figure 35-15. At a given instant, the volume contains a total energy ΔU that consists of electric field energy ΔU_E and magnetic field energy ΔU_B:

$$\Delta U = \Delta U_E + \Delta U_B \quad (35\text{-}37)$$

FIGURE 35-14
In three dimensions, dipole radiation in various directions (far from the dipole) may be depicted as a doughnut-shaped pattern, where the power radiated along a particular direction is proportional to the length of a vector drawn from the center of the dipole to the surface of the figure. The radiation is a maximum at right angles to the dipole, with no radiation occurring along the axis of the dipole.

FIGURE 35-15
A plane wave carrying electromagnetic energy through the thin slab with a speed c. The electric field varies in the $\pm y$ direction and the magnetic field varies in the $\pm z$ direction.

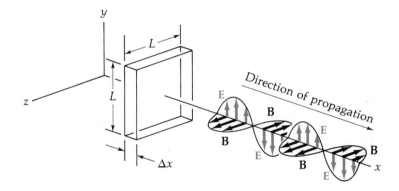

Since the volume of the slab is $L^2 \Delta x$, the energy ΔU in the slab is $L^2 \Delta x (u_E + u_B)$. Using Equations (35-35) and (35-36), we have

$$\Delta U = \frac{1}{2} L^2 \Delta x \left(\varepsilon_0 E_y^2 + \frac{1}{\mu_0} B_z^2 \right) \qquad (35\text{-}38)$$

Noting that $E_y = cB_z$, we may write Equation (35-38) so that each term contains the product $E_y B_z$:

$$\Delta U = \frac{1}{2} L^2 \Delta x \left(\varepsilon_0 c E_y B_z + \frac{1}{\mu_0 c} E_y B_z \right) = \frac{1}{2} L^2 E_y B_z \frac{\Delta x}{c} \left(\varepsilon_0 c^2 + \frac{1}{\mu_0} \right)$$

Since $c^2 = 1/\varepsilon_0 \mu_0$, we have $\qquad \Delta U = L^2 E_y B_z \dfrac{\Delta x}{c} \left(\dfrac{1}{\mu_0} \right) \qquad (35\text{-}39)$

The time Δt required for the energy ΔU to pass through the face of the volume is $\Delta t = \Delta x / c$. Designating the energy per unit time that flows through a unit area as S,

$$S = \frac{\text{(Energy)}}{\text{(Area)(Time)}} = \frac{\Delta U}{L^2 \Delta t}$$

and substituting the previous expressions we obtain

$$S = \frac{1}{\mu_0} E_y B_z \qquad (35\text{-}40)$$

Because **E** and **B** are both vectors perpendicular to the direction of propagation (Figure 35-15), we know that **E** × **B** is *along the direction of propagation*. Therefore, we may write the previous equation as

THE POYNTING VECTOR[9] $\qquad \mathbf{S} = \dfrac{1}{\mu_0} \mathbf{E} \times \mathbf{B} \qquad (35\text{-}41)$
(instantaneous value)

The vector **S** is called the **Poynting vector** in honor of its originator, John Henry Poynting (1852–1914). It is measured in SI units of *watts per square meter* (W/m^2).

---

[9] As discussed in Chapter 33, Equation (33-4), in a vacuum the *magnetic field* **H** is related to the *magnetic induction* **B** according to $\mathbf{B} = \mu_0 \mathbf{H}$. So Equation (35-41) is sometimes written as $\mathbf{S} = \mathbf{E} \times \mathbf{H}$.

The Poynting vector gives the *instantaneous* rate of energy flow per unit area in terms of E and B. For the waves we consider, these quantities vary sinusoidally, so the *instantaneous* power oscillates between zero and some maximum value. When we measure the intensity of a wave as it moves by, we measure its value averaged over many cycles of the variation. So the *average* rate of energy flow per unit area is of more practical importance. It is easy to calculate. We substitute the basic sine-wave expressions for E_y and B_z into the Poynting vector:

$$S = \frac{1}{\mu_0} E_{y0} B_{z0} \sin^2(kx - \omega t) \qquad (35\text{-}42)$$

The energy received at a given point thus varies in time as the sine squared, which repeats itself every half cycle of the basic period T. To find the average power flow, we calculate

$$S_{av} = \frac{E_{y0} B_{z0}}{\mu_0} \left[\frac{1}{(T/2)} \int_0^{T/2} \sin^2(kx - \omega t)\, dt \right] \qquad (35\text{-}43)$$

The quantity in brackets yields a factor of $\frac{1}{2}$, so

THE POYNTING VECTOR
(average value for a sinusoidal wave)

$$S_{av} = \frac{1}{2\mu_0} E_{y0} B_{z0} \qquad (35\text{-}44)$$

Thus the average power flow (in W) through a surface area A that is oriented perpendicular to the wave is

$$\int_A \mathbf{S}_{av} \cdot d\mathbf{A} = (S_{av})(A) = \left(\frac{dU}{dt}\right)_{av} = \begin{bmatrix} \text{Average power flow} \\ \text{through a surface area } A \\ \text{normal to the wave} \end{bmatrix} \qquad (35\text{-}45)$$

Energy Density

The energy densities u associated with E and B fields are also of interest. As shown previously,

$$u_E = \frac{1}{2} \varepsilon_0 E^2 \qquad \text{and} \qquad u_B = \frac{1}{2\mu_0} B^2 \qquad (35\text{-}46)$$

For a traveling electromagnetic wave, E and B are related through $E = cB$ and $c = 1/\sqrt{\varepsilon_0 \mu_0}$. Therefore, we may write

$$u_E = \frac{1}{2} \varepsilon_0 E^2 = \frac{1}{2} \varepsilon_0 c^2 B^2 = \frac{1}{2\mu_0} B^2 = u_B \qquad (35\text{-}47)$$

which shows that the instantaneous energy density in the electric field equals that in the magnetic field. *The E and B fields each contain half the total energy.* The total *instantaneous* energy density u is therefore

$$u = (u_E + u_B) = \varepsilon_0 E^2 = \frac{B^2}{\mu_0} \qquad (35\text{-}48)$$

FIGURE 35-16
Ninety-five percent of the world's current conversion of solar-to-electrical energy occurs on 1000 acres of the Mojave desert near Los Angeles, California. Here, 650 000 parabolic mirrors track the sun's motion to focus light on pipes containing synthetic oil, heating the oil to 400°C. The hot oil then flows through heat exchangers, producing superheated steam for conventional turbine generators. The peak electrical power of 196 MW is sold to the Southern California Edison Company, providing 1% of the system's peak demand of 20 000 MW. Though the process is not competitive with today's costs for conventional power plants using petroleum and coal, valuable experience in this new technology is being gained.

The *average* value of u for these sinusoidally varying fields involves a factor of $\frac{1}{2}$ [cf. Equation (35-43)] when written in terms of their peak values E_{y0} and B_{z0}. So the total average energy per unit volume u_{av} in an electromagnetic wave is

AVERAGE ENERGY DENSITY IN AN ELECTROMAGNETIC WAVE
$$u_{av} = \frac{1}{2}\varepsilon_0 E_{y0}^2 \quad \text{or} \quad u_{av} = \frac{1}{2\mu_0} B_{z0}^2 \qquad (35\text{-}49)$$

measured in SI units of joules per cubic meter (J/m³). Comparing this with Equation (35-44) and noting that $E = cB$, we conclude that

WAVE INTENSITY
$$S_{av} = u_{av} c \qquad (35\text{-}50)$$

The wave intensity in watts per square meter (W/m²) equals the average energy density (in J/m³) times the speed c.

EXAMPLE 35-3

Consider a lamp that emits essentially monochromatic green light uniformly in all directions. If the lamp is 3% efficient in converting electrical power to electromagnetic waves and consumes 100 W of power, find the amplitude of the electric field associated with the electromagnetic radiation at a distance of 10 m from the lamp.

SOLUTION

Since the lamp is 3% efficient, it emits 3.0 W of electromagnetic power, which is spread uniformly over a sphere of radius 10 m. Thus, the average power per unit area is

$$S_{av} = \frac{P}{4\pi R^2} = \frac{3.0\ \text{W}}{4\pi(10\ \text{m})^2} = \frac{0.030}{4\pi}\left(\frac{\text{W}}{\text{m}^2}\right)$$

Since the light is essentially of only one color, we can assume a single electromagnetic wave of wavelength λ and use Equations (35-49) and (35-50):

$$S_{av} = u_{av} c = \tfrac{1}{2}\varepsilon_0 c E_{y0}^2$$

Solving for E_{y0} from the outer two expressions, we obtain

$$E_{y0} = \sqrt{\frac{2 S_{av}}{\varepsilon_0 c}} = \sqrt{\frac{(2)\left(0.030\ \frac{\text{W}}{\text{m}^2}\right)}{\left(8.85 \times 10^{-12}\ \frac{\text{C}^2}{\text{N}\cdot\text{m}^2}\right)\left(3 \times 10^8\ \frac{\text{m}}{\text{s}}\right)(4\pi)}} = \boxed{1.34\ \frac{\text{V}}{\text{m}}}$$

35.6 Momentum of Electromagnetic Waves

We have shown that energy is transported in an electromagnetic wave. We will now show that the wave also possesses momentum. We begin by demonstrating that the electromagnetic wave exerts a force on a charged particle *in* the direction of the wave propagation. This is true in spite of the fact that the

35.6 Momentum of Electromagnetic Waves

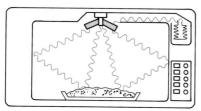

A microwave oven.

FIGURE 35-17
As a result of the development of radar during World War II, compact magnetron tubes for generating high-power microwaves became available for a variety of uses. For example, a microwave oven transfers energy by the electromagnetic radiation from a magnetron. Since microwaves tend to form standing-wave patterns within the reflecting-walls cavity of the oven, rotating metal fan blades cause a more uniform distribution of energy inside the oven. Alternatively, the food may be placed on a rotating turntable to move it through the nodes and antinodes of the standing-wave pattern. The magnetron emits at a frequency of 2.45 GHz, chosen to match a rotational vibration frequency of water molecules. The resonance absorption of microwave energy by water molecules is the mechanism of heating and cooking the food.

This magnetron has aluminum vanes that radiate thermal energy to control overheating. Small inductance coils are in series with the two wires that supply the input power, preventing microwave energy from feeding back to the power supply. The microwaves emerge through the hollow cylindrical *waveguide* at the opposite end.

electric field **E** and the magnetic field **B** are entirely *transverse* (perpendicular to the direction of propagation). Thus, if an electromagnetic wave interacts with matter (which, of course, contains electrons), it will give some momentum to the matter in the direction of the wave propagation. The details of the interaction are interesting.

Consider an electromagnetic wave traveling in the $+x$ direction, as in Figure 35-18, and striking an electron that is free to move in a sheet of resistive material. Let us suppose that, at the sheet, the electric field oscillates in the $\pm y$ direction and the magnetic field oscillates in the $\pm z$ direction.

$$\mathbf{E} = (E_0 \sin \omega t)\hat{\mathbf{y}} \quad \text{and} \quad \mathbf{B} = (B_0 \sin \omega t)\hat{\mathbf{z}} \quad (35\text{-}51)$$

The electric field will force the negatively charged electron $-e$ to move downward through the resistive material. For our purposes, we will assume that the electron moves with a drift velocity $\mathbf{v}_d$ as though it were in a viscous medium, where the electron is essentially always at its terminal velocity. Thus $\mathbf{F}_E = b\mathbf{v}_d$, where b is a constant and $\mathbf{F}_E$ is the force produced by the electric field: $\mathbf{F}_E = (-eE_0 \sin \omega t)\hat{\mathbf{y}}$. Combining these equations, we have

$$\mathbf{v}_d = -\left(\frac{eE_0}{b} \sin \omega t\right)\hat{\mathbf{y}} \quad (35\text{-}52)$$

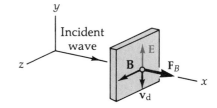

FIGURE 35-18
An electromagnetic wave exerts forces on electrons residing within a sheet of resistive material. The net average force $\mathbf{F}_B$ is in the direction of the wave propagation.

Note that the oscillating velocity of the electron is exactly 180° out of phase with the electric field oscillation.

The moving electron also experiences a magnetic force:

$$\mathbf{F_B} = -e(\mathbf{v_d} \times \mathbf{B}) \tag{35-53}$$

Substituting the expressions for $\mathbf{v_d}$ and $\mathbf{B}$ into this equation, we have

$$\mathbf{F_B} = -e\left(\frac{-eE_0}{b}\sin\omega t\right)\hat{\mathbf{y}} \times (B_0 \sin\omega t)\hat{\mathbf{z}} = \left(\frac{e^2 E_0 B_0}{b}\sin^2\omega t\right)\hat{\mathbf{x}} \tag{35-54}$$

Because the $\sin^2\omega t$ factor is always positive, *the force is always in the $+x$ direction—the direction that the electromagnetic wave travels*. The sheet of resistive material experiences the sum of all the forces on all of the electrons in the sheet.

In being forced through the resistive material, the electron absorbs energy from the electromagnetic wave to overcome the "viscous" force on the electron. The power or rate that energy is given to the electron by the electromagnetic wave is[10]

$$\frac{dU}{dt} = \mathbf{F_E} \cdot \mathbf{v_d}$$

where U is the energy absorbed by the electron. Substituting expressions for $\mathbf{F_E}$ and $\mathbf{v_d}$ in this equation, we have

$$\frac{dU}{dt} = (-eE_0 \sin\omega t)\hat{\mathbf{y}} \cdot \left(\frac{-eE_0}{b}\sin\omega t\right)\hat{\mathbf{y}} = \frac{e^2 E_{y0}^2}{b}\sin^2\omega t \tag{35-55}$$

Since $E = cB$, the rate of energy absorption can be written as

$$\frac{dU}{dt} = c\left(\frac{e^2 E_0 B_0}{b}\sin^2\omega t\right) \tag{35-56}$$

Comparing this with Equation (35-54), we have

$$\frac{dU}{dt} = cF_B = c\frac{dp}{dt} \tag{35-57}$$

where $F_B = dp/dt$, the rate of momentum change acquired by the electron in the $+x$ direction. This equation states that the rate of energy absorption dU/dt by the electron equals the speed of light times the rate of momentum change of the electron. Since both the energy absorbed and the momentum acquired by the electron were extracted from the electromagnetic wave, we apply the conservation of energy and momentum principles, integrate Equation (35-57) with respect to time, and obtain

$$\int_0^U \frac{dU}{dt} = c\int_0^p \frac{dp}{dt}$$

MOMENTUM p CARRIED BY A WAVE OF ENERGY U
$$U = cp \tag{35-58}$$

---

[10] Recall that a *magnetic* field does no work because $\mathbf{F_B}$ is always perpendicular to $\mathbf{v}$.

Because an electromagnetic wave of total energy U carries momentum p, when the wave strikes a surface perpendicularly it exerts an average force $F = dp/dt$. From Equation (35-57), this force is

FORCE OF ABSORBED RADIATION
$$F = \frac{1}{c}\frac{dU}{dt} \qquad (35\text{-}59)$$

If the radiation is totally absorbed, the force per unit area exerted on a surface is $1/c$ times the rate of energy absorbed per unit area. The force per unit area is the **radiation pressure**, or *light pressure*.[11] Since the Poynting vector is the rate of energy per unit area in the wave, we have

RADIATION PRESSURE (normal incidence)
$$\text{Pressure} = \frac{S_{av}}{c} \quad \text{(total absorption)} \qquad (35\text{-}60)$$
$$\text{Pressure} = \frac{2S_{av}}{c} \quad \text{(total reflection)} \qquad (35\text{-}61)$$

where the second expression recognizes that the momentum change for 100% reflection is twice that for total absorption.

EXAMPLE 35-4

A plane electromagnetic wave of wave intensity 6 W/m² strikes a small pocket mirror, 40 cm², held perpendicular to the approaching wave. (a) What momentum does the wave transfer to the mirror each second? (b) Find the force that the wave exerts on the mirror.

SOLUTION

(a) From Equation (35-45),

$$\frac{dU}{dt} = (S_{av})(\text{area}) = \left(6\,\frac{W}{m^2}\right)(40 \times 10^{-4}\,m^2) = 2.40 \times 10^{-2}\,\frac{J}{s}$$

In one second, the total energy U impinging on the mirror is therefore 2.40×10^{-8} J. From Equation (35-58), the momentum p transferred each second for total reflection is

$$p = \frac{2U}{c} = \frac{2(2.40 \times 10^{-8}\,J)}{(3 \times 10^8\,m/s)} = \boxed{1.60 \times 10^{-10}\,\frac{kg \cdot m}{s}} \quad \text{(each second)}$$

(b) $F = \dfrac{dp}{dt} = \dfrac{1.60 \times 10^{-10}\,kg \cdot m/s}{1\,s} = \boxed{1.60 \times 10^{-10}\,N}$

Comments on Radiation Pressure

You may be familiar with a toy called a *radiometer*, illustrated in Figure 35-19a. This device consists of vanes blackened on one side and silvered on the other. The vanes are mounted on a vertical axle in a glass bulb from which most of

---

[11] Pressure, momentum, and power all have the symbol p or P. Since all three concepts are involved here, be careful not to confuse them. For an interesting discussion of radiation pressure, see G. E. Henry, "Radiation Pressure," *Scientific American*, June 1957, p. 99.

(a) A radiometer.

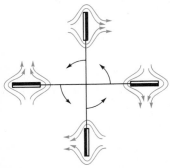

(b) The thermal creep of air around the edges of the vanes (viewed from above).

FIGURE 35-19
A radiometer turns on its axis when exposed to a moderately strong light. The torque causing it to turn is **not** produced by the pressure of the light.

the air is removed. When the radiometer is exposed to moderately strong light (or even the infrared radiation from a flatiron), the vanes rotate about the axle, with the blackened faces trailing in the motion. Being aware of radiation pressure, a person may hastily conclude that the motion is due to radiation pressure. But this conclusion is incorrect for three reasons:

(1) If the torque producing this rotation of the radiometer vanes is attributable to light pressure, the vanes are rotating in the wrong direction. (We have shown that the force exerted on the silvered side of the vane is twice the magnitude of that on the blackened side. Therefore, the silvered side should trail in the rotation.)

(2) The force exerted by the electromagnetic wave is far too small to account for the rapid angular acceleration of the vanes when the radiometer is suddenly exposed to light. (Example 35-4 indicated how small the force would be on the radiometer vanes even if the radiometer were placed close to a light bulb.)

(3) If the radiometer bulb is evacuated to an extremely low pressure, the vanes will not rotate. (The torque on the vanes due to light pressure is too small to overcome the friction on the bearings of the vane support.)[12]

The explanation of the moving vanes in a radiometer was first suggested by Maxwell in 1879. The explanation is based on the fact that air moves along the surface of an unevenly heated object toward regions of higher temperature. This phenomenon is known as *thermal creep*.[13] In the case of a radiometer vane, air flows over the edge of the vane toward the warmer blackened side. *The resulting increase in air pressure on the blackened side produces the rotation of the vanes.* In a typical radiometer, the air-pressure effect is about 10 000 times greater than the radiation pressure.

In spite of the relative smallness of the radiation pressure, in certain situations it can become a significant effect. For example, sunlight exerts a force on the earth of about 6×10^8 N (over 60 000 tons). Sunlight falling on balloon satellites circling the earth (such as the *Echo* satellite launched in the 1960s) produces noticeable alterations of the orbit. Spacecraft that have extended vanes of solar cells to capture sunlight will experience a rotation if the forces due to radiation pressure produce a net torque about the center of mass of the spacecraft.

Some comets have two tails, one composed of ionized atoms and molecules and the other of dust particles, Figure 35-20. The "nucleus" of a comet is believed to be composed of a mixture of ices, dust grains, and particles. As a comet nears the sun, thermal radiation evaporates a thickness of a meter or so from its surface. *Radiation pressure* from the sun pushes the dust particles into a curved, diffuse tail. The evaporated atoms and molecules of the ices, however, are accelerated to faster speeds (up to 100 km/s) by the *solar wind*: streams of ions (mostly electrons and protons) that are ejected more or less steadily by the sun.

Calculations show that with sufficiently large "sails," space vehicles might feasibly by propelled away from the sun through interplanetary space by radiation pressure from the sun. The method will not work for *interstellar* journeys, however, because the spacecraft moves too far away from the source of radiation.

---

[12] If the vanes are suspended by a thin quartz fiber, and if the air pressure is extremely low, then the true radiation-pressure effect can be demonstrated. If just one of the vanes is illuminated, the vanes can be turned through an angle in opposition to the restoring torque of the fiber.

[13] Experiments establishing the thermal-creep explanation of radiometers are described in E. H. Kennard, *Kinetic Theory of Gases*, McGraw-Hill, 1938.

FIGURE 35-20
The comet Mrkos, photographed in 1957, is traveling toward the left in these pictures. The curving, diffuse tail, which extends almost at right angles to the path, is formed from dust particles "blown away" by *radiation pressure* from the sun. The straight, ragged tail is composed of ionized atoms and molecules pushed away with greater speeds by the *solar wind*, a stream of ions and electrons ejected by the sun.

August 22   August 24   August 26   August 27

Summary

The following concepts and equations were introduced in this chapter:

Displacement current:

$$I_d = \varepsilon_0 \frac{d\Phi_E}{dt}$$

Maxwell's equation (for a vacuum):

$\oint \mathbf{E} \cdot d\mathbf{A} = \dfrac{q}{\varepsilon_0}$	$\oint \mathbf{B} \cdot d\mathbf{A} = 0$
$\oint \mathbf{B} \cdot d\boldsymbol{\ell} = \mu_0 \left(I + \varepsilon_0 \dfrac{d\Phi_E}{dt} \right)$	$\oint \mathbf{E} \cdot d\boldsymbol{\ell} = -\dfrac{d\Phi_B}{dt}$

Wave equation (for a plane wave traveling in the $+x$ direction):

$$\frac{\partial^2 E_y}{\partial x^2} = \mu_0 \varepsilon_0 \frac{\partial^2 E_y}{\partial t^2} \quad \text{and} \quad \frac{\partial^2 B_z}{\partial x^2} = \mu_0 \varepsilon_0 \frac{\partial^2 B_z}{\partial t^2}$$

For a sinusoidal *electromagnetic wave* traveling in the $+x$ direction:

$$E_y = E_{y0} \sin(kx - \omega t) \quad \text{and} \quad B_z = B_{z0} \sin(kx - \omega t)$$

where **E** and **B** are perpendicular to each other, so that $\mathbf{E} \times \mathbf{B}$ is the direction of the wave velocity.

Speed of light in a vacuum: $\quad c = \dfrac{1}{\sqrt{\mu_0 \varepsilon_0}}$

E *and* **B** *fields in electromagnetic waves:*

$$E_y = cB_z$$

Average energy density in an electromagnetic wave:

$$u_{av} = \frac{1}{2}\varepsilon_0 E_{y0}^2 \quad \text{or} = \frac{1}{2\mu_0} B_{z0}^2$$

Rate of energy flow in electromagnetic waves (in W/m²):

Instantaneous: $\quad \mathbf{S} = \dfrac{1}{\mu_0}(\mathbf{E} \times \mathbf{B})$

Average: $\quad S_{av} = \dfrac{1}{2\mu_0} E_{y0} B_{z0}$

where **S**, the *Poynting vector*, has units of watts per square meter and is in the direction of the electromagnetic wave propagation. It is also called the *wave intensity*.

Wave intensity: $\quad S_{av} = u_{av} c$

The average power flow (in W) across an area A:

$$(\text{Power})_{av} = \int \mathbf{S}_{av} \cdot d\mathbf{A}$$

Momentum carried by electromagnetic waves: An object acquires a momentum p in the absorption of electromagnetic energy U according to

$$U = cp$$

Radiation pressure: A pressure P is exerted on an object absorbing the radiant energy flux S_{av}:

$$P = \frac{S_{av}}{c} \quad \text{(total absorption)}$$

$$P = \frac{2S_{av}}{c} \quad \text{(total reflection)}$$

Questions

1. What kind of simple apparatus would be needed to demonstrate that a changing magnetic field produces an electric field? Similarly, what simple apparatus would be required to show that a changing electric field produces a magnetic field?

2. In her laboratory a physicist creates a magnetic field that is directed upward and increasing. When she directs a beam of electrons upward (along the direction of **B**), the beam is deflected in a certain direction. What causes the deflection? What information about the extent of the magnetic field does this provide?

3. A parallel-plate capacitor in series with a resistor is charged by a battery. How would the displacement current between the plates of the capacitor depend on the dielectric material?

4. Does the magnitude or direction of an electric field that is induced by a changing magnetic field give any information about the instantaneous direction or magnitude of the magnetic field?

5. The behavior of magnetic dipoles and quadrupoles is consistent with Maxwell's equations. Is it possible to construct a magnetic tripole (two north poles and one south pole, for example) that also has properties consistent with Maxwell's equations?

6. At a given point in space, there is an instant when both the electric and the magnetic fields associated with an electromagnetic wave are zero. How can the wave propagate from that point if no fields exist there?

7. Straight-wire radio receiving antennae are designed to detect the electric field variation of an electromagnetic wave rather than the magnetic field variation. Explain.

8. A directional radio receiving antenna is in the form of a circular coil of wire. Is such an antenna sensitive to the magnetic field variation of the transmitted electromagnetic wave or to the electric field variation? How should this antenna be oriented with respect to a straight vertical radio transmitter antenna?

9. Design an electrical apparatus by which, in principle, the speed of light could be determined through the measurement of time-varying forces alone.

10. Since the measured values of ε_0 and c are related by the defined constant μ_0, what form would Maxwell's equations take if μ_0 or ε_0 did not appear explicitly?

11. In what ways does the radiation from a light bulb differ from the radiation from a radio transmitter antenna?

12. Identify what is wrong with the following statement: "The electric field associated with the electromagnetic wave is much greater than the magnetic field because $E = cB$."

13. An electromagnetic wave transports energy in its electric and magnetic fields. Which, if either, of the fields contains the greater amount of energy?

14. Does a detector of a monochromatic electromagnetic wave experience a continuous or pulsating flow of momentum and energy? If pulsating, what is the frequency of the pulses?

15. Explain what is inappropriate about the way the following question is worded: "What fraction of the total electromagnetic spectrum does visible light represent?" What would be a better way to ask the question?

16. In what ways is an electromagnetic wave similar to a stream of particles?

17. A Crooke's radiometer turns so that the white sides of the vanes advance forward and the black sides recede. This is opposite to the direction of rotation expected if light "pressure" were causing the effect. Can you think of a way, without tampering with the radiometer, to cause the vanes to rotate in the opposite direction? [See Frank S. Crawford, "Running Crooke's Radiometer Backwards," *American Journal of Physics* **53**, 11 (1985).]

18. An ideal battery charges a capacitor to a potential difference V. All the wires and circuit elements are made of superconducting materials so that there is zero resistance in the circuit. The battery loses a charge Q at a potential difference V, so the battery loses energy QV. The capacitor gains energy $\frac{1}{2}QV$. Where did the other half of the energy go?

Problems
35.2 Displacement Current and Maxwell's Equations

35A-1 Find the distance in centimeters that light travels in one nanosecond.

35B-2 A parallel-plate capacitor consists of circular plates 10 cm in diameter and separated by 1 mm. Calculate the magnitude of the magnetic field between the plates at their outer edge while the potential difference on the capacitor is changing at the rate of 1000 V/s. (Neglect fringing of the electric field.)

35B-3 Show that the displacement current defined by $i_d = \varepsilon_0 \, d\Phi_E/dt$ has the units of amperes.

35B-4 A 0.5-μF parallel-plate capacitor is being charged through a resistance of 100 Ω by a 9-V battery. Calculate the displacement current in the capacitor 50 μs after the charging is initiated.

35B-5 Show that the displacement current i_d between the plates of a parallel-plate capacitor may be expressed by $i_d = C \, dV/dt$, where C is the capacitance of the capacitor and dV/dt is the rate of voltage change across the capacitor.

35B-6 Consider the region between the plates of a *charging* parallel-plate capacitor that has circular plates. Make a qualitative plot of the magnitude of the magnetic field as a function of the distance from the axis of the capacitor. Include the region beyond the edge of the plates. (Neglect the fringing of the electric field at the edge of the plates.)

35B-7 A parallel-plate capacitor with circular plates of radius R has a capacitance C. The potential across the capacitor is increasing at the constant rate dV/dt. Assuming that there is no fringing of the electric field, show the expressions for the magnetic field at distances radially away from the center of the capacitor are (in SI units) the following: for $r < R$: $(2rC/R^2)\,dV/cdt \times 10^{-7}$; for $r > R$: $(2C/r)\,dV/dt \times 10^{-7}$.

35.3 Electromagnetic Waves

35A-8 An electromagnetic wave in a vacuum has a magnetic field amplitude of 3×10^{-8} T. (a) Calculate the amplitude of the associated electric field. (b) When the electric field is in the $-y$ direction, what direction is the magnetic field if the propagation of the wave is in the $-x$ direction?

35A-9 Show that the equation $E = cB$ balances dimensionally in SI units.

35A-10 The electric field component of a plane electromagnetic wave has a peak value of 25 V/m. (a) Find the amplitude of the associated magnetic field. (b) If the wavelength is 2.80 m, what is the frequency? (c) Write a numerical equation in SI units for the electric component of the wave of this form: $E = E_m \sin(kx - \omega t)$.

35B-11 The ratio $\mu_0 E/B$ has dimensions of an impedance. For a traveling electromagnetic wave in a vacuum, this ratio is called *the characteristic impedance of free space*. Show that in SI units it does have units of ohms, and calculate its numerical value.

35.5 Energy in Electromagnetic Waves

35B-12 A typical value of the earth's magnetic field is 50 μT. Calculate the average wave intensity of an electromagnetic wave that would have a similar magnetic field amplitude.

35A-13 The electric field oscillations received at an FM radio antenna have an amplitude of 5×10^{-5} V/m. (a) Calculate the amplitude of the associated magnetic field oscillations. (b) Calculate the wave intensity of the radiation.

35B-14 Standard wire tables indicate that 12-gauge copper wire has a diameter of 0.080 81 in. and a resistance of 1.588 Ω/1000 ft (note units). When the wire carries an AC current of 20 A (peak), find (a) E_0, (b) B_0, and (c) S_{av} just outside the surface of the wire. (At any instant, the current is uniform throughout the volume of the wire.)

35B-15 The electric field associated with an electromagnetic wave traveling in the $+x$ direction is described in SI units by $\mathbf{E} = 6 \sin(kx - 10^{16}t)\hat{\mathbf{y}}$. (a) Write the corresponding expression for the magnetic field. (b) Calculate the wavelength of the radiation. (c) Calculate the average energy density in the radiation.

35B-16 Using the value of S_{av} obtained in Problem 35B-14, verify numerically that $\oint \mathbf{S}_{av} \cdot d\mathbf{A} = I^2 R$ for a 1000-ft length of 12-gauge copper wire.

35B-17 A pulsed laser produces a flash of light 4 ns in duration, with a total energy of 2 J, in a beam 3 mm in diameter. (a) Find the spatial length of the traveling pulse of light. (b) Find the energy density in joules/meters^3 within the pulse. (c) Find the amplitude E_0 of the electric field in the wave.

35B-18 A monochromatic light source emits 100 W of electromagnetic power uniformly in all directions. (a) Calculate the average electric-field energy density one meter from the source. (b) Calculate the average magnetic-field energy density at the same distance from the source. (c) Find the wave intensity at this location.

35B-19 Show that, for a sinusoidal electromagnetic wave, the average value of the Poynting vector $|\mathbf{S}_{av}|$ is related to the root-mean-square value of the electric field by $E_{rms} = \sqrt{\mu_0 c S_{av}}$.

35B-20 A cube, each edge 1 m long, is aligned so that the edges are parallel to a rectangular coordinate system. A plane sinusoidal electromagnetic wave propagates through the cube in the $+y$ direction with a peak electric field $E_0 = 600$ V/m. The wavelength λ is so long that at any instant the field has (essentially) the same value throughout the cube. (a) Calculate the maximum instantaneous electric-field energy within the cube. (b) When $\mathbf{E} = E_0\hat{\mathbf{x}}$, what are the magnitude and direction of $\mathbf{B}$? (c) Using the Poynting vector, calculate the average power flow through each face of the cube.

35.6 Momentum of Electromagnetic Waves

35A-21 An inflated mylar balloon 50 m in diameter orbits the earth at an altitude of approximately 1000 km. Calculate the maximum force on the balloon due to the direct electromagnetic radiation from the sun, assuming that the radiation is totally absorbed.

35A-22 A 100-mW laser beam is reflected back upon itself by a mirror. Calculate the force on the mirror.

35A-23 On a clear day, sunlight at the earth's surface delivers 840 W/m^2 on a surface oriented perpendicular to the incoming radiation. If the surface is perfectly reflecting, what pressure does this radiation exert?

35B-24 (a) Assuming that the earth absorbs all the sunlight incident upon it, find the total force that the sun exerts on the earth due to radiation pressure. (b) Compare this value with the sun's gravitational attraction.

35B-25 A 15-mW helium–neon laser ($\lambda = 632.8$ nm) emits a beam of circular cross-section whose diameter is 2 mm. (a) Find the maximum electric field in the beam. (b) What total energy is contained in a 1-m length of the beam? (c) Find the momentum carried by a 1-m length of the beam.

35B-26 Radiation with an intensity of 50 W/m^2 falls perpendicularly on the surface of a plane object that absorbs 10% of the radiation and reflects the rest. Calculate the pressure exerted upon the object by the radiation.

Additional Problems

35C-27 By means of a wire attached to a small metal sphere, the sphere is alternately charged positive and negative according to $q = (4 \text{ pC}) \sin \omega t$, where $\omega = 2\pi f$. (a) Find the displacement current $i_d(t)$ existing in one octant of the empty space surrounding the sphere if the frequency f of the charge variation is 60 Hz. (b) Repeat for a frequency of 60 MHz.

35C-28 A plane electromagnetic wave propagates along the x axis. At a time and point on the axis, the electric field associated with the wave is 7.5 V/m and changing at a rate of 2.8×10^{16} V/m·s. (a) Show that this is a reasonable value for a typical electromagnetic wave in the optical portion of the spectrum, green light: $\lambda = 500$ nm. (b) Calculate $\partial B_z/\partial x$ at the same time and place.

35C-29 Show by direct substitution that the function $E = E_0 e^{k(x-ct)}$ satisfies the wave equation $\partial^2 E/\partial x^2 = (1/c^2)\partial^2 E/\partial t^2$. (Any function of the form $f(x \pm ct)$ satisfies the wave equation.)

35C-30 Monochromatic light with a wavelength of 500 nm and an intensity of 60 μW/m^2 propagates along the $+x$ axis. At a particular instant the Poynting vector has zero magnitude at the origin. At that instant, what are the magnitudes of the electric and magnetic fields at a distances of two-thirds of a wavelength along the x axis?

35C-31 Show that $E = E_0 f(x \pm ct)$, where $f(x \pm ct)$ is an arbitrary function, satisfies the wave equation $\partial^2 E/\partial t^2 = c^2 \partial^2 E/\partial x^2$.

35C-32 A microwave transmitter utilizing a parabolic reflector emits an electromagnetic wave into a solid angle of 10^{-2} steradians. At 2 km from the transmitter, the amplitude of the electric field associated with the radiation is 8 V/m. Calculate the output power of the transmitter.

35C-33 A very long line source of radiation emits monochromatic electromagnetic waves at the rate of 20 watts per meter length of the source. Find the amplitude of the electric field of this radiation 5 m from the line source.

35C-34 For the previous problem, find the energy density in the radiation 5 m from the line source.

35C-35 A dust particle in outer space is attracted toward the sun by gravity and repelled by the radiation from the sun. Suppose that a particle is spherical, with a radius R and density $\rho = 2$ g/cm^3, and that it absorbs all the radiation falling on its surface. (a) Determine the value of R such that the gravitational and radiation forces are equal. Obtain the necessary constants from the appendices. (b) Explain why the distance from the sun is irrelevant.

35C-36 A parallel-plate capacitor is composed of circular plates with a radius of 15 cm separated by a distance of 0.1 mm. The capacitor is charged by being connected in series with a 120-V battery and a 5-MΩ resistor. Consider a point between the plates 8 cm from the axis of the plates. One millisecond after the charging starts, calculate the magnitudes of (a) the magnetic field, (b) the electric field, and (c) the instantaneous Poynting vector.

35C-37 Figure 35-21 shows the charging of a parallel-plate capacitor by a current i. As the electric field is increasing, (a) show that the Poynting vector **S** is toward the axis everywhere throughout the volume between the plates. (Ignore fringing of the electric field.) (b) The integral of the Poynting vector over the cylindrical surface surrounding the volume between the plates represents the energy flow into the volume. Show that this energy flow equals the rate of increase of energy stored in the electric field between the plates. (In this view, the energy stored in a capacitor does not come through the wires carrying the current, but flows in from the surrounding space.)

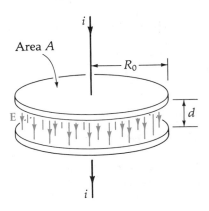

FIGURE 35-21
Problem 35C-37.

35C-38 A plane electromagnetic wave varies sinusoidally at 90 MHz as it travels along the $+x$ direction. The peak value of the electric field is 2 mV/m and it is directed along the $\pm y$ direction. (a) Find the wavelength, the period, and the peak value B_0 of the magnetic field. (b) Write expressions in SI units for the space and time variations of the electric field and of the magnetic field. Include numerical values as well as subscripts to indicate coordinate directions. (c) Find the average power per unit area that this wave propagates through space. (d) Find the average energy density in the radiation (in units of J/m^3). (e) What radiation pressure would this wave exert upon a perfectly reflecting surface at normal incidence?

35C-39 A long cylindrical resistor of radius a, made of material of resistivity ρ, carries an alternating current. (a) Show that the Poynting vector **S** is radially inward (at all times) everywhere on the surface of the resistor. (b) Integrate S_{av} over the surface for a length ℓ of the resistor to show that it equals the $I_{rms}^2 R$ losses within that length. Note that, at the surface, **E** is parallel to the axis of the cylinder. (This calculation implies that the thermal energy developed in the resistor originates not inside the resistor that carries the current, but from the space surrounding the resistor.)

35C-40 A totally reflecting "solar sail" can be used to propel a spacecraft by the radiation pressure exerted on it by solar radiation. Consider a spacecraft located midway between the orbits of Earth and Mars. (a) Using data from Appendix L, find the solar power incident normally on a square meter at this location. (b) Suppose that a flat rectangular solar sail, 900 m × 1200 m, is attached to the spacecraft and oriented with its plane perpendicular to the sun's radiation. The sail is a perfect reflector. If the total mass of the spacecraft plus sail is 2900 kg, calculate the acceleration of the spacecraft. (c) If the orientation of the sail is changed so that its normal makes an angle of 28° with the incoming radiation, find the magnitude

FIGURE 35-22
This portable probe uses a new method for locating underground gas pipes. An 80-kHz AC signal is applied at one point to the pipe where it emerges from the ground, causing the entire pipe system to act as a giant antenna, which radiates electromagnetic waves. The two cross-pieces on the probe contain identical pick-up coils that detect these waves. As the detector is moved along the ground, the strongest signal occurs when the detector is directly over a pipe, while the difference in the signal strengths from the two coils enables the depth of the pipe to be calculated automatically by a small computer in the handle. See Problem 35C-43.

and direction of the spacecraft's acceleration. (Note: for incidence at an oblique angle on a surface, the angles of incidence and reflection are equal.

35C-41 An astronaut, stranded in space "at rest" 10 m from his spacecraft, has a mass (including equipment) of 110 kg. He has a 100-W light source that forms a directed beam, so he decides to use the beam of light as a photon rocket to propel himself continuously toward the spacecraft. (a) Calculate how long it will take him to reach the spacecraft by this method. (b) Suppose, instead, he decides to throw the light source away in a direction opposite to the spacecraft. If the mass of the light source is 3 kg and, after being thrown, moves with a speed of 12 m/s *relative to the recoiling astronaut*, how long will the astronaut take to reach the spacecraft?

35C-42 (a) Derive the relationship between the radiation pressure on a nonreflecting surface with the energy density associated with radiation incident normally just outside the surface. (b) Explain why the relationship does not depend on whether or not the surface is nonreflecting. (c) Is the relationship the same for non-normal incidence? Explain.

35C-43 See Figure 35-22. The two horizontal pick-up coils are 50 cm apart, and the lower coil is 10 cm above the ground. The probe is held over a buried, straight section of gas pipe that has an AC voltage applied to it as described in the caption. The induced AC (effective) signals in the upper and lower coils are, respectively, 0.052 mV and 0.074 mV. How far below the ground surface is the pipe buried?

CHAPTER 36

Geometrical Optics I—Reflection

*Mirrors have one limitation: You can't
  either by hook or by crook
Use them to see how you look when you aren't
  looking to see how you look.*

PIET HEIN
(*Grooks 4*)

36.1 Introduction

When light reflects from smooth reflecting surfaces, images can be formed by the reflected rays. In this chapter we will be concerned with the *result* of the interaction; the physical details of *how* the interaction takes place will be covered later. Therefore, this chapter will contain relatively less new physics but more geometry than previous chapters. We will limit the discussion to electromagnetic radiation in the visible-light region, where all frequencies behave in a similar way. If we were to go very far beyond the visible portion of the spectrum, the interaction would change. For example, a thin sheet of aluminum foil, which reflects visible light, is essentially transparent to x-rays and gamma rays.

Visible wavelengths extend through the full range of the colors we see in a spectrum from deep violet to dark red, corresponding to wavelengths from about 400 nm to 700 nm. The **nanometer** (nm), a unit of length where 1 nm $\equiv$ 10^{-9} m, is the customary unit for specifying wavelengths in the visible region. A person with normal eyesight can barely distinguish[1] two colors with a wavelength difference of 1 nm. Another unit of length commonly used by spectroscopists (but gradually being replaced by the nanometer) is the *angstrom* (Å), where 1 Å $\equiv$ 10^{-10} m. Table 36-1 correlates colors of the visible spectrum with their approximate wavelengths.

As discussed in the last chapter, sources of electromagnetic radiation are basically *accelerated charges*. If we limit our considerations to the production of visible light, the source is often a *glowing hot body* such as the filament of an incandescent light bulb, typically at about 3000 K. Radiation produced by the thermal agitation of atoms and molecules in solids is a mixture of wavelengths, mostly in the infrared, with only a small percentage of the energy lying in the

TABLE 36-1
Wavelength and Color

Approximate Wavelength (nm)	Color
420	Violet
470	Blue
520	Green
570	Yellow
620	Orange
670	Red

---

[1] The human ear is in many ways much more discerning than the eye. While the visible spectrum covers less than one octave (a factor of two in frequency), the audible range of sounds is about 10 octaves, with the smallest discernible change in pitch of about one *cent*, where one octave contains 1200 cents.

visible range. The *carbon arc* is a particularly bright source formed when a DC electric arc is produced between two carbon rods a few millimeters apart. The intense electron bombardment of one rod produces a temperature of about 4000 K, resulting in a source of white light suitable for motion picture projectors and large searchlights. An *arc discharge* in a metal vapor contained within a glass rod produces the familiar blue-green mercury-arc light, or the yellowish sodium-vapor arc lights used for highway illumination. *Fluorescent* lights operate on an electric discharge in a mercury–argon vapor, which produces radiation mostly in the ultraviolet. The ultraviolet radiation is absorbed by a thin coating of phosphors on the interior walls of the tube. The phosphors *fluoresce*, re-emitting the energy as visible light. *Lasers*, those spectacular sources developed within the past few decades, emit a narrow beam of extremely intense radiation of nearly *monochromatic* light—light of essentially a single wavelength. Lasers will be discussed in Chapter 39.

36.2 Wavefronts and Rays

For this introductory discussion we consider a *point source* of light that emits radiation of a *single* wavelength λ. A cross-section of the spherical waves moving away from this point source is analogous to circular water waves moving away from a small object that is moved up and down on the surface of a pond. We may identify the *electric* field variations in the electromagnetic waves with the crests and troughs of the water waves, as shown in Figure 36-1.

The similarity between water waves and electromagnetic waves is more than just geometrical. They also share other properties. For example, electromagnetic waves bend around obstacles just as ocean waves bend around the end of a breakwater. If the obstacle has an opening, or *aperture*, in it, the waves will spread out as they pass through the opening, an effect called *diffraction* (Chapter 39). The amount of bending depends on the size of the aperture compared with the wavelength of the waves. (The closer this dimension is to the wavelength, the more the bending.) But we postpone these diffraction phenomena to a later chapter and treat here only those cases in which the obstacle or aperture size is very large compared with the wavelength, ignoring the bending and spreading effects. This approximation is an excellent one for analyzing mirrors, lenses, prisms, and other optical instruments such as telescopes and microscopes.

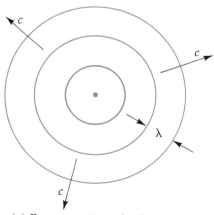

(a) Representation of spherical light waves traveling outward from a point source.

(b) Circular water waves.

FIGURE 36-1
A cross-section of the spherical waves emanating from a point source of light is geometrically similar to water waves moving outward from a localized disturbance on the surface of the water.

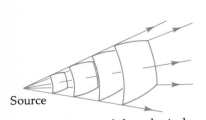
(a) A portion of the spherical wavefronts emerging from a point source.

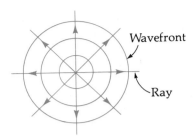
(b) The usual way of depicting spherical wavefronts from a point source.

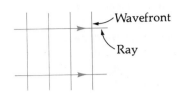
(c) A plane wave traveling toward the right.

FIGURE 36-2
Rays are perpendicular to wavefronts. The arrows on the rays indicate the direction of wavefront motion.

FIGURE 36-3
Shafts of sunlight give the impression of light rays traveling in straight lines. The shafts are actually parallel, though our perspective makes them appear to diverge.

For light waves emitted from a point source, each point on an expanding spherical surface has the *same phase* and is called a **wavefront**. In sketching diagrams we often draw wavefronts as lines, as in Figure 36-1a. However, keep in mind that wavefronts for electromagnetic waves are *surfaces*. The direction that a wavefront moves is always perpendicular to the wavefront itself. Any line drawn perpendicular to a wavefront is called a **ray**; an arrow on the ray indicates the direction of motion. Figure 36-2b illustrates some rays associated with the spherical wavefronts emerging from a point source of light. At very great distances from the source, the wavefronts become essentially *plane* because the radius of curvature is so great. Sometimes for convenience we consider wavefronts that are spaced one wavelength apart; the spacing of the rays, however, has no significance. As we will show, just two rays from a source are adequate for the analysis of an optical system.

36.3 Huygens' Principle

A useful technique for the analysis of optical systems was devised by the Dutch physicist and astronomer Christian Huygens (1629–1695). He proposed the following:

HUYGENS' PRINCIPLES **Every point on a wavefront may be considered as a point source of secondary waves, called *wavelets*. These wavelets spread outward with the speed of light. After a time t, the new position of the wavefront is the *envelope*, or *tangent surface*, to these secondary wavelets.**[2]

Figure 36-4 illustrates the procedure. Each point along a wavefront AA' is considered to be a point source, each radiating secondary wavelets. At a later time t, the envelope of these wavelets forms the new wavefront BB'. The method works for a wavefront of any arbitrary shape, not just the plane and spherical wavefronts illustrated.

---

[2] There is a degree of artificiality in this procedure. If all points on the wavefront were true point sources, the secondary wavelets would radiate not only in the forward direction of wave propagation, but also in the backward direction. Huygens ignored the backward radiation. In a more sophisticated treatment done later by Kirchhoff, it was shown that the backward radiation actually would be zero due to interference effects discussed in Chapter 38.

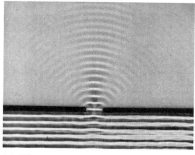

(a) Plane water waves in a ripple tank are incident upon a barrier that has a small opening whose size is comparable to the wavelength of the waves. In agreement with Huygens' principle, the opening acts as a source of secondary circular wavelets.

FIGURE 36-4

Huygens' principle. A wavefront AA' is considered to be a series of point sources for secondary wavelets. After a time t, the secondary wavelets travel forward a distance ct. The envelope

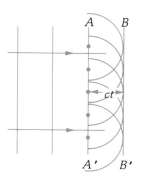

(b) Plane wavefronts.

of these secondary wavelets forms the new wavefront BB'. The arrows on the rays indicate the direction of wave propagation.

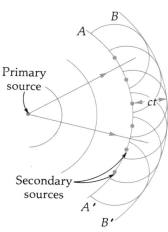

(c) Spherical wavefronts.

36.4 Reflection by a Plane Mirror

Laws describing reflection of light by mirrors were probably known as early as the time of Plato in the fourth century B.C. We now deduce these laws by two different methods, each illustrating an important principle in physics.

Using Huygens' Principle

We often speak of looking "into" a mirror in the same sense as looking into a room. We see images that certainly appear to be on the other side of the mirror, and every child has wondered what it would be like to pass through the looking glass into that other world, whose contents have a one-to-one relationship with objects in the real world. How far behind the mirror is the image of a given object? Consider a plane wave approaching a mirror as in Figure 36-5a. The rays associated with incoming wavefront AB form an angle α_1 with the surface of the mirror. As each portion of the incoming electromagnetic wave strikes the mirror, electrons in the surface of the mirror are set into oscillations. These oscillating electrons reradiate electromagnetic waves, so each becomes a source of secondary wavelets.[3]

---

[3] The idea of Huygens' wavelets originating from every point on a wavefront in *free space* does seem to be merely a "trick" that gives the right answer. However, when a material medium is present, with oscillating electrons acting as sources of reradiated waves, the idea becomes plausible and, indeed, correctly describes the mechanism of electromagnetic waves interacting with matter. In Huygens' time, it was believed that the medium that transmitted light waves—the *ether*, as it was called—was present everywhere, even in a vacuum, so it is easy to see how Huygens' principle arose. Of course, following Einstein, the present-day theory of light makes no use of the ether concept.

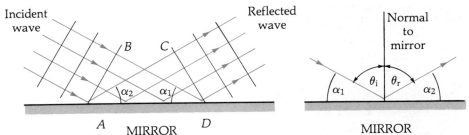

FIGURE 36-5
A plane wave reflected by a plane mirror.

(a) The reflected wavefront CD is formed by the envelope of secondary waves originating at the surface of the mirror.

(b) The angle of incidence equals the angle of reflection: $\theta_i = \theta_r$.

Let us look more closely at the reflected wavefronts shown in Figure 36-5a. As the point A on the wavefront AB strikes the mirror, a circular wavelet originating at A will proceed to a point C on the reflected wavefront CD. Meanwhile, a wavelet originating at B will proceed toward the point D on the mirror. If the time required for a wavelet to travel from A to C equals the time required for a wavelet to travel from B to D, the points C and D will be in phase, thus constituting parts of a reflected wavefront. Of course, all wavelets originating from points between A and B will be reflected to reach corresponding points between C and D. Therefore, the distances AC and BD are equal, and the right triangle ABD is congruent to the right triangle ACD. (They have the common hypotenuse AD and equal sides.) Thus angles α_1 and α_2 are equal. It follows that their complements, θ_i and θ_r, are also equal. In optics it is customary to measure angles of rays with respect to the *normal*, or perpendicular, to a surface. Therefore, in Figure 36-5b we see that the *angle of incidence* θ_i is equal to the *angle of reflection* θ_r. Moreover, if we carry out the analysis in three dimensions, it can be shown that the incident ray, the normal to the mirror, and the reflected ray all lie in the same plane.

LAWS OF REFLECTION

(1) **The angle of incidence equals the angle of reflection: $\theta_i = \theta_r$.**

(2) **The incident ray, the normal to the mirror, and the reflected ray all lie in the same plane.**

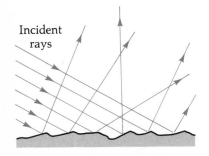

FIGURE 36-6
In the case of *diffuse* reflection, parallel rays are reflected in various directions because of surface irregularities.

If the surface is rough, as in Figure 36-6, a bundle of parallel rays will be reflected at various angles. This type of reflection, called *diffuse* reflection, is illustrated by the surface of the page you are now reading. Even though the illumination on the page is essentially parallel rays from a single study lamp, you can observe the page from any angle. Most nonluminous objects you see are observed by diffuse reflection. The difference between diffuse and *specular* (mirrorlike) reflection depends on the size of surface irregularities compared with the wavelength of the illumination. If such irregularities are small compared with the wavelength of light, specular reflection occurs. On the other hand, if such irregularities are of the order of a wavelength or larger, the reflection is diffuse. Thus the roughened surface of a piece of aluminum that has been sanded would cause diffuse reflection of visible light, but specular reflection of radar waves of 5 cm wavelength.

Using Fermat's Principle

The laws of reflection may also be deduced from Fermat's[4] principle, another important relation of physics.

FERMAT'S PRINCIPLE **In going from one point to another, a light ray travels a path that requires equal or less time in transit than the time required for neighboring paths.**

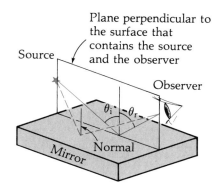

FIGURE 36-7
Fermat's principle: a light ray will be reflected in such a way that the total time in transit from the source to the observer is a minimum.

To illustrate this principle, we apply it to the situation shown in Figure 36-7. The source and the observer lie in a plane perpendicular to the surface of a plane mirror. An arbitrary path of a ray is shown as a dashed line. Clearly this path is not the shortest from source to observer, so it will not be traveled in the least time. While it seems obvious that the shortest path (the solid line) lies in the plane containing the normal to the mirror, it is also true that the angle of incidence equals the angle of reflection. Proof of the latter by Fermat's principle is left as a problem.

Let us now return to the question asked at the beginning of this section: How far behind the mirror is the image of an object? To find the answer, we trace the paths of a few rays in accordance with the laws of reflection. Figure 36-8 is a **ray diagram** that shows rays leaving the source (★) at A and being reflected by the mirror. The directions along which the reflected rays travel make them appear to come from the single point C behind the mirror, a point that is the *image* of the source. (Although we show three rays in the figure, just two rays would be sufficient to locate the point C.) We now introduce a notation that will simplify the discussion. The *object distance* p is the perpendicular distance from the object to the mirror, and the *image distance* q is the perpendicular distance from the image to the mirror. The second law of reflection ensures that the rays shown lie in the plane of the figure. The first law of reflection leads to the conclusion that the triangle ABD is congruent to the triangle CBD. (They have a common side BD, and the other two sides of the triangles form equal angles with BD.) Thus:

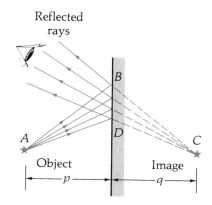

FIGURE 36-8
The image formed by a plane mirror lies behind the mirror at a distance equal to the distance the object is from the mirror.

IMAGE LOCATION IN PLANE MIRRORS **The image distance q equals the object distance p.**

This conclusion is based on a point source.

An object of finite size may be thought of as a distribution of point sources, each with its own image. Thus there is a point-to-point correspondence between an object and its image in the mirror. Because $p = q$ for each point, the object and the image are located symmetrically on opposite sides of the mirror and are the same size as shown in Figure 36-9.

An interesting feature of plane mirror images is that left and right are interchanged. For example, the image of your right hand appears as a left hand,

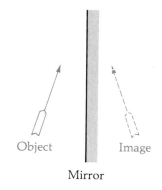

FIGURE 36-9
The image of an object formed by a plane mirror is the same distance behind the mirror as the object is in front. The image and the object are of equal size.

---

[4] Pierre de Fermat (1601–1665), a French nobleman, founded modern probability theory as a result of his interest in calculating gambling odds. In addition to Fermat's principle, he is also famous for "Fermat's last theorem," a tantalizing puzzle that still frustrates mathematicians. In a note (discovered posthumously) written on the margin of a book page, he claimed to have proved that there are no nontrivial integral solutions of $x^n + y^n = z^n$ for $n > 2$. To date, no one else has been able to prove or disprove it.

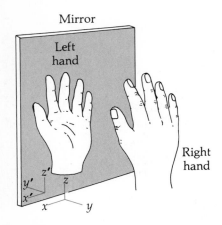

FIGURE 36-10
The images in a plane mirror have left and right interchanged. A right-handed coordinate system becomes a left-handed coordinate system in the mirror world.

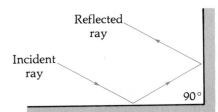

(a) Two mirrors at right angles form a two-dimensional corner reflector for rays that lie in a plane perpendicular to the mirrors. After two reflections, any incident ray is returned in an antiparallel direction back toward the source.

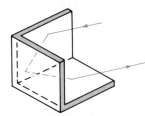

(b) A square-cube corner reflector produces a <u>retroreflection</u> for rays incident at any angle (so they reflect off all three faces), returning the light along a direction antiparallel to the incident rays. Many such small reflectors are used for highway signs and lane buttons in roadways to reflect motorists' headlights.

(c) A Laser Ranging Retroreflector (LRRR) on the moon. Three 18-in. square arrays, each containing 100 corner-cube reflectors, were placed on the moon by *Apollo* astronauts, and a fourth was deposited by a Russian spacecraft. Several earth satellites also contain corner-cube reflectors. The round-trip travel time of a laser pulse sent from the earth to these reflectors can be measured so accurately that the earth–LRRR distance is determined with an uncertainty of only a few centimeters, permitting long-term studies of subtle earth and moon motions. Continental drift is now measured directly using this technique.

FIGURE 36-12
Corner reflectors.

Figure 36-10. Also, a right-handed coordinate system has a mirror image that is a left-handed coordinate system.

Two mirrors at right angles form a two-dimensional *corner reflector*, Figure 36-12a. Any incident ray, after two reflections, is returned precisely in an antiparallel direction back toward the source (Problem 36A-1). Three mirrors forming the corner of a cube similarly act as a corner reflector in three dimensions, Figure 36-12b. Arrays of large numbers of small corner-cube reflectors are used to reflect headlights at night from road signs, safety reflectors on bicycles, etc.

36.5 Reflection by a Spherical Mirror

Much to the distress of some of us, we are greeted in the morning by a larger-than-life-size image of ourselves as we look into a shaving or makeup mirror. In most of our encounters with mirrors, the image is behind the mirror, though we will see that, under certain circumstances, images can also be formed in front

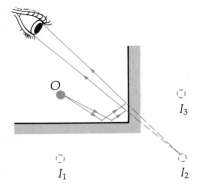

FIGURE 36-11
Two plane mirrors at right angles produce three images of an object at O. (It will be helpful to sketch the rays from the object that produce images I_1 and I_3. Each involves just one reflection.)

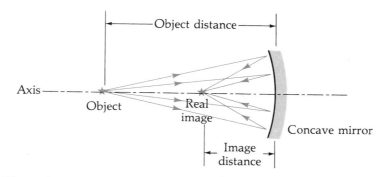

(a) All rays from this point object (★) are reflected by the concave mirror and converge to form a *real* image.

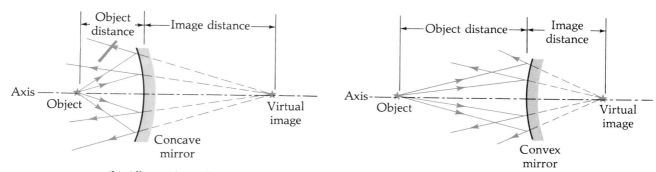

(b) All rays from these point objects (★) are reflected by a concave or a convex mirror so that they diverge to form a *virtual* image.

FIGURE 36-13
Image formation by spherical mirrors.

of the mirror. Mirrors with curved surfaces—*spherical mirrors*—may be concave or convex, depending on the type of surface curvature that the incident light rays encounter. The surface of a spherical shell approached from inside the shell is **concave**; when approached from outside the shell, the surface is **convex**.

To locate and describe an image produced by a spherical mirror, we use the technique of *ray-tracing*. A line called the **optic axis** is sketched symmetrically through the center of the mirror, perpendicular to the mirror surface. We then consider a point (★) on the axis and investigate how the mirror affects light rays that leave the object. After reflection, the rays may either converge to form a **real** image, as shown in Figure 36-13a, or diverge to form a **virtual** image, as shown in Figure 36-13b. The word *real* signifies that light rays actually converge at the image location to form an image. If we placed a screen there (without interfering with the passage of the rays), an image would appear on the screen. The word *virtual* signifies that light rays do not actually reach the image location; if a screen were placed there, no image would appear on the screen as in the case of the image in a plane mirror. In either case, if our eyes are in a position to intercept the rays after they leave the mirror, we see an image at that location. Without other clues we cannot know whether the image is real or virtual: both types of images have the same visual appearance.

To find the location of the image, we trace *two rays* whose paths we can easily determine. All reflected rays pass through the same point, so determining the paths of just two rays is sufficient to locate the image. We use the following

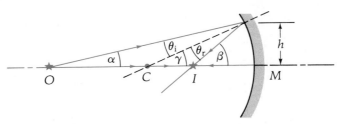

(a) Case 1. Concave mirror: real image

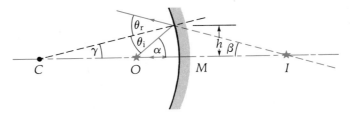

(b) Case 2. Concave mirror: virtual image

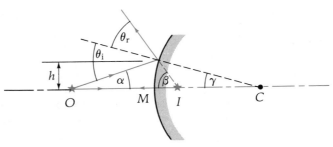

(c) Case 3. Convex mirror: virtual image

FIGURE 36-14
Ray-tracing analysis for spherical mirrors. One ray, from the object O, travels along the axis and is reflected backward along the axis. The other ray travels at an angle α to the axis and is reflected along a direction at an angle β to the axis. The image is located at the intersection of the two reflected rays.

notation, as shown in Figure 36-14. The center of curvature is at C, the object is at O, the image is at I, and the position of the mirror is at M. We restrict our considerations to those cases for which the angles involved are small enough that the tangent of the angle is approximately equal to the angle itself in radian measure. That is, $\tan \alpha \approx \sin \alpha \approx \alpha$. Such rays that lie close to the axis and are nearly parallel to it are called **paraxial rays**. Accepting this approximation, we find the results of the ray-tracing analysis to be valid for mirrors whose diameters are much smaller than the radius of curvature.[5] In all cases, we will apply the laws of reflection, and for simplicity we will drop the subscripts, so that

$$\theta_i = \theta_r = \theta$$

Let us now analyze three different cases of image formation by spherical mirrors and summarize the results in a single, convenient equation known as *the mirror equation*.

---

[5] Often this criterion is expressed by the phrase "a small-aperture mirror." That is, the aperture (diameter) of the mirror is small compared with its radius of curvature. Since this is a relative matter, an astronomical mirror 3 m in diameter may be classified as a small-aperture mirror, while a mirror 5 cm in diameter may not be.

Case 1. Concave Mirror: Real Image

In Figure 36-14a, we trace these two rays: one ray travels from the object O *along the axis* and (since it strikes the mirror perpendicularly) is reflected back along the axis. The other ray travels at *an angle α to the axis* and is reflected along a direction at an angle β to the axis. The point where these two reflected rays intersect is the image location. Because the exterior angle of a triangle is equal to the sum of the opposite interior angles, we have, for one triangle, $\beta = \gamma + \theta$, and for another triangle, $\beta = \alpha + 2\theta$. Eliminating θ, we obtain

$$\alpha + \beta = 2\gamma \qquad (36\text{-}1)$$

When we use the small-angle approximations for paraxial rays,

$$\alpha \approx \frac{h}{OM} \qquad \beta \approx \frac{h}{IM} \qquad \gamma \approx \frac{h}{CM}$$

Treating these expressions as equalities and substituting into Equation (36-1) gives

$$\frac{1}{OM} + \frac{1}{IM} = \frac{2}{CM} \qquad (36\text{-}2)$$

Note that h and θ do not appear in the expression. This implies that *all* rays emanating from the object and reflected by the mirror will converge to the image point (at least, within the validity of the small-angle approximations used in the derivation).

Case 2. Concave Mirror: Virtual Image

Referring to Figure 36-14b and proceeding as in Case 1, we have $\theta = \beta + \gamma$ and $\alpha = \theta + \gamma$. Eliminating θ gives

$$\alpha - \beta = 2\gamma$$

Identifying these angles with their tangents, we have, for paraxial rays,

$$\frac{1}{OM} - \frac{1}{IM} = \frac{2}{CM} \qquad (36\text{-}3)$$

Case 3. Convex Mirror: Virtual Image

Referring to Figure 36-14c, again we have $\theta = \alpha + \gamma$ and $2\theta = \alpha + \beta$. Eliminating θ, we obtain

$$\frac{1}{OM} - \frac{1}{IM} = -\frac{2}{CM} \qquad (36\text{-}4)$$

Note that unlike a concave mirror, which may produce either a real or a virtual image, a convex mirror *always* produces a virtual image of an object.

The ray-tracing analysis of image formation by spherical mirrors produced similar results in all three cases. Equations (36-2), (36-3), and (36-4) are identical in form, varying only in the signs of some terms. It is convenient to summarize the results by deducing a *single* equation that is valid for all cases. We do this by establishing a **sign convention** to determine the sign of the numerical values to be used in that equation. Observe that, in Figure 36-14, OM is the object distance p, IM is the image distance q, and CM is the radius of curvature R of the mirror. All object and image distances are measured along the axis to the center (M) of the mirror. Equations (36-2), (36-3), and (36-4) may then be combined in a single equation:

MIRROR EQUATION
$$\frac{1}{p} + \frac{1}{q} = \frac{2}{R} \tag{36-5}$$

where
- p = object distance
- q = image distance
- R = radius of curvature of the mirror

To use this equation, we adopt the following *sign convention*:

SIGN CONVENTION FOR MIRRORS[6]

(1) The numerical value of p is positive if the rays *approaching* the mirror are *divergent*. Otherwise p is negative.

(2) The numerical value of q is positive if the rays *leaving* the mirror are *convergent*. Otherwise q is negative.

(3) The numerical value of R is positive if the mirror is concave, and it is negative if the mirror is convex.

You should memorize this sign convention since the solutions to most problems use it in the mirror equations (36-5 and 36-7). Remember that the mirror equation is always written as shown. *Minus signs are introduced only when we substitute numerical values for the symbols.* This same procedure is followed with all general equations in physics.

In certain cases of multiple-mirror systems, the object distance p can be negative. For example, in Figure 36-15 the first mirror, M_1, acting alone, would produce a real image at I_1. In a sense, this image becomes the object for mirror 2 (with an object distance p_2). However, since mirror 2 intercepts the rays before they form the image, the rays that strike mirror 2 are *converging*. According to the sign convention, the numerical value of p_2 would therefore be negative. In such cases, the object is called a *virtual object*.

A common term applied to mirrors (and lenses) is the **focal length** f, Figure 36-16. A group of rays parallel to the axis will be reflected by a *concave* mirror so that they converge to a point a focal length f in front of the mirror.

---

[6] Several other sign conventions are in use. One version is **p, q, and f are each positive for the "standard case" of a converging mirror forming a real image of a real object.** *Any change from this standard case requires a minus sign.* This implies that the object distance p is greater than the focal length of the mirror. For mirrors, light is *reflected*, so real images are formed on the *incident-light side*. The sign for R is given by rule (3) above.

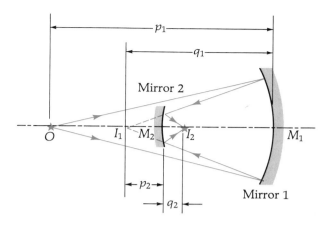

FIGURE 36-15
A multiple-mirror system in which the object distance p_2 is negative ($p_2 < 0$) according to the sign convention. This is because the rays approaching mirror 2 are converging.

The point at which they focus is the **focal point**[7] F. If parallel rays are incident on a *convex* mirror, they reflect along divergent lines that meet at a focal-length distance behind the mirror. Parallel incoming rays imply that the object is at infinity, or $p = \infty$. Substituting this value into the mirror equation, we obtain

$$\frac{1}{\infty} + \frac{1}{q} = \frac{2}{R}$$

where q then becomes equal to the focal length f. Solving for f, we get

$$f = \frac{R}{2} \tag{36-6}$$

Since the numerical value of the focal length is *positive* for *concave* mirrors and *negative* for *convex* mirrors, Equation (36-5) becomes

MIRROR EQUATION
(alternative form)
$$\frac{1}{p} + \frac{1}{q} = \frac{1}{f} \tag{36-7}$$

In this chapter it is easy to become confused in the discussions of numerous cases of mirrors and lenses in a variety of situations. However, the major content of the chapter is the single equation $1/p + 1/q = 1/f$, which is the starting point for locating images formed by both mirrors and lenses. *Knowing the sign convention is essential.* One easy way to remember the sign convention is the following: for the "standard setup" of an object situated farther from a *converging* mirror than the focal-length distance, the symbols p, q, f, and R each have *positive* numerical values. If any of these distances are on the opposite side of the mirror (compared with their locations in this standard setup), they have negative numerical values. The following two examples will illustrate the use of the mirror equation and the sign convention.

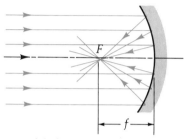

(a) Concave mirror

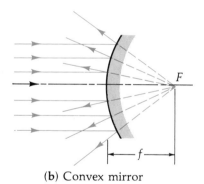

(b) Convex mirror

FIGURE 36-16
When light parallel to the axis is incident on a mirror, the image distance is called the *focal length* f of the mirror. The point F is the *focal point*. For concave mirrors, f is positive; for convex mirrors, it is negative.

---

[7] Calling F the focal point of the mirror does *not* mean that all images are formed at that location. Only for the single case of incident light *parallel to the axis* is this true; in all other cases, the image is elsewhere. It is helpful to think of F as a point that "belongs" to the mirror and that we find useful in constructing ray diagrams.

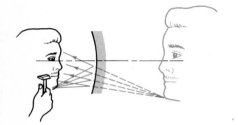

FIGURE 36-17
Example 36-1.

EXAMPLE 36-1

While holding his shaving mirror near a window, a man is able to produce the image of the sun on the wall next to the window. The mirror is 50 cm from the wall. When the man is shaving, his chin is 20 cm in front of the mirror. Find the location of the final image of his chin.

SOLUTION

Light rays from the sun are essentially parallel, so that the image of the sun is produced at the focal point of the mirror. Thus, $f = +50$ cm. (We know that f is positive from the fact that only concave mirrors are capable of producing a *real* image of an object, and according to the sign convention, concave mirrors have positive focal lengths.) Light from a point on the man's chin is diverging as it approaches the mirror, so according to the sign convention, $p = +20$ cm. Starting with the mirror equation

$$\frac{1}{p} + \frac{1}{q} = \frac{1}{f} \quad \Rightarrow \quad \frac{1}{+20 \text{ cm}} + \frac{1}{q} = \frac{1}{+50 \text{ cm}}$$

Solving for q gives
$$q = \boxed{-33.3 \text{ cm}}$$

According to the sign convention, the minus sign indicates that the light diverges from the surface as if it came from a virtual image *behind* the mirror. (Again, *virtual* implies that no rays are actually present at the image location.) So the image is 33 cm behind the mirror, Figure 36-17.

EXAMPLE 36-2

Consider the system of mirrors shown in Figure 36-18. Locate the final image of the object. Is the image real or virtual?

SOLUTION

The procedure in a multiple-mirror system is to find the image formed by each mirror acting alone in the order in which the rays are reflected. In this case

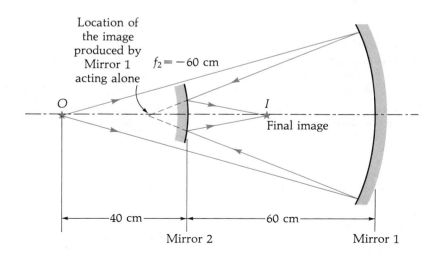

FIGURE 36-18
Example 36-2.

mirror 1 is first. Starting with the mirror equation

$$\frac{1}{p} + \frac{1}{q} = \frac{1}{f} \quad \Rightarrow \quad \frac{1}{+(40\ \text{cm} + 60\ \text{cm})} + \frac{1}{q_1} = \frac{1}{45\ \text{cm}}$$

Solving for q_1 gives

$$q_1 = +82\ \text{cm}$$

According to the sign convention, the positive value signifies that light rays are converging as they leave the first mirror. If this mirror were the only one present, a real image would be formed 82 cm in front of mirror 1. However, mirror 2 intercepts these rays 22 cm before that image can be formed. Nevertheless, we consider that hypothetical "image" to be the object for mirror 2 to work upon. Because the rays impinging on mirror 2 are converging, the object distance p_2 is negative ($p_2 = -22$ cm, according to the sign convention). The object is a **virtual object** because no light rays are actually present at the object. Substituting appropriate numerical values into the mirror equation gives

$$\frac{1}{p} + \frac{1}{q} = \frac{1}{f} \quad \Rightarrow \quad \frac{1}{-22\ \text{cm}} + \frac{1}{q_2} = \frac{1}{-60\ \text{cm}}$$

Solving for q_2 yields

$$q_2 = \boxed{34.7\ \text{cm}}$$

The positive sign indicates that the rays converge upon leaving the second mirror. The final image is thus *real* and is located 34.7 cm to the right of mirror 2, as shown in Figure 36-18. Unless otherwise intercepted, convergent rays always produce real images.

When virtual objects are involved, it is usually not easy to construct significant ray diagrams for the multiple reflections. After all, no light actually travels from the location of the virtual object to the next mirror. So in these cases, just a preliminary ray diagram may be sketched to verify the location of the first image; then the mirror equation is used to find the final image produced by the second mirror.

36.6 Ray Diagrams and Lateral Magnification

The primary function of a shaving or makeup mirror is to produce an enlarged image. In this section we will extend the ray-tracing technique to discover how an object point that is *off the axis* of the mirror is imaged. In sketches, an object that extends off the axis is usually indicated by an arrow labeled O, as in Figure 36-19. All the rays that leave *any given point* on the object and strike the mirror are brought to a focus at the *corresponding point of the image*. We usually trace rays from the tip of the arrow, recognizing that the rest of the arrow is similarly imaged. Just two rays that intersect at I are sufficient to locate the image at I. Although other choices are possible, we choose the

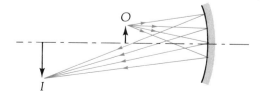

FIGURE 36-19
All rays emerging from the arrow tip that strike the mirror are brought to a focus at the image of the arrow tip.

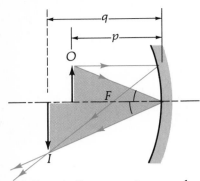

(a) Case 1. Concave mirror: real image

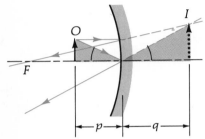

(b) Case 2. Concave mirror: virtual image

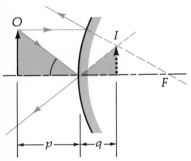

(c) Case 3. Convex mirror: virtual image

FIGURE 36-20
Magnification by spherical mirrors. Extensions of rays behind the mirrors are represented by dashed lines. Virtual images are indicated by dotted lines.

following two rays for such diagrams because they can be sketched in *all* cases, and it is easy to draw what happens to them after they strike the mirror. *Starting from the tip of the arrow:*

RAYS USED IN RAY-TRACING DIAGRAMS

(1) A ray striking the center of the mirror is reflected symmetrically. (The angle of incidence equals the angle of reflection.)
(2) A ray parallel to the axis is reflected through the focal point F.

From these two reflected rays[8] we locate the image of the arrow tip; other portions of the arrow are similarly imaged on a point-to-point basis.

As in the last section, we will treat each of the three possible cases separately. Remember that *concave* mirrors have positive focal lengths and positive radii of curvature, both located in front of the mirror. In contrast, *convex* mirrors have negative focal lengths and negative radii of curvature, both located behind the mirror. For all mirrors, $f = R/2$.

Case 1. Concave Mirror: Real Image

Referring to Figure 36-20a, the object O is the tip of the arrow located a distance p from the mirror. Two rays are drawn from the tip. One ray *strikes the center of the mirror and is reflected symmetrically* (the angle of incidence equals the angle of reflection). The other ray approaches the mirror *parallel to its axis and is reflected through the focal point F*. The intersection of these two rays locates the image I of the arrow tip.

The size of the images formed is a significant feature of optical systems. They may be larger or smaller than the object. In a ray-tracing diagram, the triangles formed by the axis, the object, and the image lead to a simple expression for the **lateral**, or **transverse**, **magnification** M (perpendicular to the axis):

$$M \equiv \frac{\text{Image size}}{\text{Object size}} \qquad (36\text{-}8)$$

Note that the shaded triangles in Figure 36-20a are similar right triangles with corresponding sides having the same ratio. Then,

LATERAL MAGNIFICATION
$$M = -\frac{q}{p} \qquad (36\text{-}9)$$

The minus sign is introduced so that a *negative* value of M indicates an *inverted* image and a *positive* value of M indicates an *erect* image. This same sign convention holds true for both mirrors and lenses.

---

[8] Two other rays also can be used for ray-tracing. Starting at the arrow-tip:

(3) *An incident ray along a mirror-radius line strikes the surface perpendicularly and is reflected back along its original path.*
(4) *A ray passing through the focal point F (or proceeding toward F) is reflected parallel to the axis.*

Sketching more than two rays is useful since it verifies your construction. Unfortunately, in some cases these additional rays are awkward to draw.

By sketching ray diagrams we can verify that if a *real* image is created by a concave mirror, it is always inverted and may be *larger* than, the *same size* as, or *smaller* than the object. It is not necessary to memorize such details for various cases, since the information is contained inherently in the sign convention and in the definition for the lateral magnification. In each case, a ray diagram verifies such characteristics.

Case 2. **Concave Mirror: Virtual Image**

If the object is placed closer than the focal point to a concave mirror, the image is virtual, as shown in Figure 36-20b. As in the first case, we use two rays: one leaves the tip of the arrow *parallel to the axis* and is then reflected through the focal point F; the other ray is *reflected symmetrically at the center of the mirror*. Unlike the first case, the rays diverge after reflection, seemingly from a point behind the mirror. This point is the image of the arrow tip. We locate it by extending the two reflected rays backward along their directions until they intersect. The point of intersection is the image point. But because no actual light rays travel along these extended lines, we draw them dashed. Also, because no actual light rays form the image, it is a *virtual* image, which we sketch with dotted lines.

The shaded triangles are again similar, so that (as before) the lateral magnification is $M = -q/p$. As we can verify by sketching ray diagrams, if a concave mirror forms a virtual image, it is always erect and always larger than the object. A shaving or makeup mirror is concave and, when held the proper distance from the face, produces an erect, virtual image behind the mirror.

The two cases just discussed differ in important ways. In Case 1, the object is *farther* than a focal-length distance from the mirror and produces an inverted, real image. In Case 2, the object is *closer* than a focal-length distance from the mirror and produces an erect, virtual image.

Case 3. **Convex Mirror: Virtual Image**

In Figure 36-20c, one ray from the tip of the arrow is reflected symmetrically at the center of the mirror. The other ray approaches the mirror parallel to the axis and is reflected in a direction *away* from the focal point. (Remember that the center of curvature as well as the focal point of a convex mirror lie behind the mirror.) The lateral magnification is $M = -q/p$. Convex mirrors always form virtual images (of real objects) and are always smaller than the object. Images seen in polished balls are of this type.

To describe an image, we specify the following:

IMAGE CHARACTERISTICS
real or virtual
erect or inverted
magnification

Do not try to memorize rules for all the types of imaging that result when objects are at various distances from converging and diverging mirrors (and lenses.) Instead, gain skill in rapidly sketching ray diagrams, which reveal the nature of the image. This approach is much simpler and enables you to deal with situations you have not seen before. *For numerical calculations, knowing the sign convention is essential.*

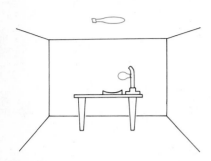

FIGURE 36-21
Example 36-3.

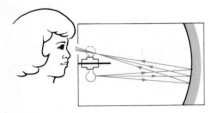

FIGURE 36-22
Here is an unusual illusion created using a concave mirror. A light bulb in a socket is mounted beneath a concealing surface and positioned so that the bulb is at the center of curvature of the concave mirror. An empty socket is mounted on top of the surface (also at the center of curvature) in clear view of the observer. When the bulb is lighted, a real image of the glowing bulb suddenly appears in the empty socket. With a good-quality mirror, the illusion is striking.

EXAMPLE 36-3

A concave mirror rests face up on a table 0.50 m below a desk lamp bulb, as in Figure 36-21. An inverted image of the bulb appears on the ceiling in clear focus and is five times the size of the bulb in the lamp. (a) How high is the ceiling above the table top? (b) Find the focal length of the mirror.

SOLUTION

(a) The situation described in this example is highly unlikely (except by chance). Ordinarily a clear image would not be produced on the ceiling because, for a given focal length, a definite relationship between the object and image distances must exist. This relationship is the mirror equation:

$$\frac{1}{p} + \frac{1}{q} = \frac{1}{f}$$

Since in this example we are given only the object distance p, we still have two unknowns: q and f. An additional relationship between p and the image distance q is needed. With the lateral magnification M known, Equation (36-9) is appropriate: $M = -q/p$. Using the numerical values of $M = -5$ (it is *negative* because the image is inverted) and $p = 0.5$ m (it is *positive* because the rays from the bulb diverge before striking the mirror), we solve for the distance v_q from the table top to the ceiling:

$$q = -Mp = -(-5)(0.5 \text{ m}) = \boxed{2.50 \text{ m}}$$

Because q is positive, we know that the rays leaving the mirror are converging (as they must to form a *real* image on the ceiling).

(b) Now that we know two of the three unknowns, we can apply the mirror equation:

$$\frac{1}{p} + \frac{1}{q} = \frac{1}{f} \quad \Rightarrow \quad \frac{1}{+0.5 \text{ m}} + \frac{1}{+2.5 \text{ m}} = \frac{1}{f}$$

Solving for the focal length f gives $\quad f = \boxed{0.417 \text{ m}}$

Summary

The *propagation of light* is characterized by *rays* and *wavefronts*:

  Rays: Imaginary lines in the direction of propagation.
  Wavefronts: Imaginary surfaces perpendicular to rays, moving in the direction of propagation; each point on a wavefront has the same phase.

  Huygens' principle: Every point on a wavefront may be considered as a point source of secondary wavelets that spread outward with the speed of light. After a time t, the new position of the wavefront is the envelope, or tangent surface, to these secondary wavelets.

  Fermat's principle: In going from one point to another, a light ray travels a path that requires equal or less time in transit than the time required for neighboring paths.

  Law of reflection: $\qquad \theta_i = \theta_r$

  We construct *ray diagrams* by tracing these two rays from the arrow tip (or from additional rays, Footnote 8). Their intersection locates the image of the tip:

(1) A ray striking the center of the mirror is reflected symmetrically.
(2) A ray parallel to the axis is reflected through the focal point F.

Images are *real* or *virtual*, *erect* or *inverted*, with *lateral magnification* according to the following:

Mirror equation

$$\frac{1}{p} + \frac{1}{q} = \frac{1}{f}$$

Lateral magnification

$$M = -\frac{q}{p}$$

where p is the object distance, q is the image distance, and f is the focal length ($=R/2$). These equations are used with the

sign convention:

(1) The value of p is *positive* if the rays that impinge on the mirror are *divergent*.
(2) The value of q is *positive* if the rays leaving the mirror are *convergent*.
(3) The sign of the focal distance $f = R/2$ is determined by the sign of R, the radius of curvature of the mirror surface. It is *positive* if the surface is *concave* and *negative* if the surface is *convex*.

The sign convention may be remembered from the fact that, for the "standard setup" of an object situated farther than a focal-length distance from a converging mirror, all the symbols in the corresponding equation are *positive*; if any of the distances are on the opposite side of the mirror, they are *negative*.

Questions

1. A plane mirror produces an image that is reversed right-for-left. Why does a plane mirror not produce an upside-down image?
2. Will convergent rays reflected by a plane mirror produce a real or a virtual image?
3. A sign painted on a store window is reversed when viewed from inside the store. When the reversed sign is viewed in a mirror, does the image of the sign appear reversed?
4. Devise a system of plane mirrors that will produce an image that is not right–left reversed as it is with a single plane mirror.
5. At one corner of a room, the ceiling and the two walls are plane mirrors. As you look into the corner, how many images of yourself can you see?
6. Under what conditions will a convex mirror produce a real image? (It can be done if a second mirror is used.)
7. Sketch a system of mirrors that would allow you to see the back of your head. Can you do it using only two mirrors such that your view is from a point on a line extending directly backward from your head? Make a ray diagram for this situation. (The image you see should be erect.) Is this image reversed left-for-right? What is the minimum number of mirrors required if the direction you look is horizontally straight ahead? Include a ray diagram. Is this image reversed left-for-right?
8. Describe the range of conditions for which a spherical mirror will form images that are (a) real, (b) virtual, (c) erect, (d) inverted, (e) enlarged, and (f) reduced. Do this for both convex and concave mirrors.
9. A very distant object is brought (along the axis) toward a concave mirror until it touches the mirror. Describe how the characteristics and location of the image change as this process occurs. Repeat for a convex mirror.
10. In some automobiles the rear-view mirror is slightly convex. Why? Do images in such a mirror appear closer or farther away than they would in a plane mirror? Do they appear to move faster or slower than they would in a plane mirror?
11. A navigator uses a sextant to measure the angle between the sun and the horizon. If the horizon is obscured by a distant fog bank, the navigator can determine the horizon angle by measuring the angle between the sun and its reflection in a pail of water and dividing this angle by 2. Using a diagram, explain why this procedure works.

Problems
36.4 Reflection by a Plane Mirror

36A-1 A light beam strikes a plane mirror and is reflected. Show that, if the mirror is rotated through an angle α about an axis in the plane of the mirror, the reflected beam moves through an angle 2α.

36A-2 As shown in Figure 36-23 (p. 840), a light ray strikes a plane mirror at a 15° angle of incidence and is reflected to a scale 3 m away. When the mirror is turned through an angle of 2°, how far along the scale will the light spot move? The scale is curved so that the reflected ray always strikes the scale perpendicularly.

36B-3 A laser beam undergoes two reflections in two right-angle mirrors as shown in Figure 36-12a. The beams and the normals to the mirrors all lie in the same plane. Show that, for all angles of incidence that result in reflections by both mirrors, the final reflected beam is always antiparallel to the incident beam.

36B-4 A woman whose eyes are 1.59 m from the floor stands before a mirror. (a) If the top of her hat is 14 cm above her eyes, find the minimum vertical dimension of a wall mirror that would enable her to see an entire image of herself (hat

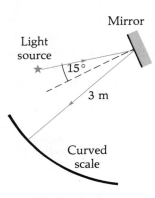

FIGURE 36-23
Problem 36A-2.

included). (b) How far from the floor is the bottom edge of the mirror?

36B-5 The edge of a plane mirror is in contact with the edge of another plane mirror, with 90° between their reflecting surfaces. One mirror is in the xy plane and the other in the yz plane, so that the joined edges are along the $\pm z$ axis. By drawing ray diagrams, locate the (x, y) coordinates of the three images of an object that is at the position $x = 30$ cm, $y = 40$ cm.

36B-6 A light ray undergoes two reflections in two mirrors as shown in Figure 36-24. All rays and normals to the two mirrors lie in the same plane. Derive an expression for β in terms of α. Verify that, for $\alpha = 90°$, $\beta = 0$.

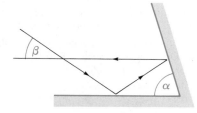

FIGURE 36-24
Problem 36B-6.

36.5 Reflection by a Spherical Mirror
36.6 Ray Diagrams and Lateral Magnification

36A-7 A spherical glass ball, 6 cm in diameter, has a mirror-like surface. The ball is at rest on a table. A fly crawls on the table toward the ball. (a) Find the distance from the ball's surface to the fly's image when the fly is 4 cm from the ball's surface. Include a ray diagram. (b) Describe the image characteristics.

36A-8 An object 2.7 cm high is placed 15 cm from a convex mirror whose radius of curvature is 29 cm. Locate and describe the final image, including its lateral magnification. Include a ray diagram.

36A-9 A concave mirror has a radius of curvature of 30 cm. (a) Where must an object be placed in front of the mirror to produce an image 15 cm behind the mirror? (b) If the mirror is convex with the same radius of curvature, where should the object be placed? Include a ray diagram for (a).

36B-10 Consider light rays parallel to the principal axis approaching a concave *spherical* mirror. According to the mirror equation, rays close to the principal axis focus at F, a distance $R/2$ from the mirror. What about a ray farther from the principal axis? Will it be reflected to the axis at a point closer than F or farther than F from the mirror? Illustrate with an accurately drawn light ray that obeys the law of reflection.

36B-11 An object placed 30 cm in front of a curved mirror produces a real image 40 cm from the mirror. Find where the object must be placed to produce the image 20 cm behind the mirror. Include ray diagrams for both cases.

36B-12 In Footnote 8, rays (3) and (4) are said to be occasionally awkward to draw in ray diagrams. Illustrate with a ray diagram for each case. (Hint: consider situations in which the object is close to F or close to the center of curvature R of the mirror.)

36B-13 An object placed 5 cm from a concave mirror produces a real image four times as large as the object. Find the radius of curvature of the mirror. Include a ray diagram.

Additional Problems

36C-14 Figure 36-25 shows a triangular enclosure whose inner walls are mirrors. A ray of light enters a small hole at the center of the short side. For each of the following, make a separate sketch showing the light path and find the angle θ for a ray that meets the stated conditions. (a) A ray that is reflected once by each of the side mirrors and then exits through the hole. (b) A light ray that reflects only once and then exits. (c) Is there a path that reflects three times and then exits? If so, sketch the path and find θ. (d) A ray that reflects four times and then exits.

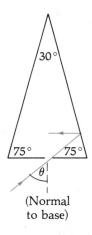

FIGURE 36-25
Problem 36C-14.

36C-15 An observer views a point source of light reflected in a mirror as in Figure 36-7 (page 827). Show by application of Fermat's principle that the angle of incidence θ_i equals the angle of reflection θ_r. Assume that the incident ray and the reflected ray lie in the same plane. However, do not assume that the source and the observer are the same distance above the mirror.

(Hint: choose the variable, upon which both θ_i and θ_r depend, to be the distance between the point on the mirror directly below the source and the point of reflection.)

36C-16 Two plane mirrors have their reflecting surfaces facing one another, with the edge of one mirror in contact with an edge of the other, so that the angle between the mirrors is α. When an object is placed between the mirrors, a number of images are formed. In general, if the angle α between the two mirrors is such that $n\alpha = 360°$, where n is an integer, the number of images formed is $n - 1$. Graphically, find all of the image positions for the case $n = 6$ when a point object is between the mirrors (but not on the angle bisector).

36C-17 The size of a real image produced by a concave mirror is doubled if the object distance is decreased from 80 cm to 50 cm. Find the radius of curvature of the mirror. Include ray diagrams for both positions.

36C-18 Figure 36-26 shows a simple vertical periscope formed by two plane mirrors placed at 45° as shown. (a) How far from the bottom mirror is the image of the arrow at O? (b) Is the final image real or virtual, erect or inverted? (c) To an observer looking into the periscope, has the image undergone a left–right reversal? Explain your reasoning. (d) Suppose you use the periscope to look around a corner at a vertical arrow by holding the periscope length horizontal. Is the image erect? Does the image undergo a right–left reversal? Explain. (e) Now view a vertical arrow by holding the periscope with its length oriented 45° to the vertical. Describe what you see.

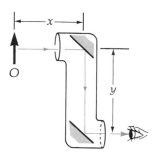

FIGURE 36-26
Problem 36C-18.

36C-19 A three-dimensional corner reflector reflects an incoming light ray so that the ray, after reflecting from the three mirrors, travels in an antiparallel direction back toward the source. Prove this statement. (Hint: consider the incoming ray that travels along the direction given by the vector $\mathbf{p} = p_x\hat{\mathbf{x}} + p_y\hat{\mathbf{y}} + p_z\hat{\mathbf{z}}$. What change in $\mathbf{p}$ occurs when the ray reflects from a mirror in the xy plane?)

36C-20 On a camping trip you and a friend communicate with one another by using signal mirrors that reflect sunlight. Each mirror is a double-sided plane mirror with a small hole through its center. You begin the aiming procedure by placing the mirror in front of your face so that the sunlight passing through the hole forms a spot of light on your face that you can see in the mirror. Next, you simultaneously view the target through the hole and tilt the mirror so that the image of the spot of light on your face appears near (or coincident with) the hole. Explain the theory of operation. (This simple device is an effective way for lost hikers, or survivors at sea, to signal a searching aircraft.)

36C-21 A concave mirror with a focal length of 25 cm produces an image 200 cm away from the object. Find the *two* object distances that produce such an object-to-image separation. Describe the image in each case.

36C-22 Complete the following table for *mirrors*. In every case assume that the diameter of the mirror is small compared with the radius of curvature of its surface. All numerical values are expressed in centimeters. Indicate the appropriate sign of the values in accordance with the sign convention.

Type of Mirror	Radius of Curvature	Focal Distance	Object Distance	Image Distance	Real?	Inverted?	Lateral Magnification
Convex	−120	−60	+30	−20	No	No	$+\frac{2}{3}$
Plane			+30				
		+10		−20			
	−100		+5				
				+100			−2
Convex			−20				$\frac{1}{4}$ (sign?)
Concave	20 (sign ?)		+100				

36C-23 Sketch ray diagrams for each of the cases given in Problem 36C-22.

36C-24 Draw ray diagrams for these three cases: (a) a concave mirror that forms a real image, (b) a concave mirror that forms a virtual image, and (c) a convex mirror that forms a virtual image. For each, sketch the *four* rays from the arrow tip as discussed in Section 36.6 (including Footnote 8).

36C-25 A man can focus clearly on objects no closer than 70 cm when he does not wear glasses. (a) When using a shaving mirror with a focal length of +75 cm, he prefers to view the image of his face at a distance of 80 cm from his eyes. Determine how far his face should be from the mirror. (b) Calculate the lateral magnification.

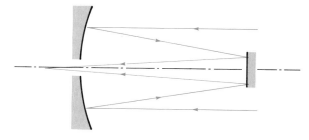

FIGURE 36-27
Problem 36C-26.

36C-26 One type of telescope is the *Cassegrain reflector* shown in Figure 36-27. Light from a distant object strikes a large concave mirror and is then reflected by a small mirror to pass through a hole in the center of the large mirror, forming an image behind the large mirror. An advantage is that it gives easy access to the image for placing photographic film or other optical instruments for analyzing the image. The radius of curvature of the large mirror is 14 m and the two mirrors are separated by 5 m. (a) Should the small mirror be concave or convex to form an image 1 m behind the surface of the large mirror? (b) Find the radius of curvature of the small mirror.

36C-27 A "floating coin" illusion consists of two parabolic mirrors, each with a focal length 7.5 cm, facing each other so that their centers are 7.5 cm apart, Figure 36-28. If a few coins are placed on the lower mirror, an image of the coins is formed at the small opening at the center of the top mirror. Show that the final image is formed at that location and describe its characteristics. (Note: a very startling effect is to shine a flashlight beam on these *images*. Even at a glancing angle, the incoming light beam is seemingly reflected off the *images* of the coins! Do you understand why?)

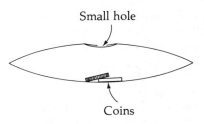

FIGURE 36-28
Problem 36C-27.

36C-28 An object is placed 50 cm in front of a concave mirror, creating a real image. As the object moves 5 cm toward the mirror, the image moves 10 cm. Find the focal length of the mirror.

CHAPTER 37

Geometrical Optics II—Refraction

Three brothers bought a cattle ranch and named it "Focus." When their father asked why they chose that name, they replied: "It's the place where the sons raise meat."

Triple pun attributed
to Professor W. B. Pietenpol, Physics Department,
University of Colorado, Boulder, Colorado

37.1 Introduction

In this chapter we continue our discussion of geometrical optics in which the paths of light rays involve only geometric considerations. *Refraction*, or the bending, of a light ray that is incident at an oblique angle at the interface between two different materials leads to the construction of *lenses* that form images. The refraction of light makes possible the cameras, telescopes, microscopes, and eyeglasses that enable us to see the tiny details of living organisms and the awesome formations in the night sky.

37.2 Refraction at a Plane Surface

The universal constant c always designates the speed of light *in a vacuum*. In matter the speed is slower. The reason is that, as light propagates through a substance, it is continually being absorbed and reradiated by atoms in the material. The incoming wave causes electrons to absorb the radiation and vibrate at the frequency of the wave. A careful analysis shows that the vibrating electrons reradiate electromagnetic waves *at a retarded phase*, depending upon the electron density of the material and their natural resonant frequencies. This retarded phase results in a slower speed for the wave in the material.[1] Thus the speed of light in matter is less than c. For air, the speed v_{air} is only about 0.03% less:

$$\frac{c}{v_{air}} = 1.000\,29 \qquad \text{(at 0°C, 1 atm)}$$

---

[1] See R. P. Feynman et al., *The Feynman Lectures in Physics* (Addison-Wesley), Vol. I, Chapter 31, for a discussion of this process.

FIGURE 37-1

"Tired" light? Some persons have speculated that perhaps light gradually slows down as it travels through space for astronomical distances. Unfortunately, it is not possible to compare the speed of light emitted from a laboratory source with that emitted from, say, a quasar located 10 billion light-years away (whose light presumably has been traveling through space for 10 billion years). Since light is absorbed and reemitted whenever it encounters atoms, we can define an *extinction length L* as the average distance light travels through matter before it is absorbed and reemitted as "reborn" light. The length L depends upon the number of electrons per unit volume. For a piece of glass, $L \approx 10^{-6}$ m; in air it is $\approx 10^{-3}$ m. Thus the light from a quasar we observe in a telescope had its actual origin within the air of the telescope! For a telescope in orbit above the earth's atmosphere, the density of interstellar gas in our galaxy is such that $L \approx 2$ light-years, so it seems that we never will have access to truly ancient light to make the test. [Adapted from John B. Schaefer, "The Unavailability of 'Old' Light," *American Journal of Physics* **57**, 3 (Mar. 1989), p. 200.]

TABLE 37-1 The Refractive Index of Some Representative Materials

Substance	Refractive Index n (for $\lambda \approx 550$ nm)
Air (0°C, 1 atm)	1.000 29
Hydrogen (0°C, 1 atm)	1.000 13
Ice	1.31
Water	1.333 ($=\frac{4}{3}$)
Fused quartz	1.46
Crown glass	1.52
Polystyrene	1.59
Flint glass	1.66
Diiodomethane	1.75
Diamond	2.42
Thallium iodide	2.78

(When three-figure accuracy is acceptable, we may take the speed of light in air to be c for ease of calculation.)

The ratio c/v is defined as the **refractive index**, or **index of refraction** n:

INDEX OF REFRACTION
$$n \equiv \frac{c}{v} = \frac{\text{Speed of light in a vacuum}}{\text{Speed of light in a medium}} \qquad (37\text{-}1)$$

Table 37-1 lists refractive indices for various substances.

We speak of materials with a high refractive index as *optically dense*. However, it is not always true that *physically* dense substances have higher values of n. For example, most oils, which float on water, have greater indices of refraction than water. Many transparent plastics have a greater index of refraction than crown glass, which is physically denser. So there is no general rule that relates physical density to optical density.

Dispersion

A complicating factor in the design of optical instruments is the fact that the glass used for making lenses does not have a constant index of refraction: the value of n varies with wavelength. Typically, the variation is about 2% over the visible spectrum. Curves displaying the refractive index as a function of wavelength are called *dispersion curves*,[2] Figure 37-2. The property of **dispersion**, *the variation of n with λ*, can be useful or troublesome, as we will see shortly. The colors of a rainbow and the brilliance of a diamond are due to dispersion.

---

[2] As far as light is concerned, electrons in atoms behave as though they were held in place by springs. For most substances, some of their electrons have natural resonant frequencies in the ultraviolet. As the frequency of light approaches a resonance from the low-frequency side, the phase retardation of the radiated wave becomes greater (see Figure 34-17b). This results in a lower propagation speed—that is, a greater index of refraction for shorter wavelengths. This is why n is greater for blue light than for red light.

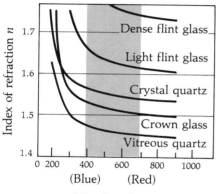

FIGURE 37-2
The *refractive index* n as a function of *wavelength* λ for various types of glass (visible-wavelength range shown shaded.) For visible wavelengths, the refractive indices can usually be measured to five significant figures. (Graph adapted from Eugene Hecht and Alfred Zajac, *Optics*, Addison-Wesley, 1974.)

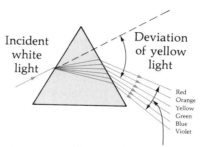

FIGURE 37-3
White light is a mixture of all wavelengths from about 400 nm to about 700 nm. Because the refractive index varies with wavelength, *dispersion* by a glass prism separates a white-light ray into a *continuous range* of wavelengths at slightly different angles. A measure of the amount of dispersion is the angle between the deviated ray for red light and that for violet light. When the dispersed beam falls on a screen, it forms a *spectrum*. If a second identical prism is placed upside-down in the dispersed beam, it will recombine the beam into a single ray of white light.

Refraction

When a light ray encounters an interface between two materials with different refractive indices, the direction of the light ray may change. The bending of a light ray in this manner is called **refraction**. In Figure 37-4, a plane wave traveling in a material of refractive index n_1 encounters a plane interface and passes into a material of higher refractive index n_2. In these two materials, the light moves with speeds

$$v_1 = \frac{c}{n_1} \quad \text{and} \quad v_2 = \frac{c}{n_2} \qquad (37\text{-}2)$$

Applying Huygens' principle to the wavefront AC, we note that, in the time t required for a secondary wavelet to move from C to D in medium 1, another secondary wavelet moves from A to B in medium 2:

$$t = \frac{CD}{v_1} \quad \text{and} \quad t = \frac{AB}{v_2} \qquad (37\text{-}3)$$

Using Equations (37-2) gives $\qquad n_1 CD = n_2 AB \qquad (37\text{-}4)$

Triangles ACD and ABD are similar right triangles with the common side AD, so

$$CD = AD \sin \theta_1 \quad \text{and} \quad AB = AD \sin \theta_2$$

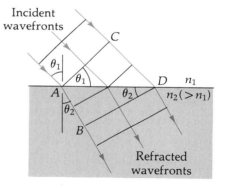

FIGURE 37-4
Refraction of a plane wave by a plane interface between two materials.

FIGURE 37-5
Light from the submerged portion of the straw is refracted at the water–air boundary, causing the straw to appear bent.

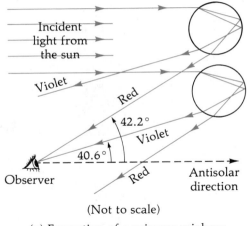
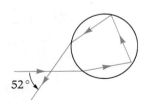

(a) Formation of a primary rainbow.

(b) A ray of green light in the secondary bow.

FIGURE 37-6
Rainbows. Because of dispersion, different wavelengths contained in the light from the sun are refracted by water droplets into different directions. In water, $n_{red} = 1.332$ and $n_{violet} = 1.343$. As shown in (a), dispersion occurs as light enters a spherical raindrop and also when it leaves the drop. The reflection at the back surface is *total internal reflection*, Section 37.3. As a result, an observer receives light of different wavelengths from slightly different directions. The outer red edge of a rainbow appears at 42.2° above the antisolar direction, while the inner violet edge is at 40.6°. These angles lie on a conical band centered on the antisolar direction. A fainter *secondary bow* is formed at angles 50.7° to 53.6° by rays that undergo two internal reflections as in (b). The spectrum of the secondary bow is reversed compared to that in the primary bow (Problem 37C-44). It is interesting that no two persons ever see precisely the same rainbow and that there is no arc of colors out there.

Substituting these relations into Equations (37-4), we obtain

SNELL'S LAW[3] FOR REFRACTION

$$n_1 \sin \theta_1 = n_2 \sin \theta_2 \qquad (37\text{-}5)$$

Note that θ_1 is the angle between the incident ray and the *normal* to the interface between the materials, and θ_2 is the angle between the refracted ray and the same normal. As in reflection, the incident ray, the normal, and the refracted ray all lie in the same plane.

The same law of refraction applies for a ray of light going from a higher refractive index to a lower, so if any light ray is reversed, it will retrace the same path in the opposite direction (provided no absorption occurs). Because the same reversibility is also true for reflection, we have the following:

PRINCIPLE OF REVERSIBILITY **If the direction of a light ray passing through any optical system is reversed, the light will retrace its original path in the opposite direction.**

---

[3] Snell's law is named after its discoverer, the Dutch physicist Willebrord Snel van Royen (1591–1626). At age 21, he succeeded his father as professor of mathematics at the University of Leyden, Holland, and his unpublished discovery in 1621 has been called one of the great moments in optics. The same relationship was probably discovered independently by the French philosopher-mathematician, René Descartes (1596–1650), who derived it using the particle theory of light and published it in his *Dioptrique*; in France the law is known as Descartes' law. In 1617, Snel determined the size of the earth by measuring the earth's curvature between Alksmaar and Bergen-op-Zoom.

Fermat's principle (page 827) applies to refracted rays just as it does to reflected rays. See Figure 37-7. That is, of all the possible paths from a point on one side of an interface between two media to a point on the other side, *light takes the path that requires the least time in transit.* In fact, Snell's law can be derived from Fermat's principle. (See Problem 37C-37.)

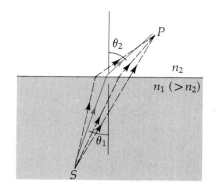

FIGURE 37-7
Fermat's principle applied to refraction. Of all the rays emanating from a point source S, only one ray will pass through the point P. This ray lies in a plane, satisfies Snell's law ($n_1 \sin \theta_1 = n_2 \sin \theta_2$), and requires the least time of transit from S to P, Fermat's principle. Even though some of the alternative paths indicated by the dashed lines are shorter in distance, *they are longer in travel time,* so no rays of light take these other routes in traveling from S to P.

EXAMPLE 37-1

A tin can 14 cm high and 12 cm in diameter is filled with an unknown liquid. An observer looking along a direction 25° above the horizontal (see Figure 37-8) can barely see the inside bottom edge of the can. Find the index of refraction of the liquid.

SOLUTION

The light ray from the bottom edge incident within the liquid on the top surface has an angle of incidence $\theta_1 = \tan^{-1}(\frac{12}{14}) = 40.6°$. The index of refraction of air is $n_2 = 1.000$ (to four significant figures). We apply Snell's law to the refraction of the ray as it emerges into the air with an angle of refraction $\theta_2 = 65°$.

$$n_1 \sin \theta_1 = n_2 \sin \theta_2$$
$$n_1 \sin 40.6° = (1.00) \sin 65°$$

Solving for n_1 gives
$$n_1 = \frac{\sin 65°}{\sin 40.6°} = \boxed{1.39}$$

Apparent Depth

When we look straight down into a pail of water resting on the floor, the bottom of the pail appears to be noticeably above the floor level. How do we visually judge distance? Human depth perception involves a variety of mechanisms. One clue is the comparison we make between the known size of an object and its perceived size. For distant landscapes, atmospheric haze provides additional helpful information. (In the absence of such haze, one can be fooled into greatly underestimating the distance of "nearby" mountains.) For objects close to us, an aid is the parallax effect that occurs when we move our head slightly. Also, we need to "aim" each eye along slightly different directions in order to match the divergence of light rays as they leave a nearby object: our minds, through experience, relate this "aiming" effect to distance estimation. The next example utilizes this last method of judging distance.

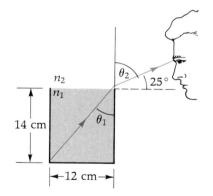

FIGURE 37-8
Example 37-1.

EXAMPLE 37-2

An observer looks straight down into the same tin can of fluid described in Example 37-1. What is the apparent depth of the fluid?

SOLUTION

In Figure 37-9a, the two rays shown coming from a point on the bottom of the can diverge as they approach the top of the fluid. As they proceed into the air, they diverge even more due to refraction. To the observer, the rays will appear to originate from a point at a depth d below the surface. To emphasize the refraction at the surface water, an exaggerated view is shown in Figure 37-9b. The

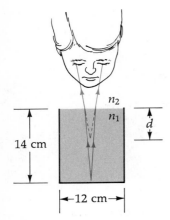

(a) Looking straight down into a can of fluid.

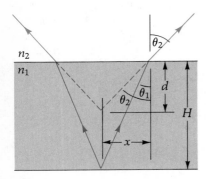

(b) An exaggerated sketch of the refraction that occurs at the liquid surface.

FIGURE 37-9
Example 37-2.

angles involved in this example are so small that we may use the small-angle approximations

$$\sin \theta_1 \approx \tan \theta_1 \approx \theta_1 \quad \text{and} \quad \sin \theta_2 \approx \tan \theta_2 \approx \theta_2$$

(For simplicity of notation, we replace the approximately equal sign with the equal sign in the discussion that follows.) From trigonometry, we have $x = d \tan \theta_2$ and $x = H \tan \theta_1$. Eliminating x between these equations and using the small-angle approximations, we obtain

$$\theta_1 H = \theta_2 d \qquad (37\text{-}6)$$

Snell's law relates θ_1 and θ_2: $\quad n_1 \sin \theta_1 = n_2 \sin \theta_2$

Since $\sin \theta \approx \theta$, we have $\quad n_1 \theta_1 = n_2 \theta_2 \qquad (37\text{-}7)$

We combine Equations (37-6) and (37-7) to obtain

APPARENT DEPTH d
(viewed perpendicularly)[4]
$$d = H \left(\frac{n_2}{n_1} \right) \qquad (37\text{-}8)$$

Thus: $\quad d = 14 \text{ cm} \left(\dfrac{1.00}{1.39} \right) = \boxed{10.1 \text{ cm}}$

37.3 Total Internal Reflection

Whenever light traveling in a medium of one refractive index encounters an abrupt transition to a medium of a different refractive index, there is always some reflection at the interface. A special case arises if the second medium has a *lower* index of refraction than the first. Under certain conditions, the reflection is 100% and *no light is transmitted through the interface*. To see this, we start with Snell's law:

$$n_1 \sin \theta_1 = n_2 \sin \theta_2$$

In Figure 37-10, as the angle of incidence increases, the angle of refraction θ_2 approaches 90°. At the "dividing-line" case of exactly 90°, $\sin 90° = 1$ and we have

$$\sin \theta_1 = \left(\frac{n_2}{n_1} \right) 1$$

For angles of incidence *larger* than this **critical angle** θ_c, *total internal reflection* occurs and there is no refracted ray.

CRITICAL ANGLE
θ_c FOR TOTAL
INTERNAL REFLECTION
$$\sin \theta_c = \frac{n_2}{n_1} \quad (\text{for } n_2 < n_1) \qquad (37\text{-}9)$$

---

[4] Because we used small-angle approximations in deriving the expression for the apparent depth, it holds true *only when the light rays from the bottom are incident almost perpendicularly on the water–air interface*. When viewed at more oblique angles, the apparent depth changes considerably. For example, note the apparently curved bottom of a (calm-water) swimming pool when viewed from the edge of the pool and how the image of the bottom changes as you walk around the pool.

(There is, of course, no critical angle of light traveling from a medium with a lower refractive index into a medium with a higher refractive index.)

Total internal reflection is used in a variety of practical applications. For example, in binoculars (Figure 37-11b) the image is made erect by several reflections at 45°. Since 45° is greater than the critical angle for glass, a 45° prism is used rather than a mirror with a silver coating (which might become tarnished with time); the reflection is 100% from the interior glass interfaces. Similarly, solid glass "corner" reflectors, whose three faces meet mutually at 90°, are used to reverse the direction of any light ray incident on the reflector. Arrays of these corner reflectors have been placed on the moon to reflect laser pulses sent from the earth. Precise timing of the round trip of a pulse enables earth–moon distances to be determined within a few centimeters, aiding studies of continental drifts, effects of tides, and numerous other phenomena. Because of total internal reflection, light is conducted along a flexible, transparent rod called a *light pipe*, provided the angle of bending is small enough to ensure that all angles of incidence are greater than θ_c. Such light pipes, only a few hundredths of a millimeter in diameter, are used for long-distance transmission of radio, TV, and telephone channels, in addition to the rapid transmission of computer data, Figure 37-12. Glass fibers a few thousandths of a millimeter

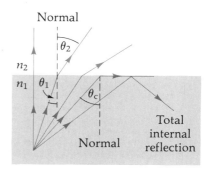

FIGURE 37-10

Total internal reflection. When a ray of light traveling in a medium of index of refraction n_1 encounters a medium of *lower* index of refraction, the ray may undergo *total internal reflection* at the interface. As the angle of incidence θ_1 increases, the angle of refraction θ_2 becomes larger. The **critical angle θ_c** is the "dividing-line" case between the refracted ray, barely emerging parallel to the boundary surface ($\theta_2 \approx 90°$), and the slightly larger incident angle, for which no light escapes and the reflection is 100%. For all angles $\theta_1 > \theta_c$, total internal reflection occurs. (Note: there is always some *reflection* (not shown) at the interface for angles less than θ_c. For a glass–air interface, it varies from about 4% at normal incidence, to $\sim 40\%$ for a refracted ray at 80°, and approaching 100% as $\theta_2 \to 90°$.)

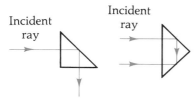

(a) Glass prisms (45°–45°–90°) reflect light rays by total internal reflection.

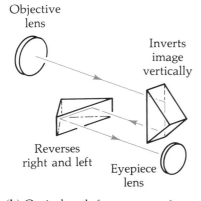

(b) Optical path for one eye of *prism binoculars*. The use of two 45° prisms oriented at right angles causes the final image to be upright and not reversed right-for-left. Because the magnification is proportional to the focal length of the objective lens, the use of prisms also "folds" the long optical path into a shorter, more convenient length.

(c) Light is transmitted through glass fibers by total internal reflection.

(d) Thousands of transparent fibers are held parallel, forming a *light pipe*, so that they transmit a true image even though the pipe is bent.

FIGURE 37-11

Examples of total internal reflection.

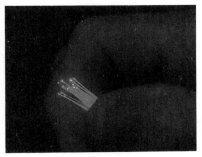

(a) Glass fibers used for long-distance transmission lines. At present, distances up to 50 km are possible before an amplifying station, or *repeater*, is necessary to compensate for absorption losses.

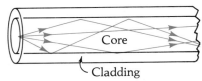

(b) Light is transmitted down the core by total internal reflection. One type of fiber in use has a diameter of 50 μm, about that of a human hair. Others have core diameters as small as 2 μm.

FIGURE 37-12
Optical fiber communication. Tremendous advances in the development of optical glass fibers for transmitting information have been made in the past 20 years. A light beam, carrying the information coded as a series of pulses (*digital modulation*), is injected into the *core* of a glass fiber. Even though the fiber bends, light is guided down the core by total internal reflections at the core walls. An outer layer of glass with a lower index of refraction, called *cladding*, protects the core surface from moisture, dust, oil, etc., that would cause light leakage by upsetting the reflections at the walls. One big advantage of optical-fiber communication is its enormous information-carrying capacity, far superior to copper-cable, or radio, systems. For example, recently a fiber the size of a human hair carried about 25 000 voice channels! Since even this impressive performance is well below theoretical limits, further advances are expected. Other advantages include small size and weight, immunity to electrical interference, security against eavesdropping, and lower cost. (See Problems 37B-14, 37C-42, and 37C-43.)

thick are sufficiently flexible that bundles of them can be used as probes, enabling physicians to see internal parts of the body, or technicians to view inaccessible parts of a mechanism. Nature has used this principle of *fiber optics* for millions of years: certain insects and crustaceans have visual sensors that consist of bundles of crystalline "light pipes" that transmit light between an array of outer corneal lenses and light-sensing elements deep within the insect body.

EXAMPLE 37-3

Find the critical angle for transparent plastic of refractive index 2.14 immersed in oil of refractive index 1.63.

SOLUTION

The critical angle is given by Equation (37-9): $\sin \theta_c = n_2/n_1 = 1.63/2.14 = 0.762$, giving

$$\theta_c = \boxed{49.6°}$$

37.4 Refraction at a Spherical Surface

Most familiar optical instruments utilize lenses rather than mirrors because of their durability and ease of combination with other elements of an optical system. As a first step in the study of lenses, we will investigate how light is refracted when it is incident on a glass surface that has a spherical curvature. This approach introduces a technique for studying lenses and also has useful applications in itself. Consider a point object on the axis of a spherical interface between two media, as in Figure 37-13a. We first take the case in which all the rays are refracted sufficiently to intersect the axis inside the medium. We will show that within the small-angle approximation all rays converge to form a real image at the point I.

Tracing a single ray, as in Figure 37-13b, we find that it intersects the axis at the image distance CI. Using the fact that the exterior angle of a triangle is equal to the sum of the opposite interior angles, we have for one triangle

$$\theta_1 = \alpha + \gamma \quad \text{and} \quad \gamma = \theta_2 + \beta \quad (37\text{-}10)$$

We eliminate θ_1 and θ_2 by multiplying these equations by the appropriate refractive indices:

$$n_1\theta_1 = n_1\alpha + n_1\gamma \quad \text{and} \quad n_2\theta_2 = n_2\gamma - n_2\beta$$

and combining them with the small-angle approximation of Snell's law (Equation 37-7): $n_1\theta_1 = n_2\theta_2$, to obtain

$$n_1\alpha + n_2\beta = (n_2 - n_1)\gamma \quad (37\text{-}11)$$

Since α, β, and γ are small, $\alpha = \tan\alpha = h/OC$, $\beta = \tan\beta = h/IC$, and $\gamma = \tan\gamma = h/RC$. Equation (37-11) then becomes

$$\frac{n_1}{OC} + \frac{n_2}{IC} = \frac{n_2 - n_1}{RC} \quad (37\text{-}12)$$

In terms of the object distance p, the small distance q, and the radius of curvature R, we have

REFRACTION AT A SINGLE SPHERICAL INTERFACE

$$\frac{n_1}{p} + \frac{n_2}{q} = \frac{n_2 - n_1}{R} \quad (37\text{-}13)$$

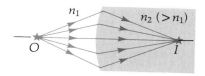

(a) A real image is formed by the convergence of rays refracted at the spherical interface.

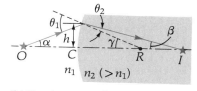

(b) Tracing a single ray from the object O to the image I.

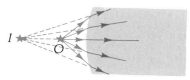

(c) If the refraction is insufficient to produce converging rays, a virtual image is formed outside the medium. (Of course, only an observer within the medium could intercept the rays after they have been affected by the interface, and thus see the virtual image.)

FIGURE 37-13
Refraction at a spherical interface between two media.

The usual sign convention applies for p and q, with R being positive for convex outer surfaces (that is, the center of curvature is inside the medium). Since h and θ do not appear in this expression, we know that *all* rays refracted by the interface will converge to the same image point I (at least within the validity of the small-angle approximations we have made). If the rays are not bent sufficiently to converge inside the medium, their diverging directions may be traced backward to a point of intersection to the left of the interface, forming a virtual image I as in Figure 37-13c. The following example illustrates this type of situation.

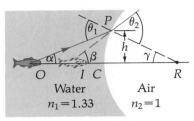

(a) A swimmer wearing a diving mask with a face plate that bulges outward. The image of a fish is much closer than the fish itself.

(b) The divergence of the light ray from the fish has been exaggerated to show more clearly the refraction at the interface between the water and air inside the diving mask.

FIGURE 37-14
Example 37-4.

EXAMPLE 37-4

A swimmer views a small fish through a face plate on her diving mask, as shown in Figure 37-14a. The face plate bulges outward, forming an outer convex surface with a radius of curvature of 0.40 m. If the actual distance to the fish is 3.0 m, find the apparent distance to the fish as viewed by the swimmer.

SOLUTION

Ignore the thickness of the face plate itself. The plate forms an interface between the water ($n_1 = 1.33$) and the air within the mask ($n_2 = 1$). If we trace a single ray from the fish as in Figure 37-14b, two triangles are formed in which $\theta_1 = \alpha + \gamma$ and $\theta_2 = \beta + \gamma$. Using Snell's law $n_1\theta_1 = n_2\theta_2$, we proceed as before and obtain

$$\frac{n_1}{OC} - \frac{n_2}{IC} = \frac{n_2 - n_1}{RC}$$

Substituting the appropriate numerical values, we have

$$\frac{1.33}{3.00 \text{ m}} - \frac{1}{IC} = \frac{1.00 - 1.33}{0.40 \text{ m}}$$

Solving for IC gives $\qquad IC = \boxed{0.788 \text{ m}}$

Instead of 3 m, the apparent distance is only 0.788 m. Obviously, a convex face plate produces large distortion of actual distances. As indicated in Example 37-2, a flat face plate would produce an image of the fish at 2.3 m, much closer to the actual location of the fish.

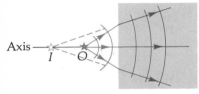
(a) Plane

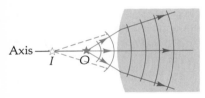
(b) Convex

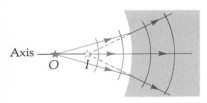
(c) Concave

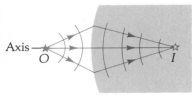
(d) Convex

FIGURE 37-15
Refraction at a single interface. Shaded regions indicate a higher refractive index. Dashed lines are extensions of the refracted rays and form virtual images.

Figure 37-15 shows examples of how rays from an object on the axis are refracted at an interface between two media. A convex surface may create either a real or a virtual image.

37.5 Thin Lenses

Most lenses have spherical surfaces, with each surface contributing some refraction. Thus, unless a ray strikes a surface at normal incidence, the ray will bend as it enters the lens and also as it leaves the lens (see Figure 37-16a). We will limit our discussion to the thin-lens case, in which the thickness of the lens is negligible compared with other dimensions. This *thin-lens approximation* means that it makes no difference whether object and image distances are measured from the front surface or the back surface. To simplify ray-tracing diagrams, we assume that all bending of a ray occurs at a plane passing through the center of the lens and that all distances are measured from this plane. We use the same notation as for mirrors: p = object distance, q = image distance, f = focal length, and R = radius of curvature of a surface. By restricting our discussion to "thin" lenses, we can use the small-angle approximations, greatly simplifying the analysis.

Figure 37-17 illustrates various types of lenses. For lenses with a refractive index greater than that of the surrounding medium, *those that are thicker in the center than on the edge are called* convergent *or* positive *lenses and have positive values of* f. *Those with a center that is thinner than the edge are called* divergent *or* negative *lenses and have negative values of* f. Centers of curvature all lie on the axis.

The analysis of refraction by a lens is a three-step process: (a) calculating the refraction of a light ray by the lens surface first encountered by the ray, (b) calculating the refraction of the ray as it emerges from the second surface, and (c) combining the results of (a) and (b) to obtain a general formula relating object distance p, image distance q, and the lens parameters. Fortunately, the thin-lens approximation makes the final result a simple expression.

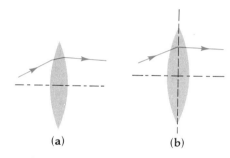

FIGURE 37-16
The thin-lens approximation. (a) In actual lenses, a ray is refracted at both surfaces (unless it happens to strike a surface at normal incidence). (b) In ray-tracing diagrams, the physical thickness of a lens is ignored, and we assume that all bending of a ray occurs at a plane passing through the center of the lens. The distances p, q, and f are all measured from this plane.

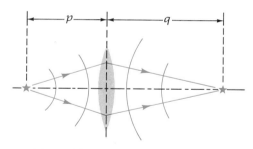

(a) Double-convex converging lens.

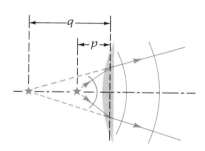

(b) Plano-convex converging lens.

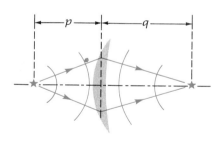

(c) Meniscus converging lens.

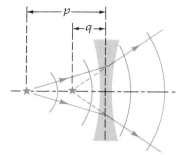

(d) Double-concave diverging lens.

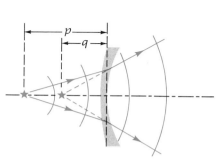

(e) Meniscus diverging lens.

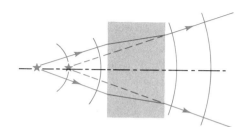

(f) A thick glass plate with parallel surfaces forms a virtual image.

FIGURE 37-17
Lenses are named according to the types of surfaces they have. Dashed lines are extensions of rays to locate virtual images. Converging lenses are always thicker at the center than at the edges; diverging lenses are always thinner at the center than at the edges. (This assumes that the index of refraction for the lens is greater than that of the surrounding medium.)

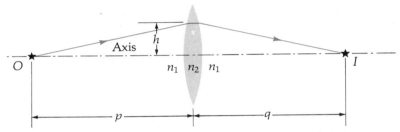

(a) An image I of an object O is formed by a lens. The refractive index of the lens is n_2, where $n_2 > n_1$.

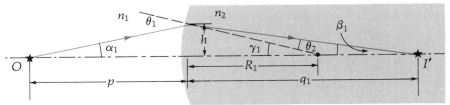

(b) The ray from the object O is bent by the first surface alone. The dashed line is the normal to the surface. If the ray remained in glass, it would arrive at the axis at I'.

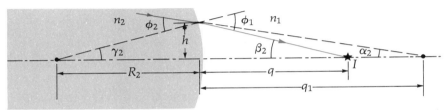

(c) The ray is bent as it emerges from inside the lens to intersect the axis at the final image distance q from the surface.

FIGURE 37-18
To determine the net effect of a lens, we must consider that in general a light ray is bent twice by the lens, once at the first surface and again at the second surface.

The First Surface

Consider the case shown in Figure 37-18a. The first surface the light ray encounters is sketched by itself in Figure 37-18b, where p is the object distance and q_1 is the image distance that would exist if the second surface were absent. This step of the analysis is the same as that done previously for a single refractive surface [Equation (37-13)]. In the notation of this case, it is

$$\frac{n_1}{p} + \frac{n_2}{q_1} = \frac{n_2 - n_1}{R_1} \tag{37-14}$$

The Second Surface

The ray is further refracted as it emerges from the second surface into the air, as in Figure 37-18c. Rather than a real image being formed at q_1, the image is formed at q. Since the exterior angle of a triangle is equal to the sum of the opposite interior angles, we have $\phi_1 = \gamma_2 + \beta_2$ and $\phi_2 = \gamma_2 + \alpha_2$. Substituting these in Snell's law for small angles, $n_1\phi_1 = n_2\phi_2$, we obtain

$$n_1\beta_2 - n_2\alpha_2 = (n_2 - n_1)\gamma_2$$

Using the small-angle approximations, $\alpha_2 \approx \tan \alpha_2 = h/q_1$, $\beta_2 \approx \tan \beta_2 = h/q$, and $\gamma_2 \approx \tan \gamma_2 = h/R_2$, we obtain

$$\frac{n_1}{q} - \frac{n_2}{q_1} = \frac{n_2 - n_1}{R_2} \qquad (37\text{-}15)$$

The Combined Result

Adding Equations (37-14) and (37-15) eliminates n_2/q_1 to give

$$\frac{n_1}{p} + \frac{n_1}{q} = (n_2 - n_1)\left(\frac{1}{R_1} + \frac{1}{R_2}\right) \qquad (37\text{-}16)$$

which we may simplify by introducing the **relative refractive index n** of the lens material, relative to the surrounding medium, n_1:

RELATIVE REFRACTIVE INDEX
$$n \equiv \frac{n_2}{n_1} \qquad (37\text{-}17)$$

Making this substitution, we have

$$\frac{1}{p} + \frac{1}{q} = (n - 1)\left(\frac{1}{R_1} + \frac{1}{R_2}\right) \qquad (37\text{-}18)$$

R is positive for convex outer surfaces and negative for concave outer surfaces (if the index of refraction of the lens is greater than that of the surroundings).

As in the case of mirrors, the *focal length f* of a lens is defined as the image distance of parallel light incident upon the lens ($p = \infty$). Substituting this value in Equation (37-18), we obtain the *lens-maker's formula*:

LENS-MAKER'S FORMULA
$$\frac{1}{f} = (n - 1)\left(\frac{1}{R_1} + \frac{1}{R_2}\right) \qquad (37\text{-}19)$$

Finally, combining Equations (37-18) and (37-19), we obtain the *thin-lens equation*[5]:

THE THIN-LENS EQUATION
$$\frac{1}{p} + \frac{1}{q} = \frac{1}{f} \qquad (37\text{-}20)$$

Our development of the lens equation was based on the analysis of a real image produced by a double-convex lens. If we analyze the other cases shown in Figure 37-14, we obtain equations similar to the lens-maker's formula,

---

[5] The equal status of p and q in the thin-lens equation led Helmholz to state a *principle of optical reversibility: if any ray is reversed, it will retrace the same path back through the optical system.*

but with various changes in the signs of the terms. However, the lens-maker's formula is valid for all cases, with the following **sign convention**[6]:

SIGN CONVENTION FOR THIN LENSES

(1) The numerical value of p is positive if the rays *approaching* the lens are *divergent*. Otherwise p is negative.
(2) The numerical value of q is positive if the rays *leaving* the lens are *convergent*. Otherwise q is negative.
(3) Assuming that the refractive index of a lens is greater than that of the surrounding medium, the radius of curvature R of the outer surface of a lens is positive if it is convex and negative if it is concave.

Note that the first two rules are identical to those used for mirrors. Regarding R, our sign convention for *both* mirrors and lens surfaces has the following consistency: *if an incident plane wave becomes a converging wave, R for that surface is positive, and vice versa.* You should memorize the sign convention since you will find it essential when using the thin-lens equation.

37.6 Diopter Power

The *strength* of a lens is a measure of its ability to alter the direction of light rays. This strength is measured in **diopters**, defined as *the reciprocal of the focal length measured in meters*:

$$D = \text{Strength (in diopters)} = \frac{1}{f \text{ (in meters)}} \qquad (37\text{-}21)$$

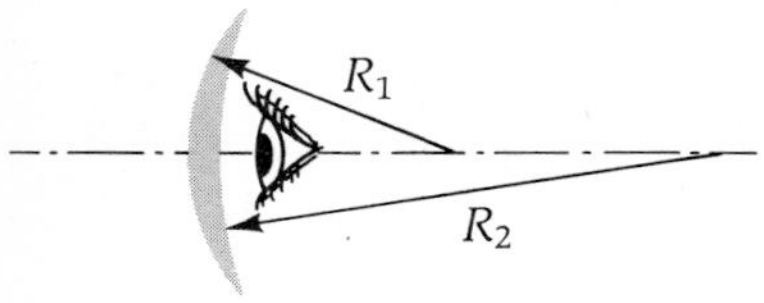

FIGURE 37-19
Example 37-5.

EXAMPLE 37-5

A converging eyeglass is constructed of crown glass ($n = 1.50$). As shown in Figure 37-19, the radii of curvature are $R_1 = 15$ cm and $R_2 = -30$ cm (minus because the outer surface is concave). Find (a) the focal length and (b) the strength of the lens. (c) Locate the image of a book held 20 cm in front of the lens.

SOLUTION

(a) For a lens in air, the relative refractive index is just that of the crown glass. Substituting numerical values into Equation (37-19), we obtain

$$\frac{1}{f} = (n-1)\left(\frac{1}{R_1} + \frac{1}{R_2}\right) \quad \Rightarrow \quad \frac{1}{f} = (1.50 - 1)\left(\frac{1}{15 \text{ cm}} + \frac{1}{(-30 \text{ cm})}\right)$$

Solving for f gives $\qquad f = \boxed{+60.0 \text{ cm}}$

---

[6] Several other sign conventions are in use. Here is a companion to the convention for mirrors in Footnote 6, Chapter 36:

> p, q, and f are each positive for the "standard case" of a converging lens forming a real image of a real object. *Any change from this standard case requires a minus sign.*

For the standard case, the object distance p is greater than the focal length of the lens. Light passes through a lens, so real images are formed on the *"far"* side of the lens. The sign of R is given by rule (3) above.

The positive sign indicates a converging lens, thicker at the center than at the edge.

(b) The strength of the lens is

$$\text{Strength} = \frac{1}{f \text{ (in meters)}} = \frac{1}{0.60 \text{ m}} = \boxed{1.67 \text{ diopters}}$$

(c) To locate the image, we use the lens equation:

$$\frac{1}{p} + \frac{1}{q} = \frac{1}{f} \quad \Rightarrow \quad \frac{1}{20 \text{ cm}} + \frac{1}{q} = \frac{1}{60 \text{ cm}}$$

The numerical sign of the object distance p is positive because rays from the object are divergent as they strike the lens. Solving for q gives

$$q = \boxed{-30.0 \text{ cm}}$$

By the sign convention, the negative sign for the image distance indicates that the rays are divergent after they leave the lens. The image is therefore *virtual*, located on the *same side* of the lens as the object, 30 cm from the lens. As we look through the lens, we see this image.

37.7 Thin Lens Ray-Tracing and Image Size

We now describe ray-tracing techniques used for locating images formed by thin lenses. They are similar to the ray-tracing methods used for mirrors. We will sometimes indicate two focal points F located equidistant on either side of the lens, called *principal foci*. This recognizes that, for thin lenses, paraxial rays incident on a converging lens *from either side* are brought to a focus at a focal-length distance on the other side of the lens (for divergent lenses, they are brought to a focus on the *same* side). Furthermore, rays from a point source at either focal point of a converging lens emerge from the lens parallel to the axis. As shown in Figure 37-20, the situation is symmetrical regarding the direction that the light passes through the lens. (This is not true for thick lenses.)

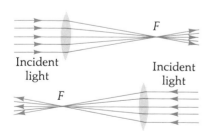

(a) Parallel light incident on opposite sides of a lens. If the lens is thin, the focal distance f is the same for each case.

(b) Light rays from a point source at either focal point F emerge from the lens in directions parallel to the axis.

FIGURE 37-20
A thin lens affects light the same way regardless of which side of the lens the incident light strikes. (This is not true for thick lenses.)

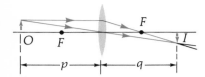

(a) Case 1. Convergent lens: real image

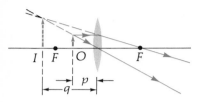

(b) Case 2. Convergent lens: virtual image

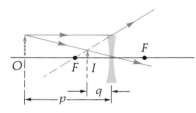

(c) Case 3. Divergent lens: virtual image

FIGURE 37-21
Image formation by ray-tracing. If the rays leaving the lens do not intersect, the virtual image is located by extending dashed lines backward from the directions of the actual rays.

As is also the case for mirrors, a focal point F is *not* the location of the image, except in the one case of incident parallel light. It is helpful to think of the focal points as *points "belonging" to a lens, which we find useful in constructing ray diagrams.*

For ray-tracing diagrams, we draw two particular rays[7] that are always easy to trace:

RAYS USED IN RAY-TRACING FOR THIN LENSES

(1) **A ray passing through the center of the lens is undeviated.** (Near the center, the lens acts as a thin piece of glass with parallel sides. The slight sideways displacement of rays not parallel to the axis can be neglected in the thin-lens approximation.)

(2) **A ray parallel to the axis is refracted so that it passes through (or extends through) the focal point F.**

Figure 37-21 illustrates the three possible cases of refraction by a thin lens (refractive index greater than the surrounding medium).

Case 1. Converging Lens: Real Image

In Figure 37-20a, the object is the tip of the arrow at O, located more than a focal-length distance from the lens. (We investigate an object that extends only *above* the axis, recognizing that, because of symmetry about the axis, an object that extends below the axis would produce the same result.) We trace two rays from the tip whose directions we can easily determine. One ray passes through the center of the lens undeviated, and the other approaches the lens parallel to the axis and is refracted so that it converges toward the focal point F of the lens. The intersection of these two rays locates the image of the arrow tip at I. (Other parts of the arrow are similarly imaged to form the complete arrow.) Since light rays actually converge to form the image, it is *real*; a screen placed at that location would have an image formed on it. The image is *inverted*, as revealed by the ray diagram.

The **linear magnification** M is defined as the ratio of the image size to the object size:

LINEAR MAGNIFICATION
$$M = \frac{\text{Image size}}{\text{Object size}} = -\frac{q}{p} \qquad (37\text{-}22)$$

The minus sign is introduced so that a *negative* value of M indicates an inverted image and a *positive* value of M indicates an *erect* image. That is, if p and q are both positive, M is negative, indicating an inverted image. This same sign convention holds true for all cases of lenses as well as mirrors. The term *magnification* is somewhat of a misnomer, since the image can be smaller than the object, in which case the absolute value of M is less than one.

---

[7] A third ray can be drawn:

(3) **A ray falling on a lens after it passes through (or extends through) the focal point F emerges from the lens parallel to the axis.**

This ray, however, is sometimes not feasible with a large object located near a focal point.

Case 2. *Converging Lens: Virtual Image*

In Figure 37-21b, the object is located closer than a focal-length distance from the lens. Again we trace two rays from the tip of the arrow. One ray passes through the center of the lens undeviated, and the other approaches the lens parallel to the axis and is refracted so that it passes through the focal point of the lens. We determine the intersection of these two rays by extending them backward along their directions until they intersect at the image location *I*. Because no actual light rays travel along these extended lines, we draw them dashed. Since no actual light rays form the image, it is *virtual*. (A screen placed at that location would *not* have an image formed on it.) The image is *erect*, as revealed by the ray diagram (and also by the fact that the numerical value of *M* is positive). You may verify that, for all cases in which the object is closer than a focal-length distance from a converging lens, the image is always larger than the object, always virtual, always erect, and on the same side of the lens as the object.

Case 3. *Divergent Lens: Virtual Image*

As shown in Figure 37-21c, the rays from the arrow tip always diverge after passing through a divergent lens, no matter how far the object is from the lens. As a consequence, a virtual image is formed that is always smaller than the object, always erect, and on the same side of the lens as the object.

Ray diagrams are very helpful in determining image characteristics since the diagram itself reveals these properties; thus there is no need to memorize the + and − sign "rules" for real or virtual, erect or inverted.

One feature of image formation should be noted. If, say, one-half of a lens is covered, the complete image is still present, though half as bright. Indeed, any fragment of a broken lens will still form complete images. The reason is clear if you remember that a lens acts on incident *wavefronts* (or on *all* the rays emanating from each point on the object), rather than on just the few rays we usually trace in ray diagrams.

37.8 Combinations of Lenses

Most optical instruments contain a system of several lenses. In many cases, the use of multiple lenses helps to correct certain image defects. In other instances, if the final image is formed in a series of steps, the overall length of the instrument is much shorter than it would be if just a single lens were used. We will limit the discussion to simple two-lens combinations.

Consider two thin lenses in contact, a common situation in many optical instruments. Figure 37-22 depicts two such lenses, which have positive focal lengths f_1 and f_2. Parallel light from the left strikes lens ① and would focus at a distance f_1 if the second lens were absent. After leaving lens ①, the rays are convergent as they strike lens ②. Therefore, according to the sign convention, the object distance p_2 for the second lens is $-f_1$. Applying the lens equation to the second lens

$$\frac{1}{p_2} + \frac{1}{q_2} = \frac{1}{f_2}$$

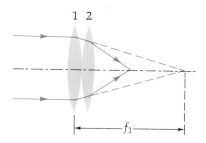

FIGURE 37-22
Two thin, converging lenses in contact.

and noting that q_2 is the resultant focal length f for the combination, we have

$$\frac{1}{-f_1} + \frac{1}{f} = \frac{1}{f_2}$$

Rearranging, we obtain

THIN-LENS COMBINATIONS (lenses in contact)

$$\frac{1}{f} = \frac{1}{f_1} + \frac{1}{f_2} \qquad (37\text{-}23)$$

or, expressed as diopters, D,

$$D = D_1 + D_2 \qquad (37\text{-}24)$$

Diopter notation is particularly convenient for lens combinations because the strength of two thin lenses in contact is merely the sum of the strengths of the individual lenses. These are general relations, valid for any combination of positive and negative lenses in contact.

EXAMPLE 37-6

Two thin lenses of focal lengths $f_1 = 20$ cm and $f_2 = 60$ cm are placed in contact. (a) Find the focal length f' of the combination. (b) Find the focal length f_3 of a third lens placed in contact with these two that would result in an overall focal length $f'' = -40$ cm.

SOLUTION

(a) Substituting numerical values in the lens-combination formula gives

Using focal lengths

$$\frac{1}{f} = \frac{1}{f_1} + \frac{1}{f_2}$$

$$\frac{1}{f'} = \frac{1}{20 \text{ cm}} + \frac{1}{60 \text{ cm}}$$

$$f' = \boxed{15.0 \text{ cm}}$$

Using diopters (D)

$$D_1 = \frac{1}{0.2 \text{ m}} = 5 \text{ diopters}$$

$$D_2 = \frac{1}{0.6 \text{ m}} = 1.67 \text{ diopters}$$

$$D' = D_1 + D_2$$
$$= (5 + 1.67) \text{ diopters}$$

$$D' = \boxed{6.67 \text{ diopters}}$$

(b) Adding one more lens in contact, we repeat the same analyses.

$$\frac{1}{f''} = \frac{1}{f'} + \frac{1}{f_3}$$

$$\frac{1}{-40 \text{ cm}} = \frac{1}{15 \text{ cm}} + \frac{1}{f_3}$$

$$f_3 = \boxed{-10.9 \text{ cm}}$$

$$D'' = \frac{1}{-0.4 \text{ m}}$$
$$= -2.5 \text{ diopters}$$
$$D'' = D' + D_3$$
$$-2.5 \text{ diopters} = 6.67 \text{ diopters} + D_3$$
$$D_3 = \boxed{-9.17 \text{ diopters}}$$

Combining positive and negative lenses together helps to correct certain image defects, as we will discuss later.

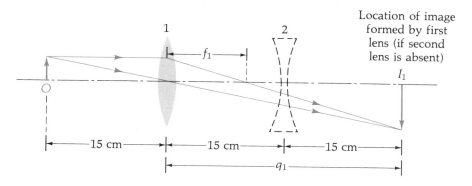

FIGURE 37-23
Example 37-7.

EXAMPLE 37-7

Consider the two lenses in Figure 37-23. An object is placed 15 cm from the convergent lens ($f_1 = 10$ cm). The divergent lens ($f_2 = -20$ cm) is placed 15 cm on the other side of the convergent lens. Locate and describe the final image formed by the two lenses.

SOLUTION

Because these two lenses are *not* in contact, we cannot use the lens-combination formula. Instead, we investigate the focusing properties of each individual lens by itself. The first step is to locate the image formed by the first lens, pretending that the second lens is absent. Applying the lens equation, we obtain

$$\frac{1}{p} + \frac{1}{q} = \frac{1}{f} \quad \Rightarrow \quad \frac{1}{15 \text{ cm}} + \frac{1}{q_1} = \frac{1}{10 \text{ cm}}$$

Solving for q_1 yields $\quad \underline{q_1 = +30.0 \text{ cm}}$

Thus the image would fall 30 cm to the right of the first lens if the second lens were absent. Because the second lens is 15 cm to the right of the first lens, the rays are still convergent when they strike the second lens (constituting a "virtual object" for the second lens), so the object distance p_2 for the second lens is -15 cm. Thus:

$$\frac{1}{p} + \frac{1}{q} = \frac{1}{f} \quad \Rightarrow \quad \frac{1}{-15 \text{ cm}} + \frac{1}{q_2} = \frac{1}{-20 \text{ cm}}$$

Solving for q_2 gives $\quad q_2 = \boxed{60.0 \text{ cm}}$

The final image is 60 cm to the right of the second lens. The image is inverted because the first lens produced a real, inverted image, and the second lens, being divergent, cannot itself produce a further inversion. Though the lens is divergent, its strength is not sufficient to change the incident converging rays to diverging rays; the *positive* value of q_2 signifies *converging* rays, forming a final real image.

The overall magnification for the two-lens system is the product

$$M = M_1 M_2 = \left(-\frac{q_1}{p_1}\right)\left(-\frac{q_2}{p_2}\right) = \left(-\frac{30 \text{ cm}}{15 \text{ cm}}\right)\left(-\frac{60 \text{ cm}}{-15 \text{ cm}}\right) = -8.00$$

As a check, we note that the negative sign signifies an inverted final image.

A word of caution. As this example illustrates, in solving multiple-lens systems we must take care to apply the sign convention correctly each step of the way. It is always helpful to sketch ray diagrams. However, a ray diagram for a *virtual* object (the case for the second lens) is not easy to draw and usually is omitted.

37.9 Optical Instruments

The Simple Magnifier

The simplest optical instrument is the single-lens magnifier or reading glass. A convergent lens is placed in front of fine print or a small object and moved closer or farther away until the best magnified image appears. Because the object is closer than a focal-length distance from the lens, the image is erect, as shown in Figure 37-24b. How much is the object magnified and how does that depend on the focal length of the lens? The answer is complicated because the image distance may be fairly short, or even infinitely far away, depending on what distance the observer finds most comfortable for viewing. Also, the observer's eye may be at various distances from the lens, so the angular size of the image the eye sees may vary. To reduce the number of possibilities, we will discuss only the case in which the observer's eye is close to the lens.

When we use a magnifier, we are interested in how much larger the image appears *with the magnifier* compared to viewing the object *with the unaided eye*. A person with so-called normal eyesight can see clearly objects located anywhere from infinity to about 25 cm from the eye. The largest angular size of an object will be when it is held as close to the eye as possible. By definition, the *"closest distance for comfortable viewing"* is taken to be 25 cm. Of course, some persons can see objects closer than this, while others cannot see objects this close; the 25 cm figure is chosen as an average value.

We define the **angular magnification** m as the ratio

ANGULAR MAGNIFICATION
$$m \equiv \frac{\left[\begin{array}{c}\text{Angle subtended by the image}\\\text{when using the magnifier}\end{array}\right]}{\left[\begin{array}{c}\text{Angle subtended by the object}\\\text{when viewed from 25 cm by}\\\text{the unaided eye}\end{array}\right]}$$

$$= \frac{\alpha}{\beta} \qquad (37\text{-}25)$$

With the unaided eye and with the object at 25 cm, the angular size is $\beta \approx h/25$ cm, Figure 37-24a. Using the magnifier with the eye close to the lens and the image at 25 cm, Figure 37-24b, we have $q = -25$ cm with an angular size $\alpha \approx h/p$. Solving the lens equation for $1/p$ and substituting $q = -25$ cm, we have

$$\frac{1}{p} = \frac{1}{f} - \frac{1}{q} = \left(\frac{1}{f} - \frac{1}{-25 \text{ cm}}\right) = \left(\frac{1}{f} + \frac{1}{25 \text{ cm}}\right)$$

When we use the magnifier, the angular size $\alpha = h/p$ is thus

$$\alpha = h\left(\frac{1}{f} + \frac{1}{25 \text{ cm}}\right)$$

Solving for the angular magnification $m = \alpha/\beta$, we obtain

ANGULAR MAGNIFICATION OF A MAGNIFIER (with image at 25 cm and the eye placed close to the lens)
$$m = \frac{25 \text{ cm}}{f} + 1 \qquad \text{(where } f \text{ is in centimeters)} \qquad (37\text{-}26)$$

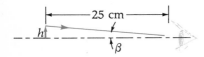

(a) Viewing the object with the unaided eye at the closest distance for comfortable viewing, 25 cm. The object subtends an angle $\beta \approx h/25$ cm.

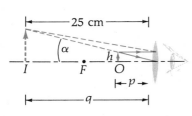

(b) With the eye close to the magnifying lens and the image at the closest distance for comfortable viewing ($q = -25$ cm), the image subtends an angle $\alpha \approx h/p$.

FIGURE 37-24
The simple magnifier.

Carrying out a similar analysis with the magnifier image at infinity (Problem 37C-53), we find that

ANGULAR MAGNIFICATION OF A MAGNIFIER (image at infinity)

$$m = \frac{25 \text{ cm}}{f}$$ (where f is in centimeters) (37-27)

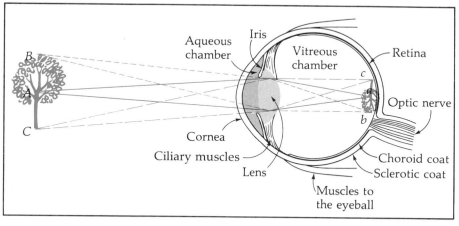

FIGURE 37-25
A simplified diagram of the human eye. The optic nerve is actually to one side of a vertical plane through the center of the eye.

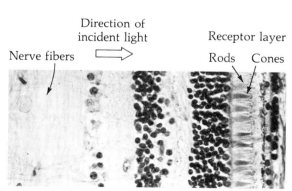

(a) A cross-section of the human retina shows the elaborate layer of nerve fibers, blood vessels, and other tissues through which the incident light must pass to reach the rods and cones. You can see this overlying layer by closing your eye and placing a tiny source of light near the lid. With practice, you can see a pattern of shadows that the blood vessels cast on the rods and cones and even the shadows of blood cells coursing through small blood vessels with each heartbeat.

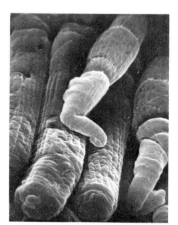

(b) Rods and cones magnified 1600 times with a scanning electron microscope.

FIGURE 37-26
(a) Cross-section of a human retina; (b) rods and cones from the retina of a salamander.

The Eye

The eye, with its linkage to that master computer—the brain—is surely one of the most remarkable of human organs.* The overwhelming majority of information input comes to us via our eyes, and the way we analyze and sort out the ever-changing pattern we see is astonishing. Basically, the eye is similar to a camera with a lens that forms images on a photosensitive surface, but there are many unique features that no camera can duplicate (see Figure 37-25). The focusing of light rays occurs primarily at the outer surface of the *cornea*, where the change in refractive index from air to the cornea is greatest. (Its power is about 40 to 45 diopters in the average person.) On the other hand, the *lens* is surrounded by fluids whose refractive indices are not too different from that of the lens material, so relatively less refraction occurs at these surfaces. (The major reason you cannot see clearly under water is that the refractive index for water, $n_{water} = 1.33$, is too close to that of the cornea, $n_c = 1.367$, to cause sufficient refraction. A face plate corrects the problem by maintaining an air contact with the eye.)

The lens is somewhat flexible, enabling the *ciliary muscles* to adjust its power from about 20 to 24 diopters. In this way, even though the lens is a fixed distance from the photosensitive surface, sharp images can be formed for varying object distances. The overall power of the cornea plus the lens is about 60 to 65 diopters. The ability of the eye to change its focal length is called *accommodation*. With age, the lens material gradually hardens, so the degree of accommodation becomes less as we grow older. The closest object distance for which sharp images can be formed is called the *near point*. For an average 10-year-old, it is around 7 cm, increasing to about 22 cm in middle age, and to about 100 cm at age 60, often requiring "reading" glasses to assist vision at closer distances.

The *retina* consists of roughly 125 million photoreceptor cells called *rods* and *cones*. An elaborate network of neurons and nerve fibers connects them to the brain via the optic nerve (Figure 37-26). As you read these words, your eye jumps abruptly from point to point, so that the center of your field of view falls on the *fovea*, a small area about 0.3 mm in diameter containing only cones packed closely together. (To get some idea of the field of view that covers the fovea, the full moon's image on the retina is about 0.2 mm in diameter.) The eye's ability to detect detail (*resolution*) is greatest in the fovea. Only the cones are sensitive to colors. Away from the fovea, the rods become relatively more numerous, and though they have no color sensitivity, they can detect very dim light. You can test the rods' sensitivity to low light levels by trying to observe a faint star. You may not see the star if you look directly at it, but shifting the direction of vision to one side a bit so the image falls on the rods makes the star's presence detectable. It is believed that some data analysis of the image occurs at the retina, particularly in certain animals with photo-

---

* *Scientific American* has many interesting articles on vision. Among them are the following: "The Visual Cortex of the Brain," David Hubel, November 1963. "Attitude and Pupil Size," E. Hess, April 1965. "Retinal Processing of Visual Images," Charles Michael, May 1969. "The Neurophysiology of Binocular Vision," John Pettigrew, August 1972. "Visual Pigments and Color Blindness," W. Rushton, March 1975. "The Resources of Binocular Perception," John Ross, March 1976.

receptors that send signals to the brain only for specific orientations of light–dark edges or for motions in certain directions.

The *iris* is an adjustable diaphragm that controls the amount of light passing into the eye. The size of the iris opening, the *pupil*, is affected not only by the amount of incident light but also by drugs and by our emotions. If something pleases us, our pupils tend to enlarge; if we are displeased, they tend to contract. Clever poker players aware of this effect claim they can sometimes discern the value of their opponents' hands by watching changes in the sizes of their pupils. Although the iris controls light intensity only by a factor of 16 or so, the retina itself has an enormously larger range of sensitivity. Light causes chemical changes in the rods and cones, reducing their sensitivity; after about half an hour in the dark, the eye becomes "dark adapted" and the greatest sensitivity is achieved. There is no completely adequate theory of color vision that explains all phenomena, though it is reasonably certain that the cones are of three types whose color sensitivities overlap somewhat; one type is most responsive to blue light, one to green light, and the third to yellow light (not red, as previously thought). About 8% of males and 1% of females have some defects of color vision, a hereditary malady that is recessive and sex linked.

In the fovea, where resolution is greatest, each cone has a separate path to the optic nerve, but near the edge of the retina several receptors may be connected to the same nerve path. The region where the optic nerve leaves the retina produces a *blind spot* in the field of vision (Figure 37-27). A portion of the nerve pathways from each eye cross over and lead to the opposite half of the visual *cortex* in the brain, a feature of the "wiring diagram" believed to be involved in depth perception and in maintaining the use of both eyes in case one side of the brain is damaged.

The eye-plus-brain combination is a surprisingly effective visual system that enables us to rapidly scan a scene, investigating interesting portions with the high-resolution fovea, sorting out the varying images, and picking up significant information on intensity, form, motion, and color to store temporarily in our memory, thereby building up a single, three-dimensional concept of our surroundings.

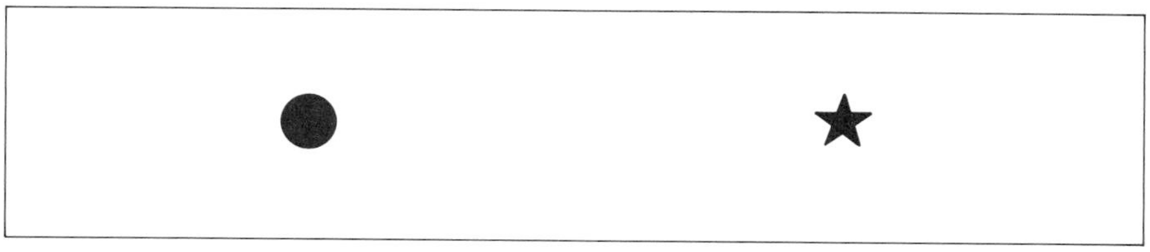

FIGURE 37-27
Diagram for revealing the *blind spot*. Close your left eye and look at the circle as you move the book closer to your eyes. When the diagram is about 20 cm away, the star will disappear. (A similar effect occurs when you close the right eye and look at the star.) The brain tends to "fill in" the missing portion of the field of view with a color and pattern similar to its surroundings. For example, if a pattern of stripes is present, the blank space seems filled with matching stripes!

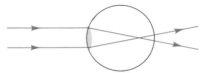
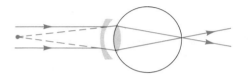

(a) In *nearsightedness* (*myopia*), a person cannot see distant objects because the eye converges incoming light too strongly and the image is formed in front of the retina.

(b) A negative lens corrects nearsightedness by forming a virtual image of a distant object closer to the eye. The eye itself can then form an image of it on the retina.

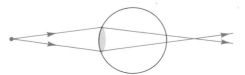

(c) In *farsightedness* (*hyperopia*), a person cannot see nearby objects because the eye does not refract the light sufficiently and the image distance is behind the retina.

(d) A positive lens corrects farsightedness by forming a virtual image of a nearby object farther from the eye. The eye itself then can form an image of it on the retina.

FIGURE 37-28
Eyeglasses can correct the visual defects caused by the eye's inability to form images on the retina.

Eyeglasses

The lens of a normal eye can slightly change its focal length (called *accommodation*) so that, even though the retina is a fixed distance from the lens, it produces sharp images on the retina for both near and distant objects. Sometimes the lens of the eye is not symmetrical and tends to produce elongated images of point sources of light, an abnormality called *astigmatism*. Both lack of accommodation and astigmatism can be corrected by eyeglasses.

See Figure 37-28. If the unaided eye is *nearsighted* and cannot form sharp images of distant objects, an eyeglass can produce a virtual image of the object at a sufficiently close distance so that the eye can look at this image and, with its own lens, focus it on the retina. Conversely, if the eye is *farsighted* and nearby objects seem blurred, an eyeglass can form a virtual image of the object farther away, so that the eye can look at it and form a sharp image on the retina. *In each case, it is easiest to think of the eyeglass as forming a virtual image in front of the observer within the distance that the eye can accommodate.* Then the eye itself looks at this image and properly focuses it on the retina. The following example illustrates these procedures.

(a) A nearsighted person's range of clear vision without eyeglasses.

(b) The person's range of clear vision with eyeglasses.

FIGURE 37-29
Example 37-8.

EXAMPLE 37-8

A nearsighted person (Figure 37-29a) can see objects easily and comfortably within the range of 15 cm to 100 cm. (a) Describe the eyeglasses that will provide a normal range of 25 cm to infinity. (b) Find the image distance these eyeglasses would produce of an object held at the convenient reading and working distance of 25 cm.

SOLUTION

(a) While this person can read easily without glasses, distant objects are out of focus. The glasses should therefore produce images of distant objects that fall within the range of 15 to 100 cm. In practice, objects at infinity should have images at the most distant point of clear vision, so that reasonably close vision is not impaired with the eyeglasses on. Thus, for this person, an object at infinity should have its image at $q = -100$ cm (negative because it is a virtual image). Starting with the lens equation,

$$\frac{1}{p} + \frac{1}{q} = \frac{1}{f} \quad \Rightarrow \quad \frac{1}{\infty} + \frac{1}{-100 \text{ cm}} = \frac{1}{f}$$

Solving for the focal length of the eyeglass, we obtain

$$f = \boxed{-100 \text{ cm}} \quad \text{or} \quad \boxed{-1.00 \text{ diopter}}$$

(b) Again, using the lens equation gives

$$\frac{1}{p} + \frac{1}{q} = \frac{1}{f} \quad \Rightarrow \quad \frac{1}{25 \text{ cm}} + \frac{1}{q} = \frac{1}{-100 \text{ cm}}$$

Solving for the image distance q yields $\quad q = \boxed{-20.0 \text{ cm}}$

This is well within the clear viewing range of the person. The eyeglasses would thus expand the range of clear vision as shown in Figure 37-29b.

Toward middle age, people often lose accommodation, so that their range of clear vision becomes smaller. (This condition is called *presbyopia*.) A person with a narrow range of clear vision near 100 cm would need *bifocals*: a convergent portion of the eyeglass for viewing nearby objects and a divergent portion for distance objects. *Trifocals* are also sometimes used.

The Astronomical Telescope

The simplest form of the astronomical telescope consists of two converging lenses: the *objective lens* (focal length f_o) and the *eyepiece lens* (focal length f_e). As shown in Figure 37-30a, the objective lens creates a real, inverted image I_1 at its focal-length distance because the object's distance is essentially infinite. In turn, the eyepiece forms a virtual image of I_1. In practice, most viewers focus the eyepiece so that the final image is at infinity. (In doing so, they can shift the eye from the eyepiece to the object without eye accommodation.) For a final image at infinity, the image I_1 must be at the first focal point of the eyepiece lens.

The angular magnification m of an astronomical telescope is the ratio of the angle subtended by the image formed by the eyepiece to that formed by the object. Thus, $m = \alpha/\beta$, where α and β are as indicated in Figure 37-30a. Ordinarily, the image formed by the objective is small compared with either the objective focal length f_o or the eyepiece focal length f_e. Then, since $h \ll f_e$, $\alpha = \tan^{-1}(h/f_e) \approx h/f_e$. Similarly, $\beta \approx h/f_o$. Thus $m = \alpha/\beta$ becomes

ANGULAR MAGNIFICATION OF AN ASTRONOMICAL TELESCOPE

$$m = \frac{f_o}{f_e} \qquad (37\text{-}28)$$

An important characteristic of a telescope is its light-gathering ability. Ideally, when we are viewing through a telescope, all of the light that enters the objective lens should ultimately enter the pupil of the eye. As shown in

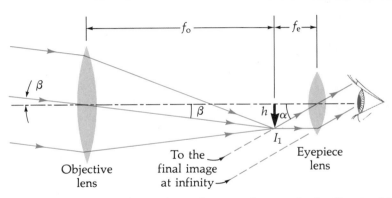

(a) The objective lens forms a real, inverted image I_1 of a distant object. The eyepiece lens forms a distant virtual image of I_1.

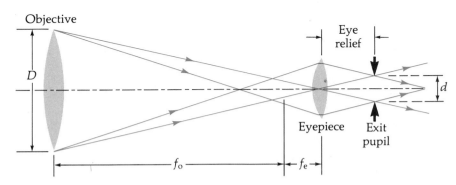

(b) The eyepiece lens forms an image of the objective lens, called the *exit pupil*. All the light that passes through the telescope goes out through the exit pupil. The pupil of the viewer's eye should be placed at the exit pupil so that the maxiumum amount of light enters the eye. The distance from the eyepiece lens to the exit pupil is the <u>eye relief;</u> it should be large enough to accommodate viewers with glasses.

FIGURE 37-30
The astronomical telescope.

Figure 37-30b, the bundle of light rays emerging from the eyepiece becomes constricted, then spreads out. The area of constriction, called the *eye ring* or *exit pupil*, is actually the image of the objective lens formed by the eyepiece lens. If the exit pupil diameter is smaller than the observer's eye pupil, the eye can capture all the light passing through the telescope.

The Simple Microscope

In the simple two-lens microscope, Figure 37-31, the object is placed just outside the focal point of the *objective* lens (whose focal length f_o is very short), producing a greatly enlarged real image I_1. This image is magnified by the *ocular* or *eyepiece* lens (focal length f_e) in the way we discussed for a simple magnifier with the final virtual image at the closest distance for comfortable viewing. The eyepiece lens has an angular magnification m [Equation (37-27)] of approximately 25 cm/f_e. (We drop the 1 in the formula because f_e is usually much shorter than 25 cm.) The objective lens has a linear magnification M, so that the total *magnifying power* of the microscope is $Mm = M(25 \text{ cm}/f)$. The linear magnification of the objective lens is ℓ/f_o, where ℓ is called the *tube length* of the microscope. A typical value of ℓ is 18 cm. Substituting these values, we have, for the magnifying power,

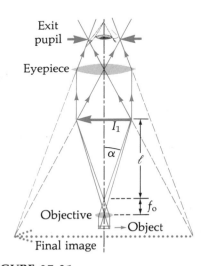

FIGURE 37-31
A simple microscope. The objective lens forms a greatly enlarged real image I_1 of the object. The eyepiece is a magnifier that creates a virtual image of I_1. The final image is usually chosen at the closest distance for comfortable viewing.

MICROSCOPE MAGNIFYING POWER

$$Mm = \left(\frac{\ell}{f_o}\right)\left(\frac{25 \text{ cm}}{f_e}\right)$$

$$= \frac{450 \text{ cm}^2}{f_o f_e} \quad \text{(where } f_o \text{ and } f_e \text{ are in centimeters)} \quad (37\text{-}29)$$

In order to achieve good exit pupil size and eye relief (see Figure 37-30b), f_e must be of the order of 1 cm. Thus, by Equation (37-41), f_o must be very short. For a magnifying power of 2000, f_o is of the order of 2 mm.

Fresnel Lens

Often it is desirable to concentrate light, as in lighthouse searchlights, certain types of solar energy collectors, and overhead projectors. Instead of a large-diameter, simple converging lens that is heavy and expensive to manufacture, we can achieve the same light concentration with a *Fresnel lens*. In Figure 37-32a, the unshaded portions of the ordinary lens do not contribute to the focusing action and are eliminated. Only the light-bending surface contour of the lens is retained in the form of a great many concentric circular ridges, greatly reducing the weight. Thin Fresnel lenses are often molded of plastic at low cost. The image quality is usually not the goal, so surface contouring is often not very precise. Fresnel lenses are used in traffic lights that are visible only in the lane intended. In addition to its light-directing uses, a large lens three feet square in sunlight can create temperatures over 3000 K at its focus, sufficient to melt a variety of metals. Smaller versions make a lightweight camp stove for hikers.

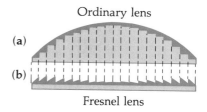

FIGURE 37-32
(a) A cross-section of an ordinary converging lens. (b) A Fresnel lens equivalent to (a) in its light-focusing properties. Concentric circular segments (many more than are shown here) contain the light-bending surface contour of the ordinary lens; the light color shaded portions in (a) have been eliminated. A negative Fresnel lens is sometimes used in the rear window of recreational vehicles to provide a wider-angle view than the window itself allows.

The Camera and f-Stops

The light-gathering ability of a lens is proportional to its area. In most cameras, the effective area can be changed by use of an *iris diaphragm*: a circular hole of variable diameter called an *aperture stop*. The aperture size is expressed by the *f/number*, or *f-stop*, defined as *the focal length divided by the diameter of the aperture*. Lenses are usually calibrated in successive *f/numbers* that change by (rounded) factors of $\sqrt{2}$. Thus, *each step corresponds to a factor of 2 in light-gathering ability*. Typical *f*-stops are $f/1.4$, $f/2$, $f/2.8$, $f/4$, $f/5.6$, $f/8$, $f/11$, $f/16$. A low *f/number* signifies a "fast" lens because its larger diameter gathers enough light to expose the film in a shorter exposure time.

EXAMPLE 37-9

The focal length of an $f/2$ camera lens is 50 mm. (a) Find the diameter of the lens. (b) If the correct exposure for photographing a scene is $\frac{1}{400}$ s at $f/2.8$, what is the correct exposure time at $f/8$?

SOLUTION

(a) Diameter $= \dfrac{\text{Focal length}}{f/\text{number}} = \dfrac{50 \text{ mm}}{2} = \boxed{25.0 \text{ mm}}$

(b) The change in aperture stop is three steps along the increasing f/number scale, or three factors of 2 *smaller* area (less light-gathering ability). Thus the exposure time should be increased by 2^3, or 8 times longer, or

$(8)(1/400) \text{ s} = \boxed{\tfrac{1}{50} \text{ s}}$

37.10 Aberrations

We used several simplifying assumptions in deriving thin-lens formulas. In particular, we employed small-angle approximations, ignored far-off-axis objects and rays, and neglected the fact that the index of refraction is not the same for all colors of light (dispersion). Consequently, every actual lens produces certain defects, or *aberrations*, in the image. Some common examples are illustrated in Figures 37-33 through 37-35. Many of these aberrations can be minimized

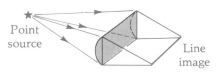

FIGURE 37-34
Astigmatism. If a lens is not perfectly spherically symmetric but has a small amount of cylindrical property, a point source forms a *line image*. (Light from off-axis sources striking a spherical lens also produce this defect.) Astigmatic vision can be helped by eyeglasses that have a compensating cylindricality whose axis is perpendicular to the axis of the corneal lens defect.

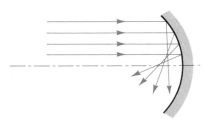 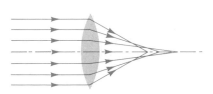

(a) Parallel rays farther from the axis of a *spherical* mirror are brought to focus closer to the mirror than are rays near the axis.

(b) Parallel rays near the edge of a spherical lens have a shorter focal length than those rays near the axis of the lens.

FIGURE 37-33
Spherical aberration. Because *spherical* surfaces are the easiest curvature to manufacture, the surfaces of most mirrors and lenses are spherical. Thus paraxial rays impinging near the edges will necessarily have different focal distances than those near the center. Consequently, the image is "smeared out" and appears out of focus. To reduce this effect, cameras often have an *iris*, or adjustable *aperture stop*, that can be "closed down" to allow only the central portion of a lens to be used. (To compensate, longer exposure times are necessary.) A *parabolic* mirror does focus parallel rays at the same point and thus has no spherical aberration. Lenses can also be ground to have special contours (expensive to fabricate) that reduce spherical aberration.

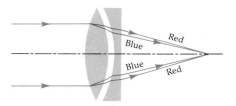

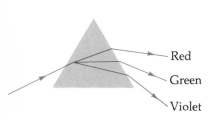 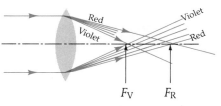

(a) A prism (dispersion exaggerated).

(b) A single convergent lens has a longer focal length for red light than for violet light.

(c) The dispersion of two specific wavelengths by a converging lens can be "undone" by a divergent lens if the divergent lens has greater dispersion than the convergent lens. (Other wavelengths, however, still focus at different points.)

FIGURE 37-35
Chromatic aberration of lenses. Because the index of refraction in glass is greater for shorter wavelengths, blue-violet light is refracted more than red light. Consequently, a lens tends to separate white-light images into a spectrum the same way a prism does. So a simple lens has different focal lengths for different colors, a defect called *chromatic aberration*. The amount of dispersion is exaggerated here. Multiple-lens systems can be designed to reduce the effect, but it can never be eliminated for all wavelengths. (Mirrors, of course, do not have chromatic aberration.)

with multiple-lens systems in which the aberrations of one lens are partially cancelled by those of another lens. The recent development of new optical-glass materials that have special index-of-refraction characteristics, and the use of high-speed computers in complex ray-tracing calculations for nonspherical surfaces, have greatly improved the design of optical systems. No lens is perfect, but (at additional expense) the most troublesome defects can be minimized.

Summary

The *index of refraction* n (which depends upon wavelength) is defined as

$$n \equiv \frac{c}{v}$$

Snell's law for the refraction at an interface between two different materials (where θ is the angle between the ray and the normal to the surface) is

$$n_1 \sin \theta_1 = n_2 \sin \theta_2$$

Total internal reflection: The *critical angle* θ_c is given by

$$\sin \theta_c = \frac{n_2}{n_1} \qquad (n_2 < n_1)$$

The thin-lens equation: $\quad \dfrac{1}{p} + \dfrac{1}{q} = \dfrac{1}{f}$

where p is the object distance, q is the image distance, and f is the focal length. The equation is to be used with the *sign convention*:

(1) The value of p is *positive* if the rays that impinge on the lens are *divergent*.
(2) The value of q is *positive* if the rays leaving the lens are *convergent*.
(3) The focal length f for a *converging* lens is *positive*; for a *diverging* lens, it is *negative*.

The lens-maker's formula: $\quad \dfrac{1}{f} = (n - 1)\left(\dfrac{1}{R_1} + \dfrac{1}{R_2}\right)$

where n is the refractive index of the lens relative to the surrounding medium and R_1 and R_2 are the radii of curvature of the lens surfaces. R_1 and R_2 are *positive* if the corresponding outer surfaces are *convex* and *negative* if they are *concave* (assuming that n of the lens is greater than that of the surrounding medium).

Ray diagrams give much information about image characteristics. We construct them by tracing two of the following rays:

(1) a ray that strikes the center of the lens and passes through undeviated;
(2) a ray that is parallel to the axis and passes (or is extended) through the focal point F;
(3) a ray that passes through (or extending through) a focal point F and then strikes the lens, emerging from the lens parallel to the axis.

Images are *real* or *virtual*, *erect* or *inverted*, with *linear magnification* $M = -(q/p)$.

The *f-stop*, or *f/number*, of a lens is the focal length divided by the diameter of the lens (or diameter of the *aperture stop*). The *diopter power* D is the reciprocal of the focal length in meters.

For *two lenses* in contact: $\quad \dfrac{1}{f} = \dfrac{1}{f_1} + \dfrac{1}{f_2}$

or, in diopters, D

$$D = D_1 + D_2$$

For two separated lenses, the image produced by the first lens acting alone is the object for the second lens.

Angular magnification:

Magnifier: $m = \dfrac{25 \text{ cm}}{f} + 1 \qquad$ (image at 25 cm, f in cm)

Telescope: $m = \dfrac{f_o}{f_e} \qquad$ (image at ∞)

Magnifying power of a microscope:

$$Mm = \left(\frac{\ell}{f_o}\right)\left(\frac{25 \text{ cm}}{f_e}\right) \quad \left[\begin{array}{l}\ell \text{ is } tube\ length \text{ (usually 18 cm)} \\ \text{and } f_o \text{ and } f_e \text{ are in centimeters}\end{array}\right]$$

Common *aberrations:* spherical aberration, astigmatism, and chromatic aberration (lenses).

Questions

1. Do we alter the focal length of a spherical mirror by immersing the mirror in water?
2. An observer walks toward a swimming pool. Why does the apparent depth of a swimming pool depend on the observer's distance from the edge of the pool?
3. Why does a straight pole penetrating the surface of a pond often appear to be bent at the point where the pole enters the water?
4. If a fisherman can see the eye of a fish in a still pond, can the fish always see the eye of the fisherman? That is, are there situations for which total internal reflection prevents either from seeing the other?
5. What does a swimmer see as she looks upward toward the smooth surface of a swimming pool? Include considerations of total internal reflection.
6. What are the optical properties of an air bubble in glass?
7. When measuring the angle between the late afternoon sun and the horizon with a sextant, a navigator must apply a correction to the observed angle. Why is a correction necessary and what is the sign of the correction?
8. Is it possible for a lens to be convergent in air and divergent in water?
9. How does the focal distance of a converging lens depend on the color of light? Is the dependence the same for a diverging lens?
10. The two focal points of a thin lens are the same distance from the lens. Can you show by sketching ray diagrams that the two focal points of a thick lens may not be the same distance from the center of the lens?
11. What is a procedure for determining the focal lengths of (a) a diverging lens and (b) a convex mirror?
12. A person's eyes appear to be smaller when he wears his glasses. Is he nearsighted or farsighted?
13. While swimming under water without a diving mask, does the swimmer become more nearsighted or more farsighted? Can she correct this by wearing eyeglasses? If so, what kind of eyeglasses?
14. A simple two-lens astronomical telescope (both converging lenses) is used to view a distant sign. Is the image simply inverted or is the lettering on the sign reversed, as in a plane mirror image?
15. Why does a person with normal vision often adjust the eyepiece of an astronomical telescope so that the image is at infinity?
16. Without asking the wearer (but being allowed to experiment with the lens), how would you determine if an eyeglass lens includes a correction for astigmatism?
17. Do two different observers see the same rainbow in exactly the same place? Explain.
18. Suppose that the top half of a lens is covered. How will this affect the image? Is the complete image still present? Are there other changes? Explain.
19. A *pinhole camera* has no lens. Instead, a tiny hole is sufficient to form images on the film at the back of the camera box. Explain how these images are formed and why the image is "sharp" *for nearby as well as distant objects.*
20. In a physics lab, a student uses a converging lens to form a real image of a window frame on a piece of paper. Should he move the paper closer to, or farther from, the lens in order to produce a sharp image of a distant tree?
21. In Figure 37-5, why does the submerged straw appear to have a smaller diameter than the straw in air?
22. How close to a converging lens can an object be placed such that the lens still produces a real image? Where is that image located?

Problems
37.2 Refraction at a Plane Surface

37A-1 A microscope may be used to measure the refractive index of a plane sheet of glass. The top surface of the glass is brought into focus by the microscope. The microscope is then lowered 2.50 mm to bring the lower surface into focus. The measured thickness of the glass is 3.80 mm. Calculate the refractive index of the glass.

37B-2 A narrow laser beam reflected from a thick glass plate produces two parallel beams, one reflected from the front surface of the plate and the other reflected from the rear surface of the plate. Assume an angle of incidence θ, a plate thickness D, and an index of refraction n for the glass plate. Derive an expression for the perpendicular distance d between the two reflected beams in terms of θ, D, and n.

37B-3 The time required for a light signal to travel vertically from the bottom to the top of an empty vessel is t_0. Show that when the same vessel is filled with a liquid ($n > 1$) the time required for a signal to travel vertically in the liquid the distance of the apparent depth of the vessel is also t_0.

37B-4 A can 12 cm deep is filled with a layer of water ($n = 1.33$) 5 cm thick and a layer of oil ($n = 1.48$) 7 cm thick that floats on the water. Calculate the apparent depth of the can when it is viewed from a point directly above the can. (Hint: use the result of Problem 37B-3.)

37B-5 A beam of light strikes a plane slab of glass at an angle of 40° with the surface of the glass. The glass is 1.5 cm thick and has a refractive index of 1.60. The beam emerging from the other side of the slab will be parallel to the incident beam but displaced laterally. Calculate the distance that the emerging beam direction is displaced sideways from the incident beam direction.

37B-6 A flat-bottomed container is filled with water ($n = 1.33$) to a depth of 8 cm. A 4-cm layer of oil ($n = 1.47$) floats on the water, Figure 37-36. A light ray in air approaches the oil at an angle of incidence $\theta = 55°$. Find the horizontal distance x.

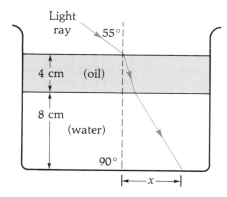

FIGURE 37-36
Problem 37B-6.

37B-7 A tin can 20 cm high and 15 cm in diameter has one end removed, and a small hole is punched in the center of the other end. When peering through the hole, you have a cone of vision that is limited by the edge of the other end of the can. (a) Calculate the maximum angle of vision away from the axis of the can. (b) Calculate the solid angle of vision in steradians (see Appendix E for definition of a solid angle). (c) The can is now filled with a clear plastic of refractive index 1.65. When you are looking through the hole, what is the maximum angle of vision (away from the axis) outside the can?

37.3 Total Internal Reflection

37A-8 A ray of light enters a 45° prism (refer to Figure 37-11a). (a) Find the minimum refractive index of the prism to produce total internal reflection, as shown. (b) If the prism is immersed in water, calculate the minimum index of refraction of the prism to produce the same result.

37B-9 Plane sheets of glass with parallel sides are used to construct a hollow prism whose cross-section is an equilateral triangle. The prism is filled with a copper sulfate solution ($n = 1.74$). As shown in Figure 37-37, a light ray is incident normally on one face (not at the center of the face). Find the maximum index of refraction of the glass that will still produce total internal reflection at the first liquid-to-glass interface.

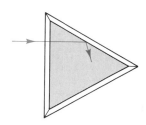

FIGURE 37-37
Problem 37B-9.

37B-10 A diamond "sparkles" more than a glass replica because the higher index of refraction of diamond causes more incident rays to be totally internally reflected back out of the upper surfaces, rather than escaping from the lower sides. In diamond, the light also undergoes more dispersion when refracted, increasing the color variation of the reflected light. Figure 37-38 shows a vertical internal ray A incident upon the side of a diamond ($n = 2.42$). (a) Show that this ray is totally internally reflected and escapes from the top of the diamond while a similar ray in a flint glass replica ($n = 1.65$) is refracted out of the lower side. (b) Find the maximum angle θ for a ray B incident upon the top of the diamond that would be totally internally reflected when striking side AC. Consider only those rays that lie in a plane passing through the central axis of symmetry.

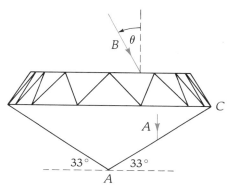

FIGURE 37-38
Problems 37B-10 and 37C-40.

37B-11 A fish at the bottom of a still pond sees the entire region above the surface of the water in a circular field of view centered on a vertical line above the fish. Calculate the solid angle (in steradians) that the circular area subtends at the fish's eye. (See Appendix E for definition of a solid angle.)

37B-12 A layer of benzene ($n = 1.501$) floats on water ($n = 1.33$). (a) Find the critical angle for total internal reflection at the interface between the two liquids. (b) Is it possible for a ray of light in air to be incident on the top surface of the benzene so that it strikes the interface at the critical angle? If so, find the angle of incidence at the top surface. If not, explain why not.

37B-13 A point source of light that emits equally in all directions is placed below the surface of a pond of water ($n = 1.33$). All of the light that reaches the surface is either totally reflected or totally transmitted. Find the fraction of the light emitted from the source that leaves the surface of the pond. (Hint: see Appendix E for definition of a solid angle.)

37B-14 See Figure 37-39 (and also Figure 37-12). Consider a light ray entering the end of an *optical fiber*. If the angle of incidence is within the *acceptance angle* θ_a, the ray is propagated down the fiber core by total internal reflections. A ray incident at a larger angle refracts into the cladding and is eventually absorbed. For a glass fiber with $n_{core} = 1.54$ and $n_{cladding} = 1.47$, find the acceptance angle for incoming light at the end of the fiber.

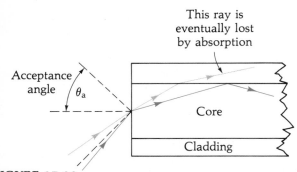

FIGURE 37-39
Problems 37B-14 and 37C-43.

37.4 Refraction at a Spherical Surface

37A-15 A small air bubble is at the center of a large glass sphere that has a refractive index n and radius R. Determine how far the air bubble appears to be from the surface of the sphere.

37A-16 A solid polystyrene sphere ($n = 1.59$) of radius 8 cm has a decorative object embedded in its interior. If a point on the object is 3 cm from the center of the sphere, how far away from the sphere's surface does it appear to an outside observer?

37B-17 A small-diameter parallel light beam is directed toward the center of a large solid sphere made of transparent plastic. The beam is brought to a focus on the opposite side of the sphere. Find the refractive index of the plastic.

37B-18 A glass rod ($n = 1.63$) with a circular cross-section has a bundle of light rays traveling parallel to the axis of the rod as shown in Figure 37-40. Find the radius of curvature R of the end of the rod that will bring the bundle of rays to a focus 12 cm from the end of the rod when the rod is immersed in water.

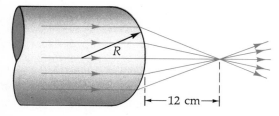

FIGURE 37-40
Problem 37B-18.

37.5 Thin Lenses
37.6 Diopter Power

37A-19 A camera has a single thin lens with a focal length of 50 mm. Determine how far and in which direction the lens must be moved relative to the film in order to change the object distance from infinity to 75 cm.

37A-20 A lens made of polystyrene ($n = 1.59$) has a power of 2 diopters. The radius of curvature of one outer convex surface is 50 cm. Calculate the radius of curvature of the other surface. Is it concave or convex?

37A-21 A pair of 1.25-diopter eyeglasses is made of glass having a refractive index 1.50. The outer surface next to the eye is concave and has a radius of curvature of 80 cm. Find the radius of curvature of the other surface of the lens.

37A-22 A lens made of glass ($n = 1.62$) has a concave outer surface with a radius of curvature of 100 cm and a convex outer surface with a radius of curvature of 40 cm. Calculate the focal length of the lens.

37B-23 The two surfaces of a double-convex converging lens have the same radius of curvature. The lens of focal length f is now cut into two equal halves by a plane through its center, perpendicular to the axis, forming two plano-convex lenses. In terms of f, find the focal length f' of each of these new lenses.

37B-24 A lens made with a material of refractive index n has a focal length f in air. When immersed in a liquid with a refractive index n_1, the lens has a focal length f'. Derive the expression for f' in terms of f, n, and n_1.

37.7 Thin Lens Ray-Tracing and Image Size

37A-25 When the full moon is viewed from the earth, its diameter subtends an angle of about 0.5°. A photograph of the full moon is obtained with a camera lens having a focal length of 50 mm. (a) Find the diameter of the moon's image on the film. (b) If the film width is 35 mm, what fraction of this width is the moon's image?

37A-26 A 6-diopter magnifying glass is held 10 cm from a printed page. Find the image size of a letter 4 mm high. Include a ray diagram.

37B-27 Figure 37-41 depicts four thin lenses made of glass ($n = 1.58$). For each lens, the two radii of curvature of the surfaces are 15 cm and 30 cm. Calculate the focal length of each lens.

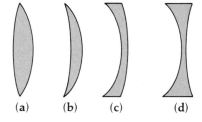

FIGURE 37-41
Problem 37B-27.

37B-28 A slide projector forms an image on a screen 5.8 m away. The image is 80 times larger in its linear dimensions than the slide. Find (a) the distance between the slide and the projection lens and (b) the focal length of the lens.

37B-29 A converging lens has a focal length of 28 cm. (a) Find the distance from the lens at which an object produces a real image twice as large as the object. (b) Repeat for a virtual image twice as large as the object. Include ray diagrams for each.

37.8 Combinations of Lenses
37.9 Optical Instruments

37A-30 One way to determine the focal length of a thin divergent lens is to place the lens in contact with a convergent lens strong enough so that the combination produces a real image of a very distant object. Suppose that a divergent lens of unknown focal length is combined with a 2-diopter converging lens to produce a real image of a distant object on a screen 75 cm away from the lenses. Calculate the focal length of the divergent lens.

37A-31 A simple telescope is constructed with two lenses of focal lengths 120 cm and 5 cm. (a) Find the angular magnification of the telescope. (b) A tower 70 m high, 2 km away, is viewed with the telescope. What is the angular size of the image (at infinity) when it is viewed through the eyepiece?

37A-32 A certain microscope has a tube length of 18 cm and an overall magnification of 800. If the eyepiece lens is 1.2 cm, find the objective lens focal length.

37B-33 A farsighted person can comfortably view objects no nearer than 2 m but can see very distant objects clearly. (a) Calculate the power of eyeglasses necessary for the person to read a book held 25 cm away. (b) Find the farthest object that the person could see comfortably while wearing these glasses, assuming that the eye cannot make more accommodation for distant vision than when unaided.

37B-34 (a) Calculate the effective focal length of the combination of two thin converging lenses, each with a focal length of 50 cm, when the lenses are separated by a distance of 5 cm. (b) Compare the result with the focal length of the two lenses in contact.

37B-35 A nearsighted person wearing eyeglasses with a power of -1.5 diopters can see clearly objects as close as 25 cm as well as very distant objects. Determine the person's range of vision without eyeglasses, assuming that no further accommodation for distant vision is possible.

37B-36 An object is located at the origin of the x axis. Two converging lenses of focal lengths 10 cm and 20 cm are placed, respectively, at $x = 15$ cm and $x = 35$ cm. (a) Locate and describe the final image. (b) Sketch a ray diagram for the first lens (acting alone).

Additional Problems

37C-37 Derive Snell's law of refraction using Fermat's principle. Use assumptions similar to those in Problem 36C-15, Chapter 36.

37C-38 A ray of light is incident upon a cube of glass ($n = 1.68$) as shown in Figure 37-42. The ray lies in a plane parallel to the plane of the diagram. (a) Find the largest angle of incidence θ_i for which total internal reflection will occur at the top face of the cube. (b) Is there an angle of incidence θ_i for which total internal reflection will also occur when the internally reflected ray strikes the right-hand face of the cube? Explain. (c) Solve part (a) for the case in which the cube is totally immersed under water.

37C-39 A convenient way to measure the index of refraction of a transparent substance is by constructing a prism of

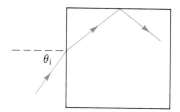

FIGURE 37-42
Problem 37C-38.

the material as shown in Figure 37-43. The total deviation angle δ of a ray incident upon a prism is the sum of the two deviation angles α_1 and α_2 as the ray passes through the prism. It is shown in advanced texts that the *minimum angle of deviation* δ_m occurs when the ray passes *symmetrically* through the prism ($\alpha_1 = \alpha_2$). Show that, for this case, the index of refraction n of the prism is given by

$$n = \frac{\sin \frac{1}{2}(\phi + \delta_m)}{\sin \frac{1}{2}\phi}$$

where ϕ is the apex angle of the prism. (Hint: show that $\theta_2 = \phi/2$ so that $\theta_1 = \theta_2 + \alpha$, and apply Snell's law for refraction.)

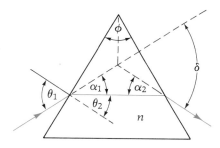

FIGURE 37-43
Problem 37C-39.

37C-40 In Figure 37-38, consider a light ray inside the diamond that is incident upon the inner conical side similar to ray A. (Limit cases to rays that lie in a plane passing through the central axis of symmetry.) Find the range of incident angles that result in the ray undergoing total internal reflection and again being totally internally reflected at the opposite side of the diamond. (Such rays usually refract out of the top surfaces, contributing to the color and brilliance of the diamond's appearance.)

37C-41 Refer to Problem 37C-39. We can find the index of refraction of a liquid by putting the liquid in a hollow prism with plane-parallel glass sides. Show that the glass sides themselves produce no effect on the deviation angle δ.

37C-42 An optical fiber shown in Figure 37-44 is made of glass ($n = 1.63$) and has a diameter of 0.060 mm. Find the smallest radius R through which the fiber could be bent and still result in total internal reflection for all rays in an incident beam parallel to the axis, spread over the cross-sectional area of the fiber.

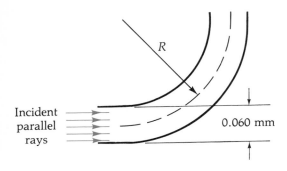

FIGURE 37-44
Problem 37C-42.

37C-43 See Figure 37-39. The core of an *optical fiber* is 50 μm in diameter, made of glass ($n = 1.58$), with a cladding layer ($n = 1.52$). Consider two rays traveling down a straight fiber 20 km long. One ray travels parallel to the axis with (essentially) no reflections, and the other ray reflects from side to side, incident always at the largest angle of incidence that produces total internal reflections. (a) What total distance within the fiber does the second ray travel? (b) How many reflections does this ray undergo in traveling to the end of the fiber? (c) If the two rays start simultaneously, find the time interval between their arrivals at the far end. (This effect smears out, or *broadens*, the light pulses, limiting the maximum rate of pulse transmission that can be used. To correct the problem, fibers with core diameters of only ~ 2 μm are used, in which the more extreme zigzag paths can be eliminated. In such small fibers, whose diameters are comparable to the wavelengths of light used, one must analyze the light as *waves* propagating through a *waveguide*, involving standing-wave patterns that eliminate certain modes of propagation.)

37C-44 Make a qualitative sketch similar to Figure 37-6b that traces a red ray and a violet ray in the secondary rainbow. Explain why the sequence of wavelengths in the observed spectrum has a reversed order compared to the spectrum in the primary bow.

37C-45 When images are projected onto a "beaded" screen, the tiny glass spheres embedded in the white surface of the screen reflect more light back toward the viewer (within about $\pm 30°$ of the projection axis) than when a plain white surface is used. The focusing action of the spheres concentrates the light on a small area at the back surface. Light from this extra-bright area is then refracted by the sphere back (approximately) toward the viewer. Consider a very narrow laser beam that is incident on a glass sphere ($n = 1.60$) along a diameter, Figure 37-45a. (a) Considering refraction at the front surface only, find the focal point for this beam in terms of the sphere radius R. Sketch a bundle of rays in the beam, showing how they strike the back surface of the sphere. (b) Consider a ray of light that approaches the sphere tangentially and does refract into the sphere. Where does this ray cross the midplane of the sphere [shown dashed in (b)]? (c) In terms of R, find the distance b from the midplane such that the incident ray strikes the center of the back surface, Figure 37-45c.

37C-46 In Figure 37-46, the small set of axes is a three-dimensional object. (a) Sketch the image formed by the converging lens, showing clearly the directions of the corresponding axes. Is the image a right-handed or a left-handed coordinate system? (b) Repeat, placing the object between F and the lens.

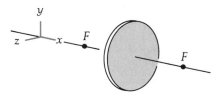

FIGURE 37-46
Problem 37C-46.

37C-47 Show that the thin-lens formula $1/p + 1/q = 1/f$ may be written in the so-called Newtonian form, $xx' = f^2$, where, for a convergent lens, x is the distance from the object to the nearest focal point and x' is the distance from the other focal point to the real image. Both x and x' are positive quantities. Describe how x and x' must be defined for a divergent lens. (This form of the thin-lens equation first appeared in Newton's *Opticks* in 1704.)

37C-48 A lens placed a distance x from a luminous object produces a clear image on a screen 30 cm from the lens. When the screen is moved 10 cm further away from the lens, the lens must be moved 1 cm closer to the object in order to restore a clear image. (a) Calculate the distance x. (b) Calculate the focal length of the lens.

37C-49 A luminous object and a screen are a distance L apart. A converging lens with a focal length f placed at either of two positions between the object and the screen will produce a real image of the object on the screen. Derive an expression for the distance between those two positions.

37C-50 In the table below, fill in the missing data. In every case assume that the diameter of the lens is small compared with the radii of curvature of its surfaces. All numerical values are expressed in centimeters. Indicate the appropriate sign of the values in accordance with the sign convention.

Type of Lens	Focal Distance	Object Distance	Image Distance	Real?	Inverted?	Magnification
Converging	+60	+20	−30	No	No	$+\frac{3}{2}$
	−40	+120				
		+50				−4
		+?	200 (sign ?)		No	+5
		+40				−5
Diverging	200 (sign ?)		−30			
		+50	−30			

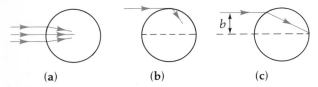

FIGURE 37-45
Problem 37C-45.

37C-51 Sketch ray diagrams for each of the cases given in Problem 37-52.

37C-52 As shown in Figure 37-2, the index of refraction of glass differs for different wavelengths. Consider a flint glass lens of focal length f (for blue light). Find the fractional change in focal length $\Delta f/f$ between light of wavelengths 434 nm (blue, $n = 1.675$) and 656 nm (red, $n = 1.644$).

37C-53 A small change in object distance Δp corresponds to the thickness (along the axis direction) of a thin object. Show that the image of the object produced by either a lens or a mirror has an apparent thickness equal to $M^2 \Delta p$, where M is the linear magnification of the lens or mirror.

37C-54 A thin double-convex lens has surfaces with radii of 40 cm and 50 cm. The index of refraction of the lens material is 1.50. The surface with the 50-cm radius is silvered so the surface forms a concave mirror. A small object is placed 60 cm from the lens on the unsilvered side. Locate and describe the image.

37C-55 An object is located 40 cm in front of a converging lens of focal length 20 cm. A plane mirror is placed behind the lens 25 cm from the lens, Figure 37-47. (a) Locate and describe the final image, including the magnification. (b) Where would you place your eye in order to view this image?

37C-56 Three lenses are lined up along the x axis as follows: the first lens, with a focal length of $+25$ cm, is at $x_1 = 40$ cm; the second lens, with a focal length of -100 cm, is at $x_2 = 55$ cm; and the third lens, with a focal length of $+40$ cm, is at

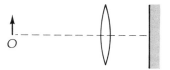

FIGURE 37-47
Problem 37C-55.

$x_3 = 70$ cm. Locate and describe the image of an object placed at the origin, $x = 0$.

37C-57 Consider the combination of lenses shown in Figure 37-23 but with the convergent ($f_1 = 0.1$ m) and the divergent ($f_2 = -0.2$ m) lenses *interchanged*. Locate and describe the final image. Compare your results with the results of Example 37-7.

37C-58 Consider a converging lens A. Suppose that a second lens B is made, with the same type glass, that is twice the diameter and twice the radii of curvature of lens A. (a) Find the focal length of lens B relative to that of lens A. (b) Lens A is used at its maximum aperture to photograph a scene correctly at $1/100$ s. If lens B is used with its maximum aperture to photograph the same scene using the same type film, what exposure time would be appropriate?

37C-59 Derive the expression for the angular magnification of a simple magnifier with the image at infinity, Equation (37-27).

CHAPTER 38

Physical Optics I—Interference

One takes up fundamental science out of a sense of pure excitement, out of joy at enhancing human culture, out of awe at the heritage handed down by generations of masters and out of a need to publish first and become famous.

LEON M. LEDERMAN, Nobel Prize 1988 (with Melvin Schwartz and Jack Steinberger). "The Value of Fundamental Science," *Scientific American* 251, 40 (Nov. 1984).

38.1 Introduction

Maxwell's electromagnetic theory describes light as waves of electric and magnetic fields that travel through space. For the next few chapters we will discuss phenomena that demonstrate these *wavelike* properties.[1] Although we developed the laws of reflection and refraction of light using a wave model, these laws can be derived just as easily with a particle model. In fact, Newton was the first to work out a particle model in some detail, explaining reflection and refraction on that basis. However, as we now discuss the interference, diffraction, and polarization of light, Newton's particle model is clearly unworkable. Only *waves* seem to make sense.

38.2 Double-Slit Interference

In 1802 and 1803, Thomas Young[2] presented papers before the Royal Society, proposing a wave model for light. It signalled the downfall of Newton's particle theory. By passing light from a single small source through two pinhole openings in an opaque screen, Young observed a series of light and dark fringes on a viewing screen. Later the effect was demonstrated using two slits to provide greater light intensity, Figure 38-1. (This figure is not to scale. Typically,

---

[1] We should mention that visible light, as well as all forms of electromagnetic radiation, possess a dual, apparently contradictory, nature. Light in transit seems to behave as a wave, but, as we will show in Chapter 42, when radiation is absorbed by matter *it always behaves as particles*. This dual nature of light—explainable as a wave in certain instances but as particles in other cases—is one of the central features of our understanding of matter and radiation.

[2] Thomas Young (1773–1829) was a brilliant English physician-scientist who contributed not only to the wave theory of light and the three-color theory of light perception, but also to Egyptology. It was largely through his efforts that the Rosetta stone, the key to Egyptian hieroglyphics, was deciphered.

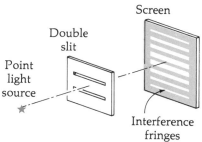 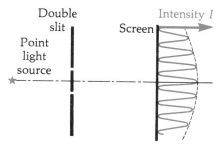 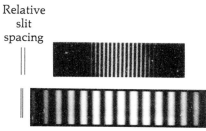

(a) A pictorial view showing the arrangement of the point light source, the double slit, and the screen.

(b) A schematic sketch that includes a graph of the intensity versus position on the screen.

(c) As the slit separation *decreases*, the distance between fringes *increases*.

FIGURE 38-1
The double-slit interference experiment.

the slit separation is about 0.5 mm and the screen is about 2 m from the slits, producing fringes about 2 mm apart.)

Unlike the phenomena of geometrical optics, the fringe pattern cannot be explained by a simple particle theory of light. The reason is the following. If we cover one of the slits, the result is a general illumination of the screen, as shown in Figure 38-2a. It does not matter which slit we cover; the screen illumination is the same broad pattern if only one slit is open. Now suppose that we uncover both slits. If light were simply a stream of particles, as Newton proposed, uncovering both slits should merely add the two individual patterns together to produce an overall intensity of twice the original value. Instead, the pattern of light and dark fringes is produced. Even more surprising, the intensity at the central axis is now *four* times the intensity of having just one slit open (instead of twice the intensity), so obviously the light passing through the two individual slits is not simply the addition of the two intensities.

The fringe pattern is explained as the superposition of two light *waves* that emerge from the slits and interfere with each other as they reach the screen. At some locations on the screen, the two waves arrive *in phase* and reinforce each other, producing an extra bright light. At other locations they arrive *out of phase* and cancel each other (see Figure 38-3). The *interference* of light waves produces the array of light and dark fringes called an *interference pattern*.

A requirement for producing an interference pattern is that the light from the two slits must be **coherent**:

COHERENCE   Two sources of light (or of any other type of waves) are coherent if they have the same wavelength and maintain a phase difference that remains constant in time.

If the two sources have different wavelengths, a constant phase difference between them is impossible. Thus, only *monochromatic* light—light of just a single frequency, and hence a precise wavelength[3]—can be coherent.

---

[3] In practice, no source of light is strictly monochromatic. But such a source can be approximated by a low-pressure, gas-discharge lamp that emits discrete colors, each involving a very narrow range of wavelengths. For example, the green line in the mercury spectrum (546.075 nm) has a wavelength range, or "line width," of about ±0.001 nm. The red line emitted by a helium–neon laser (632.8165 nm) has a line width of only about one part in 10^9.

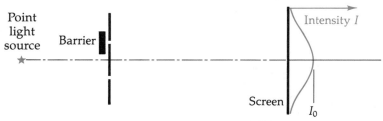

(a) With only *one slit* open, the illumination on the screen is diffuse, diminishing gradually in intensity for distances away from the center.

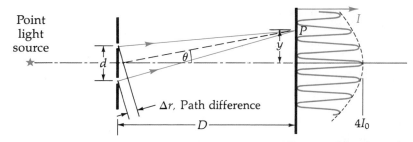

(b) With *both slits* open, the pattern on the screen is equally spaced bright and dark fringes.

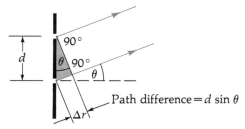

(c) If the screen distance is very much larger than y and d, we may consider the two rays as essentially parallel. With this approximation, the shaded triangle is a right triangle and the path difference Δr is equal to $d \sin \theta$.

FIGURE 38-2
A screen illuminated by coherent light from two slits produces an interference pattern.

(a) A portion of a *wave train* of infinite length.

(b) A wave train of finite length. From a low-pressure gas source, a visible-light wave train is about one meter long, containing several million wavelengths.

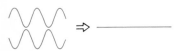

(c) Two light waves *in phase* superpose to produce an extra bright light (increased intensity). This is called *constructive* interference.

(d) Two light waves *180° out of phase* (with equal amplitudes) superpose to produce darkness (zero intensity). This is called *destructive* interference.

FIGURE 38-3
The result of the *superposition* of two coherent light waves depends strongly on the phase relation between the two waves. Other (constant) phase relations between the two cases illustrated in (c) and (d) are also possible, producing a resultant brightness between zero and maximum intensity.

As we will see in Chapter 44, light is emitted as a result of an energy transition in an atom. Each transition produces a single *wave train* of finite length, Figure 38-3b. When a single wave train illuminates *both* slits, the Huygens wavelets that emerge from the two slits are necessarily coherent since they are generated by the same wave train. The geometry of Figure 38-2 ensures that, for the wave train emitted by *each atom* in the point source, the phase difference between the two corresponding wavelets emerging from the slits is *always* zero. (With different geometry, this phase difference could be some other constant value.) *It is this constant phase difference (plus other geometry of the setup) that determines the fringe pattern on the screen.*

In contrast, if each slit had its own separate light source, there would be no interference pattern because light from the two slits would be emitted by different atoms and the light from one atom does not maintain coherence with the light from other atoms.[4]

---

[4] Laser light is different. As explained in Section 44.10, in a laser all the atoms are "locked together" in phase and frequency, so that the light in all parts of the beam is coherent.

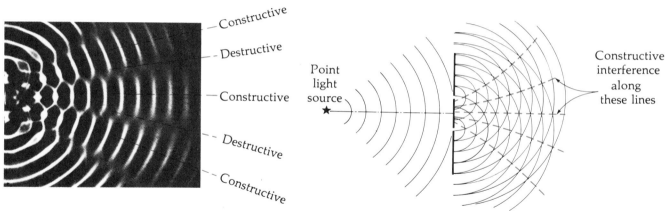

(a) Water waves spreading from two coherent sources produce a stationary pattern of constructive and destructive interference where the waves overlap.

(b) According to Huygens' principle, light waves emerging from the two slits spread out in all directions, causing interference in the region where they overlap. If the slits are located symmetrically from the source, the waves emerging from them have equal amplitudes, are in phase with one another, and are coherent.

FIGURE 38-4
Waves from coherent sources produce stationary interference patterns.

When two coherent light waves add together, they illustrate the following important principle:

PRINCIPLE OF LINEAR SUPERPOSITION[5] — **When two waves combine, the resultant wave amplitude at any given point is the sum of the instantaneous amplitudes that would be produced if each wave were present alone.**

The linear superposition principle is one of the most important principles in physical optics, as well as in other areas of physics.

Here is a summary of the criteria necessary for producing a stationary pattern of light-wave interference. We consider the superposition of two waves from the two slits in a Young's interference experiment:

CRITERIA FOR INTERFERENCE OF TRANSVERSE WAVES

(1) **The waves emerging from the two slits must be *coherent*. That is, the waves must have the same wavelength and a phase difference that remains constant in time.**
(2) **The electric field oscillations must be in the same direction so that the linear superposition principle applies.**

In the discussions that follow, we always assume that these criteria apply. A single point source behind the slits meets these criteria (Figure 38-4). If the source is equidistant from each slit, the light passing through the two slits will be *in phase* and of *equal amplitudes*, and their electric field oscillations are in the same direction.

---

[5] We do not discuss waves that add in a nonlinear fashion. Such nonlinear cases are often associated with very-large-amplitude waves.

We now investigate the details of the interference. Consider the rays leaving each of the slits shown in Figure 38-2b and arriving on the screen at the point P, a distance y from the center of the fringe pattern. The light from the lower slit will be out of phase with that from the upper slit because it travels a greater distance. As shown in Figure 38-2c the extra path distance Δr is essentially

$$\text{Path difference} \qquad \Delta r = d \sin \theta \qquad (38\text{-}1)$$

where d is the slit separation (center-to-center). This introduces a phase difference ϕ between the two waves when they arrive at the screen. The phase difference will be 2π rad for each wavelength λ in the distance Δr. That is,

$$\begin{array}{l}\text{Phase difference}\\ \text{(due to the extra} \\ \text{path length } \Delta r)\end{array} \qquad \phi = 2\pi\left(\frac{\Delta r}{\lambda}\right) \qquad \text{(in radians)} \qquad (38\text{-}2)$$

Note that ϕ is greater than 2π if Δr is greater than λ. The two light waves arriving at the point P on the screen may be represented by corresponding electric-field amplitudes,

$$E_1 = E_0 \sin \omega t \qquad (38\text{-}3)$$
and
$$E_2 = E_0 \sin(\omega t + \phi) \qquad (38\text{-}4)$$

where E_1 is the wave amplitude from the upper slit and E_2 is the wave amplitude from the lower slit. Note that E_1 and E_2 will be *in phase* for

$$\phi = m2\pi \qquad (38\text{-}5)$$

where $m = 0, 1, 2, 3, \ldots$, thus producing a resultant wave $E = 2E_0 \sin \omega t$.

For intermediate values of the phase difference ϕ, we can best obtain the resultant wave by using the mathematical technique of *phasors* (which we employed in Chapter 34 in the addition of alternating currents at an AC circuit junction). Equation (38-3) suggests that E_1 is the vertical projection of a phasor $\mathbf{E}_0$ that is rotating at an angular velocity ω, as in Figure 38-5a. Similarly, E_2 is the vertical projection of $\mathbf{E}_0$ that is rotating at the same angular velocity but is leading the phasor $\mathbf{E}_0$ of (a) by the angle ϕ. This is shown in Figure 38-5b. The sum E_3 of the *vertical projections*, E_1 and E_2, is then the sum of the waves:

$$E_3 = E_1 + E_2 = E_0 \sin \omega t + E_0 \sin(\omega t + \phi)$$

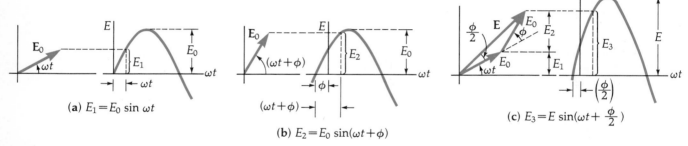

(a) $E_1 = E_0 \sin \omega t$

(b) $E_2 = E_0 \sin(\omega t + \phi)$

(c) $E_3 = E \sin(\omega t + \frac{\phi}{2})$

FIGURE 38-5
Phasor diagrams for two waves, E_1 and E_2, and their sum, $E_3 = E_1 + E_2$.

Figure 38-5c shows that the vector sum of the two rotating phasors E_0 shown in (a) and (b) produces a projection equal to E_3. The application of simple trigonometry for the sum gives (see Appendix D)

$$E_3 = \underbrace{2E_0 \cos \frac{\phi}{2}}_{\text{Amplitude}} \sin\left(\omega t + \frac{\phi}{2}\right) \qquad (38\text{-}6)$$

where $2E_0 \cos(\phi/2)$ is the projection of both phasors E_0 in the direction of E and $\omega t + \phi/2$ is the angle that E makes with the horizontal axis.

The intensity of the light is proportional to the square of the amplitude of the resultant wave.

$$I \propto \left(2E_0 \cos \frac{\phi}{2}\right)^2 = 4E_0^2 \cos^2 \frac{\phi}{2}$$

Expressed in terms of the intensity I_0 at the central maximum ($\phi = 0$), the intensity at other locations is

$$I = I_0 \cos^2 \frac{\phi}{2} \qquad (38\text{-}7)$$

Thus a maximum occurs for $(\phi/2) = m\pi$, or $\phi = m2\pi$, which is consistent with our earlier observation, Equation (38-5).

Summarizing this relation, we see that the location of the *maxima* in the intensity pattern (that is, the centers of the bright fringes) will occur when the path difference Δr is an *integral* number of wavelengths:

DOUBLE-SLIT INTERFERENCE PATTERN   *Maxima* (bright fringes) $m\lambda = d \sin \theta$ $\qquad (38\text{-}8)$
(where $m = 0, 1, 2, 3, \ldots$)

Similarly, the *minima* (the centers of the dark fringes) occur for a path difference of a *half-integral* number of wavelengths:

Minima (dark fringes) $\qquad (m + \tfrac{1}{2})\lambda = d \sin \theta \qquad (38\text{-}9)$
(where $m = 0, 1, 2, 3, \ldots$)

In practice, the distance $D \gg y$, so that $\sin \theta \approx \tan \theta = \theta$. Using the tangent approximation, we may write the above two equations as

Maxima $\qquad m\lambda = d\left(\dfrac{y}{D}\right) \qquad (38\text{-}10)$

Minima $\qquad \left(m + \dfrac{1}{2}\right)\lambda = d\left(\dfrac{y}{D}\right) \qquad (38\text{-}11)$

(small-angle approximation)

(where $m = 0, 1, 2, 3, \ldots$)

The central bright fringe is called the *zero-order* fringe ($m = 0$). As we move away on either side of the central maximum, successive bright fringes are the *first-order* fringes ($m = \pm 1$), the *second-order* fringes ($m = \pm 2$), and so on. A characteristic feature of double-slit interference is that, as the separation between the slits *decreases*, the distance between the fringes *increases*. The first experimental determination of the wavelength of light was made by Young using this double-slit method.

EXAMPLE 38-1

In a double-slit experiment using light of wavelength 486 nm, the slit spacing is 0.60 mm and the screen is 2 m from the slits. Find the distance along the screen between adjacent bright fringes.

SOLUTION

Assuming the small-angle approximation, Equation (38-10) gives the location y of the mth maximum:

$$m\lambda = d\left(\frac{y}{D}\right)$$

The separation between adjacent maxima is then

$$y_{m+1} - y_m = \frac{\lambda D}{d}[(m+1) - m]$$

$$y_{m+1} - y_m = \frac{(486 \times 10^{-9} \text{ m})(2 \text{ m})}{(0.60 \times 10^{-3} \text{ m})}[1] = 1.62 \times 10^{-3} \text{ m}$$

$$= \boxed{1.62 \text{ mm}}$$

Because this is such a small distance relative to the slit-to-screen distance, the small-angle approximation is justified.

EXAMPLE 38-2

Consider the situation shown in Figure 38-6. The source illuminates the slits with green light from a mercury lamp ($\lambda = 546$ nm). The screen is $D = 1$ m from the slits, and the slit separation d is 0.30 mm. (a) Find the intensity I of the light at a distance $y = 1$ cm from the center of the pattern relative to the intensity of the central fringe maximum I_0. (b) Find the number of bright fringes between the central fringe and the point y.

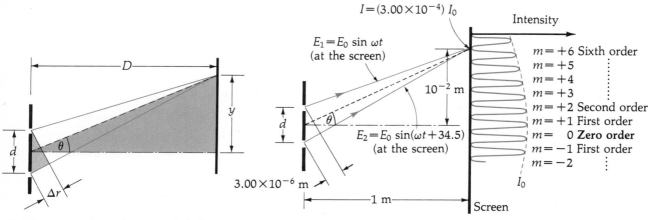

(a) For cases where $D \gg y$ and d, the two shaded triangles are similar.

(b) The numerical values. The sizes of the slit separation and fringe pattern are greatly exaggerated for clarity.

FIGURE 38-6
Example 38-2.

SOLUTION

(a) To find the difference in phase between the waves originating at the upper and lower slits shown in Figure 38-6, we first find the path difference Δr. Within the approximation of Figure 38-2c, and recognizing that $\sin\theta \approx \tan\theta$, we find that the shaded triangles in the figure are similar. Therefore, corresponding sides of the triangles are proportional. Thus, $\Delta r/y = d/D$, or

$$\Delta r = \frac{d}{D}y = \left(\frac{0.30 \times 10^{-3} \text{ m}}{1 \text{ m}}\right)(1 \times 10^{-2} \text{ m}) = 3.00 \times 10^{-6} \text{ m}$$

Equation (38-2) yields the phase angle in terms of the wavelength and the distance Δr:

$$\phi = 2\pi\left(\frac{\Delta r}{\lambda}\right) = 2\pi\left(\frac{3 \times 10^{-6} \text{ m}}{5.46 \times 10^{-7} \text{ m}}\right) = 34.523 \text{ rad}$$

Applying Equation (38-7) gives

$$I = I_0 \cos^2 \frac{\phi}{2} = \boxed{(2.98 \times 10^{-4})I_0}$$

This answer indicates that the point y lies very near a point of minimum intensity.

(b) As we move along the screen away from the central fringe, the path difference Δr increases. As Δr increases by one full wavelength, the two waves from the slits are again in phase, corresponding to moving from the central bright fringe to the adjacent fringe, and so on. How many wavelengths are there in the total path difference $\Delta r = 3.00 \times 10^{-6}$ m?

$$\frac{\Delta r}{\lambda} = \frac{3.00 \times 10^{-6} \text{ m}}{5.46 \times 10^{-7} \text{ m}} = 5.49 \text{ wavelengths}$$

Thus 5 bright fringes will exist between the central maximum and the point y. The remaining 0.49 wavelength indicates that the waves from the upper and lower slits are nearly π rad out of phase, which is consistent with the answer in part (a).

An alternative approach is to determine the number of times the phase angle ϕ is divisible by 2π. As we move away from the central fringe, each increase of 2π in the phase angle corresponds to moving from one bright fringe to the next. Thus, for $\phi = 34.5$ rad:

$$\frac{\phi}{2\pi} = \frac{34.5 \text{ rad}}{2\pi} = 5.49 \text{ multiples of } 2\pi$$

Thus, we conclude that 5 bright fringes appear between the central bright fringe and the point $y = 1$ cm. The solution to this example is summarized in Figure 38-6b.

We may also produce a phase difference between the light waves emitted from a double slit by introducing a transparent material with a different refractive index into the path of one of the waves (see Figure 38-7). Inserting a refractive material of thickness b and refractive index n increases the number of wavelengths in that path. If the wavelength in air is λ_a, a distance b (in air)

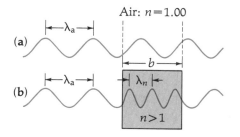

FIGURE 38-7
A phase difference may be produced by inserting a material with refractive index n in the path of a light wave.

contains b/λ_a wavelengths. In a material of refractive index n, the wavelength is shorter: $\lambda_n = \lambda_a/n$. Thus the same distance b contains b/λ_n wavelengths, or

$$\frac{b}{\left(\dfrac{\lambda_a}{n}\right)} = n\left(\frac{b}{\lambda_a}\right) \text{ wavelengths}$$

The *increase* in number of wavelengths is therefore

$$\left(\frac{nb}{\lambda_a} - \frac{b}{\lambda_a}\right) = \frac{b}{\lambda_a}(n-1) \tag{38-12}$$

Since a phase difference of 2π corresponds to each full wavelength increase, the phase difference ϕ is

Phase difference
due to inserting in one
path a material of thickness
b and refractive index n
$$\phi = 2\pi\left(\frac{b}{\lambda_a}\right)(n-1) \tag{38-13}$$
(where λ_a is the wavelength in air)

The following example illustrates an interference pattern's sensitivity to small changes in the refractive index associated with one of the light paths.

EXAMPLE 38-3

Consider the double-slit arrangement shown in Figure 38-8, where the separation d of the slits is 0.30 mm and the distance D to the screen is 1 m. A very thin sheet of transparent plastic, with a thickness of $b = 0.050$ mm (about the thickness of this page) and a refractive index of $n = 1.50$, is placed over only the upper slit. As a result, the central maximum of the interference pattern moves upward a distance y'. Find this distance.

SOLUTION

The central maximum corresponds to zero phase difference. Thus the added distance Δr traveled by the light from the lower slit must introduce a phase

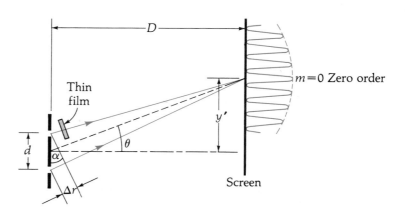

FIGURE 38-8
Example 38-3.

difference equal to that introduced by the plastic film. The *phase difference* ϕ is given by Equation (38-13):

$$\phi = 2\pi \left(\frac{b}{\lambda_a}\right)(n-1)$$

The corresponding difference in *path length* Δr is, from Equation (38-2),

$$\Delta r = \phi\left(\frac{\lambda_a}{2\pi}\right) = 2\pi\left(\frac{b}{\lambda_a}\right)(n-1)\left(\frac{\lambda_a}{2\pi}\right) = b(n-1)$$

Note that the wavelength of the light does not appear in this equation. In Figure 38-8 the two rays from the slits are essentially parallel, so the angle θ may be expressed as

$$\tan\theta = \frac{\Delta r}{d} = \frac{y'}{D}$$

Equating these expressions and solving for y' gives

$$y' = \Delta r\left(\frac{D}{d}\right) = \frac{b(n-1)D}{d}$$

$$y' = \frac{(5 \times 10^{-5}\text{ m})(1.50-1)(1\text{ m})}{(3 \times 10^{-4}\text{ m})} = 0.0833\text{ m} = \boxed{8.33\text{ cm}}$$

38.3 Multiple-Slit Interference

The mathematical technique of phasors developed in the last section for a double-slit interference pattern is easily extended to include interference of light from three or more slits. The physical arrangement of the slits is similar to that for a double-slit pattern. However, as shown in Figure 38-9, the pattern is quite different, composed of alternating *major* and *minor* maxima. (They are also called *primary* and *secondary* maxima.) The major maxima are spaced in the same way as the double-slit maxima if adjacent slits have the same separation distance d *between the centers of the slits.*

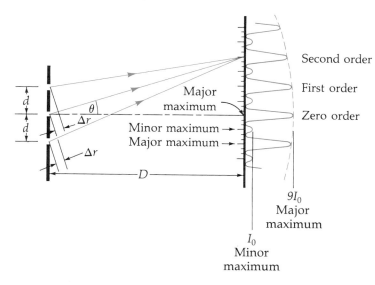

FIGURE 38-9
Triple-slit interference. On the screen, each marked interval from the center of the pattern corresponds to a phase difference between waves from adjacent slits of $\pi/3$ rad. (The slit separations and fringe pattern are greatly enlarged for clarity.)

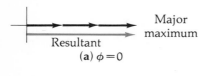
(a) $\phi = 0$ — Resultant, Major maximum

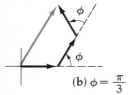
(b) $\phi = \dfrac{\pi}{3}$

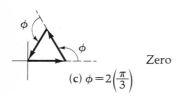
(c) $\phi = 2\left(\dfrac{\pi}{3}\right)$ — Zero

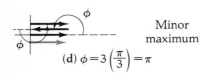
(d) $\phi = 3\left(\dfrac{\pi}{3}\right) = \pi$ — Minor maximum

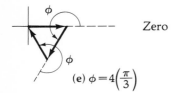
(e) $\phi = 4\left(\dfrac{\pi}{3}\right)$ — Zero

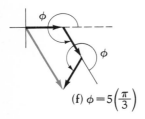
(f) $\phi = 5\left(\dfrac{\pi}{3}\right)$

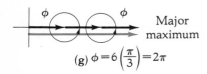
(g) $\phi = 6\left(\dfrac{\pi}{3}\right) = 2\pi$ — Major maximum

FIGURE 38-10
A series of phasor diagrams for triple-slit interference. Each successive diagram represents an additional phase delay of $\phi = \pi/3$ rad between *adjacent* phasor components.

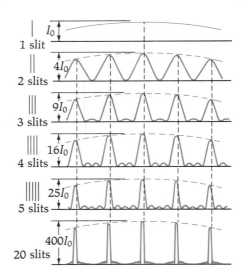

FIGURE 38-11
Multiple-slit interference patterns. As the number of slits is increased (with the slit separation kept constant), the major maxima remain fixed in position as they become narrower and more intense. The intensity of the sharp peaks increases with the square of the number of slits. (Note the changes in vertical scale.)

The development of the triple-slit interference pattern by the use of phasors is shown in Figure 38-10. The central maximum corresponds to the addition of three electric phasors, all in phase, as in Figure 38-10(a). As the distance from the central maximum increases, the phase angle ϕ between the electric phasors from adjacent slits also increases, in increments of $\pi/3$. Note that one *minor* maximum (d) occurs between each of the *major* maxima [(a) and (g)].

As the number of slits increases, the number of minor maxima between major maxima also increases, as illustrated in Figure 38-11. The number of these minor peaks is always *two less* than the total number of slits in the array. Furthermore, *as the number of slits increases, these minor peaks are suppressed in intensity, while the major maxima become much more intense and also much narrower.* Since the *positions* of the *major* maxima depend only on the slit separation d (and not on the number of slits), Equation (38-8) expresses the location of the major maxima for any number of slits.

MULTIPLE-SLIT INTERFERENCE PATTERN (major maxima)

$$d \sin \theta = m\lambda \qquad (\text{where } m = 0, 1, 2, \ldots) \qquad (38\text{-}14)$$

38.4 Interference Produced by Thin Films

We have all enjoyed a beautiful display of colors from a thin oil film on the surface of a puddle, or the colored reflection of light from the surface of a soap bubble. Both of these phenomena result from the interference of light.

Consider a thin film of refractive material such as glass. Figure 38-12 illustrates the observation of white light reflected from two different places, A and B, on the film. At both places, the light reaching the eye of the observer is a combination of light reflected from the top surface and from the lower

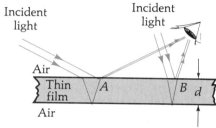

FIGURE 38-12
Interference of light reflected from a thin film.

surface. In each case, these two light waves interfere with each other, reinforcing certain wavelengths and canceling others, depending on the particular path difference between them. For example, suppose that, at A, wavelengths in the red portion of the spectrum undergo destructive interference. Then the observer will see a predominantly blue-green reflection at that location. On the other hand, if at B, where the path difference is shorter, blue wavelengths experience destructive interference, the observer will see a predominantly reddish reflection. In this manner, an entire rainbow of colors is often reflected from various portions of the film.

Light reflected from a very thin soap film also shows another interesting feature. Generally, a freshly blown soap bubble displays a swirl of reflected colors when viewed against a dark background. This is partially due to the nonuniform refractive index of the soap solution as well as the varying thickness of the film. However, if we continue to watch a soap film supported vertically, as in Figure 38-13, the various colors gradually sort themselves into horizontal rainbow stripes, slowly compressing together toward the bottom. This happens because the action of gravity drains fluid from the upper portion of the film, causing it to be thinner at the top than at the bottom. But now a surprising effect occurs. As the top part of the film becomes much thinner than a wavelength of visible light, *no light at all is reflected from the film*. It has become invisible! The reason is that light reflected from the front and back surfaces interferes *destructively* because of a phase change of π rad (180°) that occurs at one surface and not the other. A detailed analysis of the reflection of light from refractive materials shows that

FIGURE 38-13
Interference of white light reflected from a thin vertical film of soap solution. Gravity pulls the fluid downward, causing the film to become very thin near the top. If the thickness changes uniformly from top to bottom, horizontal bands of interference colors are produced as shown here. When the upper part of the film becomes sufficiently thin, the path difference between reflections from the front and rear surfaces approaches zero. Because the front reflection is shifted by 180° and the rear reflection is unshifted, with a sufficiently thin film their combination produces destructive interference for all reflected wavelengths of visible light, and the top segment of the film becomes invisible.

PHASE CHANGE UPON REFLECTION

(1) **When light traveling in a given medium reflects from another medium of *higher* refractive index, it undergoes *a phase change of π rad (180°)*.**
(2) **When light traveling in a given medium reflects from another medium of *lower* refractive index, *no phase change* occurs.**

Reflections from the front and back surfaces of the soap film are of these two different types, so the reflections alone introduce a 180° phase difference. Thus, as the film thickness shrinks toward zero, making the path differences negligible, the two reflected rays become 180° out of phase because of the different types of reflections and undergo destructive interference. If you observe reflections from a soap bubble against a dark background and watch carefully as the bubble ages, you will see the color contrasts diminish. Then, just before the film breaks, no light is reflected from the spot where the break originates.

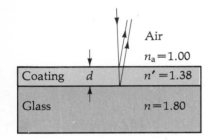

FIGURE 38-14
Example 38-4.

EXAMPLE 38-4

Nonreflecting coatings for camera lenses reduce the loss of light at various surfaces of multiple-lens systems, as well as prevent internal reflections that might mar the image. Find the minimum thickness of a layer of magnesium fluoride ($n' = 1.38$) on flint glass ($n = 1.80$) that will cause *destructive* interference of reflected light of wavelength $\lambda = 550$ nm near the middle of the visible spectrum. Consider normal incidence on the coating.

SOLUTION

In Figure 38-14, both rays reflect from a medium of higher refractive index than the medium they are traveling in, so *both* undergo a phase shift of π rad upon reflection. Therefore, the only factor contributing to a net phase shift is the extra path length of one ray. For destructive interference, the (minimum) round trip distance $2d$ should be $\lambda_{n'}/2$, where $\lambda_{n'} = \lambda_a/n'$ is the wavelength *in the coating*. Thus, $2d = \lambda_a/2n'$. Solving for d gives

$$d = \frac{\lambda_a}{4n'} = \frac{(5.50 \times 10^{-7}\text{ m})}{4(1.38)} = \boxed{99.6 \text{ nm}}$$

Though such coatings are very thin (approximately a hundred atomic diameters thick), they are easily applied by evaporating the magnesium fluoride and allowing it to condense on the glass surface. For complete destruction, the *amplitudes* of the two reflected rays must be equal. We can show that this is true only if *the refractive indices of the coating (n') are the geometric mean between the refractive indices of the materials on either side of the coating.* For air, $n_a = 1$, and we have

NONREFLECTIVE COATINGS
$$n' = \sqrt{nn_a} = \sqrt{n} \qquad (38\text{-}15)$$

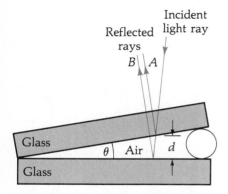

(a) A small wire (whose cross-section is greatly exaggerated) separates the glass plates at one edge.

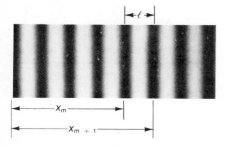

(b) The interference fringes seen by reflected light. A dark fringe occurs at the point of contact of the plates because of the 180° phase shift for one of the reflections.

FIGURE 38-15
The interference pattern produced by a wedge of air between two glass plates. The angle θ between the plates is greatly exaggerated to emphasize the variation in thickness of the air wedge.

Thin Wedges

Consider two glass plates that are in contact at one edge and separated slightly at the opposite edge by a hair or a small wire between the plates. A side view of such an arrangement is shown in Figure 38-15. Parallel, monochromatic light rays incident downward are reflected from the two surfaces of the wedge[6] back to an observer above the plates. The reflected light is thus composed of a combination of light ray A reflected from the lower surface of the top plate (no phase shift) and light ray B reflected from the upper surface of the lower plate (phase shift π). Destructive interference (dark fringes) thus occurs if the extra distance traveled by ray B ($2d$ for the round trip) is an *integral* number of wavelengths λ. That is,

Dark fringes $\qquad 2d = m\lambda \qquad$ (where $m = 1, 2, 3, \ldots$ $\qquad$ (38-16)
(air wedge) $\qquad\qquad\qquad\qquad$ and $d =$ plate separation)

---

[6] Reflections from other pairs of surfaces may be ignored. When the two surfaces are relatively far apart, or the angle becomes appreciable, the interference fringes are so close together that the eye cannot resolve them. (Exception: highly coherent parallel laser light reflected from almost parallel surfaces will produce visible fringes, even though the surfaces are far apart. The light, however, must remain coherent over the path length difference of the two rays.)

Note that this equation is the condition for *destructive* interference and includes the phase shift that occurs in one of the reflections.

If the glass plates have plane surfaces, the interference pattern is a series of equally spaced bright and dark fringes. As we proceed from one dark fringe to the next, the air wedge increases in thickness by $\lambda/2$ (making the *round-trip* path increase by λ). The separation ℓ of adjacent dark fringes is found as follows. In traveling along the plate a distance ℓ, the wedge increases by $\lambda/2$. Therefore, the tangent of the wedge angle θ is $\tan\theta = (\lambda/2)/\ell$. Since θ is ordinarily a very small angle, we may substitute $\tan\theta = \theta$ (in radians) to obtain, for ℓ,

$$\ell = \frac{\lambda}{2\theta} \quad (38\text{-}17)$$

The flatness of a glass surface is often determined by the interference pattern produced when it is placed in contact with an *optical flat*, a surface known to be flat to within a small fraction of a wavelength of light (see Figure 38-16). A *dark* fringe is located at the region where the two surfaces touch because of the 180° phase shift that occurs for (only) one of the two reflections.

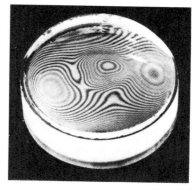

(a) A wavy fringe pattern indicates an uneven surface. Three "high" (or "low") spots are revealed by the regions of circular fringes.

(b) The surfaces are in contact at one edge and separated a small amount at the opposite edge. The regularly spaced bright and dark fringes indicate that the surface is uniformly flat.

FIGURE 38-16
We can test the flatness of a glass surface by placing it in contact with an *optical flat* and observing the interference pattern of reflected monochromatic light.

EXAMPLE 38-5

Suppose two flat glass plates 30 cm long are in contact along one end and separated by a human hair at the other end, as indicated in Figure 38-15. If the diameter of the hair is 50 μm, find the separation of the interference fringes when the plates are illuminated by green light, $\lambda = 546$ nm.

SOLUTION

The angle of the air wedge between the plates is $\theta = D/L$ (in radians), where D is the diameter of the hair and L is the length of the plates. Substituting this expression for θ into Equation (38-18), we obtain

$$\ell = \frac{\lambda}{2\theta} = \frac{L\lambda}{2D} = \frac{(0.3\text{ m})(5.46\times 10^{-7}\text{ m})}{2(5.0\times 10^{-5}\text{ m})} = 1.64\times 10^{-3}\text{ m} = \boxed{1.64\text{ mm}}$$

Newton's Rings

When illuminated from above, a plano-convex lens placed on an optical flat produces a circular interference pattern known as *Newton's rings* (Figure 38-17). The thickness d of the air wedge between the lens and the flat glass plate is related to the radius of curvature R of the lens surface and the distance r from the center of the pattern. Applying the Pythagorean theorem, we obtain $R^2 = r^2 + (R-d)^2 = r^2 + R^2 - 2Rd + d^2$. Since the radius of curvature R of the lens is much greater than the thickness d of the air wedge, we ignore the d^2 term and obtain $2d \approx r^2/R$. The condition for *destructive* interference exists when the extra (round-trip) path $2d$ for the ray reflected from the bottom surface is an *integral* number of wavelengths (because of the 180° phase change for one of the reflections):

Dark fringes (air wedge) $2d = m\lambda$ (where $m = 0, 1, 2, 3, \ldots$)

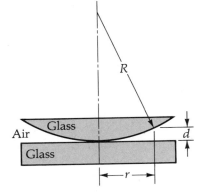
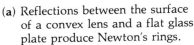
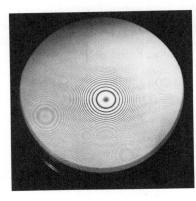

FIGURE 38-17
Newton's rings. [Note: the faint patterns in (b) are spurious.]

(a) Reflections between the surface of a convex lens and a flat glass plate produce Newton's rings.

(b) Photograph of Newton's rings obtained with monochromatic light.

Equating these two values for $2d$, we obtain an expression for r_m, the radius of the mth ring:

RADII OF NEWTON'S RINGS
$$r_m = \sqrt{Rm\lambda} \qquad (m = 0, 1, 2, 3, \ldots) \tag{38-18}$$

As we proceed outward from one dark ring to the next, the radii r_m increase in size with $\sqrt{m}$, becoming closer together. One of the most interesting aspects of this interference pattern is that the area between each of the successive circles is a constant.

38.5 The Michelson Interferometer

The Michelson[7] interferometer is an ingenious device that utilizes the interference of light to measure distances, or *changes* of distance, with great accuracy. The basic components, shown in Figure 38-18, include an *extended* light source, such as a ground-glass screen illuminated uniformly from behind with monochromatic light. (The reason a *point* source is unsatisfactory will be evident after we discuss the origin of the interference pattern.) Light from the source falls on a thinly silvered, semitransparent mirror at 45°, an angle that reflects half the light to mirror M_1 and transmits half to mirror M_2. Light reflected from M_1 and M_2 eventually merges together at the eye or other detector (minus, of course, that part further diverted by the 45° mirror). If we straighten out the several right-angle deflections caused by the 45° mirror, the situation is

FIGURE 38-18
The basic components of a Michelson interferometer. The clear glass slab C is called a *compensating plate*. It has the same dimensions and orientation as the 45° mirror in order to make the light paths in glass equal along the two arms, a condition necessary when a white-light source is used.

[7] Albert Michelson (1852–1931) was the son of Polish immigrants who were somewhat poor, and his prospects for education beyond high school were not promising. However, when his application to the U.S. Naval Academy was turned down, he shrewdly arranged to meet President Grant "by chance" while the President was walking his dog on the White House grounds. Michelson so highly impressed the President with his determination that a special appointment to Annapolis was granted. After graduating in 1873, Michelson became a physics and chemistry instructor at the Academy, where he began a lifelong interest in precision measurements of the speed of light. He then became a professor of physics at Case Institute of Technology, where he improved his earlier interferometer experiments on the ether drift, this time with a collaborator, Edward Morley, a chemist at nearby Western Reserve. Michelson was keenly disappointed in the null result: he would much have preferred to report a finite velocity through the ether, and he felt that the absence of a positive value was somehow due to an unknown defect in his method. In 1907, for his work on light, Michelson became the first American to win the Nobel prize.

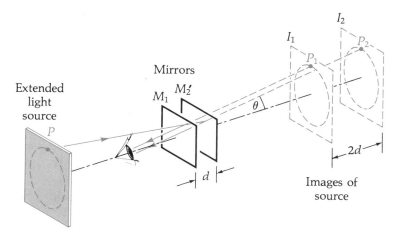

FIGURE 38-19
The origin of the circular fringes in a Michelson interferometer. In this figure, the right-angle deflections produced by the thinly silvered mirror at 45° have been straightened out. Mirror M_1 is observed directly (through the 45° mirror), while M_2' is the virtual image of M_2 produced by reflection in the 45° mirror. These mirrors form two images, I_1 and I_2, of the extended light source. Light waves from corresponding points in these images are *coherent*.

essentially as shown in Figure 38-19. The extended source is reflected by the two mirrors, forming two images of the source, I_1 and I_2. The mirrors can be aligned so that the two images are parallel. If the distance between M_1 and M_2 is d, the images are separated by a distance of $2d$.

The significant feature of these images is that *light waves from corresponding points in the images are coherent*. These waves come from a wave train emitted from a *single atom* in the source at point P. Thus, the light waves that enter the eye from the image points P_1 and P_2 are coherent and they will interfere. The phase relation between the two rays depends on their path difference, $2d \cos \theta$. The *in-phase* condition for *bright* fringes is

$$2d \cos \theta = m\lambda$$

When the two image planes are parallel, all corresponding points on a circle surrounding the central axis have the same phase relationship, producing an overall fringe pattern of concentric circles similar to Newton's rings. If one mirror is moved by $\lambda/2$, the path difference changes by λ and we are again at an in-phase condition: each fringe has moved to the position previously occupied by the adjacent fringe. This shifting of the fringe pattern enables us to observe tiny motions. (For example, if one of the interferometer arms is arranged vertically and a small mirror is attached to a mushroom, the growth rate of the mushroom can be accurately observed as fringes sweep past, usually at the rate of about one per second!) Slowly moving one mirror continuously in the same direction causes the circular fringes to shrink in size and vanish at the center (or, for the other direction, to expand from the center). As d approaches zero, the path differences for *all points* approach zero and the entire field of view thus becomes bright (or dark), depending on the *net* phase change due to reflections at the various glass surfaces. If one mirror is tilted slightly, the separation of the image planes becomes a thin wedge. In effect, this moves the center of the fringe pattern off to one side, so we now see an array of slightly curved, almost parallel bright and dark fringes that are part of the ring pattern far away from the center (Figure 38-21).

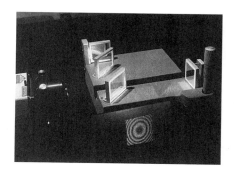

FIGURE 38-20
Light from the laser at the left passes through a small Michelson interferometer to produce the bull's-eye image on the ground glass screen in the foreground.

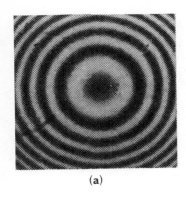

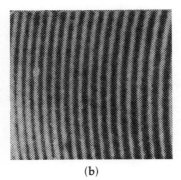

FIGURE 38-21
The interference patterns seen in a Michelson interferometer are similar to Newton's rings. In (a), the image planes are parallel (Figure 38-19) and the pattern is a series of concentric circles, whose overall size depends on the separation of the image planes. In (b), the image planes are not parallel and the pattern is a series of curved, almost parallel lines.

With monochromatic light, the fringe pattern remains sharp for path differences of 10 cm or more. However, an interferometer may also be used with *white* light, provided the path difference 2d is no more than a few wavelengths and the field of view is near the center of the pattern. With a range of wavelengths between 400 nm and 700 nm, the spacing between fringes varies for different colors; hence each bright ring for the monochromatic case becomes a spreadout rainbow of colors. Beyond about a dozen fringes from the center, the patterns overlap so much that they fade out to produce essentially white illumination. The interference is still taking place for each individual color, however, as we can verify by viewing the pattern with a filter that allows only one color to pass through. In some applications, it is necessary to have a reference position that can be found again, even though the mirror has been moved in the meantime. This can be accomplished with white-light fringes, because there is a unique, *color-free*, all-bright (or all-dark) field of view when d is exactly zero. It thus serves as a fixed reference position that can be repeatedly reached at will to within a fraction of a wavelength of visible light.

An important early use Michelson made of the interferometer was to determine the length of the then-standard meter bar in Paris in terms of the wavelengths of certain spectral lines of cadmium, counting the number of fringes that swept by as one mirror was moved along the meter bar. Based upon recent refinements in measuring the speed of light, on October 20, 1983, the Seventeenth Conférence Générale des Poids et Mesures adopted the new definition:

THE METER **The meter is the length of the path traveled by light in vacuum during a time interval of 1/299 792 458 of a second.**

Another historic use of the interferometer was in the Michelson–Morley experiment in 1887, an attempt to determine motion of the earth through the hypothetical medium, the *ether*, whose existence was believed necessary to propagate light waves. The inability to detect such motion—the famous *null result*—not only seemed exceedingly paradoxical but also was a profound blow to ether theories. The dilemmas posed by these experimental results were not resolved until Einstein presented his special relativity theory in 1905 (Chapter 41).

There is an almost endless list of applications for the Michelson interferometer, particularly when laser light is used as a source. Instruments using microwaves or other portions of the electromagnetic spectrum have also been constructed. The interferometer has proved to be a versatile and extremely precise measuring instrument, helpful in all areas of science and technology.

Summary

Coherent light waves have a phase difference that remains constant in time. When two different portions of a wave train (emitted from a single atom) are combined, they are coherent and they *interfere*. The sum of two waves

$$E_1 = E_0 \sin \omega t \quad \text{and} \quad E_2 = E_0 \sin(\omega t + \phi)$$

is

$$E_1 + E_2 = 2E_0 \cos \frac{\phi}{2} \sin\left(\omega t + \frac{\phi}{2}\right)$$

The *phase difference* ϕ may result from three effects:

(1) A *difference in path length* Δr of the waves:

$$\phi = 2\pi \left(\frac{\Delta r}{\lambda}\right) \quad \text{(in radians)}$$

(2) Placement of a material with refractive index n and thickness b in the path of one of the

waves:

$$\phi = \frac{2\pi b}{\lambda_a}(n-1) \quad \text{(in radians)}$$

(3) Reflections that the two waves may undergo:
   (a) A phase change of π rad (180°) occurs for a wave traveling in one medium when that wave is reflected from a medium of *higher* refractive index.
   (b) No phase change occurs on reflection from a medium of *lower* refractive index.

Double-slit interference:

d = slit separation (center-to-center)
D = slit-to-screen distance
y = distance along the screen from the central maximum
m = order

The *maxima* are given by

$$m\lambda = d \sin\theta \quad \text{(where } m = 0, 1, 2, 3, \ldots\text{)}$$

For small angles: $\quad m\lambda = d\left(\dfrac{y}{D}\right)$

The *minima* are spaced halfway between the bright fringes:

$$(m + \tfrac{1}{2})\lambda = d \sin\theta \quad \text{(where } m = 0, 1, 2, 3, \ldots\text{)}$$

For small angles: $\quad (m + \tfrac{1}{2})\lambda = d\left(\dfrac{y}{D}\right)$

Multiple-slit interference: For the same slit separation, the major maxima are the same as for the double-slit case.

The *maxima* are given by

$$m\lambda = d \sin\theta \quad \text{(where } m = 0, 1, 2, 3, \ldots\text{)}$$

For small angles: $\quad m\lambda = d\left(\dfrac{y}{D}\right)$

The major characteristics of multiple-slit interference may be summarized as follows:

(1) The angular separation of major maxima depends on the phase difference of waves from adjacent slits, not on the number of slits.
(2) The number of minor maxima between major maxima is two less than the number of slits.
(3) The sharpness and intensity of major maxima increases as the number of slits increases.

Interference patterns produced by thin film and wedges depend on the phase difference (upon recombination) of waves reflected from the two surfaces. Phase differences are due to the extra path length (round-trip) for one of the waves and the different types of reflections at the two surfaces.

The *Michelson interferometer* is an ingenious and versatile instrument capable of measuring distances to within a small fraction of a wavelength of light.

Questions

1. Why is it impossible for all the fringes of a double-slit interference pattern to be of exactly the same intensity?
2. Would longitudinal waves such as sound waves produce double-slit interference effects?
3. Two closely spaced parallel fluorescent light tubes, both covered with a green filter, illuminate a distant wall. Is an interference pattern produced?
4. Our discussion of double-slit interference was based on a plane light wave falling with normal incidence upon a screen containing two slits. What changes in the interference pattern would we observe if the screen containing the slits were tilted relative to the incident light? Consider tilting about an axis parallel to the slits and about an axis perpendicular to the slits.
5. Describe the interference pattern produced by two closely spaced pinholes.
6. If a pure tone is sounded in a room, a listener experiences large changes in intensity by moving his head from side to side. Is this an interference phenomenon? Why is the effect less pronounced when music is heard?
7. Suppose a double-slit experiment is immersed under water. What changes, if any, occur in the pattern of fringes on the screen?
8. In a Young's double-slit experiment, the lower halves of the two vertical slits are covered with a blue filter and the upper halves are covered with a red filter. (a) What is the appearance of the resultant interference pattern observed on a screen? (b) Suppose, instead, that one slit is covered with a blue filter and the other slit is covered with a red filter. Describe the pattern and explain the reasoning behind your conclusions.
9. How would a triple-slit interference pattern be altered if the center slit were covered by a gray filter to reduce the intensity of the light emanating from that slit?
10. An oil slick on water seems brightest where the oil film is much thinner than a wavelength of visible light. Is the refractive index of the oil greater or less than that of water?
11. A lens is coated to reduce reflection. What happens to the light energy that had previously been reflected?
12. When looking at the light reflected from a windowpane, why do we not observe an interference pattern? After all, light is reflected by both the front and the rear surfaces of the glass.
13. Why do coated camera lenses look purplish when we observe them by reflected light?

14. Suppose we use reflected white light to observe a thin, transparent coating on glass as the coating material is gradually being deposited by evaporation in a vacuum. Describe possible color changes that occur during the process of building up the thickness of the coating.

15. Consider two glass plates in contact at one edge and separated slightly at the opposite edge. In analyzing the visual appearance of the interference pattern produced by reflections from the "air wedge" between the plates, why can we ignore interference between waves reflected from the top surface of the top plate and the bottom surface of the bottom plate, even if the plates have perfectly parallel surfaces?

16. What change, if any, would occur in the pattern of Newton's rings if the space between the lens and the plate were filled with water?

17. Could an *acoustical* Michelson interferometer be used to measure the wavelength of ultrasonic sound waves? If so, how would such an interferometer be constructed and what procedure would be used in the measurement?

Problems
38.2 Double-Slit Interference

38A-1 Light of wavelength 600 nm illuminates a double slit with a slit separation of 0.30 mm. An interference pattern is produced on a screen 2.5 m from the slits. Calculate the separation of the interference fringes on the screen near the central maximum.

38A-2 Design a double-slit system that will produce fringes 2 mm apart on a screen 3 m away using light of 550 nm.

38A-3 In a double-slit experiment, sodium light ($\lambda = 589$ nm) produces fringes spaced 1.8 mm apart on a screen. Find the fringe spacing when mercury light ($\lambda = 436$ nm) is used.

38B-4 Light composed of two different wavelengths illuminates a double slit, forming two interference patterns that are superimposed on a screen. The fifth-order maximum of one color falls exactly at the location of the third-order maximum of the other color. Calculate the ratio of the two wavelengths.

38B-5 Two waves that differ only in phase are described by $E_1 = E_0 \sin(kx - \omega t)$ and $E_2 = E_0 \sin(kx - \omega t + \phi)$. Show that the linear combination of these waves produces $E_3 = E_1 + E_2 = 2E_0 \cos(\phi/2) \sin(kx - \omega t + \phi/2)$. Hint: refer to Figure 38-5.

38B-6 A double-slit interference pattern has a distance y_0 between the maxima. (a) Sketch a phasor diagram describing the wave amplitude E at a distance $y_0/4$ from the central maximum. (b) What is the intensity I at this position relative to the intensity maximum I_0 at the central peak?

38B-7 A glass plate 0.4 mm thick, with a refractive index of 1.50, is placed in a light beam ($\lambda = 580$ nm) such that the plane of the plate is perpendicular to the beam. (a) Calculate to eight significant figures the number of wavelengths of light within the glass plate. (b) Find the net phase shift in the light beam resulting from the introduction of the glass plate into the beam.

38B-8 The beam from a helium–neon laser ($\lambda = 633$ nm) is directed toward a screen. Find the number of additional wavelengths of light in the optical path from the laser to the screen when a thin slab of glass, with a thickness of 0.110 mm and a refractive index of 1.55, is inserted into the beam. The surface of the slab is perpendicular to the beam.

38B-9 Using light of wavelength 500 nm, we produce a double-slit interference pattern on a screen 1.5 m from a pair of vertical slits separated by 0.50 mm. Find the number of interference maxima that lie between the central maximum and 1.00 cm to the left of the central maximum.

38B-10 We can produce interference fringes using a *Lloyd's mirror* arrangement with a single monochromatic source S_0, as in Figure 38-22. The image S' of the source formed by the mirror acts as a second coherent source that interferes with S_0. If fringes spaced 1.2 mm apart are formed on a screen 2 m from the source S_0, 606 nm, find the vertical distance h of the source above the plane of the reflecting surface.

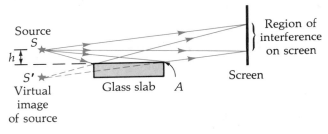

FIGURE 38-22
Problems 38B-10 and 38B-11.

38B-11 In the Lloyd's mirror setup of the previous problem, light waves are interfering in space wherever the two sets of waves pass through each other. Suppose we use a lens of high magnification to examine the interference in the vertical plane just above the edge A of the mirror. Will the fringe nearest the edge of the mirror be light or dark? Explain.

38B-12 A double slit is illuminated by light of wavelength 600 nm and produces an interference pattern on a screen. A very thin slab of flint glass ($n = 1.65$) is placed over only one of the slits. As a consequence, the central maximum of the pattern moves to the position originally occupied by the tenth order maximum. Find the thickness of the glass slab.

38.3 Multiple-Slit Interference

38B-13 The following radiations from three coherent sources combine at a point P with their electric vectors parallel (or antiparallel): $E_1 = E_0 \sin \omega t$, $E_2 = E_0 \sin(\omega t + \phi)$, and $E_3 =$

$E_0 \sin(\omega t + 2\phi)$. The resultant field is $E_p = E_r \sin(\omega t + \alpha)$. Using phasor diagrams, calculate E_r and α for (a) $\phi = 30°$, (b) $\phi = 60°$, and (c) $\phi = 120°$.

38B-14 Repeat the construction shown in Figure 38-10 for a four-slit interference pattern. Show phasor combinations corresponding to major maxima, minima, and near-minor maxima.

38.4 Interference Produced by Thin Films

38A-15 A lens is made of glass with a refractive index of 1.70 at a wavelength of 550 nm. Find (a) the minimum thickness and (b) the refractive index of a nonreflecting coating for use at this wavelength. [Hint: see Equation (38-15)].

38A-16 Find the thickness of the thinnest soap film ($n = 1.33$) that will reflect blue light of wavelength 400 nm at maximum intensity.

38A-17 In Example 38-4, the minimum thickness of a nonreflecting coating was found to be 99.6 nm. Calculate the next thicker coating that will produce the same effect.

38A-18 An air wedge is formed between two glass plates separated at one edge by a very fine wire, as was shown in Figure 38-15. When the wedge is illuminated from above by light with a wavelength of 600 nm, 30 dark fringes are observed. Calculate the radius of the wire.

38B-19 An oil film ($n = 1.45$) floating on water is illuminated by white light at normal incidence. The film is 280 nm thick. Find (a) the dominant observed color in the reflected light and (b) the dominant color in the transmitted light. Explain your reasoning.

38B-20 A glass plate ($n = 1.62$) is coated with a thin, transparent film ($n = 1.27$). Light reflected at normal incidence is observed as the wavelength is varied continuously. Constructive interference occurs for light at 680 nm, while destructive interference occurs at 544 nm (with no other such instances between these wavelengths). Find the thickness of the film.

38B-21 A film of soap solution is illuminated by white light at normal incidence and reflects bands of color, as was shown in Figure 38-13. Calculate the thickness of the film at the first green band ($\lambda = 530$ nm) below the nonreflecting portion of the film. The soap solution has a refractive index of 1.33.

38B-22 Consider the radii r_m in a Newton's-rings pattern. Show that, for $m \gg 1$, the area between successive rings is approximately equal to the constant value $\pi R \lambda$, where R is the radius of curvature of the plano-convex lens and λ is the wavelength of light.

38B-23 An air wedge is formed between two glass plates in contact along one edge and slightly separated at the opposite edge. When illuminated with monochromatic light from above, the reflected light reveals a total of 85 dark fringes. Calculate the number of dark fringes that would appear if water ($n = 1.33$) were to replace the air between the plates.

38B-24 A Newton's-rings apparatus consists of a flat plate and a plano-convex lens with a radius of curvature of 4 m. When the apparatus is illuminated from directly overhead with monochromatic light, a radial distance of 3.50 mm is measured between the tenth and thirtieth dark rings. Calculate the wavelength of the light.

38B-25 When a liquid is introduced into the air space between the lens and the plate in a Newton's-rings apparatus, the diameter of the tenth ring changes from 1.50 to 1.31 cm. Find the index of refraction of the liquid.

38.5 The Michelson Interferometer

38A-26 As the mirror M_1 of the Michelson interferometer shown in Figure 38-18 is moved through a distance of 0.163 mm, 500 bright fringes move across the field of view. Calculate the wavelength of the light illuminating the mirrors of the interferometer.

38B-27 One of the mirrors of a Michelson interferometer is attached to the growing tip of a bamboo shoot. When we use 550-nm light with a photoelectric cell to count fringes electronically, 473 bright fringes/min pass a given point in the field of view. Find how much the shoot grows in one 24-h period.

Additional Problems

38C-28 Yellow light from the mercury spectrum ($\lambda = 579$ nm) illuminates a pair of vertical slits separated by 0.20 mm. An interference pattern is produced on a screen 2.5 m from the slits. Find the intensity of the light at a distance of 1.5 cm to the right of the central maximum relative to the intensity at the central maximum.

38C-29 Show that the dashed curves representing lines of constant phase difference shown in Figure 38-4b are hyperbolas. The general form for a hyperbola in rectangular coordinates is $y^2/a^2 - x^2/b^2 = 1$.

38C-30 Monochromatic light of wavelength λ illuminates at normal incidence a pair of narrow slits separated by a distance d. The mth order interference maximum subtends an angle θ (with the incident direction) given by the equation $\sin \theta = m\lambda/d$. Derive an expression for the angular position of the mth-order interference maximum if the plane containing the slits is rotated about an axis parallel to the slits through a small angle ϕ.

38C-31 Figure 38-23 shows a *Fresnel biprism* for producing interference fringes using a single monochromatic source S_0. The biprism is two identical glass prisms joined at their bases

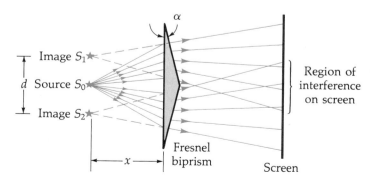

FIGURE 38-23
Problem 38C-31.

with very small vertex angles α. The prisms form two coherent virtual images, S_1 and S_2, separated a distance d. If the glass has an index of refraction n, show that $d = 2x(n - 1)\alpha$.

38C-32 Consider two coherent point sources of radiation separated by a distance of four wavelengths. In a plane containing the two sources, sketch a closed path that surrounds the two sources. As you travel once around this path, how many interference maxima do you cross?

38C-33 Consider the central peak of a double-slit interference pattern. The half-width at half maximum is twice the distance between the central maximum and the point at which the intensity I drops to $I_0/2$. Show that this half-width subtends an angle $\theta = \lambda/2d$. Make the small-angle approximations $\sin \theta \approx \tan \theta \approx \theta$.

38C-34 A double slit with a separation of 0.45 mm is illuminated by light of wavelength λ_1 and produces an interference pattern on a screen 3 m away. The tenth-order interference maximum is 4 cm from the central maximum. When light of another wavelength λ_2 also illuminates the slits, the combination of fringes overlaps such that the tenth fringe remains distinct while neighboring fringes become less clear. (a) Calculate λ_1 and (b) find the two closest values for λ_2.

38C-35 One slit of a double slit is wider than the other so that one slit emits light with three times greater amplitude than the other slit. Show that Equation (38-7) would then have the form $I = (I_0/4)(1 - 3\cos^2 \phi/2)$.

38C-36 *Pohl's interferometer.* A point source S reflected from a thin transparent film produces two coherent virtual sources S_1 and S_2 that lie on a line perpendicular to a viewing screen as shown in Figure 38-24. Derive an expression for the angular location θ of interference maxima along a vertical line. Assume appropriate small-angle approximations for θ. (Note: We can produce striking interference fringes in this manner by reflecting a diverging laser beam from a microscope slide, a plastic film, or any thin sheet that is smooth and transparent.)

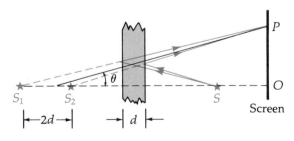

FIGURE 38-24
Problem 38C-36.

38C-37 (a) Show that, for a three-slit interference pattern, Equation (38-7) becomes $I = (4I_0/9)(1/4 + \cos \phi + \cos^2 \phi)$.

(b) Using the equation derived in part (a), verify that the first minimum occurs for $\phi = 2\pi/3$.

38C-38 Make a phasor diagram for the combination of these two parallel electric fields that have different amplitudes (in SI units) $E_1 = 2 \sin \omega t$ and $E_2 = 4 \sin(\omega t + 50°)$. Write a numerical equation for the resultant field of the form $E_r = E_0 \sin(\omega t + \alpha)$.

38C-39 The intensity distribution for a triple-slit interference pattern is given by $I = (4I_0/9)(1/4 + \cos \phi + \cos^2 \phi)$, where ϕ is the phase difference between waves from two adjacent slits. (a) In terms of the slit separation d (center-to-center) and the wavelength λ, calculate the angular half-width $\theta_{1/2}$ of the central maximum, where $\theta_{1/2}$ is the angle subtended by the central maximum I_0 and the point at which $I = I_0/2$. (b) Compare your result with the corresponding value for a double-slit pattern (see Problem 38C-33).

38C-40 In terms of the slit separation d and the wavelength λ, derive an expression for the total angular width $\Delta\theta$ of the central maximum for (a) a three-slit interference pattern, (b) a four-slit interference pattern, and (c) an N-slit interference pattern.

38C-41 A nonreflecting coating with a refractive index of 1.38 is applied to the surface of a lens of refractive index 1.90. The coating is equally nonreflecting for wavelengths of 500 nm and 600 nm. Assuming that the values of n are valid for both wavelengths, calculate the minimum thickness of the coating.

38C-42 Because of greater clarity in the interference pattern, Newton's rings are usually observed in the light that is reflected back toward the source. The light that is transmitted through the apparatus also shows an interference pattern (the "transmitted pattern"). (a) Why is the clarity, or contrast, greater in the reflected pattern? (b) Derive an expression for the transmitted pattern analogous to Equation (38-19) for the radius of the mth dark ring.

38C-43 The expression for the radius of Newton's rings, $r_m = (Rm\lambda)^{1/2}$ is the result of an approximation. Show that an exact expression is $r_m = (Rm\lambda - m^2\lambda^2/4)^{1/2}$.

38C-44 The yellow light emitted by a sodium source has two wavelengths, at 589.0 nm and 589.6 nm. Consider a Michelson interferometer used with this light. When the mirror at the end of one arm is moved continuously in one direction, the observed fringes "wash out," then reappear sharply, then wash out, and so on. (a) Explain this effect. (b) Calculate the distance between two successive positions of the mirror when the fringes are sharp.

38C-45 An air-tight tube with parallel end windows 6.0 cm apart is placed in one arm of a Michelson interferometer so that light with a wavelength of 570 nm passes through the tube, is reflected by the mirror, and again passes through the tube. When the air is withdrawn from the tube by a vacuum pump, 63 fringes pass a given point in the field of view. Calculate the refractive index of air to six significant figures.

CHAPTER 39

Physical Optics II—Diffraction

*Where the telescope ends, the microscope begins.
Which of the two has the grander view?*

VICTOR HUGO
(*Saint Dennis*)

39.1 Introduction

As we proceed into this chapter, which discusses diffraction, you will see that the phenomenon is really one of interference. There is no physical difference between interference and diffraction. In both cases, light waves interfere to produce regions of extra brightness or darkness. However, it has become customary to use the term *interference* for situations involving a finite number of point or line sources (such as multiple slits) and the term *diffraction* for the interference of waves from a single area source (essentially an infinity of neighboring point sources).

If light traveled only in straight lines, the shadows of opaque objects would have sharp edges, changing abruptly from bright to dark. The fact is, however, that light does bend somewhat around the edge of an object into the shadow region, often producing bright and dark fringes as a result of the *interference* of light waves. *The bending of light away from straight-line paths as it passes near an object* is an example of the diffraction of light, Figure 39-1.

For things we look at in our everyday experience, diffraction effects are usually quite small and therefore overlooked.[1] Another consideration is that most light sources have an *extended area*, so that diffraction patterns from one part of the source overlap with patterns from another part of the source, making them difficult to distinguish. Furthermore, each wavelength of light produces its own distinct pattern, so when many wavelengths are present, as in white light, the various patterns again overlap. It is important to understand diffraction effects because they place inescapable upper limits on the sharpness of images formed by all optical instruments. They also limit the accuracy of certain measurements.

---

[1] One way to observe diffraction is to place your hand over your eye so that you can see light from a *point* source penetrating the cracks between your fingers. A *line* source, such as a straight neon tube far away, may also be satisfactory if it is aligned in the same direction as the crack between your fingers. If the crack is narrow enough, you will observe a pattern of bright and dark fringes. They are particularly pronounced when you view a distant mercury-vapor street light, because of the dominance of only a few different wavelengths of light emitted from such a source.

(a) A magnified view of the transition from a dark shadow on the left to the bright region on the right.

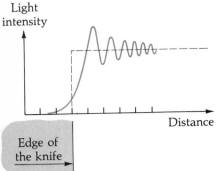

(b) A plot of the light intensity versus distance for the knife-edge diffraction pattern. If there were no diffraction, the intensity would change abruptly from dark to light as shown by the dashed line at the geometrical edge of the shadow.

FIGURE 39-1
The diffraction pattern produced by a sharp knife-edge. Note that the bright bands adjacent to the geometrical shadow are actually brighter than the uniform illumination farther to the right.

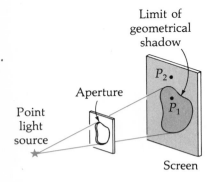

FIGURE 39-2
Diffraction: a general case. Light reaching the screen is composed of waves emanating from all parts of the wavefront as it emerges from the aperture.

Diffraction effects were known to both Newton and Huygens, but it was not until the nineteenth century that an explanation was proposed by Augustin J. Fresnel (1788–1827), a brilliant French physicist. His work, coupled with that of the British physicist Thomas Young (1773–1829), firmly established the *wave* theory of light.

Generally, diffraction effects are produced by either an *aperture* or an *obstacle* placed between a light source and a screen, as pictured in Figure 39-2. To find out what happens, we adopt Huygens' approach and imagine that *each point on the wavefront acts as a new point source of radiation*. Thus, the light falling on any given location on the screen (for example, P_1 in the directly illuminated part of the screen or P_2 in the geometrical shadow region) contains contributions from *all parts* of the wavefront passing through the aperture. The case shown in this figure is complicated for two reasons: (1) The wavefront at the aperture is *divergent* rather than *plane*. This means that as we consider different points on the wavefront, the *angle* between the normal to the wavefront and the direction to a given point P varies for different points on the wavefront. (2) The *distances* from various points on the wavefront to a given point P are all different.

Another representation of this same situation is shown in Figure 39-3a. The light diverges from a nearby point source as it moves toward the aperture. The light reaching point P on the screen is made up of Huygens wavelets that emanated from all parts of the wavefront as it emerges from the aperture. This general case is known as **Fresnel diffraction** and is quite complicated to analyze. We will mention only a few such cases, at the end of the chapter.

Figure 39-3b illustrates a situation that is easier to analyze. Rays approaching the aperture are *parallel* (with a *plane* wavefront), and rays leaving the aperture that reach a given point P on the screen are *parallel* (or essentially parallel), because the screen is so far away. This case is known as **Fraunhofer diffraction**. If large distances for the source and screen are not available, we can achieve this condition experimentally by using lenses with a nearby source and screen, as in Figure 39-3c.[2] Fraunhofer diffraction is easy to analyze because we do not have to deal with the varying angles characteristic of Fresnel diffraction.

The distinction between Fresnel and Fraunhofer diffraction patterns sometimes cannot be sharply defined. For example, if we start with a nearby source and screen and gradually move them farther away from the aperture, the Fresnel diffraction pattern gradually changes over into the Fraunhofer pattern. Thus, Fraunhofer diffraction is really just a *limiting case* of the more general Fresnel diffraction.

39.2 Single-Slit Diffraction

We will discuss two approaches to single-slit Fraunhofer diffraction. The first is a simple but useful technique of *halfwave zones* that yields the criterion for constructive and destructive interference. The second is a more detailed approach utilizing *phasors*, which yields a quantitative expression for the intensity distribution within a diffraction pattern.

---

[2] In Figure 39-3 and later figures, we draw only those rays from the wavefront at the aperture that *reach the given point P*. Of course, simultaneously there are other rays at other angles, which travel to other points on the screen.

39.2 Single-Slit Diffraction

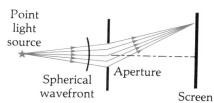
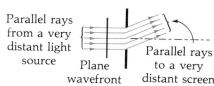
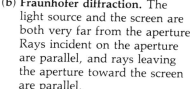

(a) **Fresnel diffraction.** The source and screen are both near the aperture. Rays from the source and rays to the screen cannot be considered parallel.

(b) **Fraunhofer diffraction.** The light source and the screen are both very far from the aperture. Rays incident on the aperture are parallel, and rays leaving the aperture toward the screen are parallel.

(c) With the use of two lenses, we can produce conditions for **Fraunhofer diffraction** using a nearby light source and a screen.

FIGURE 39-3
The distinction between *Fresnel* and *Fraunhofer* diffraction. In Fraunhofer diffraction, the light rays striking the aperture are parallel, and the light rays leaving the aperture are parallel.

Halfwave Zones

Consider the Fraunhofer diffraction apparatus illustrated in Figure 39-4. To restrict the problem to two dimensions, we analyze a slit of width a aligned perpendicular to the plane of the figure. The slit is divided into *zones* such that the path length of a ray emanating from one edge of a zone is one-half wavelength longer than that from the corresponding edge of the adjacent zone. Such zones are called *halfwave zones*. In Figure 39-4 the aperture is wide enough to contain exactly four such zones.

What happens to the rays from two adjacent zones? The rays coming from two corresponding points, such as P_1 and P_2, will differ in path length by one-half wavelength as they reach the screen. Combining similar pairs of rays for all of the two zones, we conclude that the light from one zone will interfere *destructively* with that from the neighboring zone.

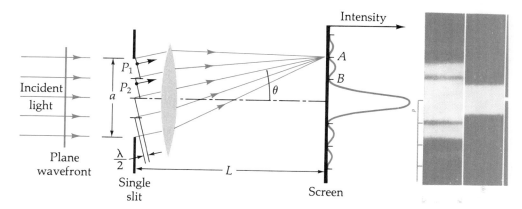

FIGURE 39-4
Fraunhofer diffraction. We can determine the criterion for destructive interference at a point on the screen by dividing the slit into *halfwave zones*. (For clarity, the width of the slit is greatly exaggerated relative to the screen distance L.) The incident light rays are parallel, forming a *plane* wavefront at the aperture.

HALFWAVE ZONE CRITERION FOR SINGLE-SLIT DIFFRACTION MINIMA A **minimum** in the diffraction pattern occurs if the slit viewed from that point on the screen contains exactly an **even** number of halfwave zones.

In reference to Figure 39-4, for point A on the screen the slit contains *four* halfwave zones, while for point B the slit contains *two* halfwave zones.

An alternative criterion for a minimum in a single-slit diffraction pattern may be based upon the total width of the slit.

ALTERNATIVE CRITERION FOR SINGLE-SLIT MINIMA A **minimum** in the diffraction pattern occurs if the path for a ray of light arriving at that point from one edge of the slit is an **integral number of wavelengths longer** than the path of a ray from the opposite edge of the slit.

Thus, in Figure 39-4 a minimum in the diffraction pattern occurs when

SINGLE-SLIT FRAUNHOFER DIFFRACTION PATTERN MINIMA

$$m\lambda = a \sin \theta \qquad \text{(minima for } m = 1, 2, 3, \ldots) \qquad (39\text{-}1)$$

Note that the central *maximum* corresponds to $m = 0$, with all other values of m designating *minima*. In most situations, the angle θ is small enough to justify the small-angle approximation: $\sin \theta \approx \tan \theta \approx \theta$. When this is true, the *central maximum and all the other minima are equally spaced*. Thus, the full width of the central maximum is *twice* the separation of adjacent minima.

Do not confuse this relation with Equation (38-8), Chapter 38:

$$m\lambda = d \sin \theta \qquad \text{(maxima for } m = 0, 1, 2, \ldots)$$

The equations have the same form, but Equation (39-1) is for the *single-slit* diffraction pattern *minima*, while the *double-slit* relation [Equation (38-8)] is for the inteference pattern *maxima*.

Phasors

The use of phasors to determine the intensity distribution in a single-slit Fraunhofer diffraction pattern is an extension of the technique used in multiple-slit interference patterns. We consider the slit to be divided into small incremental zones, Δy wide, as illustrated in Figure 39-5. Each of these zones, or strips, may be considered a source of radiation contributing an incremental electric field amplitude ΔE at the point P on the screen. The total field amplitude E_θ at the point P will be the sum of such increments from all of the zones. However, depending on the angle θ, the incremental field amplitudes will be slightly out of phase with one another. Since

$$\frac{\text{Path difference}}{\lambda} = \frac{\text{Phase difference}}{2\pi}$$

we have

$$\frac{\Delta y \sin \theta}{\lambda} = \frac{\Delta \phi}{2\pi}$$

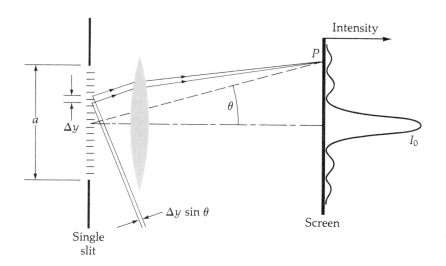

FIGURE 39-5
Fraunhofer diffraction. The electric field at P is the sum of incremental fields emanating from incremental zones Δy wide at the aperture.

where λ is the wavelength and $\Delta\phi$ is the phase difference of the electric field increments from adjacent zones. Rearranging, we have

$$\Delta\phi = \left(\frac{2\pi}{\lambda}\right) \Delta y \sin\theta \qquad (39\text{-}2)$$

Figure 39-6b shows the difference in phase between electric field increments from three adjacent zones at the top of the slit shown in Figure 39-5. If θ is small, all of the incremental field elements may be considered equal in amplitude. The angle ϕ between the first incremental zone at the top of the slit and the last zone at the bottom is shown in Figure 39-6a. The sum of all the incremental phasors is then $\mathbf{E}_\theta$, the base of the isosceles triangle with equal sides R. From trigonometry,

$$E_\theta = 2R \sin\left(\frac{\phi}{2}\right) \qquad (39\text{-}3)$$

where, from Equation (39-2), $\quad \phi = \left(\dfrac{2\pi}{\lambda}\right) a \sin\theta \qquad (39\text{-}4)$

We can obtain the value of R by letting the incremental phasor amplitude approach zero as the number of increments approaches infinity. In this limit, the sum of increments forms the arc of a circle with radius R. The length of the arc is simply the incremental phasor sum when all of the increments are in phase. This occurs for light rays parallel to the axis ($\theta = 0$) forming the central peak of the diffraction pattern. Thus the amplitude E_0 of the central maximum is

$$E_0 = R\phi \qquad (39\text{-}5)$$

Combining Equations (39-3) and (39-5), we have

$$E_\theta = \frac{2E_0 \sin\left(\dfrac{\phi}{2}\right)}{\phi} = E_0 \left(\frac{\sin\alpha}{\alpha}\right) \qquad (39\text{-}6)$$

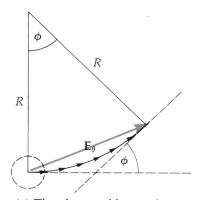

(a) The phasor addition of incremental electric fields $\Delta\mathbf{E}_n$ to produce the total field $\mathbf{E}_\theta$.

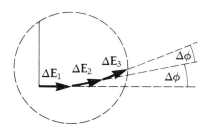

(b) A magnified view of the first three increments.

FIGURE 39-6
Phasor addition to determine the total electric field amplitude E_θ at a point P on the screen.

where $\alpha \equiv \phi/2$. From Equation (39-4) we thus obtain

$$\alpha = \left(\frac{\pi}{\lambda}\right) a \sin\theta \qquad (39\text{-}7)$$

Mathematically, $(\sin\alpha/\alpha)$ approaches unity as α approaches zero. Therefore, E_θ approaches E_0 as θ approaches zero. Recall that the intensity I of light is proportional to the *square* of the amplitude of the electric field strength ($I \propto E^2$), so the single-slit diffraction relations are as follows:

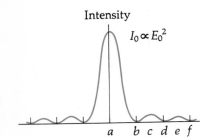

SINGLE-SLIT FRAUNHOFER DIFFRACTION INTENSITY

$$I_\theta = I_0 \left(\frac{\sin\alpha}{\alpha}\right)^2 \qquad (39\text{-}8)$$

where
$$\alpha = \left(\frac{\pi}{\lambda}\right) a \sin\theta \qquad (39\text{-}9)$$

Minima occur when $\quad \alpha = m\pi \quad$ (where $m = 1, 2, 3, \ldots$)

Combining this with Equation (31-7), we have $(\pi/\lambda) a \sin\theta = m\pi$, or

SINGLE-SLIT FRAUNHOFER DIFFRACTION MINIMA

$$m\lambda = a \sin\theta \qquad \text{(minima for } m = 1, 2, 3, \ldots) \qquad (39\text{-}10)$$

which is the same equation derived using halfwave zones.

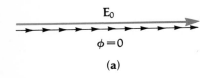

The mathematical form of Equation (39-8) makes it difficult to determine the exact relative amplitude of diffraction *maxima* and their locations. However, we can obtain approximate relative amplitudes by assuming that the maxima lie halfway between the minima. That is, since we know that the minima occur when $\alpha = m\pi$, an approximate maximum[3] occurs when

$$\alpha = (m + \tfrac{1}{2})\pi \qquad \text{(where } m = 1, 2, 3, \ldots) \qquad (39\text{-}11)$$

Substituting this into Equation (39-8), we obtain

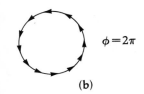

$$\frac{I_\theta}{I_0} = \left[\frac{\sin(m+\tfrac{1}{2})\pi}{(m+\tfrac{1}{2})\pi}\right]^2 \qquad \text{(where } m = 1, 2, 3, \ldots)$$

or
$$\frac{I_\theta}{I_0} = \frac{1}{(m+\tfrac{1}{2})^2 \pi^2} \qquad \text{(approximate maxima for } m = 1, 2, 3, \ldots)$$

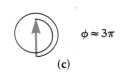

Thus, if I_0 is the intensity at the central peak, for $m = 1$, $I_\theta = 0.045 I_0$; for $m = 2$, $I_\theta = 0.016 I_0$; and for $m = 3$, $I_\theta = 0.0083 I_0$. Clearly, almost all of the light in a diffraction pattern falls within the central maximum peak.

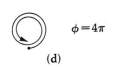

A graphical approach to diffraction pattern intensities is enlightening. In Figure 39-7a, at the central maximum the sum of the incremental electric-field

FIGURE 39-7
Phasor-addition diagrams corresponding to the maxima and minima of a single-slit diffraction pattern. For clarity, the arcs shown in (c) through (f) are drawn as spirals instead of circles.

---

[3] We find the actual maximum values of I_θ by setting $dI_\theta/d\alpha = 0$, which leads to the relation $\tan\alpha = \alpha$. The first four values of α satisfying this relation are 4.4934, 7.7253, 10.9041, and 14.0662. See Problem 39C-32.

phasors ΔE_i forms a straight electric field phasor E_0. As we move away from the central maximum (that is, as θ increases), the sum of incremental phasors forms an ever-tightening arc that closes around on itself to form successive minima, with maxima occurring between closures. In the limit of negligibly small increments, the length of the arc remains constant as it winds up. In Figures 39-7c and 39-7e we see that the lengths of the resultant phasor are a maximum just *slightly* before $\phi = 3\pi$ and 5π because, as the arc tightens, the diameter of the circle becomes smaller. However, this difference is very small, justifying the approximation used in deriving Equation (39-10).

As you look at Figures 39-8 and 39-9, it will be helpful to remember the following general characteristics of a single-slit diffraction pattern (when θ is small):

(1) The minima are *equally spaced* from one another.
(2) The full width of the central peak is *twice* the spacing between all other minima.
(3) The maxima of other peaks are relatively faint and approximately midway between the minima. (Actually, they are displaced slightly toward the central peak.)
(4) As the width of the slit is made smaller, the diffraction pattern becomes larger.
(5) As the wavelength is made smaller, the diffraction pattern becomes smaller.

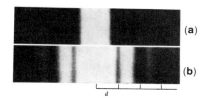

FIGURE 39-8
Two photographs of the same single-slit diffraction pattern. Ninety percent of the light passing through the slit falls in the central peak. In (b), the exposure time has been greatly increased to bring out the faint maxima on either side. (This greatly overexposes the central peak.) The scale indicates the minima.

Aperture orientation

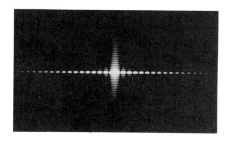

FIGURE 39-9
The diffraction pattern of a rectangular aperture. Along the horizontal axis, the minima are spaced *farther apart* than along the vertical axis because the aperture width is *narrower* along the horizontal direction.

EXAMPLE 39-1

The width of the central maximum in the diffraction pattern is often of particular interest. Suppose that a slit 3×10^{-4} m wide is illuminated by a yellow-green light ($\lambda = 500$ nm). Find the total width of the central maximum on a screen 2 m from the slit.

SOLUTION

The total width of the central maximum is the distance between the first minima on either side of the peak. We obtain the value of θ shown in Figure 39-10 by using Equation (39-1):

$$a \sin \theta = m\lambda$$

where $m = 1$ for the first minimum. Substituting values for a and λ, we obtain

$$\sin \theta = \frac{(1)\lambda}{a} = \frac{5.00 \times 10^{-7} \text{ m}}{3 \times 10^{-4} \text{ m}} = (\tfrac{5}{3}) \times 10^{-3}$$

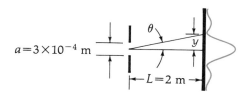

FIGURE 39-10
Example 39-1.

The distance y is half the width of the central maximum: $y = L \tan \theta$. Since $\sin \theta \approx \tan \theta$ for small angles, we have

$$y = (2 \text{ m})(\tfrac{5}{3} \times 10^{-3}) = 3.33 \times 10^{-3} \text{ m}$$

The total width $2y$ of the central maximum is thus

$$2y = \boxed{6.67 \text{ mm}}$$

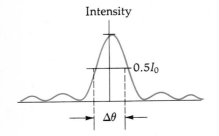

FIGURE 39-11
Example 39-2.

EXAMPLE 39-2

(a) Find the angular width $\Delta \theta$ of the half-maximum intensity within the central maximum for the situation described in Example 39-1. (b) Find the width on the screen of the central peak at half maximum (see Figure 39-11).

SOLUTION

(a) The intensity distribution is given by Equation (39-8):

$$\frac{I_\theta}{I_0} = \left(\frac{\sin \alpha}{\alpha}\right)^2$$

For $I_\theta/I_0 = 0.5$, we have $\left(\dfrac{\sin \alpha}{\alpha}\right)^2 = 0.5$

Because this equation cannot be solved algebraically, advanced optics texts contain tables of values for $(\sin \alpha)/\alpha$ versus α. However, we may use successive approximations to find α. An angle that appears as a factor must always be expressed in *radians*, so a good first approximation is that $\alpha = \pi/2$ (since the first minimum is $\alpha = \pi$). We then obtain

$$\left[\frac{\sin(\pi/2)}{(\pi/2)}\right]^2 = 0.405$$

Because this quantity increases as we decrease α, as a better approximation we try $\alpha = 1.40$ rad ($\approx 80°$):

$$\left(\frac{\sin 80°}{1.40}\right)^2 \approx 0.500$$

The angle θ is obtained from Equation (39-7): $\alpha = (\pi/\lambda)a \sin \theta$. Rearranging gives

$$\sin \theta = \left(\frac{\lambda}{a\pi}\right)\alpha = \frac{(5 \times 10^{-7} \text{ m})(1.40)}{(3 \times 10^{-4} \text{ m})\pi} = 7.43 \times 10^{-4}$$

The full angular width $\Delta \theta = 2\theta$ is thus

$$\Delta \theta = 2 \sin^{-1}(7.43 \times 10^{-4}) = \boxed{1.49 \times 10^{-3} \text{ rad}}$$

(b) On the screen, the full width of the central peak at half maximum is

$$\text{Width} = L\,\Delta \theta = (2 \text{ m})(1.49 \times 10^{-3} \text{ rad}) = 2.98 \times 10^{-3} \text{ m}$$

$$= \boxed{2.98 \text{ mm}}$$

39.3 Diffraction by a Circular Aperture

Diffraction effects impose a serious limitation on the resolving power of microscopes, telescopes, and other instruments used in all regions of the electromagnetic spectrum. Most instruments employ a circular aperture such as a lens or the circular "dish" of a radio antenna. The analysis of this diffraction pattern is more complicated than that for a single slit, though the result is similar to the minima in a single-slit pattern ($a \sin \theta = m\lambda$). For a circular aperture of diameter D, the minima are located at

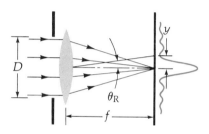

CIRCULAR APERTURE
FRAUNHOFER
DIFFRACTION MINIMA
$$D \sin \theta = p_m \lambda \quad (39\text{-}12)$$

where $p_1 = 1.220$, $p_2 = 2.233$, $p_3 = 3.238$, $p_4 = 4.241$, $p_5 = 5.243$, etc. Figure 39-12 shows the pattern. The central spot is called the *Airy disk*, after Sir George Airy, who first analyzed the pattern in 1835. The Airy disk contains 84% of the light passing through the aperture, while 91% is contained within the central spot plus the first diffraction ring.

FIGURE 39-12
The Fraunhofer diffraction pattern of a distant point source produced by a circular aperture. The size of the pattern is always larger than the diameter D of the hole. Also, the smaller the hole, the larger the pattern. The location of the first diffraction minimum determines the minimum angle of resolution θ_R. (This photograph is somewhat overexposed to bring out the faint rings surrounding the bright central spot.)

Rayleigh's criterion for barely resolving two, equal-intensity point sources is that *the peak of one diffraction pattern falls on the first minimum of the other pattern.* See Figure 39-13. Since the angles are small, $\sin \theta_R \approx \theta_R$, giving

MINIMUM ANGLE OF RESOLUTION
θ_R FOR A CIRCULAR APERTURE
(Rayleigh's criterion)
$$\theta_R = \frac{1.22\lambda}{D} \quad (39\text{-}13)$$

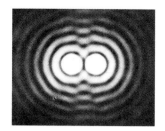

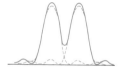

(a) The angular separation of the two patterns is clearly large enough to reveal two sources.

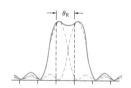

(b) The patterns overlap according to the Rayleigh criterion. The resulting pattern is barely discernible as two overlapping diffraction patterns.

FIGURE 39-13
Superimposed Fraunhofer diffraction patterns associated with the images of two distant, incoherent point sources.

(a) The world's largest radio telescope antenna is the 305-m-diameter, fixed-dish reflector at Arecibo, Puerto Rico. Its movable overhead antenna near the focus can collect signals within ±20° from the vertical. At the time of construction, its 20-acre surface was greater than the combined area of all other telescopes ever built.

(b) The Very Large Array (VLA) system in New Mexico employs 27 steerable dishes, each 26 m in diameter, arranged in a movable array in the shape of a "Y" extending over a 27-km baseline. The signals are simultaneously analyzed with an interferometric technique known as *aperture synthesis* by a large computer at the center of the "Y". The angular resolution depends on the baseline distance and is comparable to the 1-arcsecond resolution of visible-light observations from large telescopes.

(c) The Very Long Baseline Array (VLBA) to be completed in 1992 is a series of 10 antennae, each 25 m in diameter, extending 8000 km across the Northern Hemisphere. This photograph shows the antenna at Los Alamos, New Mexico, USA. The array will be operated by remote control from the Array Operations Center (AOC) at Socorro, New Mexico. Information from the magnetic data tape from each antenna will be recorded and synthesized in a computer at the AOC that can perform 10^{12} multiplications per second. This process will achieve the same resolving power as a single radio telescope 8000 km in diameter — equivalent to sitting in New York while reading a newspaper that is located in San Francisco.

FIGURE 39-14
Radio telescopes.

where D is the diameter of the circular aperture and λ is the wavelength. Applied to the human eye, this relation gives a resolving power of roughly 20 seconds of arc. In practice, the resolving power of the average eye is slightly worse due to the finite size of the receptors in the retina. On the other hand, careful analyses of photographic images can routinely achieve somewhat better results than the Rayleigh limit. The signals from two or more *radio telescopes* (Figure 39-14) can be combined to give an effective resolution comparable to that of a single instrument with a diameter equal to the baseline distance separating the telescopes (but with far less energy-gathering ability).

EXAMPLE 39-3

The world's largest operating refracting telescope is the University of Chicago's Yerkes telescope. The objective lens of the telescope is 1.02 m (40 in) in diameter and has a focal length of 18.9 m. Find the total width of the central peak (Airy disk) of the diffraction pattern at the image of a star, assuming an average wavelength of 500 nm.

SOLUTION

The total angular width of the central diffraction peak is $2\theta_R$, where θ_R is given by Equation (39-13):

$$2\theta_R = \frac{(2)(1.22)\lambda}{D} = \frac{(2)(1.22)(500 \times 10^{-9}\text{ m})}{(1.02\text{ m})}$$
$$= 1.20 \times 10^{-6}\text{ rad}$$

For an objective lens of focal length f, the linear width y corresponding to this small angular width (Figure 39-12) is given by

$$y = (f)(2\theta_R) = (18.9\text{ m})(1.20 \times 10^{-6}\text{ rad}) = \boxed{2.27 \times 10^{-5}\text{ m}}$$

39.4 The Diffraction Grating

A *diffraction grating* is a multiple-slit device in which the slits are extremely narrow and very closely spaced. The first gratings constructed by Fraunhofer were simply arrays of closely spaced, fine, parallel wires or threads. We can form a typical modern grating by making parallel scratches on glass or metal; such a grating often has the equivalent of more than five or ten thousand slits per centimeter. Because it is desirable to have as many slits per centimeter as possible, the width of each slit is very small, producing wide-angle diffraction effects. Thus the diffraction grating involves a combination of two phenomena: *multiple-slit interference* and *single-slit diffraction*. Diffraction gratings are used to make very accurate measurements of wavelengths of light.

Look again at Figure 38-11 in the previous chapter. As the number of slits in an array increases, the major maxima become much more narrow and intense. Indeed, for several thousand slits, the major peaks will be extremely narrow, and the intensities of the minor maxima in between become truly negligible. Figure 39-15 shows a source of monochromatic light passing through a slit S (aligned parallel to the grating slits). A *collimating* lens L_1 makes the light parallel as it strikes the diffraction grating (to achieve the Fraunhofer condition). The different orders of diffracted light ($m = 0, \pm 1, \pm 2, \ldots$) are emitted at various angles θ_m, where they are collected by a movable telescope that may be rotated around to the appropriate angles. Lens L_2 brings the parallel light to a line focus (really just an image of the slit S), where it is further magnified by a lens L_3 for examination by the eye. Once the device is calibrated and the grating space d is known, the angle θ permits a determination of the wavelength λ.

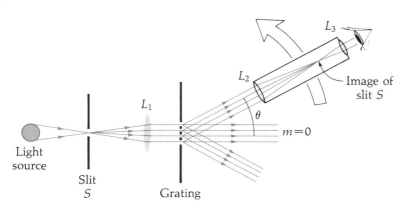

FIGURE 39-15
A simple grating spectroscope used to determine wavelengths of light from a source.

(a) Astronomer R. S. Richardson displays a 40-ft record of the sun's spectrum obtained by a grating spectrograph.

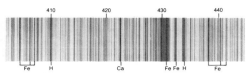

(b) A segment of the visible spectrum of the sun. Wavelengths (in nanometers) are listed above the spectrum, while elements responsible for certain strong lines are shown below.

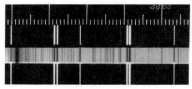

(c) A laboratory (bright-line) spectrum of iron is placed above and below the sun's absorption spectrum for comparison, indicating the presence of iron in the sun.

FIGURE 39-16
Fraunhofer was the first person to investigate the spectrum of sunlight with a diffraction grating and in so doing observed thousands of dark lines (the "Fraunhofer lines"). He noted that some lines fell in the same positions as known bright lines in the spectra of certain elements he had studied in the laboratory, but he was unable to explain the mechanism that produced the dark lines. More than half a century later, Kirchhoff gave the correct explanation that the cool atmosphere of gas atoms above the sun's glowing surface *absorbed* the characteristic wavelengths of those atoms from the continuous spectrum of the sun. The element helium (a Greek word meaning "the sun") was first discovered in the Fraunhofer lines of sunlight, as were several other elements.

If the source emits several different discrete wavelengths, then instead of a single line at each order position there will be a cluster of lines spread out at various angles, one for each wavelength present. The greater the number of lines per centimeter in the grating, the more this cluster will be spread out, allowing very precise measurements of wavelengths. If a source emits a continuous spectrum, the full distribution of wavelengths is displayed over a range of angles. Unfortunately, sometimes the spectrum of one order will overlap a portion of the spectrum of an adjacent order, a possible source of confusion that must be taken into account. Many instruments record the spectrum photographically or analyze the light with a sensitive photocell. Such devices are called *grating spectroscopes,* Figure 39-16.

Gratings are made of a series of parallel lines scratched with a diamond stylus onto a clear glass plate (forming a *transmission* grating) or onto a flat metal plate (forming a *reflection* grating, in which the interference effects are viewed by reflected light).[4] Because a good grating is so difficult to manufacture, most gratings in use are replicas formed by pouring a thin layer of a transparent collodion solution on the grating, allowing it to harden, and

---

[4] Making a series of parallel scratches sounds like a simple task. However, in practice, the procedure is full of unexpected difficulties. An interesting discussion of one of the most precise mechanical devices ever invented can be found in A. G. Ingalls, "Ruling Engines," *Scientific American,* June 1952. Most modern gratings are made on a thin layer of aluminum evaporated on a glass plate "optically flat" to within a fraction of a wavelength of light. Low-cost replica gratings made on acetate film are often used in school laboratories.

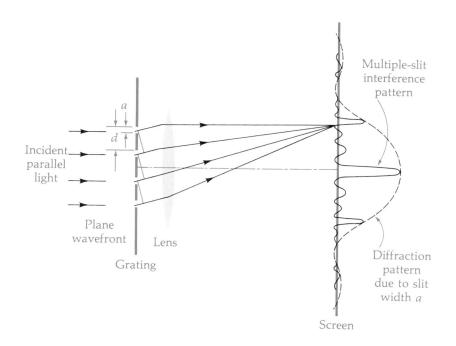

FIGURE 39-17
A four-slit grating. As the slits become narrower and the number of slits increases to that in a typical diffraction grating, the interference major maxima become sharper and more intense, while the diffraction envelope becomes broader.

peeling it off, producing a transmission grating. The collodion sheet is then mounted on glass or supported in a rigid frame. This transparent plastic replica contains a series of ridges where the scratches were, separated by undisturbed clear strips. In an overly simplified picture, we may think of such a transmission grating as allowing light to transmit through the clear strips, which therefore act as slits, while the somewhat irregular ridges scatter the light in all directions and are thus effectively opaque.

We begin a discussion of the theory of gratings by analyzing a transmission grating with just four slits, as shown in Figure 39-17. Parallel light is incident, so that as the plane wavefront passes through the grating the slits act as a series of coherent light sources. Unlike the double-slit interference discussion in Chapter 38, in which we ignored the diffraction occurring at each slit, here we take diffraction into account. Note that the slit widths a are comparable to the center-to-center slit separation d. The parallel light rays that leave the slit at an angle θ are brought to a focus on the screen as a line image perpendicular to the plane of the diagram. (Of course, diffraction causes light rays to leave the slit at other angles, too, which are similarly brought to a focus at other points on the screen; we show just one particular angle θ on the diagram.) The lens enables us to use the Fraunhofer single-slit diffraction theory we developed in the previous section.

Before proceeding further, we point out that, because of the lens, the diffraction peaks (at a given angle) produced by *all* of the slits in the grating superimpose at the same place on the screen. That is, a maximum formed by a slit at one edge of the grating falls in precisely the same place as that formed by a slit at the opposite edge. (All parallel rays entering a lens converge at the same focal point.)

In Section 38.2 we showed that the following equation describes the condition for the major maxima in a multiple-slit interference pattern:

MULTIPLE-SLIT INTERFERENCE (major maxima)
$$m\lambda = d \sin \theta \quad \text{(major maxima at } m = 0, 1, 2, \ldots\text{)} \quad (39\text{-}14)$$

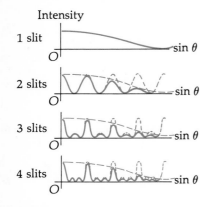

FIGURE 39-18
The diffraction and interference patterns for 1, 2, 3, and 4 slits. The resultant intensity is shown with the solid line.

where d is the center-to-center slit separation, θ is the angle between the central maximum ($m = 0$) and other major maxima, and λ is the wavelength. When we derived this equation, diffraction effects were ignored. However, if we now take into account the slit width a, we can interpret the distance d to be the *distance between corresponding points within adjacent slits*.

This equation specifies the angular location of the major interference peaks. But the overall intensity of the pattern is reduced by the single-slit diffraction effects of Equation (39-8):

$$I_\theta = I_0 \left(\frac{\sin \alpha}{\alpha} \right)^2$$

where $\alpha = (\pi/\lambda)a \sin \theta$ and $a =$ slit width. Figure 39-18 shows the net result of combining the single-slit and multiple-slit effects. The diffraction due to the *slit width a* (shown dashed) determines the upper limit of intensity of the overall pattern. Within this pattern, the multiple-slit effects further reduce the intensity at various locations determined by the slit *spacing distance d*.

EXAMPLE 39-4

For the multiple-slit pattern depicted in Figure 39-18, find the ratio of the slit width a to the slit separation d (center-to-center).

SOLUTION

Note that the *first* diffraction minimum falls at the *fourth* interference maximum. From Equation (39-14), the major interference *maxima* for multiple-slits are given by $m\lambda = d \sin \theta$. Rearranging and substituting $m = 4$ gives

$$\sin \theta = \frac{4\lambda}{d} \qquad (39\text{-}15)$$

Equation (39-10) gives the single-slit diffraction *minima*: $a \sin \theta = m\lambda$. Rearranging and substituting $m = 1$, we get

$$\sin \theta = \frac{\lambda}{a} \qquad (39\text{-}16)$$

Combining Equations (39-15) and (39-16), we obtain

$$\frac{4\lambda}{d} = \frac{\lambda}{a} \qquad \text{or} \qquad \boxed{\frac{a}{d} = \frac{1}{4}}$$

In this example, the fourth-order interference peak is missing because the first diffraction minimum occurs at that location. By noting which orders are missing in a multiple-slit pattern, one can determine the ratio of slit width to slit separation.

Until now we have concerned ourselves with a single wavelength λ. However, when we are observing light that includes several wavelengths, the spectrum is dispersed through a range of angles for each value of m. This spectral pattern is repeated for other orders of diffraction. That is, $m = 1$ corresponds to the *first-order* pattern, $m = 2$ corresponds to the *second-order* pattern, and so on. As the next example illustrates, in certain cases the pattern of one order can overlap that of another order.

EXAMPLE 39-5

A diffraction grating disperses white light so that the red wavelength $\lambda = 650$ nm appears in the second-order pattern at $\theta = 20°$. (a) Find the so-called *grating constant*—that is, the number of slits per centimeter. (b) Determine whether or not visible light of the third-order pattern appears at $\theta = 20°$.

SOLUTION

(a) Equation (38-14) gives the multiple-slit interference maxima: $m\lambda = d \sin \theta$. Rearranging, we have

$$d = \frac{m\lambda}{\sin \theta} = \frac{(2)(650 \text{ nm})}{(\sin 20°)} = 3800 \text{ nm}$$

The number of slits per centimeter ($\mathcal{N}$) becomes

$$\mathcal{N} = \frac{1}{d} = \frac{1 \text{ slit}}{3800 \times 10^{-9} \text{ m}} \underbrace{\left(\frac{1 \text{ m}}{100 \text{ cm}}\right)}_{\text{Conversion ratio}} = \boxed{2630 \text{ slits/cm}}$$

(b) Again, Equation (38-14) is appropriate: $m\lambda = d \sin \theta$. Solving for λ and substituting the given values, we get

$$\lambda = \frac{d \sin \theta}{m} = \frac{(3800 \text{ nm})(\sin 20°)}{3} = \boxed{433 \text{ nm}}$$

A wavelength of 433 nm is a faintly visible violet. Thus, for this grating visible portions of the second and third orders do overlap.

Dispersion

Diffraction gratings are often used rather than prisms in the analysis of spectra, because gratings are capable of spreading the spectrum over a wider range of angles, enabling more precise measurements of λ to be made. The **dispersion** D expresses the ability of a grating or prism to spread a range of wavelengths $d\lambda$ over an angular spread of $d\theta$.

DISPERSION
$$D \equiv \frac{d\theta}{d\lambda} \tag{39-17}$$

The greater the dispersion, the greater the angular separation of two lines that are close together in wavelength. As shown in Table 39-1, the dispersion is greater than that produced by a prism and greater for larger values of the order m.

TABLE 39-1 Comparison of the Dispersion of a Prism and a Diffraction Grating

λ (nm)	60° Flint-Glass Prism	4500 Rulings/cm Gratings	
		$m = 1$	$m = 3$
670.8	50.51°	17.57°	64.90°
656.3	50.61°	17.18°	62.38°
589.3	51.17°	15.38°	52.71°
546.1	51.64°	14.26°	47.50°
486.1	52.58°	12.64°	41.01°
404.7	54.83°	10.49°	33.12°

For the prism: $\Delta\theta = 4.32°$. For the grating: $\Delta\theta = 7.08°$ ($m=1$), $\Delta\theta = 31.78°$ ($m=3$).

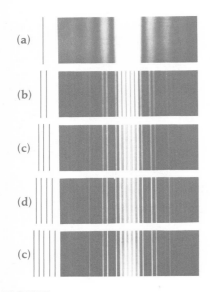

FIGURE 39-19
(a) The Fraunhofer diffraction patterns for a single slit. Multiple-slit patterns are shown in (b) through (e) for the slit systems shown at left.

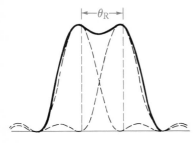

FIGURE 39-20
The Rayleigh criterion for the resolution of two diffraction maxima. The peak of one pattern falls on the first minimum of the other pattern.

Since the relationship between θ and λ for a diffraction grating is $d \sin \theta = m\lambda$, we have

$$\sin \theta = \left(\frac{m}{d}\right)\lambda$$

We obtain the dispersion D by differentiating with respect to λ

$$\cos \theta \frac{d\theta}{d\lambda} = \frac{m}{d}$$

and rearranging:

$$\frac{d\theta}{d\lambda} = \frac{m}{d \cos \theta} \tag{39-18}$$

DISPERSION OF A GRATING

$$D = \frac{m}{d \cos \theta} \tag{39-19}$$

The units of D are radians per meter (or, more commonly, degrees per nanometer).

Resolving Power

While dispersion is an important consideration in diffraction-grating design, the ability of a grating to separate perceptibly, or to *resolve*, two spectral lines of nearly the same wavelength is also important. The **resolving power** R of a diffraction grating or prism is defined as

RESOLVING POWER

$$R \equiv \frac{\lambda}{\Delta\lambda} \tag{39-20}$$

where λ is the average wavelength of two spectral lines with a wavelength difference of $\Delta\lambda$. Referring to Figures 39-19 and 39-20, we see that the principal maxima become sharper as the total number of slits N increases. The relationship between the sharpness of a principle maximum (other than the central maximum) and the total number of slits N can be shown to be

$$\theta_R = \frac{\lambda}{Nd \cos \theta} \tag{39-21}$$

The "sharpness" of the central peak is measured by θ_R, the angular separation between the center of the peak and the *first* minimum. The symbol d is the separation of the slits, and θ is the diffraction angle of the peak. The notation θ_R we adopt for this angle comes from its use in the criterion proposed by Lord Rayleigh for the minimum resolvable angular separation for two overlapping diffraction patterns. According to **Rayleigh's criterion**, *two closely spaced, equal-intensity patterns are acceptably "resolved"* (that is, one can decide they are definitely due to two point sources instead of one) if

RAYLEIGH'S CRITERION FOR MINIMUM RESOLUTION OF TWO EQUAL-INTENSITY PATTERNS

The peak of one diffraction pattern is located at the *first* minimum of the other pattern.

Figure 39-20 illustrates the criterion. For a diffraction grating, the angle θ_R is exceedingly small, corresponding to a wavelength difference $\Delta\lambda$, as given by Equation (39-18):

$$\theta_R = \frac{m}{d\cos\theta}\Delta\lambda \qquad (39\text{-}22)$$

Combining Equations (31-21) and (31-22), we have

$$\frac{m}{d\cos\theta}\Delta\lambda = \frac{\lambda}{Nd\cos\theta}$$

from which we obtain, for the *resolving power of a grating* $R = \lambda/\Delta\lambda$,

RESOLVING POWER OF A GRATING

$$R = Nm \qquad (39\text{-}23)$$

where N is the total number of slits in the grating and m is the order.

The distinction between *dispersion* and *resolving power* becomes obvious from Table 39-2, which compares data for three different gratings. As illustrated in Figure 39-21, gratings A and B have the same dispersion D (they separate two given wavelengths by the same angular distance), while gratings A and C have the same resolving power R (the ability to distinguish two wavelengths very close together, limited only by the *width* of each diffraction peak). Note that grating B has the highest resolving power, while grating C has the highest dispersion.

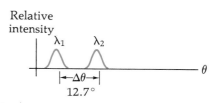

Grating A

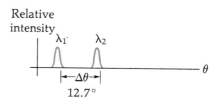

Grating B

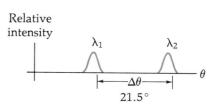

Grating C

FIGURE 39-21
The relative intensity patterns produced by the gratings of Table 39-2 for two wavelengths, λ_1 and λ_2, near 550 nm.

TABLE 39-2 The First-Order Spectrum ($m = 1$) for Light near Wavelength $\lambda = 550$ nm

Grating	N_t	d (nm)	θ	R	D (10^{-2} degrees/nm)
A	10 000	2500	12.7°	10 000	2.35
B	20 000	2500	12.7°	20 000	2.35
C	10 000	1500	21.5°	10 000	4.11

> **EXAMPLE 39-6**
>
> A sodium-vapor lamp emits a yellow light corresponding to two wavelengths, 589.00 nm and 589.59 nm. How many rulings must a grating have to barely resolve this sodium doublet in the first order?
>
> *SOLUTION*
>
> The required resolving power is given by Equation (39-20)
>
> $$R \equiv \frac{\lambda}{\Delta\lambda}$$
>
> where $\lambda = \dfrac{589.00 \text{ nm} + 589.59 \text{ nm}}{2} = 589.30 \text{ nm}$
>
> and $\Delta\lambda = 589.59 \text{ nm} - 589.00 \text{ nm} = 0.59 \text{ nm}$
>
> Thus: $R = \dfrac{\lambda}{\Delta\lambda} = \dfrac{589.3 \text{ nm}}{0.59 \text{ nm}} = 1000$
>
> The resolving power for a diffraction grating is [Equation (39-23)]: $R = Nm$. For the first order ($m = 1$), the number of rulings is thus equal to the resolving power R:
>
> $$N = \boxed{1000 \text{ rulings}}$$
>
> Since a typical diffraction grating has approximately 5000 rulings per centimeter, we can easily resolve the sodium doublet without resorting to either very fine rulings or large gratings.

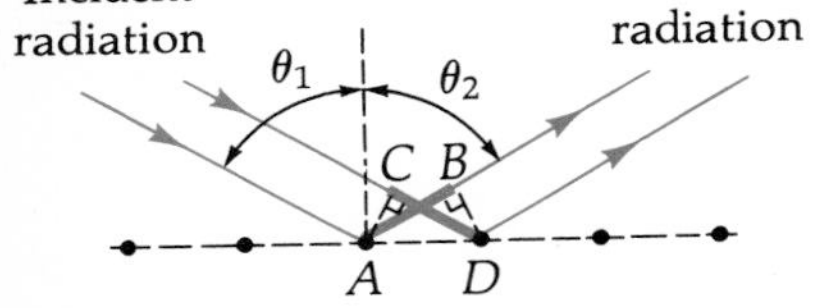

39.5 X-Ray Diffraction

In 1912, the German physicist Max von Laue (1879–1960) first suggested that a crystalline array of atoms might act as a *three-dimensional* "diffraction grating" for x-rays of wavelengths comparable to the atomic spacing in the crystal (~ 0.1 nm). The incoming radiation would be absorbed by electrons and, according to the Huygens theory, each electron would reradiate expanding wavelets in a process called *scattering*. Thus, just as the slits in a diffraction grating act as coherent sources of radiation, the three-dimensional array of scattering centers would act as coherent sources. Because electrons are concentrated near the atoms, each *atom* is effectively a scattering center. In certain directions, the scattered waves will be in phase, producing a high intensity of scattered radiation in that direction. For certain other directions, the waves will be out of phase, resulting in destructive interference and no scattering.

Consider the line of scattering centers in Figure 39-22. For radiation incident at an angle θ_1 as shown, the scattered waves will be in phase if the two distances AB and CD are equal. By symmetry, this happens when the scattering angle θ_2 equals the incident angle θ_1. Sir William Bragg[5] noted the similarity to optical reflection ("the angle of incidence equals the angle of reflection"), so he proposed an alternative explanation involving "Bragg reflection"

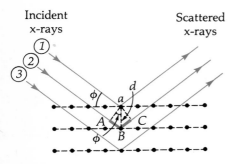

FIGURE 39-22
A line of equally spaced scattering centers.

FIGURE 39-23
A diagram illustrating the Bragg "reflection" of x-rays from planes of atoms near the surface of a crystal. In x-ray work, the angle of the incident radiation is traditionally measured with respect to the plane rather than to the normal.

---

[5] The British father–son team W. H. and W. L. Bragg received the Nobel Prize in 1915 for their studies of crystal structures by x-ray diffraction; this was just a year after von Laue received the Nobel Prize for his basic discovery of the diffraction of x-rays.

from atomic planes. Though this "reflection" is an incorrect picture of the scattering process, it is a simple and useful way of thinking about the phenomenon. Adopting this simple view, look at Figure 39-23, wherein the incident radiation strikes a cubical array of atoms, which form atomic planes spaced a distance d apart. Consider rays ① and ②. The lines aA and aC are drawn perpendicular to the incident and reflected ray ②, so that the distance ABC is the extra path length traveled by ray ②. Thus, for incident radiation at an angle ϕ with respect to the atomic planes (not to the *normal to the plane*, as in optical reflection), the extra path length is

$$\text{Path length difference} = 2(d \sin \phi)$$

When this path difference is an *integral* number of wavelengths $m\lambda$, the scattered rays will be *in phase*. (Similar relations also apply to other rays, such as ③, scattered from deeper regions.) The relation for constructive interference is called the *Bragg scattering condition*.

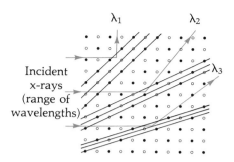

FIGURE 39-24
The lattice of atoms in a crystal may be grouped into parallel planes at various angles, each with its own spacing. The wavelength that matches the Bragg condition will be reflected from its corresponding set of planes. Three different sets of planes are illustrated.

BRAGG SCATTERING CONDITION

$$m\lambda = 2d \sin \phi \qquad (39\text{-}24)$$

where $m = 1, 2, 3, \ldots$ (the *order* of scattering)

$\phi =$ the *glancing* angle between the incident ray and the *plane* (not between the ray and the normal, as in optical reflection)

$d =$ atomic plane spacing

The scattered radiation is very sharply "peaked" at these angles. As for a plane diffraction grating with a great many slits, the three-dimensional "grating" has an enormous number of scattering centers, which causes the major maxima to become extremely narrow and intense, while suppressing all minor maxima. If a *continuous spectrum* of radiation containing all wavelengths (called "white" radiation) is incident, we may also consider the simultaneous Bragg reflections from other planes in the crystal (Figure 39-24). The various sets of parallel planes will have different spacings between them, depending on the geometric positions of the atoms. Since the incident radiation contains all wavelengths, there will be some radiation at the correct values of $m\lambda$ that match all the various Bragg conditions. A photograph of the scattered spots (Figure 39-25) is called a *Laue diffraction pattern*. The positions of the spots

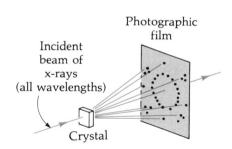

(a) Experimental arrangement for making a Laue-spot diffraction pattern.

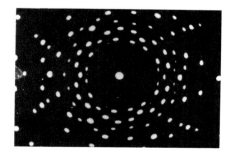

(b) A Laue diffraction pattern for a single quartz crystal.

FIGURE 39-25
X-ray diffraction patterns. The positions of the spots correlate with the configuration of atoms in the crystal. For unknown crystals, by

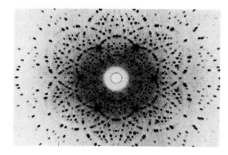

(c) Complex crystals produce strikingly beautiful Laue-spot patterns.

"working backward" from the diffraction pattern one can determine the atomic configurations.

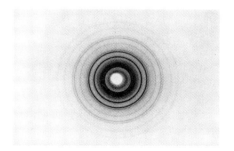

(d) A *powder pattern* of the diffraction of monochromatic x-rays from polycrystalline aluminum. By chance some microscopic crystals will be oriented properly to "reflect" x-rays from important planes, forming concentric circles.

FIGURE 39-26
A Fresnel diffraction pattern of a circular aperture, made with monochromatic light. As the screen is moved to different distances, the spot at the center changes alternately from dark to light and the number of diffraction rings changes.

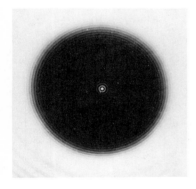

FIGURE 39-27
A diffraction pattern due to a penny reveals the Poisson bright spot at the center. To produce this pattern, a penny was placed midway between a monochromatic point source of light and a screen 40 m away.

correlate with the configuration of the atoms in the specimen. For a crystal whose atomic structure is unknown, one can "work backward" from the Laue pattern to figure out the locations of the atoms. For example, in 1962 James Watson, Francis Crick, and Maurie Wilkins[6] received the Nobel Prize in biology for discovering the double-helix structure of DNA using x-ray diffraction methods.

39.6 Fresnel Diffraction—Circular Apertures and Obstacles

When parallel light passes through a small circular hole in an opaque plate and falls on a nearby screen,[7] a surprising pattern results. Not only will the pattern be larger than the hole and contain diffraction rings, there may even be a *dark* spot in the center, Figure 39-26. This is quite unexpected since one would anticipate that the straight-through direction from the center of the opening to the screen would be bright, not dark!

Another startling result is the diffraction pattern for a small circular obstacle, such as a ball bearing with free space around it. Careful inspection reveals that there is *always a bright spot in the center*,[8] as if the ball bearing had a tiny hole! Figure 39-27 shows the bright spot in the shadow of a penny. These patterns are examples of **Fresnel diffraction**, in which light reaching a given point on the screen comes *at various angles* from different parts of a wavefront as the wavefront emerges from the aperture or passes around an obstacle, Figure 39-28. The origin of these effects is discussed in the next section.

39.7 The Fresnel Zone Plate

Consider parallel light passing through a small circular hole and falling on a screen. The point P at the center of the diffraction pattern receives light from all parts of the plane wavefront in the aperture (Huygens' principle). Let us divide that wavefront into *circular zones* by the procedure of Figure 39-29. The central circle, called a *half-period zone*, contains light reaching P that differs in phase only from 0 to π rad. Light from the next half-period zone arrives at P with phases from π to 2π rad, the next zone with phases 2π to 3π rad, and so on. The *net* contributions from any two adjacent zones are *one-half wavelength out of phase* and therefore, upon their arrival at P, tend to cancel each other by destructive interference. If $+\mathbf{E}_1$ is the net electric vector for light from the first zone, then the net electric vector for light from the second

---

[6] For a fascinating story of a scientific quest, see James D. Watson, *The Double Helix*, Atheneum Press, New York, 1968. Also see Horace Judson, *The Eighth Day of Creation*, Simon & Schuster, 1979.

[7] Because the screen is nearby, light falls at a given point on the screen at various angles, a situation called *Fresnel diffraction*—in contrast to *Fraunhofer diffraction*, in which only parallel light falls on the screen.

[8] There is an interesting anecdote about the spot. In response to a Prize Essay competition, Fresnel submitted his wave theory of diffraction to the French Academy in 1818. Poisson, a member of the judging committee and a firm believer in the corpuscular theory of light, strongly ridiculed Fresnel's theories. To clinch his objections, and hoping to deal a death blow to the wave theory of light, Poisson told a committee member, Arago, that (as Fresnel had not realized) the theory unrealistically predicted the existence of a bright spot at the center of the shadow of a circular obstruction—clearly an absurd prediction! Arago immediately tried the experiment and rediscovered the bright spot, which actually had been found 85 years earlier by Miraldi but had been long forgotten. The bright spot's existence gave a big boost to Fresnel's wave theory. Poisson, however, stubbornly clung to the Newtonian particle model for light until his death 22 years later. Ironically, today the spot is usually called *Poisson's* bright spot, ignoring the true heroes in the story: Miraldi, Fresnel, and Arago. Fresnel ultimately did win first prize for his essay.

39.7 The Fresnel Zone Plate 919

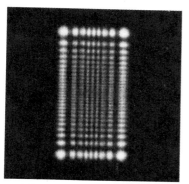

(a) A rectangular aperture.

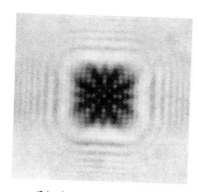

(b) An opaque square.

(c) The diffraction pattern of an opaque disk with a source consisting of an illuminated transparent portrait of Woodrow Wilson. The opaque disk acts as a sort of lens, since for every point in the source there is a Poisson bright spot in the image.

(d) Three opaque circular disks. Note the bright spot in the center of each disk.

(e) The shadow of a small screw supported by a wire.

(f) A longer exposure of (e) to bring out the faint diffraction pattern within the shadow.

(g) A magnified photograph of (f), the diffraction pattern of the head of a screw.

FIGURE 39-28
Fresnel diffraction patterns. The bright bands just outside the shadow of an opaque object are actually brighter than the unobstructed uniform illumination farther from the shadow.

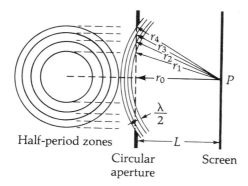

FIGURE 39-29
The geometrical construction of Fresnel zones. Within the aperture, the spheres centered at P have radii r_0, $r_0 + \lambda/2$, $r_0 + 2\lambda/2$, $r_0 + 3\lambda/2$, etc. The intersections of these spheres with the plane in the aperture form the circular half-period zones.

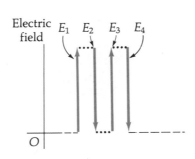

FIGURE 39-30
The average electric field vectors for the light from each of the four zones. They add together to zero. (For clarity, the vectors have been displaced sideways from one another.)

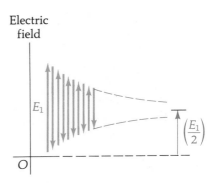

FIGURE 39-31
When the electric field vectors for a very large number of zones are added together, the resultant amplitude approaches *half* that due to the first zone acting alone.

zone is $-\mathbf{E}_2$, that from the third zone is $+\mathbf{E}_3$, and so on. Each zone has approximately the same area (Problem 39C-38), so each vector $\mathbf{E}_n$ has approximately the same amplitude,[9] but they alternate in sign, Figure 39-30. The number of zones for a given geometry depends upon the diameter of the hole, the distance L, and the wavelength λ. If there are an *even* number of zones in a small aperture, the net $\mathbf{E}$ is essentially zero, causing a dark spot at P. If the screen is moved either toward or away from the opening so that an *odd* number of zones fills the aperture, then the point P will become bright. This explains why the center spot alternates between light and dark when the screen is moved in Figure 39-26. If we block out some zones at the center, all the remaining zones will still contribute some light at P, explaining the Poisson bright spot in the shadow of a penny, Figure 39-27.

Suppose that we now make every other zone opaque (either the odd ones or the even ones). Because all the light from the transparent zones is now in phase, the electric vectors all add in the same direction and a lot more light reaches point P. A transparent film with alternate zones blocked out is called a **Fresnel zone plate**, Figure 39-32. Isn't it interesting that, by making half the area of an aperture opaque, we can dramatically increase the light transmitted to the center of the pattern? As shown in Footnote 8, the light from the entire unobstructed wavefront equals approximately half the contribution from the first zone: $E_1/2$. If we construct a zone plate that passes only the first 20 odd zones, then the electric field at P is $E = E_1 + E_3 + E_5 + \cdots + E_{39}$. Each of these terms is approximately equal to the others. Without the zone plate, the field at P is approximately $E_1/2$, but with the zone plate it is $20E_1$. Therefore, the zone plate increases the light intensity (proportional to E^2) by a factor of 1600! *Thus the zone plate acts as a sort of lens, diffracting incident parallel light so that it converges to a real point image a distance L away*.[10]

A Fresnel zone plate would be nothing more than an amusing gadget were it not for the fact that it provides an easy explanation of how a hologram generates its eerie three-dimensional image.

FIGURE 39-32
Alternate half-period zones are made opaque to form a *Fresnel zone plate*. The negative of this pattern (with the central zone opaque) is also a Fresnel zone plate. In most cases, the area of each zone ($\pi \lambda L$) is quite small. For example, for $L = 1$ m and $\lambda = 500$ nm, each area would be about 1.6 mm².

---

[9] Because light from higher-number zones has a bit farther to travel, the inverse-square law reduces their amplitudes slightly. An obliquity factor also enters in. The net result is shown in Figure 39-31.

[10] There are other point images along the axis, both real and virtual. However, the image at L that we have discussed is the brightest real image.

39.8 Holography

Everyone is familiar with holographic images, those fascinating ghostlike images that have full three-dimensional properties, formed without the use of lenses by passing coherent light through a flat sheet of film. In the old-fashioned stereoscopic image, each of a pair of almost identical pictures is viewed separately by each eye, producing the mental impression of a three-dimensional image *as viewed from a fixed perspective*. In contrast, when we view a holographic image with the unaided eye by looking through the hologram as through a window, the image has a true three-dimensional property and we can easily see behind an object in the foreground by merely changing our position. In fact, 360° holograms in the form of a cylinder have been made, allowing the viewer to move completely around the image, seeing all sides.

The principles of *holography* (Greek, meaning "whole writing") were first presented in 1948 by Dennis Gabor, who was awarded the Nobel Prize in 1971 for his theories. We can explain the basic principle of holography simply by using the idea of a zone plate. Consider Figure 39-33, in which two sets of monochromatic coherent waves impinge on a photographic film. One set, the *reference beam*, consists of plane waves. The other set is the light scattered from a point object. At the film, the interference of these two sets of coherent waves produces a pattern of light and dark rings. Upon development, the film will have opaque and clear regions, forming a *Gabor zone plate*, similar to a Fresnel-zone-plate pattern. If the developed film (called a *hologram*) is then illuminated with coherent monochromatic light, an observer located properly to receive the diffracted light coming through the hologram will see a virtual point image, as in Figure 39-34. (The real image is also present on the viewer's side of the hologram.)

Suppose that a small *extended* object is used instead of a point object. Then each point of the object forms its own zone-plate pattern, which superimposes with patterns for all the other points. The resulting hologram is a very complicated array of fringes (Figure 39-35) that contains the full information about the zone-plate pattern *for each point on the object*. When the hologram is illuminated with coherent monochromatic light, the diffracted light reconstructs a full virtual image of the object. In practice, to make about

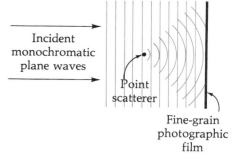

FIGURE 39-33
Plane monochromatic waves and the waves scattered coherently from the point object produce an interference pattern on the photographic film. When the film is developed, the resulting hologram is a series of light and dark concentric circles similar to a Fresnel zone plate.

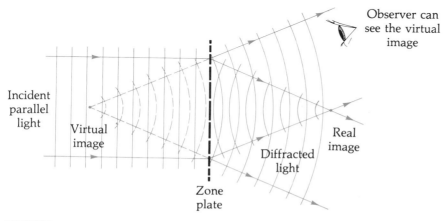

FIGURE 39-34
When parallel light is incident on a zone plate, real and virtual point images are formed on opposite sides of the plate. [Additional point images (not shown) for other orders of diffraction are also located along the central axis.]

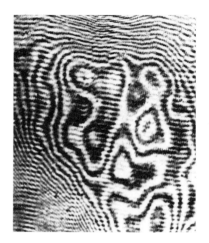

FIGURE 39-35
A highly magnified portion of a hologram, showing the complicated pattern of interference fringes.

922   39 / Physical Optics II—Diffraction

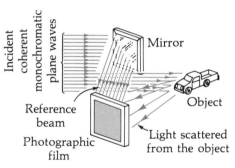
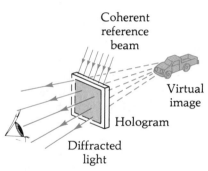

FIGURE 39-36
A common arrangement for making and viewing a hologram.

(a) One arrangement for making a hologram. Light from the reference beam combines with light scattered from the object to produce a complicated interference pattern at the film surface. When the film is developed, the recorded pattern is called a hologram.

(b) To view the hologram, we use a coherent reference beam to illuminate the hologram at the same angle as the reference beam used in making the hologram. The diffracted light forms a virtual image of the object at its original location. The image is a true three-dimensional image; we can see hidden parts by moving the eye to a new location.

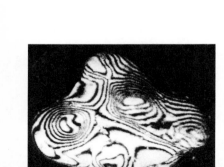

(a) A holographic contour map of a fossil badger tooth (8 mm long). Two holograms of the specimen were recorded on the same photographic plate using two slightly different wavelengths. When the resulting hologram is reconstructed using one wavelength, the interference between the two images creates fringes in the form of height contours.

(b) Details of the spark detonation of acetylene gas inside a transparent cylinder are made visible by a double exposure using a pulsed ruby laser that illuminates the scene from the rear through a ground-glass diffuser. The first exposure was made prior to ignition. The second exposure was recorded 10 ms after ignition. Upon reconstruction, three-dimensional patterns are formed by the interference of the two holographic images.

(c) A photograph of a "time-averaged holographic interferogram" of a loudspeaker vibrating at 3000 Hz. The hologram was a time exposure over several thousand cycles. Only the nodal lines and the stationary portions of the scene reconstruct brightly.

FIGURE 39-37
A few applications of holography.

equal intensities for the two sets of waves and thus achieve maximum contrast in the interference pattern, the arrangement diagrammed in Figure 39-36 is often used.[11]

Since no lenses are used at any stage of the procedure, the troublesome Rayleigh limit of resolution is avoided. The detail in the reconstructed image can actually be better than that produced by any conventional photography using lenses. Each small fragment of a hologram contains information about the entire object (as seen from that vantage point) and will reconstruct the entire image.

The applications of holography are impressive (see Figure 39-37). One limitation on making a hologram is that the incident reference beam and the scattered light (both portions of the same wave train) must not differ in optical path by more than a coherence length when they arrive at the film. Ordinary lasers produce light with coherence lengths of several meters, though special techniques can push the upper limit of laser-light coherence to $\sim 10^5$ m.

Summary

Fraunhofer single-slit diffraction:

a = slit width
θ = angle measured from the center line
λ = wavelength
m = order

A diffraction *minimum* occurs when

$$m\lambda = a \sin \theta \quad \text{(where } m = 1, 2, 3, \ldots\text{)}$$

The intensity distribution I_θ (in units of W/m²) is given by

$$I_\theta = I_0 \left(\frac{\sin \alpha}{\alpha}\right)^2 \quad \left(\text{where } \alpha = \left(\frac{\pi}{\lambda}\right) a \sin \theta\right)$$

The *maxima* of the diffraction pattern fall *approximately* halfway between the minima.

The diffraction grating:

d = separation of the slits (center-to-center)
θ = angle measured from the center line
λ = wavelength
m = order

A diffraction *maximum* occurs when

$$m\lambda = d \sin \theta \quad \text{(where } m = 0, 1, 2, 3, \ldots\text{)}$$

The *dispersion D* of a diffraction grating is defined as

$$D \equiv \frac{d\theta}{d\lambda} = \frac{m}{d \cos \theta}$$

The *resolving power R* of a diffraction grating is defined as

$$R \equiv \frac{\lambda}{\Delta \lambda} = Nm$$

where N is the total number of rulings in the grating.

In Fraunhofer diffraction by a circular aperture, the angle θ_R between the center of the diffraction pattern and the *first* minimum (measured from the center of the aperture) is

$$\sin \theta_R = \frac{(1.22)\lambda}{D}$$

where D is the diameter of the aperture and λ is the wavelength.

The Rayleigh criterion for the minimum angle of resolution θ_R is that two adjacent point sources are distinguishable if the *central peak* of one diffraction pattern falls on the *first minimum* of the diffraction pattern of the other. Thus, for a telescope aperture of diameter D (or other optical instrument with a circular aperture),

$$\theta_R = \frac{(1.22)\lambda}{D}$$

X-ray diffraction:

d = atomic plane spacing
ϕ = the *glancing* angle between the incident ray and the plane (not between the ray and the normal, as in optical reflection)
λ = wavelength
m = order

---

[11] For a method of making a hologram that can be viewed from all sides, see W. R. Schubert and C. R. Throckmorton, "Making a 360° Hologram," *The Physics Teacher* **13**, 310 (1975).

For the *Bragg scattering condition*, a diffraction *maximum* occurs at

$$m\lambda = 2d \sin \phi \quad \text{(where } m = 1, 2, 3, \ldots)$$

Fresnel diffraction occurs when either the source of light or the observing screen (or both) lies at a *finite* distance from the diffracting aperture or obstacle. Fresnel's method of analysis employs *half-period zones*, for which the average light from any two adjacent zones is out of phase by one-half wavelength.

A *Fresnel zone plate* is a special screen in which alternate half-period zones are made opaque. The zone plate has lens-like focusing properties (with multiple focal lengths).

Holography is a two-step process in which an object illuminated by coherent light produces a complicated diffraction pattern on a photographic film. When the developed film, a *hologram*, is illuminated with coherent light, diffraction effects produce a three-dimensional image in which true differences in perspective occur if the viewer's position is changed. Since no lenses are used, the conventional Rayleigh resolution limits are avoided (though other limits eventually are present).

Questions

1. A small hole illuminated by monochromatic light produces a diffraction pattern on a screen. The edges of the hole are poorly defined. If a lens is properly placed between the hole and the screen, the diffraction effects seem to disappear and the edges of the hole are well defined. Explain.
2. What happens to a Fraunhofer single-slit diffraction displayed on a screen if water replaces air in the space between the slit and the screen?
3. Since interference and diffraction effects depend on the addition of the electric fields associated with electromagnetic waves, why isn't it necessary to have a light source in which all electric field variations are polarized in the same direction?
4. Rather than a long narrow slit producing a diffraction pattern, suppose that a "slit" only twice as long as it is wide is used to produce the pattern. Qualitatively, what is the appearance of the pattern?
5. Two diffraction gratings, one larger than the other, are of the same quality and have the same number of rulings per centimeter. What are the advantages of using the larger of the two gratings?
6. Light from a slit is collimated by a lens, then passes through a diffraction grating whose rulings are parallel to the slit. What happens to the diffraction pattern on a distant screen as the grating is tilted about an axis parallel to its rulings?
7. Suppose a grating or a prism is used in the spectral analysis of light containing a mixture of wavelengths. Under what circumstances is resolving power more important than dispersion and vice versa?
8. Describe the diffraction pattern produced by two crossed diffraction gratings.
9. What are the advantages, if any, of a diffraction grating versus a prism in displaying the spectral components of a light source? What are the disadvantages, if any?
10. At night distant road signs are easier to read if they are painted in green and white rather than red and white. Why?
11. A diffraction grating produces a continuous spectrum when illuminated by white light. How does a crystal produce a discontinuous array of dots ("Laue spots") when illuminated by "white x-rays" containing a range of wavelengths?
12. What are the similarities and differences between Fraunhofer and Fresnel diffraction?
13. The shadows of objects cast by the sun seem to have a fuzzy edge. Is this a diffraction phenomenon in which fringes are not evident because of the mixture of wavelengths in sunlight? If this is not a diffraction phenomenon, what is the cause of the fuzziness?
14. If you peer at a distant light source through very small cracks between your fingers, you will see light and dark fringes. Is this a diffraction phenomenon? If so, is it an example of Fraunhofer or Fresnel diffraction?
15. Why is it necessary to have a very nearly *circular* obstacle in order to observe Poisson's bright spot?
16. To describe the diffraction of sound waves, how would our development of light-diffraction analysis have to be modified? Remember that sound waves are longitudinal pressure waves.
17. In what way is a Fresnel zone plate like a converging lens? In what ways is it dissimilar?

Problems
39.2 Single-Slit Diffraction

39A-1 Light of wavelength 550 nm passes through a single slit and forms a diffraction pattern on a screen 3 m away. The distance between the third minima on opposite sides of the central maximum is 25 mm. Find the width of the slit.

39A-2 A single slit is illuminated by light of wavelength 550 nm and produces a diffraction pattern on a screen 3 m from the slit. Find the total width of the central maximum for a slit width of (a) 0.2 mm and (b) 0.4 mm.

39B-3 A single slit 0.20 mm wide is illuminated by monochromatic light of wavelength 600 nm, producing a diffraction pattern on a screen 1.5 m away. Find the distance between the first and fifth diffraction minima.

39B-4 In a Young's double-slit experiment, green light (520 nm) produces a pattern of bright fringes spaced 1.5 mm apart on a screen 1.8 m away. (a) Find the distance between the centers of adjacent slits. (b) As one counts fringes away from the central ($m = 0$) bright fringe, every sixth bright fringe is missing. Calculate the width of each slit.

39B-5 A single slit is illuminated by light composed of two wavelengths, λ_1 and λ_2. The diffraction patterns produced overlap such that the first minimum created by the light of wavelength λ_1 falls at the second minimum of the pattern produced by light of wavelength λ_2. (a) Calculate the ratio λ_1/λ_2. (b) At what other places in the combined diffraction pattern will the minima coincide?

39B-6 Suppose that the photographs in Figure 39-8 are exact-size reproductions of the diffraction pattern produced by a slit 0.150 mm wide on a screen 1.25 m from the slit. Measure the photographs to determine the wavelength of the light producing the pattern.

39B-7 A double-slit diffraction pattern is produced by slits that are one-third as wide as the separation of their centers. Calculate the ratio of the intensity of the first-order maximum of the double-slit pattern relative to the central maximum.

39B-8 A vertical single slit 0.25 mm wide is illuminated by light with a wavelength of 600 nm, and a pattern is produced on a screen 2.5 m from the slit. (a) Find the intensity relative to the central maximum intensity I_0 at a point 2 cm left of the central maximum position. (b) Describe the position in terms of the nearest minimum location.

39B-9 Two slits, each with a width of 0.150 mm, are separated by a distance of 9 mm. Calculate the number of interference maxima that are observed (a) within the central diffraction maximum and (b) within one of the first-order diffraction maxima.

39.3 Diffraction by a Circular Aperture

39A-10 Calculate the diameter of a reflecting-telescope mirror that by the Rayleigh criterion can resolve two point sources whose angular separation is $\frac{1}{4}$ s. Assume a wavelength of 550 nm.

39A-11 A person observing the taillights of an automobile as it recedes in the distance at night can barely distinguish them as separate sources of light. Assuming that the lights are 1.5 m apart and emit at an average wavelength of 640 nm, estimate the distance between the observer and the automobile. The pupil size of the observer's eye is 6 mm in diameter. (Note: refraction effects in patches of air with different densities cause blurring, so the actual distance is shorter than calculated.)

39A-12 A parabolic microwave antenna has a diameter of 1.5 m and is designed to receive "x-band" microwave signals ($\lambda = 3$ cm). Calculate the minimum angular separation (in degrees) of two microwave sources that can be resolved by this antenna.

39A-13 An American standard television picture is composed of about 485 horizontal lines of varying light intensity. Assume that your ability to resolve the lines is limited only by the Rayleigh criterion and that the pupils of your eyes are 5 mm in diameter. Calculate the ratio of minimum viewing distance to the vertical dimension of the picture such that you will not be able to resolve the lines. Assume that the average wavelength of the light coming from the screen is 550 nm.

39A-14 Using the Rayleigh criterion, find the minimum angle of resolution (in degrees) for these two astronomical instruments: (a) the 200-in.-diameter telescope at Mt. Palomar at a wavelength of 500 nm and (b) the 1000-ft-diameter radio telescope at Arecibo, Puerto Rico, at a wavelength of 80 cm.

39B-15 A helium–cadmium laser emits a beam of light containing two wavelengths, 325 nm (in the ultraviolet) and 442 nm (blue). The beam emerges from a circular opening 3 mm in diameter, resulting in two superimposed diffraction patterns on a very distant screen. Find the distance to the screen for which the first diffraction minima for the two wavelengths are separated by 2 cm.

39B-16 The Post-Impressionist painter Georges Seurat perfected a technique known as "pointillism," whereby paintings were composed of small, closely spaced dots of pure color, each about 2 mm in diameter. The illusion of color mixing is produced in the eye of the viewer. Estimate the minimum distance away a viewer should be in order to see a blending of the color dots into a smooth variation of color. Assume that the level of illumination causes the viewer to have a pupil diameter of about 2 mm.

39B-17 A telescope with an objective aperture of 10 cm has a focal length of 80 cm. A distant point source emitting radiation with a wavelength of 550 nm produces a diffraction pattern at the focal plane of the telescope. Calculate the diameter of the ring formed by (a) the first diffraction minimum and (b) the second diffraction minimum.

39B-18 A circular radar antenna on a navy ship has a diameter of 2.1 m and radiates at a frequency of 15 GHz. Two small boats are located 9 km away from the ship. How close together could the boats be and still be detected as *two* objects?

39.4 The Diffraction Grating

39A-19 A speed-control radar system transmits microwave radiation at a wavelength of 3 cm. A wide beam of this radiation strikes a fence formed of vertical rods, 5 cm apart. Find the angle between the direction of the incident beam and the direction of the first diffraction minimum beyond the fence.

39A-20 When illuminated with monochromatic light, a certain diffraction grating produces a pattern in which the third, sixth, ninth, etc., orders are missing. Determine the ratio of slit width to slit separation for this grating.

39A-21 A diffraction grating is 2.5 cm square and has a grating constant of 5000 rulings/cm. Calculate (a) the dispersion and (b) the resolving power of this grating in the second order for a wavelength of 600 nm.

39B-22 A diffraction grating with 2500 rulings/cm is used to examine the sodium spectrum. Calculate the angular separation of the sodium yellow doublet lines (588.995 nm and 589.592 nm) in each of the first three orders.

39B-23 A certain grating has 20 000 slits spread over 5.5 cm. Find the wavelength of light for which the angle between the two second-order maxima is 60°.

39B-24 A diffraction grating has a ratio of slit separation to slit width of 10:1. Calculate the ratio of the first-order intensity maximum to the central ($m = 0$) maximum intensity.

39B-25 A certain diffraction grating has a dispersion of 2.5×10^{-2} deg/nm and a resolving power of 10^4 in the first order. Calculate the angular separation of two spectral lines near 550 nm that can barely be resolved in accordance with the Rayleigh criterion.

39.5 X-Ray Diffraction
39.7 The Fresnel Zone Plate

39A-26 Monochromatic x-rays incident upon a crystal produce first-order Bragg reflection at a glancing angle of 20°. Calculate the expected angle for the second-order reflection.

39A-27 X-rays of wavelength 0.30 nm produce a first-order reflection from a crystal of NaCl when the glancing angle of incidence is 30°. Calculate the lattice spacing that corresponds to this reflection.

39A-28 X-rays of wavelength 0.188 nm are incident on the cubic crystal LiF. The first-order scattering maximum occurs at a grazing angle of 27.9°. (a) Find the lattice spacing of the LiF crystal. (b) At what angle would second-order scattering occur?

39A-29 Let d be the spacing between adjacent atoms of a cubic crystal. Show that x-rays of wavelength greater than $d\sqrt{2}$ cannot satisfy the Bragg scattering condition for any of the three scattering planes illustrated in Figure 39-24.

39B-30 Figure 39-27 shows the diffraction pattern of a penny. The diameter of a penny is 19 mm. Using the data in the figure caption and assuming a wavelength of 546 nm, estimate the number of Fresnel half-period zones that the penny obscures when viewed from the center of the pattern on the screen.

39B-31 Light of 490-nm wavelength passes normally through a circular aperture 1 cm in diameter. At the center of a screen 6 m away, how many half-period zones are within the aperture?

Additional Problems

39C-32 The condition for the angular position θ of single-slit diffraction maxima is given by $\tan \alpha = \alpha$, where $\alpha = (\pi/\lambda)a \sin \theta$, with slit width a. This equation is most easily solved by successive approximations made with a pocket calculator. Assuming that $a = 20\lambda$, (a) show that the angular position of the first-order maximum does not lie exactly midway between the first and second minima and (b) find its value. (Hint: as a first approximation for α, use an angle θ_{av} that is the average value of θ_1 and θ_2, the first two minima. Then try slightly smaller values of α until you "zero in" on the correct value.) See Footnote 3 on page 904.

39C-33 Using the information in the previous problem, find the angular position (in radians) of the second-order diffraction maximum when $a = 20\lambda$.

39C-34 A pair of vertical slits, each of width 0.150 mm, whose centers are separated by 0.9 mm, are illuminated perpendicularly by light of wavelength 550 nm. A combined interference–diffraction pattern is produced on a screen with the double-slit maxima spaced 1 mm apart. (a) Sketch the pattern on the screen. (b) Find the slit-to-screen distance L. (c) Find the intensity of the $m = 3$ double-slit peak in terms of the central peak intensity I_0.

39C-35 In Equation (39-8), we obtain the maximum values of I_θ by setting $dI_\theta/d\alpha = 0$. (a) Show that this leads to the relation $\tan \alpha = \alpha$. (b) Using successive approximations, find (to five significant figures) the first three values of α that satisfy that relation, and compare them with the approximate values of α, using Equation (39-11).

39C-36 A diffraction grating illuminated perpendicularly by light with a wavelength λ produces an interference pattern on a large screen parallel to the grating. Calculate the ratio of the maximum slit width of the grating slits to the wavelength so that no diffraction minima are present no matter how close the screen is to the grating.

39C-37 Show that the ability of a grating to resolve two spectral lines that differ in frequency by Δf is given by $\Delta f = c/Nm\lambda$, where c is the speed of light, N is the number of grating rulings, m is the order, and λ is the wavelength.

39C-38 The dispersion D of a diffraction grating depends upon the diffraction order m, the slit separation d, and the wavelength λ. Derive an expression for D in terms of m, d, and λ.

39C-39 Consider the Bragg planes indicated in the two-dimensional lattice shown in Figure 39-24. (a) Show that the spacing of these planes can be represented by $d = a(n^2 + 1)^{-1/2}$, where $n = 1, 2, 3, \ldots$, and a is the atomic spacing in the lattice. (b) In general, $d = a(n^2 + m^2)^{-1/2}$, where both n and m are integers. Make a sketch similar to Figure 39-24, showing the planes separated by $d = a(13)^{-1/2}$.

39C-40 Show that the area of each zone in a zone plate is approximately $\pi\lambda L$, where L is the primary focal length of the zone plate.

39C-41 As shown in Figure 39-38, a beam of parallel monochromatic light passes through a large, circular hole to produce a spot of light on the screen. Consider a point P, away from the straight-through beam, located where no light strikes the screen. Now suppose that an opaque disk with an arbitrary opening in it (object A) is placed in the hole, causing some diffracted light to reach P. Next, suppose that we replace object A with its "complement," object B, which is opaque where object A is transparent, and vice versa. (If both objects were present at the same time, the combination would be completely opaque.) According to Babinet's principle, *the diffracted light reaching P is exactly the same in the two cases.* Prove the theorem using superposition concepts.

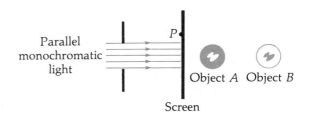

FIGURE 39-38
Problem 39C-41.

CHAPTER 40

Polarized Light

We can scarcely avoid the inference that light consists in the transverse undulation of the medium which is the cause of electric and magnetic phenomena.

JAMES CLERK MAXWELL (1856)

40.1 Introduction

Maxwell's equations describe electromagnetic radiation as a transverse wave of oscillating electric and magnetic fields. It is called *transverse* because the **E** and **B** fields are represented by vectors that lie in a plane perpendicular to the direction of propagation. By convention, the direction of the *electric* vibration is called the **direction of polarization** of the **linearly polarized wave**. Figure 40-1 shows methods of depicting linearly polarized light rays in diagrams. The

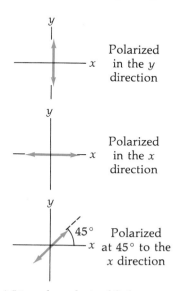

(a) Linearly polarized light rays approaching the viewer (that is, the rays are perpendicular to the plane of the paper).

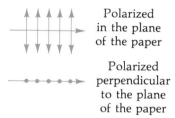

(b) Linearly polarized light rays traveling in the plane of the paper. The direction of polarization is indicated by an array of short arrows or a series of dots.

FIGURE 40-1
Ways of indicating the direction of polarization of the electric field for *linearly polarized* light rays.

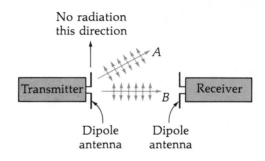

FIGURE 40-2
Radiation from a dipole antenna is polarized. The direction of polarization is perpendicular to the direction of propagation and lies in the plane containing the dipole antenna.

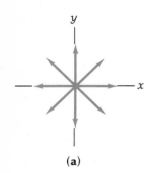

(a)

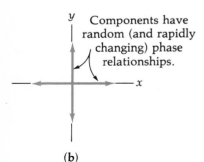

(b)

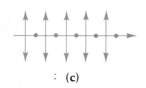

(c)

FIGURE 40-3
(a) and (b) show two ways of depicting *unpolarized* transverse waves approaching the viewer. (c) An unpolarized ray traveling in the plane of the paper; the arrays of short arrows and dots are separated in space to emphasize that there is no fixed phase relationship between the two components, which vibrate incoherently.

arrays of short arrows indicate the direction of polarization; if you wish, you may instead think of them as oscillating electric field vectors. Other words are also used to describe polarized waves. The terms *plane of polarization* and *plane-polarized* waves are common, but these have possible ambiguities. For example, in Figure 40-2 the two rays A and B have the same plane of polarization (the plane of the paper), but their directions of polarization are different.

Radio waves and microwaves emitted from antennae are polarized in directions related to the direction of the accelerated charges in the antenna wires (Figure 40-2). A receiving dipole oriented parallel to the direction of polarization will absorb energy from the waves because the alternating electric field causes electrons in the receiving dipole to accelerate back and forth along the wires, producing an oscillating potential difference between the dipole halves. However, if the receiving dipole is oriented perpendicular to the direction of polarization, the two halves of the dipole remain at the same potential and the waves are not detected by the receiver.

The fact that electromagnetic waves can be polarized is conclusive evidence that they are *transverse* waves. Interference and diffraction give evidence of their wave nature, but these effects do not differentiate between longitudinal and transverse waves. Sound waves, for example, are longitudinal and do show interference, but they cannot be polarized. Only transverse waves can be polarized.

Visible light emitted by ordinary sources, such as light bulbs and glowing hot objects, has its origin in excited atoms and molecules. Classically, each atom or molecule emits a short burst of electromagnetic waves lasting about 10^{-8} s and containing a few million vibrations, thereby sending out a *wave train* that extends up to a meter or so along the direction of propagation. Because the atoms emit light *independently* of one another, the resultant light is a superposition of many wave trains whose electric vectors are oriented randomly in all possible directions perpendicular to the direction of propagation. We call such light **unpolarized**. As shown in Figure 40-3, there are two customary ways of depicting unpolarized light in diagrams. In (a), the light ray is approaching the viewer, and the array of arrows represents the superposition of many wave trains plane-polarized with random orientations. Since an electric field at any arbitrary direction in the xy plane may be resolved into components along the x and y axes, an equivalent representation is shown in (b). Here, the electric field of each individual wave train has been resolved separately; when summed along the x and y axes, the two net components are equal in (average) magnitude. One important characteristic should be noted. Since the *phases* of the wave trains are completely random (the light from the various atoms is *incoherent*), there is no fixed phase relationship in the net components. In fact, the components have a random and rapidly changing phase relationship. However, their *time average* is the same in each direction. Consequently, our choice for the orientation of the x and y axes about the direction of propagation makes no difference for unpolarized light: in each case the (average) components at right angles are equal.

40.2 Polaroid

The human eye is not very sensitive to the direction of polarization.[1] However, polarized light can be produced and analyzed easily with a commercial material called Polaroid.[2] Ideally, a "perfect" polarizing sheet would transmit 50% of an incident unpolarized beam intensity and absorb 50%. However, in practice the transmission is about 40% or less because of reflection at surfaces and some unwanted absorption. As shown in Figure 40-4, if two Polaroid sheets are "crossed" so that their transmission axes are at an angle of 90°, approximately 90% of the light intensity is absorbed.

When two polarizing sheets are used together, the first is called the *polarizer* and the second, which is used to determine the direction of polarization of light coming from the first, is called the *analyzer*. Consider two polarizing sheets whose transmission axes are at angle θ with respect to each other, as in Figure 40-5. If light coming from the first polarizer has an electric field amplitude E_0, the analyzer (assumed "ideal") will transmit only the component $E_0 \cos \theta$ parallel to its transmission axis. Since the intensity I is proportional to the *square* of the amplitude, the transmitted intensity varies with the angle θ as

FIGURE 40-4
When polarizing sheets are *crossed*, their transmission axes are at right angles. Each individual sheet appears gray because it absorbs approximately half of the incident unpolarized light intensity.

MALUS'S LAW $\qquad I = I_0 \cos^2 \theta \qquad$ (in W/m²) $\qquad$ (40-1)

where I_0 is the intensity of the polarized light incident on the analyzer, whose transmission axis is at an angle θ with that of the polarizer. Equation (40-1) is named after its discoverer, Captain Etienne Malus, a military engineering officer in Napoleon's army (see Footnote 3). When several polarizing sheets at various angles are used in series, Equation (40-1) is applied to each successive sheet.

---

[1] The unaided eye can sometimes detect the direction of polarization through a faint pattern known as *Haidinger's brush*, which some, but not all, people can observe. For a description of this effect and other interesting features of polarized light, see the Science Series paperback *Polarized Light*, by W. A. Shurcliff and S. S. Ballard (D. Van Nostrand Co., 1965).

[2] "Polaroid" was invented in 1928 by Edwin H. Land while he was a 19-year-old undergraduate at Harvard. The modern version of Polaroid is made by the heating and stretching of a plastic sheet that contains long-chain molecules of polyvinyl alcohol. The stretching process aligns the molecules parallel to one another. The sheet is then dipped into an iodine solution, which causes iodine atoms to attach themselves to the alcohol molecules, forming chains of their own that apparently act as microscopic conducting wires. An incident electromagnetic wave that has a component of **E** parallel to the chains will drive conduction electrons along them, absorbing essentially all the energy of that component of the wave. On the other hand, if the E field is perpendicular to the chains, only a small absorption takes place and most of this component passes through. This property is called *selective absorption*. The **transmission axis** of a sheet of Polaroid is thus perpendicular to the direction the film was stretched. Sheets as large as 1 m by 30 m (or longer) are available. For protection and strength, the material is usually laminated between thin sheets of cellulose or glass. The way a sheet of Polaroid affects light has its macroscopic counterpart in a grid of parallel conducting wires. The grid affects an unpolarized beam of radio waves or microwaves in exactly the same fashion, *transmitting* only the component whose electric vector is *perpendicular* to the direction of the wires, provided that the separation between wires is somewhat less than the wavelengths of the radiation.

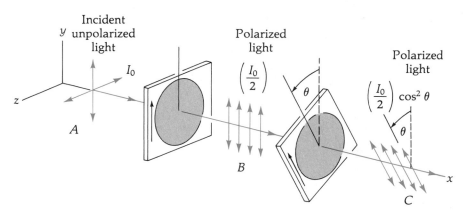

FIGURE 40-5
Two parallel polarizing sheets, one rotated so that its transmission axis is at an angle θ with respect to the other. Unpolarized light traveling along the x axis at A becomes linearly polarized in the y direction at B and polarized at an angle θ with respect to the y axis at C. If the intensity at B is I_0, the intensity at C is $I_0 \cos^2 \theta$ (for "ideal" polarizing materials).

EXAMPLE 40-1

Unpolarized light of intensity I_0 is incident upon two (ideal) polarizing sheets whose transmission axes are at an angle of 35° with respect to each other. Find the intensity I of the light emerging from the second sheet in terms of I_0.

SOLUTION

After passing through the first sheet, the light intensity is reduced to $(I_0/2)$. The second polarizing sheet further reduces the intensity by a factor of $\cos^2 \theta$. So the final beam intensity is

$$I = \left(\frac{I_0}{2}\right)\cos^2 \theta = \left(\frac{I_0}{2}\right)\cos^2 35° = \boxed{0.336 I_0}$$

40.3 Polarization by Reflection and Scattering

Another way to obtain polarized light is by *reflection* from a nonconducting surface such as glass, water, or a glossy painted surface. The reflected beam may be partially, or wholly, polarized depending on the angle of incidence. Sir David Brewster, a Scottish physicist, investigated the reflection from glass in 1812 and found that the reflected wave was 100% polarized when the refracted and reflected waves at the surface of the glass were at right angles. This relation becomes plausible when we think of the incident unpolarized light as made up of two (incoherent) E-field components at right angles (Figure 40-6). As the light is refracted into the material, it causes electrons to vibrate along these right-angle directions. However, since accelerating electrons cannot radiate energy along the direction of acceleration, the electron vibration component in the plane of the diagram in the material cannot reradiate in the direction of the reflected beam. Only the vibration component perpendicular to the plane of the paper radiates in that direction, producing a reflected beam that is 100% polarized, as shown.

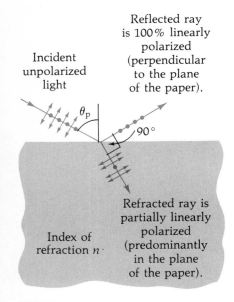

FIGURE 40-6
When light is incident at the *polarizing angle* θ_p, the reflected and refracted rays are at right angles.

Letting θ_p be the *polarizing angle* of incidence that produces this right-angle condition, we have $\theta_p + \theta_2 = 90°$. Combining this equation with Snell's law for refraction, $n_1 \sin \theta_p = n_2 \sin \theta_2$, we can derive the following relation, found in 1812 by Sir David Brewster (Problem 40B-10):

BREWSTER'S LAW
(for 100% polarization
of light by reflection $$\tan \theta_p = n \qquad (40\text{-}2)$$
from dielectric materials)

where $n = n_2/n_1$, the index of refraction of the material relative to that of the surrounding medium. The phenomenon works only for *dielectric* materials. (The process of reflection by *conducting* surfaces is more complex, and we will not take up those cases. In general, metallic surfaces reflect all components of polarization with varying degrees of effectiveness, depending on the angle of incidence.)

Sunglasses made of polarizing sheets make use of the fact that glare reflections from water surfaces, roadways, and other horizontal surfaces are (at least partially) linearly polarized; such reflections can therefore be reduced if the transmission axis of the sunglasses is oriented correctly. (What direction is correct?)

EXAMPLE 40-2

What is the polarizing angle for light incident on water (index of refraction = 1.33)?

SOLUTION

From Equation (40-2), $\tan \theta_p = 1.33$. Therefore:

$$\theta_p = \tan^{-1}(1.33) = \boxed{53.1°}$$

Polarization by Scattering

Scattered sunlight from the clear sky is partially polarized. The incident (unpolarized) sunlight sets electrons in molecules in the air into oscillations that are perpendicular to the direction of the sunlight. As explained in Figure 40-7, these vibrating electrons reradiate the light, with polarizations related to the directions of the accelerations of the charges. Vibrations along the line-of-sight do not radiate energy in that direction; *only the component of vibrations at right angles to that direction contributes to the observed scattering.* Look overhead at the clear sky through Polaroid sunglasses near sunrise or sunset, when the sun's rays are at right angles to the line-of-sight. Rotating the sunglasses reveals that the scattered radiation is polarized, but only partially because some of the light is scattered more than once before it reaches the eye. Photographers sometimes use polarizing filters to darken the sky, or to reduce light scattered by haze and unwanted reflections. Honeybees, ants and certain other insects have polarizing lenses in their eyes and are believed by some biologists to use the polarization of skylight as an aid in navigation.

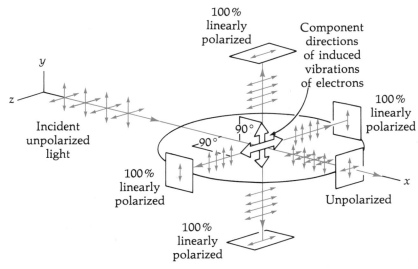

FIGURE 40-7
Scattering of unpolarized light by molecules. The transverse oscillating electric fields of the incident light set electrons into vibration in all directions in the yz plane, shown here resolved into y and z component directions. Each component radiates like a dipole antenna (see Figure 35-14). For an observer in the $\pm y$ directions, only the $\pm z$ component of electron motion radiates in that direction, so the scattered radiation is 100% polarized. (An antenna does not radiate along the direction of its length.) Similarly, only component motions of electrons in the $\pm y$ directions contribute to radiation in the $\pm z$ direction; hence it also is 100% linearly polarized. Scattering in the forward and backward directions is unpolarized; at other angles the scattered radiation is partially polarized.

FIGURE 40-8
Crossed polarizing sheets. The left-hand sheet is darker, indicating that sky light is partially polarized.

40.4 Birefringence

In a few crystalline substances, the atoms are arranged in arrays that have high degrees of symmetry. As a result, they have just a single index of refraction, which is independent of the polarization direction of the incident light. Most gases, liquids, and amorphous solids such as unstressed glass or plastic also behave this way. They are called *optically isotropic*. However, many crystalline substances and stressed amorphous materials have considerable asymmetries in their basic molecular structures. As a result, they have *two* indices of refraction, depending on the direction of polarization of the incident light. These *doubly refractive*, or *birefringent*, substances are *optically anisotropic*. The reason for the two indices of refraction is straightforward. If the crystal lattice of atoms is not symmetrical, the binding force on the electrons is also not symmetrical. That is, electrons displaced from their equilibrium positions along one direction have a greater effective "spring constant" than when displaced along another direction. Because the propagation of electromagnetic waves through materials is a process of electrons absorbing and then reradiating this energy, the fact that electrons respond differently along one direction than along another causes the waves to be transmitted with different speeds in different directions.

    Calcite, quartz, and ice are examples of birefringent materials. Figure 40-9 shows that an unpolarized ray incident on calcite splits into two polarized components: an *ordinary* ray (called the "*o*-ray"), which obeys Snell's law of refraction, and an *extraordinary* ray (the "*e*-ray"), which does not. Within the

40.4 Birefringence

(a) A calcite crystal forms a double image.

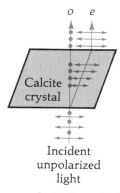

(b) An unpolarized ray incident perpendicularly on the face of a calcite crystal splits into two polarized rays. The o-ray continues in the same straight line; the e-ray is at an angle inside the crystal, emerging parallel to the o-ray but displaced to one side. (This is one case in which the direction of the ray is not normal to the wavefronts.)

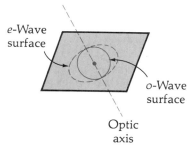

(c) A point light-source inside a calcite crystal generates two different Huygens' wave surfaces. The o-wave surface (solid line) is a sphere. The e-wave surface (dashed line) is an ellipsoid of revolution formed by rotation of an ellipse about the axis that passes through the two points where the circle and ellipse shown in the figure are in contact. This axis is called the **optic axis** and is the direction in which both the o-wave and the e-wave propogate with the same speed. In the plane perpendicular to the optic axis, the e-wave also propagates in the same direction as the o-wave, but with greater speed. (Note that the optic "axis" is a *direction*, not a line.)

FIGURE 40-9
Some optical properties of calcite, a birefringent crystal.

crystal, the extraordinary ray generally does not propagate in the same direction as the incident ray. To observe this effect, place a crystal of calcite on a piece of paper with a black dot on it; two images of the dot can be seen. Rotating the crystal causes one image to remain stationary, while the other image revolves around it. Furthermore, the two images are linearly polarized with their directions of polarization at right angles.[3] Magnetized plasmas also exhibit

---

[3] The polarization of these two images was the effect that enabled Malus to discover that reflected light could be polarized. The Paris Academy had offered a prize for a theory of double refraction. In 1808, Malus was standing at a window of his house examining a calcite crystal, hoping to learn something about double refraction. By chance, he happened to look through the crystal at the image of the setting sun reflected in the windows of the nearby Luxembourg Palace, and he was surprised to see one of the two images disappear as he rotated the calcite. Serendipity had struck again. Not only did Malus have the good fortune to have a natural polarizer in his hand, but he was lucky enough to be suitably aligned at the Brewster angle to the palace window! He spent the rest of the night experimenting with candlelight reflected at various angles from water and glass surfaces. This was about forty years before light was understood as a transverse electromagnetic wave, so the effects of polarization were truly a mystery.

FIGURE 40-10
Safety glass used for the *front* windshields of automobiles is usually formed of a transparent plastic sheet glued between two glass sheets so that if the windows shatter the glass fragments will be held together. The side and rear windows, however, are often a single sheet of glass heat-treated in a way that purposely introduces mechanical strains into the glass as it cools. If broken, the entire window then crumbles into relatively safe, gravel-sized fragments rather than shattering into large shards as ordinary glass does. The strains make the glass birefringent. As shown here, you can see this strain pattern when partially polarized sky light is reflected at the Brewster angle from the rear window. Although no polarizing filters were used for this photograph, viewing the reflection with polarizing sunglasses makes the strain pattern even more pronounced.

birefringence, and the effect is a useful tool in astronomical studies of magnetic fields in distant clouds of ionized gases.

A few crystalline substances are natural polarizers in that they absorb one component of polarization while being transparent to the other component. Tourmaline, a semiprecious stone often used in jewelry, is an example. This property of *selective absorption* is called *dichroism* (from the Greek *di*, meaning "two," and *chros*, meaning "skin" or "color"), because when viewed by transmitted light along two different directions these crystals usually exhibit two different colors. Unfortunately, natural dichroic crystals are generally very small.

40.5 Wave Plates and Circular Polarization

As mentioned previously, a birefringent material has two indices of refraction, one each for the *o*-ray and the *e*-ray. Light therefore travels with two different speeds through the material, depending on the direction of polarization of the incident light. (In calcite, the *e*-ray is faster; in some other materials, the *o*-ray is faster.) Suppose that we cut a piece of calcite into a thin slab[4] such that for a given wavelength of light the *o*-ray emerges from the slab just half a wavelength behind the *e*-ray. The two rays are thus out of phase by 180°. Such a slab is called a *halfwave plate*. It has interesting properties.

In Figure 40-11, a beam of light traveling along the x axis is linearly polarized at 45° with respect to the y axis. We can represent its electric field as two electric field components along the y and z axes that *vibrate in phase with each other*. (Do not confuse this representation with that of Figure 40-3b, in which components of *unpolarized* light have random and changing phase

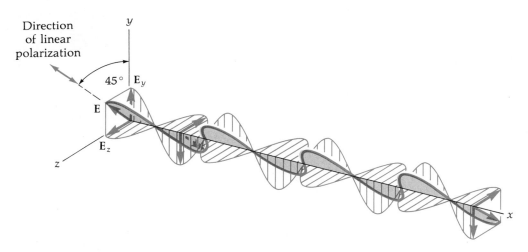

FIGURE 40-11
A linearly polarized light wave travels along the x axis, with its direction of polarization at 45° with respect to the y axis. The electric field **E** is resolved into two equal-amplitude components along the y and z directions, respectively.

---

[4] In general, a ray of light incident on a birefringent material is split into two distinct beams traveling *at an angle* to each other. However, there are certain directions in which the rays travel *along the same direction* at different speeds, along the direction of the "optic axis." The slab is constructed with this direction perpendicular to the front and back surfaces of the plate so that the two rays do not get out of alignment in traveling through the slab.

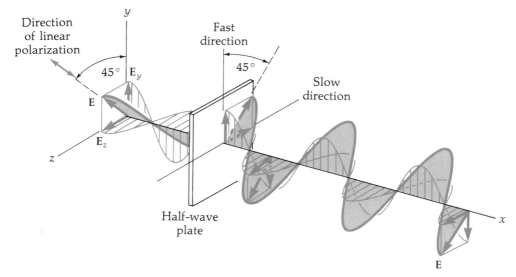

FIGURE 40-12
The halfwave plate is oriented so that E_y is along the "fast" direction and E_z is along the "slow" direction. As they emerge from the halfwave plate, the E_z vibrations have been retarded a half-wavelength (180°) behind the E_y vibrations, shifting the direction of polarization by 90°.

relationships.) Now allow this polarized light to enter a halfwave plate oriented so that these two components become the *o*- and *e*-rays in the plate. As they pass through the plate, the *o*-ray is retarded slightly relative to the *e*-ray. When they emerge, the components will be exactly 180° out of phase, shifting the direction of polarization 90°, as shown in Figure 40-12.

Figure 40-13 shows the effect of a *quarterwave plate* (that is, the slow and fast rays become out of phase by 90°). The two components of the electric field emerge 90° out of phase, producing **circularly polarized** light. If you look toward the source of such a wave as it approaches you, its electric field vector **E** will rotate at an angular frequency $\omega = 2\pi f$ (where f = light frequency).

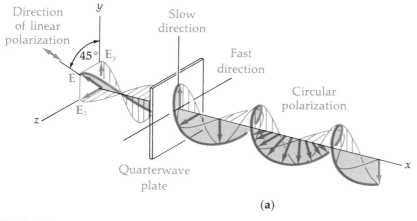

(a)

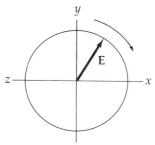

(b) When we look along the negative *x* direction, the **E** vector of the approaching wave rotates clockwise.

FIGURE 40-13
The quarterwave plate is oriented so that E_y is along the "fast" direction and E_z is along the "slow" direction. As these two equal components emerge from the quarterwave plate, the E_z vibrations have been retarded a quarter-wavelength (90°) behind the E_y vibrations, producing a *circularly polarized* wave.

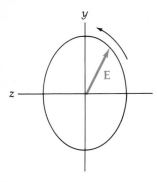

FIGURE 40-14
If a quarterwave plate is oriented at some arbitrary angle with respect to the direction of polarization of the incident light, the electric field components are *unequal* in magnitude (but 90° out of phase) and the emerging light is *elliptically polarized*.

Depending on which component lags behind, the direction of rotation will be *clockwise* or *counterclockwise*, corresponding to the two possible states of circularly polarized light.[5] If the direction of polarization is at an angle other than 90° with the fast and slow axes, the y and z components of the electric field are unequal (but still 90° out of phase), producing **elliptically polarized** light (Figure 40-14). The general name for halfwave plates, quarterwave plates, and so on, is *retardation plates*.[6]

EXAMPLE 40-3

What minimum thickness of calcite will make a halfwave plate for yellow light of wavelength $\lambda = 589.3$ nm? The indices of refraction for the o- and e-rays are $n_o = 1.6584$ and $n_e = 1.4864$, respectively.

SOLUTION

The times t_o and t_e for the o- and e-rays to travel through a plate of thickness d are $t_o = d/v_o$ and $t_e = d/v_e$. The respective velocities in the plate are $v_o = c/n_o$ and $v_e = c/n_e$. The time difference, $\Delta t = (t_o - t_e)$, is thus

$$\Delta t = \frac{d}{c}(n_o - n_e) \qquad (40\text{-}3)$$

To form a halfwave plate, we want the emerging e-ray to travel (in air) a half-wavelength before the o-ray finally emerges from the plate. The time required to do this is therefore

$$\Delta t = \frac{(\lambda/2)}{c} \qquad (40\text{-}4)$$

Combining Equations (40-3) and (40-4) gives $\lambda/2 = d(n_o - n_e)$. Solving for the plate thickness d and substituting numerical values yields

$$d = \frac{\lambda}{2}\left(\frac{1}{n_o - n_e}\right) = \frac{(5.893 \times 10^{-7} \text{ m})}{2(1.6584 - 1.4864)} = \boxed{1.713 \times 10^{-6} \text{ m}}$$

Note that a phase difference between the o- and e-rays of three-halves of a wavelength ($3\lambda/2$) would also produce a "halfwave" plate. In this case, the calcite would be three times as thick, or 5.139×10^{-6} m. Similarly, $5\lambda/2$, $7\lambda/2$, etc., would also act as "halfwave" plates.

---

[5] As discussed in Chapter 35, a light beam carries *linear* momentum, so that when it strikes an absorber it imparts a force against the absorber. It can be verified experimentally that, in addition to this force, a *circularly* polarized light beam has *angular* momentum and therefore also exerts a *torque* on the absorber. It is interesting that in the particle, or *photon*, model for light every individual photon is circularly polarized and carries one unit of angular momentum $L = h/2\pi$, where h is Planck's constant. The conservation of angular momentum requires that an atom that emits a photon must therefore itself undergo a change of angular momentum by one unit in a sense of rotation opposite to that of the photon. Plane-polarized light is actually an equal mixture of photons, with clockwise and counterclockwise senses of rotation.

[6] You can make fairly good retardation plates for amateur experimentation from certain kinds of cellophane tape, or by stretching transparent plastic food-wrap films. For interesting experiments with these plates, see the "Amateur Scientist" section of *Scientific American*, December 1977.

40.6 Optical Activity

Just as certain materials transmit linearly polarized light with two different speeds, some substances transmit circularly polarized light with two different speeds, depending on the sense of rotation of the electric vector. As will be explained, this has the interesting effect of causing a shift in the direction of *linearly* polarized light. For example, if a sugar solution is placed between a polarizer and an analyzer, the solution will rotate the direction of polarization, as shown in Figure 40-15. Such substances are called *optically active*. The amount of rotation is proportional to the distance traveled; in solutions, it is also proportional to the concentration of the optically active substance.

We can explain the mechanism causing the rotation by recognizing that linearly polarized light may be considered as the sum of two circular polarizations rotating in opposite directions, Figure 40-16. In optically active substances, one of the rotating components of light travels through the material faster than the other. This causes the two rotating components to gradually change their phase with respect to each other, so they add to a resultant vector along a different direction. Consequently, the direction of linear polarization gradually changes to a new direction as the light travels through the substance.

The shift may be in a clockwise or a counterclockwise sense, depending on the arrangement of atoms in the molecules. For example, sugar comes in two different forms with the same chemical formula, but with atoms arranged as mirror images of each other. Such pairs are called *stereoisomers*, Figure 40-17.

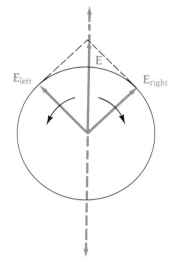

FIGURE 40-16
A *linearly* polarized light wave travels OUT of the paper toward the reader. The electric vector **E** oscillates up and down along the dashed line (the direction of polarization). We may represent the vector **E** as the sum of two *circularly* polarized components, E_{left} and E_{right}, rotating in opposite senses as shown. As they rotate, they add together to form the oscillating vector **E**.

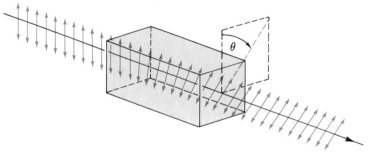

(a) As a linearly polarized wave travels through an *optically active* substance, the plane of polarization gradually rotates about the direction of propagation.

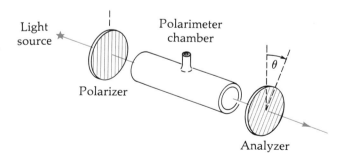

(b) A *polarimeter* measures the angle θ through which the direction of polarization is rotated. Often, different wavelengths are rotated by different amounts, causing color changes as the analyzer is rotated. To standardize measurements, usually light of a single specified wavelength is used.

FIGURE 40-15
Optically active substances cause a shift in the direction of linearly polarized light.

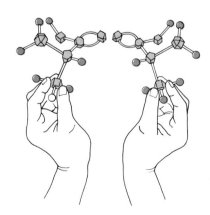

FIGURE 40-17
Stereoisomers have the same chemical composition, but the physical configurations of their atoms are mirror images.

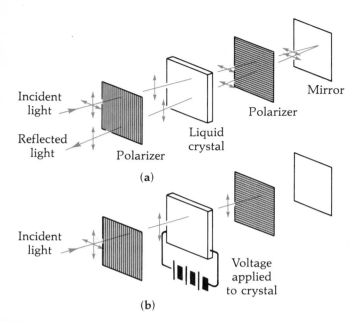

(c) A strip of Polaroid reveals the direction of polarization of the light reflected from the LC display on this hand calculator.

FIGURE 40-18

Liquid crystal displays. An interesting application of optical activity is the *liquid crystal display* (LCD) used on wristwatches, lap-top computer screens, calculators, the gallon- and dollar-displays on some gas pumps, and many other items. The molecules of a liquid crystal are more ordered than in a liquid, but not as ordered as in a crystal. They have interesting properties. Certain types of LCs have the ability to rotate the direction of polarization of polarized light and to lose that ability in the presence of a small electric field. In Figure 40-18a, a thin layer of LC that causes a 90° rotation is placed between crossed polarizing sheets and backed by a mirror. When light is incident from outside, the polarized light falling on the LC is rotated 90°, reflected, and rotated 90° again so that it passes through the front polarizer and appears bright in reflected light. When a voltage is applied to the crystal, Figure 40-18b, the plane of polarization is not rotated and no light reaches the mirror; hence that region appears dark. Electrical circuitry is formed by transparent conducting electrodes evaporated on the surface. With a grid of closely spaced conducting rows and columns, pictures can be formed as patterns of extremely small dots. LCDs require very little power because the ambient room illumination is used as the light source.

The molecules of one type form long twisted chains that rotate clockwise as you travel along the axis, while the other type forms a counterclockwise helix. The sugar called *dextrose* (from the Latin *dextro*, meaning "right") causes the direction of linear polarization to revolve clockwise as seen by an observer toward whom the light is moving, a sense of rotation called *right-handed*. The sugar called *levulose* (from the Latin *levo*, meaning "left") causes a counterclockwise or *left-handed* rotation.[7] A *saccharimeter* measures the amount of optical rotation to determine the sugar concentration in commerical syrups, wines, and so on, and in urine samples to test for suspected diabetes.

40.7 Interference Colors and Photoelasticity

If a sheet of birefringent cellophane is folded randomly several times and placed between polarizing sheets, the transmitted light shows a pattern of vividly colored areas. These colors arise because certain layers of cellophane may act

---

[7] Note that a given helix has the *same* sense of rotation as you travel along the axis in either direction— so it doesn't matter how the molecules are oriented in the solution. You can easily observe optical activity in ordinary transparent corn syrup (dextrose), which causes about 12°/cm rotation. Turpentine causes a counterclockwise rotation of $-3.7°$/cm. *Liquid crystals*, a class of organic compounds that can flow yet maintain molecular orientations, have helical molecules that produce extremely large *rotatory powers*, on the order of 40 000°/mm.

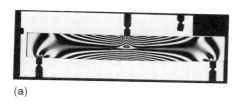
(a)

(b)

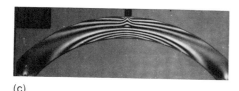
(c)

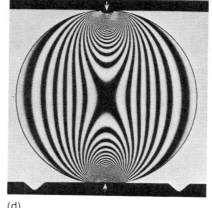
(d)

(e)

FIGURE 40-19
Models of mechanical structures are made of special photoelastic plastic and placed between polarizing sheets. When forces are applied to the models, they become birefringent, producing patterns that indicate the stress distribution within the models.

as a quarterwave plate for red light, while also acting as a halfwave plate for blue light, and so forth. Thus the direction of polarization of the light striking the cellophane may be rotated different amounts for different wavelengths, allowing some wavelengths to pass the analyzer while others are blocked. The emerging light is therefore deficient in certain portions of the spectrum, producing striking color effects. Rotating the cellophane or either polarizing sheet produces changing colors that are beautiful to see.

This aesthetically pleasing effect has practical uses. Transparent scale models of mechanical structures such as I-beams and arches are made from special *photoelastic* plastics. When the models are placed between a polarizer and an analyzer and "loaded" by having forces applied to them, the plastic becomes birefringent in amounts proportional to the applied stress. The resultant patterns of light and dark (and colors) give a map of the regions of mechanical stress within the model, Figure 40-19. Similar photoelastic colors can be observed if you use polarizing glasses to view light reflected from plastic boxes, plastic T-squares used in drafting, and other transparent objects, Figure 40-20. Some plastics are not strongly birefringent, so you may have to search to find those that show marked effects.

Polarized light is useful in numerous other applications. For example, atoms in the presence of magnetic fields emit polarized light (the Zeeman effect); this polarization is used in the investigation of magnetic fields near sunspots and in distant stars. Also, magnetic fields in far regions of our galaxy cause elongated dust grains present in interstellar gas and dust clouds to align parallel to one another. Light from nearby stars scattered by these clouds is partially polarized, so by analyzing the percentage and direction of polarization of this scattered starlight, we can obtain information about these distant magnetic fields. Much information about crystal structure, biological specimens, and other materials is obtained by analysis with polarized light.

FIGURE 40-20
A plastic template placed between sheets of Polaroid produces rainbow-colored strain patterns. When the template was manufactured, stresses caused the plastic to become birefringent.

Summary

Transverse waves are *linearly polarized* if all the vibrations associated with the waves are parallel to the direction of a fixed line in space. The *direction of polarization* of an electromagnetic wave is the direction of the *electric* field vector.

When a single (free) atom undergoes a transition from a higher to a lower energy state, it emits a *wave train* of radiation, which for visible light is of the order of 1 to 3 m long in the direction of propagation. *Unpolarized* light is the superposition of many wave trains whose electric vectors are oriented randomly in all possible directions.

Certain transparent materials such as Polaroid selectivity absorb some directions of polarization more than other directions, so they transmit electromagnetic waves that are partially or completely linearly polarized. If the direction of polarization of incident *polarized* light (intensity I_0) has an angle θ with respect to the *transmission axis* of an ideal polarizer, the transmitted intensity I (proportional to E^2) is

MALUS'S LAW $\qquad I = I_0 \cos^2 \theta$

Unpolarized light becomes 100% polarized when reflected from dielectric materials at an angle of incidence called the **Brewster angle** θ_p, for which the reflected and refracted rays are at right angles.

BREWSTER ANGLE $\theta_p \qquad \tan \theta_p = n$

where n is the index of refraction of the material relative to that of the surrounding medium. At other angles, the reflected light is partially polarized.

Birefringent substances have two indices of refraction, depending on the direction of polarization of the incident light. *Retardation plates* (or *wave plates*) are constructed of birefringent materials so that the *ordinary* (o) and *extraordinary* (e) waves emerge out of phase. When a polarized wave passes through a *quarterwave plate*, one component is shifted 90° relative to the other component; when a polarized wave passes through a *halfwave plate*, one component is shifted 180° relative to the other. *Circularly polarized* light is composed of o and e components of equal amplitude that are out of phase by 90°. *Optically active substances* (sugar solution, for example) transmit circularly polarized light with two different speeds, causing a shift in the direction of incident *linearly* polarized light.

Interference colors are produced from white light when various thicknesses of birefringent films are placed between polarizing sheets (a *polarizer* and an *analyzer*). The colors arise because certain layers of the film act as a halfwave plate for, say, blue light, but as a quarterwave plate for red light, thus shifting the direction of polarization more for certain wavelengths than for others and allowing some wavelengths to pass the second polarizer while others are blocked. Therefore, the emerging light has certain portions of the spectrum missing; the remaining portions produce the color effects.

We can analyze mechanical structures by constructing transparent models from *photoelastic* materials. When a model is placed between polarizing sheets and mechanically stressed, the material becomes birefringent, producing fringe patterns that reveal the stress conditions within the structure.

Questions

1. Can longitudinal waves such as sound waves be polarized? If so, how?
2. Which phenomenon, polarization or interference, provides the most convincing evidence for the wave nature of light?
3. What aspect of the wave nature of light do polarization phenomena reveal that interference does not?
4. A radio-telephone transmitter in an automobile uses an antenna that is straight and vertical. Is the electromagnetic radiation from such an antenna vertically or horizontally polarized? Explain.
5. A grid of closely spaced vertical wires is opaque to vertically polarized microwaves. Why?
6. Light is not transmitted through crossed polarizers. However, if a third polarizer is placed between the crossed polarizers, some light may be transmitted. Explain.
7. How can a stack of polarizing sheets be used to rotate the plane of polarization of polarized light?
8. One form of a variable-density light filter consists of two polarizing sheets placed together such that the orientations of their transmission axes may be rotated relative to each other. Does a small rotation produce a greater change in transmitted intensity when the axes are nearly aligned, nearly crossed, or at some angle in between?
9. One sheet of polarizing material is removed from a stack of randomly oriented polarizing sheets. As a result, the light transmitted through the stack decreases. How could this happen?
10. An ideal polarizing sheet transmits only half of the incident unpolarized light. What happens to the other half?
11. Many fishermen use polarized sunglasses while fishing. Why?
12. Can light be polarized by reflection at an interface between two transparent media if the light is traveling toward the interface from the region of higher refractive index?
13. How would you determine whether a beam of light is unpolarized, plane-polarized, or circularly polarized?
14. In some situations a photographer uses a polarizing filter over the lens of his or her camera. What would be some of these situations?
15. A beam of plane-polarized light may be represented by the superposition of two circularly polarized beams of opposite rotation. What is the effect of changing the relative phase of the two beams?

16. A fascinating device consists of a pair of polarizing sheets, each of which has a quarterwave plate laminated to it. Light is transmitted when one of the pair is placed over the other, but is not transmitted when the order of the pair is interchanged. What are the details of their construction and why do they behave as they do?

17. If one slit of a double-slit interference apparatus were covered by a polarizing sheet with its axis perpendicular to the slit, while the other slit were covered by a polarizing sheet with its axis parallel to the slit, would an interference pattern be produced? Explain.

18. Photoelastic plastic models of mechanical structures placed between polarizers exhibit stress by producing colored bands, as shown in Figure 40-16. How can the spacing of the bands be interpreted?

Problems

40.2 Polaroid

40A-1 Unpolarized light passes through two (ideal) polarizing sheets. If the angle between the transmission axes of the sheets is 60°, determine the fraction of the incident light intensity absorbed by the sheets.

40A-2 Two ideal polarizing sheets are placed together so that there is an angle θ between their transmission axes. Find the angle θ such that the sheets transmit 45% of the incident unpolarized light intensity.

40B-3 Two polarizing sheets are placed together with their transmission axes crossed so that no light is transmitted. A third sheet is inserted between them with its transmission axis at an angle of 45° with respect to each of the other axes. Find the fraction of incident unpolarized light intensity that will be transmitted by the combination of the three sheets. (Assume that each polarizing sheet is ideal.)

40B-4 Unpolarized light falls upon three ideal polarizing sheets. The transmission axis of the second sheet is rotated 30° with respect to that of the first sheet, and the transmission axis of the third sheet is rotated 30° with respect to that of the second sheet. Calculate the fraction of the incident light intensity transmitted by the three sheets.

40.3 Polarization by Reflection and Scattering

40A-5 A beam of unpolarized light is incident upon a sheet of glass at the polarizing angle of 58°. Find the angle of the refracted beam inside the glass.

40A-6 For a particular wavelength, the index of refraction is 1.50 for a sample of glass. Calculate the Brewster angle θ_p for this refractive index. In general, does the Brewster angle increase or decrease as the wavelength of incident light increases?

40A-7 The Brewster angle of a plate of glass is 57° when the plate is in air. Calculate the Brewster angle for the glass plate when the plate is under water ($n = 1.33$).

40A-8 An unpolarized light beam reflected from the surface of water is plane polarized for a reflection angle of 53°. (a) Calculate the index of refraction for the water. (b) Show that the angle that the refracted beam makes with the normal to the surface is the complement of 53°.

40B-9 The critical angle for total internal reflection in a dielectric material is θ_c. Derive an expression for the Brewster angle θ_p in terms of θ_c for the material.

40B-10 Derive Brewster's law for polarization by reflection, Equation (40-2).

40.4 Birefringence
40.5 Wave Plates and Circular Polarization
40.6 Optical Activity

40B-11 Quartz is birefringent, with indices of refraction of 1.553 and 1.544 for incident light of wavelength 589 nm. Find the minimum thickness of quartz that acts as a quarterwave plate at this wavelength.

40B-12 A beam of circularly polarized light is incident upon a polarizing sheet. Explain why the light is transmitted equally well for all orientations of the sheet.

40B-13 (a) Show that when a beam of circularly polarized light is incident upon a quarterwave plate the emerging light is plane-polarized. (b) Show that if the rotation sense of the circularly polarized light is reversed the direction of polarization of the emerging light is changed by 90°.

40B-14 A retardation plate made of quartz ($n_e = 1.544$, $n_o = 1.553$) is cut so that the optic axis lies in the plane of the plate. Calculate the minimum thickness of the plate such that it will be a fullwave plate for light with a wavelength of 500 nm and a halfwave plate for light with a wavelength of 600 nm.

40B-15 A concentration of one gram of cane sugar (sucrose) in one cubic centimeter of water rotates linearly polarized light 66.8° for 10 cm of path length. An unknown sucrose solution in a saccharimeter 35 cm long produces a 16° rotation. Find the concentration of the solution.

40B-16 A halfwave plate is inserted between two polarizing sheets whose directions of polarization are parallel. The halfwave plate is oriented with respect to the first sheet as in Figure 40-11. (a) Explain why *no* light passes through the combination. (b) If, instead, the two polarizing sheets are crossed at 90°, explain why *all* the light transmitted by the first sheet passes through the second sheet. (c) In part (a), explain qualitatively the nature of the light emerging from the combination as the halfwave plate is slowly rotated through 360°.

Additional Problems

40C-17 A variable transmission filter is composed of two polarizing sheets, one of which can be rotated relative to the other. Determine the angle between the transmission axes at which an incremental change in rotation $d\theta$ produces the greatest fractional change dI/I_0 in the intensity of the transmitted light.

40C-18 Two ideal polarizing sheets are placed together with their transmission axes at 90°. A third sheet is inserted between the sheets so that its transmission axis is at an angle θ with respect to that of the sheet closest to an incident beam of unpolarized light. Derive an expression for the fraction I/I_0 of the incident light intensity I_0 that is transmitted through the three sheets as a function of θ.

40C-19 A stack of polarizing sheets will rotate the direction of polarization of incident, linearly polarized light if each successive sheet is oriented at an angle θ (in the desired direction) with respect to the previous sheet. Using 10 ideal sheets to produce a 90° rotation, determine the maximum percentage of the incident, polarized light intensity I_0 that will be transmitted through the tenth sheet.

40C-20 Light composed of both linearly polarized and unpolarized light passes through an ideal polarizing sheet. As the sheet is rotated, the transmitted light varies from a maximum intensity to one-third maximum intensity. Calculate the fraction of the incident light intensity that is linearly polarized.

40C-21 Figure 40-8b shows that the extraordinary ray does not conform to Snell's law for refraction. Utilizing Huygens' principle show, by a sketch of Huygens' wavelets within the calcite, how the refraction shown in the figure is possible.

40C-22 Figure 40-21 shows a calcite prism made with its optic axis perpendicular to the plane of the paper. A beam of yellow sodium light ($\lambda = 589$ nm) is incident normally on the top face as shown. For this wavelength, the indices of refraction are $n_o = 1.658$ and $n_e = 1.486$. (a) Find the minimum prism angle θ such that the ordinary ray will be totally internally reflected. (b) Show that the extraordinary ray will emerge from the slant face and thus be 100% linearly polarized. What is this direction of polarization? (c) Find the direction of the extraordinary ray after it emerges.

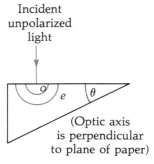

FIGURE 40-21
Problem 40C-22.

40C-23 A *Babinet compensator* consists of two quartz wedges, A and B, in contact with one another as shown in Figure 40-22. The optic axis of wedge A is vertical in the plane of the paper and that of wedge B is perpendicular to the plane of the paper. Thus, the extraordinary ray (refractive index n_2) in wedge A becomes the ordinary ray (refractive index n_1) in wedge B, and

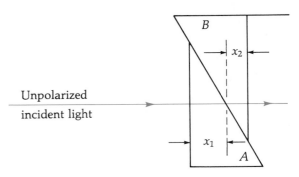

FIGURE 40-22
Problem 40C-23.

vice versa. By sliding wedge B along wedge A, we vary the difference between distances x_1 and x_2. Show that a phase difference $\Delta\phi$ between the two emergent rays can be varied according to the equation $\Delta\phi = ((2\pi)/\lambda)(n_2 - n_1)(x_1 - x_2)$.

40C-24 Unpolarized light of wavelength λ is incident upon a thin slab of birefringent material. The slab thickness is b, and the indices of refraction are n_o for the ordinary ray and n_e for the extraordinary ray. In terms of the given constants, derive an expression for the phase difference ϕ between the two emerging rays from the slab.

40C-25 The minimum thickness of the halfwave plate described in Example 40-3 is too thin to be practical. Determine an exact thickness near 0.1 mm that will produce the same effect as the minimum thickness.

40C-26 A quartz plate 0.610 mm thick is cut so that the optic axis lies in the plane of the plate. Polarized light incident on the plate has its direction of polarization at an angle of 45° with respect to the optic axis of the plate. Calculate the wavelength(s) between 600 nm and 700 nm that will produce an emergent light that is linearly polarized. (Assume that $n_e = 1.544$ and $n_o = 1.553$ for all wavelengths.)

40C-27 Show that the angle of rotation for polarized light passing through an optically active medium is exactly half of the phase shift between the right and left circularly polarized components.

40C-28 As shown in Figure 40-16, linearly polarized light may be considered as the sum of two circularly polarized components, rotating in opposite senses. In some optically active substances, circularly polarized waves travel with two different speeds, v_L and v_R, respectively, for left and right senses of rotation. As a consequence, linearly polarized light incident on a slab of this substance will emerge with its direction of polarization rotated through an angle θ. Derive an expression for the angle θ in terms of the thickness d of the slab, the two indices of refraction n_L and n_R, and the wavelength λ.

40C-29 Colorless corn syrup (from the grocery store) is mixed with 3 times its own volume of water. A 20-cm path length of this solution rotates linearly polarized light 59°. Find the rotation produced in a 10-cm path length of pure corn syrup.

40C-30 Describe the appearance of a liquid crystal display (Figure 40-18) whose polarizers are oriented with their axes parallel.

CHAPTER 41

Special Relativity

It was Einstein who made the real trouble. He announced in 1905 that there was no such thing as absolute rest. After that there never was.

STEPHEN LEACOCK

Newton, forgive me.

ALBERT EINSTEIN

41.1 Introduction

Two revolutions in physics occurred in the early part of the twentieth century that radically changed our concepts of the universe. One was the work of several people over a period of decades: the development of quantum mechanics. The other was the theory of special relativity,[1] published by Albert Einstein in 1905. Einstein's theory not only led to apparent paradoxes that seemed to violate common sense in the most radical way, but it completely changed our basic understanding of space and time. As far as is known today, special relativity unquestionably describes the way the world "is."

The main difficulties in understanding special relativity are not mathematical ones. Rather, the challenge comes from our reluctance to discard deeply ingrained ideas about space and time. We grow up using Newtonian concepts to explain physical phenomena, and it is disturbing to have cherished beliefs overthrown. Furthermore, the structure of our language reflects these common-sense classical notions, so this adds to the difficulty of gaining a new perspective. Of course, the classical way of thinking cannot be completely wrong, since it does serve admirably to explain everyday experiences. But scientists exploring the fine details of natural phenomena must abandon classical concepts and deal with a more modern theory.

The basic question that relativity asks is this:

If a given phenomenon is viewed from two different frames of reference that have uniform relative motion with respect to each other, how do the two measurements of the phenomenon compare?

---

[1] The *special* theory deals with frames of references that have constant motion in a straight line relative to each other. The *general* theory, published in 1916, treats accelerated frames of reference (see Section 41.16).

Einstein points out that making a measurement involves determining *where* and *when* something happens in *space* and *time*. In particular, we seek four quantities about an **event** that happens at a given point in space and at a given time:

A point event (x,y,z,t)

These four quantities (x,y,z,t) are measured in some inertial frame of reference that we will call the S frame, assumed to be "at rest." Another frame of reference, the S' frame, moves with constant velocity **V** along the $+x$ direction of S. For convenience, we align the two frames so that their origins and respective axes are coincident at the time $t = 0$ (see Figure 41-1). Unprimed quantities designate measurements made in the S frame, while primed (') quantities are for the S' frame. Each frame is equally valid, and measurements made in either frame correctly measure space and time for that frame. Both frames of reference are *inertial frames*, since neither has acceleration. Relativity shows that it is only the *relative* velocity that is important, not which system is imagined to be "at rest." We could equally well assume that S' is at rest and S moves with a velocity $-$**V** (in the negative x' direction). The basic conclusions of relativity would be exactly the same.

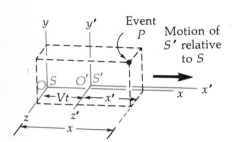

FIGURE 41-1
The two coordinate systems S and S'. The S' frame has velocity **V** in the $+x$ direction relative to S. The frames are coincident at the time $t = t' = 0$. At a later time t, their origins O and O' have moved a distance Vt apart when the event P occurs.

How to Make Measurements

Basically, all measurements reduce to determining the four quantities associated with a point event: (x,y,z,t). Einstein suggests that, in principle, a meter-stick framework be extended throughout the frame of reference and an "observer" be stationed at every location within the frame. Each observer has a clock that has been synchronized with all other clocks in the frame. *Every event is to be measured by a "local" observer, situated where the event occurs.* The spatial coordinates (x,y,z) of the event are found by reference to the meter-stick framework in that vicinity, and the time (t) is given by the observer's clock. Because all measurements are of *local* events, one does not have to take into account the transit times that would be involved for light signals to travel from some distant event to the observer. Even if such cases were allowed, however, the conclusions of relativity would be the same.

41.2 The Galilean Transformation

Classically, we assume that measurements of spatial intervals and time intervals are the same for observers in all inertial frames. Indeed, this is the basic assumption upon which Newtonian mechanics is founded. It agrees with our common sense. For relatively slow velocities, Newtonian mechanics is sufficiently valid in all inertial frames of reference (as anyone who has flown in a smoothly moving airplane will testify). Stated another way, there is no mechanical effect by which observers in the S and S' frames could determine which frame is "truly" moving and which is "at rest." This fact is known as the **Galilean relativity principle**: *the laws of Newtonian mechanics are the same in all inertial frames*.

If these assumptions are true, how do we express the relationship between an event as measured in the S frame and the measurement of the *same* event as made in the S' frame?[2] For the event P depicted in Figure 41-1, simple

---

[2] Note that *events happen in space and time*. They do not happen in a particular frame of reference. Any inertial frame can be used for determining the four coordinates of an event, relative to that frame.

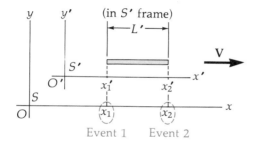

FIGURE 41-2
In the S frame, the two events that determine the location of the ends of the moving stick at x_1 and x_2 are *simultaneous* events.

geometry reveals the equations relating the two sets of measurements. They are called **the Galilean transformation**.

THE GALILEAN TRANSFORMATION

$$\begin{aligned} x &= x' + Vt' & x' &= x - Vt \\ y &= y' & y' &= y \\ z &= z' & z' &= z \\ t &= t' & t' &= t \end{aligned} \qquad (41\text{-}1)$$

The transformation equations are a sort of foreign-language dictionary that translates the description of an event as measured in one frame (x,y,z,t) into the description of the same event as measured in the other frame (x',y',z',t'), and vice versa. Note that in the left-hand set primed quantities appear only on the right side, while in the other set they appear only on the left side. Many of our basic assumptions about the nature of space and time are contained in the transformation equations. For example, the fact that we write $t = t'$ implies a universal (or absolute) time scale that is valid for all frames of reference. Similarly, the equations imply that the space in which events happen is the same in both frames. The difference in the x-coordinate descriptions clearly has its origin in the relative motion of the frames; it does not imply that *space itself* is different in the two frames. These classical ideas about space and time are so strongly ingrained in experience that is seems impossible to imagine they are not correct. Indeed, for centuries philosophers have accepted them without question. This makes all the more remarkable the great revolution Einstein brought about when his relativity theory showed these classical ideas to be wrong.

To illustrate the use of the transformation equations, consider a rod in the S' frame and determine its length as measured in both the S' and the S frame.[3] The rod is oriented along the x' axis as shown in Figure 41-2. We define the length of the rod to be $L' = x'_2 - x'_1$. *Since the locations of the ends do not change with time*, we measure these values in the usual way by placing a ruler next to the rod and locate the ends in terms of two point events:

Event 1, locating the left end: (x'_1, y'_1, z'_1, t'_1)
Event 2, locating the right end: (x'_2, y'_2, z'_2, t'_2)

The length L' of the rod depends only on x'_2 and x'_1. In particular, the times t'_2 and t'_1 are not involved.

However, in the S frame, the rod is moving. Let us investigate the quantity $x'_2 - x'_1$ as expressed in terms of measurements in S. Applying the

---

[3] In reading about relativity, it is essential that you keep in mind at all times which frame of reference is being discussed. Those primes are of prime significance!

Galilean transformation, Equation (41-1), we have

$$x'_2 - x'_1 = (x_2 - Vt_2) - (x_1 - Vt_1)$$

or, rearranging the right-hand side,

$$x'_2 - x'_1 = (x_2 - x_1) - V(t_2 - t_1) \qquad (41\text{-}2)$$

The quantity $(x_2 - x_1)$ is the length of L of the rod as measured in the S frame. Obviously, it would not make much sense to locate the left end of the rod at t_1 and later, after the rod has moved, to locate the other end at a different time t_2. So, in the S frame we adopt the reasonable procedure of *determining the locations of the ends simultaneously*, when $t_2 = t_1$. Thus, Equation (41-2) becomes

$$x'_2 - x'_1 = x_2 - x_1$$
$$L' = L$$

or

The ends of the moving object are located *simultaneously*, so the length as measured in S (where the object is moving) corresponds to the length as measured in S' (where the object is at rest). In Galilean relativity, these two length measurements give the same answer. Einstein showed that this conclusion is incorrect.

Velocity Addition in Galilean Relativity

At the instant the S and S' frames coincide at $t = t' = 0$, assume that a particle passes the origin moving along the $+x$ (and $+x'$) direction with a constant speed u' as measured in the S' frame. We obtain the velocity addition relation by considering the following two events:

		In S	In S'
FIRST EVENT:	*Particle at origin*	(0,0,0,0)	(0,0,0,0)
SECOND EVENT:	*Particle away from origin*	(x,0,0,t)	(x',0,0,t')

In the S' frame, during the time t' the particle moves a distance x' with constant speed u', so $v' = x'/t'$. The S' frame itself moves along the $+x$ direction with constant speed V relative to the S frame. This motion is the second velocity V that we will add to the particle velocity u' to give the velocity u of the particle as measured in the S frame. In S, the particle's speed is $u = x/t$. We make use of the Galilean transformation to express this speed in terms of the primed measurements:

$$u = \frac{x}{t} = \frac{x' + Vt'}{t'} = \frac{x'}{t'} + Vt'$$

Because $x'/t' = u'$, we have

GALILEAN VELOCITY ADDITION $\qquad u = u' + V \qquad (41\text{-}3)$
(for velocities along
the $\pm x$ direction)

Here, u is the velocity of the particle as measured in S, u ⎓ of the particle as measured in S', and V is the velocity of S' relative to S. If any velocities are in the $-x$ (or $-x'$) direction, minus signs are used with the corresponding numerical values. This relation is the same as the one that we derived previously in Section 9.4.

EXAMPLE 41-1

A child on a moving train rolls a ball along the aisle with a speed of 2 m/s toward the front of the train. (a) If the train moves along a straight track at a constant speed of 5 m/s, what is the speed of the ball as measured in the earth's frame of reference? (b) If the child rolls the ball toward the rear of the train, what velocity does the ball have in the earth's frame?

SOLUTION

(a) We choose the train as the S' frame and the earth as the S frame, with the $+x$ and $+x'$ axes in the direction of the train's motion. In S', the ball's velocity is $u' = 2$ m/s. Thus:

$$u = u' + V = 2 \text{ m/s} + 5 \text{ m/s} = \boxed{7 \text{ m/s}}$$

(b) In this case, $u' = -2$ m/s so

$$u = u' + V = -2 \text{ m/s} + 5 \text{ m/s} = \boxed{3 \text{ m/s}}$$

The numerical value is positive, so in S the ball's velocity is in the $+x$ direction.

To obtain the transformation relation for accelerations, we differentiate Equation (41-3) with respect to time:

$$\frac{du}{dt} = \frac{du'}{dt} + \frac{dV}{dt}$$

Since V is constant, we obtain

$$a = a' \tag{41-4}$$

The acceleration of a particle is thus the same in all inertial frames of reference in relative motion. In classical physics, the mass m of a particle is not affected by motion, so $ma = ma'$, which leads to the conclusion that $\Sigma F = ma$ and the rest of Newtonian laws are valid in both S and S', and therefore are valid in all inertial frames of reference. As we have shown in prior chapters, the fundamental conservation relations for energy and momentum are direct consequences of Newton's laws, so we conclude that *all the laws of mechanics are the same in all inertial frames of reference*. This statement is called the **Galilean relativity principle**. True, the velocity, momentum, and kinetic energy of a particle will have different values in different frames that are in uniform relative motion. But the fundamental laws of mechanics will be the same in all inertial systems. We express this property by saying that "the fundamental laws of mechanics are *invariant* under the Galilean transformation."

To the degree of accuracy normally required, classical mechanics provides an excellent description of the motions of objects. Engineers and scientists have used it for centuries and will continue to do so. However, after Maxwell developed electromagnetic theory in the 1860s, certain puzzles emerged that

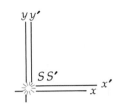

(a) At $t = t' = 0$, a flashbulb is set off at the coincident origins O and O'.

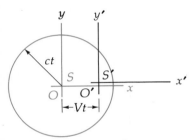

(b) In the S frame at a later time t, the expanding spherical wavefront is centered on the origin O. The S' frame has moved in the $+x$ direction a distance Vt.

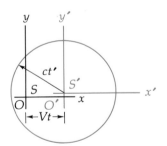

(c) In the S' frame at a later time t', the expanding spherical wavefront is centered on the origin O'. The S frame has moved in the $-x$ direction a distance Vt'.

FIGURE 41-3
The so-called "paradox" of the expanding light sphere. Observers in *each* frame of reference, measuring the *same* expanding wavefront, find that it is a sphere centered on their own origin. This is not a paradox in the context of the new space and time of special relativity.

were not resolved until Einstein took a bold new approach. The major difficulty arose because Maxwell's equations predicted a specific speed for light: $c = 1/\sqrt{\mu_0 \varepsilon_0} = 3 \times 10^8$ m/s. In the late nineteenth century, light was believed to be an electromagnetic wave propagated through a medium called the *ether*. But if this were true, the speed of light would certainly not have the same value in frames of reference that moved relative to the ether.[4] This lack of invariance of Maxwell's equations with respect to the Galilean transformation profoundly disturbed Einstein. For philosophical reasons, Einstein felt deeply that a relativity principle ought to apply to *all* the laws of physics, not just the laws of mechanics. In fact, it would be bizarre if mechanics were separated from the rest of physics in this respect.

41.3 The Fundamental Postulates of Special Relativity

Einstein based his theory of relativity on two assumptions:

BASIC POSTULATES OF SPECIAL RELATIVITY

(1) **All the laws of physics have the same form in all inertial frames.** (*The Principle of Relativity*)

(2) **The speed of light in a vacuum has the same value c in all inertial frames.** (*The Principle of the Constancy of the Speed of Light*)

The entire theory of special relativity is derived from just these two postulates. Their simplicity and generality are characteristic of Einstein's genius. As a consequence, Einstein showed that Newtonian mechanics is only approximately correct, usable in cases in which velocities are small compared with the speed of light. In fact, Einstein's relativistic mechanics approaches Newtonian mechanics when $v \ll c$.

The first postulate appears quite reasonable and can be accepted without qualms. However, the implications of the second postulate seem absurd. For example, suppose that at the instant the two frames are coincident, a flashbulb is set off at the coincident origins O and O', Figure 41-3. If the speed of light is c in all frames of reference, at a later time observers in each frame would detect a symmetrically expanding sphere of light that is centered on their respective origin. *Though each set of observers measures the same expanding wavefront, each finds it to be an expanding sphere, centered on the observers' own origin!* In this chapter we will convince you that this, indeed, is the true situation, and that it is not a paradox.

We now examine some important conclusions of relativity, particularly with respect to space and time. By themselves, these conclusions seem paradoxical and contrary to common sense. But if we consider all the conclusions of special relativity together, *and manage to give up our Newtonian concepts of absolute space and time,* they form a coherent and satisfying theory—one that has been verified experimentally an overwhelming number of times. And, of course, experiment is the ultimate test of any theory. Einstein commented upon the fact that relativity disagrees radically with our common sense by saying, "Common sense is that layer of prejudices laid down in the mind prior to the age of eighteen."

---

[4] For example, motion through the ether (or, equivalently, an ether "wind" blowing past the observer) should result in a different speed for light along two right-angle paths: parallel to the motion and at right angles to the motion. This is the effect sought (but not found) in the Michelson–Morely experiment, Section 38.5. An analogy to this situation is treated in Problem 9B-23.

41.4 Setting Clocks in Synchronism

Einstein points out that to make measurements of events, a "local" observer, situated where the event occurs, determines the coordinates (x,y,z) by comparison with the meter-stick framework, and the time (t) by comparison with the observer's local clock, which has been synchronized with all other clocks in the frame of reference. In principle, *all measurements are to be made in this fashion.* We now discuss the procedure for synchronizing a system of clocks that are stationed at various points throughout the frame of reference. This matter of synchronization is the source of many of the "paradoxes" of relativity, so it has greater significance than might be suspected at first glance.

We cannot synchronize clocks when they are together, then move them to their respective positions. Because of an effect called *time dilation*, which we will discuss shortly, to transport clocks in this fashion would cause them to get out of synchronization. Instead, Einstein proposed the method illustrated in Figure 41-4, which utilizes the constant speed of light c in its procedure. When two clocks are placed at their appropriate locations, a flash-bulb situated at the point midway between the clocks is set off, sending light pulses in opposite directions. The light pulses take the same time to traverse equal distances. *The clocks are set so that they indicate the same times at the arrivals of the pulses.* This is the basic synchronizing procedure that, in principle, could be extended, to all other clocks in the frame of reference one by one. The entire array of synchronized clocks establishes a time scale by which the simultaneity of events separated in space is judged in that frame. This gives the speed of light a more fundamental significance than that of being just one of the constants of nature. In particular, it is intimately related to our concepts of time and of simultaneity.

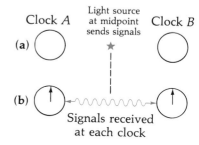

FIGURE 41-4
One method of synchronizing two clocks that are separated in space. A flashbulb at the midpoint sends light signals to each clock. If the clocks are set to read the same times when the signals arrive at each clock, they are correctly synchronized.

41.5 The Lorentz Transformation

Einstein derived a new set of transformation equations that replaced the Galilean transformation. They have the same mathematical form as an earlier transformation developed by H. A. Lorentz, so they are called the **Lorentz transformation.**[5] However, Einstein derived them using reasoning different from that of Lorentz, and the interpretation of the equations is vastly different from the meaning Lorentz attached to them. The derivation is based upon the second postulate of relativity and certain assumptions about the homogeneity of space and time, for example, that as far as physical experiments are concerned, all points in space are equivalent. To simplify the mathematical form, we define $\beta \equiv V/c$, where V is the relative speed along the x axis of the two frames of reference.

THE LORENTZ TRANSFORMATION (where $\beta \equiv V/c$)

$$x = \frac{x' + Vt'}{\sqrt{1 - \beta^2}} \qquad x' = \frac{x - Vt}{\sqrt{1 - \beta^2}}$$
$$y = y' \qquad y' = y$$
$$z = z' \qquad z' = z \qquad (41\text{-}5)$$
$$t = \frac{t' + Vx'/c^2}{\sqrt{1 - \beta^2}} \qquad t' = \frac{t - Vx/c^2}{\sqrt{1 - \beta^2}}$$

---

[5] Appendix I presents a simplified derivation of the Lorentz transformation.

Note that the two sets of equations "turn the crank" in opposite directions. We obtain either set from the other by interchanging primed and unprimed quantities and changing the signs of V and β. (To observers in S, the other moving frame has a velocity $+V$; but in S', the other frame has a velocity $-V$. Hence there is a sign change.)

The Lorentz transformation has an interesting characteristic. If the velocity of the moving frame is much smaller than c, then the factor β approaches zero and the Lorentz transformation becomes identical to the Galilean transformation. *So classical relativity is just a special case contained within Einstein's more comprehensive special relativity theory.*

We now discuss specific details. Note that the only novel features of the derivations are the use of Einstein's two postulates (as expressed by the Lorentz transformation). All the surprising conclusions that follow are contained implicitly in these two assumptions. Their justification rests on the tremendous successes that special relativity has had in explaining physical phenomena.

FIGURE 41-5
Einstein enjoying a moving frame of reference in 1936.

Albert Einstein

Between 1900 and 1927, there were two great revolutions in physics: quantum mechanics and relativity. The former grew from contributions by many physicists (including Einstein), but relativity was the creation of Einstein alone, a stunning accomplishment ranking easily with the achievements of Newton.

Albert Einstein was born in Ulm, Germany, in 1879, the year of Maxwell's death. His father owned a small electrochemical shop. Einstein did not speak at all before the age of three, nor fluently until he was almost nine. He particularly disliked the rigid discipline and authoritarian teaching methods common in German schools. His relatives predicted he would never amount to much, and his high school teachers considered him a "disruptive influence," asking him to leave school, which he did at age 15. Yet during this time he was intensely interested in geometry, algebra, and calculus; these he studied diligently on his own. After a year of roaming about in Northern Italy, at age 16 (two years younger than most applicants) he took the entrance examination for admission to the Federal Institute of Technology in Zurich, a renowned engineering school. He failed the test because of deficiencies in modern languages, zoology, and biology. After returning to high school to earn his diploma and doing some extra studying with the help of a friend, he took the exam again and was admitted. He seemed an indifferent student, uninspired by the old-fashioned nature of the curriculum, attending classes sporadically, and spending considerable time in the local cafes. But he also thought a great deal about physics and during this time taught himself Maxwell's theory of electromagnetism. He graduated in 1900 with no particular distinction.

Perhaps it was Einstein's middling academic record that prevented him from obtaining the immediate teaching position he desired. After an unsatisfactory interval of trying to earn a living by tutoring poor students, he obtained a job in the Swiss patent office in Bern through the aid of a friend. It was an undemanding position with modest pay, but it left a great deal of spare time for his absorbing intellectual pursuits. During the next eight years, Einstein made remarkable contributions to physics. Though isolated from the ferment and stimulation of an academic environment, he completed his doctoral thesis and published several papers on statistical mechanics and molecular motions. The year 1905 was truly a banner period, in which he published four short papers on the photoelectric effect, Brownian motion, and the special theory of relativity. In spite of Einstein's questionable background, the scientific community began to recognize the value of his accomplishments. He was offered numerous professorial positions in various universities; he accepted those in Zurich and Prague, and he finally took a prestigious appointment at the University of Berlin which left him entirely free from specified duties.

In 1916, Einstein published his general theory of relativity. Its abstract, mathematical nature made acceptance slow until one of its predictions—the bending of starlight in the strong gravitational field of the sun—was experimentally verified by a group of English physicists in 1919. After that, Einstein's reputation soared in academic circles and with the general public (for whom he became the perfect symbol of the absent-minded brilliant professor, whose theories, it was reputed, "only seven people in the world could understand"). In 1921, he was awarded the Nobel Prize in physics—not for relativity (!), but for his explanation of the photoelectric effect.

Einstein was noted for his warm, generous personality and his gentle sense of humor. He was a fairly accomplished musician, playing his violin or piano frequently. Mozart and Bach were his favorites. He had a dogged persistence in intellectual pursuits, repeatedly seeking simplicity and unity in describing nature. This fondness for simplicity, for eliminating all but essentials, was also evident in his personal life: in his clothes and in his behavior.

Unfortunately, political events—World War I, increasing nationalism, and the rise of the Nazis—had considerable impact on Einstein's life. His invited lectures in France and England were occasionally boycotted by some professors whose nationalistic feelings apparently overwhelmed their scientific interests. Being a Jew and a confirmed pacifist who refused to support the German war effort, Einstein became the target of Nazi anti-Semitism. His prestige protected him for a time, but in 1933 he decided to emigrate to the United States, settling after a few years at the Institute for Advanced Study at Princeton. He continued to work on a unified field theory in which he attempted (unsuccessfully) to combine gravitation and electromagnetism into a single theoretical structure.

In his later years, Einstein devoted much attention to pacifist ideas, the Zionist movement, world government, and similar social and political issues. He became a passionate and fearless spokesman for causes of human freedom. Some persons considered him naive, but all believed in his sincerity. He frequently was perplexed and saddened by the contradictions of people and politics. In 1939, concerned about the rising fury in Europe and aware of German research in uranium fission, he lent his name to a letter to President Roosevelt urging immediate investigation into the possibility of a nuclear bomb. After the war, in response to criticism in a Japanese journal reproaching him for this involvement, he wrote: "There are circumstances in which I believe the use of force is appropriate—namely, in the face of an enemy bent on destroying me and my people."

Einstein died in 1955. His most famous legacy—the truly brilliant insight of relativity—gives a new unity and clarity to our understanding of the universe.

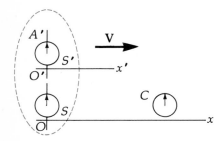

(a) *The first event.* The moving clock A' is coincident with the stationary clock B in the S frame. (For convenience, we set all clocks to read zero at this instant.)

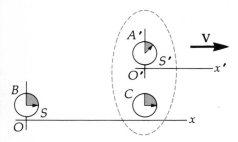

(b) *The second event.* The moving clock A' is coincident with the stationary clock C in the S frame.

FIGURE 41-6
The two events, *measured in the S frame*, by which we compare the rate of a moving clock A' with synchronized clocks B and C at rest in S.

41.6 Comparison of Clock Rates

How do we determine the rate at which a moving clock runs? We cannot just compare a moving clock with a stationary clock at a single instant of time; the procedure necessarily involves measuring a time interval between *two* events. Figure 41-6 illustrates the procedure in "our" frame of reference, the S frame. *Three* clocks are involved. The clock A' is at rest in the S' frame. At one instant, it is adjacent to clock B, which is at rest in the S frame. This coincidence is *the first event*. In each frame, observers located next to the clocks record the readings on the clocks at this time: t_1 and t'_1. The *second event* occurs later, when the moving clock A' is coincident with clock C, which is at rest in the S frame. Again, *local* observers *situated where this event occurs* record clock readings for this event: the times t_2 and t'_2. The time interval between these two events is

$$T = t_2 - t_1 \quad \text{(in the } S \text{ frame)}$$

and

$$T' = t'_2 - t'_1 \quad \text{(in the } S' \text{ frame)}$$

The two time intervals are not the same. Using the Lorentz transformation, Equation (41-5), we find how the time intervals are related.

$$T = t_2 - t_1 = \frac{\left(t'_2 + \dfrac{Vx'_2}{c^2}\right)}{\sqrt{1-\beta^2}} - \frac{\left(t'_1 + \dfrac{Vx'_1}{c^2}\right)}{\sqrt{1-\beta^2}}$$

In the S' frame, the two events occur at the same location, $(x'_2 = x'_1)$, so the above expression becomes

$$T = \frac{t'_2 - t'_1}{\sqrt{1-\beta^2}} = \frac{T'}{\sqrt{1-\beta^2}}$$

The time interval $T' = t'_2 - t'_1$ is measured by a *single* clock in S' (in contrast to the time interval T, which is measured by *two different* clocks in S). As we will point out in Section 41.9, this has a special significance. Since single-clock readings may occur in *either* frame, instead of a prime we will use a zero subscript to signify this type of measurement.

TIME DILATION
$$T = \frac{T_0}{\sqrt{1-\beta^2}}$$
(where T_0 must be a time interval measured by a *single* clock) (41-6)

Because the factor $\sqrt{1-\beta^2}$ is always less than unity, the time interval T is always larger than T_0. We conclude that *moving clocks run slower than clocks at rest*. The effect is called **time dilation**. The moving clocks run slower not because the motion somehow deforms them so that they show an incorrect time; rather, it is time itself that is different for a moving frame compared with the time scale in a "stationary" frame. All clocks in a frame of reference show the correct time for that frame.

An even more startling feature of time dilation is that since either frame may be considered "at rest," observers in S' who carry out the procedure described above would find that clocks in S run more slowly than their own S' clocks. The effect is entirely symmetrical: *observers in each frame find that the other "moving" clocks run slower than clocks at rest in their own frame.* All measurements depend on the frame of reference of the observer, and each frame has its own scale of time, which does not necessarily agree with the time scale in other frames. We can properly answer the question "Do moving clocks *really* run slower?" by pointing out that according to all measurements made on moving clocks, yes, they certainly *do* run slower than clocks in our own frame of reference. It is not an illusion. All clocks show the correct time in their own frame of reference. There is simply no absolute time scale, valid in all frames. By itself, this conclusion may seem paradoxical. But when all aspects of special relativity are taken together, they form a most logical and impressive structure that agrees completely with experimental evidence. This unusual behavior is a basic feature of our universe.

EXAMPLE 41-2

A clock at rest in the S' frame gives a "tick" once each second. Thus, as measured in the S' frame, the time interval between ticks is $T_0 = 1$ s. If the S' frame has a velocity of $0.80c$ relative to the S frame, what is the time between ticks as determined in the S frame?

SOLUTION

Since $T_0 = 1$ s and $\beta = 0.80$, we have

$$T = \frac{T_0}{\sqrt{1-\beta^2}} = \frac{(1\text{ s})}{\sqrt{1-0.64}} = \frac{(1\text{ s})}{\sqrt{0.36}} = \boxed{1.67\text{ s}}$$

The moving clock thus runs slower than our own clocks at rest.

41.7 Comparison of Length Measurements Parallel to the Direction of Motion

In Section 41.2, we compared the length L' of a rod at rest in the S' frame with a measurement L of the rod made in the S frame (in which the rod is moving with speed V parallel to the x axis). We now follow the identical procedure, but instead of the Galilean transformation, we will use the Lorentz transformation. Consider a meter stick at rest in the S' frame, aligned along the direction of relative motion of the two frames, the x and x' axes. See Figure 41-7. As measured in S', the stick's length is $L' = x'_2 - x'_1$. Applying the Lorentz transformation yields

$$L' = x'_2 - x'_1 = \frac{x_2 - Vt_2}{\sqrt{1-\beta^2}} - \frac{x_1 - Vt_1}{\sqrt{1-\beta^2}}$$

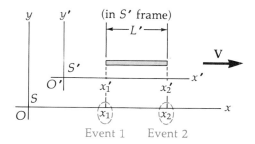

FIGURE 41-7
In the S frame, the two events that determine the location of the ends of the moving stick at x_1 and x_2 are *simultaneous* events.

In the S frame, the two events of locating the ends of the stick *must occur simultaneously* ($t_2 = t_1$), so $t_1 = t_2$, and the above expression reduces to

$$L' = \frac{x_2 - x_1}{\sqrt{1 - \beta^2}} = \frac{L}{\sqrt{1 - \beta^2}}$$

The length L' is made in the S' frame, in which the object is *at rest* (in contrast to L being measured in the S frame, in which the object is moving). Since measurements of an object at rest may occur in either frame, instead of a prime we will use a zero subscript to signify this type of measurement.

LENGTH CONTRACTION $L = L_0 \sqrt{1 - \beta^2}$ (where L_0 must be a measurement in a frame in which the object is at rest)   (41-7)

Because the factor $\sqrt{1 - \beta^2}$ is always less than unity, the length L is always less than L_0. Consequently, we conclude that the length of a moving object is less along the direction of its motion than it is when the object is measured at rest. (Distances perpendicular to the direction of motion are unchanged.) The effect is called *length contraction*.

As with time dilation, this, too, is a symmetrical effect. *Observers in each frame measure the other meter stick to be shorter than theirs.* There is no paradox, since the two sets of measurements are made in different frames of reference. The length of an object is not some attribute possessed by that object. Rather, it is the result of a measurement. We can properly answer the question "Is the moving stick *really* shorter?" by pointing out that all measurements made of the moving stick show that, yes, it certainly *is* shorter than meter sticks at rest in our own frame of reference. Because there is no absolute space and no absolute time, measurements in one frame do not necessarily agree with those made in another frame. Nevertheless, measurements made in each frame are equally valid. Martin Gardner[6] makes an interesting analogy: if two people stand on opposite sides of a huge reducing lens, each sees the other as smaller. But that is only to say that, *in each person's frame of reference*, the other person is smaller. It is not the same as making the paradoxical statement that each person actually *is* smaller than the other.

EXAMPLE 41-3

A meter stick moving with speed $0.60c$ is oriented parallel to the direction of motion. Find the length of this meter stick as measured by an observer at rest.

SOLUTION

Since $L_0 = 1$ m and $\beta = 0.60$, we have

$$L = L_0\sqrt{1 - \beta^2} = (1 \text{ m})\sqrt{1 - 0.36} = (1 \text{ m})\sqrt{0.64} = \boxed{0.800 \text{ m}}$$

Thus the moving meter stick is shorter than a meter stick at rest.

---

[6] Martin Gardner, *The Relativity Explosion*, Vintage Books (1976).

41.8 Proper Measurements

Observers in different frames of reference may find different answers for measurements of lengths and time intervals. It is customary to use a special name for the measurements made, as follows:

Proper length: A length determination made in a frame of reference in which the object is at rest.

Proper time interval: A time interval between two events when measured in a frame of reference in which the events occur at the same location. Proper time is measured by only a *single* clock. All clocks indicate the proper time at their respective locations.

The use of the word *proper* does not imply that other measurements are somehow improper or incorrect. The adjective is used in the sense of "naturally belonging to" or "characteristic of." Although one can always find the proper length of an object, there are situations in which the concept of a proper time interval does not apply. Note that a proper time interval is measured *by only a single clock*. Thus, if two events occur apart in space, but so close together in time that a frame of reference cannot move fast enough to enable a *single* clock to be located where each event occurs (without traveling at the speed of light, or faster), then the concept of a proper time interval does not apply. It is important to remember that the symbols T_0 and L_0 in the expressions for time dilation and length contraction are *proper* measurements, regardless of which frame of reference is designated the primed frame.

Because classical ideas of space and time are so deeply ingrained in our thought processes, it is surprisingly easy to be led astray in solving relativity problems. For this reason, it is prudent always to think in terms of *point events* and to make careful sketches of these events *as measured in a particular frame of reference*.

41.9 Relativistic Momentum

Thus far, our discussion of special relativity has been restricted to kinematics. We now develop relativistic dynamics using the same basic concept that forms the foundation of Newtonian mechanics: *the conservation of momentum*. If we apply a constant force to an object, Newton's second law ($F = dp/dt$, where $p = mv$) places no limit on the speed that an object may acquire. Experimentally, however, the momentum of an object approaches infinity as its speed approaches the speed of light, so there is a relativistic upper limit to the maximum attainable speed.

To investigate this effect, we analyze an elastic collision between two identical particles and require that momentum conservation hold true in all frames of reference, in accordance with Einstein's first postulate. Consider two railroad flatcars, the S and S' frames, approaching each other on parallel tracks with equal speeds in opposite directions. Figure 41-8a shows the situation in the earth's frame of reference. Observers in both frames have balls whose masses m are equal. The two observers launch their balls perpendicular to the direction of motion with equal speeds u (as measured in their respective frames). After traveling the same distance y perpendicular to the motion of the cars, the balls collide elastically and rebound the same distance y before each ball is caught. We use the following notation: in S, ball A of mass m travels a total

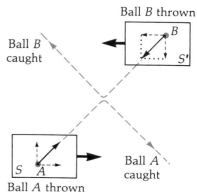

(a) In the earth's frame, the elastic collision is completely symmetrical.

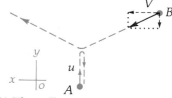

(b) The collision as measured in the S frame (observer A).

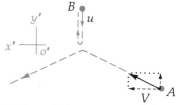

(c) The collision as measured in the S' frame (observer B).

FIGURE 41-8
A hypothetical experiment involving an elastic collision of two identical balls. In the earth's frame of reference, the collision is entirely symmetrical.

distance $2y$ at the speed u. In S', ball B of mass m travels a total distance $2y'$ at the speed u. Since lengths in the y direction are not contracted, $y = y'$. The situation is symmetrical.

An unusual feature emerges, however, when the collision is analyzed in one of the moving frames, Figure 41-8b. The y component of B's velocity is the distance $2y$ ($=2y'$) divided by the time interval between throwing and catching B. In S', this time interval is a *proper time* T_0 since the two events (throwing and catching) occur at the same location in S and thus are measured by a *single* clock. But in S these same two events (throwing and catching B) occur at two different locations. The time T measured in S is related to T_0 according to the time-dilation relation: $T = T_0/\sqrt{1-\beta^2}$. Consequently, even though $y = y'$, the times are different and the y component of ball B's speed (as measured in S) is not the same as the y component of ball A's speed (as measured in S).

Speeds of A and B in S
$$(u_B)_y = \frac{2y}{T} = \frac{2y}{T_0}\sqrt{1-\beta^2} = u\sqrt{1-\beta^2} \quad (41\text{-}8)$$
$$(u_A)_y = \frac{2y}{T_0} = u \quad (41\text{-}9)$$

Momentum conservation requires that the change of y momentum of one ball must equal the (negative) change of y momentum of the other ball. But if momentum is defined as (mass)(velocity), these changes do *not* have equal magnitudes:

In S: $\quad (\Delta p_A)_y = -2mu \quad$ and $\quad (\Delta p_B)_y = 2mu\sqrt{1-\beta^2}$

(The sign difference is due to our choosing the positive y axis as "up" in Figure 41-8). So we conclude that momentum, defined as $p = $ (mass)(velocity), apparently is not conserved in relativity!

Momentum conservation is so important in physics that we seek some way of rescuing this fundamental principle. The trouble arose because in our analysis the y component of momentum depended upon the x component of the velocity associated with the moving frame of reference. Here is a possible alternative. Let us define the velocity in terms of *the proper time $\Delta\tau$ of the moving object itself*, that is, the time measured by a clock attached to the moving object. Then the quantity $\Delta y/\Delta\tau$ is the same for all observers. This proper time is related to the observer's time Δt by

$$\Delta\tau = \Delta t\sqrt{1-\frac{u^2}{c^2}} \quad (41\text{-}10)$$

Thus: $\quad$ Velocity $= \dfrac{\Delta y}{\Delta t}\dfrac{1}{\sqrt{1-\dfrac{u^2}{c^2}}} \quad (41\text{-}11)$

Therefore, for all frames moving at constant velocity along the $\pm x$ direction, the y component of the velocity of a particle is the same in all frames:

$$[y\text{ component of velocity}] = \frac{u}{\sqrt{1-\dfrac{u^2}{c^2}}}$$

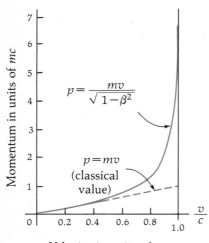

FIGURE 41-9
As the velocity increases, relativistic momentum deviates greatly from its classical value of mv. As the speed approaches c, the momentum approaches infinity.

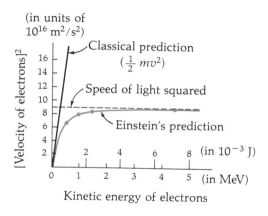

FIGURE 41-10
The experimental points show evidence for the speed of light as a limiting velocity for any particle that has mass. (Adapted from W. Bertoozi, *American Journal of Physics* **32** (1964), p. 555, with permission of the American Journal of Physics.)

We generalize this equation and *define* relativistic momentum as

RELATIVISTIC MOMENTUM p
$$\mathbf{p} = \frac{m\mathbf{u}}{\sqrt{1 - \frac{u^2}{c^2}}} \qquad (41\text{-}12)$$

Note that this definition does not involve the relative speed of frames of reference. Instead, the speed u is the particle velocity as measured in a frame of reference. With this definition, momentum conservation holds true for relativistic situations. It also reduces to the familiar classical value $\mathbf{p} = m\mathbf{u}$ for $u \ll c$. Figure 41-9 shows how the relativistic momentum varies with velocity.

Because of the relativistic momentum increase, c is the upper limit to the velocity attainable by any particle that has a rest mass. As the momentum increases, an increasingly larger force is required to further accelerate the particle. It would take an infinite amount of energy to achieve the speed c. Thus the speed of light is truly an upper limit.[7] Figure 41-10 shows convincing experimental evidence for this limiting velocity. Here, the square of the speed of an electron is plotted versus its kinetic energy. *On the scale of this graph*, electrons emerging from Stanford's three-kilometer accelerator (Figure 41-11) would be a point plotted 188 m to the right (more than the length of two football fields!). These electrons have measured velocities that are essentially c, but, of course, they still have not achieved a speed of precisely c—a remarkable discrepancy from the classical prediction for that energy.

EXAMPLE 41-4

Electrons emerging from Stanford's three-kilometer linear accelerator are traveling at 99.999 999 97% the speed of light. Find their momentum in terms of mc.

SOLUTION

Their momentum is not $mv \approx mc$ as classical theory predicts, but instead is given by Equation (41-12). Because β is extremely close to unity, we may use the

FIGURE 14-11
The Stanford three-kilometer linear accelerator for electrons at the Stanford Linear Accelerator Center (SLAC). (An interstate highway passes over the accelerator.) The operation of the accelerator verifies all aspects of special relativity. Electrons emerging from the accelerator differ from the speed of light by only about 5 parts in 10^{11}. If classical (Galilean) relativity were correct and the relativistic momentum increase did not occur, the accelerator would need to be only a few inches long to achieve this speed.

---

[7] It has been proposed that particles called *tachyons*, which always travel faster than c, might exist. For them, the speed of light would be a lower limiting velocity. The existence of such particles is consistent with special relativity; approached from either side, c remains an impenetrable barrier. So far, experiments to detect them have been unsuccessful, and they may not exist. For more information, see G. Feinberg, "Particles That Go Faster Than Light," *Scientific American* **223**, 2 (Feb. 1970), p. 69.

approximation (see Appendix E)

$$1 - \beta^2 = (1 + \beta)(1 - \beta) \approx 2(1 - \beta)$$

For $\beta = 0.999\,999\,999\,7$, the factor $(1 - \beta)$ is equal to 3×10^{-10}. Hence, $\sqrt{1 - \beta^2} \approx \sqrt{2(1 - \beta)} = \sqrt{6 \times 10^{-10}}$, and we have

$$p = \frac{mv}{\sqrt{1 - \beta^2}} \approx \frac{mv}{\sqrt{6 \times 10^{-10}}} \approx \boxed{4.08 \times 10^4 mc}$$

This agrees with the experimentally measured momentum for the electrons when they are deflected by a magnetic field as they emerge from the accelerator. Although it is common to speak of these electrons as having a relativistic mass 4×10^4 greater than their rest mass, we emphasize again that this change occurs because of the unusual properties of space and time, not because of any peculiar changes in the mass itself. (See the next section.)

EXAMPLE 41-5

A baseball moves at 30 m/s. By what fraction does its true relativistic momentum differ from the classical value of mv?

SOLUTION

Here the velocity is very small compared with c, so we use the following approximation, valid for $\beta^2 \ll 1$ (see Appendix E):

$$(1 \pm \beta^2)^n \approx 1 \pm n\beta^2 \qquad (\text{for } \beta \ll 1)$$

Therefore,
$$\frac{1}{\sqrt{1 - \beta^2}} = (1 - \beta^2)^{-1/2} \approx 1 + \frac{\beta^2}{2}$$

The fraction we seek is

$$\frac{[\text{Difference}]}{mv} = \frac{p - mv}{mv} = \frac{p}{mv} - 1 = \frac{mv(1 - \beta^2)^{-1/2}}{mv} - 1 \approx 1 + \frac{\beta^2}{2} - 1 \approx \frac{\beta^2}{2}$$

The numerical value of β is
$$\beta = \frac{v}{c} = \frac{30 \text{ m/s}}{3 \times 10^8 \text{ m/s}} = 1 \times 10^{-7}$$

so
$$\frac{\beta^2}{2} = \frac{1 \times 10^{-14}}{2} = \boxed{5 \times 10^{-15}}$$

Thus the relativistic correction is negligible for speeds we usually encounter in everyday experience.

Note that for such problems as Examples 41-3 and 41-4, which involve speeds of $v \ll c$ and $v \approx c$, the approximation formulas help to avoid awkward procedures such as directly calculating $\sqrt{1 - (0.999\,999\,999\,7)^2}$, an operation beyond the capability of most pocket calculators. If you find yourself tangled in such unwieldy operations, you have not made the appropriate approximations before substituting numerical values.

41.10 A Note about Rest Mass

Sometimes Equation (41-12) is interpreted to mean that as a particle's speed increases, its mass increases. In the following discussion, m_0 refers to the *rest mass* and m_{rel} designates the so-called *relativistic mass*.

$$p = \frac{m_0 u}{\sqrt{1 - \frac{u^2}{c^2}}}$$

$$m_{rel} u = \frac{m_0 u}{\sqrt{1 - \frac{u^2}{c^2}}}$$

Dividing by u gives
$$m_{rel} = \frac{m_0}{\sqrt{1 - \frac{u^2}{c^2}}} \qquad (41\text{-}13)$$

Some authors use this definition since it leads to a few expressions that are similar to classical formulas, such as the relativistic momentum, $p = m_{rel} v$, and the convenient formula for the total energy, $E = m_{rel} c^2$. However, other classical equations are incorrect when m_{rel} is substituted for m. F does not equal $m_{rel} a$, nor does the relativistic kinetic energy equal $\frac{1}{2} m_{rel} v^2$. Further misunderstanding occurs if we make the claim that "mass increases with speed," ignoring the fact that the square root factor in Equation (41-9) comes into the derivation in connection with a *velocity* measurement (involving space and time) when the momentum is determined. *Thus the square root factor is a consequence of the transformation properties of space and time, not those of mass.* Relativity changes our ideas about space and time, and it affects dynamical quantities like velocity and momentum. But it does not affect the intrinsic properties of fundamental particles, such as charge and mass. **In this text, m always refers to the invariant rest mass**—a notation also preferred in advanced treatments of relativity.

41.11 Relativistic Velocity Addition

Suppose that a particle has a speed ω along the x' direction in the S' frame of reference. The S' frame itself has the speed V along the $+x$ direction relative to the S frame. See Figure 41-12. At the instant the two frames coincide at $t = t' = 0$, the particle passes the origin O (and O'). We obtain the relativistic velocity addition relation by following the identical procedure we used to obtain the classical velocity addition [Equation (41-3)], except that we use the Lorentz transformation rather than the Galilean transformation. Consider again the following two events:

		In S	In S'
FIRST EVENT:	*Particle at origin*	(0,0,0,0)	(0,0,0,0)
SECOND EVENT	*Particle away from origin*	(x,0,0,t)	(x',0,0,t')

In the S' frame, the particle has the speed $u' = x'/t'$. The S' frame itself moves along the $+x$ direction with constant speed V relative to S. This motion is the second velocity V that we will add to the particle velocity u' to obtain the velocity u of the particle as measured in the S frame. In the S frame, $u = x/t$.

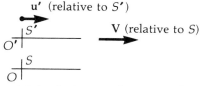

(a) At the instant the S and S' frames coincide, at $t=t'=0$, the particle passes the origin moving in the $+x$ (and $+x'$) direction.

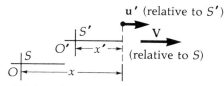

(b) At a later time, the particle is at x in the S frame and at x' in the S' frame. The particle has the speed $u=x/t$ in S and $u'=x'/t'$ in S'.

FIGURE 41-12
A hypothetical experiment to investigate velocity addition. The S' frame has the velocity V relative to the S frame.

Making use of the Lorentz transformation, Equation (41-5), we have

$$u = \frac{x}{t} = \frac{\frac{x' + Vt'}{\sqrt{1-\beta^2}}}{\frac{t' + Vx'/c^2}{\sqrt{1-\beta^2}}} = \frac{t'(u' + V)}{t'(1 + u'V/c^2)}$$

RELATIVISTIC
VELOCITY ADDITION
(for velocities along
the $\pm x$ direction)

$$u = \frac{u' + V}{1 + \left(\dfrac{u'V}{c^2}\right)} \qquad (41\text{-}14)$$

For speeds much less than c, this expression reduces to the classical velocity addition relation $u = u' + V$. If any velocities are in the $-x$ (or $-x'$) direction, minus signs are used with the corresponding numerical values.

What happens if both of the velocities, u' and V, are close to the speed of light? Can this result in a velocity greater than c? No. The successive addition of *any number* of such velocities less than c, all in the same direction, still results in a final velocity less than c.

EXAMPLE 41-6

Suppose that two stars, A and B, recede from the earth in opposite directions, with speeds as shown in Figure 41-13a. Find the speed that star B would have for observers on star A.

SOLUTION

In terms of the notation we have developed, star A is the S frame, while the earth (S' frame) is the moving frame ($V = 0.7c$), in which star B is observed to have the speed $u' = 0.8c$ relative to the earth (Figure 41-13b). Using the relativistic velocity addition formula, Equation (41-14), we have

$$u = \frac{u' + V}{1 + \left(\dfrac{u'V}{c^2}\right)} = \frac{(0.8c + 0.7c)}{1 + \left[\dfrac{(0.8)(0.7)c^2}{c^2}\right]} = \boxed{0.962c}$$

Note that this is less than the speed of light. [The Galilean velocity addition relation would give the incorrect value $u = u' + V = (0.8c + 0.7c) = 1.5c$.)]

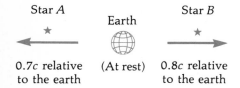

(a) As seen in the earth's frame of reference (the S' frame)

(b) As seen in star A's frame of reference (the S frame)

FIGURE 41-13
Example 41-6. Two stars move away from the earth in opposite directions with relativistic speeds. What would observers on star A measure for the speed of star B?

EXAMPLE 41-7

To push velocity addition to its limit, suppose that a spaceship (S' frame) passes the earth (S frame) at an extremely fast speed, say, $V = 0.9999c$. A rider aboard the spaceship sets off a flashbulb at the rear of the ship and measures the speed of the light pulse progressing toward the nose of the ship to be c in the S' frame. Using the relativistic velocity addition relation, find the speed of the *same* light pulse as measured in the earth's frame of reference.

SOLUTION

Given: $u' = c$ and $V = 0.9999c$. Substituting in Equation (41-14), we get

$$u = \frac{u' + V}{\left(1 + \frac{u'V}{c^2}\right)} = \frac{c + v}{\left(1 + \frac{cV}{c^2}\right)} = \frac{(c + V)c}{(c + V)} = \boxed{c}$$

We should not be surprised at the answer. Regardless of the numerical value of V, the Lorentz transformation was developed specifically to guarantee that the speed of light is c in all frames of reference.

41.12 Relativistic Energy

In Section 6.6 we derived the work–energy relation, which says that the work done on a particle by the net force F equals the change in the kinetic energy ΔK. We now carry out the same calculation using relativistic ideas. We assume that the particle starts at rest, so that $K_0 = 0$. For one-dimensional motion,

$$K = \int_0^x F\,dx \qquad (41\text{-}15)$$

It turns out to be simpler to evaluate this integral using the relativistic momentum p as a variable of integration. We thus substitute $F = dp/dt$ and $dx = v\,dt$. From Equation (41-9),

$$p = \frac{mv}{\sqrt{1 - v^2/c^2}}$$

Solving for v gives

$$v = \frac{p/m}{\sqrt{1 + (p/mc)^2}} \qquad (41\text{-}16)$$

Substituting these relations into Equation (41-15), we have

$$K = \int_0^t \frac{dp}{dt} v\,dt = \int_0^p v\,dp = \int_0^p \frac{p/m}{\sqrt{1 + (p/mc)^2}}\,dp \qquad (41\text{-}17)$$

After a bit of messing about, the result is

$$K = mc^2 \left[\sqrt{1 + \left(\frac{p}{mc}\right)^2} - 1\right]$$

Substituting the relativistic momentum $p = mv/\sqrt{1 - \beta^2}$, we obtain

RELATIVISTIC KINETIC ENERGY

$$K = \frac{mc^2}{\sqrt{1 - \frac{v^2}{c^2}}} - mc^2 \qquad (41\text{-}18)$$

where m is the rest mass. For low speeds, this expression reduces to the familiar Newtonian kinetic energy: $K = \frac{1}{2}mv^2$. To show this, we expand the square

TABLE 41-1 Mass-Energies of Some Common Particles (rounded 1986 CODATA values)

Particle	Symbol	mc^2 (in MeV)	m (kg)
Electron (or positron)	e or e^- (e^+)	0.511	$9.109\,390 \times 10^{-31}$
Muon	$\mu^\pm$	105.658	$1.883\,533 \times 10^{-28}$
Pi meson (neutral)	π^0	134.964	$2.405\,95 \times 10^{-28}$
Pi meson (charged)	$\pi^\pm$	139.569	$2.488\,05 \times 10^{-28}$
Atomic mass unit	u	931.494	$1.660\,540 \times 10^{-27}$
Proton	p	938.272	$1.672\,623 \times 10^{-27}$
Neutron	n	939.565	$1.674\,929 \times 10^{-27}$
Deuteron	d or ^{2}H	1875.613	$3.343\,586 \times 10^{-27}$
Alpha particle	α or ^{4}He	3727.380	$6.644\,653 \times 10^{-27}$

root term using the binomial formula (see Appendix E):

$$K = mc^2 \left[1 + \frac{1}{2}\frac{v^2}{c^2} + \frac{3}{8}\left(\frac{v^2}{c^2}\right)^2 + \cdots - 1 \right]$$

For small speeds, the v^4/c^4 term may be neglected compared with v^2/c^2, so the expression becomes

$$K_{\text{classical}} = \tfrac{1}{2}mv^2 \qquad (\text{for } v/c \ll 1)$$

In Equation (41-18) the term mc^2 is called the **rest energy** E_0:

REST ENERGY E_0 $\qquad\qquad E_0 = mc^2 \qquad\qquad$ (41-19)

This implies that there is an equivalence between a *mass m* and an *energy E_0*. Because of the large numerical value of c, the energy equivalent to even a small mass is most impressive. The generation of electrical power in nuclear reactors is one example in which the uranium nucleus undergoes fission, usually resulting in two lighter nuclei plus two or three neutrons. The total mass of the products is less than the mass of the initial uranium nucleus by an amount Δm. The kinetic energy of all the fragments is exactly equal to $(\Delta m)c^2$. This kinetic energy is then used to heat steam for the conventional generation of electricity. The daily needs of a city the size of San Francisco could be met by the conversion to energy of a mass approximately half that of a penny. The reverse process—the conversion of energy to mass—is possible. For example, in the *pair-production* process (Section 42.6) the energy of a "particle of light"—a *photon*—is converted into the creation of the masses of an electron and a positron, plus their kinetic energies.

It is common practice to express the masses of particles in the energy units[8] of *electron volts*, a usage particularly convenient for calculations, Table 41-1. Similarly, momentum is conveniently expressed in terms of MeV/c. Note: when we say that "a particle has an energy of 2 MeV," we usually mean that its *kinetic* energy is 2 MeV.

---

[8] Of course, *mass* does not equal *energy*. These quantities are related only through the factor c^2.

EXAMPLE 41-8

A penny has a mass of about 3 g. Compute the energy that would be released if this mass were entirely converted into energy.

SOLUTION

$$E = mc^2 = (0.003 \text{ kg})(3 \times 10^8 \text{ m/s})^2 = \boxed{2.70 \times 10^{14} \text{ J}}$$

This is about equal to the maximum energy output of Hoover Dam for $2\frac{1}{2}$ days.

The sum of the rest energy mc^2 and the kinetic energy K equals the total energy E of a system:

TOTAL RELATIVISTIC ENERGY E

$$E = mc^2 + K \tag{41-20}$$

and

$$E = \frac{mc^2}{\sqrt{1 - \beta^2}} \tag{41-21}$$

This leads to a new conservation principle, *the conservation of mass-energy*, which unites the two separate conservation principles of classical physics—the conservation of energy and the conservation of (classical) mass (as in chemical reactions).

The internal energy U of a system of particles is part of the rest energy $E_0 = mc^2$ of the system. For example, if we stretch a spring, thereby giving it positive internal potential energy U_{sp}, the rest energy of the spring increases slightly (though by an amount far too small to measure directly). An example having negative internal energy is the bound system of a proton and a neutron, which forms the stable particle called a *deuteron* (the nucleus of the isotope ^{2}H). To pull the proton and the neutron apart against the attractive force that holds them together, we must do work on the system. In other words, the internal *binding energy* is negative (relative to *zero* potential energy when the particles are separated at rest), and the rest energy of the deuteron is slightly less than the combined rest energies of the free proton and neutron.

EXAMPLE 41-9

A deuteron is composed of a neutron and a proton bound together. Referring to Table 41-1, calculate how much energy would be required to break up the deuteron into a proton and a neutron.

SOLUTION

The combined rest energies of a proton and a neutron are 938.280 MeV + 939.573 MeV = 1877.853 MeV. The rest energy of a deuteron, 1875.628 MeV, is subtracted from this to yield $\boxed{2.22 \text{ MeV}}$, the binding energy of the deuteron.

In the above example, to supply the energy required to break apart the deuteron, we could bombard the deuteron with another particle or with an energetic photon (symbol γ, for gamma ray). Such a *photo-induced reaction* is written

$$\gamma + d \to n + p$$

The inverse reaction is the combination of a proton and a neutron to form a deuteron, releasing a photon having 2.22 MeV of energy to account for the change in the rest energies of the particles.

$$n + p \to d + \gamma$$

See Chapter 45 for a more detailed discussion of nuclear reactions.

We usually think of "particles" as having mass greater than zero. However, there are three types of particles that are believed to have zero mass: photons, neutrinos, and (as yet unobserved) gravitons.[9] From the relation $E\sqrt{1-\beta^2} = mc^2$, we conclude that particles with zero mass must travel only at the speed of light, $v = c$, in order to make the square-root factor zero.

Combining equations for E, K, and p, we can obtain the following useful relations:

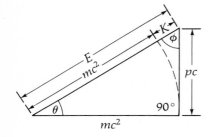

FIGURE 41-14
To help you remember relations between E, K, and p, this right triangle and the Pythagorean theorem illustrate that $E^2 = (pc)^2 + mc^2$. Also, note that $E = mc^2 + K$. It is also easy to show that $\sin \theta = \beta$ and $\sin \phi = \sqrt{1 - \beta^2}$.

ADDITIONAL RELATIVISTIC ENERGY AND MOMENTUM RELATIONS

$$E^2 = (mc^2)^2 + (pc)^2 \tag{41-22}$$

$$p = \frac{1}{c}\sqrt{K^2 + 2mc^2 K}$$

$$= \sqrt{2mK\left(1 + \underbrace{\frac{K}{2mc^2}}_{\text{Relativistic correction term}}\right)} \tag{41-23}$$

$$\frac{p^2}{2m} = \frac{(pc)^2}{2mc^2} = K\left(1 + \underbrace{\frac{K}{2mc^2}}_{\text{Relativistic correction term}}\right) \tag{41-24}$$

$$v = \frac{pc^2}{E} = c\sqrt{1 - \left(\frac{E_0}{E}\right)^2} \tag{41-25}$$

When the total energy E is much greater than the rest energy mc^2, the first term of Equation (41-22) may be neglected, giving the useful relation

HIGH-ENERGY APPROXIMATION

$$E \approx pc \quad \text{(for } E \gg mc^2\text{)} \tag{41-26}$$

---

[9] In the detection of Supernova 1987A, the time of arrival of the neutrino burst relative to the light flash suggests that one form of neutrino—the *electron antineutrino*—may have a small mass, no greater than ~ 14 eV/c^2. This conclusion relies on how well we understand the details of supernova explosions (still somewhat controversial), so all neutrino masses may be truly zero. The *graviton* is a zero-mass particle proposed in current theories of gravitation.

EXAMPLE 41-10

Find (a) the momentum and (b) the speed of a proton whose kinetic energy equals its rest energy.

SOLUTION

(a) From Equation (41-23) we have

$$p = \frac{1}{c}\sqrt{K^2 + 2mc^2 K} = \frac{K}{c}\sqrt{1 + 2\frac{mc^2}{K}}$$

For $K = mc^2$, we obtain

$$p = \frac{mc^2}{c}\sqrt{1 + 2} = \frac{(938 \text{ MeV})}{c}\sqrt{3} = \boxed{1625 \frac{\text{MeV}}{c}}$$

(b) For $K = mc^2$, we have $E = mc^2 + K = 2mc^2$. Thus, from Equation (41-24),

$$v = \frac{pc^2}{E} = \frac{\left(1625 \frac{\text{MeV}}{c}\right)(c^2)}{(2)(938 \text{ MeV})} = \boxed{0.866c} \quad \text{or} \quad \boxed{2.60 \times 10^8 \frac{\text{m}}{\text{s}}}$$

EXAMPLE 41-11

Find (a) the total energy E, (b) the kinetic energy K, and (c) the momentum p of an electron moving with speed $v = 0.6c$.

SOLUTION

(a) $E = \dfrac{mc^2}{\sqrt{1 - \beta^2}}$

Since $\sqrt{1 - \beta^2} = \sqrt{1 - (0.6)^2} = 0.8$, and $mc^2 = 0.511$ MeV, we have

$$E = \frac{0.511 \text{ MeV}}{0.8} = \boxed{0.639 \text{ MeV}}$$

(b) The kinetic energy is

$$K = E - mc^2 = 0.639 \text{ MeV} - 0.511 \text{ MeV} = \boxed{0.128 \text{ MeV}}$$

(c) The momentum is $p = mv/\sqrt{1 - \beta^2}$. Multiplying numerator and denominator by c^2, we have

$$p = \frac{mc^2 v}{\sqrt{1 - \beta^2}\, c^2} = \frac{(0.511 \text{ MeV})(0.6c)}{(0.8)(c^2)} = \boxed{0.383 \frac{\text{MeV}}{c}}$$

> **EXAMPLE 41-12**
>
> Protons emerge from an accelerator with a kinetic energy of 500 GeV ($= 5 \times 10^5$ MeV). (a) By how much does β differ from 1 for these protons? (b) Find their momentum in units of GeV/c.
>
> **SOLUTION**
>
> (a) Because the kinetic energy of these protons is more than 500 times their rest energy, we may use the approximation suitable for the extreme relativistic case (cf. Example 41-4):
>
> $$E = \frac{mc^2}{\sqrt{1-\beta^2}} \cong \frac{mc^2}{\sqrt{2(1-\beta)}}$$
>
> Rearranging gives
>
> $$\sqrt{2(1-\beta)} = \frac{mc^2}{E} = \frac{938 \text{ MeV}}{5 \times 10^5 \text{ MeV}} = 1.876 \times 10^{-3}$$
>
> $$(1-\beta) = \frac{(1.876 \times 10^{-3})^2}{2} = \boxed{1.76 \times 10^{-6}}$$
>
> (b) From Equation (41-26), we have
>
> $$p = \frac{E}{c} = \boxed{\frac{500 \text{ GeV}}{c}}$$
>
> Note the obvious convenience of expressing momentum in units of GeV/c.

41.13 The Nonsynchronism of Moving Clocks

A system of clocks, properly synchronized in the moving S' frame, will appear not to be properly synchronized when viewed from the S frame of reference. This effect is in addition to the time dilation phenomenon and is perhaps relativity's greatest jolt to our commonsense ideas. The effect is the source of most of the so-called "paradoxes" of special relativity.

Recall the procedure for synchronizing two clocks, A' and B', at rest in the S' frame (Section 41.3). As seen in the S' frame, a light flash at the midpoint between the clocks sends light signals in opposite directions. When a pulse arrives at a clock, that clock is set to indicate $t' = 0$. In this manner, A' and B' are correctly synchronized in the S' frame.

Now let us view this procedure from the S frame of reference, Figure 41-15. Since clock A' is moving toward its light signal, it will intercept the light pulse first and be set to read $t'_A = 0$. At some *later* time (as seen in the S frame), the other light signal reaches clock B', which has been moving away from its light signal, and clock B' is set to read $t'_B = 0$. Thus, according to observers in the S frame, the "chasing" clock is set to read a *later* time than the "leading" clock. As measured at a given instant in the S frame, the clocks are not in synchronism.

As usual, the situation is a symmetrical one. Observers in S' similarly find that clocks in S are not properly set in synchronism. Yet, the synchronizing of clocks establishes a time scale *by which the simultaneity of events is judged in that particular frame of reference*. There is no reason, however, to prefer one

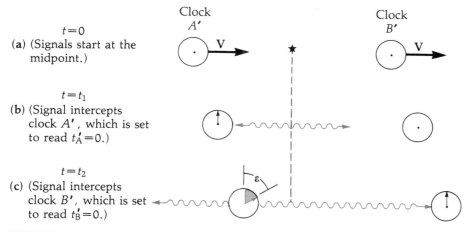

(a) (Signals start at the midpoint.) $t=0$

(b) (Signal intercepts clock A', which is set to read $t'_A = 0$.) $t=t_1$

(c) (Signal intercepts clock B', which is set to read $t'_B = 0$.) $t=t_2$

FIGURE 41-15
As measured in the S frame, the procedure for synchronizing two clocks in S' results in their being out-of-synchronism by an amount ε. Of course, *in the S' frame* the clocks have been correctly synchronized, since the procedure illustrated in Figure 41-4 was followed. There is no absolute "scale" of simultaneously, valid in all frames.

sense of simultaneity over another. Thus, events (separated in space) that appear simultaneous in one frame are not necessarily simultaneous in another frame. The amount of nonsynchronism is directly related to the (Vx'/c^2) term in the Lorentz transformation for time. Consequently, the time t' depends not only on t and V, but also on the space coordinate x. Space and time are truly interdependent in relativity. It can be shown that the discrepancy ε between two moving clocks is

NONSYNCHRONISM OF MOVING CLOCK SYSTEMS *Two clocks, separated a distance $\Delta x'$ and correctly synchronized in the S' frame, are incorrectly synchronized to observers in the S frame by an amount ε:*

$$|\varepsilon| = \frac{V \Delta x'}{c^2} \quad (41\text{-}27)$$

*The "chasing" clock indicates a **later** time than the "leading" clock.*

Only moving clocks located along the $\pm x'$ direction are out-of-synchronism in the S frame; a line of clocks along the y' or z' direction (in S') is correctly synchronized in both frames of reference.

Another feature of nonsynchronism is illustrated if we consider *three* frames of reference, each with a line of several clocks along the direction of motion, Figure 41-16. We consider the S frame to be at rest, the S' frame moving in the $+x$ direction, and the S'' frame moving in the $-x$ direction. For simplicity, we assume all of the center clocks read zero at the instant depicted in the S frame. For a line of moving clocks, each individual clock reads a *later* time than its predecessor. Now suppose that at the instant sketched, two lightning bolts, A and B, strike the left and right groups of clocks, respectively. These two events would be judged simultaneous in the S frame because clocks in that frame indicate the same time. However, as measured in S', the clock readings indicate that B occurs before A and, in the S'' frame, that A occurs before B. There is no such thing as absolute simultaneity.

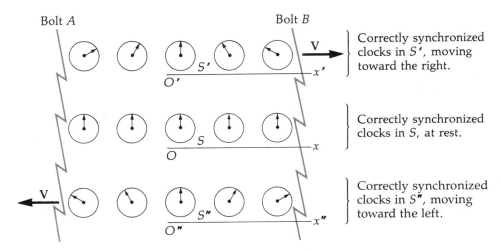

AS MEASURED IN THE S FRAME

FIGURE 41-16
A comparison of clocks in three frames of reference *as measured at a given instant in the S frame*. Each set of clocks is correctly synchronized in its own frame. However, when measured at a given instant in the S frame, the moving clock systems are found not to be in synchronism. To simplify the comparison, we suppose that the clocks located at the respective origins read zero at the instant the origins coincide.

Does this reversal of the time sequence of events imply that in some frame of reference an "effect" might occur before a "cause"? Could the arrow hit the target before the bowstring is released? No. A careful analysis reveals that only those events that could *not* conceivably be causally related in any way can occur in a reversed time sequence in some frame of reference. So the important principle of *cause and effect* is still preserved in relativity theory.

It should be emphasized that this lack of agreement regarding the time sequence of certain events is *not* due to the fact that light signals from a distant event take a finite time to reach an observer (and thus the observer may visually see one event after the other). Even after all corrections for finite transit times of light signals are made, the same peculiarities of simultaneity (or the lack thereof) still remain. Of course, within any given frame, the concept of simultaneity is clearly defined; it just does not agree with the scale of simultaneity in other frames. *All the so-called "paradoxes" of special relativity are traceable to the lack of absolute simultaneity.*

One possible misunderstanding about relativity should be clarified. The message of relativity is not that "everything is relative." True, we must discard absolute space, absolute time, and a few other "absolute" concepts. But the major significance of Einstein's theory (aside from being the theory that agrees best with experimental facts) is this:

THE "MESSAGE" OF RELATIVITY
When correctly expressed, the laws of nature are the <u>same</u> for all observers.

What a chaotic situation it would be if each frame had its own fundamental laws of nature, which would not agree with laws valid in other frames. (This is actually the situation if one clings to Newtonian concepts.) By devising a model for the universe in which nature behaves exactly the same way for all frames of reference, Einstein made a great unifying simplification to our understanding of the universe.

41.14 The Twin Paradox

The so-called *twin paradox* has generated more controversy than any other topic in relativity.[10] Briefly stated, the paradox is as follows. Two twins live on the earth. One decides to take a relativistic trip to a distant star and return. According to relativity, upon his return the traveling twin will be younger than the brother who remained on earth. The paradox arises when one asks why the traveling twin cannot claim that, in *his* frame of reference, his earth brother moved away from him and returned, and that the earth twin (not the traveling twin) is therefore the younger upon their reunion. After all, does not relativity tell us that absolute motion is a fiction? Cannot either twin be considered the stationary one and thus the situation be symmetrical? No. Because the traveling twin must accelerate in some fashion to change his velocity for the return trip, acceleration is involved with only the traveling twin's frame of reference. Acceleration is an absolute, not a relative, matter, so the situation is not a symmetrical one. The consequences are laborious to straighten out, but the conclusion is inescapable: the traveling twin really would be younger upon his return compared with the twin who stayed home.

We can analyze the twin paradox effect using just special relativity by imagining a straight-line trip in which the turnaround time involving acceleration is negligibly short compared to other time intervals. The acceleration times in starting and stopping are also assumed to be negligible.[11] Consider a trip to the star Alpha Centauri, 4 light-years away. One twin, in the S' frame, travels at a constant velocity $V = 0.8c$ to the destination, turns around in a negligibly short time, and returns to the earth at the same constant speed. His twin brother remains on the earth, the S frame. In the earth's frame, the round-trip distance is 8 light-years. (It is convenient to write this as $8\ c\cdot\text{yr}$, because the unit c may cancel in equations just as other units do.) The time to make the journey at a constant speed of $0.8c$ is $t = x/v = (8\ c\cdot\text{yr})/(0.8c) =$ <u>10 years in the earth's frame</u>. In the traveling twin's frame, the distance is contracted to $L = L_0\sqrt{1 - \beta^2} = (8\ c\cdot\text{yr})\sqrt{1 - (0.8)^2} = (8\ c\cdot\text{yr})(0.6) = 4\ c\cdot\text{yr}$. The relative velocity is $0.8c$. Therefore, it takes the time $t' = x'/v = (4.8\ c\cdot\text{yr})/(0.8c) =$ <u>6 years in the traveling twin's frame</u>.

To further verify the elapsed times for each twin, suppose that the journey starts on January 1st. To notify the other twin of the elapsed time, each twin agrees to send the other a New Year's message via radio waves on January 1st of each year during the journey. These signals travel with the speed of light c and are emitted at a frequency f_0 of 1 per year according to the sender's local time scale. Figure 41-17 shows a diagram of the journey as drawn in the *earth* frame of reference. Here, we plot distance (in units of light-years) on the horizontal axis and time (in years) on the vertical axis.

We now make use of a well-verified effect known as the *relativistic Doppler shift* for light (similar to the Doppler shift for sound discussed in Section 18.10). The effect describes how light signals (or any electromagnetic wave) received from a moving source are shifted in frequency f. (Remember, however, that the *speed* of light received from a moving source is always c.) When the light

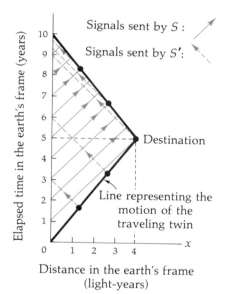

AS DRAWN IN THE EARTH'S FRAME OF REFERENCE (S)

FIGURE 41-17
A diagram of the twin-paradox example as drawn in the earth's frame of reference. The traveling twin moves in a straight line at constant speed $v = 0.8c$. (The time intervals for starting, stopping, and turnaround are assumed to be negligibly short.) Each twin sends a radio signal to the other twin on every January 1st, <u>local time.</u> These radio signals travel at the speed c and thus are drawn at 45° with respect to the x-axis.

---

[10] An excellent source of information about relativity is *Resource Letter SRT-1 (Selected Reprints: Special Relativity Theory)*, published by the American Institute of Physics, 335 East 45th St., New York, NY 10017. For an interesting discussion of the historical origins of relativity, see G. Holton, *American Journal of Physics* **28**, 627 (1960). A comprehensive discussion of the twin paradox is L. Marder's *Time and the Space Traveller*, University of Pennsylvania Press, 1971.

[11] It has been experimentally verified (via the Mössbauer effect) that accelerations up to the order of $10^{16}\ g$ produce no effect in clock rates. Only relative *velocities* alter clock rates. See C. W. Sherwin, *Physical Review* **120**, 17 (1960).

source is *receding* along the line of sight with a speed $V = \beta c$, the received frequency f is *lower* than the frequency f_0 emitted by the source. When the source is *approaching* along the line of sight, the received frequency is *higher* than f_0.

	Light source moving away	Light source approaching	
RELATIVISTIC DOPPLER SHIFT FOR LIGHT[12]	$f = f_0 \sqrt{\dfrac{1-\beta}{1+\beta}}$	$f = f_0 \sqrt{\dfrac{1+\beta}{1-\beta}}$	(41-28)

For the twin paradox example, the radio signals are sent with a frequency f_0 of 1 pulse per year. The speed of the source is $\beta = 0.8$. Putting these values into the Doppler shift formulas, we calculate the rate of signals received for the two cases:

When separating: $\quad f = f_0 \sqrt{\dfrac{1-\beta}{1+\beta}} = f_0 \sqrt{\dfrac{1-0.8}{1+0.8}} = \tfrac{1}{3} f_0$

When approaching: $\quad f = f_0 \sqrt{\dfrac{1+\beta}{1-\beta}} = f_0 \sqrt{\dfrac{1+0.8}{1-0.8}} = 3 f_0$

It is clear that 10 years elapses in the earth's frame (S) for the journey. The most puzzling feature is that only 6 years elapse in S'. Here is how the twins can verify this result using the Doppler shift formula and the rate at which radio signals are received in each frame of reference. They *calculate* the elapsed time as follows:

Calculated in S'
: The twin in S' receives signals at the rate of $\tfrac{1}{3}$ per year for half the journey and 3 per year for half the journey. The average rate of receiving signals during the entire journey is thus

$$(\tfrac{1}{2})(\tfrac{1}{3}) + (\tfrac{1}{2})(3) = \tfrac{5}{3} \text{ per year}$$

Ten signals are received altogether, so the total time for S' is $10/(\tfrac{5}{3}) = 6$ years.

Calculated in S
: The twin in S receives signals at the rate of $\tfrac{1}{3}$ per year for 9 years and 3 per year for 1 year. The total number of signals received by S is thus $(\tfrac{1}{3})(9) + (3)(1) = 6$, signifying that 6 years has elapsed in S'.

Thus, both twins conclude that the elapsed time in S' is 6 years. Although both twins have aged during the trip, after they are reunited the space traveler is 4 years younger than the twin who remained on earth.

A more detailed analysis reveals that the turnaround of the S' frame is the crucial feature. This acceleration does not alter clock *rates*, but it does dramatically change the *scale of simultaneity* for that frame (cf. Figure 41-15). You

---

[12] If there is an angle θ between the line of sight and the velocity of the source, the equation is

$$f = f_0 \left(\frac{\sqrt{1-\beta^2}}{1 + \beta \cos \theta} \right)$$

For $\theta = 90°$, the shift in frequency is just the time dilation effect.

may enjoy the challenge of working out the details. (Hint: sketch arrays of clocks in the two frames at various instants during the turnaround, being careful to depict only *point events* as measured at a given instant in a frame of reference. Remember that events that are simultaneous in one frame are not necessarily simultaneous in another frame.[13]

The twin-paradox effect has been amply verified experimentally. For example, radioactive particles that have a very short half-life have been placed in "storage rings" associated with high-energy accelerators. More of these particles survive one round trip than we would predict for identical particles at rest in the laboratory because a shorter time elapses in the "traveler's" frame of reference. In one experiment, the discrepancy was a factor of 30, exactly in accordance with the predictions of relativity. The first direct experiment using macroscopic clocks was made in 1971, when four cesium clocks were flown on commercial jets around the world, two eastward and two westward.[14] The results confirm the twin-paradox effect. It does seem odd that two clocks initially synchronized, *both of which always show the proper time*, will disagree after being separated and then brought together in this manner. Nevertheless, this is the essence of the twin paradox. It is merely a consequence of the fact that there is no absolute time and no absolute simultaneity.

As a final comment, a startling example of the twin paradox is a hypothetical straight-line trip in which travelers on a spaceship undergo constant acceleration g throughout, accelerating the first half of the outward journey, decelerating the second half, and coming to rest at the destination. The return trip is made in a similar fashion. Such constant acceleration of g would be comfortable for the travelers, since it simulates earth-gravity conditions. For a round trip to Andromeda galaxy, 2 million light-years away, the elapsed time in the spaceship would be only 59 years. Yet the earth would be more than 4 million years older upon the travelers' return. For a similar round trip lasting 78 years in the spaceship's frame, it would be possible to reach a destination 500 million light-years away, returning to find the earth more than one billion years older. Such trips are essentially impossible, however, because of practical engineering difficulties (not because of any limitations in the laws of nature).[15]

41.15 Relativity and Electromagnetism

Consider a single electric charge q at rest in the S' frame of reference. To observers in S', there is an electric field surrounding the charge. However, to observers in the S frame the charge is in motion, so there is not only an electric field but also a magnetic field: the moving charge constitutes an electric current, and currents generate magnetic fields. Thus, electric and magnetic fields are viewed differently in frames of reference that have relative motion. Interestingly, this phenomenon was the subject of Einstein's original paper on special relativity: "On the Electrodynamics of Moving Bodies," *Annalen der Physik*, Volume 17, 1905. All of the startling ideas about space and time for which special relativity is famous emerged unexpectedly from one man's delving into a question about charged objects in motion.

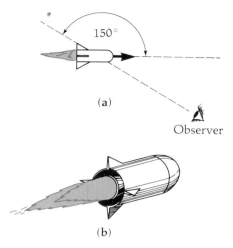

FIGURE 41-18
The Terrell effect. In 1959, James Terrell showed (surprisingly) that if a snapshot is taken of an object in rapid motion at a relatively large distance away, the object will appear to have undergone *rotation*, not contraction. As mentioned in the reference cited below, Terrell considers a relativistic rocketship approaching an observer with speed $v/c = 0.98974$, viewed in a direction at 150° from the flight direction, as sketched in (a). As shown in (b), a snapshot, or the *visual* appearance to the observer, will show the rocketship approaching almost tail-end first! This unusual effect results partly from the fact that, in a snapshot, the camera captures light that arrives simultaneously at the camera. Thus light from more distant parts of the object must have left earlier than the light from closer parts of the object because it had farther to travel. Other unexpected shear distortions occur if a finite solid angle of viewing is considered or if a pair of stereoscopic photos are obtained. This example emphasizes that the data acquired in an experiment depend crucially on the method of measurement employed. [See Letter to the Editor, "The Terrell Effect," James Terrell, *American Journal of Physics* **57**, 9 (Jan. 1989).]

---

[13] The consequences of this are explained in an article by E. S. Lowry, *American Journal of Physics* **31**, 59 (1963). Good discussions of the twin paradox will also be found in articles by G. David Scott, *American Journal of Physics* **27**, 580 (1959) and A. Schild, *American Mathematical Monthly* **66**, 1 (1959).

[14] See two consecutive articles: J. C. Hafele and R. E. Keating, "Around-the-World Atomic Clocks," *Science* **177**, 14 July 1972, pp. 166–70. Later experiments have verified the effect to better than 1%.

[15] For an interesting discussion of the practical difficulties of extended space travel, see S. von Hoerner, "The General Limits of Space Travel," *Science* **137**, (1962), pp. 18–23.

FIGURE 41-19
One situation, viewed from two different frames of reference. Whether the force on the electron is magnetic, electrostatic, or a combination of both depends upon the frame of reference. In the S frame, the electron e moves to the right with a speed v and the wire is stationary. The force is *entirely magnetic*. In the S' frame, the wire moves to the left with a speed v and the electron e is stationary. The force is *entirely electrostatic*.

THE S FRAME

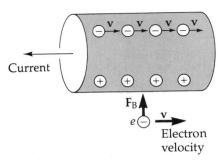

Conditions:

(1) The $\oplus$ charges are stationary.

(2) The $\ominus$ charges are moving to the right with a speed v. Therefore the separation of the $\ominus$ charges is Lorentz contracted.

(3) The net linear charge density on the wire is *zero* (because the Lorentz-contracted distance between the moving $\ominus$ charges is the same as the distance between the $\oplus$ charges at rest).

(4) **The force on the electron e is entirely magnetic.**

(a)

THE S FRAME

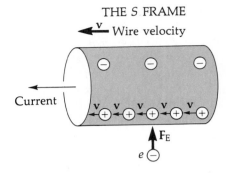

Conditions:

(1) The $\ominus$ charges are stationary.

(2) The $\oplus$ charges are moving to the left with a speed v. Therefore the separation of the $\oplus$ charges is Lorentz contracted.

(3) The net linear charge density on the wire is *positive* (because the distance between the $\ominus$ charges at rest is greater than the Lorentz-contracted distance between the moving $\oplus$ charges).

(4) **The force on the electron e is entirely electrostatic.**

(b)

We now describe a situation that shows how an electric field in one frame of reference will be viewed as a magnetic field in the other frame of reference. The situation is somewhat artificial, but it does clarify the interesting relation between electromagnetism and relativity.

Suppose that a single electron e is moving parallel to a current-carrying wire, as shown in Figure 41-19a. For simplicity, the electrons in the wire are shown separated from the positive ions and moving in straight-line motion with the drift velocity $\mathbf{v}$. We will assign the same velocity $\mathbf{v}$ to the electron outside the wire and assume that the wire has no net charge. The current in the wire produces a magnetic field out of the plane of the diagram at the moving (negative) electron, resulting in a magnetic force $\mathbf{F_B}$ toward the wire. The electron is accelerated toward the wire.

Now consider the same situation viewed from the moving charge, as shown in Figure 41-19b. Here the wire moves to the left and the electron e is stationary. The wire is still observed to carry a current, because although the electrons are at rest the positive charges are now moving toward the left. But since the electron is not moving, there is no magnetic force $\mathbf{F_m} = q\mathbf{v} \times \mathbf{B}$ on the electron. Obviously, if the electron accelerates toward the wire in one frame of reference, it must also do so when viewed from any other frame. What (if not a *magnetic* force) is the origin of a force that could produce such an acceleration?

The theory of special relativity provides the answer. Viewed in frame S (Figure 41-18a), the positive ions are at rest and the electrons in the wire are moving to the right with a velocity **v**. The electrons appear closer together along the wire than their "proper" separation by the Lorentz length-contraction factor $\sqrt{1-v^2/c^2}$. However, the contracted separation is just equal to the separation of the stationary positive ions *because the wire has no net charge*. Viewed from the S' frame (which is moving with the charge e), the situation is quite different: the electrons in the wire are at rest and thus are more widely separated than they were in frame S. At the same time, the positive ions now appear closer together by the Lorentz contraction factor $\sqrt{1-v^2/c^2}$. The net effect is that the *wire now has a net positive charge*. Therefore, the electron e viewed in S' is attracted to the wire by an *electrostatic* force. A detailed analysis shows that *the magnetic force viewed in the S frame is exactly equivalent to the electrostatic force viewed in the S' frame*.

The validity of this analysis is based on the supposition that the magnitude of the electronic charge does not vary with relative motion between a charge and the observer. A variety of experiments indicates that this is true. For example, when a block of metal is heated, the thermal motion of the electrons increases much more than that of the positive ions. Yet the net charge on the block does not change.

Recall Example 28-1, which showed that the drift speed of electrons in a typical current-carrying wire is only on the order of 0.1 mm/s. How astonishing that the relativistic length contraction effect for speeds this low accounts for the magnetic field!

41.16 General Relativity

Up to this point, we have sidestepped a curious puzzle. There are two, seemingly different, properties of mass: a *gravitational attraction* for other masses and an *inertial* property that resists acceleration. These two attributes are apparently distinct. To designate them, we will use the subscripts g and i and write

Gravitational property	$W = m_g g$
Inertial property	$F = m_i a$

The numerical value for the gravitational constant G was chosen to make the magnitudes of m_g and m_i numerically equal. But regardless of how G is chosen, the strict *proportionality* of m_g and m_i has been established experimentally to an extremely high degree: a few parts in 10^{12}. Thus it appears that gravitational and inertial mass may be indeed exactly proportional.

But why? They seem to involve two entirely different attributes: a force of mutual gravitational attraction between masses, and the resistance a single mass has regarding acceleration. This puzzled Newton and other physicists until Einstein published his theory of gravitation known as *general relativity* in 1916. It is a mathematically complex theory, and thus we will be able to only hint at the elegance and insight Einstein achieved.

In Einstein's view, the remarkable coincidence that m_i and m_g seem to be exactly proportional was evidence for a very intimate and basic connection between the two concepts. He pointed out that no *mechanical* experiment (such as dropping a mass) could distinguish between the two different situations sketched in Figure 41-20 (a and b). In each case, if the observer released a mass from his hand, it would undergo a downward acceleration of g relative to the floor of the box.

Einstein carried this idea further to propose, as one of two fundamental postulates in his general theory of relativity, that *no* experiment, mechanical *or otherwise*, could distinguish the difference between the two cases. This

(a) An observer at rest in a uniform gravitational field where the acceleration due to gravity is g.

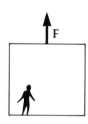

(b) An observer in a region where gravity is negligible, but whose frame of reference is accelerated through space (by the external force **F**) with an acceleration equal to g.

(c) If (a) and (b) are truly equivalent, as Einstein proposed, then a ray of light would be bent in a gravitational field. Such an effect has been experimentally verified by light and radio signals that pass close to the strong gravitational field of the sun.

FIGURE 41-20
According to Einstein, these (a) and (b) frames of reference are equivalent in every way. No experiment of any sort could distinguish any difference.

extension to include all phenomena (not just mechanical ones) has interesting consequences. For example, suppose that a pulse of light is sent horizontally across the box in Figure 41-19b. The pulse of light would have a trajectory that bends downward toward the floor as the box accelerates upward to meet it. Therefore, proposed Einstein, in case (a) a beam of light should be bent downward by the gravitational field. (No such bending is predicted in Newton's theory of gravitation.)

The two postulates of Einstein's **general relativity** are

POSTULATES OF GENERAL RELATIVITY

(1) **All the laws of nature may be stated so that they have the same form for observers in any space-time frame of reference, whether accelerated or not.** (This is the *principle of covariance*.[16])

(2) **In the neighborhood of any given point, a gravitational field is equivalent in every respect to an accelerated frame of reference in the absence of gravitational effects.** (This is the *principle of equivalence*.)

The second postulate implies that gravitational mass and inertial mass are completely *equivalent*, not just proportional. What were thought to be two different types of mass are actually, in a basic sense, identical.

One interesting effect predicted by general relativity is that time scales are altered by gravity. A clock in the presence of gravity runs more slowly than one situated where gravity is negligible. Consequently, spectral lines emitted by atoms in the presence of a strong gravitational field are *red-shifted* to lower frequencies when compared with the same spectral emissions in a weak field. This gravitational red shift has been detected in spectral lines emitted by atoms in massive stars. It has also been verified on the earth by comparisons of the frequency of gamma rays emitted from nuclei separated vertically by about 20 m.[17]

The second postulate suggests that a gravitational field may be "transformed away" at any point if we choose an appropriately accelerated frame of reference—a freely falling one. Einstein developed an ingenious way of describing the exact amount of acceleration necessary. He specifies a certain quantity, the *curvature of spacetime*, that describes the gravitational effect at every point. In fact, the curvature of spacetime completely replaces Newton's gravitational theory.[18] According to Einstein, there is no such thing as a gravitational force. Rather, the presence of a mass such as the sun causes a curvature of spacetime in its vicinity, and this curvature dictates the spacetime path that all freely moving objects follow. As one physicist says: "Mass tells spacetime how to curve; curved spacetime tells mass how to move."

If the concentration of mass becomes very great, as is believed to occur when a large star exhausts its nuclear fuel and collapses to a very small volume, a **black hole** may be formed. Here, the curvature is so extreme that, within a certain distance from the center, all matter and light become trapped.

---

[16] An equation that has the same form after transformation to another frame of reference is *covariant* with respect to the transformation.

[17] See R. V. Pound and J. L. Snider, "Effects of Gravity on Gamma Radiation," *Physical Review B* **140**, 788 (1965). For another test, see R. F. C. Vessot et al., "Test of Relativistic Gravitation with a Space-Borne Hydrogen Maser," *Physical Review Letters* **45**, 2081 (1980).

[18] For an introduction to curved spacetime, see J. J. Callahan, "The Curvature of Space in a Finite Universe," *Scientific American*, August 1976, pp. 90–100.

Summary

Special relativity compares measurements of *events* made in two different frames of reference (S and S') that have uniform relative velocity V with respect to each other.

A point event:
$$\begin{cases} (x,y,z,t) & \text{(in the } S \text{ frame)} \\ (x',y',z',t') & \text{(in the } S' \text{ frame)} \end{cases}$$

Each event is to be measured by a *local* observer, situated where the event occurs and equipped with a clock that has been synchronized with other clocks in the frame.

Postulates of special relativity:

(1) *All the laws of physics have the same form in all inertial frames* (the principle of relativity).
(2) *The speed of light in a vacuum has the same value c in all inertial frames* (the principle of the constancy of the speed of light).

From these two postulates, the following relations are derived. (Note: $\beta \equiv V/c$ and $\gamma \equiv 1/\sqrt{1-\beta^2}$.)

Time dilation:

$T = \dfrac{T_0}{\sqrt{1-\beta^2}}$ where T_0 must be a time interval measured by a *single* clock) $\qquad T = \gamma T_0$

Length contraction:

$L = L_0\sqrt{1-\beta^2}$ (where L_0 must be a measurement made in a frame in which the object is at rest) $\qquad L = \dfrac{L_0}{\gamma}$

Relativistic momentum:

$\mathbf{p} = \dfrac{m\mathbf{v}}{\sqrt{1-\beta^2}} \qquad \mathbf{p} = \gamma m \mathbf{v}$

Relativistic velocity addition (for velocities along the $\pm x$ direction):

$v = \dfrac{v' + V}{1 + \left(\dfrac{v'V}{c^2}\right)}$

Kinetic energy:

$K = \dfrac{mc^2}{\sqrt{1-\beta^2}} - mc^2 \qquad K = mc^2(\gamma - 1)$

Rest energy: $\qquad$ Total energy:

$E_0 = mc^2 \qquad E = mc^2 + K \qquad E = \gamma mc^2$

When $E \gg mc^2$, then $\qquad E \approx pc$

[Also see Equations (41-22) through (41-25).]

If an amount of mass Δm disappears when particles combine into a bound system, the equivalent energy $\Delta E = (\Delta m)c^2$ is called the *binding energy* of the system.

The *twin paradox*: If one twin goes on a relativistic round-trip journey, that twin will be younger upon returning than the twin who remained at home.

The nonsynchronism of moving clock systems. Two clocks, separated a distance $\Delta x'$ and correctly synchronized in S', are incorrectly synchronized to observers in the S frame by an amount $|\varepsilon| = V\Delta x'/c^2$. The "chasing" clock indicates a <u>later</u> time than the "leading" clock.

The message of relativity: *When correctly expressed, the laws of nature are the **same** for all observers.*

General relativity. Experiment shows that two different attributes of mass are exactly proportional:

m_g = *gravitational mass* (the property of attraction for other masses)

m_i = *inertial mass* (the property of resisting acceleration)

The value of G, the universal gravitational constant, is chosen so that there is numerical equivalence for the units of m_g and m_i. Einstein generalized his theory of relativity to include accelerated frames of reference as well as the inertial frames of special relativity.

Postulates of general relativity:

(1) *All the laws of nature have the same form for observers in any frame of reference, accelerated or not* (the principle of covariance).
(2) *In the neighborhood of any given point, a gravitational field is equivalent in every respect to an accelerated frame of reference in the absence of gravity* (the principle of equivalence).

In place of Newtonian gravitational forces, the curvature of spacetime determines the trajectories that freely moving objects follow.

Questions

1. What were Galileo's contributions to special relativity?
2. Explain how it is possible for the moving spot on an oscilloscope screen to move across the screen faster than the speed of light without violating relativity.
3. Discuss what life would be like if the speed of light were, say, 100 km/h.
4. List several quantities whose measured values would be different in two inertial frames in relative motion. Other than the speed of light, what quantities would have the same values in these two frames?
5. Under what circumstances would you be older than your parents?
6. Interestingly, there is nothing in special relativity that forbids speeds faster than c as long as such particles *always* travel faster than c. As a particle approaches the speed of light from either side, the speed c seems to be an effective barrier that cannot be "penetrated" from either direction. It is proposed that particles that always travel faster than c be called *tachyons*, after the Greek word *tachos*, meaning "speed." Experiments have been performed to detect them, without success. What might be some properties of tachyons? Could they have a rest mass? What would be some consequences for fundamental ideas about causality? (See Bilaniuk and Sudarshan, "Particles Beyond the Light Barrier," *Physics Today*, May 1969, p. 43.)
7. Explain why it has been suggested that the "theory of relativity" could equally well be called the "theory of absolutism."
8. In a famous science fiction story, aliens kidnap several people and take them away in a spaceship. One person remarks, "We are traveling at the speed of light—look at your watches." Someone does and exclaims, "My God! My watch has stopped!" What blunder has the author made in writing this incident?

Problems

41.6 Comparison of Clock Rates
41.7 Comparison of Length Measurements
41.8 Proper Measurements

41B-1 The speeds of electrons emerging from Stanford's linear electron accelerator differ from the speed of light by about 5 parts in 10^{11}. Find this difference in centimeters per second.

41B-2 In 1849, H. L. Fizeau experimentally determined the speed of light by sending a light beam through the slots of a rotating toothed wheel to a distant mirror 8633 m away, Figure 41-21. Upon return of the reflected light pulses, if the rotation speed of the wheel was just right the light pulses could again pass through the slot openings between the teeth and thus be seen by the experimenter. On the other hand, with a different rotation speed the teeth interrupted the return light pulses, so no light was observed. Thus, as the wheel was speeded up, the observer would see a gradual progression from brightness to darkness to brightness, and so on, depending on whether the return pulses met a tooth or a slot on the rotating wheel. Fizeau reported that as the wheel was speeded up, the first "eclipse" of the return pulses occurred when the speed of rotation was 12.6 rev/s. The wheel had 720 teeth and 720 slots, all of the same width. Using these data, find the speed for light that Fizeau must have calculated. (His value was somewhat larger than more accurate determinations made later.)

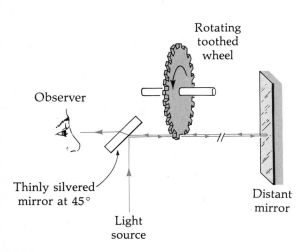

FIGURE 41-21
Problem 41B-2.

41A-3 According to his wristwatch, an astronaut takes 2 min to eat a chocolate bar. (a) If the astronaut is traveling with a speed of $0.5c$ relative to the earth, determine the amount of time that elapses in the earth's frame of reference during this time interval. (b) Find the distance in the earth's frame that the spaceship travels during this time.

41A-4 Though the Shinkansen "Bullet Train" in Japan can travel safely at 260 km/hr, its cruising speed is limited to 210 km/hr to keep the loudness level down to 75 phons. At this slower speed, by how much is the moving train's length (in the earth's frame) shorter than its rest length of 230 m?

41A-5 Alpha Centauri is a star about 4 light-years away. A rocketship travels at constant speed from the earth to this star in one day as measured by the rocketship's occupants. (a) Find the speed of the rocketship relative to the earth. Express your answer as the amount by which β differs from 1. [Hint: because β is so nearly equal to 1, use the convenient approximation $1 - \beta^2 = (1 + \beta)(1 - \beta) \approx 2(1 - \beta)$.] (b) In the rocketship's frame, how far away is the star at the beginning of the trip?

41B-6 Two spaceships, A and B, pass close to each other as they travel in opposite directions. Each ship has a proper length of 300 m. In ship A's frame of reference, it takes 2×10^{-6} s for the nose of ship B to pass the full length of ship A. A clock located in the nose of ship A reads exactly zero as the nose of B passes close by. Find the reading on this clock as the tail of B passes close by.

41B-7 The half-life of a given sample of radioactive particles is the time it takes for half the initial number of particles to undergo a disintegration. A group of radioactive particles moving at a speed of $0.8c$ travels through the laboratory a distance of 30 m. Half the particles survive the trip. Find the half-life of the particles in their own frame of reference.

41B-8 A beam of π^+ pions has a speed of $0.7c$. When at rest, the pions have an average lifetime of 2.6×10^{-8} s before disintegrating. (a) In the laboratory frame of reference, how long, on the average, will the moving pions live before disintegrating? (b) On the average, how far will they travel through the laboratory in this time?

41B-9 If you travel on a jet plane from New York to Los Angeles (4000 km air distance), at an average speed of 1000 km/h, how much younger are you on arrival than you would have been had you remained in New York during the time it took the plane to make the journey? (Hint: note that T, the time that would have been spent in New York, is extremely close to T_0, the time spent on the plane.)

41B-10 An astronaut wishes to visit the Andromeda galaxy (2 million light-years away) in a one-way trip that will take 30 yr in the spaceship's frame of reference. Assuming that his speed is constant, how fast must he travel relative to the earth? Express your answer as the amount by which β differs from 1.

41B-11 A spaceship has a proper length of 100 m. It travels close to the earth's surface with a constant speed of $0.8c$. Observers on earth decide to measure the length of the ship by erecting two towers, A and B, that coincide with the ends of the ship simultaneously (in the earth's frame) as it passes by. Tower A is at the tail of the ship, and tower B is at the nose of the ship. (a) How far apart do the earthmen build the towers? (b) How long do the earthmen say it takes for the nose of the ship to travel from tower A to tower B? (c) How long, according to measurements in the spaceship frame, does it take for the nose of the ship to travel from tower A to tower B? (d) As measured by the space travelers, how far apart are the towers? (e) Find the proper time interval between event 1, in which the nose of the ship coincides with tower A, and event 2, in which the nose of the ship coincides with tower B.

41B-12 Refer to the previous problem. (a) In the spaceship frame, how long does it take a beam of light to travel from the front to the rear end of the spaceship? (b) How long, according to earthmen, is required for a beam of light to travel from the front to the rear end of the moving spaceship? (c) A projectile is fired from the rear of the spaceship toward the front end with a speed of $0.6c$ as measured by the space travelers. Find the speed of the projectile in the earth frame of reference. (d) Find the earth speed of the projectile if it had been fired in the opposite direction with the same speed relative to the spaceship.

41.9 Relativistic Momentum
41.11 Relativistic Velocity Addition

41A-13 A certain type of meson decays at rest into two equal-mass particles, which are ejected in opposite directions with speeds of $0.8c$. Suppose that the meson is traveling through the laboratory with a speed $v = 0.6c$ when the decay particles are emitted along the line of motion (in opposite directions). Find the speeds of the two decay particles as measured in the laboratory frame.

41A-14 A meter stick, oriented parallel to the direction of motion, and a 1-kg object are on board a spaceship that has a speed $v = 0.6c$ relative to the earth. Find (a) the length of the meter stick and (b) the momentum of the object as measured in the earth's frame of reference. (c) If it takes an astronaut 6 h to do her physics homework, calculate the time it takes her as measured in the earth's frame of reference. (d) According to observers on earth, how far (in $c \cdot$hr) does the spaceship travel during this time?

41A-15 An astronomer observes that two distant galaxies are traveling away from the earth in opposite directions, each with speed $v = 0.9c$. What would an observer in one galaxy measure for the speed of the other galaxy?

41A-16 A certain quasar recedes from the earth with a speed $v = 0.87c$. A jet of material is ejected from the quasar toward the earth with a speed of $0.55c$ relative to the quasar. Find the speed of the ejected material relative to the earth.

41B-17 A particle of mass M moving at $v_1 = 0.6c$ collides head-on with and sticks to another particle of mass m moving at $v_2 = 0.8c$ in the opposite direction. After the collision, the combined mass is at rest with respect to the laboratory. Find the ratio M/m of the masses.

41B-18 A mass m moving with an initial speed v_0 has a head-on *elastic* collision with a mass $3m$, which is initially at rest. Nonrelativistically, the mass m rebounds with a speed $v_0/2$ while the mass $3m$ moves in the forward direction, also with a speed $v_0/2$. Relativistically, however, the final speeds cannot be equal. Letting $v_0 = 0.8c$, show that if the final speeds are each assumed to be $0.4c$, then momentum is not conserved.

41.12 Relativistic Energy

41A-19 Determine an object's speed if its kinetic energy equals its rest energy.

41A-20 A proton moves with speed $0.8c$. Find, in units of MeV, (a) the proton's total energy E and (b) its kinetic energy K. (c) Find its momentum in units of MeV/c.

41A-21 It is estimated that the total energy input (from all sources) to the U.S. economy in the year 1987 was about 8×10^{19} J. Assuming that all this energy came from nuclear reactions in which mass is converted to energy according to $E = mc^2$, determine the total mass annihilation that would be involved.

41A-22 The rest energy of a tritium nucleus, ^{3}He (two protons and a neutron), is 2808.413 MeV. Find the minimum energy required to remove one proton, resulting in a deuteron, ^{2}H, plus the proton.

41B-23 The *Stefan–Boltzmann radiation law* (Section 42.2) states that the total power R radiated per square meter by a surface at kelvin temperature T is $R = \sigma T^4$, where the *Stefan–Boltzmann constant* $\sigma = 5.672 \times 10^{-8}$ W/m²·K⁴. Calculate the rate of mass loss of the sun due to the conversion of mass to energy by nuclear reactions in the sun's core. (See Appendix G for additional data.)

41B-24 At normal incidence at the top of the earth's atmosphere, the incident solar power per unit area is about 1370 W/m². From this information (and other data from Appendix L), estimate the sun's mass loss per second ($E = mc^2$).

41B-25 Starting with fundamental definitions of E and p, derive Equation (41-22), $E^2 = (mc^2)^2 + (pc)^2$.

41B-26 (a) Determine the work required to accelerate an electron from rest to $0.8c$ according to Newtonian mechanics. (b) How much work is required according to relativity? Express your answers in terms of mc^2. (Hint: recall that the work done on an object equals the change in kinetic energy of the object.)

41B-27 Starting with fundamental definitions of E and p, derive the first relation of Equation (41-25), $v = pc^2/E$.

41B-28 A free neutron will decay into a proton, an electron, and a massless particle called an antineutrino. From the difference between the mass energies of the neutron and of the decay particles, calculate the total kinetic energy (in joules) the decay particles would have if the neutron were initially at rest.

41B-29 Starting with fundamental definitions of K and p, derive Equation (41-23), $p = \sqrt{2mK[1 + (K/2mc^2)]}$.

41B-30 An electron's kinetic energy is three times its rest energy. Find (a) the electron's total energy in electron volts and (b) its speed in terms of c.

41B-31 Starting with the fundamental definitions of E_0 and E, show that the second relation of Equation (41-25) is true: $v = c[1 - (E_0/E)^2]^{1/2}$.

41.13 The Nonsynchronism of Moving Clocks
41.14 The Twin Paradox

41A-32 A certain galaxy moves away from the earth so fast that the spectral lines in its light emission are Doppler-shifted to one-half their frequencies here on the earth. Find the galaxy's speed.

41B-33 Two clocks are located in the nose and the tail of a spaceship whose proper length is 300 m. They are correctly synchronized in the spaceship frame of reference. If the spaceship moves past the earth with a speed $V = 0.90c$, (a) find the difference in the readings of the two clocks as measured simultaneously in the earth's frame. (b) Which clock reads the earlier time?

41B-34 Imagine that the entire sun collapsed to a sphere of radius R_g such that the work required to remove a small mass m from the surface would be equal to its mass energy mc^2. This radius is called the *gravitational radius* for the sun. Find R_g. (It is believed that the ultimate fate of many stars is to collapse to their gravitational radii or smaller.)

41B-35 Refer to Problem 41B-11. As the spaceship passes the towers, the following two events are simultaneous in the earth frame:

> Event (a): *Coincidence of tower A with the tail of the ship.*
> Event (b): *Coincidence of tower B with the nose of the ship.*

(a) Make pictorial sketches (with dimensions) to show how these same two events look in the spaceship frame of reference. (b) Find the time interval, if any, between these events as measured in the spaceship frame.

41B-36 A spaceship (S' frame) passes the earth (S frame) at the times $t = t' = 0$ in their respective frames. The spaceship's velocity relative to the earth is $0.9c$. One second later as measured in the earth's frame, a radio signal is sent to the spaceship. Find the time in the spaceship's frame when the radio signal is received.

Additional Problems

41C-37 At exactly noon in our frame of reference, a clock moving with speed $v = 0.8c$ reads 12:00 (noon) as it passes the origin of our frame. (a) How far away will it be when its hands indicate 1 s after 12:00? (Leave the symbol c in the answer.) (b) When the clock face indicates 1 s after 12:00, a light signal is sent from the clock back toward the origin of our frame of reference. At what time (in our frame) does this signal arrive at our origin?

41C-38 A spaceship of proper length L travels past the earth with a speed $v = (4/5)c$. When a clock at the tail of the spaceship reads $t' = 0$ (and when earth clocks also read $t = 0$), a light signal is sent from the tail to the front of the spaceship. Determine the time at which the signal reaches the front end of the ship (a) according to spaceship clocks and (b) according to earth clocks. (c) The answers to parts (a) and (b) are *not* related according to the time dilation formula. Why not? (d) The light signal is reflected by a mirror at the front end back toward the rear. Find the time at which it reaches the rear according to rocket clocks. (e) Find the time in (d) according to earth clocks. (f) Are the answers to parts (d) and (e) related according to the time dilation formula? Explain why or why not.

41C-39 Imagine that a runner carries a mirror 1 m (in the runner's frame of reference) in front of her face to observe her own reflection as she runs, Figure 41-22. Her speed is $0.6c$ relative to the earth. She blinks. (a) In the runner's frame of reference, how much time will pass after she blinks before she sees the blink of her mirror image? (b) In the earth's frame of reference, what is this time interval? Leave the symbol c in the answers.

FIGURE 41-22
Problem 41C-39.

41C-40 Electrons in Stanford's 10 000-ft linear accelerator attain a final velocity of $(0.999\,999\,999\,7)c$. (a) In a frame of reference moving at this speed, how long is the accelerator? (Use the appropriate mathematical approximation.) (b) Traveling at this (constant) speed, how long would it take to travel this distance in the frame of reference of an electron? (c) How long would the electron's journey take as measured by a Stanford physicist?

41C-41 We could define "the length of a moving rod" to be the product of its velocity times the time interval between the instant one end of the rod passes a fixed point in our frame of reference and the instant the other end passes the same point. Show that this definition also leads to the familiar result for length contraction, $L = L_0(1 - \beta^2)^{1/2}$.

41C-42 A golf ball travels with a speed of 90 m/s. By what fraction does its relativistic momentum p differ from mv? That is, find the ratio $(p - mv)/mv$.

41C-43 One way of expressing the relativistic momentum increase is the fraction f by which the relativistic momentum p exceeds its classical value mv. That is, $f \equiv (p - mv)/mv$. Derive the following expression for the speed ratio $\beta = v/c$ in terms of f: $\beta = \sqrt{f(f+2)}/(f+1)$.

41B-44 Bandits try to stop a train (which is moving forward) by setting off explosive charges near the engine and near the caboose. The two explosions are simultaneous in the earth's frame of reference. In the train's frame, which explosion, if either, occurred first according to relativity? Does it make any difference whether the train is traveling in the $+x$ or the $-x$ direction? Justify your answers.

41C-45 Primary cosmic "rays" are high-energy protons that impinge on the earth from outer space. They collide with and break apart atomic nuclei in the upper atmosphere, creating secondary cosmic rays: a debris of electrons, positrons, neutrons, mesons, photons, etc., that shower down upon the earth's surface. (The most penetrating particles reach the deepest mines within the earth.) By recording the simultaneous arrival of particles over an area of a square mile or so at the earth's surface, one can estimate the energy of the single proton that initiated the shower of particles. Events involving a shower of perhaps 100 million particles have been measured whose total energy is about 10^{21} eV (1 eV = 1.6×10^{-19} J). Suppose that a photon (which travels always with the speed c in a vacuum) and a proton of 10^{21}-eV energy have a race to earth from the nearest star, 4 light-years away. By how much time would the proton lose the race?

41C-46 An unmanned spaceship recedes from the earth with a speed of $0.8c$. Transponder equipment aboard the spaceship sends back to earth a radio signal of exactly 1 s duration (measured in the spaceship's frame of reference) whenever an "interrogation pulse" is received from earth. Suppose that two very short interrogation pulses are sent from the earth, 10 s apart as measured in the earth's frame. (a) In the earth's frame, find the duration Δt of a single signal pulse received from the spaceship. (b) Find the time interval T between the leading edges of the two response signals as received at the earth.

41C-47 A high-energy proton has a speed of approach v relative to a proton at rest on the earth. Find the speed V relative to the earth of a frame of reference in which the two protons have equal speeds.

41C-48 A laser emits monochromatic light of wavelength λ. The laser beam is directed at normal incidence on a mirror that is moving away at speed V (relative to the laser). Show that the beat frequency between the incident light and the reflected light is approximately $2V/\lambda$. (Hint: in the mirror's frame of reference, light is received from a source moving away from the mirror. In the laser's frame, the reflected light is as if it were emitted from a receding source.)

41C-49 The total energy E of a proton from a high-energy accelerator is 5 times its rest energy E_0 (equal to mc^2). In terms of its rest energy E_0, find (a) its kinetic energy K and (b) its momentum p. (c) Find the value of β. When appropriate, leave the symbol c in the answer.

41C-50 Suppose that noted astronomers conclude that our sun is about to undergo a supernova explosion. In an effort to escape, we depart in a spaceship and head toward the star Tau Ceti, 12 light-years away. When we reach the midpoint (in space) of our journey, we see the supernova explosion of our sun and, unfortunately, at the same instant we see the explosion of Tau Ceti. (a) In the spaceship's frame of reference, should we conclude that the two explosions occurred simultaneously? If not, which occurred first? (b) In a frame of reference in which the sun and Tau Ceti are at rest, did they explode simultaneously? If not, which occurred first?

The following problems are famous "paradoxes" of special relativity. They are presented here without answers. Explaining why these situations are not paradoxical will challenge your understanding of the true nature of space and time.

41C-51 In a space war, two identical rocketships pass close to each other, traveling in opposite directions at speeds close to that of light. When the tail of ship B is adjacent to the nose of ship A, a mortar shell is fired sideways (perpendicular to the relative motion) from a gun barrel located near the tail of A in an attempt to hit ship B, Figure 41-23. Obviously, *both*

(a) As seen in the frame of reference of ship A, the length of B is contracted; thus the shell does *not* hit ship B.

As seen in Ship A's frame of reference

(b) As seen in the frame of reference of ship B, the length of A is contracted; thus the shell *does* hit ship B.

As seen in Ship B's frame of reference

FIGURE 41-23
Problem 41C-51.

statements in Figure 41-23 and their corresponding diagrams cannot be true. Ship B is either hit or it is not. Find the ambiguities in the statements of this problem, and discuss briefly what really happens and why there is no paradox. Assume that the ships pass very close to each other and that the shell's speed is very great, so that the transit time of the shell itself is not a factor in the analysis.

41C-52 The "stick-in-the-hole" paradox is one of the most puzzling paradoxes in special relativity. A 100-cm stick is moving horizontally with relativistic speed such that its length is contracted to 50 cm as seen in the earth's frame, Figure 41-24. An observer in the earth's frame has a thin board with a circular hole, 70 cm in diameter, cut out of it. As the pole passes by, the observer quickly lifts the board vertically (keeping its plane horizontal), allowing the stick to pass through the hole. Thus, at some instant, the (contracted) stick fits entirely inside the horizontal hole. Here is the paradox: "In the stick's frame of reference, the stick is 100 cm long, and the hole is only 35 cm across. How can the 35-cm opening engulf the 100-cm stick?" (Hint: focus your attention on *point events*. For example, consider four points equally spaced along the stick's length. At some instant in the earth's frame, these four points simultaneously lie in the plane of the board. What do these four events look like in the pole's frame? Are they simultaneous?)

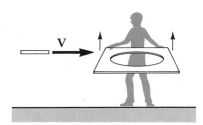

FIGURE 41-24
Problem 41C-52.

41C-53 The "pole-in-the-barn" paradox is one of the classic puzzles of special relativity. Consider a 20-ft pole carried along so fast that it is only 10 ft long as measured in the earth's frame, Figure 41-25. The pole is carried through a barn that has doors C and D on opposite walls. The barn is 12 ft long (2 ft longer than the moving pole), so both doors could be simultaneously shut for a brief period, trapping the pole inside the barn. (Door D would then be opened to permit the pole to travel on through.) On the other hand, to the runner carrying the pole, the barn is only 6 ft long because of length contraction. Here is the apparent paradox: "To the runner, how can his 20-ft pole fit inside the 6-ft barn with both doors closed?" Obviously, there is an inconsistency somewhere. Can you resolve this apparent paradox? (Hint: as with most paradoxes in relativity, the root of the problem lies in the fact that two events that are simultaneous in one frame of reference are not necessarily simultaneous in another frame.)

41C-54 Here is the famous "string paradox." Consider two frames: S is "our" frame, and S' is a "moving" frame traveling at constant speed $0.8c$ in the $+x$ direction. Two spaceships, A and B, each of proper length 100 m, are at rest in our frame, aligned in the same $+x$ direction with A in front of B and with their "noses" 200 m apart. A 300-m string (with 100 m of "slack") ties the nose of A to the nose of B. Simultaneously (in our frame) the two ships are now given identical constant accelerations along the $+x$ direction until they each reach a speed of $0.8c$, when the accelerations are simultaneously stopped. Thus each ship is finally at rest in S'.

- **In our S frame:** Because the ships started simultaneously, had identical accelerations, and stopped accelerating simultaneously, their nose-to-nose separation remains 200 m. (Also, the distance between any such pair of corresponding points on the two ships remains constant at 200 m.) Because of the Lorentz contraction, the final length of each is 60 m.
- **In the moving S' frame:** Each ship was initially contracted to 60 m long, and the initial nose-to-nose distance was contracted to 120 m. In the final rest position, each ship is 100 m long.

Here is the problem: (a) Show that, in the S' frame, the final nose-to-nose separation (at rest) is actually 333 m *and the string is therefore broken*. (b) The comments under "In our S frame" are correct, yet they seem to imply that the string should not break because the final nose-to-nose distance is only 200 m. Resolve this apparent paradox. Include diagrams for the initial and final situations in each frame.

41C-55 Consider the twin paradox. In the traveling twin's frame the earth clocks move away and come back, so as measured in that frame the moving earth clocks run *more slowly* than clocks at rest in that frame. This is true during *both the receding and the approaching motions of the earth*. Therefore, why is it that, upon his return, the traveling twin finds that *more* time has elapsed on the earth than in the traveling frame of reference?

Note: for an interesting example in which twins undergo the same acceleration for the same length of time, yet age differently, see S. P. Broughn, "The Case of the Identically Accelerated Twins," *American Journal of Physics* **57** (Sept. 1989).

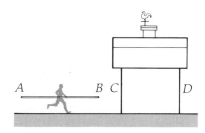

FIGURE 41-25
Problem 41C-53.

CHAPTER 42

The Quantum Nature of Radiation

All these fifty years of pondering have not brought me any closer to answering the question, What are light quanta?

EINSTEIN
(in a letter to Besso, 1951)

Physics is very muddled again at the moment; it is much too hard for me anyway, and I wish I were a movie comedian or something like that and had never heard anything about physics!

WOLFGANG PAULI
(in a letter to R. Kronig, 25 May 1925)
[*American Journal of Physics* 43, 208 (1975)]

I do not like it, and I am sorry I ever had anything to do with it.

E. SCHRÖDINGER
(on quantum mechanics)

42.1   Introduction

Toward the end of the nineteenth century, our understanding of what is now called *classical physics* had reached an impressive stage. It was believed that almost everything was known about the physical world and its interactions—at least, this was the opinion expressed by several well-known scientists at that time. *A more embarrassing misconception can hardly be imagined.* Yet, considering the widespread success of Newtonian mechanics in explaining the motion of all kinds of objects from baseballs to the solar system, and the fact that these same ideas also brought all heat phenomena under the rules of mechanics, it seemed reasonable that we had, at last, found a great unifying theory that explained all phenomena. There were also radio waves, light, and thermal radiation, which were obviously apart from mechanics, but these, too, were brought together in another unifying theory: Maxwell's electromagnetism. Together these two theories seemed to complete our understanding of all natural phenomena in terms of *particles* and *waves*.

However, a few surprises began to surface. In 1895 Wilhelm Konrad Roentgen discovered x-rays; the next year Antoine Becquerel discovered nuclear radioactivity; and the year after that J. J. Thomson's measurements of e/m for electrons showed that they were a fundamental component of all atoms, so the model of an atom needed revision. In addition, there were a few well-known phenomena that still remained a mystery. For example, the spectral distribution of wavelengths emitted by hot, glowing bodies had no satisfactory theoretical explanation. And the fact that ultraviolet light could eject electrons from metals had some very puzzling aspects. But most scientists felt that these were merely a few isolated instances that sooner or later would also be explained by the two "complete" theories of the day, Newton's mechanics and Maxwell's electromagnetism. If this had been true, the future activity for physicists would have been quite dull—merely applying these theories to the few remaining puzzles and determining the next decimal places in the fundamental constants of nature (the charge on the electron, the speed of light, Avogadro's constant, and so on).

We now tell the story of how the few minor cracks in the foundations of physics widened and brought the smug complacency of the nineteenth century tumbling down. In the process, physics itself expanded rapidly and became greatly strengthened. The revolution that occurred—the *quantum revolution*—was even more troubling and difficult to accept than Einstein's theory of relativity was a few years later. In a sense, relativity is considered part of classical physics (prequantum, that is) because the fundamental concepts of mass, momentum, energy, and the way systems interchange energy remain essentially unchanged. Einstein's revolution was to change completely the structure of space and time within which measurements are made and to extend classical concepts so that physical laws would be correct for high velocities. The quantum revolution revised classical concepts so that they were correct for very small distances. The new physics of both relativity and quantum mechanics includes classical physics as special cases. But the quantum revolution was perhaps the more revolutionary because it altered our most basic concepts of *particles* and of *electromagnetic waves*—the only "stuff" physicists in those days believed the universe was made of. The new quantum physics demonstrated that these classical ideas were inadequate and often led to profound contradictions, both in disagreeing with experiment and in challenging basic philosophical issues about the nature of matter and our perception of it.

42.2 The Spectrum of Cavity Radiation

One outstanding unsolved puzzle in physics in the late nineteenth century was the spectral distribution of so-called **cavity radiation**, also referred to as **blackbody radiation**. It was shown by Kirchhoff that the most efficient *radiator* of electromagnetic waves was also the most efficient *absorber*. A "perfect" absorber would be one that absorbs all incident radiation; since no light would be reflected, it is called a *blackbody*.

To investigate the nature of radiation, it seems best to construct the most efficient radiator of all. How does one make a blackbody? The nearest practical approach to an ideal blackbody is a tiny hole in a cavity with rough walls (Figure 42-1). Any radiation that enters the hole has negligible chance of being reflected out through the opening: it is essentially 100% absorbed. As the walls of the cavity absorb this incoming radiation, their temperature rises and they begin to radiate. They continue to radiate until *thermal equilibrium* is reached, at which time they radiate electromagnetic energy at the same rate they absorb it. The radiation inside is then called *blackbody radiation*, or *cavity radiation*, and the tiny amount that manages to leak out the hole can be studied. *The hole itself is the blackbody.*

In 1879, the Austrian physicist J. Stefan first measured the total amount of radiation emitted by a blackbody at all wavelengths and found that it varied as the fourth power of the absolute temperature. This was later explained

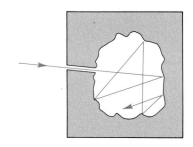

FIGURE 42-1
A practical approximation of an ideal blackbody is a hole that leads to a cavity with rough walls. The hole itself is the blackbody, since essentially all of the radiation incident on the hole is absorbed. The radiation inside the cavity is called *blackbody radiation* or *cavity radiation*.

through a theoretical derivation by L. Boltzmann, so the result became known as the *Stefan–Boltzmann radiation law*.

STEFAN–BOLTZMANN RADIATION LAW
$$R = \sigma T^4 \quad (42\text{-}1)$$

where the **total emittance** R is the total energy at all wavelengths emitted[1] per unit time and per unit area of the blackbody, T is the kelvin temperature, and σ is the *Stefan–Boltzmann constant*, equal to 5.672×10^{-8} W/m$^2 \cdot$ K^4.

In examining the spectral distribution of cavity radiation (the amount of energy at various wavelengths), researchers made a startling discovery. *The spectral distribution does not depend on the material of the cavity, but only on the absolute temperature T.* No matter what the cavity is made of, the spectral distribution is the same for a given temperature. Whenever physicists discover a phenomenon that is independent of the material involved, there is a strong probability that the effect involves a very basic interaction. So it is important to understand the effect thoroughly.

42.3 Attempts to Explain Cavity Radiation

Many capable physicists tried to develop a theory based on classical ideas that could predict the spectral distribution of cavity radiation. The goal was to derive the **spectral energy density** (in joules/meter^3) for the cavity radiation between wavelengths λ and $\lambda + d\lambda$. This is defined in terms of a mathematical function, $f(\lambda, T)$, that depends on both the wavelength λ and the absolute temperature T. Figure 42-2 shows experimental curves for three different temperatures. Note that as the temperature increases, the wavelength at the peak of each curve is displaced toward shorter wavelengths. The German physicist W. Wien obtained an empirical relationship for this feature, known as *Wien's displacement law*.

WIEN'S DISPLACEMENT LAW
$$\lambda_m T = \text{constant} \quad (42\text{-}2)$$

where λ_m is the wavelength at the maximum of the spectral distribution, T is the absolute temperature, and the constant is experimentally found to be 2.898×10^{-3} m$\cdot$K.

The total energy density at all wavelengths is the area under the curve:

$$\text{Total energy density} \atop \text{(all wavelengths)} = \int_0^\infty f(\lambda, T)\, d\lambda \quad (42\text{-}3)$$

According to the Stefan–Boltzmann law, the total energy radiated is proportional to the fourth power of T, so the area under the curve for $T = 6000$ K is 16 times that for $T = 3000$ K. Also note that the fraction of the radiation that falls within the *visible* range[2] is not uniform. At low temperatures, there

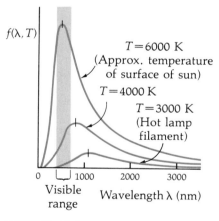

FIGURE 42-2
The *spectral distribution* curves for cavity radiation at three different equilibrium temperatures. The function $f(\lambda, T)$ describes how the intensity of the radiation inside the cavity depends upon wavelength. It is in units of energy per unit volume per unit wavelength interval between wavelengths λ and $\lambda + d\lambda$. The small vertical lines on the peaks of the curves show that as the temperature becomes hotter the wavelength at the peak becomes shorter, according to Wien's displacement law.

---

[1] As pointed out in Question 4 at the end of the chapter, the *surfaces* of materials have a total emittance somewhat less than that of an ideal blackbody. The emittance depends upon the physical conditions of the surface and is different for different materials. A surface coated with lampblack (carbon soot) is close to an ideal emitter and absorber.

[2] It is interesting that the visual sensitivity of our eyes centers on the peak of the sun's radiation distribution. If there are sensing beings on planets around a star with a different temperature, perhaps, through evolution, their "eyes" evolved to respond to a different portion of the electromagnetic spectrum.

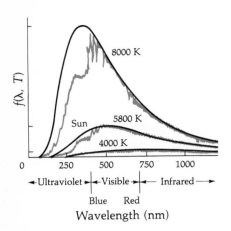

FIGURE 42-3
Most exposed surfaces are not perfect blackbody radiators, though they are often close enough to the Planck curves so that temperatures can be accurately estimated. These curves are the best fit to the spectral distribution from the surfaces of three different stars. (Absorption by the earth's atmosphere, particularly in the ultraviolet, greatly distorts the spectral distributions obtained by earth-based telescopes.) Our sun, at 5800 K, looks yellowish. The 8000-K star emits more blue light than our sun and appears bluish-white. The 4000-K star is reddish, emitting most of its radiation in the invisible infrared. (From W. M. Protheroe, E. R. Capriotti, and G. H. Newsom, *Exploring the Universe*, 2nd ed., Charles E. Merrill Publishing Company, 1981.)

is relatively more energy radiated at long wavelengths (red) than at shorter wavelengths (blue). As the temperature increases, this changes to relatively more radiation in the blue, which explains the color changes that occur as a solid is heated: it first begins to glow with a dull red, progressing through orange, yellow-white, and finally, at very high temperatures, blue-white.

EXAMPLE 42-1

The wavelength at the peak of the spectral distribution for a blackbody at 4300 K is 674 nm (red). At what temperature would the peak be 420 nm (violet)?

SOLUTION

From the Wien displacement law, we have (for the wavelengths at the maximum)

$$\lambda_1 T_1 = \lambda_2 T_2$$

Substituting numerical values gives

$$(674 \times 10^{-9} \text{ m})(4300 \text{ K}) = (420 \times 10^{-9} \text{ m})(T_2)$$

Finally, $T_2 = \boxed{6900 \text{ K}}$

Wien's Theory

The search for a theoretical basis for the radiation formula is one of the most fascinating chapters in the history of physics. We will mention just a few high points here. In 1884, Boltzmann used a thermodynamic approach to the problem. He assumed that the cavity radiation was in a cylinder with a movable piston, and he calculated the results of a Carnot-cycle process. (The radiation exerts a force on the walls, so ideas of work came into the analysis.) In 1893, Wien expanded on this result and, from considerations of the Doppler shift upon reflection from the moving piston, derived the fact that some function of the product (λT) was involved. Making certain assumptions about the emission and absorption of radiation, Wien derived an expression for the spectral distribution curve $f(\lambda, T)$. It is usually written as du_λ, the **spectral energy density** (in joules/meter^3) for wavelengths between λ and $\lambda + d\lambda$:

WIEN'S RADIATION LAW
$$du_\lambda = f(\lambda, T)\, d\lambda = \frac{c_1 \lambda^{-5}}{e^{(c_2/\lambda T)}}\, d\lambda \qquad (42\text{-}4)$$

The unknown constants c_1 and c_2 are chosen to make the "best fit" to the experimental data. The curve fits the data well at short wavelengths, but as more experimental points at long wavelengths were obtained, the disagreement became obvious. Figure 42-4 shows how the Wien curve falls below the experimental points at long wavelengths.

The Rayleigh–Jeans Theory

The thermodynamic derivation of Boltzmann and Wien was a helpful step in revealing that some function of λ and T (and nothing else!) was involved. However, since thermodynamics is based on very general principles that apply

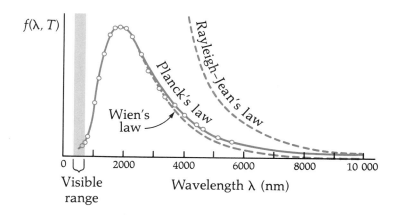

FIGURE 42-4
The circles are experimental points for cavity radiation at 1600 K. Curves for three different theories are shown for comparison.

to *all* systems, often thermodynamic arguments do not give insight into the particular processes involved in a given system. Perhaps more success would come if one focused on the *source* of the cavity radiation, the actual process of electromagnetic radiation and absorption by the walls.

Rayleigh (1900) approached the problem from this viewpoint. He considered a rectangular cavity with metallic walls and assumed that the electric charges in the walls were the source of the radiation. They behaved as simple harmonic oscillators and could radiate as well as absorb radiation, each with its "characteristic" natural frequency of oscillation. For any sufficiently large enclosure, there was such an extremely great number of oscillators that the resulting negligible differences between adjacent frequencies caused the radiation to appear continuous over all wavelengths. At a given temperature T, constant operation of the oscillators means that *standing waves* would be set up in the enclosure. With perfectly conducting walls, the standing waves must have nodes at each wall. The total number[3] of such standing waves (per unit volume) turned out to be $8\pi\lambda^{-4}$.

The *equipartition theorem* (Chapter 21) states that, on the average, $\frac{1}{2}kT$ of energy is associated with each variable required to specify the energy of a system in thermal equilibrium at absolute temperature T. For electromagnetic waves[4] there are two variables (the two directions of polarization), so the total energy associated with each is

AVERAGE ENERGY OF A CLASSICAL SHM OSCILLATOR
(in a system at thermal equilibrium)

$$E_{av} = kT \quad (42\text{-}5)$$

where the Boltzmann constant k is equal to 1.381×10^{-23} J/K. Multiplying the number of standing waves by the average energy of each gives the

---

[3] Rayleigh made a trivial error of a factor of 8 in the derivation. After the result was published, Sir James Jeans pointed out the obvious mistake, so the corrected formula became known as the Rayleigh–Jeans law. In this instance, considerable fame resulted from a rather minor contribution. Of course, Jeans also made a great many other contributions to physics.

[4] One could also apply the equipartition theorem to the SHM oscillators in the walls. In SHM, two variables are required: one for the kinetic energy and one for the potential energy. Therefore, each SHM oscillator has an average energy of kT.

Rayleigh–Jeans law for the *spectral energy density* (in joules/meter^3) between wavelengths λ and $\lambda + d\lambda$:

RAYLEIGH–JEANS RADIATION LAW
$$du_\lambda = f(\lambda, T)\, d\lambda = 8\pi k T \lambda^{-4}\, d\lambda \qquad (42\text{-}6)$$

where k is the Boltzmann constant.

The theory fits the data at extremely long wavelengths, but as shown in Figure 42-4 it was in drastic error everywhere else. The curve never "bent over": as the wavelength approached zero, the curve continued to increase toward infinity. Since the discrepancy was greatest at short wavelengths, it became known as the *ultraviolet catastrophe*. And a catastrophe for classical physics it was. The Rayleigh–Jeans derivation was based on classical concepts of thermodynamics and statistical mechanics, which had been completely successful in every other application. Each step of the derivation seemed so plausible that it was extremely disturbing to find the result so inaccurate. Where was the error in thinking?

42.4 Planck's Theory

In 1900, the German physicist Max Planck stumbled upon a solution to the difficulties. He first found it by some purely mathematical reasoning, then tried to figure out the physical implications of the mathematical trick he employed. Even though he obtained a radiation law that agreed with the experimental data, the physical implications were so startling that for many years Planck himself did not want to accept them as describing what the "real world" was like. The quantum ideas were just too radical.

Planck's stratagem was the following. In the Rayleigh–Jeans derivation, an important step in the procedure was to find the average energy of a SHM oscillator by integrating over all possible energies the oscillator might have. Classically, such an oscillator (as, for example, a mass on a spring) could vibrate with any amplitude from zero on up. Since the energy is proportional to the square of the amplitude, the oscillator could have any of a *continuum* of energy states, a range of values that varied smoothly from 0 to ∞. The trouble was that *integrating* over a *continuous* range of energies from 0 to ∞ made the function become infinite as $\lambda \to 0$. Planck was a good enough mathematician to realize that if, instead, he made a *summation* over a *discrete* range of energies from 0 to ∞, the result was a function that "turned over" and approached zero as $\lambda \to 0$, just like the experimental radiation curves. As it turned out, the curve Planck obtained matched the experimental points exactly. This put Planck in a position similar to that of a student who has looked in the back of the book to find the right answer to a problem, but is then faced with finding out how to get there from the given facts. What was it about nature that made a summation of *discrete* energy states the proper approach?

Planck decided on a bold step. Although it disagreed with all classical theories, he assumed that a SHM oscillator with a natural frequency f was "allowed" to have only one of a discrete series of energies: 0, hf, $2hf$, $3hf$, ..., where h is a constant.

ALLOWED ENERGIES FOR A QUANTIZED SHM OSCILLATOR
$$E_n = nhf \qquad (\text{where } n = 0, 1, 2, 3, \ldots) \qquad (42\text{-}7)$$

Planck first determined the constant by fitting experimental data to the expression for $f(\lambda, T)$ that evolved from his theory. He obtained a value very close to the currently accepted value:

PLANCK'S CONSTANT $\quad h = \begin{cases} 6.626 \times 10^{-34}\ \text{J·s} \\ 4.136 \times 10^{-15}\ \text{eV·s} \end{cases}$

Figure 42-5 compares *energy-level* diagrams for the classical and the quantum cases.

As a second assumption, Planck proposed that the only amount of energy ΔE an oscillator could emit or absorb was a **quantum**[5] of energy:

$$\Delta E = hf \qquad (42\text{-}8)$$

With these assumptions, Planck found the average energy for a collection of oscillators in thermal equilibrium at absolute temperature T to be

$$E_{av} = \frac{hf}{(e^{hf/kT} - 1)} \qquad (42\text{-}9)$$

Transforming the variable from frequency f to wavelength λ through $f\lambda = c$, we have

AVERAGE ENERGY OF A
QUANTIZED SHM OSCILLATOR
(in a system at thermal
equilibrium)

$$E_{av} = \frac{\left(\dfrac{hc}{\lambda}\right)}{(e^{hc/\lambda kT} - 1)} \qquad (42\text{-}10)$$

This is quite different from the classical value of $E_{av} = kT$ [Equation (42-5)]. However, as $h \to 0$, this relation does reduce to the classical value (see Problem 42C-39).

If the oscillators had this average energy, then it must also be the average energy of the waves in the cavity (because the walls and the radiation are in thermal equilibrium). Multiplying this average energy by the Rayleigh–Jeans calculation for the number of standing waves, $8\pi\lambda^{-4}$, Planck obtained his expression for the spectral distribution $f(\lambda, T)$. The Planck *spectral energy density* (in joules/meter³) for cavity radiation between wavelengths λ and $\lambda + d\lambda$ is

PLANCK'S
RADIATION LAW $\quad du_\lambda = f(\lambda, T)\, d\lambda = \dfrac{8\pi hc\lambda^{-5}}{(e^{hc/\lambda kT} - 1)}\, d\lambda \qquad (42\text{-}11)$

As you can see from Figure 42-4, the Planck theory fits the experimental points beautifully. For short wavelengths, the Planck equation approaches the Wien expression, which was correct in that region. For long wavelengths, the Planck equation approaches the Rayleigh–Jeans law, correct for long wavelengths. Planck effectively built a bridge between the two classical radiation theories. However, to do so, he had to make a radical break with all previous ideas about the energy a system could possess. If nature really behaved this way and all systems had quantized energy states, why wasn't it discovered long ago? The following example will explain why.

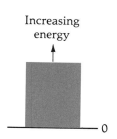

(a) According to classical mechanics, the possible energy states from a *continuous* distribution.

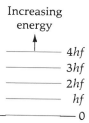

(b) According to quantum mechanics, the possible energy states form a *discrete* distribution.

FIGURE 42-5
Energy-level diagrams for a SHM oscillator of natural frequency f.

---

[5] The word *quantum* comes from the Latin word *quantus*, meaning "how much." Planck originally proposed that quanta could have integral multiples of hf, but Einstein and others later showed that only single units of hf were permissible.

> **EXAMPLE 42-2**
>
> A 5-g mass is hung from a string 10 cm long and is set into motion so that, at extreme positions, the string makes an angle of ± 0.1 rad with the vertical. Because of friction with the air, the amplitude gradually decreases. Can we detect the quantum jumps in energy as the amplitude decreases?
>
> **SOLUTION**
>
> The frequency of oscillation f is obtained from Equation (15-21):
>
> $$f \approx \frac{1}{2\pi}\sqrt{\frac{g}{\ell}} = \frac{1}{2\pi}\sqrt{\frac{9.8 \text{ m/s}^2}{0.1 \text{ m}}} = 1.58 \text{ s}^{-1}$$
>
> The energy of the pendulum is equal to the gravitational potential energy at an extremity:
>
> $$E = mg\ell(1 - \cos\theta)$$
> $$E = (0.005 \text{ kg})(9.8 \text{ m/s}^2)(0.1 \text{ m})(1 - \cos 0.1) = 2.45 \times 10^{-5} \text{ J}$$
>
> The quantum jumps in energy would be
>
> $$\Delta E = hf = (6.63 \times 10^{-34} \text{ J}\cdot\text{s})(1.58 \text{ s}^{-1}) = 1.05 \times 10^{-33} \text{ J}$$
>
> The ratio is $\Delta E/E = 4.28 \times 10^{-29}$. Therefore, in order to detect the quantized nature of the energy states, we would have to measure energy to better than 4 parts in 10^{-29}, a sensitivity far beyond the capability of any experimental technique.

As the example shows, the quantization of energy states is undetectable for *macroscopic* mechanical systems. The "graininess" of energy transfers is usually not noticed in everyday phenomena because of the smallness of h. If h were bigger, we would see quantum effects all around us. Quantum effects are always present, but they become noticeable only for *microscopic* systems on an atomic scale, that is, for cases in which ΔE is of the order of E. This condition is what makes blackbody radiation (at high frequencies) behave in an unusual way, traceable to quantum effects. It is interesting that if we let $h \to 0$, all quantum equations turn into the corresponding classical expressions. Thus the new quantum mechanics is a more general theory that contains classical mechanics as a special case.

42.5 The Photoelectric Effect

Today it is hard to realize the magnitude of the break with classical thinking that Planck initiated. Planck himself, who strongly resisted giving up the continuity of possible energy states, spent much effort in trying (unsuccessfully) to find an alternative solution to the ultraviolet catastrophe within the framework of classical physics. Though he grudgingly came to accept the idea that oscillators could have only quantized energy states and emit or absorb radiation in units of hf, he held to the classical view of radiation: electromagnetic waves were not quantized. But soon even this link to classical physics was broken.

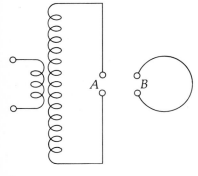

FIGURE 42-6
The experimental apparatus used by Hertz to detect electromagnetic waves.

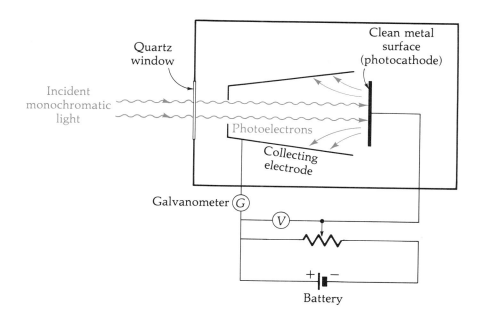

FIGURE 42-7
An experimental arrangement for investigating the photoelectric effect. The quartz window passes wavelengths in the ultraviolet that would be stopped by ordinary glass. The variable voltage V applied to the electrodes can be reversed by a switching arrangement (not shown).

Heinrich Hertz was the first (in 1887) to experimentally produce the electromagnetic waves predicted by Maxwell's equations. Using an *induction coil* (a step-up transformer with a great many turns on the secondary) attached to two small metal spheres as shown in Figure 42-6, he initiated an *oscillating*[6] spark across gap A. A nearby metal ring with a gap B would respond by sparking across its gap, verifying that electromagnetic energy had traveled from A to B. Quite by accident, Hertz discovered that the spark at B could be initiated much more easily if the gap were illuminated by ultraviolet light. Ten years later Thomson discovered the electron, and it was then verified that the ultraviolet light ejected electrons from the gap electrodes, making the spark easier to form. The phenomenon of electron ejection by light is called the *photoelectric effect*.

Figure 42-7 shows an experimental apparatus for investigating the effect. At any one time, monochromatic light is used. According to classical wave theory, the electric field of the incident light could transfer some of its energy to electrons in the surface of the metal, allowing them to acquire sufficient energy to escape. If the intensity of the light is increased, the ejected *photoelectrons* should acquire greater kinetic energy because of the stronger electric field of the light. The frequency of the light, however, should not make any difference at all. *Both of these deductions from classical theory disagree with experimental data.*

Figure 42-8 shows experimental curves for the *photocurrent* resulting when light of (essentially) a single wavelength is incident. The *stopping potential* is the negative voltage V_0 applied to the collecting electrode such that the kinetic energy of the most energetic electrons will be converted to potential energy at the collector. That is, the voltage V_0 barely stops the most energetic photoelectrons from reaching the collector. The relation is

$$eV_0 = \tfrac{1}{2}mv_{max}^2 \tag{42-12}$$

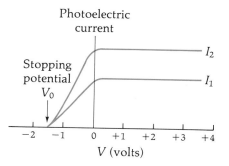

FIGURE 42-8
The photoelectric current versus the potential V of the collecting electrode with respect to the photocathode. Curves for monochromatic light of two different intensities are shown. Both have the same stopping potential.

---

[6] The frequency of the oscillation was determined by the capacitance of the spheres and the inductance of the induction coil.

The above two features of the photoelectric effect that are contrary to predictions of classical theory, along with a third feature, are summarized as follows:

Classical prediction	Experimental fact
(1) As the intensity of light is increased, the electric field E becomes larger. Since the force on an electron is eE, increasing the intensity should increase the kinetic energy acquired by the electrons.	(1) Figure 42-8 shows that even though the light intensity is increased, the maximum[7] kinetic energy of the photoelectrons remains the same.
(2) The *frequency* of light should not affect the kinetic energy of the ejected photoelectrons. Only the *amplitude* of the electric field should change their energies.	(2) As the frequency of the light is reduced, a *threshold* frequency is reached, below which *no* photoelectrons are produced, regardless of the light intensity (see Figure 42-9).
(3) Assuming that a single electron in the metal surface could absorb energy over an "effective target area" about the size of an atom, very dim light should require a longer time before the electron absorbs sufficient energy to escape.	(3) No appreciable time delay has ever been observed (though *fewer* electrons are produced by the dimmer light). The upper limit on measurements of the time delay is $<10^{-9}$ s; the actual time delay may be much less than that.

Clearly, classical ideas just do not come up with the correct predictions. The following example illustrates one of the discrepancies.

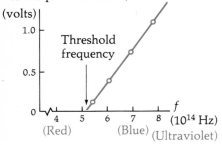

FIGURE 42-9
The frequency of the incident light determines the maximum kinetic energy of the photoelectrons. Below a certain "cut-off" value called the *threshold frequency*, no photoelectrons are ejected regardless of the intensity of the incident light. (The values are for a cesium surface.)

EXAMPLE 42-3

A cesium surface is 2 m from a point light source of 1 μW power that emits light uniformly in all directions. The area is perpendicular to the incident light. Assume that a single electron can absorb energy over a circular area of one atom (radius $\approx 10^{-10}$ m). The minimum energy required to extract an electron from the surface is 2.14 eV. Estimate how much time is required, according to classical theory, for the electron to absorb this amount of energy.

SOLUTION

The effective target area is $\pi r^2 \approx 3 \times 10^{-20}$ m². According to classical theory, the energy from the point source is spread uniformly over a spherical wavefront of radius $R = 2$ m. Therefore, the power per meter² at the cesium surface is

$$\frac{P}{4\pi R^2} = \frac{10^{-6} \text{ W}}{16\pi \text{ m}^2} \approx 2 \times 10^{-8} \frac{\text{W}}{\text{m}^2}$$

---

[7] The photoelectrons emerge with a spectrum of energies from zero to the maximum value indicated in Equation (42-12). Presumably, many come from varying depths, just within the surface, where they must expend some energy to make their way through the lattice of metal atoms as well as to overcome the attractive forces at the surface.

> The power that falls on the area of one atom is
>
> $$\left(2 \times 10^{-8} \frac{W}{m^2}\right)(3 \times 10^{-20} \text{ m}^2) = 6 \times 10^{-28} \frac{J}{s}$$
>
> The minimum energy needed to escape the surface is
>
> $$(2.14 \text{ eV})\underbrace{\left(\frac{1.60 \times 10^{-19} \text{ J}}{1 \text{ eV}}\right)}_{\text{Conversion ratio}} = 3.42 \times 10^{-19} \text{ J}$$
>
> and the time required to absorb this much energy is then
>
> $$t = \frac{3.42 \times 10^{-19} \text{ J}}{6 \times 10^{-28} \frac{J}{s}} = \boxed{5.71 \times 10^8 \text{ s}}$$
>
> This is about 18 years! Yet, experimentally, the upper limit to any possible time delay is less than 10^{-9} s—a discrepancy of a factor of $\sim 10^{17}$!

From electromagnetic theory a plausible argument can be made that an electron might absorb energy over a larger target area of the order of λ^2, where λ is the wavelength of the incident radiation. For visible light ($\lambda \approx$ 500 nm), this improves the situation by a factor of only $\sim 10^8$, still leaving a factor of $\sim 10^9$ unaccounted for. There are not many experiments that disagree with theory so drastically!

In 1905, Einstein[8] proposed a solution to the photoelectric dilemma. Though Planck was reluctant to accept the possibility that electromagnetic waves were quantized, Einstein saw that if one assumed that radiation was actually well-localized "bundles" or **quanta** (later called **photons**), then the photoelectric effect could be simply explained. Einstein proposed the following:

EINSTEIN'S ASSUMPTION OF THE QUANTIZATION OF RADIATION	The emission and absorption of radiation of frequency f always occur in quanta (or *photons*) of energy: $E = hf$. The photon remains localized in space as it moves away from the source with a velocity c.

If photons remain well localized, then, Einstein reasoned, in the photoelectric process the photon could be completely absorbed by a single electron. After gaining an energy hf, the electron would use part of this energy in escaping from the surface, and its remaining energy would appear as kinetic energy of the electron. The *minimum* energy required to barely escape from a surface is called the **work function** w_0. (Typical values for metals are about 2 to 6 eV. Visible photons have energies of around 2 eV in the red to somewhat above 3 eV for blue. For this reason, some materials exhibit a photoelectric effect for only the more energetic photons of ultraviolet light.) Applying conservation of energy to the process, Einstein proposed that the maximum kinetic

---

[8] It was an incredible year for 26-year-old Einstein. Volume 17 of *Annalen der Physik* (1905) included his revolutionary paper on special relativity, a treatise on Brownian motion that enabled Perrin to determine Avogadro's number, and his article on the photoelectric effect. It was this last article that led to Einstein's Nobel Prize in 1921. (See the chronology of quantum theory development at the end of Chapter 43.)

energy K_{max} of the electrons would be related to the photon energy hf according to

EINSTEIN'S PHOTOELECTRIC EQUATION
$$hf = K_{max} + w_0 \qquad (42\text{-}13)$$

This simple idea immediately explained the three baffling features of the photoelectric effect mentioned above:

(1) Since K_{max} depends on only the frequency of the light, and not on its intensity, dim light has the same stopping potential as bright light (Figure 42-7).
(2) For certain materials, the photon energy at a given wavelength may be less than the work function. Therefore, there is a threshold frequency, below which no photoelectrons would be produced.
(3) Since the photon energy is localized in space (rather than spread uniformly over a wavefront), its total energy can be transferred to an electron in a single step, ejecting the electron with negligible time delay no matter how dim the illumination. (Of course, the number of photoelectrons depends on the light intensity.)

This close agreement with experiment in another area, distinct from blackbody radiation, seemed to force acceptance of the photon's existence. However, as we will discuss shortly, it was a large pill to swallow.

Photoelectric experiments yield a great deal of important information. For example, combining Equations (42-12) and (42-13) and rearranging, we have

$$V_0 = \left(\frac{h}{e}\right)f - \left(\frac{w_0}{e}\right) \qquad (42\text{-}14)$$

This is a straight-line function for the stopping potential V_0 as a function of frequency f (Figure 42-9). The slope of the line is h/e, which furnishes another experimental method of determining Planck's constant h. These values agree with those found previously from the completely different phenomenon of blackbody radiation. It is reassuring that separate pieces of evidence lock together like this to form an overall coherent picture. Another feature of Equation (42-14) is shown in Figure 42-10. The intercept of the straight line on the horizontal axis is the threshold frequency, and the intercept with the vertical axis is the work function w_0.

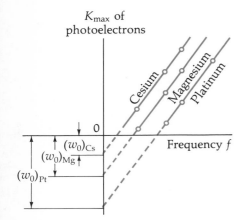

FIGURE 42-10
Photoelectric data for various substances produce straight lines whose slope is h/e. The lines intersect the horizontal axis at the threshold frequencies and the vertical axis at the respective work functions.

EXAMPLE 42-4

If the 1-μW light source of Example 42-3 emits only monochromatic light of wavelength $\lambda = 550$ nm, (a) find the number of photons per second incident normally on a circular target area 1 cm in diameter and located $R = 2$ m from the source. (b) Find the maximum kinetic energy (in electron volts) of the photoelectrons. (c) Find the threshold frequency for cesium.

SOLUTION

(a) The fraction of the energy output of the source that falls on a circular target area ($r = 5$ mm) that is 2 m away is

$$\frac{\pi r^2}{4\pi R^2} = \frac{\pi(0.005 \text{ m})^2}{4\pi(2 \text{ m})^2} = 1.56 \times 10^{-6}$$

The power incident on the target is therefore

$$\left(1 \times 10^{-6} \frac{J}{s}\right)(1.56 \times 10^{-6}) = 1.56 \times 10^{-12} \frac{J}{s}$$

Each photon has an energy of

$$hf = h\left(\frac{c}{\lambda}\right) = \frac{(6.63 \times 10^{-34}\ J \cdot s)(3 \times 10^8\ m \cdot s^{-1})}{550 \times 10^{-9}\ m} = 3.62 \times 10^{-19}\ J$$

The number of photons per second striking the target is therefore

$$\frac{\left(1.56 \times 10^{-12} \frac{J}{s}\right)}{3.62 \times 10^{-19}\ J} = \boxed{4.33 \times 10^6 \frac{\text{photons}}{\text{second}}}$$

(b) The photon energy in electron volts is

$$(3.62 \times 10^{-19}\ J)\underbrace{\left(\frac{1\ eV}{1.6 \times 10^{-19}\ J}\right)}_{\text{Conversion ratio}} = 2.26\ eV$$

The work function w_0 for cesium is (from Example 42-3) 2.14 eV. The maximum kinetic energy, K_{max}, of the photoelectrons is given by Equation (42-13):

$$hf = K_{max} + w_0$$

Solving for K_{max} gives
$$K_{max} = hf - w_0$$
$$= 2.26\ eV - 2.14\ eV = \boxed{0.120\ eV}$$

(c) At the threshold frequency, f_{th}, the photon energy $hf_{th}(= hc/\lambda_{th})$ equals the work function w_0. Solving for λ_{th}, and substituting numerical values, we obtain

$$\lambda_{th} = \frac{hc}{w_0} = \frac{(4.136 \times 10^{-15}\ eV \cdot s)(3 \times 10^8\ m/s)}{2.14\ eV}$$
$$= 5.80 \times 10^{-7}\ m = \boxed{580\ nm}$$

This is in the orange-yellow portion of the spectrum, so shorter wavelengths of visible light (toward the green-blue) will eject photoelectrons from cesium.

The photoelectric effect has many practical applications. Most light meters for determining proper exposures in photography use the photocurrent produced by incident light for operating the meter. A *photocell* is the "electric eye" that opens a door, or sets off an alarm, when a beam of light is interrupted. It is also used to detect holes in punched cards or paper tape. An instrument widely used in nuclear physics experiments is the *scintillation counter*, shown in Figure 42-11. A typical detector uses certain materials that emit tiny flashes of light, or scintillations, when energy is absorbed from photons or charged particles. This light, in turn, falls on a *photocathode surface*,

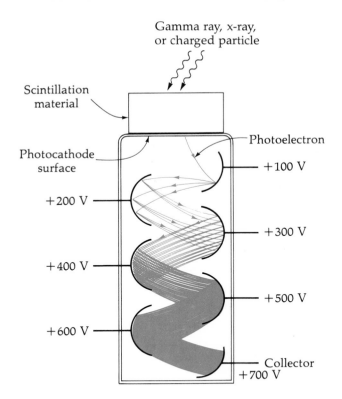

FIGURE 42-11
A scintillation counter uses a scintillation material with a photomultiplier tube to produce a large electrical pulse at the collector when a gamma ray, an x-ray, or a charged particle is absorbed in the scintillator.

ejecting photoelectrons that subsequently strike a series of *dynodes*. If the impact velocity is high enough, a single electron striking a dynode will eject one or more additional electrons in a process called *secondary emission*. Typical multiplication factors are from 2 to 5 or more. In a *photomultiplier* with 10 dynodes and a multiplication factor of 4 at each impact, a single photoelectron that starts down the chain produces 4^{10} ($\approx 10^6$) electrons at the collector, sufficient to produce an electrical pulse that can be easily amplified. Many photomultipliers have gains as high as 10^9 or more.

42.6 The Compton Effect and Pair Production

An additional piece of evidence for the existence of photons was presented by A. H. Compton in 1923. Directing a monochromatic beam of x-rays at a thin slab of carbon, he observed that the x-rays that were scattered from the carbon at various angles had a *longer* wavelength λ' than the incident wavelength λ_0. Figure 42-12 shows the experimental arrangement, and Figure 42-13 shows

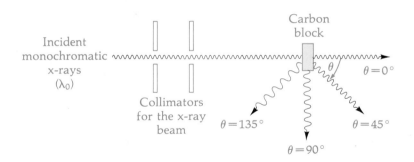

FIGURE 42-12
X-rays scattered at various angles have longer wavelengths λ' than the incident wavelength λ_0.

the experimental data. The amount of wavelength shift, $\Delta\lambda = \lambda' - \lambda_0$, was the same regardless of the target material, implying that it is an effect involving electrons rather than the atom as a whole. *Classical wave theory cannot explain this result.* According to classical theory, the oscillating electric field of the incoming wave would set electrons in the target material into oscillations. These vibrating electrons would then reradiate electromagnetic waves, *but necessarily at the same frequency of the incident wave,* contrary to what was observed.

Compton invoked the photon model to explain the results in a simple way. From Einstein, the energy of a photon is $E = hf$. According to relativity, energy and mass are related by $E = mc^2$. Combining these equations gives

$$hf = mc^2 \qquad (42\text{-}15)$$

If photons travel with a speed c, their momentum is $p = mc$, which, from Equation (42-15), becomes

MOMENTUM p OF A PHOTON
$$p = \frac{hf}{c} = \frac{h}{\lambda} \qquad (42\text{-}16)$$

It should be noted that even though photons have momentum, *they have zero mass*. This is seen from the relativistic relation [Chapter 41, Equation (41-22)] between energy E, momentum p, and mass m:

$$E^2 = c^2p^2 + (mc^2)^2 \qquad (42\text{-}17)$$

Since the momentum of a photon is $p = hf/c = E/c$, it becomes clear that the mass term in Equation (42-17) must be zero.

Compton viewed the interaction as a billiard-ball type of "collision" between the incoming photon and an (essentially) "free" electron[9] at rest. Figure 42-14 sketches the process. Conservation of energy and of momentum applies in the collision. Since the scattered electron acquires some energy, the scattered photon must have *less* energy than the incident photon. Applying relativistic equations for the conservation of energy and momentum, Compton derived the following expression for the shift in wavelength:

COMPTON SHIFT
$$\lambda' - \lambda_0 = \frac{h}{mc}(1 - \cos\theta) \qquad (42\text{-}18)$$

COMPTON WAVELENGTH
$$\lambda_C \equiv \frac{h}{mc} = 0.002\,43 \text{ nm} \qquad (42\text{-}19)$$

Because Compton shifts are of this order, the effect is noticeable only for photons of comparably short wavelengths (x-rays and gamma rays).

Equation (42-18) agrees with the experimental data of Figure 42-13. The presence of the *unshifted* line at λ_0 is the result of scattering from inner-shell electrons, which are firmly bound to the atom, so that the atom as a whole recoils. Because of its relatively great mass, the atom acquires negligible energy in the collision (see Problem 9C-50). The success of the photon model in ex-

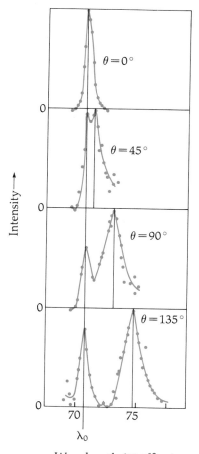

FIGURE 42-13
Experimental data for Compton scattering. The intensity of the x-rays scattered at various angles is plotted versus the wavelength. The presence of the peak at λ_0 is due to scattering from the atom as a whole. Using the atomic mass rather than the electronic mass in Equation (42-18) produces a wavelength shift of only about 10^{-16} m, a negligible amount on this scale.

---

[9] The bonds that hold outer electrons to atoms have energies of only a few electron volts. The x-rays Compton used had energies thousands of times greater, so the outer electrons were essentially "free" in their interactions with the incoming photons.

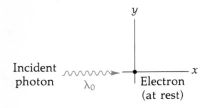

(a) Before

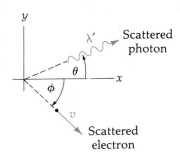

(b) After

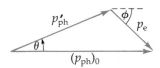

(c) A momentum vector diagram for the conservation of momentum.

FIGURE 42-14
In a Compton scattering process, a photon of wavelength λ_0 undergoes a particle-like collision with an electron initially at rest. The scattered photon has a longer wavelength λ'.

plaining Compton scattering further reinforced belief in the particle-like nature of radiation.

EXAMPLE 42-5

Write equations for the conservation of energy and momentum in the Compton scattering process of Figure 42-14. Outline the derivation of Equation (42-18).

SOLUTION

We use relativistic expressions for energy and momentum. From $E^2 = (mc^2)^2 + (pc)^2$, we note that since a photon has no mass, the incident photon energy is $p_{ph}c$ and the final photon energy is $p'_{ph}c$. The electron's initial energy is mc^2, and its final energy E is given by the above expression.

Conservation of energy: $\quad E_0 = E$

$$p_{ph}c + mc^2 = p'_{ph}c + E_e$$

Solving for E and squaring, then substituting $E_e^2 = (mc^2)^2 + (p_e c)^2$, we get

$$(p_{ph}c - p'_{ph}c + mc^2)^2 = (mc^2)^2 + (p_e c)^2 \quad (42\text{-}20)$$

We eliminate p_e from this equation by using the vector diagram (Figure 42-14c) representing *momentum conservation*: $(\mathbf{p}_{ph})_0 = \mathbf{p}'_{ph} + \mathbf{p}_e$. From the law of cosines for this triangle, we have

$$p_e^2 = (p_{ph})_0^2 + p'^2_{ph} - 2(p_{ph})_0 p'_{ph} \cos\theta \quad (42\text{-}21)$$

This value for p_e^2 is now substituted into Equation (42-20), and the left-hand side is multiplied out. After simplification, we obtain

$$\left[\frac{mc}{p_{ph}'} - \frac{mc}{(p_{ph})_0} \right] = 1 - \cos\theta \quad (42\text{-}22)$$

Substituting $(p_{ph})_0 = h/\lambda$ and $p_{ph} = h/\lambda'$, we obtain the Compton scattering relation, Equation (42-18).

Pair Production

Another interaction in which a photon behaves as a particle is the process called **pair production**. If a photon of sufficient energy passes close to a nucleus, the photon can disappear and create an *electron positron pair*, $\gamma \rightarrow e^+ + e^-$. The rest energy of the pair is $2m_e c^2 = 1.022$ MeV (twice that of a single electron), so the photon must have at least this much energy. Any additional photon energy appears as kinetic energy of the electron and positron. Electric charge is conserved in the reaction because of the equal and opposite charges of the pair. Momentum is conserved by the presence of the nucleus (which absorbs usually a negligible amount of kinetic energy). Problem 42C-49 shows that pair production *cannot* occur in empty space because both momentum and energy conservation cannot be simultaneously satisfied.

PAIR PRODUCTION $\quad hf = 2m_e c^2 + K_1 + K_2 \quad (42\text{-}23)$

42.7 The Dual Nature of Electromagnetic Radiation

Up to this point, we have reviewed some of the experimental evidence for the particle-like behavior of radiation. No doubt the photon model now appears logical and straightforward. However, its acceptance was a slow and painful process for most physicists. Robert Millikan, the noted American physicist, expressed his reluctance thus (in 1916):

> *I spent ten years of my life testing the 1905 equation of Einstein's and contrary to all my expectations, I was compelled in 1915 to assert its unambiguous experimental verification in spite of its unreasonableness since it seemed to violate everything that we knew about the interference of light.*

The reasons for the reluctance are as follows. All interference and diffraction phenomena seem to furnish ample evidence that radiation is a *wave*. If we accept the photon model, can we interpret an effect such as double-slit interference on the basis of *photons*? You will recall that we explained the light and dark fringes as an interference between two coherent *waves* that spread out as they emerge from the slits. What happens if we assume that the incident light is a stream of *photons*?

First, we can clearly associate the light intensity pattern on the screen with the varying numbers of photons that arrive at different locations. Each individual photon arrival is a *localized* "point event," perhaps knocking an electron off a silver-halide molecule in a photographic emulsion, causing the molecule to deposit a silver grain during the development process. If a photon is, indeed, a localized particle small enough to interact with a single electron, it certainly should go through just one of the slits at a time. Therefore, it should not make any difference if we close one of the slits for half the exposure time, then open it and close the other slit for the other half of the exposure time. *Yet if we do that experiment, we do not obtain the double-slit pattern.* As shown in Figure 42-15, the light pattern is just a superposition of two *single*-slit patterns, due to each slit acting alone. Apparently the photon, even though it is a well-localized particle, "knows" whether or not the other slit is open.

How do photons cause interference effects? Could one photon pass through one slit and interfere with another photon going through the other slit? No. Experiments have been performed using extremely dim light, which

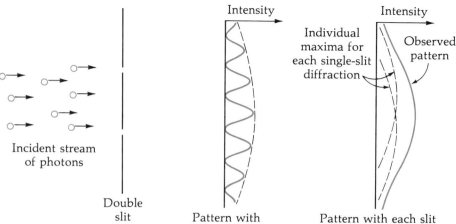

FIGURE 42-15
An attempt to interpret a double-slit interference experiment in terms of photons.

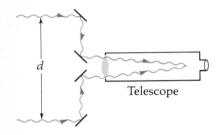

FIGURE 42-16
A stellar interferometer. Mirrors at 45° reflect light from a distant star into a telescope, causing certain interference effects in the image. Essentially, the stellar interferometer is a double-slit apparatus, in which the slit separation d may be as large as 10 meters.

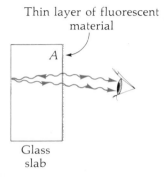

FIGURE 42-17
One face of a slab of glass has a thin layer of fluorescent material that glows when illuminated by ultraviolet light. Consider the light from a single atom at A. (Because of coherence requirements for forming an interference pattern, light from a single atom interferes *only with itself*, not with light from other atoms.) The part of the light that reflects from the rear surface of the glass slab interferes with the light traveling directly to the eye, and the observer sees a pattern of light and dark rings similar to Newton's rings (Figure 38-17). This effect is easily understandable in terms of spherical *wavefronts* that expand outward from the atom and eventually come together to interfere, forming the pattern. But in the photon model for light, the atom emits a *single photon*. Does this photon start to travel outward simultaneously in two opposite directions? In this experiment, thinking in terms of photons clearly leads to perplexities.

guarantees that (on the average) only one photon at a time passes from the source to the screen. In one such case, the experimenter started the exposure in an interference experiment, then went on a sailing trip for a few months. Upon his return, he developed the photographic film and found the usual fringe pattern, even though only one photon at a time had passed through the apparatus. **Each photon interferes only with itself.**[10]

Does this imply that the photon is "smeared out" so that part of it goes through each slit? This is hard to imagine when we consider an instrument known as the *stellar interferometer*, Figure 42-16. Basically this instrument is a double-slit apparatus with the two slits separated by up to 10 m. Both slits must be open simultaneously for the correct interference effect to be obtained. But if we try to imagine a photon as spread out so much that parts of it can go through both slits simultaneously, we must keep in mind that the photon must also be capable of giving up all its energy to a single electron should the photon, instead, just happen to undergo a photoelectric process. Such a scenario is certainly inconsistent. We do run into serious difficulties if we try to imagine photons as spread out in space. Figure 42-17 shows another experiment that cannot be explained using a photon model.

The behavior of photons in such an experiment is understandable only in a probabilistic way. It is not possible to predict where a single photon will hit the screen. Only the average distribution of a statistically large number of photon impacts is predictable. The observed distribution is the same as the intensity distribution calculated from the wave theory of light. Here we have an important clue to a new way of thinking about light, described in the next chapter. *The probability that a photon arrives at a given location is proportional to the intensity of the light wave at that location.*

When interpreting light phenomena, apparently one has to become an expert in "double-think." For some experiments, a wave model for light gives us insight into what is occurring; for another class of experiments, only a particle model makes sense. Are there any hints we can find for choosing a model? One clue is the following. If the dimensions of the apparatus (slit widths, apertures, and so on) are of the order of λ, then the *wave* nature of the radiation is usually most important because of interference and diffraction. On the other hand, when significant dimensions are $\gg \lambda$ (as in Chapters 36 and 37, "Geometrical Optics"), we are usually not interested in the wave characteristics, so we can assume that light rays do not bend around edges but travel in straight lines—as *particles*, if we wish. Another clue is that if the energy and momentum of a photon are comparable with other energies and momenta in the system, then we must treat the photon as a *particle* (as in the photoelectric effect and Compton scattering). However, all of these clues are only rule-of-thumb considerations. We must use care: for example, in a stellar interferometer we think of *waves* passing through the apparatus, but *photons* arriving at the photographic plate.

At this stage in the development of physics (the early 1900s), light seemed to develop a split personality. Even today, for most applications we still think of light in terms of *waves* or *particles*. But one significant fact should be noted. *Whenever we detect light experimentally, it always involves a particle-like interaction, not a wave-like one.* We need the wave model to understand such effects as interference and diffraction, *yet we never physically detect light in those*

---

[10] Photon-photon interactions do take place under certain circumstances, but they are rare and of no consequence here.

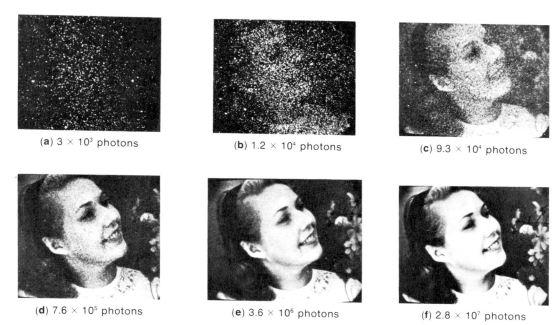

FIGURE 42-18
A great many photons are needed to form a complete image. The number of photons involved is indicated below each picture.

regions where we think of it as waves. If light interacts with matter, we must always use a particle model. The formation of an interference pattern is the result of a very large number of photons that *statistically* sort themselves out to gradually form the pattern of light and dark fringes. This statistical behavior of photons is present in all image formation (see Figure 42-18). It is ironic that we need the wave model to understand the propagation of light *only through that part of the system where it leaves no trace!*

Perhaps the moral of the story is that we should not take either the particle model or the wave model too seriously. They are useful, but inherently contradictory: particles are *localized,* waves are *spread out.* Conceptually, we cannot blend them together. The modern resolution to this paradoxical duality is revealed in the next chapter.

Summary

One of the characteristics of *blackbody* (or *cavity*) radiation is that the *total emittance* R at all wavelengths (in watts/meter^2) is proportional to the fourth power of the Kelvin temperature T.

Stefan–Boltzmann radiation law $\qquad R = \sigma T^4$

where the *Stefan–Boltzmann constant* is $\sigma = 5.672 \times 10^{-8}$ W/(m$^2 \cdot$ K^4). Another characteristic is that as the Kelvin temperature increases, the wavelength λ_m at the maximum of the spectral distribution becomes shorter according to

Wien displacement law $\qquad \lambda_m T = $ constant

The classical radiation laws of Wien and Rayleigh–Jeans are approximations that are correct only for short and long wavelengths, respectively. Planck derived the correct expression by making the following assumption:

Planck assumption of quantization

> SHM oscillators (with a natural frequency f) can exist only in quantized energy states:
>
> $$E_n = nhf$$
>
> where $n = 0, 1, 2, 3, \ldots$ and $h = 6.626 \times 10^{-34}$ J·s, known as *Planck's constant*. The oscillators emit or absorb only quanta of energy: $\Delta E = hf$. (Planck originally proposed nhf, but it was shown later that only $n = 1$ occurs.)

Planck's radiation law
$$du_\lambda = f(\lambda, T)\, d\lambda = \frac{8\pi hc\lambda^{-5}}{e^{hc/\lambda kT} - 1}\, d\lambda$$

where $f(\lambda, T)\, d\lambda$ is the energy per unit volume from wavelength λ to $\lambda + d\lambda$, and the *Boltzmann constant* k equals 1.381×10^{-23} J/K. As $h \to 0$, quantum mechanical expressions approach the corresponding classical expressions.

Einstein explained the *photoelectric effect* by assuming that radiation is quantized as *photons* that remain localized in space as they travel with speed c (in a vacuum) and that have zero rest mass.

Photon energy $\qquad E = hf$

Photon momentum $\qquad p = \dfrac{h}{\lambda}$

Einstein's photo-electric equation $\qquad hf = K_{\max} + w_0$

where f is the frequency of the illumination, $K_{\max}$ is the maximum kinetic energy of the photoelectrons, and w_0 is the *work function* of the surface.

The *Compton shift* for the scattering of photons by free electrons is

$$\lambda' - \lambda_0 = \frac{h}{m_0 c}(1 - \cos\theta)$$

where m is the mass of the electron. The quantity h/mc is the *Compton wavelength* $\lambda_C = 0.002\,43$ nm.

As a result of the dual nature of electromagnetic radiation, we use both *wave* and *particle* models. The wave model enables us to predict interference and diffraction effects, but all interactions of radiation with matter that are experimentally detected are particle-like interactions.

Questions

1. As the power input to an incandescent bulb is reduced, the brightness decreases. Why does the color of the emitted light also change?
2. Metal objects put into a heat-treating furnace are heated to incandescence. When we peer through a small hole in the oven door, the objects seem to have almost disappeared. Why?
3. What is meant by the adjective *black* in "blackbody radiation"?
4. Materials that are heated radiate energy from their surfaces in accordance with a modified form of the Stefan–Boltzmann radiation law, $R = e\sigma T^4$, where e (called the *emissivity*) is equal to one for a blackbody and less than one for other materials. Two different metals at the same temperature may thus glow at different intensities. On a thermodynamic basis, how is the emissivity of a material related to the material's ability to absorb radiation?
5. At a temperature that causes metals to become incandescent, glass does not even glow. Why not?
6. Steam radiators in buildings are sometimes painted with aluminum paint, or their surfaces are polished bare metal. Explain why they would be more efficient if painted with a nonglossy black paint.
7. Discuss the assumptions that Planck made in his theory of blackbody radiation.
8. Doesn't an inconsistency exist in ascribing an energy $E = hf$ to a *photon*, when f refers to the frequency of a *wave*?
9. Could a faint star be visible to the eye if the light from the star were not corpuscular in nature?
10. Roughly how large would Planck's constant have to be in order for the unaided eye to be able to observe the quantum effects in simple mechanical oscillators?
11. Discuss the assumptions that Einstein made in his theory of the photoelectric effect.
12. Why is the maximum kinetic energy with which photoelectrons leave the surface of a metal independent of the intensity of the light falling on the surface?
13. In a photocell, electrons emanating from a photosensitive cathode are drawn to an anode that is normally at a higher potential than the cathode. How does the electron current through the photocell depend on the intensity of the light falling on the cathode and upon the potential difference between the cathode and the anode?
14. An isolated sheet of zinc exposed to ultraviolet light emits photoelectrons when first exposed to the light, then seems to stop. Why? (Hint: does the zinc sheet become charged?)
15. The photoelectric effect occurs in gases as well as on the surfaces of solids. When a gas target is used, is there a threshold wavelength? If so, is it the same as for the same substance in solid form? Explain.
16. Discuss the assumptions that Compton made in his theory of the Compton effect.
17. Why is it reasonable that the Compton shift in wavelength of the scattered photons is independent of the scattering material?

18. Why is the Compton effect not readily observable for visible light?
19. In what way does the Compton effect reinforce the photoelectric effect in substantiating the quantum theory of radiation?
20. What is wrong with the following explanation of the Compton effect? Electromagnetic radiation is only a wave phenomenon. The wave interacts with electrons, causing the electrons to recoil due to the momentum carried by the wave as well as causing the electrons to oscillate at the frequency of the incoming electric wave. The frequency shift observed is simply a Doppler shift of radiation produced by the oscillating electrons, which are also moving under the recoil.
21. A photon and an electron have the same momentum. Which has the greater total energy (including rest-mass energy)?

Problems

42.2 The Spectrum of Cavity Radiation

42A-1 A 200-W tungsten-filament light bulb operates with a filament temperature of 2200 K. Assuming that the filament radiates as an ideal blackbody, calculate its surface area.

42B-2 (a) Assuming that the sun's surface is an ideal blackbody emitter at 5780 K, find the total power radiated from the sun. (b) Find the incident power of sunlight at the earth (above the atmosphere) on a square meter of surface area oriented perpendicular to the incident radiation.

42B-3 Suppose that a small area on the surface of a person's skin increases to 37.5°C above the normal surface temperature of 37.0°C. Assuming blackbody radiation, calculate $\Delta R/R$, the fractional increase in the rate of radiation per unit area for the warmer area compared to the normal rate of radiation. (Such slight differences can be revealed by *thermography*, an infrared or microwave photographic technique that is useful in detecting tumors and other diseases located a few centimeters below the surface of the skin.)

42B-4 An insulated oven operating at a temperature of 500°C has a peephole with a diameter of 2 cm. Calculate the net amount of energy per second that is transferred through the peephole into a room at 30°C. (Hint: consider both the room and the oven as ideal blackbody radiators.)

42.3 Attempts to Explain Cavity Radiation

42A-5 Find the wavelength at the maximum of the blackbody radiation curve for a room temperature of 27°C.

42A-6 As a result of the Big Bang and the expansion of the universe, interstellar space contains a background radiation at a temperature of about 2.7° K. Find (a) the wavelength and (b) the frequency at which this radiation is a maximum.

42A-7 The sensitivity of the human eye is greatest at a wavelength of about 555 nm. Find the temperature of blackbody radiation that produces the maximum spectral output at this wavelength.

42B-8 The radius of our sun is 6.96×10^8 m and its total power output is 3.86×10^{26} W. (a) Assuming that the sun's surface emits as an ideal blackbody, calculate its surface temperature. (b) Using the result of part (a), find the wavelength at the maximum of the spectral distribution of radiation from the sun.

42.4 Planck's Theory

42A-9 Find the wavelength of a photon that has an energy equal to the rest energy of an electron (0.511 MeV).

42A-10 An FM radio station emits 80 kW of power at a frequency of 92.4 MHz. How many photons per second does it emit?

42A-11 A useful relation between the energy E of a photon and its wavelength λ is $E\lambda = 1.240 \times 10^{-3}$ MeV·nm. Derive this expression.

42A-12 A He–Ne laser emits light at a wavelength of 632.8 nm. (a) In what portion of the electromagnetic spectrum is this light? (b) How many photons per second are emitted by a He–Ne laser whose beam power is 2 mW?

42B-13 Experiments indicate that a dark-adapted human eye can detect a single photon of visible light. Consider a point source that emits 2 W of light of wavelength 555 nm in all directions. How far away would this source have to be for, on the average, one photon per second to enter an eye whose pupil is 6 mm in diameter?

42B-14 For small amplitudes, a simple pendulum behaves like a simple harmonic oscillator. Consider a 50-g mass suspended by a string (of negligible mass) of length 40 cm. (a) According to Planck, what is the smallest nonzero energy that this pendulum may have? (b) What is the amplitude of oscillation of the pendulum bob at this minimum energy? (The answer reveals why quantization is not observable for macroscopic motions.)

42.5 The Photoelectric Effect

42A-15 The work function for sodium is 2.75 eV. Find the threshold wavelength for the photoelectric effect in sodium.

42A-16 Bismuth exhibits photoelectron emission only for ultraviolet wavelengths shorter than 294 nm. Calculate the work function (in electron volts) for bismuth.

42B-17 Ultraviolet light ($\lambda = 384$ nm) illuminates a clean calcium surface whose work function is 2.87 eV. Calculate (a) the maximum speed of the emitted photoelectrons and (b) the threshold wavelength.

42B-18 Light of wavelength 410 nm is incident upon a metallic surface. The stopping potential for the photoelectric effect is 0.83 V. Find (a) the maximum kinetic energy (in electron volts) of the ejected photoelectrons, (b) the work function for the metal, and (c) the threshold wavelength.

42.6 The Compton Effect and Pair Production
42.7 The Dual Nature of Electromagnetic Radiation

42A-19 Find the change in wavelength of a photon that is "back-scattered" at 180° by an electron initially at rest. Does this change depend upon the wavelength of the incident photon?

42A-20 In a Compton scattering process, a photon undergoes a wavelength increase of 4.1 pm. At what angle was the photon scattered by the electron?

42A-21 A high-energy photon can create a proton–antiproton pair in a pair production process. A 2.10-GeV photon creates such a pair, with the proton having a kinetic energy of 95 MeV. Find the kinetic energy of the antiproton.

42B-22 A gamma-ray photon with an energy equal to the rest energy of an electron (511 keV) collides with an electron that is initially at rest. Calculate the kinetic energy acquired by the electron if the photon is scattered 30° from its original line of approach.

42B-23 A pair-production process, $\gamma \to e^+ + e^-$, can occur only near a nucleus in order to conserve momentum. Show that even though the nucleus absorbs all of the initial momentum of the photon, it absorbs very little of the energy. (Hint: find the ratio of the final kinetic energy of the nucleus, $\frac{1}{2}Mv^2$, to the initial energy of the photon, and show that this ratio is truly negligible. Consider photon energies less than ~10 MeV for which nonrelativistic equations are sufficiently accurate.)

42B-24 The nucleus of a radioactive isotope of chlorine (^{38m}Cl) decays by the emission of a 660-keV photon. (The symbol m indicates a *metastable* state. Instead of decaying immediately, the nucleus exists in this excited state a relatively long time.) If the nucleus is initially at rest, determine the ratio of the kinetic energy acquired by the nucleus to the energy of the emitted photon. The mass–energy equivalent of the ^{38m}Cl nucleus is 35.4 GeV.

42B-25 A 2-W helium–neon laser beam (632 nm) is completely absorbed when it strikes a target. Find (a) the number of photons striking the target each second and (b) the momentum of each photon. (c) Using these data, find the force that the laser beam exerts on the target.

Additional Problems

42C-26 A person whose skin area is 1.70 m² sits naked in a sauna that has a wall temperature of 61°C. The person's skin temperature is 37°C. Assuming blackbody radiation, find the *net* rate at which the person absorbs heat by radiative transfer. (b) The latent heat of evaporation of sweat is essentially the same as that of water at 37°C: 2427 kJ/kg. At what rate must sweat evaporate to compensate for this heat absorption?

42C-27 The net power radiated from an object at absolute temperature T in surroundings at absolute temperature T_0 is proportional to $(T^4 - T_0^4)$. Show that if the temperature difference is small, then *Newton's law of cooling* holds true: *the rate of cooling of a body is approximately proportional to the temperature difference between the body and its surroundings.*

42C-28 Show that, for short wavelengths, Planck's radiation law, Equation (42-11), approaches Wien's radiation law, Equation (42-4).

42C-29 Show that, for long wavelengths, the Planck radiation law, Equation (42-11), approaches the Rayleigh–Jeans law, Equation (42-6). (Hint: expand the exponential term in a power series.)

42C-30 By differentiating Planck's radiation law, Equation (42-11), to find the peak value, show that it agrees with Wien's displacement law, Equation (42-2).

42C-31 A point source of monochromatic light ($\lambda = 550$ nm) emits 2 W of light uniformly in all directions. Calculate the distance from the light source at which the average volume density of photons is one photon per cubic centimeter.

42C-32 A 10-g mass oscillates with an amplitude of 3.0 cm under the influence of a spring whose force constant is 0.01 N/m. Find the decrease in amplitude of oscillation corresponding to the loss of a single quantum of energy.

42C-33 A parallel beam of uniform, monochromatic light of wavelength 546 nm has an intensity of 200 W/m². Find the number of photons in 1 mm³ of this radiation.

42C-34 The dark-adapted human eye can barely detect green light (500 nm) that delivers 1.7×10^{-18} W to the retina. Assume that the incoming light is parallel so that it is focused on a single receptor. (a) Find the average number of photons per second arriving at the receptor. (b) If the pupil of your dark-adapted eye is 8 mm in diameter, at what distance would you barely be able to detect a point source that emits 10 W of 500-nm light uniformly in all directions? Only about 20% of the light incident on the eye reaches the retinal receptors; the other 80% is absorbed by the layer of nerve fibers, blood vessels, and other tissues overlaying the receptors.

42C-35 Show that the average energy E_{av} of a quantized SHM oscillator, Equation (42-10), approaches the classical value kT as λ becomes very large.

42C-36 In the Planck law for cavity radiation, Equation (42-11), change the variable from λ to f and obtain the spectral energy density du_f between frequencies f and $f + df$:

$$du_f = f(f,T)\,df = \frac{8\pi h f^3}{c^3(e^{hf/kT} - 1)}\,df$$

42C-37 (a) By integrating the result of Problem 42C-36 over all frequencies, find the total energy density u of cavity radiation: $u = \int_0^\infty f(f,T)\,df$. (Hint: change to the variable $x = hf/kT$. An integral you will encounter is $\int_0^\infty x^3(e^x - 1)^{-1}\,dx = \pi^4/15$.) (b) Find the numerical value of u for $T = 300$ K (room temperature).

42C-38 Show that Planck's radiation law, Equation (42-11), integrated over all wavelengths, is consistent with the Stefan–Boltzmann law. That is, show that $\int_0^\infty f(\lambda,T)\,d\lambda = \alpha T^4$, where α is a constant. (Hint: make the change of variable, $x = hc/\lambda kT$, and note the integral given in the previous problem.)

42C-39 Show that the average energy of a quantized SHM oscillator [Equation (42-10)] reduces to the classical value (Equation 42-5) as Planck's constant h approaches zero.

42C-40 An electron initially at rest recoils from a head-on collision with a photon. Show that the kinetic energy acquired by the electron is given by $2hf\alpha/(1 + 2\alpha)$, where α is the ratio of the photon's initial energy to the rest energy of the electron.

42C-41 A metal target is placed in a beam of 662-keV gamma rays emitted by a radioactive isotope of cesium (^{137}Cs). Find the energy of those photons that are scattered through an angle of 90°. The electrons in the target may be considered as essentially free electrons.

42C-42 The table below shows data obtained in a photoelectric experiment. (a) Using these data, make a graph that plots as a straight line. From the graph, determine (b) an experimental value for Planck's constant (in joules per second) and (c) the work function (in electron volts) for the surface. (Two significant figures for each answer are sufficient.)

Wavelength (nm)	Maximum Kinetic Energy of Photoelectrons (eV)
588	0.67
505	0.98
445	1.35
399	1.63

42C-43 A low-energy photon ($E \ll$ electron rest-mass energy) collides head-on with a free electron initially at rest. The photon is scattered backward along the line of approach. Show that the ratio of the scattered photon energy to the kinetic energy acquired by the electron is approximately c/v, where v is the speed of the electron. (Hint: this is a nonrelativistic Compton scattering problem.)

42C-44 A 200-MeV photon is scattered at 40° by a free proton initially at rest. (a) Find the energy (in mega electron volts) of the scattered photon. (b) What kinetic energy (in mega electron volts) does the proton acquire?

42C-45 Following the suggestions in Example 42-5, derive the Compton shift relation $\lambda' - \lambda_0 = (h/mc)(1 - \cos \theta)$.

42C-46 Show that a photon colliding with a moving electron cannot be totally absorbed by the electron because to do so violates the relativistic conservation laws. For simplicity, consider a one-dimensional collision.

42C-47 A photon of initial energy E_0 undergoes a Compton scattering at an angle θ by a free electron (mass m) initially at rest. Using relativistic equations for energy and momentum conservation, derive the following relation for the final energy E of the scattered photon: $E = E_0[1 - (E_0/mc^2)(1 - \cos \theta)]^{-1}$.

42C-48 A photon strikes a free *proton* initially at rest in a Compton type of collision. Find the minimum energy of the photon that will give the proton a kinetic energy of 4 MeV.

42C-49 Figure 42-19 shows momentum considerations in a pair-production process, $\gamma \to e^+ + e^-$, occurring in empty space. Show that this is impossible (without the presence of a nucleus to conserve momentum) because energy and momentum conservation cannot both be true. (Hint: using the figure, write equations for momentum conservation in the x and y directions and an equation for energy conservation. Divide the momentum equations by c and the energy equation by c^2. Square and add the momentum equations, and compare with the square of the energy equation. Show that they are inconsistent.)

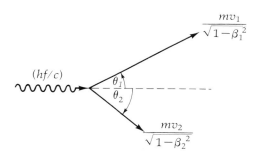

FIGURE 42-19
Problem 42C-49.

CHAPTER 43

The Wave Nature of Particles

... one may feel inclined to say that Thomson, the father, was awarded the Nobel prize [in 1906] for having shown that the electron is a particle, and Thomson, the son, for having shown that the electron is a wave [in 1937].

MAX JAMMER
(commenting on J. J. Thomson and G. P. Thomson)
The Conceptual Development of Quantum Mechanics,
McGraw-Hill (1966)

43.1 Introduction

The discovery of the dual nature of radiation was a fascinating revelation in its own right, but in the 1920s, an equally startling development occurred when particles of matter were found to exhibit wave-like behavior. This rounded out the physicist's "picture" of nature in a particularly symmetrical and satisfying way. Radiation *and* matter exhibit particle-like characteristics as well as wave-like characteristics. To place this discovery in context, we will describe some related developments that set the stage for this important step.

43.2 Models of an Atom

At the turn of the century it was believed that atoms were made of just two components: positive charges and electrons. But how were these components put together so they formed stable atoms? What configuration of charged particles could produce the extraordinary complexity of atomic spectral lines observed when we excite a gas by passing an electrical current through it (Figure 43-1)? These spectra had been studied and catalogued carefully, and many attempts were made to discover some mathematical relationship between wave-

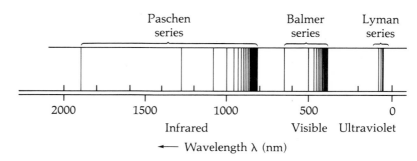

FIGURE 43-1
When hydrogen gas is heated by having an electrical current pass through it, the gas emits light consisting of a series of spectral lines called a *bright-line spectrum,* or an *emission spectrum* (indicated here by dark lines).

lengths that might reveal a clue to the atom's structure. Also, the cyclic variation in chemical properties of atoms in the periodic table was another clue to the puzzle. As a starting point, atoms were assumed to be spherical with radii $\sim 10^{-10}$ m. This could be calculated from the density, the atomic mass, and Avogadro's number.

The Thomson Model

One notable attempt to devise an atomic model was that of the British physicist J. J. Thomson at Cambridge University. In 1898, he suggested a kind of fluid of positive charge Ze (where Z is the atomic number) that contained most of the mass of the atom. The electrons were embedded within this positive fluid somewhat like plums in a plum pudding (Figure 43-2a). Supposedly, the electrons could then vibrate in various modes of oscillation and thereby (according to classical theory) emit radiation at these natural frequencies of oscillation. Unfortunately, quantitative agreement with observed spectral frequencies was lacking.

The Rutherford Model

Before 1910, physicists had made many attempts to discover the secrets of atomic structure by observing how incident particles and radiation were scattered from atoms. X-rays, electrons, and alpha particles were the main projectiles. A former student of Thomson's, Professor Ernest Rutherford,[1] was conducting experiments at the University of Manchester in England on the scattering of alpha particles by matter. An alpha particle was known to have a positive charge twice the magnitude of the electronic charge and a mass about four times that of hydrogen. Alpha particles were a convenient projectile since they were emitted with several million electron volts of energy by certain naturally radioactive elements. Rutherford wanted a very thin target because he hoped to observe the scattering by just a single atom rather than the multiple scattering by many atoms; multiple encounters would tend to obscure the characteristics of the single collision he wished to investigate. Although several different elements were investigated, gold was a particularly convenient target substance because it could be hammered to extremely thin foils, only a few hundred atoms thick. As shown in Figure 43-3, the scattered alpha particles struck a small screen coated with zinc sulfide, causing tiny flashes of light that

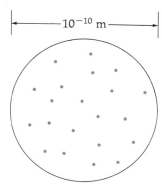

(a) Thompson's "plum pudding" model, with electrons embedded in a sphere of positively charged fluid.

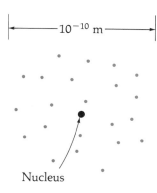

(b) Rutherford's nuclear model, with all the positive charge (and most of the mass) concentrated in a very small region at the center. Electrons surround the nucleus in an unknown way.

FIGURE 43-2
Classical models of the atom.

---

[1] Rutherford received the Nobel prize in *chemistry* in 1908 for discovering that the radiation from uranium consisted of at least two types he called *alpha* and *beta* radiation. He later showed that alpha "radiation" actually consisted of particles, being nuclei of helium atoms.

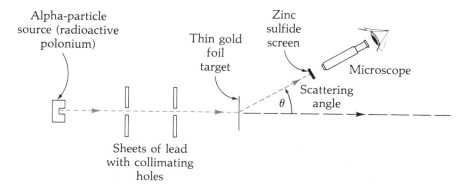

FIGURE 43-3
The Rutherford alpha-scattering experiment. The zinc sulfide detector can be moved to record scattering at various angles. The apparatus is placed within an evacuated chamber.

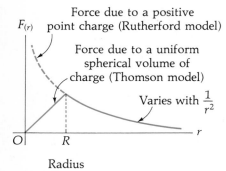

FIGURE 43-4
The force on an alpha particle due to a positive charge in two different configurations: a point charge and a uniform spherical volume of charge. (See Figure 25-16, Chapter 25.)

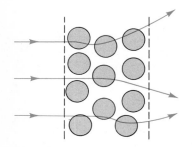

(a) According to the Thomson model, multiple scattering could occur if the alpha particle penetrates more than one atom. (The scattering is greatly exaggerated.)

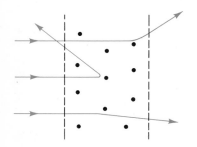

(b) According to the Rutherford model, a single close encounter with a nucleus could produce a large-angle scattering.

FIGURE 43-5
Scattering of alpha particles by a thin foil. The target foil is typically several hundred atoms thick.

were observed by watching the screen with a microscope. It was tedious work, requiring well-dark-adapted eyes. Rutherford's assistants were Dr. Hans Geiger[2] and an undergraduate student, Ernest Marsden.

Early data for small-angle scattering of 1° or 2° seemed to confirm the Thomson model. Wishing to start Marsden on a research project of his own, Rutherford suggested he look for scatterings in the backward direction (>90°), though Rutherford personally felt that the chance of a fast alpha particle being scattered backward by a Thomson atom was truly negligible. Much to everyone's amazement, many alphas were back-scattered. The reason for surprise is clear from the estimates of the scattering probabilities. The mass of an alpha is about 8000 times the mass of an electron, so electrons have negligible effect on the scattering: all the scattering occurs from the massive positive charge. In a Thomson atom, the positive charge is spread uniformly throughout a spherical volume, so the maximum force a single atom could exert on an alpha particle was limited (Figure 43-4), causing a deflection of just a few hundredths of a degree at most. Thus, thousands of scatterings would have to take place, *with a majority adding up in the same direction*, to cause a net deflection of 90° or more. The chance of a backward scattering by Thomson atoms in the foil used in one experiment was calculated to be incredibly small—about 1 in 10^{3500}. Yet Geiger and Marsden found roughly 1 in 10^4! Undoubtedly this discrepancy of a factor of $\sim 10^{3496}$ takes the all-time prize for the greatest disagreement between theory and experimental results ever encountered. Rutherfold later wrote of his reaction:

> *It was quite the most incredible event that has ever happened to me in my life. It was almost as incredible as if you fired a 15-inch shell at a piece of tissue paper and it came back and hit you.*

Recognizing that a single scattering at large angles could occur only if the forces were extremely strong, in 1911 Rutherford proposed his nuclear model of an atom. In it, the massive positive charge was concentrated in a region he called the **nucleus**, no bigger than 10^{-14} m, since to create a force big enough to scatter the alpha particle backward, the alpha would have to approach at least that close to the point charge Ze. Figure 43-5 compares the two situations, and Figure 43-6 shows experimental points for one experiment.

Though the Rutherford model was clearly superior to the Thomson model, there were still some troublesome aspects. For example, what held the positive charges in the nucleus together? And what held the negatively charged electrons away from the positively charged nucleus? They presumably could not rotate around the nucleus in a "solar system" motion because Maxwell's equations predicted that accelerated charges radiate electromagnetic waves. Indeed, such radiation was observed in every instance in which electrons were accelerated. According to classical physics, if you started electrons moving in circular orbits, they would radiate energy and spiral into the nucleus in less than 10^{-8} seconds. Obviously atoms did not do this, so what was wrong?

The Bohr Model

As shown in Figure 43-7, the spectrum of a hydrogen atom—the simplest atom of all—had a baffling complexity and regularity. How could just a proton and an electron interact to produce this series of spectral lines? A Swiss high

---

[2] To avoid the painstaking and boring method of data taking, Geiger later invented an electronic gadget for detecting charged particles: the "Geiger" counter, widely used today.

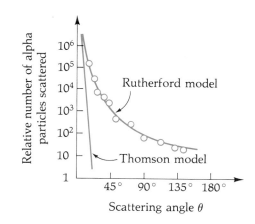

FIGURE 43-6
Typical data by Geiger and Marsden for the scattering of alpha particles by gold foils. The solid lines are theoretical curves based on the Thomson and Rutherford models.

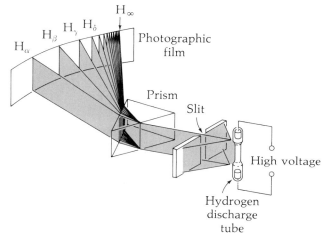

(a) A prism spectrometer. Light from the hydrogen discharge tube is refracted by the prism to form the line spectrum on the photographic film.

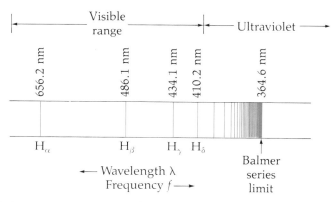

(b) The Balmer series is a group of an infinite number of spectral lines whose spacings regularly converge toward the short-wavelength limit of 364.6 nm.

FIGURE 43-7
The Balmer series emission spectrum of hydrogen.

school teacher of descriptive geometry, J. Balmer, had found by trial and error an *empirical* formula that agreed almost exactly with the observed wavelengths.

THE BALMER SERIES IN HYDROGEN

$$\lambda = (364.56 \text{ nm}) \left(\frac{n^2}{n^2 - 2^2} \right) \quad \text{(where } n = 3, 4, 5, \ldots \text{)} \quad (43\text{-}1)$$

But how the hydrogen atom produced this mathematically simple series of lines remained a nagging puzzle.

In 1913, the Danish physicist Niels Bohr proposed his famous model of the hydrogen atom. Bohr was young (age 28) and fearless. His theory contained radical ideas that were clearly contrary to classical physics, but his model predicted all observed lines almost exactly. It was based on the following assumptions, known as the **Bohr Postulates**:

(1) The electron travels in circular orbits around the proton, obeying the classical laws of mechanics. (The Coulomb force of attraction is the centripetal force.)

(2) Contrary to classical theory, the electron can move in certain *allowed* orbits of radius r_n *without radiating*. Since the energy E_n is constant in such orbits, the electron is said to be in a *stationary state*.

(3) The allowed orbits are those for which the angular momentum mvr of the electron (mass m) is an integral multiple of Planck's constant divided by 2π (notation[3]: $\hbar \equiv h/2\pi$).

$$mvr = n\hbar \quad \text{(where } n = 1, 2, 3, 4, \ldots \text{ and } \hbar = 1.0546 \times 10^{-34} \text{ J·s)} \qquad (43\text{-}2)$$

(4) *Transitions* between stationary states are possible when the electron somehow "jumps" from one allowed orbit to another. Electromagnetic radiation is emitted or absorbed by the atom, and the difference in the two energy states is the energy hf of the radiation emitted or absorbed.

$$hf = E_{\text{final}} - E_{\text{initial}} \qquad (43\text{-}3)$$

Bohr's proposal was a peculiar mixture of classical and quantum physics. Thanks to the classical Coulomb force, the electron moved in circular orbits according to classical mechanics. Contrary to classical physics, it did not radiate. Also, Planck's quantum constant h entered the picture in two ways: in the energy hf associated with the radiation and in an entirely new way by quantizing the angular momentum, a parameter that had previously been nonquantized.

The allowed radii and energy states are calculated as follows. Applying Newton's second law to the circular motion of the electron of charge e and mass m about a nucleus[4] of charge Ze (Figure 43-8), we have

$$\Sigma F = ma$$
$$\left(\frac{1}{4\pi\varepsilon_0}\right)\frac{(Ze)(e)}{r^2} = m\left(\frac{v^2}{r}\right) \qquad (43\text{-}4)$$

The quantum restriction on the angular momentum is

$$mvr_n = n\hbar \qquad (43\text{-}5)$$

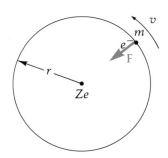

FIGURE 43-8
The Bohr model for a one-electron atom. The electron of charge $-e$ travels in a circular orbit around a fixed nucleus of charge Ze. The Coulomb force **F** is the centripetal force on the electron.

---

[3] For convenience, $h/2\pi$ is often written as $\hbar$, pronounced "h-bar."
[4] If we consider a charge Ze in the nucleus (Z = atomic number), the analysis also applies to a singly ionized helium, doubly ionized lithium, and so on. The equations obtained predict the observed spectra for all these cases very well.

FIGURE 43-9
Niels Bohr (1885–1962, facing camera) was a Dutch physicist who received the Nobel Prize in 1922 for his model of the hydrogen atom. He is shown here with his physicist son, Aage Bohr (b. 1922), who succeeded his father as Director of the Niels Bohr Institute in Copenhagen. Niels Bohr made many later contributions to the liquid-drop model of the nucleus and to theories of nuclear fission. His son also received the Nobel Prize in 1975 (with Ben Mottleson and James Rainwater) for theoretical studies on nuclear structure. (Photo courtesy of AIP Niels Bohr Library, Margrethe Bohr Collection.)

Combining the two equations to eliminate v, we obtain the radii r_n for the allowed orbits:

RADII OF BOHR ORBITS FOR HYDROGEN
$$r_n = \frac{\varepsilon_0 h^2 n^2}{\pi m Z e^2} \qquad (n = 1, 2, 3, 4, \ldots) \qquad (43\text{-}6)$$

Substituting numerical values ($Z = 1$) gives
$$r_n = (0.0529 \text{ nm})n^2 \qquad (43\text{-}7)$$

The allowed radii are thus proportional to n^2.

The energy state E of the atom is found from $E = K + U$. Defining the zero reference for $U \equiv 0$ when the electron is infinitely far from the nucleus, we have

$$U = -\left(\frac{1}{4\pi\varepsilon_0}\right)\frac{(Ze)(e)}{r}$$

Therefore:
$$E = \frac{1}{2}mv^2 - \left(\frac{1}{4\pi\varepsilon_0}\right)\frac{(Ze)(e)}{r} \qquad (43\text{-}8)$$

Substituting values of v and r from Equations (43-5) and (43-6), we obtain

ENERGY STATES OF THE BOHR HYDROGEN ATOM
$$E_n = -\frac{mZ^2 e^4}{8\varepsilon_0^2 h^2 n^2} \qquad (n = 1, 2, 3, 4, \ldots) \qquad (43\text{-}9)$$

Substituting numerical values ($Z = 1$) gives
$$E_n = -\frac{13.6 \text{ eV}}{n^2} \qquad (43\text{-}10)$$

The allowed hydrogen energy states are thus *negative* and proportional to $1/n^2$ (see Figure 43-10). Each series of spectral lines is characterized by the common *final* state involved in the transitions. Problem 43C-35 shows that the Balmer series, Equation (43-1), can be obtained from this expression.

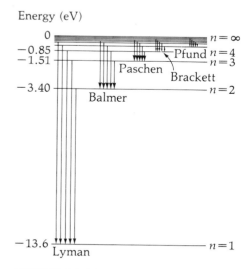

FIGURE 43-10
Energy states of a hydrogen atom. Between the levels $n = 4$ and $n = \infty$, there are an infinite number of energy levels. Transitions from higher to lower energy states result in emission of radiation of energy hf. The names of the experimenters who investigated the different spectral series are shown. Only a portion of the Balmer series is in the visible range of wavelengths.

43.3 The Correspondence Principle

Every new revolution in physics introduces concepts radically different from the older, established theories. For example, in relativity the equations appropriate for high speeds are quite different from those of Newtonian mechanics. Similarly, the quantum ideas of radiation are radically different from the classical Maxwell equations. Yet, physically, the transition between cases in which classical equations apply and in which the newer ideas must be used cannot be an abrupt one; there must be a smooth transition in the "overlap" region from one theory to the other.

In quantum physics, the relation between the new and old theories was pointed out by Bohr in a statement he called the *correspondence principle*. According to classical electromagnetic theory, the frequency emitted by an electron traveling in a circular orbit is just the orbital frequency of revolution f_0. From Equations (43-5) and (43-6) we obtain, for this orbital frequency for hydrogen ($Z = 1$),

$$f_0 = \frac{me^4}{4\varepsilon_0^2 h^3 n^3} \tag{43-11}$$

In the newer, Bohr theory, the frequency f emitted in a transition between adjacent energy states is intermediate between the two orbital frequencies, given by Equation (43-3):

$$hf = E_{\text{final}} - E_{\text{initial}}$$

From Equation (43-9),
$$hf = \frac{me^4}{8\varepsilon_0^2 h^2}\left[\frac{1}{n^2} - \frac{1}{(n+1)^2}\right] \tag{43-12}$$

The factor in brackets may be written as

$$\left[\frac{1}{n^2} - \frac{1}{(n+1)^2}\right] = \left[\frac{n^2 + 2n + 1 - n^2}{n^2(n+1)^2}\right]$$

When n becomes very large, we have

$$\lim_{n \gg 1}\left[\frac{2n+1}{n^2(n+1)^2}\right] = \frac{2}{n^3} \tag{43-13}$$

So for large n the frequency of emission is

$$f \approx \frac{me^4}{4\varepsilon_0^2 h^3 n^3} \tag{43-14}$$

Comparing this equation with Equation (43-11), we see that *in the limit of large n, the quantum expression agrees with the classical expression*. This illustrates Bohr's **correspondence principle**.

BOHR'S CORRESPONDENCE PRINCIPLE   Any new theory must reduce to the classical theory to which it corresponds when applied to situations appropriate to the classical theory.

This means that the new theory must contain the old theory as a special case. In Bohr's model of the hydrogen atom, if n becomes very large the system approaches a *macro* system (not a *micro* system) and, as shown above, the classical equations become an adequate description. Of course, electrons do not change their behavior for large n—they *always* obey quantum mechanics (as does every other object in the universe). But for $n = 10\,000$, say, the very large values of the radius, energy, and angular momentum make the small quantum differences from the $n = 10\,001$ values essentially negligible, and the behavior of the system approaches that described by classical equations. As another example, we have seen that for slow speeds Einstein's special relativity reduces to Newtonian mechanics—so special relativity also illustrates Bohr's correspondence principle. This principle provides a valuable check on the validity of new theoretical developments.

43.4 De Broglie Waves

Bohr's model for the hydrogen atom was a great triumph. It agreed very closely with wavelengths of the Balmer series, and it correctly predicted the spectrum of other series outside the visible range. Yet small but unmistakable discrepancies were still present. The reason for part of these discrepancies originated in the fact that energies were calculated on the basis of a *fixed* nucleus (which is equivalent to assuming that the proton has an infinitely large mass compared with the electron mass). Agreement with experimental data was improved by consideration of the proton's motion about the CM of the rotating proton–electron system. Still further improvements were made by A. Sommerfeld, who considered elliptical as well as circular orbits and included relativistic effects for the electron's motion.

In some respects, this improved theory was still not completely satisfactory. What was the reason for the strange quantum restriction on angular momentum? It implied, for example, that a top could spin only with certain discrete values of angular velocity ω instead of with any arbitrary value among a smooth continuum of possible velocities. As experiments continued, more puzzles were uncovered. Some individual spectral lines are apparently multiple lines at the same frequency, because subjecting the atom to an electric or magnetic field "splits" the lines into a cluster of two or more lines spaced closely together. One spectral line of dysprosium, for example, splits into 137 closely spaced lines!

Other questions were also disturbing. Why does the orbiting electron not radiate as classical laws of electromagnetism say it should—those very same laws that provide the central force for these orbits? Why do atoms undergo transitions? Why was the Bohr theory a failure in calculating the spectrum of atoms with more than one electron? All of these drawbacks were eliminated in the new quantum theory that emerged in the next decade. We will now trace these developments step by step, culminating in the next chapter with a quantum mechanical description of atomic structure.

A crucial step toward understanding these mysteries was made by a graduate student in physics at the University of Paris, Prince Louis Victor de Broglie (1892–1987). While studying for his doctor's degree in physics, de Broglie began to think that perhaps the wave–particle duality applied not only to radiation but also to particles of matter. It would, indeed, form a grand sort of symmetry in nature if particles showed wave-like characteristics just as waves have particle-like characteristics. In his doctoral thesis (1924), de Broglie proposed the following ideas (somewhat simplified here). Since photons of

FIGURE 43-11
Louis Victor de Broglie was a member of an old, aristocratic French family that pronounces its last name to rhyme approximately with the English word *troy*. He originally majored in medieval history at the Sorbonne, specializing in Gothic cathedrals. However, he later became interested in physics, switched majors, and received his first degree in physics in 1913. De Broglie's novel proposal that a wave was associated with a moving particle was soon developed by Erwin Schrödinger into the quantum mechanical theory known as *wave mechanics*. Disturbed by the probabilistic nature of quantum mechanics, de Broglie made great (unsuccessful) efforts to find a causal, rather than probabilistic, interpretation of wave mechanics. He was awarded the 1929 Nobel Prize in physics. For a summary of the development of de Broglie's ideas, see H. Medicus, "Fifty Years of Matter Waves," *Physics Today*, Feb. 1947.

electromagnetic radiation have momentum p according to

Photons $$p = \frac{h}{\lambda} \qquad (43\text{-}15)$$

de Broglie proposed that a wavelength λ is also associated with any particle having momentum mv according to

Particles $$mv = \frac{h}{\lambda} \qquad (43\text{-}16)$$

DE BROGLIE WAVELENGTH
(for a particle having momentum p)
$$\lambda = \frac{h}{p} \qquad (43\text{-}17)$$

Just as for electromagnetic radiation, the question as to *what* it is that is "waving" (if anything) requires a long explanation. It definitely is *not* electromagnetic waves. De Broglie called them *matter waves*, or *phase waves*, since he believed there might be interference between the phase of the waves as there is for light waves.

EXAMPLE 43-1

A particle of mass 1 g moves at a speed of 1 mm/s. Calculate the de Broglie wavelength associated with this particle.

SOLUTION

The associated de Broglie wavelength is

$$\lambda = \frac{h}{mv} = \frac{6.63 \times 10^{-34} \text{ J} \cdot \text{s}}{(0.001 \text{ kg})(0.001 \text{ m/s})} = \boxed{6.63 \times 10^{-28} \text{ m}}$$

This is an impossibly small wavelength to measure, since a single proton is about 10^{13} times larger. Indeed, de Broglie waves are of little consequence for macroscopic particles. However, for microscopic particles such as electrons, neutrons, and atoms, interference effects due to these waves are clearly evident and lead to some surprising effects.

De Broglie showed that if one assumes there are matter waves for electrons, there is a reasonable explanation for the Bohr quantum condition on angular momentum that originally seemed so baffling. According to de Broglie, it is simply a case of a *standing-wave pattern* for the electron's motion. This requires that, for stationary states, only an integral number of wavelengths can fit around the circular orbit, as in Figure 43-12: $n\lambda = 2\pi r$. Substituting the de Broglie relation for the wavelength, $\lambda = h/mv$, and rearranging, we obtain

$$mvr = n\left(\frac{h}{2\pi}\right)$$

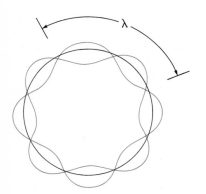

FIGURE 43-12
De Broglie waves for the orbiting electron in the Bohr model for hydrogen form a standing-wave pattern. The distance between adjacent nodes is $\lambda/2$. This illustration is for the energy state $n = 4$.

This is just the Bohr condition for allowed orbits. Thus the *arbitrary* assumption Bohr made for no reason other than that it led to the right answer could now be derived in a plausible way if one assumed that only the motions of the electron that make *standing-wave patterns* represent *stationary states* of the atom. Note that what is involved is interference between different parts of the de Broglie wave associated with a *single* electron. (This is similar to the case of light, for which it is the interference between different parts of the electric field wave of a *single* photon that is significant, not the interference between waves of one photon and waves of another.)

De Broglie's proposal did not win immediate acceptance. While it was recognized as a worthy exercise in theoretical physics, it was treated more as a curious hypothesis that might turn out to have some validity but on the other hand might not. During the oral examination for his doctoral degree, de Broglie was asked how one might detect these waves. He suggested that perhaps a beam of electrons impinging on a crystal would exhibit interference effects, since the crystal lattice of atoms would provide the necessary close spacing of the order of λ that was required to bring out the interference behavior of the waves. The first experiment to detect de Broglie waves did not succeed because of a variety of experimental difficulties. But three years after de Broglie presented his thesis, a dramatic confirmation of matter waves occurred in the United States.

43.5 The Davisson–Germer Experiments

The experiments that first verified de Broglie waves began in 1921, when an American physicist, Clinton Davisson, was investigating the reflection of electrons by metal surfaces for the Western Electric Company (now the Bell Telephone Laboratories).[5] Some of the results he obtained were puzzling. Instead of being scattered uniformly at all angles, the electrons seemed to be scattered at certain angles more than at others. Davisson published the results, but could give no satisfactory explanation for the unusual scattering. He continued the experiments with an assistant, Lester Germer.

In 1925, Davisson was using a target of pure nickel metal in the usual metallic form: innumerable microcrystals with random orientations. An accidental explosion in the laboratory shattered the glass enclosure that kept the apparatus in a vacuum. The exposure to air oxidized the surface of the nickel making it unusable for the experiment. To remove the layer of oxide, Davisson and Germer rebuilt the vacuum enclosure and then heated the target, inadvertently heating it so much that the nickel melted and recrystallized into just a few large crystals at the spot where the electrons struck the target. When they resumed the experiment, the data showed unmistakable peaks in the scattering distribution when the velocity of the electrons was adjusted to certain values. They traced the difference to the fact that the target now consisted of just a few large crystals rather than being in a polycrystalline state. However, unaware of de Broglie's ideas, they proposed an incorrect origin for the peculiar scattering. They felt that the crystal lattice planes somehow "channeled" electrons in certain directions. In 1927, after Davisson attended a physics meeting at Oxford University and learned that *matter waves* might be responsible, he checked de Broglie's theory with the data and found an excellent agreement. Figure 43-13 shows results from an experiment using a single large crystal.

---

[5] The General Electric Company had brought a patent suit against Western Electric over a vacuum-tube design. This experimental work on the scattering of electrons was undertaken to obtain evidence with which to fight the suit. Western Electric won.

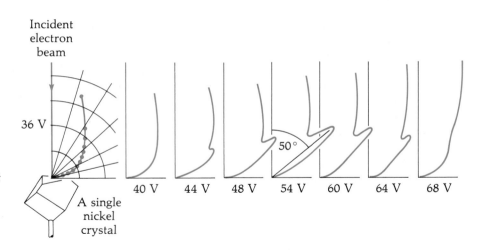

FIGURE 43-13
The scattering of electrons in a Davisson–Germer experiment. Each plot is a polar graph for the number of scattered electrons as a function of angle. Several different values of the accelerating voltage are shown.

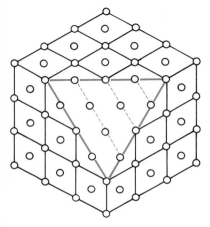

FIGURE 43-14
A detailed view of the cleaved nickel crystal, showing the arrangement of atoms on its surface.

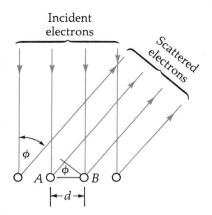

FIGURE 43-15
An edge-on view of the cleaved surface shown in Figure 43-14.

Electrons are scattered from the surface of a metallic crystal in preferred directions. The wave-like character of the electrons causes them to interact with the regular array of atoms on the surface to produce interference effects, similar to the way that light impinging on a diffraction grating produces interference. Let's examine the crystal in Figure 43-13 in greater detail. Figure 43-14 is an enlarged view of the crystal, showing the arrangement of atoms on its surface. We need not be concerned with the arrangement of atoms within the crystal, because low-energy electrons do not penetrate the surface of the crystal to any significant degree. A nickel crystal is composed of basic units called *face-centered cubic* units. Figure 43-14 shows 27 such units with a corner cleaved off. This cleaved surface reveals rows of atoms, indicated by the dashed lines in the figure. These rows of atoms are separated by a distance $d = 0.215\,79$ nm. The scattering of electrons from these rows produces interference of the wave-like electrons. Figure 43-15 is an edge-on view of the cleaved surface. Consider electrons incident normal to the cleaved surface that are scattered at an angle ϕ relative to the normal to the surface. More specifically, consider electrons impinging on rows A and B shown in Figure 43-15. These electron waves will interfere constructively when scattered if the path difference is a multiple of the wavelength associated with the electrons. That is,

$$m\lambda = d \sin \phi \tag{43-18}$$

where $m = 1, 2, 3, \ldots$ (the order of scattering)
λ = wavelength associated with the electrons
d = distance between the rows of atoms
ϕ = angle between the scattered beam of electrons and the normal to the surface

We obtain the relationship between the energy of the electrons and their de Broglie wavelength by recognizing that electrons accelerated through a potential difference V less than 100 V have nonrelativistic kinetic energies given by

$$eV = \tfrac{1}{2}mv^2$$

Solving for v yields
$$v = \sqrt{\frac{2eV}{m}}$$

    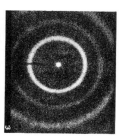

(a) X-rays on NaCl (b) Neutrons on NaCl (c) 0.071-nm x-rays (d) 600-eV electrons (e) 0.057-eV neutrons

FIGURE 43-16
Diffraction patterns produced by electromagnetic waves and by particles. (a) and (b): Laue-spot patterns demonstrate the wave nature of photons and of neutrons. (c), (d), and (e): Diffraction rings produced by scattering from polycrystalline metal samples.

The momentum $p = mv$ is thus

$$p = m\sqrt{\frac{2eV}{m}} = \sqrt{2meV} \qquad (43\text{-}19)$$

and the de Broglie wavelength $\lambda = h/p$ is

$$\lambda = \left(\frac{h}{\sqrt{2me}}\right)\frac{1}{\sqrt{V}} \qquad (43\text{-}20)$$

Substituting numerical values, we obtain the useful relation

DE BROGLIE WAVELENGTH FOR ELECTRONS (nonrelativistic)

$$\lambda = \frac{1.226 \text{ nm}}{\sqrt{V}} \quad \text{(where } V \text{ is in volts)} \qquad (43\text{-}21)$$

In Figure 43-13, the most prominent peak at 50° occurred for 54-eV electrons with a de Broglie wavelength of $\lambda = (1.226 \text{ nm})/\sqrt{54 \text{ eV}} = 0.167$ nm. Comparing this value of λ with that predicted by Equation (43-18) and assuming $m = 1$, we have

$$\lambda = d \sin \phi = (0.215\ 79 \text{ nm}) \sin 50° = 0.165 \text{ nm}$$

which is in excellent agreement with the de Broglie wavelength.

The experiments of Davisson and Germer in 1925–1927, and similar studies by G. P. Thomson in Scotland, were the first experimental confirmation of the de Broglie wave properties of particles.[6] Essentially all of the interference and diffraction effects of electromagnetic waves were later duplicated with particles (see Figures 43-16 and 43-17).

---

[6] Davisson and Thomson shared the Nobel Prize in 1937 for demonstrating the *wave* properties of electrons. Thirty-one years earlier, Thomson's father, J. J. Thomson, received the Nobel Prize for investigating the conduction of electricity by gases, a phenomenon involving the *particle* properties of electrons.

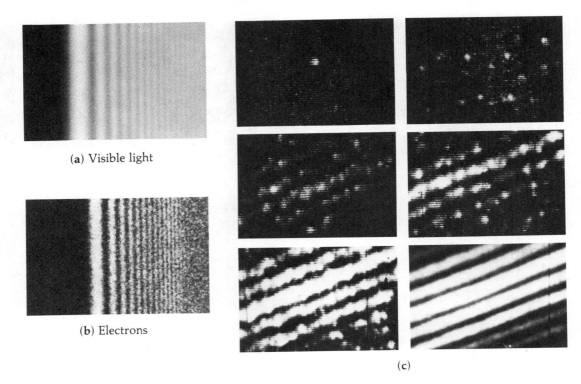

FIGURE 43-17
Fringes formed in the shadow of a straightedge by visible light and by electrons. In (b), the fringes were recorded with the aid of an electron microscope. (c) An interference pattern produced by electrons is the sum of many independent events. As the number of events increases, the pattern becomes more distinct.

43.6 Wave Mechanics

Before matter waves were experimentally verified, two physicists used de Broglie's ideas in 1925–1926 to develop a theory called *wave mechanics*, or *quantum mechanics*, which describes what happens when a force acts on a de Broglie wave. The two theories are vastly different in mathematical form. The German physicist Werner Heisenberg used sophisticated matrix methods, while the Austrian physicist Erwin Schrödinger devised a differential equation approach.[7] Shortly after the theories were published, it was discovered that they were entirely equivalent; either could be derived from the other. Since matrix methods are usually treated in more advanced mathematics courses, we will discuss only the Schrödinger theory here.

For all but the simplest cases, the theory is mathematically difficult to apply. Perhaps the most troublesome aspect of the theory is that its concepts are foreign to our everyday experience and common sense. Yet it has proved to be the only correct way of analyzing the microphysical world. In fact, in its complete relativistic form, known as *quantum electrodynamics* ("Q.E.D."), there is *no* discrepancy with experimental data (at least, to the date of this publication).

---

[7] Nobel prizes were awarded to Heisenberg in 1932 and to Schrödinger (along with P. A. M. Dirac) in 1933 for their accomplishments in developing quantum mechanics.

The central idea of quantum mechanics is contained in a differential equation called "the Schrödinger equation." (Its counterpart in classical mechanics is the differential equation of Newton's second law: $m\, d^2x/dt^2 = F$.) A rigorous derivation would lead us too far astray, so we will give just a plausibility argument here for its origin.

The (nonrelativistic) kinetic energy K of a particle may be written in terms of the momentum p as

$$K = \frac{1}{2} mv^2 = \frac{p^2}{2m} \tag{43-22}$$

If the potential energy is U, the total energy $E = K + U$ becomes

$$E = \frac{p^2}{2m} + U$$

Solving for p gives

$$p = \sqrt{2m(E - U)} \tag{43-23}$$

If we put this value into the de Broglie relation $\lambda = h/p$, we obtain

$$\lambda = \frac{h}{\sqrt{2m(E - U)}} \tag{43-24}$$

As developed in Chapter 18, a solution to the classical wave equation for a wave traveling in the $+x$ direction [Equation (18-16)] is

$$y = A \sin(kx - \omega t) \qquad \left(\text{where } k = \frac{2\pi}{\lambda} \text{ and } \omega = \frac{2\pi}{T}\right)$$

The kx term gives the space variation of y, while the ωt term gives the time variation that causes y to vary in amplitude at the angular frequency ω. We will discuss only the space variation. If we take partial derivatives with respect to x, we obtain the following:

$$\frac{\partial^2 y}{\partial x^2} + \left(\frac{2\pi}{\lambda}\right)^2 y = 0 \tag{43-25}$$

where

$$y = A \sin\left(\frac{2\pi x}{\lambda}\right) \tag{43-26}$$

This relation describes *any* type of mechanical wave—sound waves, waves on a stretched rope, and so on.

Schrödinger put the value of λ from Equation (43-24) into Equation (43-25) to obtain the *time-independent Schrödinger wave equation*:

SCHRÖDINGER'S TIME-INDEPENDENT WAVE EQUATION (one dimension)

$$\frac{\partial^2 \psi}{\partial x^2} + \left(\frac{2m(E - U)}{\hbar^2}\right)\psi = 0 \tag{43-27}$$

where

$$\psi = \psi_{\max} \sin\left(\frac{2\pi x}{\lambda}\right) \tag{43-28}$$

The Schrödinger equation is used in the following way. To find the effect of applying a force to a particle, we substitute into the Schrödinger equation the potential energy function U that is associated with the force. Solutions to the differential equation then express the behavior of the matter wave for the particle. For example, if we put the Coulomb potential $U(r) = -(1/4\pi\varepsilon_0)(qq'/r)$ into the Schrödinger equation for three dimensions, we obtain the ψ functions that represent the matter waves for the stationary states of the electron in a hydrogen atom. (We do this in the next chapter.)

But what does ψ itself represent? We have called it a "matter wave," but naming it does not give us much insight. Since waves are inherently spread out in space, does this mean, for example, that an electron in a hydrogen atom is somehow "smeared out" in space in a way described by the value of ψ? Schrödinger originally proposed this interpretation, but it did not gain much support. The difficulties arose in the complete time-dependent theory, in which the *wave packet* representing a free electron gradually spreads out in space as time goes on. Interpreting this to mean that the charge and mass of an electron in free space similarly spread out seemed impossible for most physicists to accept.

In 1926, a more reasonable interpretation for ψ was proposed by Max Born, a professor at the University of Göttingen. Born noted that Einstein had put forth a new interpretation of the amplitude of the electric field E for electromagnetic waves. Since the square of the amplitude is proportional to the intensity of the wave, Einstein suggested that E^2 is proportional to the *probability* of finding a photon near that location. Thus the light and dark fringes on a photographic film (which can be predicted from wave interference) may be interpreted as the probability of a photon arriving near that particular location on the film. Born extended this idea to the wave function ψ. He proposed that ψ^2 represents the probability that the particle is located near that region of space. This interpretation gave back to the electron its status as a particle rather than a smeared-out entity. *Only our ability to predict the electron's location becomes spread out.*

Generally ψ is a *complex* mathematical function (that is, it involves $\sqrt{-1}$). Because only mathematically real numbers correlate with physically real objects, Born removed the complex characteristics of ψ by suggesting that the *square of the absolute value* of ψ be used. In particular,

BORN'S PROBABILITY INTERPRETATION OF ψ
$$|\psi|^2 \Delta V = \begin{bmatrix} \text{The probability of being found} \\ \text{within the volume element } \Delta V \end{bmatrix}$$

The **probability density function** P is defined as

$$P = |\psi|^2 \tag{43-29}$$

Then, the **probability** $\mathscr{P}$ of finding an electron in a given volume V is

$$\mathscr{P} = \int_V P \, dV \tag{43-30}$$

where the integral is evaluated over the volume V. In order to identify ψ with a probability, we recognize that the probability of finding the electron somewhere is a certainty. That is, when we integrate the probability density function P over all space, it must equal 1. This imposes the following *normalization condition* on the wave function:

NORMALIZATION OF ψ
$$\int_{\substack{\text{all} \\ \text{space}}} |\psi|^2 \, dV = 1 \tag{43-31}$$

Particle in a Box

To illustrate the connection between the wave function ψ and the probability $\mathcal{P}$, consider the case of an electron moving in one dimension between rigid walls a distance D apart, Figure 43-18. The electron confined in this "box" is described as a standing-wave pattern of de Broglie waves that must have nodes at each wall. That is, we fit an integral number n of half-wavelengths within the distance D. Therefore,

$$n\left(\frac{\lambda}{2}\right) = D \quad \text{or} \quad \lambda = 2D/n \qquad (43\text{-}32)$$

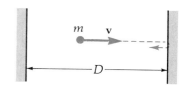

FIGURE 43-18
A particle is confined to move in a one-dimensional box, bouncing elastically at each wall.

The solution to the Schrödinger equation [Equation (43-28)] becomes $\psi = \psi_{max} \sin(2\pi x/\lambda) = \psi_{max} \sin[2\pi x/(2D/n)]$, or

WAVE FUNCTION FOR A PARTICLE IN A BOX
$$\psi(x) = \psi_{max} \sin\left(\frac{n\pi x}{D}\right) \qquad (43\text{-}33)$$

where x is measured from one wall. Before proceeding further, we *normalize* the wave function by integrating Equation (43-33) over all of the space that is available to the electron, from $x = 0$ to $x = D$, and set the result equal to 1:

$$\int_0^D |\psi|^2 \, dx = 1$$

Substituting Equation (43-33) gives

$$\int_0^D (\psi_{max})^2 \sin^2\left(\frac{n\pi x}{D}\right) dx = 1$$

Evaluating this integral, we obtain

$$\psi_{max} = \sqrt{\frac{2}{D}} \qquad (43\text{-}34)$$

The normalized wave function is then

NORMALIZED WAVE FUNCTION FOR A PARTICLE IN A BOX
$$\psi(x) = \sqrt{\frac{2}{D}} \sin\left(\frac{n\pi x}{D}\right) \qquad (43\text{-}35)$$

and the probability density function $P = |\psi|^2$ is

$$P(x) = \left(\frac{2}{D}\right) \sin^2\left(\frac{n\pi x}{D}\right) \qquad (43\text{-}36)$$

Figure 43-19 illustrates the wave functions and probability density functions for the first three quantum states corresponding to $n = 1$, 2, and 3. Note that the probability of finding the electron in a small region at the center between the walls is a maximum for $n = 1$ and a minimum for $n = 2$. The total area under each of the probability density function curves is equal to 1 because each of the wave functions is normalized.

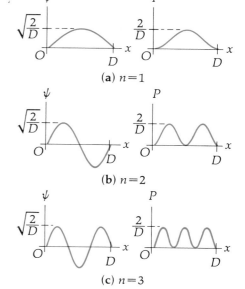

FIGURE 43-19
The first three quantum states for an electron confined to one-dimensional motion between rigid walls a distance D apart.

EXAMPLE 43-2

An electron is confined to one-dimensional motion between two rigid walls separated by a distance D. (a) What is the probability of finding the electron within the interval $x = 0$ to $x = D/3$ from one wall if the electron is in its $n = 1$ state? (b) Compare this value with the classical probability.

SOLUTION

(a) The probability $\mathscr{P}$ of finding the electron in an interval Δx along a line is given by the one-dimensional version of Equation (43-30):

$$\mathscr{P} = \int_x^{x+\Delta x} P(x)\, dx$$

where $P(x)$ is given by Equation (43-36). Substituting this value in the above equation gives

$$\mathscr{P} = \frac{2}{D} \int_0^{D/3} \sin^2\left(\frac{n\pi x}{D}\right) dx$$

$$\mathscr{P} = \frac{2}{D}\left[\frac{x}{2} - \sin\left(\frac{2n\pi x}{D}\right)\Big/\left(\frac{4n\pi}{D}\right)\right]_0^{D/3} = \boxed{0.196} \quad \text{(for } n = 1\text{)}$$

The wave-mechanical probability of finding the electron somewhere between $x = 0$ and $x = D/3$ is thus about $1/5$ for the $n = 1$ state.

(b) To illustrate the correspondence principle, as $n \to \infty$ we note that $\lim_{n\to\infty} [(\sin an)/bn] = 0$. Therefore, the classical probability becomes $\lim_{n\to\infty} \mathscr{P} = 1/3$. Viewed classically, the electron moves back and forth with constant speed between the walls, so the probability of finding it in one-third of the available space is, indeed, $1/3$.

Energy States of a Particle in a Box

The energy of each of the quantized states that the electron may have as it moves between the walls is found from $E = U + K$. Here, $U = 0$ (Why?). K may be written in terms of the momentum $p = h/\lambda$, so we have

$$E = \frac{p^2}{2m} = \frac{h^2}{2m\lambda^2} \tag{43-37}$$

For stationary energy states (a standing-wave pattern), there must be an integral number of *half*-wavelengths within the distance D between the walls (in contrast to an integral number of *whole* wavelengths around a circle for standing waves in the hydrogen atom):

$$n\left(\frac{\lambda}{2}\right) = D \quad \text{(where } n = 1, 2, 3, \ldots\text{)} \tag{43-38}$$

Substituting this value in Equation (43-37), we obtain

ENERGY STATES OF A PARTICLE IN A BOX
$$E_n = \left(\frac{h^2}{8mD^2}\right) n^2 \quad (n = 1, 2, 3, \ldots) \tag{43-39}$$

where the number n refers to the nth quantum state of the electron.

> **EXAMPLE 43-3**
>
> An electron with an energy of approximately 6 eV moves between rigid walls exactly 1 nm apart. (a) Find the quantum number n for the energy state that the electron occupies. (b) Find the exact value for the electron's energy.
>
> **SOLUTION**
>
> The relationship between the quantum number and the energy is given by Equation (43-37). Solving this equation for n yields
>
> $$n = \left(\frac{2D}{h}\right)\sqrt{2mE}$$
>
> Substituting numerical values for $E = (5 \text{ eV})(1.6 \times 10^{-19} \text{ J/eV}) = 8 \times 10^{-19}$ J, we obtain
>
> $$n = \frac{2(10^{-9} \text{ m})}{(6.626 \times 10^{-34} \text{ J} \cdot \text{s})}\sqrt{(2)(9.1 \times 10^{-31} \text{ kg})(8 \times 10^{-19} \text{ J})} = 3.642$$
>
> Since n must equal an integer, we try $n = 4$ in Equation (43-37), which gives $E = 6.017$ eV. For $n = 3$, we obtain $E = 3.384$ eV. Because the value for $n = 4$ is closer to "approximately 6 eV," we conclude that
>
> and (a) $n = 4$
> (b) $E = 6.02$ eV

We have given only the briefest introduction to quantum mechanics. Numerous innovations and additions to the theory were made by many physicists, most notably the British physicist P. Dirac (1928), who developed the *relativistic* wave equation that accounts for the splitting of spectral lines in the presence of a magnetic field and predicts the existence of *antimatter*.[8]

43.7 Barrier Tunneling

One fascinating conclusion of quantum mechanics is that the wave function for a particle may penetrate into a region forbidden by classical theory. Suppose that we repeatedly throw a grain of sand at a piece of paper held fixed in space. If the kinetic energy of the sand grain is insufficient to break through the paper, our expectation is that we would never find the particle traveling at the same speed on the other side, *with the paper intact*. But in the analogous situation of an electron approaching a potential "wall" with kinetic energy less than the height U of the potential barrier, the electron wave function can penetrate the barrier and have a finite amplitude on the far side of the wall. This means that occasionally we would find that the electron has quantum-mechanically "tunneled" through the barrier to appear on the other side where,

---

[8] Antimatter—antielectrons, antiprotons, antineutrons, and so forth—is another form of matter, created in high-energy interactions of photons and particles. An antiparticle has the same mass and the same spin (see Section 44.4) as its ordinary matter counterpart, but it has opposite electric charge and the alignment between its spin and magnetic moment is opposite to that of the particle. If an antiparticle comes in contact with a particle of the same type, they mutually annihilate, forming an equivalent amount of energy (mc^2) in photons. Since matter and antimatter are always experimentally formed in equal amounts, one of the problems to be solved in cosmology is why we live in a universe that seems dominated by matter rather than antimatter.

classically, it could never be found, Figure 43-20. The probability of tunneling is essentially zero for macroscopic objects, but for particles on a quantum-mechanical scale it becomes important. Problem 43B-23 calculates the probability of such tunneling.

Barrier tunneling has many practical applications. A very slight change in the height of the barrier causes a very great change in the probability of penetration. For example, in the *tunnel diode*, the flow of electrons between oppositely charged regions can be rapidly varied by tiny changes in the potential of the thin barrier wall separating the two regions. Another interesting device is the *scanning tunneling microscope*.[9] In this device, an extremely sharp metal needle that (ideally) terminates in a single atom is brought to within about 2 atomic diameters of the surface of a conductor. With a low potential difference, electrons cannot classically move across the gap. However, barrier tunneling does occur. If the gap between the tip and the surface increases, the current decreases. The needle is now moved across the sample's surface, while the height of the tip is constantly adjusted to keep the current constant. Thus the vertical motions of the tip plot a sort of topographic map of the peaks and valleys, revealing the locations of atoms on the surface. Successive lines are scanned, forming a complete picture. The individual lines are "smoothed" by a computer program to form Figure 43-21. Differences as small as one-hundredth of an atomic diameter can be detected, in contrast to the lesser precision of a light microscope, whose resolution is ~ 2000 atomic diameters.

43.8 The Uncertainty Principle

Wave mechanics replaces the precise trajectories of particles with a "cloud" of probability estimates spread out in space. This is a profound change in the way we deal with nature. The most all-inclusive theory we have—quantum mechanics—is not based upon the kind of *physical* models that all previous theories were. It does not tell us exactly where the electron is or how it moves, but only how to *estimate the probability of finding it within a certain region traveling within a certain range of velocities*. But the nagging question remains: The electron must be *somewhere*. Can't we improve our measuring technique to pin down its location exactly and find out precisely how it moves from one place to another?

In 1927, Heisenberg pointed out that there is a *fundamental* limit, inherent in *all* measurements, that prevents us from doing this. No amount of cleverness or refinement of our measuring apparatus will get around this basic obstacle, because *the limitation is a consequence of the wave–particle duality of nature*, and we cannot change that.

The uncertainty principle can be illustrated in the following way. Suppose that we wish to determine the position of an electron along the x axis with a very powerful microscope, Figure 43-22. Because of diffraction effects due to the lens diameter D, the image of a (point) electron will be a *diffraction pattern* whose central peak has an angular size θ_R according to Equation (39-22):

$$\theta_R = \frac{(1.22)\lambda}{D}$$

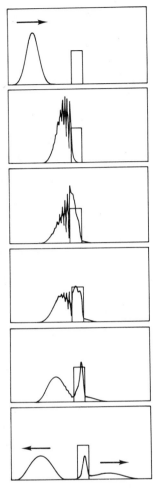

FIGURE 43-20
Frames from a computer-animated film of the probability density function $|\psi|^2$ for a particle approaching a rectangular potential barrier with not quite enough energy (classically) to pass through the barrier. After the impact, however, two "wave packets" of probability travel away on either side of the barrier, showing that there is a finite chance that the particle has tunneled through the classically forbidden region. This does not mean that the particle splits into two parts; the fact that $|\psi|^2$ is finite to the right of the barrier only means that the *chances of finding it there* after the impact are finite, not zero as classical physics predicts. (From the film by A. Goldberg, H. M. Schey, and J. L. Schwartz, "Scattering in One Dimension," described in *American Journal of Physics* **35**, 177 (1967).]

---

[9] See Gerd Binnig and Heinrich Rohrer, "The Scanning Tunneling Microscope," *Scientific American* **253**, (Aug. 1985) p. 50. The article explains how such tiny, controlled movements of the needle tip are achieved. The 1986 Nobel Prize in physics was awarded to these authors for their invention (shared with Ernst Ruska for his earlier invention of the electron microscope).

where λ is the wavelength of light used and D is the lens diameter. This minimum angle of resolution θ_R may also be written as $\Delta x/d$. It implies that the electron's position is known only within an uncertainty $\pm \Delta x$.

$$\frac{\Delta x}{d} = \frac{(1.22)\lambda}{D}$$

Rearranging gives
$$\Delta x = \frac{(1.22)\lambda}{(D/d)}$$

If 2α is the angle of the cone of light from the object the lens gathers, then $\tan \alpha = (D/2)/d = \frac{1}{2}(D/d)$. For an order-of-magnitude estimate, we may replace $\tan \alpha$ with the approximation $\sin \alpha$ (not an overwhelmingly good approximation, but it is still in the ballpark).

$$\Delta x \approx \frac{(1.22)\lambda}{2 \sin \alpha}$$

Finally, in the same spirit of estimation, we drop the factor 1.22/2 to obtain

$$\Delta x \approx \frac{\lambda}{\sin \alpha} \qquad (43\text{-}40)$$

This is the inherent uncertainty in determining the x coordinate of the position of the electron. It is due to the fact that we used a lens of diameter D. If we used a lens with a smaller diameter, the uncertainty would be greater (because $\sin \alpha$ would be smaller).

Perhaps we could try to improve matters by using light of shorter wavelength, say, in the x-ray region. But, unfortunately, a photon of shorter wavelength has a greater momentum $p = h/\lambda$ and would give the electron a harder "kick" as it scatters off the electron into the microscope lens. The scattered photon can enter the lens *anywhere* within an angle 2α. We do not know the exact direction because we do not detect the photon until after it travels through the lens to reach the image location. All we know is that it went through the lens at *some* point. As the photon scatters off the electron in a Compton interaction, its x component of momentum can vary anywhere from $+(p_x \sin \alpha)$ to $-(p_x \sin \alpha)$. And, by the conservation of momentum, this uncertain amount is transferred to the electron. So the uncertainty in the x component of the electron's momentum becomes

$$\Delta p_x \approx 2p \sin \alpha \approx 2\left(\frac{h}{\lambda}\right) \sin \alpha \qquad (43\text{-}41)$$

Combining these uncertainties in position and momentum, we have

$$\Delta x \Delta p_x \approx \frac{\lambda}{\sin \alpha} 2\left(\frac{h}{\lambda}\right) \sin \alpha \approx 2h \qquad (43\text{-}42)$$

As the uncertainty in position is reduced, inevitably the uncertainty in momentum increases, and vice versa. Note that this uncertainty is not due in any way to lack of refinement in our measuring instruments. Even with the most ideal apparatus imaginable, the fundamental limitation still remains; *this limitation is traceable to the wave–particle aspects of both matter and radiation.*

FIGURE 43-21
A computer-processed image of data obtained with a scanning tunneling microscope. Each ring-shaped image is an hexagonal array of the six carbon atoms in a benzene molecule. The molecules have been deposited on a rhodium metal surface.

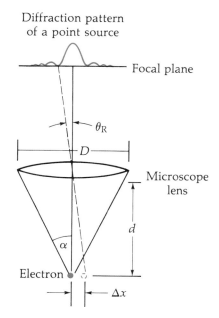

FIGURE 43-22
Observing the position of an electron with a microscope. The central peak of the diffraction pattern is within $\pm \theta_R$ of the axis.

A more rigorous statement of the **Heisenberg uncertainty relation** is

HEISENBERG UNCERTAINTY RELATION
$$\Delta x \, \Delta p_x \gtrsim \hbar \qquad (43\text{-}43)$$

In a simultaneous measurement of the position and momentum of a particle, the product of the uncertainties is equal to or greater than a number of the order of $\hbar \equiv h/2\pi$.

No amount of ingenuity or improvement in measurement techniques can outwit this limitation. Because of the wave–particle aspects of matter and radiation, the very act of measurement itself inevitably disturbs the system under investigation in an *unknown* way that cannot be avoided. It is a built-in limitation in nature. The uncertainty principle underscores the fact that classical models of atomic phenomena are bound to be misleading.

Note, however, that there is no limit on determining the position (only) of a particle to any desired degree of accuracy, or the momentum (only). But as we narrow down the uncertainty in position ($\Delta x \to 0$), inevitably the uncertainty in the *simultaneous* determination of the momentum of the particle becomes larger and larger ($\Delta p_x \to \infty$), and vice versa. The precise relation between Δx and Δp_x depends on how one defines the limits of uncertainty in a particular case. The product may vary somewhat in the range of $2h$ down to about $\hbar$. Similar relations also apply in the y and z directions.

$$\Delta y \, \Delta p_y \gtrsim \hbar \qquad (43\text{-}44)$$
$$\Delta z \, \Delta p_z \gtrsim \hbar \qquad (43\text{-}45)$$

Different sets of variables are also related in the same way. It can be shown that

$$\Delta E \, \Delta t \gtrsim \hbar \qquad (43\text{-}46)$$

where ΔE is the uncertainty in the measurement of energy E and Δt is the time interval for determining the energy. The principle also applies to angular measurements. For example, if we wish to determine where the electron is located in the orbit of a Bohr-model hydrogen atom, the uncertainty in the angle ϕ measurement is related to the uncertainty in the angular momentum L_ϕ:

$$\Delta \phi \, \Delta L_\phi \gtrsim \hbar \qquad (43\text{-}47)$$

This form of the uncertainty principle essentially leads to the destruction of the planetary view of the Bohr model, in which the electron occupies a well-defined position in an orbit. Consider the following example:

EXAMPLE 43-4

Estimate the uncertainty in the angular position $\Delta\phi$ of the electron in a Bohr orbit.

SOLUTION

The value $\Delta\phi$ is related to the electron's uncertainty in angular momentum ΔL_ϕ by Equation (43-47): $\Delta\phi \, \Delta L_\phi \gtrsim \hbar$. Since L_ϕ is quantized according to one of the

Bohr-model postulates, it has *discrete values only*, with no uncertainty in any of the Bohr orbits:

$$\Delta L_\phi = 0$$

Equation (43-47) then states that $\Delta\phi$ must have no finite value, which is equivalent to stating that ϕ is *completely uncertain*. The electron is equally likely to be anywhere in the orbit all of the time. Thus, it is meaningless to speak of the electron as moving from point to point along its orbit.

EXAMPLE 43-5

An electron ($m = 9.11 \times 10^{-31}$ kg) and a bullet ($m = 0.02$ kg) each have a speed of 500 m/s, accurate to within 0.01%. Within what limits could we determine the position of the objects?

SOLUTION

(a) The electron's momentum is $p = mv = (9.11 \times 10^{-31}$ kg$)(500$ m/s$) = 4.56 \times 10^{-28}$ kg·m/s. The uncertainty Δp_x in this momentum measurement is given as 0.01%. Thus:

$$\Delta p_x = (4.56 \times 10^{-28} \text{ kg·m/s})(0.0001) = 4.56 \times 10^{-32} \text{ kg·m/s}$$

From the Heisenberg uncertainty relation [Equation (43-43)], the uncertainty Δx in position is of the order of

$$\Delta x \approx \frac{\hbar}{\Delta p_x} = \frac{h}{(2\pi)\Delta p_x} = \frac{6.63 \times 10^{-34} \text{ J·s}}{(2\pi)\left(4.56 \times 10^{-32} \frac{\text{kg·m}}{\text{s}}\right)}$$

$$\approx 0.00231 \text{ m}, \quad \text{or} \quad \boxed{\approx 2.31 \text{ mm}}$$

This is an unbeatable lower limit on the uncertainty with which we could determine the electron's position. A model of an electron as a small point mass is not valid for this situation.

(b) The bullet's momentum is $p = mv = (0.02$ kg$)(500$ m/s$) = 10.0$ kg·m/s. The uncertainty Δp_x in this momentum measurement is given as 0.01%, or

$$\Delta p_x = (10.0 \text{ kg·m/s})(0.0001) = 10^{-3} \text{ kg·m/s}$$

From the Heisenberg uncertainty relation [Equation (43-43)], the uncertainty Δx in position is of the order of

$$\Delta x \approx \frac{\hbar}{\Delta p_x} = \frac{h}{(2\pi)(\Delta p_x)} = \frac{6.63 \times 10^{-34} \text{ J·s}}{(2\pi)\left(10^{-3} \frac{\text{kg·m}}{\text{s}}\right)} \approx \boxed{1.00 \times 10^{-31} \text{ m}}$$

This uncertainty in position is far below any conceivable possibility of measurement (an atomic nucleus is about 10^{-15} m in size), so for macroscopic objects under everyday circumstances, we may confidently treat them as *classical* particles.

The uncertainty principle has profound philosophical consequences. Just as Einstein showed that absolute space, absolute time, and absolute simultaneity are inherently unmeasurable and therefore meaningless concepts that should be eliminated from our theories, Heisenberg pointed out that precise knowledge of the position and momentum of an electron at a given instant is inherently limited.

This is in contrast to the situation in classical physics, in which *any* measurements could, in principle, be made with increasing precision without limit. The uncertainty principle denies this. It points out the impossibility of making a measurement without disturbing the object *by an unknown amount*, thereby reducing our knowledge of some related quantity. This is true even with "perfect" measuring instruments that have no technical imperfections because the uncertainties do not originate in defects in the equipment or in the measuring techniques. *The uncertainties originate in the wave–particle duality of matter and radiation.* Since we can never experimentally determine the exact behavior of particles at the atomic level, we should not speak of their motions in classical terms.

It now becomes clear why the paradoxes arose in the analysis of the double-slit interference effect in terms of classical trajectories for photons (or particles) as they pass through the slit system. In an experiment in which a beam of electrons incident on two slits whose spacing is of the same order of magnitude as the de Broglie wavelength, the usual two-slit interference pattern results. Even if we send only one electron at a time to the slit, the interference fringes are still formed (statistically) if enough electrons are used. However, if one slit at a time is covered alternately during the exposure time, we do not get the two-slit fringe pattern, but just the single-slit diffraction pattern. Thus we must conclude that with both slits open each electron somehow interacts simultaneously with both slits, in spite of our classical model of an electron as a well-defined particle that could go through only one slit at a time. As far as we can experimentally verify, electrons are not classical particles with well-defined trajectories, so we should not talk as if they were. This is the essence of the *positivist* philosophy that gained a strong foothold in physics, first through Einstein's relativity (which rejected the idea of an ether because it was unmeasurable) and later through quantum mechanics (which rejected precise classical descriptions of atomic phenomena as unmeasurable). In its place, quantum mechanics sometimes offers only probability estimates. If a series of identical measurements is made of a property of a system, quantum mechanics can predict *precisely* the average value of these measurements, yet it can give only a *probability estimate* for any single measurement.

This probabilistic interpretation of quantum mechanics is associated with the *Copenhagen* school of thought, so-named because of its main architect, the Danish physicist Niels Bohr. The majority of physicists today accept this interpretation. However, there are some notable exceptions. Einstein, for example, never accepted the abandonment of the strict causality on which classical physics is based. "God does not play dice with the universe," he said, and felt there must be some underlying causal relations that produce the statistical behavior we observe. He had faith that some future theory could reveal a strict causality at a deeper level. A few good theorists have devoted years in attempting to devise such a "hidden parameter" theory. None has succeeded to date.

Finally, one point deserves emphasis.[10] All observations are described in the language of classical physics because we ultimately record measurements

---

[10] The following remarks are adapted from Herman Feshbach and Victor F. Weisskopf, "Ask a Foolish Question...," *Physics Today*, Oct. 1988. The April 1989 issue contains Letters to the Editor that express other viewpoints with lively enthusiasm.

with macroscopic instruments. However, this does not imply that measuring instruments and other large-scale objects obey classical laws instead of quantum laws. *Every object obeys quantum laws.* It is only because macroscopic objects are so large that we can describe their behavior using classical concepts with negligible error. But when we analyze atomic phenomena, only quantum physics gives correct predictions.

Quantum mechanics makes certain predictions with extreme precision. For example, it gives the ground-state energy for hydrogen to one part in 10^{12}. Yet for certain other questions quantum mechanics gives only a probability distribution rather than a definite answer. As Feshbach and Weisskopf point out:

> *The Heisenberg uncertainty relations are the signposts saying, "You are allowed to use [certain pairs of] classical . . . variables up to here, but go no further. The use of such variables beyond this limit is inappropriate. If you ask an inappropriate question you get a probability distribution as a response." On the other hand, if an appropriate question is asked, quantum mechanics gives a crisp, precise answer such as the energy of a hydrogen atom in its ground state.*

These authors clarified their use of the word *inappropriate*: "Observations are formulated in the language of classical physics. . . . But classical physics concepts are not always appropriate for the description of atomic situations." They did not mean that such "inappropriate" questions should not be asked. The meaning of quantum mechanics remains a continuing, heated debate among certain physicists and philosophers.

43.9 The Complementarity Principle

We have described how physicists came to believe in a certain symmetry in nature involving particles and waves. But this new unity came at the price of new conceptual difficulties. The best theory we have—quantum electrodynamics—does not allow us to picture the motions and interactions of microscopic objects as we did in classical physics. They are neither particles nor waves, yet on occasion they show more strongly one or the other of these attributes. An experiment designed to bring out the *wave* aspects (such as double-slit interference) cannot be dealt with in terms of particles. Similarly, an experiment that brings out *particle* aspects (such as Compton scattering) cannot be visualized in terms of waves. Bohr (1928) recognized this essential characteristic of nature by suggesting a **principle of complementarity** at the atomic level.

BOHR'S COMPLEMENTARITY PRINCIPLE   **In the quantum domain, wave and particle aspects complement each other. Though the choice of one description precludes the simultaneous choice of the other, *both* are required for a complete understanding.**

In explaining this principle, Bohr suggested an analogy: both sides of a coin must be included for a complete description of the coin, yet we cannot see both sides simultaneously. As with the Copenhagen interpretation of quantum mechanics, a few physicists and philosophers still seek an alternative view. Nevertheless, Bohr's principle of complementarity does seem to express in general terms why we find ourselves in a dilemma when we try to cling to classical ideas at the atomic level.

Our concepts, our modes of thought and language—indeed, what we call common sense—all originate in our experiences. Classical physics is the crowning achievement of this common sense. In the 1920s, however, our experiences in the microworld and in the relativistic domain began to include observations that violated classical ideas, so our "common sense" had to be enlarged and changed to include these new types of experiences. Nature continues to challenge us with new mysteries. What surprising concepts will we need to accept in the future in order to unravel them?

43.10 A Brief Chronology of Quantum Theory Development

1900 Explanation of blackbody radiation by energy quantization.
Max Planck (Nobel Prize 1918).

1900 Discovery that the energy of electrons emitted by the photoelectric effect was independent of the light intensity.
Philip von Lenard (Nobel Prize 1905).

1905 Explanation of the photoelectric effect.
Albert Einstein (Nobel Prize 1921).

1905 The theory of special relativity.
Albert Einstein (Nobel Prize 1921).

1907–1911 Explanation of the specific heats of solids by energy quantization.
Albert Einstein (Nobel Prize 1921).

1911 Observation of the nuclear atom.
Ernest Rutherford (Nobel Prize, Chemistry, 1908)

1913 First quantized model of the hydrogen atom.
Niels Bohr (Nobel Prize 1922).

1916 Experimental studies of the photoelectric effect.
Robert Millikan (Nobel Prize 1923).

1923 Discovery and explanation of the collisions between light quanta and electrons.
Arthur Compton (Nobel Prize, with C. T. Wilson, 1927).

1924 Proposal that electrons have an associated wavelength $\lambda = h/p$.
Prince Louis Victor de Broglie (Nobel Prize 1929).

1925 Mathematical theory of wave mechanics.
Erwin Schrödinger (Nobel Prize, with P. Dirac, 1933).

1925 Mathematical theory of matrix mechanics.
Werner Heisenberg (Nobel Prize 1932).

1925 The Exclusion Principle.
Wolfgang Pauli (Nobel Prize 1945).

1926 Statistical interpretation of the wave function.
Max Born (Nobel Prize 1954).

1927 The Uncertainty Principle.
Werner Heisenberg (Nobel Prize 1932).

1927 Observation of electron-wave diffraction by crystals.
Clinton Davisson (Nobel Prize, with G. P. Thompson, 1937).

1928 Relativistic theory of quantum mechanics and the prediction of the positron.
Paul Dirac (Nobel Prize, with E. Schrödinger, 1933).

1932 Observation of the positron.
Carl Anderson (Nobel Prize, with Victor Hess, 1936).

1948 Completion of the theory of quantum electrodynamics.
Sin-Itiro Tomanaga, Julian Schwinger, and Richard Feynman (Nobel Prize 1965).

Summary

The *Bohr model for hydrogen* assumes the following:

(1) The electron travels in circular orbits about the proton. The Coulomb force is the centripetal force.
(2) There exist allowed energy states E_n for which the electron moves without radiating.
(3) The allowed energy states are those for which

$$mvr = n\hbar$$

(4) Transitions between allowed energy states involve the emission or absorption of photons of energy hf, where

$$hf = E_{\text{final}} - E_{\text{initial}}$$

The orbital radii and the energy of allowed energy states in the Bohr model are

$$r_n = \frac{\varepsilon_0 h^2 n^2}{\pi m Z e^2} = (0.0529 \text{ nm})n^2$$

$$E_n = -\frac{mZ^2 e^4}{8\varepsilon_0^2 h^2 n^2} = -\frac{13.6 \text{ eV}}{n^2} \qquad (n = 1, 2, 3, \ldots)$$

Bohr's *correspondence principle*:

Any new theory must reduce to the corresponding classical theory when applied to situations appropriate to the classical theory.

Under certain circumstances, particles exhibit wave characteristics with a *de Broglie wavelength*

$$\lambda = \frac{h}{p}$$

where p is the momentum of the particle. For electrons accelerated from rest through a potential difference V,

$$\lambda = \frac{1.226 \text{ nm}}{\sqrt{V}} \qquad \text{(where } V \text{ is in volts)}$$

Wave mechanics, or *quantum mechanics*, is a theory developed by Erwin Schrödinger (and independently by Heisenberg in a different mathematical format) that includes the wave and particle characteristics for both matter and radiation. It is a differential equation for an amplitude ψ. In Born's interpretation,

$$|\psi|^2 = \begin{cases} \text{the probability that the particle will} \\ \text{be found within the region } \Delta x \end{cases}$$

Heisenberg's uncertainty relation places a fundamental limit on the accuracy with which certain pairs of variables can be measured simultaneously. The product of the uncertainties is $\gtrsim \hbar$. Following is a partial list of these variables:

Position and momentum: $\quad \Delta x \, \Delta p_x \gtrsim \hbar$

Energy and time: $\quad \Delta E \, \Delta t \gtrsim \hbar$

Angular position and
angular momentum: $\quad \Delta \phi \, \Delta L_\phi \gtrsim \hbar$

SCHRÖDINGER'S TIME-INDEPENDENT WAVE EQUATION (one dimension)

$$\frac{\partial^2 \psi}{\partial x^2} + \left(\frac{2m(E-V)}{\hbar^2}\right)\psi = 0$$

where $\quad \psi(x) = \psi_{\max} \sin\left(\frac{2\pi x}{\lambda}\right)$

In Born's interpretation,

$$|\psi|^2 \Delta x = \begin{bmatrix} \text{The probability that the particle will} \\ \text{be found within the region } \Delta x \end{bmatrix}$$

We *normalize* the wave function by determining $\psi_{\max}$ from

$$\int_{\substack{\text{all} \\ \text{space}}} |\psi|^2 \, dV = 1$$

For a particle confined in a one-dimensional box of width D with rigid walls, the normalized wave functions form standing-wave patterns with nodes at each wall:

PARTICLE IN A BOX
$$\psi(x) = \sqrt{\frac{2}{D}} \sin\left(\frac{n\pi x}{D}\right)$$
$$E_n = \left(\frac{h^2}{8mD^2}\right)n^2 \qquad n = 1, 2, 3, \ldots$$

where x is measured from one wall.

The wave function ψ for a particle may penetrate into regions forbidden by classical theory (where $E < U$), leading to *barrier tunneling*. (See Problem 43B-23 for the probability of penetrating a rectangular potential barrier.)

Bohr's *complementarity principle*:

In the quantum domain, wave and particle aspects complement each other. Though the choice of one description precludes the simultaneous choice of the other, both are required for a complete understanding.

Questions

1. How does the correspondence principle apply to Einstein's theory of special relativity?
2. What would be the observable consequences[11] if Planck's constant were on the order of 0.1 J·s?
3. What are the similarities between particle waves and electromagnetic waves? What are the dissimilarities?
4. In what ways are high-energy electrons and photons similar? In what ways are they dissimilar?
5. Do the wave-like properties of particles imply that a baseball pitched through an open door may be deflected?
6. In what ways does the wave-like concept of particles contradict Bohr's model of the hydrogen atom?
7. Attempt to clarify this statement: If a beam of electrons were used to produce a double-slit interference pattern, each of the electrons would have to pass through both slits.
8. In what way is the uncertainty principle a direct consequence of the wave-like nature of particles?
9. How is the de Broglie concept of an orbital standing wave for the electron in the hydrogen atom inconsistent with the uncertainty principle?
10. What is the role of the complementarity principle in an experiment that demonstrates electron diffraction?
11. For a particle confined in a box. Figure 43-18, the probability density may be zero at certain points. Can the particle move through these points?

Problems

43.2 Models of an Atom

43A-1 Before the Bohr model for hydrogen was developed, J. R. Rydberg obtained an empirical expression for the wavelength λ emitted when an atom undergoes a transition from the initial state n_i to the final state n_f:

RYDBERG FORMULA $$\frac{1}{\lambda} = R\left[\frac{1}{n_f^2} - \frac{1}{n_i^2}\right]$$

where R is the *Rydberg constant*. Using the fact that the Balmer series transition from $n = 3$ to $n = 2$ produces the H_α line at 656.3 nm, show that for hydrogen $R = 1.097 \times 10^7$ m^{-1}.

43A-2 When spectroscopists tabulate wavelengths, those longer than 200 nm are given as they would be in air, since that is how they are usually measured. (Wavelengths shorter than about 200 nm don't penetrate air, so these values are tabulated for a vacuum.) The H_α line (Balmer series) has a listed wavelength of 656.28 nm. Calculate its value in a vacuum to five significant figures.

43B-3 Derive the following expression for the hydrogen spectrum wavelengths emitted when the electron undergoes a transition from the n_i state to the n_f state.

$$\lambda = 91.13 \text{ nm} \left(\frac{n_i^2 n_f^2}{n_i^2 - n_f^2}\right)$$

43B-4 Consider a hydrogen atom in the ground state. Find the following quantities (in electron volts): (a) the kinetic energy of the electron, (b) the potential energy, (c) the total energy, and (d) the energy required to remove the electron completely from the proton.

43B-5 Solve the previous problem for singly ionized helium (a helium atom with one electron removed).

43B-6 Determine the longest and shortest wavelengths of light that are emitted in the Paschen series of spectral lines from atomic hydrogen.

43B-7 Consider an ideal, rigid, diatomic molecule in which two equal point masses m, separated by a (constant) distance $2a$, are rotating about an axis that is halfway between the masses and perpendicular to the line joining the masses. Assuming quantization of angular momentum as in the Bohr hydrogen atom, show that the rotational energy levels are given by $E_n = n^2 h^2 / 16\pi^2 m a^2$.

43B-8 A photon is emitted when the hydrogen atom undergoes a transition from the $n = 3$ state to the $n = 1$ state. The work function for lead is 4.25 eV. Find the maximum kinetic energy (in electron volts) that a photoelectron can have when ejected from lead by this photon.

43.4 De Broglie Waves
43.5 The Davisson–Germer Experiments

43A-9 A certain electron microscope uses 50-keV electrons. By what factor is the de Broglie wavelength of these electrons smaller than that of visible light of 500-nm wavelength?

43A-10 A 1-g particle and an electron are moving at 150 m/s each. Calculate the de Broglie wavelength of each.

43A-11 Calculate the de Broglie wavelength of an electron that has been accelerated from rest through a potential difference of 50 V.

43A-12 A moving neutron has a de Broglie wavelength of 0.2 nm. Find (a) its speed and (b) its kinetic energy in electron volts.

---

[11] For an amusing account of the strange consequences of relativity and quantum theory see George Gamow, *Mr. Tompkins in Wonderland* (Macmillan, 1940). Here, $c = 10$ mi/hr, $h = 1$ erg·s and $G = 10^{12}$ times larger than its actual value. A companion volume is *Mr. Tompkins Explores the Atom* (Macmillan, 1940). Both are currently available in *Mr. Tompkins in Paperback* (Cambridge Univ. Press, 1967).

43A-13 An alpha particle is a helium nucleus whose mass is 4 u (where u is the *atomic mass unit*: $1\,u = 1.661 \times 10^{-27}$ kg). Calculate the de Broglie wavelength associated with an alpha particle that has a kinetic energy of 2 MeV.

43A-14 An electron microscope achieves very high resolution by using electrons whose de Broglie wavelengths are usually less than 0.01 nm. Explain why we can't design a photon microscope using photons with wavelengths of this order of magnitude.

43A-15 Electron A moves such that its de Broglie wavelength is twice that of electron B. Find the ratio of their kinetic energies, K_A/K_B.

43B-16 Explain, in a quantitative way, why second-order electron scattering peaks are not evident for any of the electron energies shown in Figure 43-13.

43B-17 Consider the experimental result of the Davisson–Germer experiment shown in Figure 43-13 for incident 60-eV electrons. (a) Find the de Broglie wavelength of the incident electrons on the basis of their energy. (b) What scattering angle would you predict for this case?

43B-18 A beam of "white" x-rays (containing many different wavelengths) is incident upon a cubic crystal at a glancing angle of incidence of 35° with respect to the crystal surface. The longest wavelength of x-rays that are "reflected" symmetrically at the same glancing angle is 0.330 nm. (a) Find the spacing between adjacent planes of atoms in the crystal. (b) If a beam of electrons were substituted for the x-ray beam, what minimum energy (in electron volts) of electrons would also produce a strong "reflection" at this angle?

43B-19 Electrons are accelerated through a potential difference V and then directed at a target of powdered crystals whose largest atomic-plane separation is 0.283 nm. Find the smallest value of V for which Bragg reflection occurs when the reflected beam is deviated through an angle of 130° with respect to the incident beam direction.

43.6 Wave Mechanics
43.7 Barrier Tunneling

43B-20 The space part of the wave function describing a free electron is $\psi(x) = A\sin(7 \times 10^9 x)$ in SI units. Find (a) the de Broglie wavelength of the electron, (b) the electron's speed, and (c) its kinetic energy in electron volts.

43B-21 A particle is confined to one-dimensional motion between two rigid walls separated by a distance D. The probability density function $P = |\psi|^2$ is given by Equation (43-29). Show that the distance Δx between minima is D/n.

43B-22 A particle of dust whose mass is 80 pg floats in air, trapped between two rigid walls 0.6 mm apart. It takes the dust particle 5 min to move from one wall to the other. Considering this situation quantum-mechanically as that of a particle trapped in a one-dimensional box, find (a) the quantum number n for this energy state. (b) Explain why it is not possible to experimentally determine the quantum number for this state. (c) Now assume that this dust particle is in its lowest possible ($n = 1$) energy state. Find the time (in years) it would take the particle to travel from one wall to the other wall.

43B-23 The *transmission coefficient* T gives the probability that a particle of mass m approaching the rectangular potential barrier of Figure 43-23 may "tunnel" through the barrier:

$$T = e^{-2kD} \quad \text{where} \quad k = \sqrt{\frac{8\pi^2 m(U-E)}{h^2}}$$

Consider a barrier with $U = 5$ eV and $D = 950$ pm. Suppose that an electron with energy $E = 4.5$ eV approaches the barrier. Classically, the electron could not pass through the barrier because $E < U$. However, quantum-mechanically there is a finite probability of tunneling. Calculate this probability.

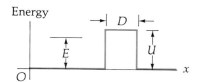

FIGURE 43-23
Problems 43B-23 through 43B-26.

43B-24 In the previous problem, calculate the probability that a 4.5-eV proton could tunnel through the barrier. Obtain a finite (though extremely small!) nonzero numerical answer.

43B-25 In Problem 43B-23, by how much would the width D of the potential barrier have to be increased so that the chance of an incident 4.5-eV electron tunneling through the barrier is one in a million?

43B-26 (a) In Problem 43B-23, calculate the de Broglie wavelength of the 4.5-eV electron as it approaches the potential barrier. (b) What fraction of this de Broglie wavelength is the barrier width of 950 pm? (c) Repeat (b) for a 4.5-eV proton.

43.8 The Uncertainty Principle

43B-27 A 9-g marble is rolling along a table at 2 m/s. (a) If its linear momentum is measured to an accuracy of 0.1%, what is the uncertainty in the simultaneous measurement of its position? (b) Repeat for an electron moving at the same speed. Comment upon the answers.

43B-28 An atom in an excited state 1.8 eV above the ground state remains in that excited state on the average 2×10^{-6} s before undergoing a transition to the ground state. Find (a) the frequency and (b) the wavelength of the emitted photon. (c) Find the approximate uncertainty in energy of the photon.

43B-29 A π^0 meson is an unstable particle that is produced in high-energy particle collisions. It has a mass–energy equivalent of about 135 MeV, and it exists for an average lifetime of only 8.7×10^{-17} s before decaying into two gamma rays. Using the uncertainty principle, estimate the fractional uncertainty $\Delta m/m$ in its mass determination.

Additional Problems

43C-30 A negative μ-meson (called a *muon*) has a charge of $-e$ and a mass of about $206.8 m_e$. Consider a hydrogenlike atom formed of a proton and a muon. (a) Assuming that the proton remains fixed, find the $n = 1$ Bohr orbit radius for this "atom." (b) What is the ground-state energy in electron volts? (c) Find the wavelength of the radiation emitted for the transition from the $n = 2$ state to the $n = 1$ state.

43C-31 An electron and a positron (same mass as an electron but with a positive electronic charge) can form a bound system known as *positronium*. The two particles revolve about their mutual center of mass, and the total angular momentum is quantized according to the Bohr condition. Derive general expressions for (a) the quantized radii r_n and (b) a numerical expression (in electron volts) for the energy states E_n. (c) Calculate the longest and shortest wavelengths of radiation emitted from positronium in transitions to the ground state.

43C-32 Consider a hypothetical atom having a neutron for a nucleus, with an electron held in orbit by the gravitational force between the neutron and the electron. Using an analysis similar to that used for the Bohr hydrogen atom, determine (a) the radii of the orbits, similar to Equation (43-7), and (b) the energy states, similar to Equation (43-10).

43C-33 As a photon is emitted from an atom, a small fraction of the energy associated with the transition appears as the recoil energy of the atom. Show that this fraction is approximately equal to $E/2mc^2$, where E is the energy of transition and m is the mass of the atom.

43C-34 A 50-kg satellite is in a circular orbit about the earth with a period of 2 h. (a) Applying the Bohr quantum condition on angular momentum, calculate the quantum number n for this orbit. (b) Find the radial distance between this orbit and the next "allowed" higher orbit. Could we experimentally detect this distance?

43C-35 Starting with Equation (43-9), derive the empirical relation for the Balmer series in hydrogen, Equation (43-1).

43C-36 A singly ionized helium atom (designated He II) has one electron and a nucleus of charge $+2e$. Apply the Bohr theory to find expressions for (a) the energies E_n and (b) the electron radii r_n for allowed states of this ionized atom. (c) Show that for every spectral line in the hydrogen spectrum, there is a line of identical wavelength in the ionized helium spectrum. What is the relationship between the corresponding n-values for these "matching" lines? (Note: these lines are identical in the original Bohr theory. Actually, they differ slightly because the Rydberg constant R has a small dependence on the nuclear mass.)

43C-37 An example of the correspondence principle is that the relativistic kinetic energy $K = mc^2[1/(1 - v^2/c^2)^{1/2} - 1]$ reduces to the classical value $K = mv^2/2$ for $v \ll c$. Prove this statement.

CHAPTER 44

Atomic Physics

If all this damned quantum jumping were really here to stay, I should be sorry I ever got involved with quantum theory.

EDWIN SCHRÖDINGER (in a heated discussion with Bohr regarding the Bohr postulates)

The great initial success of quantum theory cannot convert me to believe in that fundamental game of dice.

ALBERT EINSTEIN (in a letter to Max Born, November 7, 1944)

44.1 Introduction

As we saw in the last chapter, there is a fundamental difference between classical mechanics and quantum mechanics. Classical Newtonian mechanics describes the motion of an object under the influence of a force in terms of measurable parameters such as mass, position, velocity, and acceleration, giving us (supposedly) precise predictions for numerical values of these quantities at any instant. The results agree with our everyday experience. Quantum mechanics also describes relationships between measurable parameters, but it reveals a basic limit. Because of the uncertainty principle (whose roots lie in the fundamental wave–particle duality of matter and radiation), certain pairs of parameters cannot be measured simultaneously with unlimited accuracy. Consequently, quantum mechanics makes some of its predictions by giving a precise statement of the *probability* that a given parameter has a certain range of values about some average, rather than giving the exact value of the parameter as classical physics does.

Lest you think that quantum mechanics is not a very good substitute for the unlimited precision of classical mechanics, we point out that classical mechanics is merely an approximation of the more subtle and rich theory of quantum mechanics. The exactness—without limit—of classical mechanics is an illusion. That approach is valid for macroscopic conditions in which so many atoms are involved that the uncertainties in the average values are negligible. But for small-scale systems, quantum mechanics must be used. A particularly pleasing aspect of quantum theory is that it contains within itself the full Newtonian theory, which emerges automatically when quantum mechanics is applied to macroscopic systems. So quantum mechanics is the single best theory to date that describes most[1] of our wondrous universe. In the words of Herman Feshback and Victor F. Weisskopf[2]:

Quantum physics holds a unique position in intellectual history as the most successful framework ever developed for the understanding of natural phenomena.

In this chapter we apply quantum mechanics to the hydrogen atom and interpret the results. This example of the simplest two-particle system will dramatically illustrate the unique features of quantum mechanics.

The Bohr model of the hydrogen atom was a magnificent achievement. It predicted results that were in remarkable agreement with experimental data. Probably the greatest impact of Niels Bohr's discovery, however, was that the model raised more problems than it solved, thereby initiating a closer look into the nature of atomic structure. Among the unresolved problems were these:

(1) How, in clear contradiction to firmly established electromagnetic theory, could an electron orbit about a proton and not continually lose energy by radiation?
(2) Why, upon careful observation of the hydrogen spectrum, do we find many of the lines to be closely spaced combinations of two or more lines (*fine structure*)?
(3) How could the Bohr model account for the fact that some spectral lines are more intense than others?
(4) What is the justification for the quantization of orbits in the Bohr model?

As we pointed out in the previous chapter, in 1924 Louis de Broglie provided a rationale for quantization through the idea of matter waves. With the experimental verification of matter waves by C. J. Davisson and L. H. Germer in 1925, the stage was set for a new theory of atomic structure. The two main architects of the new theory were Erwin Schrödinger, who devised a wave-mechanical model, and Werner Heisenberg, who used mathematical matrices to represent transitions between initial and final energy states of the atom. Both theories were later found to be exactly equivalent. We discuss only the simpler wave-mechanical model.

Wave mechanics yields predictions that are in exact agreement with experimental data. But by accepting this purely mathematical model we are forced to reject the idea of electrons orbiting a nucleus in precisely defined trajectories. Instead, we can only say that the electron has a certain probability of being in *this* region of space, or in *that* region of space. It is gratifying, however, that the regions of highest probability correspond to the discrete orbits of the old Bohr atomic model.

Another success of wave mechanics is that quantization arises naturally when only *standing-wave solutions* to the wave equation are "allowed." **Allowed solutions** are those for which certain *boundary conditions* are met. As an illustration, in the previous chapter we discussed an electron moving in one-dimensional motion between rigid walls. For that case we require that $\psi = 0$ at the walls, automatically restricting solutions to standing-wave patterns between

---

[1] There are still enough unsolved puzzles in nuclear physics and fundamental particles to keep physicists challenged for a long time to come.
[2] *Physics Today*, Oct. 1988.

the walls. For the three-dimensional case of an electron in an atom, we require that the value of ψ for $\phi = 0°$ must equal ψ for $\phi = 360°$. (That is, one complete rotation brings us back to the original angular position.) Another boundary condition is that $\psi \to 0$ as $r \to \infty$. As discussed in the next section, such restrictions lead to *quantum numbers*, which designate the allowed solutions.

44.2 The Schrödinger Wave Equation

The wave-mechanical approach to the solution of the hydrogen atom is to consider the electron, influenced by the Coulomb potential U of the proton nucleus, as a de Broglie "matter" wave. As a wave, the electron must obey the wave equation. For a one-dimensional wave, the time-independent Schrödinger equation [Equation (43-27)] is

$$\frac{d^2\psi}{dx^2} + \left(\frac{2m(E-U)}{\hbar^2}\right)\psi = 0$$

which is often written as

SCHRÖDINGER WAVE EQUATION (one dimension)
$$\left[-\left(\frac{\hbar^2}{2m}\right)\frac{d^2}{dx^2} + U(x)\right]\psi = E\psi \qquad (44\text{-}1)$$

Since the electron wave is three-dimensional (analogous to the mechanical vibrational waves in a wiggly sphere of gelatin), the wave equation must be written in a three-dimensional form:

$$-\frac{\hbar^2}{2m}\left(\frac{\partial^2\psi}{\partial x^2} + \frac{\partial^2\psi}{\partial y^2} + \frac{\partial^2\psi}{\partial z^2}\right) + U(x,y,z)\psi = E\psi \qquad (44\text{-}2)$$

where the potential energy $U(x,y,z)$ is the Coulomb potential in Cartesian coordinates:

$$U(x,y,z) = -\left(\frac{1}{4\pi\varepsilon_0}\right)\frac{e^2}{(x^2+y^2+z^2)^{1/2}} \qquad (44\text{-}3)$$

Because of the spherical symmetry of $U = -ke^2/r$, however, it is advantageous to write the Schrödinger wave equation in *spherical* coordinates: r, θ, and ϕ, Figure 44-1. With these substitutions, the wave equation in spherical coordinates becomes

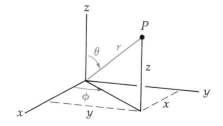

FIGURE 44-1
The point P may be specified by its *rectangular* coordinates (x,y,z) or by its *spherical* coordinates (r,θ,ϕ).

THE SCHRÖDINGER WAVE EQUATION (spherical coordinates)
$$-\frac{\hbar^2}{2m}\left[\frac{1}{r^2}\frac{\partial}{\partial r}\left(r^2\frac{\partial\psi}{\partial r}\right) + \frac{1}{r^2\sin\theta}\frac{\partial}{\partial\theta}\left(\sin\theta\frac{\partial\psi}{\partial\theta}\right) + \frac{1}{r^2\sin^2\theta}\frac{\partial^2\psi}{\partial\phi^2}\right] + U(r)\psi = E\psi \qquad (44\text{-}4)$$

where the potential energy function is simply

$$U(r) = -\left(\frac{1}{4\pi\varepsilon_0}\right)\frac{e^2}{r} \qquad (44\text{-}5)$$

Don't be alarmed at this elaborate equation—we won't be working with it directly. The complete solution of Equation (44-4) is complicated, so we will present only some important aspects of its solution that give physical insight into the nature of quantum mechanics. They are the following:

(1) The solution of the wave equation $\psi(r,\theta,\phi)$ is expressed as the product of three functions: a <u>radial</u> part $R(r)$, which is a function only of r; a <u>polar</u> part $\Theta(\theta)$, which is a function only of θ; and an <u>azimuthal</u>[3] part $\Phi(\phi)$, which is a function only of ϕ. Thus:

$$\psi(r,\theta,\phi) = R(r)\Theta(\theta)\Phi(\phi) \qquad (44\text{-}6)$$

*The central thread of the story is that for each of the spatial variables r, θ, and ϕ a **quantum number** (which designates the "allowed" solutions) arises naturally when we restrict solutions to only those that are single-valued and approach zero as r $\to \infty$. These are "standing-wave" solutions representing different quantum states of the atom.*

(2) The **radial function** $R(r)$ that satisfies the boundary condition $[\psi \to 0$ as $r \to \infty]$ exists only for integral values of a quantum number $n = 1, 2, 3, \ldots$. The number n is called the <u>principal quantum number</u> because the energy of the electron depends principally upon n in the following way:

$$E_n = -\left(\frac{me^4}{8\varepsilon_0^2 h^2}\right)\frac{1}{n^2} \qquad (44\text{-}7)$$

Note that this is the same energy function that was obtained with the Bohr model, as should be expected. After all, the Bohr model was very successful in providing energy levels.

(3) The solution of the **polar function** $\Theta(\theta)$, which satisfies the boundary conditions, gives rise to the <u>orbital</u> quantum number ℓ:

$$\ell = 0, 1, 2, \ldots, (n-1)$$

Thus, for $n = 1$, ℓ may only be 0; for $n = 3$, $\ell = 0, 1,$ or 2; and so forth. It is called the orbital quantum number because it determines the *orbital angular momentum* **L** of the electron about the proton. The discrete values of ℓ quantize the orbital angular momentum to only these values:

$$L = \hbar\sqrt{\ell(\ell+1)} \qquad (44\text{-}8)$$

(Note that this result does *not* agree with the (incorrect) Bohr quantization of angular momentum: $L = n\hbar$.)

(4) The solution of the **azimuthal function** $\Phi(\phi)$ gives rise to a third quantum number m_ℓ called the <u>magnetic</u> quantum number m_ℓ:

$$m_\ell = 0, \pm 1, \pm 2, \pm 3, \ldots, \pm \ell$$

The value of m_ℓ determines the z component of the angular momentum **L** according to the relation (see Figure 44-2)

$$L_z = m_\ell \hbar \qquad (m_\ell = 0, \pm 1, \pm 2, \ldots, \pm \ell) \qquad (44\text{-}9)$$

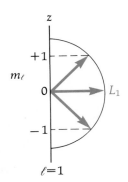
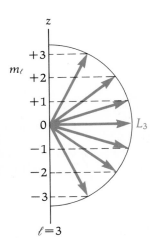

FIGURE 44-2
The allowed values of $L_z = m_\ell \hbar$ for two different values of ℓ (drawn to different scales). The values of m_ℓ are the integral numbers along the z axis, covering all possibilities from $+\ell$ to $-\ell$. The magnitude of $L = \sqrt{\ell(\ell+1)}\hbar$.

---

[3] The word *azimuth* comes from astronomy, where it designates the angular distance around the horizon, measured eastward from the north point.

The orbital angular momentum **L** of the electron is associated with a *magnetic dipole moment* $\boldsymbol{\mu}_\ell$ (see Problem 30C-41) given by

$$\boldsymbol{\mu}_\ell = -\left(\frac{e}{2m}\right)\mathbf{L} \qquad (44\text{-}10)$$

Therefore, the z component of the magnetic dipole moment $(\mu_\ell)_z$ is also quantized:

$$(\mu_\ell)_z = -m_\ell\left(\frac{e\hbar}{2m}\right) \qquad (44\text{-}11)$$

where $e\hbar/2m$ is called the **Bohr magneton**:

BOHR MAGNETON $\qquad \left(\dfrac{e\hbar}{2m}\right) = 9.27 \times 10^{-24} \text{ A}\cdot\text{m}^2 \qquad (44\text{-}12)$

Since $(\mu_\ell)_z$ is the measurable quantity, it is physically more significant[4] than μ_ℓ or **L**.

Atomic states that have the same values of n and ℓ, but different wave functions, represent *different directions* for **L**. To measure these differences experimentally, we place the atom in a weak magnetic field that is aligned along the $+z$ direction (to identify a specific direction in space). We infer the discrete orientations of **L** in Figure 44-2 by measuring the z component of μ_ℓ.

A helpful way to picture this situation is with a **vector model**. The angular momentum **L** and the magnetic moment $\boldsymbol{\mu}_\ell$ are rigidly connected together. The magnetic field exerts a torque on $\boldsymbol{\mu}_\ell$. As a consequence of the gyroscopic behavior of angular momentum, **L** and $\boldsymbol{\mu}_\ell$ precess together about the z axis. But because of Heisenberg's uncertainty principle, $(\Delta L_z)(\Delta \phi) \geq \sim \hbar$, we can never measure *where* in the precessional motion these vectors are at any instant. Our mental image of this vector model must show the precessional motion only as an average blurred cone, Figure 44-3. The *only* information that we can experimentally obtain is the *magnitude* of these vectors and their *projections along the z axis*. Nothing else!

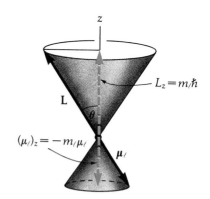

FIGURE 44-3
A *vector model* for visualizing the quantized spatial orientations of vector quantities in quantum mechanics. Here we show one of the possible orientations for **L** and its projection on the z axis. The vectors **L** and $\boldsymbol{\mu}_\ell$ precess together about the z axis.

EXAMPLE 44-1

Find the minimum angle θ between **L** and the z direction for $\ell = 3$.

SOLUTION

In Figure 44-3, $\cos\theta = (L_z)_{\max}/L = \ell\hbar/\sqrt{\ell(\ell+1)}\hbar = 3/\sqrt{12}$.

$$\theta = \boxed{30°}$$

---

[4] Your first introduction to quantum mechanics may seem confusing unless you keep certain distinctions in mind. Classically, we picture particles moving in orbits, leading to the *mechanical* concept of angular momentum **L**. Because the particles are charged, these motions also produce magnetic moments $\boldsymbol{\mu}$, an *electromagnetic* concept. The two concepts are inevitably linked together, and they are quantized. Sorting out the consequences is a major achievement of quantum mechanics. Avoiding confusion between these *mechanical* and *electromagnetic* concepts will help smooth the way to your understanding the new world of quantum mechanics.

The solution to the wave equation that we have described fails to account for the so-called *fine structure* of the spectral lines. A high-resolution spectrometer reveals that some of the lines are actually closely spaced combinations of two or more lines. As we will see in the next section, this fine structure was explained in 1925 by S. A. Goudsmit and G. E. Uhlenbeck, graduate students at Leiden University in the Netherlands, who proposed that the electron itself possesses an angular momentum, or "spin," and a related magnetic moment. Both of these characteristics are inherent properties of the electron, just like the electronic charge and mass. A simple way of visualizing the origin of these properties is to imagine that the electron is a charged sphere, spinning on its axis.[5] Thus the total angular momentum of an electron in the hydrogen atom is made up of two parts: its orbital angular momentum **L** and its spin angular momentum **S**. Aware that fine-structure lines often come in pairs, Goudsmit and Uhlenbeck proposed that the electron spin could have only two possible orientations with respect to an external magnetic field: *parallel* or *antiparallel*. Consequently, a fourth quantum number enters the picture.

(5) **Electron spin** gives rise to the *spin quantum number* m_s and is related to the z component of the *spin angular momentum* S_z:

$$S = \hbar\sqrt{s(s+1)} \quad \text{(where } s = \tfrac{1}{2}\text{)} \quad (44\text{-}13)$$

$$S_z = m_s \hbar \quad \text{(where } m_s = \pm\tfrac{1}{2}\text{)} \quad (44\text{-}14)$$

If $S_z = +\tfrac{1}{2}\hbar$, the electron's spin is said to be "up"; if $S_z = -\tfrac{1}{2}\hbar$, its spin is "down." Analogous to the case of orbital motion, there is a z component of the *spin magnetic moment* $(\mu_s)_z$ associated with s_z:

$$(\mu_s)_z = -m_s\left(\frac{e\hbar}{m}\right) \quad \text{(where } m_s = \pm\tfrac{1}{2}\text{)} \quad (44\text{-}15)$$

Referring to Equation (44-11), we note that the electron spin angular momentum seems to be twice as effective in producing a magnetic dipole moment as is the orbital angular momentum.

Here is a summary of the four quantum numbers that designate the allowed states of the hydrogen atom. The first three arise naturally in the quantum-mechanical description of an electron confined in a particular region of space by the Coulomb attraction of a proton. The fourth is due to the inherent properties of spin of the electron.

QUANTUM NUMBERS
$$\begin{cases} n = 1, 2, 3, \ldots & \text{Principal quantum number} \\ \ell = 0, 1, 2, \ldots (n-1) & \text{Orbital quantum number} \\ m_\ell = 0, \pm 1, \pm 2, \ldots \pm \ell & \text{Magnetic quantum number} \\ m_s = \pm\tfrac{1}{2} & \text{Spin quantum number} \end{cases} \quad (44\text{-}16)$$

---

[5] Though it appeals strongly to our imagination, this spinning-sphere model must not be taken literally. Even if somehow we could put a mark on an electron and experimentally attempt to follow its motion as the electron spins, the uncertainty principle forbids such a procedure. *Quantum mechanics does not say that an electron is a spinning sphere!*

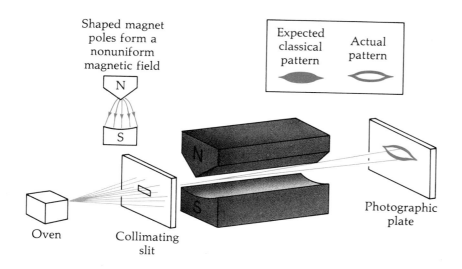

FIGURE 44-4
The Stern–Gerlach experiment (1922) demonstrates the spatial orientation of spin magnetic moments in a magnetic field. A beam of neutral silver atoms is sent through a nonuniform magnetic field. The magnetic moment of the silver atom is due solely to the single valence electron, which has zero *orbital* magnetic moment ($\ell = 0$); only the *spin* magnetic moment for that electron is present. Classically, a single smeared pattern is expected since the magnetic moments of the atoms in the beam should be able to have any orientation as they pass through the field. Instead, the beam splits into two distinct lines, verifying the spatial orientation of spin magnetic moments in a magnetic field. The spin magnetic moments align either *parallel* or *antiparallel* to the field direction, and the nonuniform field then pushes them either up or down to form the double-line pattern.

44.3 Electron Spin and Fine Structure

In the 1920s, the development of atomic theory provided a scenario that would rival that of a good mystery story. The discrepancies between theory and experimental evidence began to grow in the early part of the decade. Two notable problems were that spectral lines had a *fine structure* and that neutral atoms passing through a nonuniform magnetic field were deflected either in one direction or in the opposite direction, Figure 44-4. These phenomena could not be explained by the existing theory. In 1925 Goudsmit and Uhlenbeck made two proposals that did lead to correct predictions. They suggested (1) that an electron behaves as though it is a spinning ball of charge with quantized angular momentum and (2) that in the presence of a magnetic field the magnetic dipole moment can assume only two orientations: *parallel* or *antiparallel* to the field. But such a literal picture of a spinning electron did not fit into the current framework of wave mechanics and thus was not a completely satisfactory explanation. A welcome solution to the spinning-electron mystery came in 1928, when P. A. M. Dirac introduced relativity to the wave-mechanical treatment of the electron. The concepts that previously had to be accepted only because they led to the right answers now emerged as the natural consequences of applying relativity to wave mechanics. Indeed, *the Dirac theory, using only the electron charge and mass as given data, predicts all the other intrinsic properties of electrons, including spin and the existence of anti-electrons (positrons)!* It is justifiably considered one of the major triumphs of theoretical physics. Dirac received the 1933 Nobel Prize (with Schrödinger).

44.4 Spin–Orbit Coupling

As mentioned in the last section, the *fine structure* of spectral lines is due to the interaction of two magnetic dipole moments: the one associated with electron *spin* and the other associated with the *orbital motion* of the electron. This interaction, or "coupling," is called **spin–orbit** or **L–S coupling**. Because of the abstractions of the purely mathematical description of the atom, we often think of the visual picture of a spinning electron orbiting a nucleus.[6] Such a view, although incorrect, does help visualize spin–orbit coupling. Thinking classically, we note that in the electron's frame of reference the proton circulates around the electron, Figure 44-5. This motion is equivalent to a current

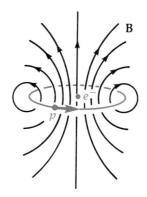

FIGURE 44-5
In the electron's frame of reference, the proton circulates around the electron. This motion is equivalent to a current loop, producing a magnetic field **B** at the location of the electron.

---
[6] The earth–sun system is an analogy: the orbital motion of the earth about the sun produces angular momentum, while the earth's rotation about its own axis adds additional angular momentum.

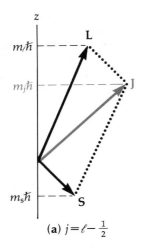

(a) $j = \ell - \tfrac{1}{2}$

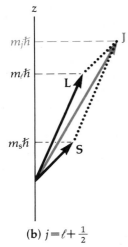

(b) $j = \ell + \tfrac{1}{2}$

FIGURE 44-6
In L–S coupling, the orbital and spin angular momenta may add in two ways to form the total angular momentum **J**. Because of the unusual quantized values for their magnitudes, $L = \sqrt{\ell(\ell+1)}$, $S = \sqrt{s(s+1)}$, and $J = \sqrt{j(j+1)}$, these vectors can add only at certain discrete angles.

loop, producing a magnetic field **B** at the location of the electron. The magnetic moment $\boldsymbol{\mu}_s$ of the electron orients itself either parallel or antiparallel to **B**, with a corresponding potential energy $U = -\boldsymbol{\mu}_s \cdot \mathbf{B}$ for this interaction. (The magnetic field **B** is calculated in the electron's frame of reference.)

In quantum mechanics we must combine quantized vectors in the manner discussed previously, so we extend those ideas to spin–orbit coupling. Since the magnetic dipole moment is associated with angular momentum, we may couple dipole moments by combining angular momenta. The angular momentum of orbital motion **L** [defined by Equation (44-8)] and the angular momentum of spin **S** [defined by Equation (44-13)] are added *vectorially* to produce the total angular momentum **J**:

SPIN–ORBIT
COUNPLING $\qquad\qquad$ **J = L + S** $\qquad\qquad$ (44-17)
(or L–S)

The magnitude of **J** is quantized in a manner similar to **L** and **S** by the relation

$$J = \hbar\sqrt{j(j+1)} \qquad (44\text{-}18)$$

The vector addition of **L** and **S** to form **J** is shown in Figure 44-6. The symbol j is the <u>inner</u> *quantum number* that tells how ℓ and s combine. Since $s = \tfrac{1}{2}$, we have only these values:

$$j = \ell \pm \tfrac{1}{2} \qquad (44\text{-}19)$$

The projection of **J** on the z axis is quantized in the same way that **L** has a quantized projection $m_\ell \hbar$. Thus, the z component of **J** is

$$J_z = m_j \hbar \qquad (44\text{-}20)$$

where m_j may have the values

$$m_j = j, (j-1), (j-2), \ldots, -(j-2), -(j-1), -j \qquad (44\text{-}21)$$

Therefore, there are $2j + 1$ values of m_j.

The way that the angular momentum vectors **L** and **S** add vectorially determines how the corresponding dipole moments add. The vector addition is shown in Figure 44-6. The magnitudes of all three vectors **L**, **S**, and **J** are quantized, so the angles between these vectors can have only certain discrete values. The energy difference of the doublet levels is the difference between the electron being in the $(\ell + \tfrac{1}{2})$ "up" state and the $(\ell - \tfrac{1}{2})$ "down" state.

At this point it may appear that, by introducing spin–orbit coupling, we have added to the list of quantum numbers required to define the state of the electron. As we will see later, the state of the electron can be described *either* by the quantum numbers n, ℓ, m_ℓ, and m_s or by the quantum numbers n, ℓ, j, and m_j. *No more than four quantum numbers are needed.*

ALTERNATE
QUANTUM
NUMBERS
(for L–S
coupling)
$\begin{cases} n = 1, 2, 3, \ldots & \text{Principal quantum number} \\ \ell = 0, 1, 2, \ldots, (n-1) & \text{Orbital quantum number} \\ j = \ell \pm \tfrac{1}{2} & \text{Inner quantum number} \\ m_j = j, (j-1), \ldots, -(j-1), -j & (z\text{ component of } j) \end{cases}$ (44-22)

44.5 Quantum States of the Hydrogen Atom

We now show how to describe the various possible energy states of the electron in the hydrogen atom. We have pointed out that the state of the electron is specified by four quantum numbers. The following example illustrates the procedure for determining all of the possible states.

EXAMPLE 44-2

List the possible quantum energy states that an electron may have for the $n = 1$ and $n = 2$ states. Derive the list from both (a) the system of quantum numbers n, ℓ, m_ℓ, and m_s and (b) the quantum numbers n, ℓ, j, and m_j.

SOLUTION

(a) Quantum numbers can have only the following values:

$$\ell = 0, 1, 2, \ldots, (n-1)$$
$$m_\ell = 0, \pm 1, \pm 2, \ldots, \pm \ell$$
$$m_s = \pm \tfrac{1}{2}$$

Applying these rules, we form the table of unique states, Table 44-1. There are a total of 2 states for $n = 1$ and a total of 8 states for $n = 2$.

(b) Using the quantum numbers n, ℓ, j, and m_j, again we can have only certain values, which are described by

$$\ell = 0, 1, 2, \ldots, (n-1)$$
$$j = \ell \pm \tfrac{1}{2} \quad \text{(where } j > 0\text{)}$$
$$m_j = \pm j, \pm(j-1), \pm(j-2), \pm \ldots$$

See Table 44-2. Again, we have a total of 2 states for $n = 1$ and 8 states for $n = 2$.

TABLE 44-1 States Based on the Quantum Numbers n, ℓ, m_ℓ, and m_s.

n	ℓ	m_ℓ	m_s
1	0	0	$+\tfrac{1}{2}$
1	0	0	$-\tfrac{1}{2}$
2	0	0	$+\tfrac{1}{2}$
2	0	0	$-\tfrac{1}{2}$
2	1	0	$+\tfrac{1}{2}$
2	1	0	$-\tfrac{1}{2}$
2	1	$+1$	$+\tfrac{1}{2}$
2	1	$+1$	$-\tfrac{1}{2}$
2	1	-1	$+\tfrac{1}{2}$
2	1	-1	$-\tfrac{1}{2}$

Spectroscopic Notation

Rather than list the quantum numbers for a particular quantum state, we often use **spectroscopic notation** to simplify the description of a state. An example of this notation is $3d_{5/2}$. The number preceding the letter is the principal quantum number n. The letter corresponds to the orbital quantum number ℓ according to the following scheme:

ℓ Value	0	1	2	3	4	5
Letter	s	p	d	f	g	h

The letters s, p, d, and f were originally derived from the visual appearance of spectral lines, in which s implies *sharp*, p implies *principal*, d implies *diffuse*, and f implies *fundamental*. The subscript is the inner quantum number j. The following example illustrates the use of spectroscopic notation.

TABLE 44-2 States Based on the Quantum Numbers n, ℓ, j, and m_j.

n	ℓ	j	m_j
1	0	$\tfrac{1}{2}$	$+\tfrac{1}{2}$
1	0	$\tfrac{1}{2}$	$-\tfrac{1}{2}$
2	0	$\tfrac{1}{2}$	$+\tfrac{1}{2}$
2	0	$\tfrac{1}{2}$	$-\tfrac{1}{2}$
2	1	$\tfrac{3}{2}$	$+\tfrac{3}{2}$
2	1	$\tfrac{3}{2}$	$-\tfrac{3}{2}$
2	1	$\tfrac{3}{2}$	$+\tfrac{1}{2}$
2	1	$\tfrac{3}{2}$	$-\tfrac{1}{2}$
2	1	$\tfrac{1}{2}$	$+\tfrac{1}{2}$
2	1	$\tfrac{1}{2}$	$-\tfrac{1}{2}$

EXAMPLE 44-3

Among the following electron states, some are not allowable. Identify those states and tell why they are incorrect.

(a) $1p_{3/2}$ (b) $1s_{1/2}$ (c) $2p_{5/2}$ (d) $4d_{3/2}$ (e) $5f_{5/2}$ (f) $6f_{3/2}$

SOLUTION

Only (b), (d), and (e) are possible because $j = \ell \pm \frac{1}{2}$ and $\ell < n - 1$. (a) is incorrect because $\ell > n - 1$, (c) is incorrect because $j > \ell + \frac{1}{2}$, and (f) is incorrect because $j < \ell - \frac{1}{2}$.

Shell Notation

Although not used in spectroscopic notation, the value of the principal quantum number n is sometimes indicated by a capital letter according to the following notation used in x-rays:

n Value	1	2	3	4	5	6	7
Letter	K	L	M	N	O	P	Q

Thus, a $3p$ electron is said to be in the M shell and the p subshell.

44.6 Energy Level Diagram for Hydrogen

The Bohr theory of the hydrogen atom presented in Chapter 43 produced a single quantum number n, which identified the energy state. Consequently, a simple energy-level diagram could be drawn (Figure 43-10). The wave-mechanical model with spin–orbit coupling adds a fine structure to the energy levels, making the diagram considerably more complicated. A portion of such a diagram is shown in Figure 44-7. The spin–orbit energy dependence is greatly exaggerated in the figure. Actually, the $3p_{3/2}$ level is above that of the $3p_{1/2}$ level by only about 1/1000 of the energy difference between the $3p_{1/2}$ and $3s_{1/2}$ levels.

Transitions do not occur between all possible higher energy states and lower energy states; only "allowed" transitions that obey certain **selection rules** normally take place. If the atom gets rid of the energy difference by emitting a photon, the photon carries off one unit of angular momentum.[7]

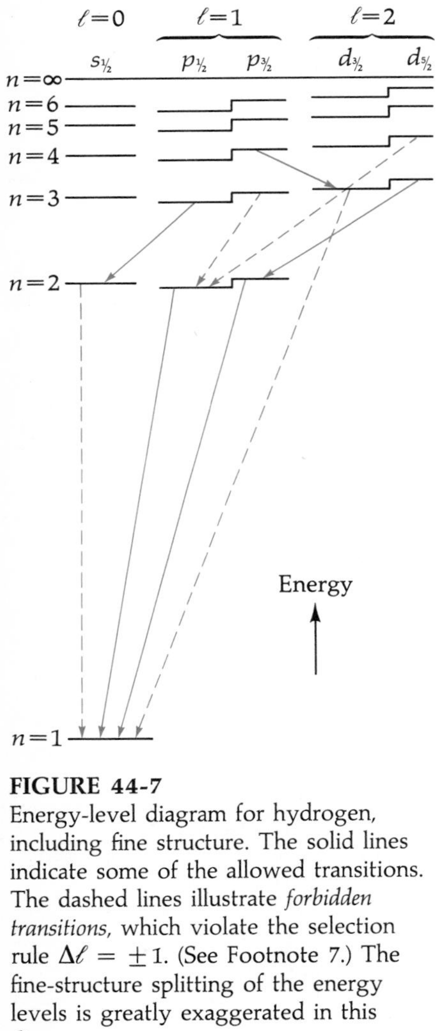

FIGURE 44-7
Energy-level diagram for hydrogen, including fine structure. The solid lines indicate some of the allowed transitions. The dashed lines illustrate *forbidden transitions*, which violate the selection rule $\Delta\ell = \pm 1$. (See Footnote 7.) The fine-structure splitting of the energy levels is greatly exaggerated in this diagram.

---

[7] Each photon has an angular momentum of one unit of $\hbar$. (The classical analogue is a circularly polarized electromagnetic wave.) The requirement of conserving angular momentum is expressed by the *selection rules for "allowed" transitions*:

SELECTION RULES $\Delta\ell = \pm 1$
("allowed" transitions) $\Delta m_\ell = 0, \pm 1$

Though the selection rules forbid them, "forbidden" transitions do occur rarely because of certain effects not discussed here. In any case, the conservation of angular momentum is never violated.

44.7 The Hydrogen Atom Wave Functions

The solutions to the wave equation, Equation (44-4), always have a constant multiplier that is not initially determined. For example, in the lowest (1s) state of hydrogen, the solution is of the form

$$\psi = Ae^{-(r/a)}$$

where A is an arbitrary constant. As will be shown, the symbol a is the **Bohr radius**, the radius of the ground-state orbit in the Bohr model:

BOHR RADIUS $$a \equiv \frac{\varepsilon_0 h^2}{\pi m e^2} = 0.0529 \text{ nm} \qquad (44\text{-}23)$$

Choosing a suitable value for the constant A is called **normalization**, discussed first in Section 43-6. (Here, the hydrogen wave functions are three-dimensional, while in Section 43-6 we dealt with a one-dimensional situation.) The physical significance of ψ is that it provides information about where the electron is likely to be relative to the nucleus (in contrast to the Bohr theory, which states that, for example, in the ground state the electron is *precisely* at a radial distance of 0.0529 nm). The **probability density function** P, as before, is defined to be

$$P = |\psi|^2 = \psi\psi^* \qquad (44\text{-}24)$$

where ψ may be complex.[8]

For the ground state (1s) of hydrogen, the solution of the wave equation is $\psi = Ae^{-(r/a)}$, so

$$P = |\psi|^2 = \psi\psi^* = A^2 e^{-(2r/a)} \qquad (44\text{-}25)$$

The **probability** $\mathscr{P}$ of finding the electron within the volume dV is

PROBABILITY $\mathscr{P}$ OF FINDING THE PARTICLE DESCRIBED BY ψ WITHIN THE VOLUME $\int dV$
$$\mathscr{P} = \int |\psi|^2 \, dV \qquad (44\text{-}26)$$

We now recognize that the probability $\mathscr{P}$ of finding the electron *somewhere* between $r = 0$ and $r = \infty$ is 1. Since the wave function for the ground state is symmetrical about the nucleus, the probability does not depend upon the coordinates θ or ϕ. So the volume differential dV is chosen as a spherical shell with area $4\pi r^2$ and thickness dr, giving

$$\mathscr{P} = \int_0^\infty A^2 e^{-(2r/a)} 4\pi r^2 \, dr = 1$$

Evaluating the integral yields $\quad 4\pi A^2 (a^3/4) = 1$

Solving for A, we get $\quad A = (\pi a^3)^{-1/2}$

---

[8] Complex numbers involve $i = \sqrt{-1}$. The *complex conjugate* of a number, denoted by (*), replaces i with $-i$. Thus if $\psi = Ae^{i\phi}$, then $\psi^* = Ae^{-i\phi}$. The product $\psi\psi^* = A^2 e^{i\phi} e^{-i\phi} = A^2 e^0 = A^2$, always a real number.

The *normalized wave function* for the ground state (1s) is thus

$$\psi = (\pi a^3)^{-1/2} e^{-(r/a)} \tag{44-27}$$

The normalized wave functions for the lowest two states of hydrogen are given in Table 44-3. The value of the constant a is the radius of the Bohr orbit for hydrogen in its ground state. Note that in the 1s and 2s states the wave functions do not depend upon either θ or ϕ, which indicates that in these states the wave functions are spherically symmetric about the nucleus.

TABLE 44-3 Normalized Wave Functions* of the Hydrogen Atom

	n	ℓ	m_ℓ	ψ
K Shell	1	0	0	$(\pi a^3)^{-1/2} e^{-(r/a)}$
L Shell	2	0	0	$(32\pi a^3)^{-1/2}(2 - r/a)e^{-(r/2a)}$
	2	1	0	$(32\pi a^5)^{-1/2} r\, e^{-(r/2a)} \cos\theta$
	2	1	$+1$	$(64\pi a^5)^{-1/2} r\, e^{-(r/2a - i\phi)} \sin\theta$
	2	1	-1	$(64\pi a^5)^{-1/2} r\, e^{-(r/2a + i\phi)} \sin\theta$

* Note: $a = \varepsilon_0 h^2 / \pi m e^2 = 0.0529$ nm, the Bohr radius, and $i = \sqrt{-1}$.

Where Is the Electron?

The next question that we address is this: For the hydrogen atom in its ground state, what is the probability of finding the electron at a distance r from the nucleus within a small incremental radial distance Δr? Since the wave function for the ground state is spherically symmetric, we use the volume element $dV = 4\pi r^2\, dr$ in Equation (44-26) with the ground-state wave function from Table 44-3:

$$\mathcal{P} = \int_r^{r+\Delta r} \left(\frac{1}{\pi a^3}\right) e^{-(2r/a)} 4\pi r^2\, dr.$$

We need not perform the integration if we assume that the value of r is essentially constant over the incremental distance Δr. The value of $\mathcal{P}$ then becomes

$$\mathcal{P} = \left(\frac{4r^2}{a^3}\right) e^{-(2r/a)} \Delta r \tag{44-28}$$

In this case, the *radial probability density*, $P(r)$, is defined as

$$P(r) = \left(\frac{4r^2}{a^3}\right) e^{-(2r/a)} \tag{44-29}$$

Then,

$$\mathcal{P} = \int P(r)\, dr \tag{44-30}$$

It is important to make the distinction between the *probability density function*, P, and the *radial probability density function*, $P(r)$. $P\Delta V$ is the probability of finding the electron within a small *volume* element ΔV, whereas $P(r)\Delta r$ is the probability of finding the electron within a small *radial* distance Δr.

EXAMPLE 44-4

For the ground state (1s) of the hydrogen atom, determine the distance r (from the proton) near which the electron is most likely to be found.

SOLUTION

The electron is most likely to be found near the distance corresponding to the maximum value of the radial probability density function, that is, near the value of r for which

$$\frac{dP(r)}{dr} = 0$$

$$\frac{d}{dr}\left[\left(\frac{4r^2}{a^3}\right)e^{-(2r/a)}\right] = 0$$

Eliminating constants gives
$$\frac{d}{dr}[r^2 e^{-(2r/a)}] = 0$$

$$2re^{-(2r/a)} - \left(\frac{2r^2}{a}\right)e^{-(2r/a)} = 0$$

which yields
$$\boxed{r = a}$$

In the ground state, the most probable distance from the nucleus is the Bohr radius for $n = 1$. Note that the result of this example is *not* in agreement with the Bohr model. The Bohr model defines a precise orbital radius, while wave mechanics describes only the likelihood of finding the electron within various radial increments from the center. The next example emphasizes this point.

EXAMPLE 44-5

For the ground state of hydrogen, what is the probability of finding the electron closer to the nucleus than the Bohr radius corresponding to $n = 1$?

SOLUTION

The probability, $\mathcal{P}$, of finding the electron within the Bohr radius is given by Equation (44-30):

$$\mathcal{P} = \int_0^a P(r)\,dr$$

Substituting $P(r)$ given by Equation (44-29), we obtain

$$\mathcal{P} = \left(\frac{4}{a^3}\right)\int_0^a r^2 e^{-(2r/a)}\,dr = 1 - 5e^{-2} = \boxed{0.323}$$

The electron is likely to be within the Bohr radius about one-third of the time. The Bohr model indicates *none* of the time.

We can better understand the results of the last two examples by examining Figure 44-8, which is a graph of the radial probability density function,

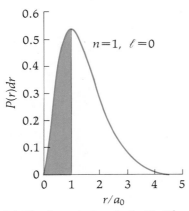

(a) The 1s state ($n=1$, $\ell=0$). The shaded portion indicates that there is about a 32% probability of the electron being inside the classical Bohr radius a_0 (the peak of the curve) and a 68% probability of its being farther from the nucleus.

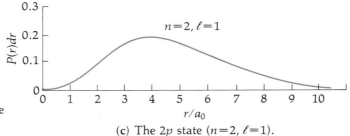

(b) The 2s state ($n=2$, $\ell=0$).

(c) The 2p state ($n=2$, $\ell=1$).

FIGURE 44-8
The radial probability density function $P(r)$ for the three lowest states of hydrogen.

$P(r)$, for the 1s state of hydrogen. As shown in Example 44-4, the maximum of the curve in Figure 44-8a occurs at $r = a_0 = 0.0529$ nm (the Bohr $n = 1$ radius). The shaded portion is 32.3% of the total area under the curve, indicating that during this fraction of its time, the electron is closer to the nucleus than the Bohr radius.

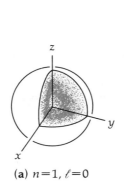

(a) $n=1$, $\ell=0$

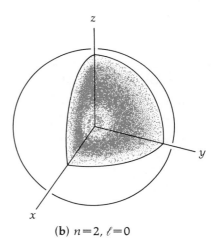

(b) $n=2$, $\ell=0$

FIGURE 44-9
One way of representing the probability density for the 1s, 2s, and 2p states of the hydrogen atom. (We have drawn rather artificial boundaries to the distributions; the probability of finding an electron outside the boundary of a cloud is less than about 10%.) In each case, the nucleus is at the coordinate origin. The greater the cloud density, the greater the probability of finding the electron in that region.

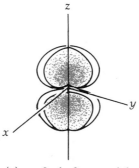

(c) $n=2$, $\ell=0$, $m_\ell = \pm 1$

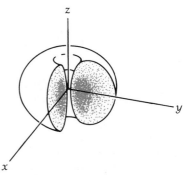

(d) $n=2$, $\ell=1$, $m_\ell = 0$

While it is relatively easy to visualize the definite orbitlike states of the Bohr model of the hydrogen atom, the visualization of the wave-mechanical model requires not only a three-dimensional perspective but also a way of showing the most probable locations of the electron. One way of picturing this is shown in Figure 44-9. The figure shows cross-sectional views of probability "clouds." The greater the density of the cloud, the greater the probability of finding the electron there. The rather artificial boundaries of the clouds shown in the figure are such that the probability of finding the electron outside the boundary is less than about 10%. Do not confuse this probability density representation with the *radial* probability density $P(r)$. Consider, for example, Figure 44-9a. Even though the cloud is most dense near its center, the electron will spend little time there because the volume for a given Δr at small r is much less than it is for a large r. The combination of both a high-volume probability density and high incremental volume makes it most probable for the electron to be found at 0.0529 nm from the center (the Bohr $n = 1$ orbital distance). Another way of representing the volume probability density is shown in Figure 44-10. Here, the cloud density is represented by the height of the bumps on the cross-sectional slice through the nucleus.

44.8 The Pauli Exclusion Principle and the Periodic Table of the Elements

In this section, we show how electrons in multi-electron atoms are distributed among the possible energy states. The lowest possible energy state of the electrons within the atoms account for such things as the chemical and electrical properties of certain elements. Thus, the periodic table of chemical behavior established by Mendeléef in 1870 can be explained on a physical basis.

Suppose that we build atoms by adding electrons (and, of course, adding corresponding positive charges to the nucleus to preserve the overall electric neutrality of the atom). We begin with hydrogen in its ground state, $1s^1$, where the superscript indicates the number of electrons in the $1s$ state. As we add electrons, they seek the lowest possible energy state. Thus, when helium is formed, both of its electrons will be in the $n = 1$, $\ell = 0$ state, and will have opposite spins. The ground state of helium is written as $1s^2$. Proceeding to lithium, no more $n = 1$ states exist, so the third electron goes to the $n = 2$ shell. The ground state of lithium is $1s^2 2s^1$. As we add more electrons, we find ground-state configurations as shown in Table 44-4.

Apparently electrons do not always seek the lowest energy state. If they did, they would all be in the $1s$ state. An inspection of the ground-state configurations shown in Table 44-2 reveals that only two electrons can occupy the $n = 1$ state and that only eight electrons can occupy the $n = 2$ state. In Example 44-1 we discovered that there are only two possible states for the electron in the $n = 1$ state (the K shell) and only eight possible states for $n = 2$ (the L shell). The connection between possible energy states and ground-state configurations was stated by Wolfgang Pauli in 1925 as follows:

| THE PAULI EXCLUSION PRINCIPLE | In the same atom, no two electrons can have the same set of values for the four quantum numbers $[n, \ell, m_\ell, m_s]$ or $[n, \ell, j, m_j]$. |

This rule was later *derived* as a consequence of a more sophisticated version of quantum theory. At the time, it was of great help in our understanding of the characteristics of atoms and the regularities of the periodic table. The

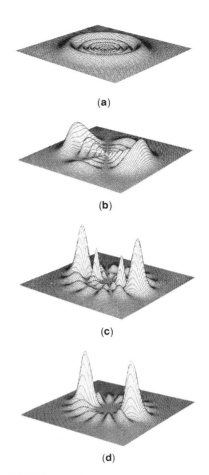

FIGURE 44-10
Representations of the probability-density distributions for highly excited states ($n = 8$) of the hydrogen atom that have different values of angular momentum. The nodal lines are either concentric circles or straight lines passing through the nucleus. The true three-dimensional distributions may be visualized by imagining that the graph is rotated about a horizontal line passing through the nucleus, forming nodal surfaces that are spherical shells or cones. In these excited states, the hydrogen atom is much larger than it is in its lowest energy state. The distance from the nucleus to the edge of these graphs is 380 times the Bohr radius for $n = 1$.

TABLE 44-4 Ground-State Configuration of the Elements

Element	Number of Electrons	Ground State*	n Value
H	1	$1s^1$	K shell
He	2	$1s^2$	($n = 1$)
Li	3	$1s^2 2s^1$	
Be	4	$1s^2 2s^2$	
B	5	$1s^2 2s^2 2p^1$	
C	6	$1s^2 2s^2 2p^2$	L shell
N	7	$1s^2 2s^2 2p^3$	($n = 2$)
O	8	$1s^2 2s^2 2p^4$	
F	9	$1s^2 2s^2 2p^5$	
Ne	10	$1s^2 2s^2 2p^6$	
Na	11	$1s^2 2s^2 2p^6 3s^1$	
Mg	12	$1s^2 2s^2 2p^6 3s^2$	M shell
Al	13	$1s^2 2s^2 2p^6 3s^2 3p^1$	($n = 3$)

* A shorthand notation is often used for closed inner shells. Thus, lithium may be written [He] $2s^1$; aluminum, [Ne] $3s^2 3p^1$; and so forth. The symbol in brackets designates the closed-shell configuration for that atom.

following example illustrates how to determine the number of states corresponding to a particular shell.

EXAMPLE 44-6

Determine the number of electrons that can occupy the $n = 3$ shell.

SOLUTION

Following the procedure in Example 44-1, we enumerate the quantum states for $n = 3$.

For $\ell = 2$ there are *five* values for m_ℓ, each with *two* values of m_s, producing a total of *ten* $\ell = 2$ states, or $3d^{10}$.

For $\ell = 1$ there are *three* values for m_ℓ, each with *two* values of m_s, producing a total of *six* $\ell = 1$ states, or $3p^6$.

For $\ell = 0$ there is only *one* value for m_ℓ, with *two* values for m_s, producing a total of *two* $\ell = 0$ states, or $3s^2$.

Thus, the configuration for the filled $n = 3$ shell would be $3s^2 3p^6 3d^{10}$, a total of *eighteen* states.

Table 44-3 suggests a pattern of simply filling the $n = 2$ shell, the $n = 3$ shell, and so on. It is a bit more complicated than that. Because the energy level depends not only on n but also on ℓ (and on the particular L–S coupling), the energy states in some higher shells begin to overlap those of an inner shell, disrupting the orderly sequence of adding electrons. See Table 44-5. Nevertheless, quantum theory does explain these exceptions. *The ground state con-*

TABLE 44-5	Shells					
X-Ray Notation	K	L	M	N	O	P
n	1	2	3	4	5	6
S U B S H E L L S	1s	2s	3s	4s	5s	6s ...
		2p	3p	4p	5p	6p ...
			3d	4d	5d	6d ...
				4f	5f	6f ...
					5g	6g ...
						6h ...

TABLE 44-5
Paschen's triangle, an array that organizes shells and subshells in a convenient pattern. The arrows indicate the sequence of energy levels for adding electrons. (There are a few exceptions in heavy atoms.) The electronic configuration for cobalt is thus $_{27}$Co: $1s^2 2s^2 2p^6 3s^2 3p^6 4s^2 3d^7$, also written [Ar] $4s^2 3d^7$. In another common notation, the sequence is listed in order of increasing n, so the last two terms could be interchanged: [Ar] $3d^7 4s^2$.

figuration of an atom places electrons in the lowest possible energy state without violating the Pauli exclusion principle.

For a filled shell, $\mathbf{S} = 0$, $\mathbf{L} = 0$, and $\mathbf{J} = 0$. Thus electrons in closed shells combine to give zero net angular momentum and zero net magnetic moment. The chemical properties[9] of an atom are principally determined by the atom's outermost electrons. Thus all the "filled-shell-plus-one" configurations have similar chemical properties that depend mainly on just the extra electron that is relatively loosely bound and "located" far outside the inner closed shells. This group forms the highly reactive *alkali metals* (lithium, sodium, potassium, rubidium, cesium, and francium). These atoms readily give up their extra electron to certain other atoms, forming an ion of $+1$ charge, while the other atom becomes an ion of -1 charge. The "filled-shell-minus-one" group are the *halogens* (fluorine, chlorine, bromine, iodine, and astatine)—atoms that strongly seek an extra electron to form a closed shell. One type of chemical bonding that joins atoms to form molecules is the *ionic bond*. For example, in the formation of sodium chloride, the sodium atom gives up its 3s electron to fill the 3p subshell of chlorine, forming Na^+Cl^-; the two ions are held together by their mutual Coulomb attraction. There are other types of chemical bonds, including those in which the atoms share more than one electron. In a *covalent bond*, there is a more or less equal sharing of one or several electrons by two or more atoms. The hydrogen molecule H_2 is an example of a molecule with the covalent type of bonding. Finally, "filled-shell" configurations are the *inert gases* (helium, argon, krypton, xenon, and radon); they have little tendency to gain or lose an electron so do not normally form molecules with other atoms.[10] See Figure 44-11.

---

[9] Some physicists get a kick out of telling their chemist friends that all of chemistry is contained in the Schrödinger equation—an exaggeration that does contain some truth. The chemists usually respond by pointing out that the Schrödinger equation cannot be *exactly* solved for any atom containing more than one electron! Approximation methods must be used. Of course, in *all* of the sciences (including chemistry) only approximations are achieved. Absolute certainty is claimed only by a few disciplines outside science.

[10] They are also called the "noble" gases, signifying their aloofness in associating with other, less royal atoms!

FIGURE 44-11
The *ionization energy* of an atom is the minimum energy (in electron volts) required to remove an electron from the atom in its ground state. The peaks at the inert gases are for atoms whose electron subshells are all complete. The next added electron must go into the next higher shell, farther from the nucleus, so notably less energy is required to remove it from the atom; these form the alkali metals. As more electrons are added (and, of course, more protons to the nucleus), the binding of the electrons becomes progressively stronger, until the next shell is complete. Thus, each period in the periodic table starts with a strongly reactive alkali metal, and ends with an inert noble gas. The numbers of elements in these periods are 2, 8, 8, 18, 18, and 32.

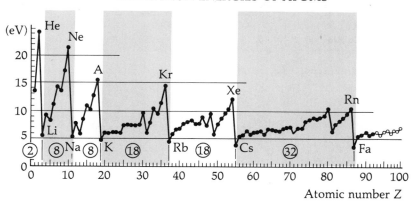

IONIZATION ENERGIES OF ATOMS

44.9 X-Rays

When energetic electrons strike a metal target, they produce **x-rays**—photons of very short wavelengths from roughly 0.001 to 10 nm. Figure 44-12 shows a modern x-ray tube and typical x-ray spectra. Two different processes occur when the electrons strike the target. The rapid deceleration of the electrons produces a smooth, **continuous spectrum** of photon wavelengths called *bremsstrahlung* (German: *bremse*, brake, and *strahlung*, radiation). The spectrum has a short-wavelength limit λ_{min} that depends upon the voltage across the tube, that is, upon the kinetic energy of the bombarding electrons. From the conservation of energy,

$$[K \text{ of electron}] = [\text{Maximum photon energy}]$$

$$Ve = hf_{max} = \frac{hc}{\lambda_{min}} \qquad (44\text{-}31)$$

Thus, the cutoff wavelength λ_{min} depends on only the accelerating voltage, not the target material.

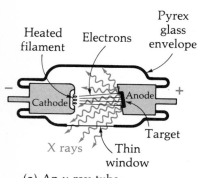

(a) An x-ray tube.

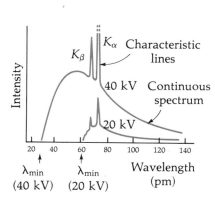

(b) The x-ray spectrum of a metal target obtained at two different accelerating voltages.

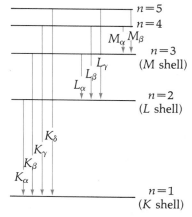

(c) The x-ray series are named for the shell vacancy that the electron fills.

FIGURE 44-12
X-rays.

The **characteristic line spectra** are sharp peaks superposed on the continuous spectrum. They are produced when the bombarding electrons have sufficient energy to knock out an electron from one of the inner shells of the target atoms, creating a vacancy in the K shell, the L shell, or another inner shell. When one of the outer electrons falls to fill this vacancy, the atom emits an x-ray photon whose energy equals that lost by the electron in making the transition. If an electron falls from the L shell to the K shell (from $n = 2$ to $n = 1$) it produces the K_α line; a transition from the M shell to the K shell (from $n = 3$ to $n = 1$) produces the K_β line, and so forth. There are also L series, M series, etc., named for the shell vacancy the electron *fills* (not the shell from which it came). These lines are all "characteristic" of the particular element used for the target material. K electrons, being close to the nucleus, are very sensitive to the nuclear charge Ze. Electrons in higher shells "feel" a smaller nuclear charge because inner electrons neutralize, or "screen," a portion of that charge from those outer electrons.

In 1913, the British physicist H. G. J. Moseley (1887–1915) investigated the characteristic x-ray spectra using a variety of different elements as targets. He found an interesting *straight-line* relationship by plotting the square root of the K_α frequency vs. the atomic number Z, Figure 44-13, establishing the atomic *number* Z (rather than atomic *weight*) as the true "signature" of an atom.[11] This *Moseley diagram* showed that a few elements fell off the line unless their positions in the atomic table were interchanged with those of a neighbor. The reason was that prior to Moseley's work the periodic table was ordered on the basis of atomic *weights*. But for a few elements, different isotopic abundances (Section 45.2) caused their masses to be out of line with their neighbors. For example, nickel formerly came before cobalt in the periodic table. Moseley's diagram, however, clearly showed that the atomic number Z of cobalt was smaller than that of nickel and that the atomic number Z was the best basis for ordering the periodic table. A few other discrepancies were

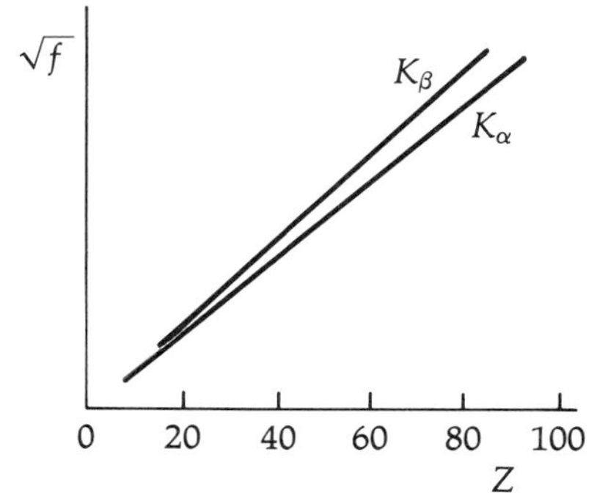

FIGURE 44-13
This *Moseley diagram* plots the square root of the frequency $\sqrt{f}$ vs. the atomic number Z of the target element for two lines of the K series. [Later plots used $(Z - 1)$ instead of Z, still obtaining a straight line. See Problem 44C-38.]

[11] Moseley later plotted $\sqrt{f}$ vs. $[Z - 1]$, rather than Z. He justified this by pointing out (correctly) that a vacancy in the K shell still leaves one K electron, which effectively shields one nuclear charge from the other electrons.

FIGURE 44-14
Harry G. J. Moseley (1887–1915). Immediately after graduating with honors from Oxford, Moseley began to work in Rutherford's laboratory in Manchester, the same year that Bohr was developing his atomic model. Moseley's work was especially valuable since it provided the first experimental link between the chemist's periodic table and the physicist's new model of the atom. Moseley was an ingenious experimenter and a tireless worker. For example, to solve the problem of the longer-wavelength x-rays being absorbed by the glass wall of the vacuum tube, Moseley cut a hole in the glass and covered it with the thin membrane from the large intestine of an ox. It worked—between frequent ruptures! Moseley traveled to Australia to report his research at a meeting of the British Association for the Advancement of Science, arriving the day that England declared war on Germany in 1914. Returning home a few weeks later, he volunteered his services to the government. Though offered a job in a research laboratory, he preferred more active service and accepted a commission in the Royal Engineers. His brilliant career was cut short when he was killed at Gallipoli at age 28.

similarly corrected, resulting in better agreement with the known chemical properties of the elements. At the time, vacancies in the sequence soon led to claims of discoveries of new elements—claims that were validated by the K_α line in their spectra (or, more often, were clearly shown by this test to be false!). Moseley's simple method also sorted out the rare-earth elements of atomic numbers 57 through 71—a confusing group whose similar chemical properties made Z determination difficult.

44.10 The Laser

The word **laser** is an acronym for the phrase "**L**ight **A**mplification by **S**timulated **E**mission of **R**adiation." Einstein was the first to predict the effect in a 1916 lecture (published the following year). Consider an atom that can undergo a transition from an excited state E_2 to a lower state E_1, emitting a photon of energy $hf = (E_2 - E_1)$ in the process. Suppose that while the atom is in the excited state, a photon with exactly the energy hf passes nearby. This photon can stimulate the excited atom to decay and emit a photon of energy hf. The intriguing aspect to this process is that we now have *two* photons with the *same energy*, traveling in the *same direction* with the *same phase* and the *same polarization* as the original photon. The two photons can, in turn, stimulate emission by other excited atoms, in a sort of chain reaction. The light is *coherent* (Section 38.2) and can build up to very great intensities.

The trick is to keep more atoms in state E_2 than in E_1. After all, an atom in state E_1 can *absorb* a photon of energy hf in a resonance process, raising the atom to state E_2. This causes photons to disappear—an unwanted event. In a collection of atoms in thermal equilibrium, the number of atoms N in various energy states follows the Boltzmann equation, $N = Ce^{-E/kT}$, where E is the energy of a state, k is Boltzmann's constant, and T is the kelvin temperature (C is a constant). Thus, normally the ratio of two state populations is

$$\frac{N_2}{N_1} = e^{-(E_2 - E_1)/kT} \tag{44-32}$$

Higher-energy states are *less* populated. We must create a **population inversion** that reverses the normal condition, so that the production of photons by stimulated emission occurs more often than the absorption of photons. To maintain the population inversion, we need to "pump" atoms up to an excited state continuously; this state is always a *metastable*[12] state to aid in prolonging the inversion condition. Pumping can be done in many ways: by an intense flash of light (pulsed ruby laser), by an electrical discharge (argon laser), by a

FIGURE 44-15
The He–Ne gas laser. The windows W at each end are tilted at the Brewster angle to reflect light of the unwanted polarization out of the laser.* The other polarization component transmits essentially 100% through the windows. (With windows perpendicular to the beam, each reflection would have $\sim 4\%$ loss—an intolerable situation.) Focussing mirrors M reflect light back and forth about 100 times, while the right-hand mirror permits a tiny fraction ($\sim 1\%$) to pass through, forming the external beam.

* This does *not* mean that the laser loses half its power. After only *one pass* through the tube, the Brewster reflection removes that component, so it doesn't build up much energy in the tube in the first place.

[12] The mean lifetime for excited atomic states is roughly $\sim 10^{-8}$ s. However, certain *metastable* states exist, on the average, for up to $\sim 10^{-3}$ s before decaying, because spontaneous transitions to lower states are "forbidden" (Section 44.6, Footnote 7).

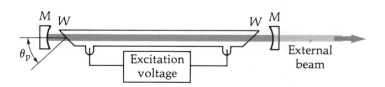

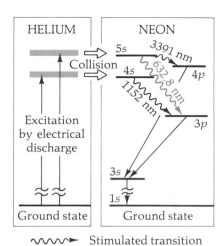
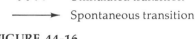

FIGURE 44-16
Significant energy levels in the He–Ne gas laser.

FIGURE 44-17
The top left photograph is a highly magnified view of a defective hypodermic needle—note the tiny hook at the tip. When illuminated by a laser, that tip produces the diffraction pattern below. A perfect needle produces the diffraction pattern at the right, a difference that enables rapid, automatic quality control in the manufacturing process.

FIGURE 44-18
The bar code scanner in supermarket checkout counters uses a narrow beam from a He–Ne laser to sweep across the Universal Product Code (UPC) symbol that identifies the item. The reflection from the light and dark bars is detected by a photocell, and the information is sent to a central computer. If the number is listed in the computer memory, the scanner sounds a beep, prints out the product information and price on the sales slip, and tallies the sale of that item in the store's inventory records. Some stores analyze daily records to make smart marketing decisions, taking advantage, for example, of the evidence that candy bar sales zoom when the bars are placed near the checkout counter, or that sales of bean dip increase if taco chips are on sale. Daily record analysis can also measure the effectiveness of a newspaper advertisement vs. an in-store display. In the code above, the initial zero identifies a grocery item, the next five digits the manufacturer (The Campbell Soup Co.), the next five digits the specific item ($10\frac{3}{4}$ oz., reduced-salt, condensed tomato soup), and the final digit the weight or volume.

chemical reaction (CO_2 laser), or by atomic collisions (He–Ne laser). The popular helium–neon laser, Figure 44-15, contains a gaseous mixture of those elements. Pumping is accomplished by an electrical discharge through the gas, which excites helium atoms to two upper levels that, by a lucky coincidence, are very close to two excited metastable states of neon, Figure 44-16. The excited helium atoms collide inelastically with ground-state neon atoms, transferring their internal energies to the neon. This raises neon to the metastable $5s$ and $4s$ levels, where they form a population inversion with the lower $4p$ and $3p$ states. (Transitions from $5s$ to $4s$ are forbidden.) The main transitions are by stimulated emission in the infrared (1152 and 3391 nm) and the familiar bright red (632.8 nm). The fact that stimulated emission is between two *upper* levels is most advantageous: the p states immediately drain off to the $3s$ state, maintaining the population inversion. (If the lower state were the ground state, it would rapidly fill up to become the most populated.) Furthermore, we can easily create the inversion condition without having to half-empty the greatly populated ground state.

Laser technology is leaping ahead furiously, and any list of the most powerful laser, the smallest laser, or the most unusual laser application would be out-of-date by the time you read it here. Lasers are now bounced off the moon to determine continental drift; they spot-weld detached retinas; they play hi-fi and TV discs; they store 10 billion "bits" of information (the contents of about 250 books, each the size of an Encyclopaedia Brittanica volume) on an ultra-high-density computer disc 1 ft in diameter; they guide milling machines and missiles; they provide a surgical "knife" that automatically cauterizes the cut as it removes cancerous growths; they make holograms; they generate fusion by imploding tiny spheres of deuterium–tritium; they alter genes; they simultaneously carry hundreds of TV and telephone signals in optical fibers; and they are used in myriad other ways that continue to amaze us all.

Summary

The *time-independent Schrödinger equation* (one dimension) is

$$\left[-\left(\frac{\hbar^2}{2m}\right)\frac{d^2}{dx^2} + U(x)\right]\psi = E\psi$$

For a hydrogen atom, E is the total energy and U is the Coulomb potential energy

$$U = -\left(\frac{1}{4\pi\varepsilon_0}\right)\frac{e^2}{r}$$

Because of spherical symmetry, the wave equation in three dimensions is most conveniently solved in spherical coordinates: r, θ, and ϕ. The requirements that solutions be single-valued and approach zero as $r \to \infty$ restrict "allowed" solutions to only those characterized by *four quantum numbers*: n, ℓ, m_ℓ, and m_s.

The *principal* quantum number n is identified with the total energy E_n:

ALLOWED ENERGIES OF THE HYDROGEN ATOM
$$E_n = -\left(\frac{me^4}{8\varepsilon_0^2 h^2}\right)\frac{1}{n^2}$$
$$= -\frac{13.6 \text{ eV}}{n^2} \quad (n = 1, 2, 3, \ldots)$$

The *orbital* quantum number ℓ is identified with the angular momentum L of the electron about the nucleus:

$$L = \hbar\sqrt{\ell(\ell+1)}$$

The *magnetic* quantum number m_ℓ is identified with the projection of the angular momentum on the z axis:

$$L_z = \hbar m_\ell$$

Associated with the angular momentum and its projection on the z axis is a *magnetic dipole moment* $\boldsymbol{\mu}$ such that

$$\boldsymbol{\mu} = -\left(\frac{e}{2m}\right)\mathbf{L}$$

and
$$\mu_z = -\left(\frac{e}{2m}\right)L_z = -\left(\frac{e\hbar}{2m}\right)m_\ell$$

The constant $e\hbar/2m$, called the *Bohr magneton*, has the value

BOHR MAGNETON
$$\left(\frac{e\hbar}{2m}\right) = 9.27 \times 10^{-24} \text{ A}\cdot\text{m}^2$$

The electron behaves as if it were spinning on its axis, producing an additional quantum number m_s, called the *spin quantum number*. It is associated with the projection of the spin angular momentum S on the z axis:

$$S = \hbar\sqrt{s(s+1)}$$
$$S_z = \hbar m_s$$

where $s = \frac{1}{2}$ and $m_s = \pm\frac{1}{2}$. The associated *spin magnetic dipole moment* μ_s has a z component:

$$(\mu_s)_z = m_s\left(\frac{e\hbar}{m}\right)$$

The *quantum state* of an electron in an atom is described by the set of four *quantum numbers* n, ℓ, m_ℓ, and m_s, where

$n = 1, 2, 3, \ldots$	Principal quantum number
$\ell = 0, 1, 2, 3, \ldots, (n-1)$	Orbital quantum number
$m_\ell = 0, \pm 1, \pm 2, \pm 3, \ldots, \pm\ell$	Magnetic quantum number
$m_s = \pm\frac{1}{2}$	Spin quantum number

The interaction of the magnetic dipole moments due to the electron's orbital motion and its spin produces a *spin–orbit*, or **L–S**, coupling. The coupling slightly alters the energy levels of the electron to produce a *fine structure* in the hydrogen spectrum. As a consequence, an alternate set of quantum numbers may be used to describe the quantum state of an electron. They are

$n = 1, 2, 3, \ldots$	Principal quantum number
$\ell = 0, 1, 2, 3, \ldots, (n-1)$	Orbital quantum number
$j = \pm\frac{1}{2}$	Inner quantum number
$m_j = \pm j, \pm(j-1), \pm(j-2), \ldots$	(z component of j)

A few hydrogen atom wave functions ψ are listed in Table 44-3, page 1044. They are *normalized* such that

$$\int |\psi|^2 \, dV = 1$$

where, for radial functions, $dV = 4\pi r^2 \, dr$.

The *probability density function* P is

$$P = |\psi|^2$$

The probability $\mathscr{P}$ of finding the electron in the *volume* dV is

$$\mathscr{P} = \int P \, dV$$

The probability of finding the electron in the *radial* element dr is

$$\mathscr{P} = \int P(r)\, dr$$

where $P(r)$ is the *radial probability density function*.

The *periodic table of the elements* can be constructed from two principles:

(1) The *Pauli exclusion principle*. No two electrons in an atom may have the same set of four quantum numbers $[n, \ell, m_\ell, m_s]$ or $[n, \ell, j, m_j]$.
(2) Electrons tend to seek the lowest possible energy level without violating the Pauli exclusion principle.

Spectroscopic notation identifies the energy level of an electron as illustrated by the example, $4d_{5/2}$, where the number preceding the letter is the principal quantum number n, the subscript is the value of the quantum number j, and the letter corresponds to the orbital quantum number ℓ according to the following scheme: the ℓ values of 0, 1, 2, 3, 4, and 5 correspond to the letters s, p, d, f, g, and h, respectively.

The *ground-state configuration* may be expressed in spectroscopic notation as illustrated by the example, $1s^2 2s^2 2p^1$, where the superscript indicates the number of electrons in a particular n, ℓ state.

X-rays. When a beam of energetic electrons strikes a target and causes vacancies in the inner shells of the target atoms, x-rays are produced when outer electrons fill these vacancies. The resulting *characteristic sharp-line spectra* are designated by the common lower energy level of the transitions. Thus we have the K series, the L series, etc. A *continuous spectrum*, called *bremsstrahlung*, is also produced by the abrupt deceleration of the incident electrons. The continuous spectrum has a minimum *cutoff wavelength* $\lambda_{\min}$, produced when the initial K of an incident electron produces a single photon of energy $hf_{\max} = hc/\lambda_{\min}$. Moseley showed that the x-ray spectrum of an atom is a true "signature" of an element when he obtained a straight line by plotting $\sqrt{f_{K_\alpha}}$ vs. the atomic number $[Z - 1]$ (rather than the atomic weight A). Such a plot is called a *Moseley diagram*.

Lasers generate beams of coherent light of high intensity by stimulating radiative transitions from an excited metastable state to a lower state. The populations of the two states must be *inverted* through "pumping," so that stimulated emissions from the higher state occur more frequently than the absorption of photons by the lower state.

Questions

1. Why must the wave function ψ describing a three-dimensional situation always have the dimensions $[L]^{-3/2}$?
2. In the wave-mechanical model of the atom, how can there be uncertainty in the position and velocity of an electron, yet precise values for angular momenta?
3. For all of the wave functions listed in Table 44-3, what is implied by the fact that $|\psi|^2$ (or $\psi\psi^*$) is independent of ϕ for each?
4. In the wave-mechanical view of the hydrogen atom, is the electron a point charge, a ball of charge, a charge distributed around the nucleus, or something else?
5. Consider the assumptions that Bohr made in devising his model of the hydrogen atom. Which assumptions are consistent with classical theory and which are not?
6. A student (incorrectly) writes the ground-state configuration of sodium as $1s^2 2s^2 2p^6 2d^1$. Discuss the error.
7. In the Stern–Gerlach experiment, Figure 44-2, what would happen if singly ionized atoms of silver (rather than neutral atoms) were sent through the apparatus? Since this experiment reveals the spatial orientation of the magnetic moments of electrons in a magnetic field, why couldn't a beam of electrons be used (rather than neutral silver atoms)? Why is it necessary to use a nonuniform magnetic field rather than a uniform field?
8. What is meant by the phrase "allowed solutions" to the Schrödinger equation?
9. Discuss how our mental picture of the hydrogen atom differs in the Schrödinger theory from that in the Bohr theory.
10. Consider the spin angular momentum and the magnetic moment of an electron. Why are these vectors in opposite directions?
11. In three-dimensional space, three parameters—for example, the three rectangular components—are required to describe a vector. Yet we use only two quantum numbers to describe the vector angular momentum of an electron in the hydrogen atom. Explain.
12. Define these terms and explain the differences between them: (a) *wave function*, (b) *probability density function*, and (c) *radial probability density function*.
13. Explain why it takes *more* energy to remove an electron from an argon atom $(Z = 18)$ than from a potassium atom $(Z = 19)$, which has a higher positive charge in the nucleus.
14. Why must the electrons in the ground state of the helium atom have opposite spins?
15. About 5 eV are required to remove an electron from a potassium atom. Would you expect the energy required to remove an additional electron to be more, less, or about the same? Explain.
16. The ionization energies of the first five alkali atoms are highest for lithium (5.39 eV), dropping fairly uniformly

to the lowest value for cesium (3.89 eV). Which of these two atoms would you expect to be the most chemically active? Why? In terms of atomic structure, explain why these values drop monotonically as we progress toward higher-Z atoms.

17. Do the characteristic x-ray lines for the L series have longer or shorter wavelengths than those for the K series?
18. The fact that only specific orientations of certain quantum-mechanical vectors are allowed is called *spatial quantization*. Is *space* quantized? If not, what is?

Problems
44.3 Electron Spin and Fine Structure
44.4 Spin–Orbit Coupling

44A-1 We can observe the effects of the magnetic quantum number m_ℓ by placing the atom in a magnetic field (the Zeeman splitting of energy levels). Into how many levels will the $\ell = 3$ state split? Include a freehand sketch of the orientations of the magnetic dipole moment $\boldsymbol{\mu}$ with the field direction (the $+z$ axis).

44B-2 All objects, large and small, behave quantum-mechanically. (a) Estimate the quantum number ℓ for the earth in its orbit about the sun. (b) What energy change (in joules) would occur if the earth made a transition to an adjacent allowed state?

44B-3 In the presence of a magnetic field, an electron orients its magnetic moment $\boldsymbol{\mu}_s$ "parallel" or "antiparallel" to the field direction (the z axis). Actually, $\boldsymbol{\mu}_s$ makes a finite angle θ (not 0°) with the field because of the way such vectors must project on the z direction. Determine the two values of θ.

44B-4 The following constants appear often in atomic physics theories:

Bohr radius: $a_0 \equiv \dfrac{\varepsilon_0 h^2}{\pi m_e e^2}$

Compton wavelength: $\lambda_C \equiv \dfrac{h}{m_e c}$

Classical electron radius: $r_e \equiv \dfrac{e^2}{4\pi\varepsilon_0 m_e c^2}$

Fine structure constant: $\alpha \equiv \dfrac{e^2}{2\varepsilon_0 hc}$

By direct calculation, find the numerical value of each (including units) in the SI system.

44B-5 The magnitude of the total angular momentum $\mathbf{J}$ that an electron in hydrogen may have is $J = \sqrt{j(j+1)}\hbar$. The possible projections of $\mathbf{J}$ on the z axis are given by $J_z = m_j\hbar$. Find the allowed angles between $\mathbf{J}$ and the $+z$ axis for $j = \frac{5}{2}$.

44B-6 The magnitude of the orbital angular momentum $\mathbf{L}$ that an electron in hydrogen may have is $L = \sqrt{\ell(\ell+1)}\hbar$. The possible projections of $\mathbf{L}$ on the z axis are given by $L_z = m_\ell\hbar$. Find the allowed angles between $\mathbf{L}$ and the $+z$ direction for $\ell = 2$.

44.5 Quantum States of the Hydrogen Atom
44.6 Energy Level Diagram for Hydrogen
44.7 The Hydrogen Atom Wave Functions

44B-7 List all the quantum states of the hydrogen atom for $n = 4$ in a manner similar to that of Example 44-2.

44A-8 A hydrogen atom is in a state for which $\ell = 3$. What are the possible values for n, m_ℓ, and m_s?

44B-9 In interstellar space, atomic hydrogen produces the sharp spectral line called the **21-cm radiation**, which astronomers find most helpful in detecting clouds of hydrogen between stars. This radiation is useful because interstellar dust that obscures visible wavelengths is transparent to these radio wavelengths. The radiation is not generated by an electron transition between energy states characterized by n. Instead, in the ground state ($n = 1$), the electron and proton spins may be *parallel* or *antiparallel*, with a resultant slight difference in these energy states. (a) Which condition has the higher energy? (b) The line is actually at 21.11 cm. What is the energy difference between the states? (c) The average lifetime in the excited state is about 10^7 yr. Calculate the associated uncertainty in energy of this excited energy level.

44B-10 Consider hydrogen in its ground state. Using the approximation of Equation (44-26), estimate the probability of the electron being within the range of distance $(1 \pm 0.01)a$ from the nucleus. The distance a is the Bohr radius ($n = 1$).

44B-11 Consider the hydrogen atom in its ground state. Using the approximation of Equation (44-28), estimate the ratio of (1) the probability of finding the electron within a distance $\Delta r = (1 \pm 0.01)a$ from the nucleus to (2) the probability of finding it within a distance $\Delta r = (4 \pm 0.01)a$ from the nucleus. The distance a is the Bohr radius ($n = 1$).

44.8 The Pauli Exclusion Principle and the Periodic Table of the Elements

44A-12 Identify the elements corresponding to the following electron configurations: $1s^2 2s^2 2p^1$ and [Ar] $3d^{10} 4s^2 4p^6$.

44B-13 Using Table 44-4, write the electronic configuration for the ground state of an atom whose "last" electron is in the $4p^2$ state. What is the element?

44B-14 Consider an atom whose M shell is completely filled (with no additional electrons). (a) Identify the atom. (b) List the number of electrons in each of its subshells.

44B-15 Show that the number of quantum states in the nth shell is $2n^2$.

44B-16 All atoms are roughly the same size. (a) To show this, estimate the diameters for aluminum, with molar atomic mass = 27 g/mole and density 2.70 g/cm^3, and uranium, with molar atomic mass = 238 g/mole and density 18.9 g/cm^3. (b) What do the results imply about the wave functions for inner-shell electrons as we progress to higher and higher atomic weight atoms? (Hint: the molar volume is roughly proportional to $D^3 N_A$, where D is the atomic diameter and N_A is Avogadro's number.)

44.9 X-Rays
44.10 The Laser

44A-17 The wavelength of the K_α line from silver is 56.3 pm. If we are using a silver target, what minimum accelerating voltage on an x-ray tube must we exceed to (barely) make this line appear in the emitted spectrum?

44A-18 Find the cutoff wavelength for an x-ray tube operated at 45 kV.

44B-19 In the Bohr model of an atom of fairly high atomic number, a K-shell electron moves in a hydrogenlike orbit under the Coulomb force between the nuclear charge Ze and the electron's charge $(-)e$. Adapt Equation (43-12) in Chapter 43 to this situation, and derive the following relation between the frequency f of the K_α x-ray line and the atomic number Z (Moseley's law) (This expression ignores the screening effects of inner electrons.):

$$\sqrt{f} = \left[\frac{e^2}{\varepsilon_0} \sqrt{\frac{3m}{32h^3}} \right] Z$$

44B-20 The same x-ray tube that was used to obtain the graph of Figure 44-12b is operated at 15.5 kV. Make a freehand sketch of its x-ray spectrum, including the numerical value of λ_{min}.

44B-21 For a photon traveling along the axis of a typical He–Ne laser, the amplification due to stimulated emissions is $\sim 0.7\%$ per meter of path. Find the average number of additional photons that the original photon generates while traveling the 1-m length of the tube 200 times.

44B-22 A high-power, pulsed laser delivers 30 kJ of energy in 4 ns. (a) What is the power in this pulse? (b) What is the physical length of the pulse as it travels through space? (c) Find the impulse delivered to a target that completely absorbs this radiation.

44B-23 A pulsed ruby laser emits light at 694.4 nm. For a 14-ps pulse containing 3 J of energy, find (a) the physical length of the pulse as it travels through space, and (b) the number of photons in the pulse. (c) The beam has a circular cross section of 0.6 cm diameter. Find the number of photons per cubic millimeter in the beam.

44B-24 A Nd:YAG laser used in eye surgery emits a 3-mJ pulse in 1 ns, focussed to a spot 30 μm in diameter on the retina. (a) Find (in SI units) the power per unit area at the retina. (This quantity is called the *irradiance*.) (b) What energy is delivered to an area of molecular size, say a circular area 0.6 nm in diameter?

Additional Problems

44C-25 In a Stern–Gerlach experiment, a beam of silver atoms of mass M and magnetic moment μ has a most probable speed v as it travels a distance x through a magnetic field whose gradient is dB/dz. Derive an expression for the distance d between the two subbeams as they emerge from the magnetic field. Note that quantum-mechanically, the vector $\boldsymbol{\mu}$ has an angle θ with respect to the field direction. (See Problem 44B-3.)

44C-26 Consider a classical model of an electron as a spinning sphere of uniform mass m_e, with a radius r_e as given in Problem 44B-4. From the known spin angular momentum $s_z = \frac{1}{2}\hbar$, calculate the speed of rotation at the equator.

44C-27 A hypothetical one-electron atom emits radiation with wavelengths of 160 nm, 120 nm, 100 nm, 90 nm, and 85 nm, with a series limit of 80 nm. (a) Assuming that all of the radiation results from transitions to the lowest energy ($n = 1$) state, calculate the three lowest energy levels of the atom. (b) Show that the energy levels cannot be represented by $E_n = E_1/n^2$. (c) Calculate the wavelength of radiation corresponding to a transition from the $n = 3$ state to the $n = 2$ state.

44C-28 Consider an electron in the lowest (classical) Bohr orbit, moving in a circle in the xy plane around a stationary proton at the origin. The direction of motion is such that the electron's angular momentum is in the $-z$ direction. From the electron's frame of reference, the proton moves in a circle around the electron. (a) Find the magnitude and direction of the magnetic field $\mathbf{B}$ at the electron's location due to this circular motion by the proton. (b) In which case does the magnetic potential energy $U = -\boldsymbol{\mu}_s \cdot \mathbf{B}$ have a positive value (relative to $U \equiv 0$ for 90° orientation): when $j = \ell + \frac{1}{2}$, or when $j = \ell - \frac{1}{2}$? (Hint: relative to the angular momentum direction due to spin, in what direction is the electron's spin magnetic moment? Remember that the electron has a negative charge.) (c) Calculate the energy difference (in electron volts) between these closely spaced doublet states.

44C-29 The ground-state configuration of an element is $1s^2 2s^2 2p^6 3s^2 3p^6 4s^1$. (a) Identify the element. (b) The following configurations are excited states of this element. From which of these are transitions directly to the ground state possible according to the *selection rules* of Footnote 7: [ ] $3p^6 4p^1$, [ ] $3p^5 4s^2$, [ ] $3p^6 4d^1$, and [ ] $3p^5 4p^2$?

44C-30 A hydrogen atom in its ground state is in a magnetic field of 0.3 T. Find (in electron volts) the magnitude of the magnetic interaction energy $E = -\boldsymbol{\mu} \cdot \mathbf{B}$ between the field and the spin of the electron. (b) What is the difference in energy between the parallel and antiparallel orientations of the spin magnetic moment? (c) Find the wavelength of an incident photon that would induce a "resonance" transition from the parallel to the antiparallel state. (See Problem 44B-3.)

44C-31 Consider excited states of hydrogen. (a) Show how a $3p \to 2s$ transition results in a fine structure of two closely spaced lines. (b) How many lines are in the fine structure for a $4d \to 3p$ transition? Indicate any forbidden transitions between these levels. (Hint: see Figure 44-4 and the *selection rules* of Footnote 7.)

44C-32 The average, or mean, value of the distance r that the electron is from the hydrogen nucleus is given by $r_{av} = \int_0^\infty r P(r) \, dr$. Find the value of r_{av} in terms of the $n = 1$ Bohr radius a for hydrogen in the ground state. (Hint: consult Appendix G-III, Equation 1.)

44C-33 Verify that the hydrogen wave function for $n = 2$, $\ell = 0$, and $m_\ell = 0$ (Table 44-3) is normalized. That is, show that $\int |\psi|^2 \, dV = 1$, where $dV = 4\pi r^2 \, dr$.

44C-34 Consider the hydrogen atom in its ground state. For $r = a$, calculate the values of (a) ψ, (b) $|\psi|^2$, and (c) $P(r)$. What physical meanings do we associate with these values?

44C-35 The wave function for the ground state of hydrogen is independent of the polar angle θ and the azimuthal angle ϕ. Show by direct substitution that the wave function for the ground state (1s) in Table 44-3 satisfies the Schrödinger wave equation, Equation (44-4).

44C-36 For hydrogen in the 1s state, what is the probability of finding the electron farther than $2.50a$ from the nucleus?

44C-37 The 2p state of hydrogen is described by the three wave functions $\psi_{2,1,0}$, $\psi_{2,1,+1}$, and $\psi_{2,1,-1}$ listed in Table 44-3. All of these wave functions are for the same energy state. Suppose that an electron with this energy is described by each of these wave functions one-third of the time. The probability density P for this energy state would then be

$$|\psi_{2,1}|^2 = \tfrac{1}{3}|\psi_{2,1,0}|^2 + \tfrac{1}{3}|\psi_{2,1,+1}|^2 + \tfrac{1}{3}|\psi_{2,1,-1}|^2$$

(a) Calculate $|\psi_{2,1}|^2$. Note that the result is independent of θ and ϕ, indicating spherical symmetry. (b) Determine the radial probability density function $P(r)$. (c) By calculating $dP(r)/dr = 0$, find the most probable radial distance for the electron in the 2p state. Express the answer in terms of a, the Bohr radius for $n = 1$.

44C-38 Accepting the Bohr model as correct for atoms, prove that a Moseley diagram will be a *straight* line, regardless of the amount of screening that is assumed; that is, plotting $\sqrt{f}$ vs. Z or vs. $(Z - 1)$ or vs. $(Z - k)$ where $k = $ constant, will in each case result in a straight line.

44C-39 (a) Find the normal population ratio (without "pumping"), N_{5s}/N_{4s}, for the two excited states of neon that produce the 632.8-nm red light from a He–Ne laser. The gas inside the laser is at 27°C. (b) For lasing to be achieved, a population inversion must occur. That is, $N_2/N_1 > \tfrac{1}{2}$. At what temperature would the gas (at equilibrium) have $N_2/N_1 = \tfrac{1}{2}$?

44C-40 A high-power CO_2 laser operates continuously, producing 200 kW at 10.6 μm (sufficient to cut through a 1-in.-thick steel plate in a few seconds). (a) If the beam coming from the laser has a diameter of 4 mm, find the average power per square millimeter of cross-sectional area of the beam. This beam now passes through an ideal lens of focal length 6 cm. At the image plane, there is a circular diffraction pattern as shown in Figure 39-12. The central spot contains 84% of the beam energy. Find (b) the diameter of this central spot and (c) its average power per square millimeter.

44C-41 The conditions of a population inversion are sometimes referred to as a state of *negative absolute temperature*. (a) Explain why this term is appropriate. (b) For an inversion population ratio $N_2/N_1 = 1.09$, what would be the equivalent negative kelvin temperature of an argon laser that emits 514.5 nm?

CHAPTER 45

Nuclear Physics

The energy produced by breaking down the atom is a very poor thing. Anyone who expects a source of power from the transformation of these atoms is talking moonshine.

ERNEST RUTHERFORD, 1933
(five years before fission
was accidentally discovered
by the German physicists
Hahn and Strassman)

Consider $E = mc^2$. Hitler, Stalin, Churchill, and FDR had only the dimmest notion of what it means. Yet this simple equation is the product of a theory as beautiful as a Mozart concerto, more useful to humanity in the long run than the stock market, more revolutionary than the Communist party. And this theory, the theory of relativity, was something a funky mathematician, kicked out of Germany because the practical men who were running the Fatherland couldn't stand Jews, made up in his head. If Hitler had understood the formula he might not have lost the war.

MARTIN GARDNER, *Order and Surprise*
(Prometheus Books, 1983, page 299)

45.1 Introduction

We now turn our attention to the properties and behavior of the atomic nucleus. Our knowledge of the nucleus has been gained over a period of about one century, with a tremendously accelerated growth over the last 50 years. In no other period in human history has physics had a more awesome and profound impact on the world.

In the 1930s, the nucleus was believed to consist of neutrons and protons, with electrons and photons completing the list of basic building blocks for constructing everything in the physical universe. But with new data from more powerful accelerators built after World War II, many additional particles were soon discovered—currently we know of over 200 so-called "elementary" particles and antiparticles. One of the major frontiers of physics today is the quest for a theory that brings this huge zoo of elementary particles into what we hope is a simple, conceptual unity with all other phenomena. In this chapter we discuss the structure and behavior of nuclei, radioactivity, nuclear reactions, and nuclear power, and we conclude with some comments about elementary particles.

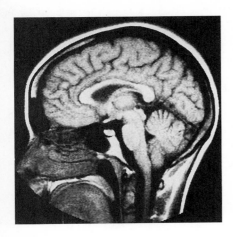

FIGURE 45-1
A cross section of a human head, obtained by *magnetic resonance imaging* (MRI).* Like an electron, a proton has spin $\frac{1}{2}\hbar$, so its magnetic moment can assume either of two quantized orientations with respect to an external magnetic field: "spin up" or "spin down." The two states differ slightly in energy, and normally most protons are in the lower energy state. Because of gyroscopic action, their magnetic moments precess about the field direction with a frequency f (see Figure 44-3). If a short pulse of an alternating electromagnetic field of the same frequency f is now applied, resonant transitions to the upper state can be induced. As these excited states decay down to the lower state, the "spin flips" can be detected externally. The interesting feature of this process is that the frequency f at which a transition occurs depends on the precise magnetic field in the proton's vicinity, and this field is affected slightly by surrounding electrons and nuclei. Thus hydrogen atoms in different chemical compounds will have slightly different resonant frequencies, allowing discrimination between different organic materials. Computer processing of the data can produce an image of a cross section of the body, revealing clear differences between various organs and soft tissues. The imaging technique is similar to *computerized tomography* (CT) scanning using x rays. However, MRI has the unique advantages of showing many details not revealed by x rays and of discriminating sensitively between healthy and diseased tissues. Furthermore, it is a non-invasive technique and does not subject the patient to the physiological hazards of x-ray dosages. [See Ian L. Pykett, "NMR Imaging in Medicine," *Scientific American* **246**, 78 (May 1982). For a simple explanation of the remarkable technique of computerized tomography, see Margaret Stautberg Greenwood, "X-Ray CT-Scan Analogy," *The Physics Teacher* **23**, 94 (Feb. 1985).]

* Formerly called *nuclear magnetic resonance* (NMR) imaging. The name was changed when it was realized that the word *nuclear* caused some patients undue apprehension.

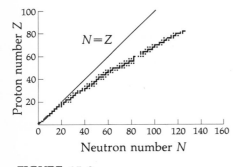

FIGURE 45-2
The isotopes found in nature. The black squares represent isotopes that are completely stable (nonradioactive). The colored squares represent radioactive isotopes, with half-lives greater than 100 000 yr [with the exception of ^{14}C (5730 yr) and ^{226}Ra (1600 yr)].

45.2 A Description of the Nucleus

The nucleus is composed of protons and neutrons, each of which is called a *nucleon*. The combination is called a *nuclide*. The proton has a positive charge equal in magnitude to the charge on the electron, 1.602×10^{-19} C, and a mass of 1.673×10^{-27} kg. The neutron has no charge and a mass of 1.675×10^{-27} kg. Both the neutron and the proton have a "radius" of about 10^{-15} m, or one *femtometer* (1 fm). Each has a spin of $\frac{1}{2}\hbar$. Every element is characterized by an **atomic number** Z, the number of protons in the nucleus. The nucleus may also have a number of neutrons, designated by the **neutron number** N. Each element has the same number of protons but may have a variety of *isotopes*, each with a different number of neutrons. For example, the oxygen nucleus has eight protons but has three isotopes found in nature: 99.785% of these have eight neutrons, 0.038% have nine neutrons, and 0.204% have ten neutrons. The total number of protons and neutrons is the **mass number** A. Thus:

MASS NUMBER A $\qquad\qquad A = Z + N \qquad\qquad$ (45-1)

The isotope of oxygen that has nine neutrons in its nucleus is identified by the notation $^{17}_{8}$O. The superscript preceding the letter is the atomic mass number A, and the subscript preceding the letter is the atomic number Z— in general, $^{A}_{Z}$X. Since the letter and the subscript both represent the element (here oxygen), the subscript is often omitted. The three naturally occurring isotopes are then designated by ^{16}O, ^{17}O, and ^{18}O.

The nucleus is bound together by a very strong attractive force, which has a limited range. This force, called the *strong nuclear force*, acts on both protons and neutrons alike. The "stable" isotopes found in nature are indicated by the squares in Figure 45-2. For low Z, the stable isotopes have roughly equal numbers of neutrons and protons ($N = Z$). But as Z increases, the

Coulomb repulsion also increases, so it is reasonable that additional numbers of neutrons (which experience only the *attractive* nuclear force) are required for stability. Beyond $Z = 82$, there are no completely stable nuclides; here, apparently, additional neutrons are unable to overcome the very large Coulomb repulsion. The number of naturally occurring isotopes for a given element varies. Tin ($Z = 50$) has ten such isotopes, while gold ($Z = 79$) has only one. The elements technetium ($Z = 43$) and promethium ($Z = 61$) have not been found in nature; when artificially produced, all their isotopes have relatively short half-lives.

It is easy to show why a strong nuclear force must exist. The gravitational force between nucleons is far too weak to counteract the Coulomb repulsive force between protons, as the following example illustrates.

EXAMPLE 45-1

Find the ratio of the repulsive Coulomb force to the attractive gravitational force between two protons.

SOLUTION

Coulomb repulsion

$$F_C = \left(\frac{1}{4\pi\varepsilon_0}\right)\frac{e^2}{r^2}$$

Gravitational attraction

$$F_G = G\frac{m_p^2}{r^2}$$

The ratio of the two forces is

$$\frac{F_C}{F_G} = \frac{1}{4\pi\varepsilon_0 G}\left(\frac{e}{m_p}\right)^2$$

$$= \left[\frac{9 \times 10^9 \text{ N·m}^2/\text{C}^2}{6.67 \times 10^{-11} \text{ N·m}^2/\text{kg}^2}\right]\left[\frac{1.602 \times 10^{-19} \text{ C}}{1.673 \times 10^{-27} \text{ kg}}\right]^2 = \boxed{1.24 \times 10^{36}}$$

The fact that all nuclei have about the same density leads us to conclude that the strong attractive nuclear force has a very short range. If the nuclear force were to extend very far beyond the nearest neighbors of a nucleon, the cumulative effect would be to draw all of the nucleons closer together, thus increasing the density of the nucleus as the atomic mass number increases. Scattering experiments with high-energy electrons indicate that most nuclei are approximately spherical with a radius R given by

RADIUS R OF THE NUCLEUS $\qquad R = R_0 A^{1/3} \qquad$ (45-2)

where A is the mass number and R_0 is a constant[1] equal to about 1.2 fm. Since the volume of a sphere is proportional to R^3, this suggests that all nuclei have

---

[1] The "size" of the nucleus depends upon the particular interaction used to probe the nucleus. Values of R_0 range from about 1.0 to 1.5 fm. Data from the scattering of high-energy electrons (which feel only the Coulomb force) give ~ 1.2 fm, while data from scattering of neutrons and protons (which respond to the nuclear force) give somewhat larger values, probably because the nuclear force field extends a bit beyond the nucleus.

The unit *femtometer* is sometimes called the *fermi*, in honor of Enrico Fermi (1901–1954), the brilliant Italian physicist who made important contributions to both theoretical and experimental physics. In addition to his studies of β decay and nuclear fission, Fermi received the Nobel Prize in 1938 for the production of new isotopes by neutron bombardment.

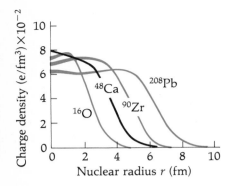

FIGURE 45-3
How is positive charge distributed within a nucleus? Because of Coulomb repulsion, is it mainly near the surface like a hollow sphere? Or concentrated near the center? Or uniformly spread throughout the interior? These charge density distributions show some experimental results from the elastic scattering of high-energy electrons at the Stanford Linear Accelerator Center (SLAC). The thickness of the line represents the experimental uncertainty. For many nuclei, the charge densities in the interior are known to an accuracy of 1%, an order-of-magnitude better than any current theory predicts. Note the tendency of the central charge distribution to diminish slightly as A increases. (From Bernard Frois and Costas N. Papanicolas, "Electron Scattering and Nuclear Structure," *Annual Review of Nuclear and Particle Science* **37** (1987), pp. 133—176.)

approximately the same density. Each individual nucleon thus behaves somewhat like a small hard sphere; the "spheres" combining to form a nucleus is analogous to the formation a drop of liquid in which the density does not depend upon the size of the drop. Indeed, the *liquid-drop model* has proven to be a useful representation. The following example indicates the enormity of the nuclear density.

EXAMPLE 45-2

Determine the density of nuclear matter.

SOLUTION

The density ρ of the nucleus is $\rho = M/V = Am/V$, where the volume $V = \frac{4}{3}\pi R^3$. From Equation (45-2), the radius $R = R_0 A^{1/3}$. Thus, $V = \frac{4}{3}\pi R_0^3 A$, and

$$\rho = \frac{Am}{V} = \frac{Am}{\frac{4}{3}\pi R_0^3 A} = \frac{3m}{4\pi R_0^3}$$

$$\rho = \frac{3(1.67 \times 10^{-27} \text{ kg})}{4\pi (1.2 \times 10^{-15} \text{ m})^3} = \boxed{2.31 \times 10^{17} \frac{\text{kg}}{\text{m}^3}}$$

A sphere with this density and a mass of the earth would have radius of only 184 m.

45.3 Nuclear Mass and Binding Energy

The mass-energy equivalence formulated by Einstein plays the central role in all nuclear reactions, ranging from radioactive decay to nuclear power reactors. As an illustration, consider the hydrogen isotope ^{2}H, called *deuterium*. The proton and the neutron are bound together in the nucleus by a very strong nuclear force, and work must be done on the nucleus to separate them. The work done appears as increased mass of the separated nucleons. Conversely, when a proton and a neutron combine to form deuterium, some mass disappears and an equivalent amount of energy is released in the form of an emitted photon.

In order to apply mass-energy conversions in a quantitative way, we establish a mass unit appropriate for nuclear masses. This unit is the *unified atomic mass unit* (u), defined as exactly $\frac{1}{12}$ of the mass of atomic ^{12}C, including the mass of the six electrons in the atom. The mass and energy equivalents are

$$\text{UNIFIED ATOMIC MASS UNIT, u} \quad 1 \text{ u} = \begin{cases} \frac{1}{12} \text{ the mass of atomic } ^{12}\text{C} \quad \text{(including electrons)} \\ 1.660\,540 \times 10^{-27} \text{ kg} \\ 931.494 \text{ MeV}/c^2 \\ 1.492\,42 \times 10^{-10} \text{ J}/c^2 \end{cases} \quad (45\text{-}3)$$

where the speed of light c is defined as exactly $2.997\,924\,58 \times 10^8$ m/s. Table 45-1 lists the mass of some selected isotopes. Since mass-energy is given by $E = mc^2$, we have also expressed the mass of fundamental particles in units of MeV/c^2.

TABLE 45-1 Selected Particles and Elements*

Particle	Charge	kg	u	(MeV/c^2)
Proton	e	1.6726×10^{-27}	1.007 277	938.272
Neutron	0	1.6749×10^{-27}	1.008 665	939.566
Electron	$-e$	9.1094×10^{-31}	5.4858×10^{-4}	0.511

Element ($_Z$symbol, name)	A	Atomic Mass (including electrons) (u)	BE/Nucleon (MeV)	Half-Life[†]	Decay Mode[‡]
$_1$H Hydrogen	1	1.007 825	—	stable	—
	2	2.014 102	1.11	stable	—
$_2$He Helium	3	3.016 029	2.57	stable	—
	4	4.002 603	7.07	stable	—
$_3$Li Lithium	6	6.015 121	5.33	stable	—
	7	7.016 003	5.61	stable	—
$_4$Be Beryllium	9	9.012 182	6.46	stable	—
$_5$B Boron	10	10.012 937	6.48	stable	—
$_6$C Carbon	12	12 (exactly)	7.68	stable	—
	14	14.003 241	7.52	5730 yr	β^-
$_7$N Nitrogen	14	14.003 074	7.48	stable	—
$_8$O Oxygen	16	15.994 915	7.97	stable	—
	17	16.999 131	7.75	stable	—
	18	17.999 160	7.76	stable	—
$_{10}$Ne Neon	22	21.991 383	8.08	stable	—
$_{11}$Na Sodium	22	21.994 434	7.92	2.601 yr	β^+, EC
$_{26}$Fe Iron	56	55.934 939	8.79	stable	—
$_{37}$Rb Rubidium	90	89.914 811	8.63	4.26 min, 3.03 min	β^-
$_{55}$Cs Cesium	137	136.907 073	8.38	30.17 yr	β^-
	143	142.927 220	8.24	1.78 s	β^-
$_{56}$Ba Barium	137	136.905 812	8.39	stable	—
$_{79}$Au Gold	197	196.966 543	7.92	stable	—
	198	197.968 217	7.77	2.693 d	β^-
$_{80}$Hg Mercury	198	197.966 743	7.91	stable	—
$_{82}$Pb Lead	206	205.974 440	7.88	stable	—
$_{84}$Po Polonium	210	209.982 848	7.83	138.38 d	α
$_{85}$At Astatine	210	209.987 126	7.81	8.1 h	α, EC
$_{90}$Th Thorium	232	232.038 054	7.61	1.4×10^{10} yr	α
$_{91}$Pa Proactinium	233	233.040 242	7.60	27.0 d	β^-, γ
$_{92}$U Uranium	233	233.039 628	7.60	1.59×10^5 yr	α, γ
	235	235.043 924	7.59	7.04×10^8 yr	α, γ, SF
	236	236.045 562	7.59	2.342×10^7 yr	α, γ, SF
	238	238.050 784	7.57	4.468×10^9 yr	α, γ, SF
$_{93}$Np Neptunium	239	239.052 933	7.56	2.35 d	β^-, γ
$_{94}$Pu Plutonium	239	239.052 157	7.56	2.411×10^4 yr	α, γ, SF

* Particle data (rounded) from *CODATA Bulletin*, No. 63, Nov. 1986. Atomic data from *CRC Handbook of Chemistry and Physics*, 66th ed., CRC Press, 1985–86. Additional data in Appendix J, *Periodic Table of the Elements*.
[†] s, second; min, minute; h, hour; d, day; yr, year.
[‡] EC = orbital electron emission; SF = spontaneous fission.

> **EXAMPLE 45-3**
>
> Calculate the amount of work required (in MeV) to separate the neutron and the proton in the nucleus of deuterium.
>
> **SOLUTION**
>
> The amount of work done, or the energy added to the system, is equivalent to the increase in mass in transforming ^{2}H into ^{1}H plus a neutron. From Table 45-1, we have
>
> After separation $\begin{cases} ^1\text{H mass} = 1.007\,825 \text{ u} \\ \text{Neutron mass} = 1.008\,665 \text{ u} \\ \text{Total mass} = 2.016\,490 \text{ u} \end{cases}$
>
> Before separation $\quad ^2\text{H mass} = 2.014\,102 \text{ u}$
>
> The mass difference $\Delta m = 0.002\,388$ u. The energy equivalent of this mass, $\Delta E = (\Delta m)c^2$, is
>
> $$0.002\,388 \text{ u}\left(\frac{931.5 \text{ MeV}/c^2}{1 \text{ u}}\right) = \boxed{2.22 \text{ MeV}}$$

The above example illustrates an important procedure for calculating mass differences in nuclear reactions. Note that both ^{2}H and ^{1}H atoms have a single extranuclear electron. *We thus may use <u>atomic</u> masses for such calculations because the same number of electrons appear in both the "before" and "after" masses; the mass difference is not affected.*[2] (This simplification ignores the energy that binds electrons to atoms. But this energy is on the order of ~ 10 eV, so it may be neglected compared with the usual energies involved in nuclear reactions.)

Binding Energy

The nucleons are more tightly bound in some nuclei than in other nuclei. The strength of the nuclear bonding is characterized by the **binding energy per nucleon** (BE/nucleon). The more tightly bound the nucleons are, the more stable the nucleus becomes. For a nucleus of the atom ^A_ZX, we determine the BE/nucleon by calculating the total mass of $Z\,^1_1$H atoms and $(A - Z)$ neutrons and then subtract the mass of the ^A_ZX atom. This calculates the mass *increase* upon separation of the atom into individual nucleons. Dividing this by the number of nucleons A and converting to energy units gives

$$\left[\frac{\text{BE}}{\text{nucleon}}\right] = \frac{1}{A}[Zm_\text{H} + (A - Z)m_\text{n} - m_x]\left[\frac{\text{Energy equivalent}}{\text{Mass}}\right]$$

where m_H is the atomic mass of $^1_1$H, m_n is the neutron mass, and m_x is the atomic mass of the ^A_ZX atom. Obtaining the result in MeV/nucleon, we have

$$\left[\frac{\text{BE}}{\text{nucleon}}\right] = \frac{1}{A}[(1.007\,825 \text{ u})(Z) + (1.008\,665 \text{ u})(A - Z) - m_x]\left[\frac{931.494 \text{ MeV}/c^2}{1 \text{ u}}\right] \tag{45-4}$$

---

[2] There is one exception. See the discussion of *positron decay* leading to Equation (45-23).

EXAMPLE 45-4

Calculate the binding energy per nucleon for (a) ^{2}H, (b) ^{4}He, (c) ^{56}Fe, and (d) ^{238}U.

SOLUTION

(a) We can calculate the binding energy per nucleon for ^{2}H using the results of Example 45-3 or Equation (45-3). We will use the former. The total binding energy corresponds to a mass increase of 0.002 388 u or (1/2)(0.002 388 u) = 0.001 194 u per nucleon. Converting to the appropriate energy units, we obtain

$$\text{BE/nucleon} = (0.001\ 194\ \text{u}) \left[\frac{931.5\ \text{MeV}/c^2}{1\ \text{u}} \right] c^2 = \boxed{1.11\ \text{MeV}}$$

For the remaining parts of this example we refer to Table 45-1 for the atomic masses. *Atomic* masses are valid for these calculations because the same number of electrons appear in the "before" and "after" calculations and thus do not affect the mass *differences*.

(b) For ^{4}He we apply Equation (45-3), using appropriate units:

$$\text{BE/nucleon} = \tfrac{1}{4}[(1.007\ 825)2 + (1.008\ 665)(4-2) - 4.002\ 603](931.5)$$
$$= \boxed{7.07\ \text{MeV}}$$

(c) Similarly, for ^{56}Fe we obtain

$$\text{BE/nucleon} = \tfrac{1}{56}[(1.007\ 825)26 + (1.008\ 665)(56-26) - 55.934\ 939](931.5)$$
$$= \boxed{8.79\ \text{MeV}}$$

(d) And for ^{238}U we obtain

$$\text{BE/nucleon} = \tfrac{1}{238}[(1.007\ 825)92 + (1.008\ 665)(238-92) - 238.050\ 784](931.5)$$
$$= \boxed{7.57\ \text{MeV}}$$

In this example, we have shown that the binding energy per nucleon varies considerably from one nuclide to another. Figure 45-4 is a plot of the binding energy per nucleon *vs.* atomic mass number. Except for rather erratic behavior for atomic mass numbers less than about 20, the curve is fairly smooth, with a peak value at about $A = 63$. Of particular importance is the fact that ^{4}He lies considerably above the curve joining most of the points for low mass numbers, as do ^{8}Be, ^{12}C, ^{16}O, and ^{20}Ne. This suggests that an alpha particle (^{4}He—2 protons plus 2 neutrons) is a particularly stable combination of nucleons. Another indication is that when a nucleus spontaneously decays, an alpha particle is often ejected intact from the nucleus. We discuss this mode of decay in Section 45.5. We will also discover that because the binding energy per nucleon of the heavy nuclides is lower than that of nuclides near $A \approx 50-80$, nuclear energy can be obtained through *fission* (breaking apart) of heavy nuclei. Similarly, the fact that very light nuclides have a lower binding energy per nucleon compared to that of ^{4}He and ^{8}Be accounts for obtaining nuclear energy by *fusion* (combining together) of very light nuclei such as ^{2}H, ^{3}He, and ^{6}Li.

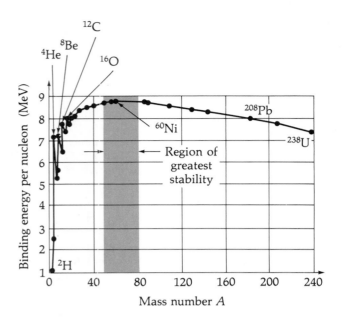

FIGURE 45-4
Binding energy per nucleon. We plot this graph by calculating the BE/nucleon for some representative isotopes, plotting the points, and connecting the points by straight lines. Note that ^{4}He lies well above the general trend of points for low mass numbers.

45.4 Radioactive Decay and Half-Life

Not all combinations of neutrons and protons form stable nuclei. Most of the approximately 1960 currently known nuclei decay to form other nuclei. Such nuclei are said to be *radioactive*. Only 279 of the naturally occurring nuclei are considered stable or *nonradioactive*. That is, they exhibit very little or no tendency to decay. The tendency for a nucleus to decay is indicated by its **half-life** $T_{1/2}$:

HALF-LIFE $T_{1/2}$ The average amount of time required for half the nuclei in a given large sample to decay to other nuclei.

Radioactive decay is a purely random process: any given radioactive nucleus may decay in the next second or one year from now. Thus, only by considering a very large number of nuclei can we make the concept of half-life meaningful. For a very large number of nuclei N, the rate of decay dN/dt, called the **activity**, is proportional to the number N of nuclei present:

ACTIVITY $$\frac{dN}{dt} = -\lambda N \qquad (45\text{-}5)$$

where λ is a positive constant of proportionality indicative of the stability of the nucleus. The larger the value of λ, the *less* stable the nucleus. The minus sign indicates that the number of nuclei decreases with increasing time. The constant λ is called the **decay constant**. Equation (45-4) may be rewritten in the form

$$\frac{dN}{N} = -\lambda\, dt$$

Integrating both sides of the equation, we obtain

$$\int_{N_0}^{N} \frac{1}{N}\, dN = -\lambda \int_{0}^{t} dt \qquad (45\text{-}6)$$

where N_0 is the number of nuclei at $t = 0$ and N is the number at time t. Performing the integration, we obtain

$$\ln\left(\frac{N}{N_0}\right) = -\lambda t$$

RADIOACTIVE DECAY
(using decay constant λ)
$$N = N_0 e^{-\lambda t} \qquad (45\text{-}7)$$

The half-life $T_{1/2}$ is related to λ in the following way. Equation (45-6) gives

$$\frac{N_0}{2} = N_0 e^{-\lambda T_{1/2}}$$

Taking the natural logarithm of both sides, we obtain

$$\ln 2 = \lambda T_{1/2}$$

or
$$T_{1/2} = \frac{\ln 2}{\lambda} \qquad (45\text{-}8)$$

A convenient form of Equation (45-6) then becomes

RADIOACTIVE DECAY
(using half-life $T_{1/2}$)
$$N = N_0 e^{-(\ln 2/T_{1/2})t} \qquad (45\text{-}9)$$

The following example illustrates the statistical nature of radioactive decay and its relationship to half-life.

EXAMPLE 45-5

Suppose that we have a huge number of dice. In order to make our calculations simple and avoid rounding-off difficulties, let us start with 279 936 000 dice at noon on April 1st. Suppose that at noon on April 2nd we throw all the dice and extract those that have only one dot uppermost. At noon on April 3rd, we again throw the remaining dice, extracting those with one dot uppermost. We continue the process day after day. When would only about half of the original number of dice remain?

SOLUTION

Since each side of a die is equally probable to be uppermost, approximately one-sixth of the dice will fall with one dot uppermost. The greater the number of dice, the more valid this assumption becomes. Thus, on April 2nd we extract (279 936 000)/6, leaving only 233 280 000 dice. We continue this process each noon and tabulate the results in Table 45-2. Because $\ln(N/N_0) = -\lambda t$, a *semilog plot* is appropriate to display the results,[3] Figure 45-5. Interploating between the

TABLE 45-2

Date	Dice Remaining
April 1	279 936 000
April 2	233 280 000
April 3	194 400 000
April 4	162 000 000
April 5	135 000 000
April 6	112 500 000
April 7	93 750 000
April 8	78 125 000
April 9	65 104 167
April 10	54 253 472

---

[3] Because it is easiest to curve-fit data points that plot as a straight line, exponential functions are usually plotted on *semilog graph paper* on which exponential curves plot as straight lines. On semilog graph paper, the spacings of divisions along the vertical axis are logarithmic. This causes the vertical distance Δy to be the same between all pairs of numbers having the same ratio (for example, 4 and 2, 6 and 3, or 20 and 10). Thus, a given distance Δy between points implies a fixed fractional ratio $\Delta N/N$ between numbers. Formally, if $N = N_0 e^{-\lambda t}$, then $\log N = \log N_0 - \lambda t$. Defining $\log N \equiv y$, we have $y = y_0 - \lambda t$, a straight line.

FIGURE 45-5
Example 45-5. The data points given in Example 45-5 are plotted on a semilog graph to produce a straight line.

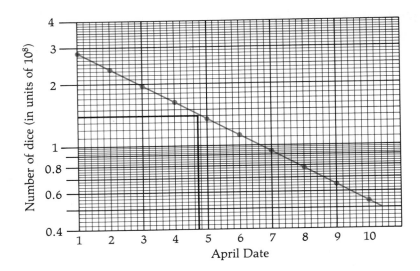

plotted points, we find that half of the original number of dice would remain at the end of about 4.7 days.

Another approach to the solution is to utilize Equation (45-4),

$$\frac{dN}{dt} = -\lambda N$$

where λ is one-sixth per day. The half-life of the dice remaining is then

$$T_{1/2} = \frac{\ln 2}{\lambda} = \frac{\ln 2}{(1/6)} = 4.16 \text{ days}$$

The discrepancy between the two results arises from the fact that throwing the dice once a day is not a continuous process as implied by Equation (45-4).

An interesting extension to this example is to continue the dice-throwing process until only one-fourth of the dice remain. Whereas the discrepancy between the two values obtained above is about 10%, the extension to two half-lives is less than 4%. (See Problem 45B-10.) The discrepancy is smaller because the dice-throwing process over a two–half-life period more nearly approximates a continuous process.

The rate dN/dt at which an isotope decays may be expressed in terms of the *becquerel* (Bq), defined as one disintegration per second. A more traditional unit is the *curie* (Ci):

THE CURIE $\qquad 1 \text{ Ci} = 3.71 \times 10^{10} \dfrac{\text{disintegrations}}{\text{second}}$

The equivalence is 1 Ci = 37.1×10^9 Bq = 37.1 GBq. (See Problem 45B-9 for the origin of the curie.)

EXAMPLE 45-6

A radioactive isotope has an initial activity of 5 mCi. Forty-eight hours later, the observed activity is 4 mCi. (a) Determine the half-life of the isotope. (b) Determine the initial number of nuclei in the sample of the isotope.

SOLUTION

(a) From Equation (45-9) we have

$$N = N_0 e^{-(\ln 2/T_{1/2})t}$$

Since by Equation (45-5) $N = -(1/\lambda)(dN/dt)$, we may write

$$\frac{dN}{dt} = \left(\frac{dN}{dt}\right)_0 e^{-(\ln 2/T_{1/2})t}$$

Setting $(dN/dt)_0 = A_0$ as the initial activity, we have

ACTIVITY A
$$A = A_0 e^{-(\ln 2/T_{1/2})t} \qquad (45\text{-}10)$$

Taking the logarithm of both sides of this equation and solving for $T_{1/2}$, we have

$$T_{1/2} = \frac{(\ln 2)t}{\ln\left(\dfrac{A}{A_0}\right)} = \frac{(\ln 2)48\text{ h}}{\ln\left(\dfrac{5\text{ mCi}}{4\text{ mCi}}\right)} = \boxed{149\text{ h}}$$

(b) To obtain the number of radioactive nuclei corresponding to a given half-life and activity, we eliminate λ between Equation (45-5)

$$\frac{dN}{dt} = -\lambda N$$

and Equation (45-8)
$$T_{1/2} = \frac{\ln 2}{\lambda}$$

and obtain
$$N = -\left(\frac{dN}{dt}\right)\frac{T_{1/2}}{\ln 2}$$

Noting that the initial activity, $5\text{ mCi} = (5 \times 10^{-3})(3.71 \times 10^{10}) = 1.86 \times 10^8$ disintegrations per second, corresponds to -1.86×10^8 nuclei per second and that the half-life must be expressed as $(149\text{ h})(3600\text{ s/h}) = 5.36 \times 10^5$ s, we obtain

$$N = -(-1.86 \times 10^8)\frac{5.36 \times 10^5}{\ln 2} = \boxed{1.44 \times 10^{14}\text{ nuclei}}$$

45.5 Modes of Radioactive Decay

Certain isotopes may spontaneously decay to form other isotopes. Though a given isotope will choose only one (or, rarely, more) mode of decay, the process may happen in many different ways:

(1) α decay
(2) β decay
(3) γ decay
(4) internal conversion
(5) electron capture
(6) spontaneous fission

We will discuss each in turn. What determines whether a given isotope may spontaneously decay? The criterion is this:

The mass of the reaction products must be less than the mass of the original isotope.

An important measure of this difference is the *"Q" of the reaction*. If a mass Δm disappears in the reaction, an amount of energy, $(\Delta m)c^2 \equiv Q$, appears as energy of the products.

Q OF A REACTION (Original mass)c^2 = (Product mass)c^2 + Q (45-11)

In the following discussions, note how we always turn our attention first to the *mass differences*.

(1) Alpha Decay

The decay of a nuclide by **alpha decay** produces a new element. Such a process, called *transmutation*, is written as

$$^{A}_{Z}X \longrightarrow \,^{A-4}_{Z-2}Y + \,^{4}_{2}He \qquad (45\text{-}12)$$

Note that this representation shows the conservation of the number of nucleons because the superscript on the left side of the equation equals the sum of those on the right. Similarly, charge is conserved because the number of protons represented by the atomic number Z, as well as the atomic electrons, also balances. The nuclide $^{A}_{Z}X$ is called the **parent** nuclide, while $^{A-4}_{Z-2}Y$ is called the **daughter** nuclide.

The high binding energy per nucleon for ^{4}He shows that it is a particularly tightly bound configuration (see Figure 45-4). This accounts for the likelihood of alpha decay among the heavy radioactive nuclides. In fact, alpha decay rarely takes place for elements lighter than osmium, ^{186}Os, because such a process would produce an isotope above the curve of stable isotopes in Figure 45-1. A particular mode of decay is enhanced if the Q defined by Equation (45-11) is large. The following example illustrates the relative likelihood of alpha decay.

EXAMPLE 45-7

Calculate the energy associated with the alpha decay of ^{210}Po.

SOLUTION

The equation for the alpha decay of ^{210}Po is

$$^{210}_{84}Po \longrightarrow \,^{206}_{82}Pb + \,^{4}_{2}He$$

We identify the daughter nuclide by balancing the atomic mass numbers and atomic numbers on both sides of the equation. The masses associated with these nuclides are obtained from Table 45-1:

	Parent atom	^{210}Po	209.982 848 u
Decay Products {	Daughter atom	^{206}Pb	205.974 440 u
	Helium atom	^{4}He	4.002 603 u
	Decay product sum:		209.977 043 u

The mass of the parent atom *exceeds* the mass of the decay products by 0.005 805 u, indicating that alpha decay can (and does) occur. Converting the mass difference to energy, we have

$$E = (\Delta m)c^2 = (0.005\ 805\ \text{u})\left(\frac{931.5\ \text{MeV}/c^2}{1\ \text{u}}\right) = \boxed{5.41\ \text{MeV}}$$

(Note that, as usual, we use *atomic* masses here; when we calculate Δm, the masses of the extranuclear electrons cancel.) This energy appears as kinetic energy of the products. Because the α particle mass is much less than the mass of the daughter nucleus, momentum conservation requires that the α particle recoil with much greater velocity than the daughter nucleus. The lighter particles always receive most of the kinetic energy in nuclear decays.

Quantum Mechanical Tunneling in Alpha Decay The mechanism of alpha decay has an interesting quantum mechanical explanation. The nucleons that constitute an alpha particle are strongly bound to the nucleus by an attractive nuclear force of short range and are repelled by the Coulomb force of the other protons in the nucleus. Figure 45-6 shows the net potential energy that results from the combination of these two forces. The nucleons forming the alpha particle have the total energy E shown in the figure. Classically, the alpha particle would be bound forever within the nucleus because of the Coulomb potential barrier. Recall that the total energy E is equal to the sum of the kinetic energy K and the potential energy U; consequently, within the shaded region between R and R_1 the kinetic energy of the alpha particle would have to be negative—classically impossible. However, quantum-mechanically the alpha-particle wave function extends beyond the boundary of the nucleus to where the kinetic energy is positive, and therefore it has a finite probability of being found outside R_1. In effect, the alpha particle repeatedly "knocks on the door" of the barrier until it *quantum-mechanically tunnels* through it to appear outside the nucleus (cf. Section 43.7). The dashed line of Figure 45-6 shows one possible wave function for the alpha particle.

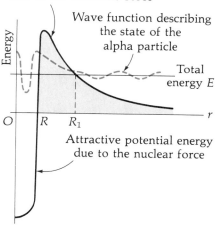

FIGURE 45-6
The total potential energy of the alpha-particle–nucleus interaction versus the separation distance r.

(2) Beta Decay

Certain nuclei emit electrons, $_{-1}^{0}e$ (or β^-), or positrons, $_{+1}^{0}e$ (or β^+), in a process called **beta decay**. These reactions involve the *"weak" interaction*—one of the four basic interactions in nature: electromagnetic, strong, weak, and gravitational.

BETA DECAY PROCESSES

Electron (β^-) decay $\quad _{Z}^{A}X \longrightarrow\ _{Z+1}^{A}Y + _{-1}^{0}e + \bar{\nu}$ (45-13)

Positron (β^+) decay $\quad _{Z}^{A}X \longrightarrow\ _{Z-1}^{A}Y + _{+1}^{0}e + \nu$ (45-14)

As will be discussed shortly, the process also includes the emission of a *neutrino*[4] ν or an *antineutrino* $\bar{\nu}$. Since β particles are emitted from a nucleus,

---

[4] There are three kinds of neutrinos and their corresponding antineutrinos: the *electron neutrino* ν_e (emitted during beta decay), and two other neutrinos emitted in other processes—the *muon neutrino* ν_μ, and the *tauon neutrino* ν_τ. Neutrino masses are postulated to be zero, though they may have a small mass. Experiments to measure neutrino masses can place only an *upper limit* to the mass. To date (1989), the ν_e mass is believed to be less than ~ 20 eV/c^2 from ^{3}He decay and less than ~ 14 eV/c^2 from the delay between the arrival times of the neutrino pulse and the light pulse from Supernova 1987A. (This does not rule out the possibility of truly zero mass.)

it is reasonable to ask, "Do electrons and positrons actually reside inside a nucleus?" As Problem 45B-24 shows, because of the uncertainty principle, β particles do not exist as a separate entity inside nuclei; they are created during the beta-decay process itself.

β^- Decay The β^- decay process is equivalent to the transformation of a neutron within the nucleus into a proton, an electron, and an antineutrino, which together escape from the nucleus, Figure 45-7. To predict whether a given nucleus may undergo such a process, we calculate the Q value, Equation (45-11). The atomic mass of the parent nuclide M_X is equal to the atomic mass of the nucleus m_X plus Z atomic electrons, each with a mass m_e:

$$M_X = m_X + Zm_e \tag{45-15}$$

Similarly, the mass of the daughter nuclide is

$$M_Y = m_Y + (Z + 1)m_e \tag{45-16}$$

The Q of the reaction, Equation (45-11), becomes

$$Q = [m_X - (m_Y + m_e)]c^2 \tag{45-17}$$

indicating that although the mass of the beta particle is involved in the mass loss, the Z atomic electrons are not. Using Equations (45-15) and (45-16), we have

$$Q = (M_X - Zm_e)c^2 - [M_Y^Y - (Z + 1)m_e + m_e]c^2$$

Q FOR β^- DECAY
$$Q = (M_X - M_Y)c^2 \tag{45-18}$$

Thus, if the parent atomic mass is at all greater than the daughter atomic mass, β^- decay can occur. Conversely, if the atomic mass of the parent nuclide is less than that of the daughter, β^- decay cannot occur, as the following example illustrates.

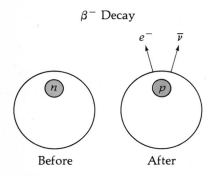

FIGURE 45-7
In a β^- decay, the number of nucleons remains the same. The net effect is that a neutron within the nucleus disintegrates into a proton, which remains in the nucleus, and an electron e^- plus an antineutrino $\bar{v}$ escape. (In β^+ decay, a proton in the nucleus transforms into a neutron plus a positron and a neutrino.)

EXAMPLE 45-8

Show that β^- decay of ^{210}Po cannot occur.

SOLUTION

We first identify the daughter nuclide by writing the reaction equation

$$^{210}_{84}\text{Po} \longrightarrow {}^{210}_{85}\text{At} + {}^{0}_{-1}e + \bar{v}$$

The daughter nuclide must have an atomic number Z one unit greater than that of polonium, with the atomic mass A unchanged. This is the nuclide *astatine* ($A = 210$, $Z = 85$). From Table 45-1, we obtain the atom mass numbers ^{210}Po, 209.982 848 u; and ^{210}At, 209.987 126 u. The daughter nuclide is *more* massive than the parent nuclide. Therefore, β^- decay of ^{210}Po cannot occur.

Figure 45-8 shows how the available kinetic energy is distributed among the beta particles. This energy distribution reveals that essentially all of the emitted electrons have *less* kinetic energy than the Q of the reaction provides. (Because the daughter nucleus is so massive compared to the electron's mass, the recoil nucleus carries negligible kinetic energy.) Where is the missing energy? Furthermore, when the trajectories of the beta particle and the recoiling nucleus are determined, they almost never have exactly opposite directions, so linear momentum is not conserved. Also, for reasons beyond the scope of this discussion, angular momentum is not conserved. Following Wolfgang Pauli's suggestion (in 1930) that another uncharged particle participated in the decay, Enrico Fermi developed a new theory in 1934. Fermi proposed that a neutral particle that escaped detection shared some of the kinetic energy. He coined the name "neutrino" for this unseen particle. *By proposing that the neutrino had the properties of zero charge, no rest mass, and spin $\frac{1}{2}$, Fermi could thereby preserve the three important principles of the conservation of energy, linear momentum, and angular momentum.* A worthy achievement!

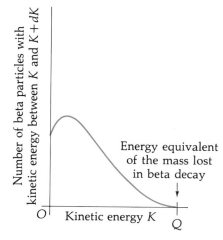

FIGURE 45-8
The kinetic energy distribution among the emitted beta particles in beta decay. Most particles have considerably less than the available energy Q for the reaction. The "missing" energy is mainly taken up by the *antineutrino* in β^- decay (or the *neutrino* in β^+ decay), which is emitted simultaneously with the beta particle. A small amount of energy is taken up by the recoil of the daughter nucleus.

β^+ Decay β^+ (positron) decay is a rarer occurrence than β^- decay because a much greater mass difference between parent and daughter nuclides is necessary. The reaction equation for positron decay is

$$^A_Z X \longrightarrow {}^{A}_{Z-1}Y + {}^{0}_{+1}e + \nu \tag{45-19}$$

where $^{0}_{+1}e$ is the positron and ν the neutrino. Following the procedure we used for β^- decay, we seek an expression for the Q of the reaction. The mass of the parent nuclide is

$$M_X = m_X + Zm_e \tag{45-20}$$

where M_X is the atomic mass, m_X is the nuclear mass, and Zm_e is the total mass of the atomic electrons. For the daughter nuclide, we have

$$M_Y = m_Y + (Z-1)m_e \tag{45-21}$$

From Equation (45-11), we obtain the equation for Q

$$Q = m_X c^2 - (m_Y + m_e)c^2 \tag{45-22}$$

where m_e is the positron mass (equal to the electron mass). Again the Z *atomic* electrons are not involved in the reaction. Using Equations (45-20) and (45-21), we obtain

$$Q = (M_X - Zm_e)c^2 - [M_Y - (Z-1)m_e + m_e]c^2$$

Q FOR β^+ DECAY
$$Q = (M_X - M_Y - 2m_e)c^2 \tag{45-23}$$

Thus, for β^+ decay, *the parent nuclide mass must exceed the daughter nuclide mass by at least two electron masses.* (This is an exception to the general procedure of using *atomic* masses only in nuclear reaction equations.)

> **EXAMPLE 45-9**
>
> Find the maximum energy of the positrons emitted from ^{22}Na.
>
> **SOLUTION**
>
> As in the previous example, we identify the daughter nuclide by writing the reaction equation:
>
> $$^{22}_{11}\text{Na} \longrightarrow {}^{22}_{10}\text{Ne} + {}^{0}_{+1}e + \nu$$
>
> From Table 45-1, we obtain the atomic masses for ^{22}Na, 21.994 434 u; ^{22}Ne, 21.991 383 u; and m_e, 0.000 549 u. Substituting into Equation (45-23), we have
>
> $$Q = [21.994\ 434\ \text{u} - 21.991\ 383\ \text{u} - 2(0.000\ 549\ \text{u})]c^2 \left(\frac{931.5\ \text{MeV}/c^2}{1\ \text{u}}\right)$$
>
> $$= \boxed{1.82\ \text{MeV}}$$

(3) Gamma Decay

Gamma rays emitted from the nucleus are high-energy photons of electromagnetic radiation emitted during energy-state transitions within the nucleus. (They are analogous to the photons emitted from atoms when atomic electrons move from a higher energy state to a lower state.) Since **gamma decay** does not alter the atomic mass number or the atomic number, transmutation to another element does not take place.

Excited nuclear states occur most often in the daughter nuclide of another decay reaction. A simple illustration is the β^- decay of ^{137}Cs. When observing the decay of ^{137}Cs, we find that electrons and gamma rays are emitted essentially simultaneously. However, the gamma rays are emitted by the daughter nuclide. The reaction equations for this decay are

$$^{137}_{55}\text{Cs} \longrightarrow [^{137m}_{\ 56}\text{Ba}^*] + {}^{0}_{-1}e + \bar{\nu} \qquad (45\text{-}24)$$

$$[^{137m}_{\ 56}\text{Ba}^*] \longrightarrow {}^{137}_{\ 56}\text{Ba} + \gamma \qquad (45\text{-}25)$$

where the asterisk (*) represents an excited state of barium. The state is called *metastable* because it exists for a relatively long time before undergoing a transition to a lower energy state—long enough for the state's half-life to be directly measurable. The half-life of the beta decay shown in Equation (45-24) is 30.17 yr, while the gamma decay shown in Equation (45-25) has a half-life of 2.55 min.

Energy levels within a nucleus may be inferred from the energies of the gamma rays emitted. In the case of ^{137}Cs decay, an energy level within the ^{137}Ba nuclide is determined by reference to the energy-level diagram shown in Figure 45-9. From Equation (45-24) we obtain a Q value of 1.17 MeV. This corresponds to the sum of the maximum energy of the beta decay electron, 0.512 MeV, and the observed gamma-ray energy of 0.662 MeV. Thus, ^{137}Ba has a nuclear excited energy state of 0.662 MeV above the ground state. Most decay processes are accompanied by the emission of gamma rays with many discrete energies. For example, the beta decay of ^{140}Cs produces gamma rays with twenty different energies, indicating a very complicated nuclear-energy-level structure of the daughter nuclide, $[^{140}\text{Ba}^*]$.

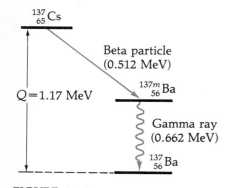

FIGURE 45-9
The energy-level diagram associated with the beta decay of $^{137}_{55}$Cs to the metastable state of barium, $[^{137m}_{\ 45}\text{Ba}^*]$, followed by gamma decay to the ground state of barium.

(4) Internal Conversion

Instead of undergoing gamma decay, an excited nucleus can get rid of its excess energy in a process called **internal conversion**. The wave functions for atomic electrons penetrate the nucleus slightly, permitting a direct interaction between the nucleus and an atomic electron. The nucleus transfers its excess energy to the electron, ejecting it with a kinetic energy equal to the nuclear transition energy minus the Coulomb energy that bound the electron to the atom. Though internal conversion occurs most frequently with K-shell electrons, it can also occur with electrons in other shells. These internal-conversion electrons appear as spikes of discrete energies superposed upon the continuous beta-decay spectrum of Figure 45-7. The ejection of such electrons leaves vacancies in low-lying atomic energy levels, and outer electrons falling into these vacancies produce x-rays. For example, for the decay of Equations (45-24) and (45-25), an x-ray of energy 0.032 MeV is produced.

(5) Electron Capture

As we have shown, β decay is essentially the transformation of either a proton p or a neutron n within the nucleus. Such reactions are

THE BASIC BETA-DECAY PROCESS

$$\beta^- \text{ Decay} \quad {}^1_0n \longrightarrow {}^1_1p + {}^{\ 0}_{-1}e + \bar{\nu}$$

$$\beta^+ \text{ Decay} \quad {}^1_1p \longrightarrow {}^1_1p + {}^{\ 0}_{+1}e + \nu$$

The question arises, "Can an electron originating from outside a nucleus be captured by a proton within the nucleus, transforming the proton into a neutron?" The answer is yes. The process is

ELECTRON CAPTURE

$$ {}^{\ 0}_{-1}e + {}^1_1p \longrightarrow {}^1_0n + \nu \qquad (45\text{-}26)$$

Electron capture (EC) is a very common decay process. As discussed in Chapter 44, atomic electrons in an atom have wave functions that extend to the nucleus and overlap it slightly, so there is a finite probability that such electrons could be captured by a proton in the nucleus. Because the wave functions for K-shell electrons have a larger amplitude near $r = 0$ than those for electrons in outer shells, capture of K-shell electrons is most probable. For this reason, electron capture is often called *K capture*. The reaction equation for electron capture is

$$ {}^A_Z X \longrightarrow {}^{\ \ A}_{Z-1}Y^+ + \nu \qquad (45\text{-}27)$$

where the superscript "plus" sign indicates that the daughter nuclide has lost one of its atomic electrons, thus becoming a positive ion.

Electron capture is possible only if there is a mass loss in the reaction. That is, the Q of the reaction must be positive. We derive an expression for the Q of the reaction as follows:

$$Q = (m_X - m_Y)c^2$$

where m_X and m_Y are nuclear masses. The atomic masses are

$$M_X = m_X + Zm_e \qquad (45\text{-}28)$$
$$M_Y = m_Y + (Z-1)m_e - m_e \qquad (45\text{-}29)$$

TABLE 45-3 Radioactive Decay Processes

Process	Daughter Nucleus	Q
β^- emission	One Z higher	Positive
β^+ emission	One Z lower	Positive ($> 2m_e c^2$)
Gamma emission	Same Z	—
Internal conversion	Same Z	—
Electron capture (K capture)	One Z lower	Positive

Substituting values of m_X and m_Y from Equations (45-28) and (45-29), we have

$$Q = (M_X - Zm_e)c^2 - [M_Y - (Z-1)m_e - m_e]c^2$$

Q FOR ELECTRON CAPTURE

$$Q = (M_X - M_Y)c^2 \qquad (45\text{-}30)$$

Electron capture and positron emission are competing processes [see Equation (45-23)], and each leads to a nuclide one unit lower in Z. For each, the atomic mass of the parent exceeds that of the daughter. However, if the Q for the reaction is less than the equivalent of two electron masses ($2m_e c^2 = 1.02$ MeV), only electron capture can occur. Table 45-3 summarizes the various processes by which an excited nucleus gets rid of its excess energy. The characteristic γ rays and x-rays identify the Z of the daughter nucleus, thus distinguishing between the processes.

(6) Spontaneous Fission

Many heavy nuclides above $Z = 90$ experience spontaneous fission, splitting into two unequal fragments plus two or three neutrons. Figure 45-10 shows the distribution of the fragments from ^{236}U. The fragments generally lie along a straight line (shown dashed) joining the parent nuclide and the origin. From Figure 45-3, the binding energy per nucleon of the fragments is greater than that of the original nuclide, so *the total mass of the fragments is always less than that of the parent*, releasing roughly 200 MeV in each fission process. Both fragments have excess neutrons, and they immediately ($< 10^{-15}$ s) emit two or three so-called *prompt neutrons*. (A few other *delayed neutrons* may be emitted later.) The following example describes one of the probable fission reactions of ^{236}U.

EXAMPLE 45-10

Determine the Q associated with the spontaneous fission of ^{236}U into the fragments ^{90}Rb and ^{143}Cs.

SOLUTION

The reaction equation must be written in order to determine the number of neutrons involved in the fission:

$$^{236}_{92}\text{U} \longrightarrow {}^{90}_{37}\text{Rb} + {}^{143}_{55}\text{Cs} + 3{}^{1}_{0}n \qquad (45\text{-}31)$$

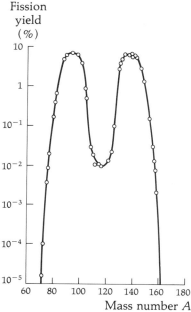

(a) The distribution of the yields of fission fragments indicates that fission is usually asymmetrical, with the most probable fission yielding two nuclei having mass numbers around 96 and 140. [Adapted from J. M. Siegel et al., "Plutonium Project Report on Nuclei Formed in Fission," *Review of Modern Physics* **18**, 538 (1946).]

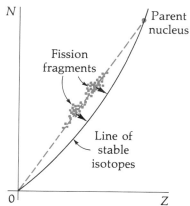

(b) The line of stable isotopes is derived from Figure 45-1. The fission fragments initially "land" on the dashed line, and they usually move diagonally downward by β^- emission to reach the *stability curve* of stable isotopes.

FIGURE 45-10
The fission of ^{236}U after the absorption of a thermal neutron by ^{235}U. The kinetic energy of each fragment is roughly 90 MeV. The fragments subsequently decay by ejecting neutrons, β particles, γ rays, and neutrinos, yielding an additional 20 MeV or so, for a total energy release of roughly 200 MeV per fission.

We deduce the fact that three neutrons are products of the reaction by balancing the mass numbers and atomic numbers on both sides of the reaction equation. From Table 45-1 we obtain the atomic masses ^{236}U, 236.045 562 u; ^{143}Cs, 142.927 220 u; ^{90}Rb, 89.914 811 u; and n, 1.008 665 u. It can be shown (see Problem 45A-16) that the $Q = (\Delta m)c^2$ for this fission reaction is

$$Q = (M_U - M_{Rb} - M_{Cs} - 3m_n)c^2$$
$$Q = [236.045\ 562\ u - 89.914\ 811\ u - 236.045\ 562\ u - 3(1.008\ 665\ u)]c^2$$
$$= (0.1775\ u)c^2$$

Converting to the conventional units of MeV, we have

$$Q = (0.1775\ u)c^2 \left[\frac{931.5\ \text{MeV}/c^2}{1\ u} \right] = \boxed{166\ \text{MeV}}$$

TABLE 45-4
Distribution of Energy in Fission

	Energy (MeV)
Kinetic energy of	
fission fragments	165 ± 5
neutrons	5 ± 0.5
γ rays (prompt)	7 ± 1
Delayed γ rays	6 ± 1
Delayed β particles	7 ± 1
Delayed neutrinos	10
Total	200 ± 6

As shown in Table 45-4, if we include the additional energy from the decay of fission fragments, the total energy of this fission process is approximately 200 MeV. This very high energy is typical of fission reactions but, as we will discuss later, for several reasons it is not the ideal source of energy.

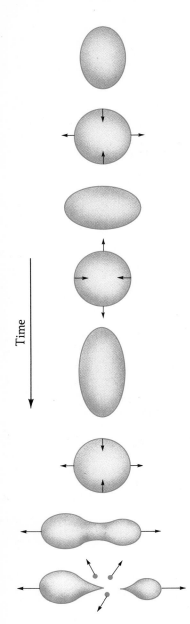

FIGURE 45-11
The liquid-drop model of an excited ^{236}U nucleus undergoing fission. The excess energy causes the charged drop to undergo rapid oscillations between flattened and elongated shapes (the arrows indicate motions of the drop's surface). Finally, the drop elongates sufficiently to form a "neck," and electrostatic repulsion breaks the nucleus into two unequal-size fragments with the emission of a few prompt neutrons.

We can form a mental picture of the fission process by considering the parent nucleus to behave as a liquid drop with a "surface tension" arising from the strong nuclear forces. If energy is added to a nucleus by the absorption of a neutron (or other energy transfers) this excitation energy sets up oscillations of the drop, distorting the shape alternately into a football-shaped ellipsoid or a flattened-doorknob shape as shown in Figure 45-11. Surface tension forces tend to pull the drop back into a spherical shape, while the excitation energy tends to distort the drop even further. When it distorts into a dumbbell shape with a neck, the Coulomb repulsion of the two ends can split the drop into two fragments,[5] with a few energetic neutrons emitted immediately in the process.

Radioactive Decay Series

The radioactive decay of one nuclide may result in successive decays of a series of isotopes until a stable nuclide terminates the sequence of decays. Prominent among these *radioactive decay series* is one that originates with ^{238}U and terminates with ^{206}Pb, as shown in Figure 45-10. The half-life for each decay process is shown in the figure. Notice that the half-life of the first decay, from ^{238}U to ^{234}Th (4.5×10^9 yr), is so much longer than the others that it is essentially the half-life of the entire series that transforms ^{238}U to ^{206}Pb. This long half-life enables geologists to determine the age of certain rocks. As molten rocks crystallize, there is often a natural separation of minerals because of their different melting-point temperatures. Thus, when initially formed, the rocks contain known proportions of different elements. If one of the elements is radioactive, the composition of the mineral changes as time passes, forming a sort of geological calendar. The following is an example of such a process.

EXAMPLE 45-11

A specimen of uranite (a uranium-bearing mineral) contains five times as many atoms of ^{238}U as of ^{206}Pb. Assume that all of the lead originated from the radioactive decay sequence shown in Figure 45-12 and that the uranite contained uranium and no lead when it was formed. (Other isotopes of uranium form series not terminating in ^{206}Pb.) Calculate the number of years that have passed since the formation of the uranite.

SOLUTION

The total number of nuclei in a given sample does not change, so

$$N_0 = N_U + N_{Pb}$$

where N_0 is the original number of ^{238}U nuclei, N_U is the current number of ^{238}U nuclei, and N_{Pb} is the current number of ^{206}Pb nuclei. Then

$$\frac{N_0}{N_U} = 1 + \frac{N_{Pb}}{N_U} = 1 + \frac{1}{5} = \frac{6}{5}$$

The current number of uranium nuclei N relative to the original number N_0 is given by Equation (45-9):

$$N = N_0 e^{-(\ln 2/T_{1/2})t}$$

---

[5] If the excitation energy is not large enough to cause fission, the drop can get rid of the excess energy by gamma radiation.

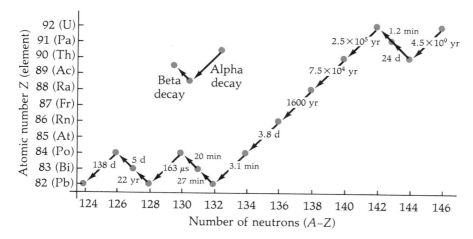

FIGURE 45-12
The primary sequence in the ^{238}U-to-^{206}Pb decay series. The arrows represent the steps in the decay process. The half-life of each step is shown along the arrow. This figure corresponds to the upper right-hand portion of Figure 45-2.

Solving for t gives
$$t = \frac{\ln(N_0/N)T_{1/2}}{\ln 2}$$

With the half-life of the series essentially 4.5×10^9 yr and the ratio $N_0/N = N_0/N_U = 6/5$, we obtain

$$t = \frac{\ln(6/5)(4.5 \times 10^9 \text{ yr})}{\ln 2} = \boxed{1.18 \times 10^9 \text{ yr}}$$

For time periods much shorter than a billion years, an interesting fact is that one gram of ^{238}U decays to produce 1.33×10^{-10} g of ^{206}Pb per year. (See Problem 45B-21.)

45.6 Nuclear Cross Section

The likelihood of an interaction between two particles depends upon their mutual "sphere of influence." A convenient way to picture the situation is to imagine that the incoming particles are point projectiles and that each nucleus presents a *projected target area* called the **cross section** σ to the incoming particles, Figure 45-13. A reaction occurs only if a particle strikes a target area.[6] The cross section has little relation to the actual physical size of the interacting particles. Indeed, a give nucleus can have widely different cross sections for different nuclear reactions. For example, the cross section σ_s for scattering the incoming particle may have two parts: the cross section σ_e for *elastic* scattering (involving no kinetic energy loss) and a different cross section σ_i for *inelastic* scattering. Furthermore, σ can depend strongly on the speed of the incoming particle, as shown in Figure 45-14. The unit for measuring cross sections is the *barn* (b):

THE BARN $\qquad 1 \text{ b} \equiv 10^{-28} \text{ m}^2 = 10^{-24} \text{ cm}^2 \qquad (45\text{-}32)$

Not all of the particles incident upon a target foil necessarily interact with target nuclei—some may pass through the foil without interacting. To

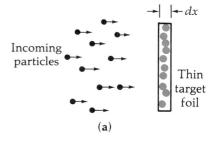
(a)

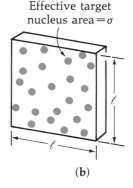
(b)

FIGURE 45-13
A beam of point particles incident upon a thin foil of target nuclei. Each target nucleus presents an effective projected area σ to the incoming beam. A reaction occurs if a particle strikes an area σ.

---

[6] Another analogy is throwing darts at a wall on which inflated balloons are attached at various points. The chance that a randomly thrown dart will strike a balloon depends upon the projected area σ that each balloon presents to the incoming dart, the number of balloons, and the total area of the wall.

find the probability that an incoming particle will interact with a nucleus, we calculate the total *effective interaction area* that an incoming particle "sees" as it approaches the foil. This target area is the product of the number of nuclei in the foil and the cross-sectional area σ of each nucleus. For a square foil of area ℓ^2 and thickness dx, this equals $n\sigma\ell^2\,dx$, where n is the number of nuclei per unit volume, $\ell^2\,dx$ is the volume, and σ is the cross section. The ratio of the number of collisions dN to the number of incident nuclei N equals the ratio of the total area of target nuclei to the area ℓ^2 of the foil. Thus:

$$-\frac{dN}{N} = \frac{n\sigma\ell^2\,dx}{\ell^2} = n\sigma\,dx \tag{45-33}$$

(The minus sign indicates that particles are being removed from the beam.)

Integrating,
$$\int_{N_0}^{N} \frac{dN}{N} = -\int_0^x n\sigma\,dx$$

$$\ln N \Big|_{N_0}^{N} = -n\sigma x \Big|_0^x$$

$$\ln\left(\frac{N}{N_0}\right) = -n\sigma x \tag{45-34}$$

we obtain
$$N = N_0 e^{-n\sigma x} \tag{45-35}$$

Thus the number of incoming particles that penetrates a distance x into a target material without interacting decreases exponentially with the distance x.

EXAMPLE 45-12

Control rods made of cadmium are often used to capture excess slow neutrons in a fission reactor because of the very high slow-neutron capture cross section of the cadmium isotope ^{113}Cd, Figure 45-14. This cross section of 1.99×10^4 b is the largest of any known nuclide. Calculate the approximate thickness (in centimeters) of a sheet of natural cadmium that will absorb half the slow neutrons falling upon its surface. The cadmium isotope ^{113}Cd is 12.22% abundant in natural cadmium. The density of natural cadmium is 8.65 g/cm³, and its molecular weight is 112.41 g/mole.

SOLUTION

Equation (45-34) gives the absorption of a beam of neutrons as it passes through a thickness x of material whose nuclear cross section for absorption is σ. The number n of cadmium nuclei per unit volume is

$$n = \frac{\rho N_A}{\text{(mol. wt)}}$$

where ρ is the density, N_A is Avogadro's number, and mol. wt is the molecular weight. Of this number, only 12.22% are ^{113}Cd nuclei. Thus the number n of ^{113}Cd nuclei per cubic centimeter is

$$n = (0.1222)\frac{\rho N_A}{\text{(mol. wt)}} = \frac{(0.1222)(8.65\text{ g/cm}^3)(6.022 \times 10^{23}\text{ molecules/mole})}{(122.41\text{ g/mole})}$$

$$n = 5.66 \times 10^{21}\text{ nuclei/cm}^3$$

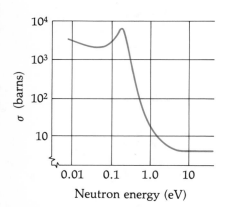

FIGURE 45-14
The cross section for neutron capture by natural cadmium $_{48}$Cd as a function of neutron energy. Note the logarithmic scales. The large cross section for slow neutrons is due almost entirely to the isotope ^{113}Cd. (Adapted from Donald J. Hughes and Robert B. Schwartz, *Neutron Cross Sections*, Brookhaven National Laboratory, July 1, 1958.)

Using this value in Equation (45-34) to find the distance x in which the incident number of neutrons is reduced to one-half, we get

$$\ln\left(\frac{N}{N_0}\right) = -n\sigma x$$

$$x = \frac{-\ln(N/N_0)}{n\sigma}$$

$$= \frac{-\ln(0.5)}{(5.66 \times 10\ 10^{21}\ \text{nuclei/cm}^3)(1.99 \times 10^4\ \text{b})(10^{-24}\ \text{cm}^2/\text{b})}$$

$$x = 0.00615\ \text{cm} = \boxed{6.15 \times 10^{-3}\ \text{cm}}$$

The concept of a cross section is useful in describing various nuclear reactions, the scattering of alpha particles by nuclei, neutron activation analysis (Section 45.7), and many other types of interactions.

45.7 Nuclear Reactions

There is a large class of nuclear reactions in which a particle is incident upon a target nucleus (initially at rest). This forms a "compound nucleus" in an excited state, which immediately decays by emitting one or more particles. Such a reaction was first observed by Rutherford in 1919 when he found that alpha particles ($_2^4$He) passing through nitrogen gas produced protons. The reaction is written as

$$_2^4\text{He} + {}_7^{14}\text{N} \rightarrow [{}_9^{18}\text{F}^*] \longrightarrow {}_8^{17}\text{O} + {}_1^1\text{H} + Q$$

We infer the presence of the compound nucleus by balancing the total mass numbers and charge numbers on both sides of the reaction. Other examples (omitting the compound state) are

Reaction	Notation
$_0^1 n + {}_7^{14}\text{N} \longrightarrow {}_5^{11}\text{B} + {}_2^4\text{He}$	$_7^{14}\text{N}\ (n,\alpha)\ {}_5^{11}\text{B}$
$_1^2 d + {}_3^6\text{Li} \longrightarrow {}_3^7\text{Li} + {}_1^1 p$	$_3^6\text{Li}\ (d,p)\ {}_3^7\text{Li}$
$_1^1 p + {}_6^{13}\text{C} \longrightarrow {}_7^{13}\text{N} + {}_0^1 n$	$_6^{13}\text{C}\ (p,n)\ {}_7^{13}\text{N}$
$_0^1 n + {}_{16}^{32}\text{S} \longrightarrow {}_{16}^{33}\text{S} + {}_0^0 \gamma$	$_{16}^{32}\text{S}\ (n,\gamma)\ {}_{16}^{33}\text{S}$

Of course, the conservation of energy and momentum applies to such reactions. Consider the general case of an incident particle x striking a target nucleus X (initially at rest). The reaction products are y and Y:

$$x + X \longrightarrow y + Y \qquad (45\text{-}36)$$

Letting K represent kinetic energies, we have for the conservation of mass-energy

$$E_0 = E$$
$$K_x + m_x c^2 + M_X c^2 = K_Y + K_y + m_y c^2 + M_Y c^2 \qquad (45\text{-}37)$$

We limit our discussion to low-energy reactions for which the kinetic energies and momenta may be considered classically instead of relativistically. The Q value for a reaction is the difference between the initial and final mass-energies of the particles. From the above equation, we see that it also equals the difference in kinetic energies:

Q VALUE FOR A REACTION
$$\begin{cases} Q = (\Delta m)c^2 = (m_x + M_X - m_y - M_Y)c^2 \\ Q = (K_y + K_Y - K_x) \end{cases} \quad (45\text{-}38)$$

If some mass disappears in the reaction, Q is *positive* and the kinetic energy of the products is greater than the initial kinetic energy; some mass-energy has been transformed into kinetic energy. This is called an *exoergic* reaction— one that releases some mass-energy. If Q is *negative*, some mass has been created at the expense of the output kinetic energy—an *endoergic* reaction.

When Q is negative, not all of the kinetic energy of the incident particle is available for the reaction because a portion of it is tied up in the energy associated with the motion of the center of mass (Section 9.5). Thus the incident particle must have a kinetic energy larger than $-Q$ to make the reaction "go." As shown in Problem 45C-37, from the conservation of energy and momentum we find that the minimum kinetic energy that will cause the reaction is the *threshold energy*[7] E_{th}:

THRESHOLD ENERGY (when $Q < 0$)
$$E_{th} = -Q\left(\frac{m_x + M_X}{M_X}\right) \quad \text{(nonrelativistic)} \quad (45\text{-}39)$$

EXAMPLE 45-13

Calculate the minimum kinetic energy that an alpha particle must have to produce the following (endoergic) reaction that Rutherford investigated:

$${}^{4}_{2}\text{He} + {}^{14}_{7}\text{N} \longrightarrow {}^{1}_{1}\text{H} + {}^{17}_{8}\text{O}$$

SOLUTION

Using the values of atom mass units given in Table 45-1, we have

Before		After	
${}^{14}\text{N}$	14.003 074 u	${}^{17}\text{O}$	16.999 131 u
${}^{4}\text{He}$	4.002 603 u	${}^{1}\text{H}$	1.007 825 u
Total	18.005 677 u	Total	18.006 956 u

Because the final mass is greater than the initial mass, the change of mass is $\Delta m = -0.001\,279$ u. So the Q of the reaction is

$$Q = (\Delta m)c^2 = (-0.001\,279 \text{ u})\left(\frac{931.5 \text{ MeV}}{1 \text{ u}}\right) = -1.19 \text{ MeV}$$

---

[7] The initial kinetic energy must also be at least large enough to overcome the Coulomb repulsion so that the nuclei can get close enough together to interact.

The threshold energy, from Equation (45-39), is

$$E_{th} = -Q\left(\frac{m_x + M_X}{M_X}\right) = -(-1.19 \text{ MeV})\left(\frac{18.01 \text{ u}}{14.00 \text{ u}}\right) = \boxed{1.53 \text{ MeV}}$$

Most alpha particles from naturally radioactive isotopes have energies in excess of 4 MeV. So the alpha particles that Rutherford used were sufficiently energetic to make the reaction occur. The excess energy appears as kinetic energy of the reaction products.

After Rutherford's experiment, many attempts were made to accelerate other charged particles to energies high enough to cause nuclear reactions. In 1930, J. D. Cockcroft and E. Walton succeeded in accelerating protons to an energy of 0.3 MeV, which was more than sufficient to induce the reaction

$$^{1}_{1}\text{H} + ^{7}_{3}\text{Li} \longrightarrow [^{8}_{4}\text{Be}^*] \longrightarrow 2 \, ^{4}_{2}\text{He} \qquad (45\text{-}40)$$

The excited state of beryllium decays with a half-life of about 10^{-16} s into two alpha particles. This reaction has historical interest since it was one of the first quantitative verifications of Einstein's mass-energy relationship, $\Delta E = (\Delta m)c^2$.

With the possibility of inducing transmutations artificially, physicists attempted to realize the alchemists' dream of producing gold from metals of less value. In 1936, gold was produced by a transmutation performed by J. M. Cork and E. O. Lawrence. The reactions involved were

$$^{196}_{78}\text{Pt} + ^{2}_{1}\text{H} \longrightarrow ^{197}_{78}\text{Pt} + ^{1}_{1}\text{H}$$

followed by

$$^{197}_{78}\text{Pt} \longrightarrow ^{197}_{79}\text{Au} + ^{0}_{-1}e + \bar{\nu}$$

In the words of J. M. Cork, "the luster of the achievement was somewhat dulled by the fact that the parent element in the reaction was platinum."

Neutrons play a central role in nuclear technology because their lack of charge allows easy penetration to the nuclear surface, resulting in high reaction cross sections with nuclei. In contrast, as Figure 45-15 shows, an incoming proton must overcome the Coulomb barrier, so only very energetic charged particles reach the nucleus. This is dramatically demonstrated in the analytical technique known as *neutron-activation analysis*, in which a minute quantity of an unknown substance is subjected to a high concentration of neutrons. The neutrons are absorbed by the nuclei, often forming radioactive isotopes, which in turn decay. By recognizing the characteristic gamma-ray energy spectrum associated with the decay, we can then identify the elements in the unknown substance.

An example of neutron-activation analysis is the detection of arsenic through the reaction (see Figure 45-14)

$$^{1}_{0}n + ^{75}_{33}\text{As} \longrightarrow [^{76}_{33}\text{As}^*] \longrightarrow ^{76}_{34}\text{Se} + ^{0}_{-1}e + \bar{\nu} + 2\gamma \qquad (45\text{-}41)$$

Extremely small amounts of arsenic can be detected by this method. At some airport check-in gates, neutron-activation detectors are now used to discover the presence of explosives.

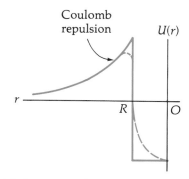

(a) An approaching proton must overcome the Coulomb repulsion to reach the nuclear surface.

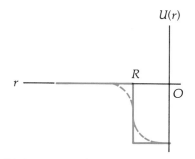

(b) An approaching neutron "sees" no barrier, thus allowing easy penetration of the nuclear surface.

FIGURE 45-15
Using a simplified *square-well* nuclear potential, we can make the difference between the potential energy of an approaching proton or neutron clear. (The actual potential energy is more like the dashed curves.)

The Discovery of the Neutron

Early researchers found that when an alpha emitter was placed in contact with some of the light elements, such as boron or beryllium, a very penetrating type of radiation resulted. Originally the reaction was thought to produce gamma rays. But in 1932 J. Chadwick found that after he placed paraffin in the path of the radiation, an unusually large number of protons emerged from the paraffin, a result inconsistent with incident gamma rays. After a variety of experiments, Chadwick concluded that the unknown radiation must be *uncharged* particles with about the *same mass* as the proton. If such a particle hits a proton in the paraffin "head-on," it can transfer all its energy in one collision. He named the particles "neutrons." Because of their lack of charge, high-energy neutrons are highly penetrating, capable of passing through several centimeters of lead.

A common source of energetic neutrons is a mixture of powdered beryllium and an alpha-emitter such as plutonium. The alpha particles from the plutonium have a high cross section for the production of neutrons by the following reaction:

$$_2^4\text{He} + {_4^9}\text{Be} \longrightarrow {_6^{12}}\text{C} + {_0^1}n + Q \qquad (45\text{-}42)$$

Because most of the kinetic energy of the products resides in the lighter-mass neutron, the kinetic energy of the neutron is essentially the Q value plus the kinetic energy of the incident alpha. In this case, the maximum neutron energy is about 6 MeV.

Detection of Charged Particles

In addition to the *scintillation detector* described in Chapter 42, Figure 42-11, various methods of detecting radiation and particles all rely on the ionization produced when a photon or a charged particle passes through matter. The familiar *Geiger counter*, Figure 45-16, is a metal tube filled with a gas at low pressure. A wire along the axis is maintained at a high positive potential ($\sim 10^3$ V) with respect to the outer tube. When a charged particle passes through the chamber, it produces a trail of ionized gas atoms. The high field accelerates the freed electrons, which in turn ionize other atoms, producing an avalanche of electrons that causes a voltage pulse when the electrons arrive at the wire. *Semiconductor detectors*, such as a silicon crystal, similarly record a pulse of conduction electrons generated by the passage of a charged particle. Since semiconductors are solids, they are particularly useful where a detector of small size is required.

Neutrons have no charge, so they do not directly produce appreciable ionization when passing through matter. One type of neutron detector utilizes the very high cross section for neutron absorption by boron, which produces energetic alpha particles that create the necessary ionization for detection. Often a tube filled with boron trifluoride gas (BF_3) is used, producing the reaction

$$_0^1 n + {_5^{10}}\text{B} \longrightarrow {_2^4}\text{He} + {_3^7}\text{Li} + Q \qquad (45\text{-}43)$$

Radioactive Dating

A neutron reaction is involved in the radioactive dating of ancient organic materials. Neutrons resulting from interactions of cosmic radiation with the upper atmosphere are absorbed by the common nitrogen isotope ^{14}N to pro-

(a) A simple end-window Geiger tube.

(b) A portable battery-operated Geiger counter measures the activity of pitchblende, a radioactive mineral containing uranium.

FIGURE 45-16
A Geiger–Müller (GM) counter, often called simply a *Geiger counter*.

duce the radioactive isotope of carbon ^{14}C. The reaction is

$$_0^1n + {}^{14}_7N \longrightarrow {}^{14}_6C + {}^1_1H + Q \qquad (45\text{-}44)$$

Because the production of ^{14}C has been going on much longer than its half-life of 5730 years, the rate of production of ^{14}C equals the rate of disintegration of ^{14}C. Thus an equilibrium concentration of ^{14}C exists in the atmosphere: about one ^{14}C atom for every 10^{12} stable atoms of ^{12}C and ^{13}C. The radioactive atoms interact chemically in the same way as other carbon atoms, so carbon is ingested by living plants and animals in this definite proportion. When the living organism dies, it no longer ingests carbon, and the proportion of ^{14}C to the stable carbon gradually decreases as the radioactive isotope ^{14}C decays according to

$$^{14}_6C \longrightarrow {}^{14}_7N + {}^{\;\;0}_{-1}e + \bar{\nu} \qquad (T_{1/2} = 5730 \text{ yr}) \qquad (45\text{-}45)$$

As time passes, the organic material becomes less radioactive, making *radiocarbon dating* possible. The very low energy of the beta decay of ^{14}C and the low activity of samples limit the accuracy of radiocarbon dating to about ± 50 years in 5000 years. The validity of the technique depends upon whether the rate of ^{14}C production has been constant over time. Comparing data for objects whose ages are accurately known from historical records shows that radiocarbon dating is a valid technique.

EXAMPLE 45-14

An archeologist obtains ashes from an ancient fire pit. Carbon derived from the ashes proves to be only one-third as radioactive as an equal mass of carbon derived from recent plants. Calculate the date of the ancient fire.

SOLUTION

From Equation (45-10), the activity A of a decaying radioactive material is given by

$$A = A_0 e^{-(\ln 2/T_{1/2})t}$$

Solving for t, we obtain

$$t = \ln\left(\frac{A_0}{A}\right)\frac{T_{1/2}}{\ln 2} = \frac{(\ln 3)(5730 \text{ yr})}{\ln 2} = \boxed{9082 \text{ yr}}$$

45.8 Nuclear Power

The extent of energy extraction from nuclear reactions is based upon the binding-energy-per-nucleon curve shown in Figure 45-4. Nuclides corresponding to atomic mass numbers near the peak are *less* massive per nucleon than those on either side of the peak. Consequently, any reaction that results in nuclides nearer the peak results in a mass-to-energy conversion. Thus the **fission**, or splitting apart, of a very massive nuclide into less massive nuclides results in a large energy release. Similarly, the **fusion**, or combining together, of low-mass nuclides into a heavier nuclide also results in a mass-to-energy conversion. The mechanisms of fission and fusion processes are so different that we will discuss them separately.

Nuclear Fission

All of the naturally occurring isotopes of uranium as well as most of the heavier nuclides undergo spontaneous fission. The tremendous energy associated with such fission is illustrated in Example 45-10. While a nuclear power fission reactor may utilize a number of reactions, currently the most common reactor uses a "fuel" consisting primarily of ^{235}U. This isotope accounts for only 0.72% of natural uranium, while ^{238}U constitutes 99.27% and ^{234}U only 0.005%. Even though it is possible to fuel a reactor with natural uranium, a greater concentration of ^{235}U is more practical. A costly gas-diffusion process is commonly used to "enrich" natural uranium to a ^{235}U concentration as high as 90%.

The first step in the energy-extraction process is to produce ^{236}U by the neutron capture by ^{235}U. The ^{236}U immediately undergoes fission such as that shown in Example 45-10. Since each fission reaction produces two or three neutrons, each of which is capable of initiating another fission, a violent chain reaction will occur unless excess neutrons are removed from the reactor. Excess neutrons are those not required to maintain the desired rate of fission production. In a reactor operating at a constant power level, for every fission only one of the neutrons produced is used to initiate another fission. Excess neutrons escape the reactor, decay radioactively,[8] or are captured by *control rods* made of cadmium or other elements with a high neutron-capture cross section. Happily, not all of the neutrons emitted in fission are "prompt"—about 1% are "delayed" because they originate in the neutron-rich fission fragments with lifetimes of a fraction of a second to a few minutes. The presence of these delayed neutrons enables the relatively simple mechanical insertion of the cadmium rods for control purposes. The fuel consumption of uranium for power generation may seem small, yet the plants are rapidly depleting the world supply of easily obtainable uranium.

EXAMPLE 45-15

A typical nuclear fission power plant produces about 1000 MW of electrical power. Assume that the plant has an overall efficiency of 40% and that each fission produces 200 MeV of thermal energy. Calculate the mass of ^{235}U consumed each day.

---

[8] The *free* neutron is an unstable particle that decays as follows, with a mean lifetime of 900 s:

Free neutron $\qquad n \longrightarrow p + e^- + \bar{\nu} \qquad (45\text{-}46)$

The reason that neutrons can be stable in nuclei is a consequence of the Pauli exclusion principle. In the ground state of a nucleus, the lowest nuclear energy states are filled. The proton produced in the reaction must therefore go into one of the higher vacant states. For most nuclei, the proton does not have sufficient energy to do this. Thus the exclusion principle restricts the decay of neutrons in stable nuclei. Only free neutrons decay.

Some recent theories suggest that a *free* proton may also be unstable, with a lifetime of $\sim 10^{30}$ yr or longer. Several reactions have been proposed, with a probable one being

Free proton $\qquad p \xrightarrow{?} e^+ + \pi^0 \qquad (45\text{-}47)$

Experimental evidence is difficult to obtain, but to date lifetimes shorter than $\sim 10^{32}$ y appear unlikely. If the proton is unstable, fortunately its mean lifetime is more than 10^{22} *times* the age of the universe since the Big Bang, so we are not in imminent danger of annihilation.

SOLUTION

If the electrical power output of 1000 mW is 40% of the power derived from fission reactions, the power output of the fission process is

$$\frac{1000 \text{ MW}}{0.40} = 2500 \text{ MW} = \left(2.5 \times 10^9 \, \frac{\text{J}}{\text{s}}\right)\left(\frac{86\,400 \text{ s}}{\text{d}}\right) = 2.16 \times 10^{14} \, \frac{\text{J}}{\text{d}}$$

The number of fissions per day is

$$\left(2.16 \times 10^{14} \, \frac{\text{J}}{\text{d}}\right)\left(\frac{1 \text{ fission}}{200 \times 10^6 \text{ eV}}\right)\left(\frac{1 \text{ eV}}{1.602 \times 10^{-19} \text{ J}}\right) = 6.74 \times 10^{24} \text{ d}^{-1}$$

This also is the number of ^{235}U nuclei used, so the mass of ^{235}U used per day is

$$\left(6.74 \times 10^{24} \, \frac{\text{nuclei}}{\text{d}}\right)\left(\frac{235 \text{ g/mole}}{6.02 \times 10^{23} \text{ nuclei/mole}}\right) = 2631 \text{ g/d} = \boxed{2.63 \text{ kg/d}}$$

In contrast, a coal-burning steam plant producing the same electrical power uses more than 6×10^6 kg/d of coal.[9]

The Nuclear Reactor Many problems need to be solved in designing a reactor. For example, the neutrons emitted during fission are "fast" neutrons with energies from ~1 MeV to ~15 MeV. The cross section for absorption by ^{235}U is large only for "slow" neutrons with energies just a fraction of 1 eV. So fission neutrons must be slowed down by the use of a *moderator*—atoms whose masses are close to that of neutrons so that the average energy loss per elastic collision is large. Unfortunately $^{1}_{1}$H (whose mass is ideal) does absorb some neutrons, so it is not the best material to use. Deuterium, $^{2}_{1}$H, is the next best choice, and its neutron capture cross section is low. Thus *heavy water*, formed by replacing ^{1}H atoms in water with ^{2}H, is a feasible moderator. Very pure carbon ^{12}C is another alternative. Purity is essential since many other elements absorb neutrons, including the fission products themselves. Some neutrons escape from the surface of the reactor, further reducing the overall neutron supply. The minimum amount of fissionable material that will maintain a chain reaction is called the *critical mass*. It depends on the type of nuclear fuel, the degree of *enrichment*, the moderator, and the *geometry* of arranging lumps or rods of fuel spaced apart by the moderator. Pure ^{235}U with ordinary water as a moderator has a critical mass of about 3 kg.

Since only one neutron per fission is utilized in sustaining a chain reaction, some of the excess neutrons may be used to convert ordinary ^{232}Th and ^{238}U into the fissionable isotopes ^{233}U and ^{239}Pu. Such an arrangement is called a *breeder reactor*; it not only produces enough fuel to maintain the operating level of the reactor, but also generates additional fissionable fuel for another reactor

---

[9] In 1986, nuclear power plants generated 16.6% of the electric power consumed in the United States; coal-burning plants generated 55.7%. (The remaining power was produced by petroleum, natural gas, hydro-electric, and other sources.) An interesting way to illustrate the magnitude of this energy usage is the following. The **daily** (1986) U.S. consumption of coal for electric power generation would fill a railroad train of coal cars 197 miles long! In the same year, the **daily** (total) U.S. use of petroleum would fill a railroad train of oil tank cars 301 miles long!

in about 10 years.[10] The use of breeder reactors to produce plutonium has been curtailed primarily for two reasons. First, plutonium is extremely dangerous, both biologically and radioactively. Second, plutonium produced by breeders (and other nuclear fuels) may be stolen by terrorist groups, either for ransom or for the construction of rather simple but devastating explosive devices—a possibility of grave concern as the worldwide use of reactors increases.

As a power source, all reactors generate heat, which is then used in the conventional way to operate steam turbines that drive electric generators. Some special problems are the intense radiation bombardment that structural members of the reactor must withstand and the containment of the hot fluids that transfer heat from the reactor core to the turbine. Any rupture within the core could release dangerous radioactivity. Also, the safe disposal of long-lived, highly radioactive fission products is a serious problem. The lifetime of a reactor is limited to about three decades because of the radiation weakening of the structure. (Note that decommissioning a large nuclear power plant is not cheap!) Reactors used for research are sources of high-intensity neutron beams and gamma rays that are valuable tools in many scientific investigations. They also produce useful radioactive isotopes for "tracer" studies in biological and medical research.

Nuclear Fusion

The *fusion* of light nuclei is the source of energy emitted by the sun and other stars. A sequence of fusion reactions called the *proton–proton cycle* is believed to be the main source of stellar energy in the sun and other stars cooler than the sun. The net effect of this sequence is to combine four protons to form $^{4}_{2}\text{He}$ plus two positrons, two neutrinos, and two gammas:

$$4\,{}^{1}_{1}\text{H} \longrightarrow {}^{4}_{2}\text{He} + 2e^{+} + 2\nu + 2\gamma + Q \qquad (45\text{-}48)$$

As shown in Problem 45C-39, the total energy released is 27.7 MeV, about 6.9 MeV/nucleon, compared with an energy release in fission reactions of roughly 1 MeV/nucleon. For stars hotter than the sun, the *carbon cycle* (Problem 45B-17) is believed to be the principal source of energy.

The prospect of achieving a practical fusion reactor for power generation is very appealing. In contrast to fission, fusion involves much less of a radioactive-waste-disposal problem. No weapons-grade materials are involved, and there is no danger of a runaway nuclear accident. In addition, the fuel cost could be extremely low if the naturally occurring deuterium $^{2}_{1}\text{H}$ in seawater and lakes could be utilized. A possible sequence of reactions is

$$\begin{aligned}
{}^{2}_{1}\text{H} + {}^{2}_{1}\text{H} &\longrightarrow {}^{3}_{1}\text{H} + {}^{1}_{1}\text{H} & (Q = 4.0 \text{ MeV}) & \qquad (45\text{-}49)\\
{}^{2}_{1}\text{H} + {}^{2}_{1}\text{H} &\longrightarrow {}^{3}_{2}\text{He} + {}^{1}_{0}n & (Q = 3.3 \text{ MeV}) & \qquad (45\text{-}50)\\
{}^{2}_{1}\text{H} + {}^{3}_{1}\text{H} &\longrightarrow {}^{4}_{2}\text{He} + {}^{1}_{0}n & (Q = 17.6 \text{ MeV}) & \qquad (45\text{-}51)
\end{aligned}$$

The last reaction, involving tritium ($^{3}_{1}\text{H}$), is interesting by itself, since it yields the most energy. However, tritium is radioactive (β^{-} decay, $T_{1/2} = 12.3$ yr) so it does not occur naturally in appreciable amounts. It is very expensive to produce artificially (current price about \$2 million/kg!). One intriguing proposal is to surround a fusion reactor core with molten lithium. Energetic neutrons from the reactor would be absorbed by the lithium, raising its temperature. This

---

[10] This does not imply an endless supply of fuel; the initial material to be converted to fissionable isotopes must be supplied. But breeder reactors could extend the available supply by a factor of 100 or so.

thermal energy could then be used to generate steam to operate the electric generator. As a bonus, neutrons from the reactor produce tritium in the following reactions:

$$^1_0n + ^7_3Li \longrightarrow ^3_1H + ^4_2He + ^1_0n \quad (45\text{-}52)$$
(fast) (slow)

$$^1_0n + ^6_3Li \longrightarrow ^3_1H + ^4_2He + 4.8 \text{ MeV} \quad (45\text{-}53)$$
(slow)

The tritium could then be circulated back into the reactor as fuel—a sort of breeder reaction that converts inexpensive lithium to the more valuable tritium.

Though these fusion reactions do not produce radioactive fission fragments, the copious production of radioactive ^3_1H and of neutrons that induce radioactivity in the surrounding structures does present radiological hazards. An interesting reaction that avoids such hazards uses abundant natural boron and hydrogen. Called *thermonuclear fission* because of the multiple fragments in the product, it is

$$^{11}_5B + ^1_1H \longrightarrow 3\ ^4_2He + 8.7 \text{ MeV} \quad (45\text{-}54)$$

There is hope that some method may be devised to convert the energetic alpha particles directly to electrical power without the intermediate steam–turbine–generator process. The main difficulty in achieving this reaction is attaining a temperature of about 3×10^9 K to overcome the Coulomb barrier of the boron nucleus.

In Example 45-16 we compare fusion power with the power derived from the burning of fossil fuels. While the reaction has not yet been utilized in a reactor, it clearly indicates the tremendous potential of fusion power.

EXAMPLE 45-16

Calculate the energy ideally derivable from one gallon of seawater, utilizing the deuterium that it contains in the following reaction:

$$^2_1H + ^2_1H \longrightarrow ^3_2He + ^1_0n \quad (45\text{-}55)$$

Deuterium is a stable isotope that makes up 0.015% of natural hydrogen. (The cost of extracting the water molecules containing deuterium from one gallon of water is currently less than 10 cents.)

SOLUTION

We begin by calculating the Q of the reaction. From Table 45-1, the mass loss in the reaction is

	Initial mass		Final mass
$2\ ^2H$	2(2.014 102 u)	^3He	3.016 029 u
		^1n	1.008 665 u
Total	4.028 204 u	Total	4.024 694 u

with a mass difference of 0.003 510 u, which corresponds to

$$Q = (\Delta m)c^2 = (0.003\ 510 \text{ u})\left(\frac{931.5 \text{ MeV}/c^2}{1 \text{ u}}\right)c^2 = 3.27 \text{ MeV}$$

> In joules, $Q = (3.27 \times 10^6 \text{ eV})\left(\dfrac{1.602 \times 10^{-19} \text{ J}}{1 \text{ eV}}\right) = 5.24 \times 10^{-13}$ J
>
> The number of water molecules in a gallon of water is $(6.03 \times 10^{23}$ molecules/mole$)(3785$ g/gal$)[1/(18$ g/mole$)] = 1.27 \times 10^{26}$ molecules/gal. The number of ^{2}H nuclei in a gallon of water is $(1.27 \times 10^{26}$ molecules/gal$)(2$ H nuclei/molecule$)(0.000\ 15\ ^2\text{H}/\text{H}) = 3.80 \times 10^{22}$ ^{2}H nuclei/gal. Two deuterons are involved in each reaction. The total energy derivable from the gallon of water then becomes
>
> $$\left(\dfrac{3.80 \times 10^{22}\ ^2\text{H/gal}}{2\ ^2\text{H/reaction}}\right)\left(5.24 \times 10^{-13}\ \dfrac{\text{J}}{\text{reaction}}\right) = \boxed{9.95 \times 10^9 \text{ J/gal}}$$
>
> In order to comprehend the immensity of this amount of energy, we compare this to the combustion of one gallon of gasoline, which provides about 1.3×10^8 J/gal.
>
> $$\left[\dfrac{9.95 \times 10^9 \text{ J}}{1.3 \times 10^8 \text{ J/gal}}\right] = \underline{\underline{76.6 \text{ gal}}}$$
>
> Potentially, one gallon of ordinary seawater has the fusion energy content of 76.6 gallons of gasoline!

Containment Nuclei are positively charged. To get them close enough together for the short-range attractive nuclear force to cause fusion, the nuclei must have very large kinetic energies to overcome their Coulomb repulsion (see Problem 45A-25). High-energy accelerators achieve these speeds easily. But to produce large amounts of power, very high collision rates are necessary. Thus the problem to be solved is to hold together nuclei at the highest densities possible at the fusion temperatures of roughly 200–400 million K. The British physicist John D. Lawson showed that the required conditions for a self-sustaining reaction at fusion temperatures are expressed by *Lawson's criterion*: $n\tau < \sim 10^{20}$ s·m^{-3}, where n is the interacting particle density and τ is the confinement time.

Currently, two types of confinement mechanisms show promise. In **magnetic confinement**, a neutral plasma of nuclei and electrons is contained in a "magnetic bottle." (For a discussion of the forces that confine the plasma, see Figure 30-6, Chapter 30.) Unfortunately, a shape as simple as that shown in Figure 45-17a tends to be leaky at the ends. An improved configuration, first developed in the USSR, joins the ends together, forming a toroid, Figure 45-17b. It is called a *tokamak*, an acronym for the Russian words for "torus," "chamber," and "magnetic." The hot plasma must not touch the walls of the vacuum chamber; it is not that the walls might melt, but rather that the plasma would chill below the temperature required for fusion. A variety of techniques are being investigated to heat the confined plasma to fusion temperatures: by passing a current through the electrically conducting plasma, by bombarding it with neutral particles, by compressing the plasma with a greatly increasing magnetic field, and by radiofrequency heating.

Another method of containment is called **inertial confinement**. A fuel pellet about 1 mm in diameter or smaller is suddenly imploded by simultaneous bombardment from all sides with very powerful laser beams. This produces an inwardly moving shock wave that momentarily increases the density of the material about a factor of 10^3 and heats the core of the pellet to fusion temperatures. Due to the inertia of the nuclei, these fusion conditions exist for

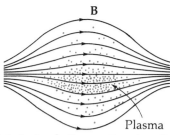

(a) A simple "magnetic-mirror" field configuration. Magnetic forces on the moving charged particles "reflect" them at each end back toward the center. Unfortunately, the ends are leaky since particles traveling parallel to the field lines experience no deflecting forces.

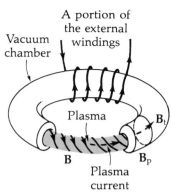

(b) In the *tokamak*, the two ends of the bottle in (a) are joined to form a toroid, or doughnut shape. A *toroidal* magnetic field B_t around the toroid is produced by current in the external windings. A *poloidal* field B_p is produced by current in the plasma itself (which also helps heat the plasma). The combination of these two fields is a net helical field **B** that improves the confinement characteristics.

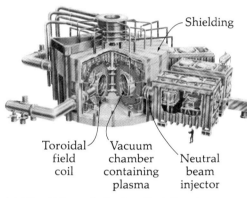

(c) The Tokomak Fusion Test Reactor (TFTR) at Princeton, New Jersey. In addition to the heating produced by the induced current in the plasma, further heating is produced by injecting high-speed neutral atoms into the plasma.

FIGURE 45-17
Schemes for confining hot plasmas of nuclei and electrons to achieve fusion.

about 10^{-11} to 10^{-9} s, after which the high pressure created within the pellet blows the pellet apart. One calculation suggests that if good efficiencies are achieved, the fusion of only 10 pellets/s would be sufficient to supply a 1000-MW power station.

To date (1989), an experimental reactor has achieved the *break-even* point at which the fusion energy produced equals the energy input needed to trigger fusion. No device has yet reached *ignition*, the conditions necessary for the plasma to sustain its own thermonuclear reactions. A great many difficult engineering problems are yet to be solved. With international cooperation, it is hoped that fusion will be achieved in the near future.

Brief History and Status of Particle Physics

Thomas Ferbel
University of Rochester

"We have sailed many months, we have sailed many weeks
  (Four weeks to the month you may mark),
But never as yet ('tis your Captain who speaks)
  Have we caught the least glimpse of a Snark!"

LEWIS CARROLL
"The Hunting of the Snark" (1891)

Introduction

The structure of matter and the laws that govern its behavior have always fascinated scientists. As energies of probing instruments have increased, allowing matter to be examined with ever finer resolution, new and at times mystifying phenomena have been revealed. Although the pace of discovery during this century has been truly breathtaking, one of the more remarkable aspects of the current evolution of modern physics is the extent to which our accepted theoretical ideas can accommodate the rich and varied spectrum of mounting experimental observations. What is emerging is that the universe is composed of a very small number of fundamental objects. These objects interact through just four forces, which seem at first glance to be quite distinctive. Yet closer examination has given us the hope that one day they will be seen to be different aspects of a single fundamental law.

As we look up into the sky, we first notice the effects of gravity in our Solar System: enormous masses attracting each other over vast distances. There is barely a hint of the presence of the far stronger electromagnetic, nuclear, or weak forces. Although the electromagnetic force falls with distance at the same rate as the gravitational force, most large objects have a very small net charge, and the overwhelming importance of electromagnetism is therefore evident only when we start looking at things on the molecular or atomic scale. Because the two remaining forces are exceedingly short-ranged, we do not notice the strong force until we penetrate the atomic nucleus, and the effect of the weak force is obvious only when we observe the radioactive decay of certain unstable nuclei and particles.

We have already learned that the atomic world, condensed matter, and chemical phenomena can be understood, at least in principle, with the help of quantum mechanics applied to electromagnetism. The nucleus is a complex object, which through its very stability implies the existence of a strong attractive force that is able to overcome the electromagnetic repulsion among its closely crowded positively charged protons. This new force seems to be independent of electric charge. But what is the origin of the nuclear force?

Does it reflect some truly basic property of matter—as Coulomb's law reflects presence of electric charge—or can it be accounted for through a *remanent effect* after the cancellation of the more fundamental attributes? For example, though the total electric charge of neutral molecules adds to zero, there is still an electromagnetic attraction between the molecules because their charge distributions are not uniform. As we are about to learn, experiments in particle physics indicate that the nuclear force is most likely a residual phenomenon.

Progress in particle physics, especially during the past thirty years, has been impressive. New, higher-energy accelerators produced a veritable zoo of hundreds of different and unexpected particles. They were given names and symbols such as the muon (μ), the pion (π), the lambda (Λ), the sigma (Σ), and so forth, until the capacity of the Greek alphabet was exhausted. All were unstable, with mean lives ranging from $\sim 10^{-23}$ s to $\sim 10^{-6}$ s. This remarkable flood of discovery was initially very confusing. But as important similarities began to be observed among different particles, theoretical schemes were suggested for grouping them into larger families that had some common characteristics. Although this brought a degree of order to the subject, it was still all pretty complex.

Eventually, particle physicists began to wonder whether the many supposedly fundamental particles might not be composed of combinations of just a few, more elementary objects. Indeed, current theory maintains exactly that. It describes most of the hundreds of particles as combinations of just a few constituents called **quarks** and certain "messengers" called **gluons**. The properties of these new constituents (e.g., *color* and *flavor*) are given somewhat fanciful names (there are six flavors, among which we have *strangeness* and *charm*). Of course, these words do not mean what their ordinary usage implies— they are just easy-to-remember names for quantum numbers that characterize the way the constituents interact.

The latest discoveries and their theoretical interpretations have led to the formulation of what is called the **Standard Model** of particle physics. Though not a complete theory, this view of particle phenomena does correlate and predict essentially all the known interactions of all elementary particles. Before delving into the consequences of the theory, we will mention three characteristics of particles that are particularly useful for sorting out their different attributes. These help us organize a large number of particles into the few categories that are the basis of the Standard Model.

1. **Spin.** Particles with *half-integral* spin (in units of $\hbar$), called *fermions*, obey the Pauli exclusion principle (Section 44.6). This means that only one fermion can occupy any given quantum state—from which it also follows that fermions *cannot* be produced one at a time. Particles with *integral* spin, called *bosons*, do not obey this principle; any number of identical bosons can occupy a given state, and any number *can* be produced at once in high-energy collisions.
2. **Fundamental interactions.** All objects, because they have energy (and energy is equivalent to mass), respond to the gravitational force. All electrically charged objects are influenced by the electromagnetic interaction. Particles that sense *both* the *weak* and the *strong* nuclear forces are called **hadrons**. Those that respond to *only* the *weak* force are called **leptons**. There are two types of hadrons—*baryons* and *mesons*—the distinction being that baryons are fermions and mesons are bosons. Protons and neutrons are the simplest baryons. All baryons carry the *baryon quantum number*; that is, all baryons are composed of the same kind of matter that exists within the nucleus of the hydrogen atom. Mesons, on the

TABLE 45-5 Types of Elementary Particles

Particles	Weak Force (Leptons)	Strong Force (Hadrons)	Produced in Collisions
Fermions	Leptons	Baryons	As antifermion–fermion pairs
Bosons*	Not observed	Mesons	One at a time†

* See Table 45-6 for the fundamental bosons.
† Unless specifically forbidden through other considerations (see *charm, strangeness, etc.*).

other hand, have baryon number zero. All fermions, and some bosons, carry a type of quantum or property that is always *conserved*. That is, it cannot be created or destroyed. For example, when new baryons are produced in high-energy collisions, it is always in the company of their *antibaryon* partners.

3. **Particle–antiparticle symmetry.** Every particle appears to have an antiparticle counterpart, which has the same spin and mass but the opposite sign of electric charge and baryon number. We designate an antiparticle by a bar over its symbol; for example, an antiproton is written $\bar{p}$. As an antiparticle passes through matter, it loses energy by ionizing atoms in its path, slows down, meets a particle, and the two annihilate, converting the rest masses of the particle–antiparticle system into particles with smaller rest masses but higher kinetic energies. Why our universe contains mostly matter rather than more equal amounts of matter and antimatter is one of the current puzzles that challenges physicists and astronomers.

Table 45-5 summarizes the nomenclature we have introduced thus far. According to the Standard Model, all interactions in nature can be described in terms of six leptons and six quarks, and their antiparticles, and six fundamental bosons.

The Beginning of the Modern Era

The onset of the modern phase of particle physics can be traced to an observation that parallels Rutherford's discovery of the dense, massive atomic nucleus. The decisive experiment involved the scattering of electrons from protons. Just as Rutherford's alpha particles were scattered by gold nuclei to larger angles (and with greater interchange of momentum) than had been anticipated, so electrons at the Stanford Linear Accelerator Center (SLAC) were scattered more often with large momentum transfers than expected from a purely uniform distribution of charge within the proton. That is, the observed distribution of the scattered electrons implied that the charge of the proton was concentrated within miniscule regions that were at least a factor of 100 smaller than the size of the proton itself.

The discovery at SLAC, along with other evidence from the then-known spectrum of elementary particles, indicated that nucleons (much like the nucleus) were composed of other constituents. Surprisingly, the quark constituents have fractional charge and fractional baryon number. Subsequent experiments revealed that, in addition to charged pointlike quarks, nucleons also contain neutral pointlike gluons. (By *pointlike* we simply mean "without any apparent substructure.")

The stable world around us, consisting of electrons and nuclei, can be described in terms of only a few fundamental constituents and forces. Electrons appear to be pointlike, and they carry one unit of negative electric charge (a definition attributed to Benjamin Franklin). The nucleus consists of positively charged protons and electrically neutral neutrons, which, in turn, are composed of quarks and gluons. Just as electric charge is known to be the source of the electromagnetic field (as well as the origin of its quanta, the photons), so a "*color*" charge, contained within both quarks and gluons, is the source of the strong (color) force (and the origin of its quanta, the gluons). Both electrons and quarks have electric charge, so both electrons and quarks feel the electromagnetic force. Gluons, on the other hand, being neutral and pointlike, do not sense electromagnetism. Electrons, which also appear fundamental, neither contain nor sense the presence of color.

Colors, Flavors, QED, and QCD

As the previous section implies, there is an important distinction between the carriers of electromagnetism (photons) and the carriers of the color force (gluons): photons cannot send any messages or couple to other photons, while gluons can communicate with other gluons. This difference arises because there is only one type of electric charge in nature (ignoring the difference between positive and negative charge), while experimental evidence is conclusive that there must be three different types of color quantum numbers for quarks and gluons. The fully quantized field theories of electromagnetism, *Quantum Electrodynamics* (QED), and of the color force, *Quantum Chromodynamics* (QCD), are based on similar principles and account for this crucial difference. Figure 45-18 shows schematically the allowed processes in terms of Feynman diagrams for scattering of two quarks, two electrons, and two gluons, and the appropriate exchanged force-carrying quanta.

Two kinds (or *flavors*) of quarks, *up* (u) and *down* (d), suffice to describe normal matter. The electric charges are $+\frac{2}{3}e$ and $-\frac{1}{3}e$ for the u and d quarks, respectively. The proton consists of two u and one d valence quarks, and the neutron of one u and two d valence quarks. The existence of antimatter suggests that antiquarks $\bar{u}$ and $\bar{d}$ must also exist. It also follows that if the properties of baryons can be accounted for by three valence quarks, then quarks must themselves be fermions.

The discovery, during the late 1940s, of the pion and the muon in cosmic rays indicated that particles intermediate in mass between that of electrons and nucleons abounded in nature. Muons appear to be pointlike objects that behave exactly like electrons, except that they have a mass 207 times greater. (We will return to their properties later.) Pions have strong interactions akin to those observed for nuclear matter, but they do not carry baryon number, and nucleons therefore cannot transform into pions (at least not yet!). The neutral pion usually decays into two photons (bosons), which suggests that, in addition to being bosons, pions must have no quark content, or, perhaps more reasonably, that they may be regarded as having a quark–antiquark substructure. Pions are the simplest members of the meson family of hadrons.

During the 1950s, cosmic-ray experiments revealed the presence of other objects with properties that could not be comprehended without the introduction of a new flavor quantum number. These "*strange*" particles, which also appear to possess the kind of properties observed for nuclear matter, suggested the need for an additional quark, the *s*-quark. Many strangeness-carrying baryons and mesons were eventually discovered, which expanded the spectrum of hadrons. During the 1970s, "*charm*" flavor was discovered (attributed to the *c*-quark), and this was followed shortly by the discovery of the "*bottom*" or *b*-quark. All these newer and more massive hadrons can be understood on

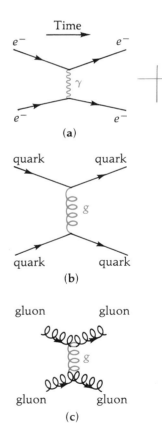

FIGURE 45-18
Feynman diagrams showing the scattering of several pointlike objects. The quanta or "messenger" particles that communicate the force between the two interacting particles are the photon [γ] in (a) and the gluon [g] in (b) and (c).

the basis of a substitution of one or more of the heavier quarks for a light one in the structure of the previously known particles.

Color Confinement

All known hadrons come in integral multiples of both electric charge and baryon number. Also, there has been no confirmed observation of a physical (or "bare") quark or gluon. This is all entirely consistent with QCD, which predicts that quarks and gluons are confined within hadrons, which, in turn, have no net color. Table 45-6 provides a summary of the properties of quarks and gluons. It contains an entry for the yet to be discovered "*top*" quark, which is the last flavor required to complete the Standard Model. Table 45-7 indicates the valence-quark composition of several well-studied hadrons.

It might seem outrageous to discuss properties of physical objects (of zero net color) in terms of their colored fundamental building blocks that we can never even hope to see. The situation is not quite as anomalous as one might first imagine. Consider, for example, the neutron. As we know, this object has no electric charge, yet it has a large magnetic moment, a fact that must be attributed to a distribution of currents within the neutron. If the neutron were structureless, there would be no way of understanding such an electromagnetic effect. As another example, consider again molecular forces. Molecules are electrically neutral, yet most properties of matter can be attributed to molecular electromagnetic interactions. Although the total charge

TABLE 45-6 The Fundamental Fermions and the Boson Force-Carriers

FUNDAMENTAL FERMIONS (spin = $\frac{1}{2}$)

Quarks*	Electric Charge (e)	Leptons	Electric Charge (e)	Lepton Mass (GeV/c^2)
up (u)	$\frac{2}{3}$	ν_e	0	$<2 \times 10^{-8}$
down (d)	$-\frac{1}{3}$	e	-1	5.1×10^{-4}
charm (c)	$\frac{2}{3}$	ν_μ	0	$<2.5 \times 10^{-4}$
strange (s)	$-\frac{1}{3}$	μ	-1	0.106
top (t)	$\frac{2}{3}$	ν_τ	0	<0.035
bottom (b)	$-\frac{1}{3}$	τ	-1	1.784

POINTLIKE BOSONS (integral spin)

Type of Force	Force Carrier	Electric Charge (e)	Spin	Mass (GeV/c^2)
Electromagnetism (QED)	Photon (γ)	0	1	0
Weak	W^+	$+1$	1	81
	Z^0	0	1	92
	W^-	-1	1	81
Strong Color (QCD)	Gluon (g)*	0	1	0
Gravity	Graviton (G)	0	2	0

* Quarks and gluons carry the color quantum number. There are three different colors for quarks, and gluons come in eight unique color–anticolor combinations. Because quarks and gluons do not appear as isolated physical objects (they are always bound within other particles), their masses cannot be defined in an unambiguous way. The u- and d-quarks, as well as gluons, are often taken to be massless. The masses of the heavier quarks turn out to be less problematic, and are approximately $m_c = 1.5$ GeV/c^2, $m_b = 4.7$ GeV/c^2, $m_t > 70$ GeV/c^2, and $m_s \sim 0.15$ GeV/c^2. The leptons and quarks are shown grouped together horizontally into three "generations." The Standard Model requires just these three generations; however, it is not known whether there are more pointlike fermions or more generations at higher masses.

TABLE 45-7 Quark Structure of Hadrons

Symbol	Name	Quark Structure	Electric Charge (e)	Mass (GeV/c^2)	Spin	Mean* Life (s)
Fermions (Baryons and Antibaryons)						
p	proton	uud	1	0.938	$\frac{1}{2}$	$>5 \times 10^{32}$
n	neutron	udd	0	0.940	$\frac{1}{2}$	898
Λ	lambda	uds	0	1.116	$\frac{1}{2}$	2.6×10^{-10}
Ω^-	omega-minus	sss	-1	1.672	$\frac{1}{2}$	0.8×10^{-10}
$\Delta^{++}(1232)$	delta	uuu	2	1.232	$\frac{3}{2}$	$\sim 6 \times 10^{-24}$
$\bar{p}$	antiproton	$\bar{u}\bar{u}\bar{d}$	-1	0.938	$\frac{1}{2}$	$>3 \times 10^{14}$
$\bar{\Lambda}$	antilambda	$\bar{u}\bar{d}\bar{s}$	0	1.116	$\frac{1}{2}$	2.6×10^{-10}
Bosons† (Mesons)						
π^+	pion	$u\bar{d}$	1	0.140	0	2.60×10^{-8}
π^0	pion	$(u\bar{u} - d\bar{d})/\sqrt{2}$	0	0.135	0	0.87×10^{-16}
K^+	kaon	$u\bar{s}$	1	0.494	0	1.24×10^{-8}
ρ^+	rho	$u\bar{d}$	1	0.770	1	$\sim 4.4 \times 10^{-24}$
D^0	d-meson	$c\bar{u}$	0	1.865	0	4.3×10^{-13}
B^+	b-meson	$u\bar{b}$	1	5.271	0	1.4×10^{-12}

* Lifetimes of $<10^{-18}$ s are inferred from the observed distribution of measured mass values of the particles. The "widths" or mass uncertainties (Δmc^2) of such distributions can be used with the help of the Heisenberg uncertainty principle to estimate lifetimes. (That is, using $\Delta E \cdot \Delta t \sim \hbar$ provides a measure of the mean life $\Delta t = \hbar/\Delta mc^2$.) These very short lifetimes signal the presence of very strong transitions, of order of nuclear reactions times ($r_p/c \sim 10^{-13}$ cm$/3.0 \times 10^{10}$ cm/s $\approx 3 \times 10^{-24}$ s), in which all flavors are conserved. Longer lifetimes involve weak processes in which quark flavors may change. Intermediate lifetimes in the 10^{-16}-s range generally involve electromagnetic transitions.

† Unlike the case with baryons, antibosons are not always distinct from bosons. In particular, the antiparticle of the π^0 is the π^0 itself. The three pions π^+, π^0, and π^- all have similar strong interaction properties and can be grouped into a pion "triplet," just like the proton and neutron that form the nucleon "doublet." However, whereas the antinucleons $\bar{p}$ and $\bar{n}$ can be distinguished from p and n, the antipion triplet is identical to the pion triplet. Mesons that carry net flavor content (beyond the u and d varieties), for example, K^0 ($d\bar{s}$), are distinct from their antiparticles.

on any molecule sums to zero, it is the spatial *distribution* of the charge (individual positions of the atomic electrons) that is important for determining the physical properties of large objects. Similarly for the case of the colorless hadrons; these objects have properties and interactions that arise from the character of the distribution of the quarks and gluons within them; the sum of all these quark and gluon color charges, just as in the case of molecular electric charge, is zero, but the residual effects of color are nevertheless strongly felt.

Electric charge and color charge are not exactly analogous, because, as illustrated in Figure 45-18, in QCD two gluons can interact with each other through the exchange of another gluon, whereas in QED two photons cannot scatter simply by exchanging another photon. This effect causes an *increase* in the force that binds color-charged objects as they are pulled *apart*, opposite to what happens in electromagnetism, where the force *decreases* as charges are moved farther *apart*. The net result is that a single quark can never be dislodged from a colorless hadron. Hadrons can be broken apart or "split" at very high energies, but this results only in the "boiling off" of other colorless particles, such as pions or photons, while the hadrons that are left remain colorless. Single quarks or gluons never seem to emerge from any interactions of the elementary particles. Nevertheless, we believe in their existence because they account so well for all observed phenomena.

Weak Processes, Generations, and Lepton Number

The interactions we have been discussing thus far, in addition to obeying the usual classical laws of energy, momentum, and angular momentum conservation, also conserve quark flavor content. We have already alluded to the fact that, in collisions of normal particles, when a baryon is produced it is accom-

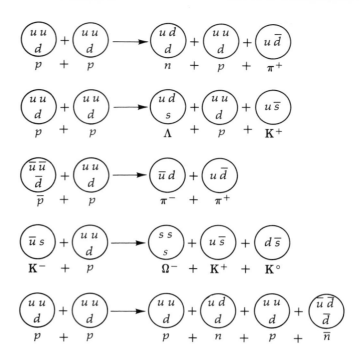

FIGURE 45-19
Examples of allowed particle collisions. In all such strong interactions, the sum of the quark flavors on the left and right sides of the arrow is the same (conserved!). When a $\bar{u}$ is produced, a u-quark must also appear to balance the reaction, etc.

panied by an antibaryon. Similarly, when a charm-flavored meson or baryon is produced, another particle is always produced that contains the $\bar{c}$-quark. This is what is referred to as *associated production*, and the principle is illustrated in several of the reactions shown in Figure 45-19. There are, however, other processes in which quark flavor is not conserved (although baryon number is). These kinds of interactions are very weak, and they do not compete favorably when other transitions are possible. All beta-decay processes (Section 45.5), for example, proceed through this *weak interaction*.

Being pointlike, having no electric charge, and carrying no color charge, the neutrino interacts only weakly. There is definite evidence for the existence of two kinds of neutrinos (and, of course, their antineutrino partners). These are the electron neutrino (v_e) and the muon neutrino (v_μ). Just as the electron neutrino arises in processes that involve electron or positron weak interactions, so the muon neutrino is found in interactions involving muons. There is another weakly interacting charged particle, the tau (τ), that also has essentially the same properties as the electron, but is even more massive than the muon. Although the existence of the tau-associated neutrino (v_τ) has yet to be proven, no one doubts that it will have an inherent property that will distinguish it from the v_e and the v_μ. These six pointlike, weakly interacting particles (and their antiparticles) constitute the lepton family of fermions. The electron, the muon, and the tau (and their partner neutrinos) carry electron, muon, and tau *lepton numbers* that are unique; and, just as flavor content is preserved in the strong interactions, lepton number (or lepton flavor) is always conserved in all particle interactions. The lepton flavor of any neutrino is defined to be the same as that of its charged partner. So, anytime a lepton is produced, it is accompanied by an appropriate antilepton, and consequently the total lepton flavor of the universe is unchanged (that is, it is conserved). As illustrated in Figure 45-20, the beta decay of a neutron, for example, involves the production of an electron and its antineutrino.

All weak processes are now thought to be mediated by the massive and pointlike W^+, W^-, and Z^0 bosons, the carriers of the weak force. These objects are the analogues of the gluons and photon for the other forces. Properties of

Neutron β Decay

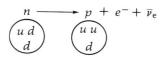

Muon Decay

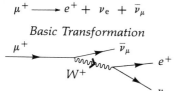

Neutrino–Proton Scattering

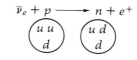

Neutrino–Electron Scattering

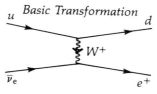

Neutrino "Elastic" Scattering

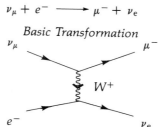

FIGURE 45-20
Examples of weak interactions. All such processes involve either $W^{\pm}$ or Z^0 bosons as the messengers of the weak force. The basic mechanisms are illustrated in terms of Feynman diagrams.

these mediators (Table 45-6) and some of their simpler reactions are illustrated in Figure 45-20. The figure shows examples of weak interactions that proceed through W exchange, in which quark flavor is not conserved. Weak transitions that conserve quark flavor involve the Z^0. There are many rules that emerge from the Standard Model that we cannot consider here for of lack of space. It is worth mentioning, however, that though W's and Z's can be emitted or absorbed directly by quarks, they do not carry color and consequently do not couple to gluons. They can originate only from the *"weak charge"* contained within leptons and quarks. W's and Z's can therefore carry messages of the weak force only among quarks and leptons.

Unification and the Future

A noteworthy point is that, except for mass, the photon and the Z^0 have many properties in common. From quantum theory, this implies that for some reactions it may not be possible to tell whether the Z^0 or the photon was the mediator. This suggests that the weak force and the electromagnetic force may be related. In fact, this, among other features, has led to the formulation of a single theory that unifies the weak and electromagnetic interactions, forming one of the underpinnings of the Standard Model. The success of this *Electroweak Theory* in predicting both the observed properties and the precise values of the masses of the W's and the Z^0, as well as other subtle effects, gives physicists hope for even unifying the strong force with the electroweak ("*Grand Unification*").

In Figures 45-18 and 45-20, we presented interactions of constituents and electrons in terms of their exchanged quanta. Is it possible that all fundamental reactions proceed through such mechanisms? There is at present no quantized field theory of gravitation, but the search for the **graviton** (G), the quantum boson thought to be the carrier of the gravitational field, has been going on for several years. Although there is as yet no convincing experimental evidence for such an object, few doubt that the graviton exists, and that it is to quantum gravity (QGD) what the photon is to QED, what the gluon is to QCD, and what the W's and the Z^0 are to the weak interaction. Also, just as interacting (or accelerating) charges of the weak, electromagnetic, and strong interactions provide the sources for their respective quanta, accelerating mass (or energy) is the source of gravitons. Gravity is, however, the weakest of the known forces, one that does not have much of an effect on elementary particles at energies available at existing accelerators. As an example, we can compare the relative strengths of the different forces for two protons that are about 10^{-13} cm apart (nucleon dimensions). If the Coulomb force is equated to a strength of unity, then the gravitational attraction would be about 10^{-36} of that, and the *residual* strong force about a factor of 20 times the electromagnetic. The weak interaction would have a value of about 10^{-7} on this scale.

The similarity of the fundamental interactions (and of quarks to leptons) has inspired theorists to try to unify all these forces under one framework. How can one theory encompass forces of such varying strengths? The answer is that the strengths of the individual forces depend on the distances between objects, and therefore, through the Heisenberg uncertainty principle, on the size of momentum and energy transfers in the interactions. For separations of about 10^{-32} cm the strengths of all the forces seem to merge. This is a fascinating concept that, although appealing, does not as yet stand on firm ground.

In fact, although the Standard Model is a remarkable achievement, there are still gaping holes in our understanding. For example, the simple question of why there are three groupings (or "*generations*") of pointlike fermions, or whether there are more Z's and W's yet to be discovered, cannot be answered. We also do not understand the origin of mass: Why is the tau lepton 3500 times heavier than the electron? Why is the photon massless while the Z^0 has a mass of 100 protons? Nor do we know whether the leptons and quarks are truly elementary and indivisible; they certainly appear to be structureless down to scales of order 10^{-16} cm. It is the hope of particle physicists that answers to these questions will be forthcoming from the next generation of accelerators, like the *Superconducting Super Collider* (SSC, proposed to be constructed in Ellis County, Texas) and the *Large Hadron Collider* (LHC, proposed for construction at the European physics laboratory CERN outside of Geneva, Switzerland); these machines have been specifically designed to probe the energy scales where many of the important issues that we have just discussed are likely to be clarified.

The Cosmic Connection[11]

Cosmologists today believe that the universe was born in a single tremendous explosion, the "*Big Bang*." The elementary particles, the fundamental forces, the chemical elements, the stars and galaxies—all trace their origin to this primordial conception. The Big Bang was not an explosion of matter and radiation into a previously empty space, but was the creation of space itself along with everything else. We measure both time and the amount of space in the expanding universe from the Big Bang instant at $t = 0$.

The earliest moments of the universe were too hot for atoms or nuclei to exist. There were only the simplest objects interacting through fundamental forces. It is thought that initially there was only one force; but when the universe was at the barely imaginable age of 10^{-42} of a second, gravity sepa-

---

[11] This section is based largely on the booklet "To the Heart of the Matter—The Superconducting Super Collider," Universities Research Association, Washington, D.C., April 1989.

rated from the unified strong–electroweak force, and therefore two types of forces became apparent. The universe expanded and cooled rapidly. When it was 10^{-35} s old, and the temperature had fallen to an equivalent of 10^{24} eV in energy, the strong force and the electroweak force became distinct. The universe continued to cool. At 10^{-12} of a second, and 10^{12} eV, the electroweak force split into two, and the four forces that we know today all became distinct. A little later, close to 10^{-6} of a second, quarks and gluons coalesced into protons and neutrons. Later still, at an age of several minutes, atomic nuclei began to condense from the sea of protons and neutrons; whole atoms started appearing only after hundreds of thousands of years. It is certainly difficult to imagine that at an age of 10^{-42} s all the matter and energy in the universe today was squeezed into a volume less than one tenth of a millimeter across!

Today, the universe is about 10^{10} years old and has a typical temperature of 2.3×10^{-4} eV (2.7 K). What astronomers observe today are the cool remnants of that dazzling initial fireball. But cosmologists have been remarkably successful in reconstructing cosmic history back to the first microsecond of the Big Bang, and particle physics is an essential ingredient in the reconstruction. Two particles colliding in an accelerator recreate an early moment from cosmic history. And the greater the collision energy, the further back in time we see. To explore the world of elementary particles is to explore the early universe.

Particle physicists and cosmologists now find they have many common goals and interests. As an example, the issue of *"dark matter"* is one region of great mutual concern. The mass of a galaxy, as measured from the motion of gas clouds about its center, turns out to be greater than the combined mass of all the observed stars, gas, and dust. The unseen matter—the dark matter—has long puzzled astronomers, but now particle physicists may be able to offer an explanation. Dark matter may consist of certain particles that have survived unseen since their production long ago, when the universe was much hotter. (The only known stable particles in the universe are protons, photons, electrons, and neutrinos, and they cannot account for the missing mass.) A discovery of such new particles at higher-energy accelerators will not only guide physicists toward unification of forces, but may also help cosmologists solve the dark matter problem.

Postscript

The W's and the Z^0, the leptons and the quarks, the gluons, and all the other mind-boggling richness we have described would not be ours were it not for the accelerator scientists who invented the machines that provided the opportunity for discovery, and the ingenious experimenters who built the detectors to "see" the new phenomena created at increasingly higher energies. At times we tend to forget that physics is an experimental science. Viki Weisskopf, one of the major theorists of this century, likes to compare our experimenters to the explorer Columbus, accelerators to the ships he sailed, and our theoretical physicists to the know-it-alls who stayed behind in Spain and convinced everyone (including Columbus) that India would be reached sailing West. The irony of the story, which might have even escaped Viki, is that to this very day the islands scattered in the Caribbean are called the West Indies! Theorists have always had great influence!

It is not possible in the space available to do justice to the beautiful technical achievements that have propelled this field. Some of the techniques have been sketched in several chapters of this book, but their scale cannot be captured. (The late Sir John Adams, a master builder of the CERN accelerators, was once overheard saying that his kind publishes in cement!) An example

FIGURE 45-21
Assembly of the 2000-ton Collider Detector at Fermilab (CDF). Sophisticated detector systems are used to measure energies and directions of all particles produced in high-energy collisions. Such detectors are built up in layers surrounding the interaction point. Each layer is designed to reveal specific information about the traversing particles. Closest to the point of collision is a *vertex detector* (not shown in the photograph) to detect any particles with exceedingly short flight paths (that is, with short lifetimes). The next layer, here shown being moved into position, is the *central tracking chamber*. This consists of sets of coaxial, cylindrically positioned planes of wire electrodes that sense and trace the paths of any charged particles by recording ionization produced along the particle trajectories. Bathing the chambers in a magnetic field allows positively charged particles to be distinguished from those of negative charge by the sense of curvature of the reconstructed paths. In CDF, a 1.5-T axial magnetic field is produced by a 5-m-long superconducting solenoidal coil that surrounds the central tracking chamber. Beyond the coil are additional layers of instrumentation. First come segmented *"calorimeter"* modules, constructed from lead or steel plates, interleaved with planes of scintillator. Photomultiplier tubes record energies deposited by particles as they pass through the calorimeter stacks, interact in the material, and produce more particles and scintillation light. Because muons deposit very little energy along their paths, they penetrate beyond the region of calorimetry, where they are measured using special outer muon-detection chambers. In the photograph, the end calorimeters are shown in their operating positions, while the wedge-shaped central calorimeters are retracted for servicing. The boxlike iron superstructure, besides providing mechanical support, also serves as the yoke for the return path of the magnetic flux. The object on the side that looks like a nose cone is one of two sets of end calorimeters that fit snugly into the front and rear apertures of CDF. When fully assembled, this multilayered, intricate, about-100,000-channel device is rolled during a one-day operation from its "garage" into the collision hall to study collisions between 1-TeV protons and 1-TeV antiprotons. Detector performance is monitored and controlled on-line through a houseful of electronic devices and computers.

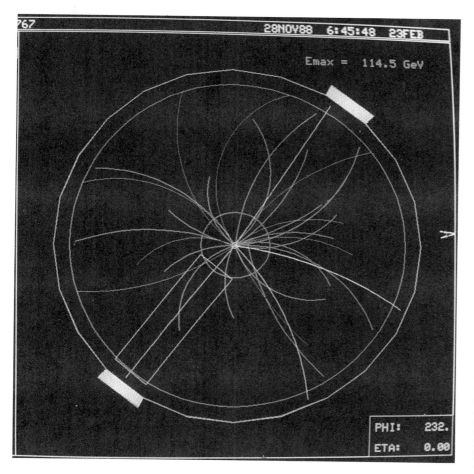

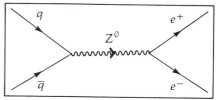

FIGURE 45-22
Production of a Z^0 boson at CDF. A computer reconstruction of particles created in a proton–antiproton collision at CDF. The curved lines represent trajectories of electrically charged particles that are deflected in the axial magnetic field. (The straighter the track, the higher the momentum of the particle.) The energies of the two high-energy, back-to-back particles, as measured in the calorimeter (represented by the light rectangles on the perimeter of the display), show that they are electrons (one e^+ and one e^-) with an effective mass of a Z^0 particle. The lifetime of the Z^0 is too short for it to interact directly with any part of the detector; its existence is inferred from its characteristic decay modes. The other particles accompanying the Z^0 have very low energy, and correspond to the remnants of the "spectator" quarks that barely participated in the collision. The Feynman diagram shows how to understand the production of the Z^0 in this $\bar{p}p$ reaction. A quark from the p and an antiquark from the $\bar{p}$ fuse to form a Z^0, which then decays into the e^+e^- pair. Overall color, like charge and momentum, must be conserved in the collision. (The rectangular box just singles out the particle with highest momentum.)

is the apparatus shown in Figure 45-21, which is used by the Collider Detector at the Fermi National Accelerator Laboratory—Fermilab, just outside of Chicago—to study interactions of quarks and gluons contained within colliding protons and antiprotons. This detector is a maze of sophistication. Its tens of thousands of separate elements track charged and neutral particles and measure all their momenta and energies. From reconstruction of the properties of collisions of the type shown in Figure 45-22, the unimagined can emerge. The Standard Model may be a magnificent achievement, but finding its limitations, and recognizing that it is only a fleeting approximation to nature, would be a truly exhilarating experience, and one that only an experimenter can stumble upon![12]

---

[12] For further reading see F. Close, M. Martin, and C. Sutton, *The Particle Explosion*, Oxford University Press, 1987, and L. Lederman and D. Schramm, *From Quarks to the Cosmos*, W. H. Freeman, 1989. We also wish to thank A. Hudson, K. Metropolis, and C. Quigg for valuable comments on this essay.

Summary

A nucleus is characterized by the *atomic number* Z (the number of protons) and the *neutron number* N. The *mass number* A is the sum $A = Z + N$. *Isotopes* have the same Z but different values of N and are written in the following notation: ^A_Z[element symbol]. (The Z is often omitted since the element symbol identifies the atomic number.)

The approximate *nuclear radius* $R = kA^{1/3}$, where k is a constant equal to ~ 1.2 fm to ~ 1.4 fm, depending on the type of interaction used to probe the nucleus. The nuclear density is approximately constant, indicating the short-range nature of the nuclear force.

Nuclear masses are measured in *unified atomic mass units* u:

$$1\ u = \begin{cases} \frac{1}{12} \text{ the mass of atomic } ^{12}\text{C} \\ 931.494 \text{ MeV}/c^2 \\ 1.660\ 540 \times 10^{-27} \text{ kg} \\ 1.492\ 42 \times 10^{-10} \text{ J}/c^2 \end{cases}$$

The *binding energy per nucleon* is the energy required to separate a nucleus into its nucleons divided by the number of nucleons.

A sample containing N_0 initial radioactive nuclei decays according to

$$N = N_0 e^{-\lambda t} \quad \text{or} \quad N = N_0 e^{-(\ln 2)t/T_{1/2}}$$

where λ is the *decay constant* and $T_{1/2}$ is the *half-life*. Radioactivity is also measured in units of the *curie* (Ci), where the *activity* A is

$$A = A_0 e^{-\lambda t} \quad 1 \text{ Ci} \equiv 3.71 \times 10^{10} \frac{\text{disintegrations}}{\text{second}}$$

A radioactive nucleus may decay by alpha emission (α), beta emission (either β^- or β^+), gamma decay (γ), internal conversion, electron capture (also called K capture), or in a few heavy nuclei, *spontaneous fission*. In each case, the Q of the reaction must be positive:

$$Q = [\text{Original mass} - \text{Product masses}]c^2$$

In the case of positron decay (β^+), the parent nuclide must exceed the daughter nuclide by at least $2m_e c^2$.

Typical *nuclear reactions* involve an incident particle x striking a target nucleus X (initially at rest), yielding the reaction products y and Y. Reactions are classified as exoergic (Q positive), in which mass-energy is released, and endoergic (Q negative), in which some initial kinetic energy is converted to mass-energy. When Q is negative, the incident particle must have at least the minimum *threshold kinetic energy* E_{th} to cause the reaction

$$E_{th} = -Q\left(\frac{m_x + M_X}{M_X}\right) \quad \text{(nonrelativistic)}$$

The *nuclear cross section* σ is a measure of the relative probability that a reaction will happen. It can be pictured as the effective area a target nucleus presents to an incoming particle (assumed to be a point); a nuclear reaction occurs only if the incident particle strikes a target area. Cross sections are measured in *barns* (1 b $\equiv 10^{-28}$ m^2 = 10^{-24} cm^2). Cross sections vary widely depending upon the type of reaction and the kinetic energy of the incoming particle.

Radioactive dating makes use of the 5730-yr half-life of ^{14}C to determine the age of carbon-based artifacts. We also can date geologic specimens by noting the ratios of certain elements that are part of radioactive decay series.

Nuclear power can be obtained from the neutron-induced *fission* of very heavy nuclei, such as ^{235}U and ^{239}Pu, or by the *fusion* of very light nuclei, such as ^{2}H and ^{3}H. Commercial power generation by fusion has not yet been achieved because *Lawson's criterion* at fusion temperatures, $n\tau > \sim 10^{20}$ s·m^{-3}, has not been reached for large-scale operations.

Particle physics. The "Standard Model" classifies particles (other than photons) into two groups: *hadrons* (which interact mainly via the strong force) and *leptons* (which interact only via the weak force). Hadrons are subdivided into *baryons*, which have half-integral spin (fermions), and *mesons*, which have integral spin (bosons). Recent experiments support the *quark* model, in which hadrons are made up of various combinations of quarks. Interactions between particles occur by means of "messenger" particles (or "force carriers") that are exchanged during the process.

Force	Force Carrier
Electromagnetism	Photon (γ)
Weak	$W^\pm$, Z^0
Strong	Gluon (g)
Gravity	Graviton (G)

Theorists seek a *unified theory* in which (at sufficiently high energies) all interactions are combined into just one universal force.

The only known *stable* particles in the universe are the *electron*, the *proton*, the *neutrino*, and the *photon*.

Questions

1. If the size of a hydrogen atom were scaled upward so that the diameter of the proton were one millimeter, what would be the approximate diameter of the atom?

2. What are the similarities and differences among electrostatic, gravitational, and nuclear forces?

3. Why is nuclear mass-density independent of the number of nucleons in the nucleus?
4. What is the qualitative comparison of the size of nuclei and the barn?
5. If radioactive decay is a random process, why can it be represented by a simple mathematical function?
6. What mode(s) of decay would you expect for ^{124}Ba? Why?
7. Why are nuclei of the form $^{2n}_nX$ (n even) particularly stable for $n \leq 10$ yet unstable for $n > 10$?
8. Why do most of the fragment nuclei of nuclear fission undergo β^- decay rather than β^+ decay?
9. An isotope of gold is $^{197}_{79}$Au. What other pair of numbers would also identify this isotope?
10. Technetium results from the decay of molybdenum, which is a common product of nuclear fission. What inference can be made about the fact that natural technetium is probably not present in the earth's crust?
11. Compare the value for the cross section for neutron capture by ^{113}Cd given in Example 45-12 with the data in Figure 45-14. Why is the former value about an order of magnitude larger than that indicated by the graph?

Problems

45.2 A Description of the Nucleus
45.3 Nuclear Mass and Binding Energy

45A-1 A rough estimate of the relative strengths of the nuclear force that holds protons in a nucleus and the Coulomb force of repulsion can be made by the following calculation. Find the ratio of the *binding energy* (BE) of one proton in the ^{4}He nucleus to the *Coulomb potential energy U* of the two protons, assuming that in the nucleus they are 1 fm apart.

45A-2 (a) Find the approximate radius of the nuclide $^{133}_{55}$Cs. (b) What is the approximate mass number A of a nucleus whose geometrical cross section is 0.8 b?

45A-3 On the Richter scale, the magnitude M of an earthquake is related to energy E released according to the relation $M = (1/1.5)\log_{10}(E/25\,000)$, where E is expressed in joules. The energy release of a one-megaton (equivalent to 10^6 tons of TNT) hydrogen bomb is 4.18×10^{12} J. Find the magnitude of an earthquake that produces the same energy as the explosion of a two-megaton hydrogen bomb.

45B-4 How much energy (in joules) would be required to separate the nuclei in one gram of $^{56}_{26}$Fe into separate nucleons?

45.4 Radioactive Decay and Half-Life
45.5 Modes of Radioactive Decay

45A-6 A mummy known as Whiskey Lil was discovered in a cave near Lake Winnemucca, Nevada, in 1955. With carbon dating methods, it was determined that 73.9% of the original ^{14}C was still present. What year did Whiskey Lil die?

45A-7 Find the time required for the activity of an isotope with a half-life of 12 min to decay to one-fifth its initial activity.

45B-8 The half-life of ^{241}Am is 432 yr. The isotope decays by alpha emission. (a) Write the reaction for this decay. (b) Find the mass of this isotope that has an activity of 1 mCi.

45B-9 Assuming that the molecular weight of radium is 226 and that its half-life is 1620 yr, find the activity of one gram of ^{226}Ra.

45B-10 Refer to Example 45-5. Of the original 279 936 000 dice, one-sixth are removed every succeeding day. (a) From the plot of data shown in Figure 45-6, determine the time elapsed for only one-quarter of the original dice to remain. (b) Using Equation 45-6, calculate the time for only one-quarter to remain. (c) Find the percent discrepancy between these two results. Explain.

45B-11 After a nuclear explosion, the resulting aggregate radioactivity does not follow the exponential decay law. Instead, for a period of about six months following the explosion, the activity decreases according to the relation

$$A = A_0(t/t_0)^{-1.2}$$

where A_0 is the activity at a time t_0 after the explosion. (After six months, the decay is more rapid, so that after 10 years the activity is only about $\frac{1}{25}$th that predicted by the above relationship.) Calculate the short-term half-life $T_{1/2}$ in terms of t_0 according to this relationship.

45B-12 The isotope $^{49}_{24}$Cr decays by positron emission with a half-life of 42 min. (a) Write the reaction equation. (b) If a sample has an initial activity of 24 mCi, what is the activity two hours later?

45.6 Nuclear Cross Section

45B-13 A beam of thermal neutrons is incident upon a slab of carbon, ^{12}C. The total capture cross section for thermal neutrons is 3.5 mbn. Calculate the thickness L of carbon that will capture 20% of the incident neutrons. The specific gravity of carbon is 2.25.

45B-14 A lead brick is mostly empty space. Assuming that the lead atoms within the brick are uniformly distributed, how thick would the brick have to be for the projected area of the geometrical cross sections of all of the nuclei on a face of the brick to be one-tenth of the area of the face of the brick? The value of R_0 is 1.3 fm.

45B-15 Supernova 1987A, located about 170 000 light-years from the earth, is estimated to have emitted a burst of $\sim 10^{46}$ J of neutrinos. Assuming an average neutrino energy of 6 MeV and a 5000-cm^2 cross-sectional area for your body, how many of these neutrinos passed through your body? [Adapted from a problem in the *Back of the Envelope* column in *American Journal of Physics* **56**, 5 (May 1988).]

45.7 Nuclear Reactions

45A-16 The isotope ^{236}U undergoes fission to produce the fission products ^{90}Rb and ^{143}Cs. Show that the Q of this reaction is given by $Q = (M_U - M_{Rb} - M_{Cs} - 3m_n)c^2$, where M_U, M_{Rb}, and M_{Cs} are the respective atomic masses and m_n is the neutron mass.

45B-17 The *carbon cycle*, believed to be the main source of energy in stars hotter than the sun, is the following sequence of reactions:

$$^{12}_{6}C + ^{1}_{1}H \longrightarrow ^{13}_{7}N + \gamma$$
$$^{13}_{7}N \longrightarrow ^{13}_{6}C + e^+ + \nu$$
$$^{13}_{6}C + ^{1}_{1}H \longrightarrow ^{14}_{7}N + \gamma$$
$$^{14}_{7}N + ^{1}_{1}H \longrightarrow ^{15}_{8}O + \gamma$$
$$^{15}_{8}O \longrightarrow ^{15}_{7}N + e^+ + \nu$$
$$^{15}_{7}N \longrightarrow ^{12}_{6}C + ^{4}_{2}He$$

(a) Show that the net effect of this sequence is the same as the proton–proton cycle (Problem 45C-39), namely, combining four protons to form He. (Note that no carbon is consumed; it merely acts as a catalyst in the sequence of reactions.) (b) Explain why the carbon cycle requires a higher temperature than the proton–proton cycle.

45B-18 Using the result of Problem 45C-41, calculate the energy of neutrons emitted at 90° with respect to an incident 0.5-MeV beam of deuterons upon a deuterium target in the ^{2}H(d,n) reaction.

45B-19 Natural gold has only one isotope, $^{197}_{79}$Au. If natural gold is irradiated by a flux of slow neutrons, β^- particles are emitted. (a) Write the appropriate reaction equations. (b) Calculate the maximum energy of the emitted beta particles.

45B-20 Write reaction equations for the following reactions (where X is to be determined); include the compound nucleus and all A and Z values: ^{9}Be$(\alpha,n)X$, $X(\alpha,p)^{31}$P, ^{7}Li$(d,2\alpha)X$; ^{11}B$(p,\alpha)X$, ^{12}C$(\gamma,\alpha)X$.

45B-21 Show that the decay of 1 g of $^{238}_{92}$U produces 1.33×10^{-10} g of $^{206}_{82}$Pb per year for time periods much shorter than a billion years. As Figure 45-12 indicates, the half-life of the chain of decays from $^{238}_{92}$U to $^{206}_{82}$Pb is essentially 4.5×10^9 yr.

45.8 Nuclear Power

45A-22 Pure ^{238}U, with ordinary water as a moderator, has a critical mass of about 3 kg. Find the diameter of a sphere of this much uranium.

45A-23 Calculate the kinetic energy (in eV) of a ^{2}H nucleus having the root-mean-square speed v_{rms} for gas particles in equilibrium at a fusion temperature of 3×10^8 K.

45B-24 Suppose that an electron resides in a nucleus with a diameter of 10^{-14} m. Determine the approximate energy (in MeV) that such an electron must have. Use the uncertainty principle and interpret the uncertainty in position as the diameter of the nucleus and the uncertainty in momentum as the momentum. Your answer should justify using the relativistic approximation $E = pc$. Since electrons emitted in beta decay seldom have energies greater than 1 MeV, the uncertainty principle supports the non-existence of electrons in the nucleus.

45B-25 Assuming that the centers of two nuclei must approach to within 10 fm of each other to cause fusion, calculate the minimum total energy E (in MeV) for fusion to occur between two nuclei in (a) the deuterium–deuterium reaction and (b) the deuterium–tritium reaction.

45B-26 (a) Find the rms speed of deuterons in a contained plasma at 200 million K. (b) About how long would it take the deuterium to escape from a sphere of 10-cm diameter if not contained?

Additional Problems

45C-27 A living specimen of organic material in equilibrium with the atmosphere contains one atom of ^{14}C (half-life = 5730 yr) for every 10^{12} stable carbon atoms. An archeological sample of wood (cellulose, $C_{12}H_{22}O_{11}$) has a mass of 21.0 mg. When the sample is placed inside a beta counter whose counting efficiency is 88%, 937 counts are accumulated in one week. Assuming that the cosmic-ray flux and the earth's atmosphere have not changed appreciably since the sample was formed, find the age of the sample.

45C-28 The rate of decay of a radioactive sample is measured at 10-s intervals, beginning at $t = 0$. The following data are obtained: 1137, 861, 653, 495, 375, 284, 215, 163. (a) Plot these data on semilog graph paper and determine the best-fit straight line. (b) From the graph, determine the half-life of the sample.

45C-29 The rate of decay of a radioactive sample is measured at 1-min intervals starting at $t = 0$. The following data (in counts per second) are obtained: 260, 160, 101, 72, 35, 24, 13, 10, 5.2, 4.0. (a) Plot these data on semilog graph paper and sketch the best-fit straight line. Determine, to two significant figures, (b) the half-life for this sample and (c) the decay constant.

45C-30 If semilog graph paper is not available, measurements of the activity vs. time for a radioactive sample will plot as a straight line on linear graph paper if the logarithm of the counting rate is plotted vs. time. Solve the previous problem by this method.

45C-31 When radioactive tracers are used in living systems, the system may expel tracer atoms through natural metabolic processes, reducing the counting rate below that due to radioactive decay alone. An exponential decay is a reasonable assumption for these biological processes. Thus there are two decay constants, the *physical* decay constant λ_p and the *biological* decay constant λ_b, leading to

$$\frac{dN}{dt} = -N(\lambda_p + \lambda_b) \qquad \text{and} \qquad N = N_0 e^{-(\lambda_p + \lambda_b)t}$$

Show that the *effective half-life* T_e that combines these two effects is given by $1/T_e = 1/T_p + 1/T_b$.

45C-32 An alpha particle is emitted from ^{226}Ra with an energy of 4.7845 MeV. Calculate the total decay energy (including the recoil of the parent nucleus).

45C-33 A pellet of ^{210}Po with an activity of 50 Ci feels warm to the touch. Find the rate at which thermal energy is produced within the pellet.

45C-34 Example 45-12 describes a reaction cross section that is very much larger than the projected area of the nucleus. At the other extreme is the tiny cross section for neutrino interactions, making neutrinos exceedingly difficult to detect. One reaction used to detect antineutrinos is $\bar{\nu} + {}_1^1\text{H} \longrightarrow {}_{+1}^{0}e + {}_0^1n$. This reaction has an approximate cross section of 10^{-19} b.
(a) Determine the thickness of water (in kilometers) required to reduce an incident neutrino flux by one part in a million.
(b) Compare your answer to the earth–sun distance.

45C-35 Refer to the previous problem. The antineutrino flux from a nuclear reactor is 10^{13} antineutrinos/cm$^2 \cdot$s. Suppose that this flux impinges uniformly on one side of a cube of water 10 cm on a side. Calculate the average number of interactions per day between the antineutrinos and the protons in the water.

45C-36 A slow neutron with negligible kinetic energy is absorbed by a boron nucleus at rest, producing the reaction of Equation (45-43). Find the kinetic energies of each of two nuclei produced in the reaction.

45C-37 Consider an endoergic reaction whose Q value is negative. Using the conservation of energy and momentum (nonrelativistic) in a one-dimensional collision, show that the *threshhold energy* E_{th} is given by Equation (45-39).

45C-38 In a particular fission of ^{235}U by a slow neutron, one of the fission fragments is ^{137}Te. No neutrons are emitted. (a) Identify the other fragment. (b) The Q for this reaction is 190 MeV. Using whole-number masses, calculate the kinetic energy of each fission fragment.

45C-39 The following sequence of fusion reactions is the *proton–proton cycle*, believed to be the main source of energy in the sun:

$$\begin{aligned}{}_1^1\text{H} + {}_1^1\text{H} &\longrightarrow {}_1^2\text{H} + e^+ + \nu \\ {}_1^1\text{H} + {}_1^2\text{H} &\longrightarrow {}_2^3\text{He} + \gamma \\ {}_2^3\text{He} + {}_2^3\text{He} &\longrightarrow {}_2^4\text{He} + {}_1^1\text{H} + {}_1^1\text{H}\end{aligned}$$

(a) Find the Q value for each reaction. (b) Noting that each of the first two reactions must occur twice to produce the two $_2^3$He nuclei for the third reaction, find the total energy released in the (net) fusion of four protons to produce one $_2^4$He nuclei by the proton–proton cycle. (c) When each of the two positrons produced in the cycle encounters an electron, a positron–electron annihilation occurs, producing two photons of equal energy emitted in opposite directions (to conserve momentum).

The energy released in an (e^+, e^-) annihilation is $2m_ec^2$, and this also contributes to the total energy release for the process. Calculate the total energy released in the p–p cycle, including positron annihilations.

45C-40 The fusion of two deuterons (^{2}H) will form an alpha particle (^{4}He). (a) Calculate the energy released in this reaction due to the decrease in mass. (b) How many such reactions must occur each second to light a 60-W light bulb? (c) Starting with one milligram of deuterium atoms, how long could we keep the light bulb burning?

45C-41 A particle of mass m and initial kinetic energy K_0 is captured by a nucleus of mass M initially at rest. The compound nucleus immediately ejects a light particle of mass M_1 at 90° to the incident-particle direction, and the recoil nucleus has a mass M_2. Show that the kinetic energy K of the ejected light particle is

$$K = \left[Q - \left(\frac{m - M_2}{M_2}\right)K_0\right]\left(\frac{M_2}{M_1 + M_2}\right)$$

where the energy equivalent of the total mass difference is Q.

45C-42 Which conservation laws, if any, are violated in the following reactions? Could these processes proceed through one of the fundamental interactions? Which ones?
(a) $\rho^+ \longrightarrow \pi^+ + \pi^0$
(b) $\rho^+ \longrightarrow K^+ + \pi^0$
(c) $\rho^+ \longrightarrow \pi^+ + \gamma$
(d) $\pi^+ \longrightarrow p + \bar{n}$
(e) $\Lambda \longrightarrow p + \pi^-$
(f) $\Lambda \longrightarrow p + e^- + \bar{\nu}_e$
(g) $\pi^+ \longrightarrow \mu^+ + \nu_\mu$
(h) $\bar{\Lambda} + p \longrightarrow K^+$
(i) $\bar{\Lambda} + p \longrightarrow K^+ + \pi^0$
(j) $\pi^+ + n \longrightarrow K^+ + \Lambda$

45C-43 A K^+ meson comes to rest and decays into $e^+ + \nu_e$. Ignoring the rest mass of the positron, find the energy carried away by the e^+. Find the energy carried away by the ν_e.

45C-44 An e^- and an e^+, each with an energy of 46 GeV, collide head-on to form the Z^0 particle. The Z^0 then decays into $\pi^+ + \pi^-$. Is this possible? Can you suggest a mechanism for this transition? (Hint: you may have to allow a quark–antiquark pair to "pop out of the vacuum.")

45C-45 A Λ particle with momentum of 1.116 GeV/c is produced in a collision. What will be its mean flight path in the laboratory before it decays?

45C-46 A neutral particle M^0 decays in flight into two photons. Writing the rest mass of the particle in terms of the two energies and momenta of the photons, show that for small angles (cos $\theta \sim 1 - \theta^2/2$), $\theta^2 = m^2c^4/E_1E_2$, where E_1 and E_2 are the photon energies, θ is the angle between the momentum vectors of the photons (small angle), and m is the rest mass of the particle. Prove that the minimum value of θ occurs when $E_1 = E_2 = E/2$, where E is the energy of the meson ($E = E_1 + E_2$ and therefore $\theta_{min} = 2mc^2/E$.) What is the observed distance between two photons from a π^0 decay ($\pi^0 \to \gamma + \gamma$) if the photons are detected 2 m from their point of production for a π^0 of 10 GeV? What about a π^0 of 100-GeV energy? (Hint: remember that a photon has no rest mass and that consequently $E = pc$ holds.)

Appendices

CONTENTS

A. SI Prefixes A-1
B. Mathematical Symbols A-1
C. Conversion Factors A-2
D. Mathematical Formulas A-4
E. Mathematical Approximations, Expansions, and Vector Relations A-6
F. Fourier Analysis A-6
G. Calculus Formulas A-8
H. Finite Rotations A-10
I. Derivation of the Lorentz Transformation A-10
J. Periodic Table of the Elements A-12
K. Constants and Standards A-13
L. Terrestrial and Astronomical Data A-14
M. SI Units A-15

APPENDIX A
SI Prefixes

Multiple	Prefix		Symbol
10^{18}	exa	(ĕk′så)	E
10^{15}	peta	(pĕt′å)	P
10^{12}	tera	(tĕr′å)	T
10^{9}	giga	(jĭ′gå)	G
10^{6}	mega	(mĕg′å)	M
10^{3}	kilo	(kĭl′ō)	k
10^{2}	*hecto	(hĕc′tō)	h
10^{1}	*deka	(dĕk′å)	da
10^{-1}	*deci	(dĕs′ĭ)	d
10^{-2}	†centi	(sĕn′tĭ)	c
10^{-3}	milli	(mĭl′ĭ)	m
10^{-6}	micro	(mī′krō)	μ
10^{-9}	nano	(năn′ō)	n
10^{-12}	pico	(pē′cō)	p
10^{-15}	femto	(fĕm′tō)	f
10^{-18}	atto	(ăt′tō)	a

In each case, the accent is on the *first* syllable.
* Rarely used.
† Generally used only as *centimeter* (cm).

APPENDIX B
Mathematical Symbols

Symbols

$=$	is equal to
$\neq$	is not equal to
$\equiv$	is identical to or by definition
$a > b$	a is greater than b
$a \gg b$	a is much greater than b
$a < b$	a is less than b
$a \ll b$	a is much less than b
$a \geqq b$	a is equal to or greater than b
$a \leqq b$	a is equal to or less than b
$a \sim b$	a is of the order of magnitude of b; i.e., a is within a factor of 10 or so of b
$\propto$	is proportional to
$\approx$	is approximately equal to
$\pm$	plus or minus (for example, $\sqrt{4} = \pm 2$
$r \to \infty$	r approaches infinity
$\Rightarrow$	implies
Σ	the sum of
$|\ |$	the absolute value of
$\|A\|$ or A	the magnitude of the vector **A**
$\oint$	a line integral around a closed path or a surface integral over a closed surface
$\cdot$	multiplication symbol
$\cdot$	(as in $\mathbf{A} \cdot \mathbf{B}$) dot product
$\times$	(as in $\mathbf{A} \times \mathbf{B}$) cross product
$\times$	(as in 3.2×10^4) multiplication symbol in scientific notation

The Greek Alphabet

Alpha	A	α	Nu	N	ν
Beta	B	β	Xi	Ξ	ξ
Gamma	Γ	γ	Omicron	O	o
Delta	Δ	δ	Pi	Π	π
Epsilon	E	ε	Rho	P	ρ
Zeta	Z	ζ	Sigma	Σ	σ
Eta	H	η	Tau	T	τ
Theta	Θ	θ	Upsilon	Υ	υ
Iota	I	ι	Phi	Φ	ϕ
Kappa	K	κ	Chi	X	χ
Lambda	Λ	λ	Psi	Ψ	ψ
Mu	M	μ	Omega	Ω	ω

APPENDIX C
Conversion Factors
The SI unit is listed first in each table.

Use of Conversion Factors

The ratio of any pair of quantities listed in a table of conversion factors is dimensionless, having the value of 1. To illustrate, consider the expression 1 mi = 5280 ft. Dividing both sides by 5280 ft, we obtain the ratio

$$\left(\frac{1 \text{ mi}}{5280 \text{ ft}}\right) = 1 \quad \left[\begin{array}{l}\text{The reciprocal is also a}\\\text{ratio that equals 1.}\end{array}\right]$$

Any quantity may be multiplied by a **conversion ratio** without changing the value of the quantity.

Example C-1

To express 44 ft/s in units of miles per hour, we make use of two conversion ratios:

$$\left(\frac{1 \text{ mi}}{5280 \text{ ft}}\right) = 1 \quad \text{and} \quad \left(\frac{3600 \text{ s}}{1 \text{ h}}\right) = 1$$

Multiplying conversion ratios and canceling units, we get

$$44 \frac{\text{ft}}{\text{s}} = \left(\frac{44 \text{ ft}}{\text{s}}\right)\underbrace{\left(\frac{3600 \text{ s}}{1 \text{ h}}\right)\left(\frac{1 \text{ mi}}{5280 \text{ ft}}\right)}_{\text{Conversion ratios}} = 30 \frac{\text{mi}}{\text{h}}$$

Thus:
$$\boxed{44 \frac{\text{ft}}{\text{s}} = 30 \frac{\text{mi}}{\text{h}}} \quad \text{(exact)}$$

To change units, we multiply by whatever conversion ratio will cancel the unwanted units.

Length

1 **m** = 39.3701 **in.** = 3.280 84 **ft** = 1.0936 **yd**
1 **km** = 0.621 37 **mi** = 0.539 96 **nautical mile (nmi)**
1 **in.** = 2.54 **cm** (exact)
1 **yd** = 0.914 40 **m** (exact)
1 **mi** = 5280 **ft** = 1.609 × 10³ **m** = 1760 **yd**
1 **nautical mile (nmi)** = 1.151 **mi**

---

[1] One **astronomical unit** (AU) is defined as the mean radius of the earth's orbit.
   One **parsec** (pc) is the distance at which one astronomical unit subtends an angle of one second of arc. Its name comes from the fact that, as seen from the earth, a star at this distance has an annual *parallax* of one *second*. Note the convenient relation (distance in parsecs) = (parallax in seconds)$^{-1}$.
   One **light-year** (ly) is the *distance* that light travels in one year in a vacuum. In numerical calculations in which the speed of light is represented by *c*, the unit may be conveniently written as (*c*·y), so that the symbol *c* may cancel in the calculation as other units do.

1 **astronomical unit**[1] (AU) = 1.4960 × 10¹¹ **m**
  = 4.8481 × 10⁻⁶ **parsec (pc)**
  = 1.5812 × 10⁻⁵ **light-year (c·y)**
1 **parsec (pc)** = 3.2616 **light-year (c·y)**
  = 3.0857 × 10¹⁶ **m**
1 **light-year (c·y)** = 9.4607 × 10¹⁵ **m**
  = 6.324 × 10⁴ **AU** = 0.3066 **pc**
  = 5.879 × 10¹² **mi**
1 **angstrom (Å)** = 1 × 10⁻¹⁰ **m** = 1 × 10⁻⁴ **μm**
  = 1 × 10⁻¹ **nm**

Area

1 **m²** = 1.196 **yd²** = 10.76 **ft²** = 1550 **in.²**
  = 1.974 × 10⁹ **circular mils**
1 **km²** = 0.3861 **mi²** = 1.076 × 10⁷ **ft²**
1 **b** = 1 × 10⁻²⁸ **m²** = 1 × 10⁻²⁴ **cm²**
1 **mi²** = 2.590 × 10⁶ **m²** = 2.788 × 10⁷ **ft²**
  = 640 **acres**
1 **in.²** = 6.452 × 10⁻⁴ **m²** = 6.452 **cm²** = $\frac{1}{144}$ **ft**
1 **ft²** = 9.290 × 10⁻² **m²** = 144 **in.²** = $\frac{1}{9}$ **yd²**
1 **acre** = 43 560 **ft²**

Volume

1 **m³** = 1 × 10³ **L** = 35.31 **ft³** = 6.102 × 10⁴ **in.³**
  = 264.2 **gal (U.S. fluid)**
1 **km³** = 0.2399 **mi³**
1 **gal (U.S. fluid)** = 3.785 × 10⁻³ **m³** = 0.1337 **ft³**
  = 231.0 **in.³**
1 **gal (British imperial and Canadian fluid)**
  = 4.546 × 10⁻³ **m³** = 1.201 **gal (U.S. fluid)**
  = 277.4 **in.³**

Time

1 **s** = $\frac{1}{60}$ **min** = $\frac{1}{3600}$ **h** = 1.157 × 10⁻⁵ **d**
  = 3.169 × 10⁻⁸ **yr**
1 **min** = 60 **s**
1 **h** = 3600 **s**
1 **d** = 8.640 × 10⁴ **s** = 1440 **min** = 24 **h**
  = 2.738 × 10⁻³ **yr** = 1.003 **sidereal days**
1 **yr** = 3.156 × 10⁷ **s** = 365.24 **d**
  = 366.24 **sidereal days**

Mass

1 **kg** = 6.852 × 10⁻² **slug** = 6.022 × 10²² **u**
1 **slug** = 14.59 **kg**
1 **unified atomic mass unit (u)**
  = 1.660 540 2 × 10⁻²⁷ **kg**
1 **kg mass weighs**[2] 2.205 **lb** (at standard *g*)
1 **ounce (oz)** = $\frac{1}{16}$ **lb**
  = weight of 2.835 × 10⁻² **kg** (at standard *g*)

1 **pound (lb)** (at standard g) has a mass of[3] 0.4536 **kg**
1 **metric ton** = 1000 **kg**
1 **ton-mass** = 907.2 **kg** = 2000 **lb-mass**

Density

$1 \text{ kg/m}^3 = 1 \times 10^{-3}$ **g/cm³** $= 1.940 \times 10^{-3}$ **slug/ft³**
1 **ft³** of water weighs[3] 62.43 **lb** (at standard g, 4°C)

Speed

1 **m/s** = 3.600 **km/h** = 3.281 **ft/s** = 2.237 **mi/h**
30 **mi/h** = 44 **ft/s** (exact)
1 **knot** (or 1 **nautical mile per hour**) = 0.5144 **m/s**
   = 1.852 **km/h** = 1.688 **ft/s** = 1.151 **mi/h**

Acceleration

1 g (standard gravity) = 9.806 65 **m/s²** (exact)
   = 32.174 **ft/s²**
1 **ft/s²** = 0.304 80 **m/s²** (exact)
1 **Gal** = 0.010 **m/s²** (exact). The *gal* (Gal) is a special unit used in geodesy and geophysics, named to honor Galileo.

Plane Angle

1 **rad** = 57.30° = 0.1592 **rev**
1 **rev** = 360° = 2π **rad**
1° = 60 **min** (') = 3600 **s** ('') = 1.745×10^{-2} **rad**
   = $\frac{1}{360}$ **rev**
1 **rev/min** = 0.1047 **rad/s**

Solid Angle

1 **sphere** = 4π **steradian (sr)**

Force

1 **newton (N)** = 1×10^5 **dynes** = 0.2248 **lb**

---

[2] It is incorrect to state that 1 kg = 2.205 lb, since there are units of *mass* on one side of the equal sign and units of *force* on the other. Furthermore, the value of the gravitational force depends on the local value of g, varying from point to point on the earth. However, with care, the fact that a mass of 1 kg weighs 2.205 lb (at standard g) can be used to change the mass of an object as expressed in one system to its weight in another system. It is generally safest to make this conversion *prior* to solving a numerical problem. In this table, we use the *avoirdupois pound* = 16 ounces = 0.4536 kg. Another system of weight used for precious metals and stones is the *troy system*:

1 **pound troy** = 12 **ounces troy** = 0.8229 **pound avoirdupois** = 0.3732 **kg**

and

1 **ounce troy** = $\frac{1}{12}$ **pound troy** = 3.1103×10^{-2} **kg**

The troy system is also called the *apothecary system* (of dry weight) used in pharmacy.

[3] From this fact, one can obtain the *weight density* (in pounds/foot³), which is dimensionally different from the *mass density* (in slugs/foot³).

Pressure[4]

1 **pascal (Pa)** = 1 **N/m²** = 10 **dynes/cm²**
   = 9.869×10^{-6} **atm**
   = 1×10^{-5} **bar** = 2.089×10^{-2} **lb/ft²**
   = 1.450×10^{-4} **lb/in.²**
   = 7.501×10^{-3} **torr**
1 **atm** = 1.013×10^5 **Pa** (or **N/m²**)
   = 1.013×10^6 **dynes/cm²**
   = 1013 **millibars** = 2116 **lb/ft²** = 14.70 **lb/in.²**
   = 76.00 **cm Hg** (0°C)
   = 29.92 **inches of mercury** (0°C)
   = 406.8 **inches of water** (4°C)
   = 33.90 **feet of water** (4°C)
1 **bar** = 1×10^5 **Pa** (or **N/m²**) = 0.9869 **atm**
   = 75.01 **cm Hg**
1 **torr** = 1 **millimeter of mercury (mm Hg)**
   = 1.333×10^2 **Pa** (or **N/m²**)

Work and Energy[5]

1 **J** = 1×10^7 **erg** = 0.7376 **ft·lb** = 0.2388 **cal**
   = 9.478×10^{-4} **Btu** = 9.872×10^{-3} **L·atm**
   = 2.778×10^{-7} **kW·h**
   = 3.725×10^{-7} **hp·h** = 6.242×10^{18} **eV**
1 **ft·lb** = 1.356 **J** = 0.3239 **cal** = 1.285×10^{-3} **Btu**
   = 3.766×10^{-7} **kW·h**
1 **cal** = 4.186 **J** (exact)
1 **Btu** = 1055 **J** = 252.0 **cal** = 2.930×10^{-4} **kW·h**
1 **kW·h** = 3.600×10^6 **J** = 2.655×10^6 **ft·lb**
   = 8.598×10^5 **cal** = 3412 **Btu**
1 **eV** = 1.602×10^{-19} **J**
1 **kg** (mc^2 equiv.) = 8.987×10^{16} **J**
1 **u** (mc^2 equiv.) = 1.492×10^{-10} **J**

Power

1 **W** = 1 **J/s** = 0.7376 **ft·lb/s** = 1.341×10^{-3} **hp**
   = 0.2389 **cal/s** = 3.413 **Btu/h**
1 **hp** = 550 **ft·lb/s** (exact) = 745.7 **W** = 178.1 **cal/s**
   = 2545 **Btu/h**

Temperature

$T = T_C + 273.15°$
$5(T_F + 40°) = 9(T_C + 40°)$
$T_C = (\frac{5}{9})(T_F - 32°)$
$T_F = (\frac{9}{5})T_C + 32°$

$\begin{cases} T \text{ is in } \textbf{kelvin (K)} \\ \text{(absolute scale); } T_C \text{ is} \\ \text{in } \textbf{degrees Celsius (°C);} \\ T_F \text{ is in } \textbf{degrees} \\ \textbf{Farenheit (°F).} \end{cases}$

continued

---

[4] Pressures in barometric units that involve the height of a column of mercury or water are measured where the acceleration due to gravity has the standard value g = 9.806 65 m/s².

[5] There are several other (slightly different) definitions of the *calorie* and the *British thermal unit*.

Magnetic Field

1 **tesla (T)** = 1 **weber per square meter (Wb/m²)**
$\quad\quad\quad\quad = 1 \times 10^4$ **gauss**

$$\sin\alpha \cos\beta = \tfrac{1}{2}[\sin(\alpha-\beta) + \sin(\alpha+\beta)]$$
$$\sin\alpha \sin\beta = \tfrac{1}{2}[\cos(\alpha-\beta) - \cos(\alpha+\beta)]$$
$$\cos\alpha \cos\beta = \tfrac{1}{2}[\cos(\alpha-\beta) + \cos(\alpha+\beta)]$$
$$\sin\alpha + \sin\beta = 2 \cos\tfrac{1}{2}(\alpha-\beta) \sin\tfrac{1}{2}(\alpha+\beta)$$

APPENDIX D
Mathematical Formulas

Pythagorean Theorem

$a^2 + b^2 = c^2$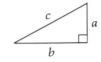

Quadratic Formula for Roots

If
$$ax^2 + bx + c = 0$$
then
$$x = \frac{-b \pm \sqrt{b^2 - 4ac}}{2a}$$

Trigonometric Functions of Angle θ

$\sin\theta = \dfrac{y}{r}\quad\quad \csc\theta = \dfrac{r}{y}$

$\cos\theta = \dfrac{x}{r}\quad\quad \sec\theta = \dfrac{r}{x}$

$\tan\theta = \dfrac{y}{x}\quad\quad \cot\theta = \dfrac{x}{y}$

Trigonometric Identities

$\sin(-\theta) = -\sin\theta \quad\quad \sin\theta = \cos(90° - \theta)$
$\cos(-\theta) = \cos\theta \quad\quad \cos\theta = \sin(90° - \theta)$
$\tan(-\theta) = -\tan\theta \quad\quad \cot\theta = \tan(90° - \theta)$

$\sin^2\theta + \cos^2\theta = 1$
$\sec^2\theta - \tan^2\theta = 1 \quad\quad \tan\theta = \dfrac{\sin\theta}{\cos\theta}$
$\csc^2\theta - \cot^2\theta = 1$

$\tan 2\theta = \dfrac{2\tan\theta}{1 - \tan^2\theta} \quad\quad \tan\dfrac{\theta}{2} = \sqrt{\dfrac{1 - \cos\theta}{1 + \cos\theta}}$

$\sin 2\alpha = 2 \sin\alpha \cos\alpha$
$\cos 2\alpha = 1 - 2\sin^2\alpha = 2\cos^2\alpha - 1 = \cos^2\alpha - \sin^2\alpha$

$\sin(\alpha \pm \beta) = \sin\alpha \cos\beta \pm \cos\alpha \sin\beta$
$\cos(\alpha \pm \beta) = \cos\alpha \cos\beta \mp \sin\alpha \sin\beta$

$\tan(\alpha \pm \beta) = \dfrac{\tan\alpha \pm \tan\beta}{1 \mp \tan\alpha \tan\beta}$

For All Plane Triangles

Sides a, b, and c, with opposite angles α, β, and γ.

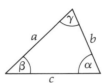

$\alpha + \beta + \gamma = 180°$

Law of sines: $\quad \dfrac{a}{\sin\alpha} = \dfrac{b}{\sin\beta} = \dfrac{c}{\sin\gamma}$

Law of cosines: $\quad c^2 = a^2 + b^2 - 2ab\cos\gamma$

Law of tangents: $\quad \dfrac{(a + b)}{(a - b)} = \dfrac{\tan\tfrac{1}{2}(\alpha + \beta)}{\tan\tfrac{1}{2}(\alpha - \beta)}$

Plane Angle θ

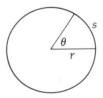

$\theta = \dfrac{s}{r} \quad \begin{bmatrix}\text{measured in}\\ \text{radians (rad)}\end{bmatrix}$

The whole plane angle (360°) is 2π radians.

Solid Angle Ω (in general)

$\Omega = \dfrac{\text{Subtended area}}{r^2} \quad \begin{bmatrix}\text{measured in}\\ \text{steradians (sr)}\end{bmatrix}$

The subtended area A on the curved surface of the sphere of radius r may have any arbitrary shape.
The whole solid angle = 4π steradians.

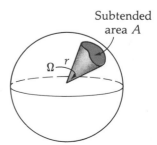

Subtended area A

Conical Solid Angle Ω

$$\Omega = 2\pi(1 - \cos\theta)$$

where $\theta = \begin{bmatrix} \text{half the vertex} \\ \text{angle of the cone} \end{bmatrix}$

Pythagorean Theorem in Three Dimensions

$a^2 + b^2 + c^2 = r^2$

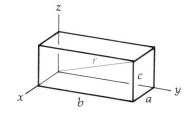

Sphere

Surface area $= 4\pi r^2$

Volume $= \frac{4}{3}\pi r^3$

Spherical Cap

Surface area $= 2\pi rh$

Volume (shaded) $= \frac{1}{3}\pi h^2(3r - h)$

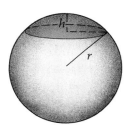

Right Circular Cone

Lateral surface area $= \pi r \ell$

Volume $= \frac{1}{3}\pi r^2 h$

Right Circular Cylinder

Lateral surface area $= 2\pi rh$

Volume $= \pi r^2 h$

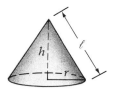

Truncated Right Circular Cone

Lateral surface area $= \pi(a + b)\ell$

Volume $= \frac{1}{3}\pi h(a^2 + ab + b^2)$

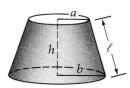

Sagitta Formula (approximation)

$$R \approx \frac{b^2}{2a} \quad \left(\text{for } \frac{a}{b} \ll 1\right)$$

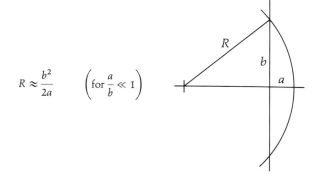

where R is the radius of the arc, a is the arc-to-chord distance, and b is the half-chord length.

Exponentials

$$e = 2.718\,281 \qquad e^0 = 1 \qquad \frac{1}{e} = 0.367\,879$$

If $y = e^x$, then $x = \ln y$

$e^{\ln x} = x \qquad \ln e^x = x$

$$e^x = 1 + x + \frac{x^2}{2!} + \frac{x^3}{3!} + \cdots$$

Logarithms

If $\log x = y$, then $x = 10^y$

If $\ln x = y$, then $x = e^y$

If $\log_b x = y$, then $x = b^y$

Change of base:

$$\ln x = (\ln 10)(\log x) = 2.3026 \log x$$

$$\log x = (\log e)(\ln x) = 0.434\,29 \ln x$$

$\ln e = 1 \qquad \ln e^x = x \qquad \ln 0 = -\infty$

$\ln 1 = 0 \qquad \ln a^x = x \ln a \qquad \log 0 = -\infty$

$\ln xy = \ln x + \ln y \qquad \ln a^x = x \ln a$

$\ln \dfrac{x}{y} = \ln x - \ln y \qquad \ln \sqrt[b]{a} = \dfrac{\ln a}{b}$

$$\ln(1 + x) = x - \frac{x^2}{2} + \frac{x^3}{3} - \cdots \qquad (\text{for } -1 < x \le 1)$$

APPENDIX E
Mathematical Approximations, Expansions, and Vector Relations

Approximations

For $x \ll 1$:

$$\sqrt{1 \pm x} \approx 1 \pm \frac{x}{2} \qquad \frac{1}{1 \pm x} \approx 1 \mp x$$

$$\frac{1}{\sqrt{1 \mp x^2}} \approx 1 \pm \frac{x^2}{2} \qquad \frac{1}{(1 \pm x^2)^{3/2}} \approx 1 \mp \frac{3x^2}{2}$$

For $x \approx 1$: $\qquad (1 - x^2) \approx 2(1 - x)$

For small θ (in radians):

$$\left. \begin{array}{l} \sin\theta \approx \theta \\[4pt] \cos\theta \approx 1 - \dfrac{\theta^2}{2} \\[4pt] \tan\theta \approx \theta \end{array} \right\} \quad \left[\begin{array}{l} <1\% \text{ discrepancy} \\ \text{for } \theta < 10° \end{array} \right]$$

Stirling's Approximation for Factorials

For large n:

$$n! \approx \sqrt{2\pi n}\, n^n e^{-n} \qquad \left[\begin{array}{l} <1\% \text{ discrepancy} \\ \text{for } n > 10 \end{array} \right]$$

Binomial Theorem

$$(1 \pm x)^n = 1 \pm \frac{nx}{1!} + \frac{n(n-1)}{2!} x^2 \pm \cdots \qquad (x^2 < 1)$$

$$(1 \pm x)^{-n} = 1 \mp \frac{nx}{1!} + \frac{n(n+1)x^2}{2!} \mp \cdots \qquad (x^2 < 1)$$

Expansions

$$e^x = 1 + x + \frac{x^2}{2!} + \frac{x^3}{3!} + \cdots$$

$$\ln(1 + x) = x - \tfrac{1}{2}x^2 + \tfrac{1}{3}x^3 - \cdots \qquad (|x| < 1)$$

$$\left. \begin{array}{l} \sin\theta = \theta - \dfrac{\theta^3}{3!} + \dfrac{\theta^5}{5!} - \cdots \\[6pt] \cos\theta = 1 - \dfrac{\theta^2}{2!} + \dfrac{\theta^4}{4!} - \cdots \\[6pt] \tan\theta = \theta + \dfrac{\theta^3}{3} + \dfrac{2\theta^5}{15} + \cdots \end{array} \right\} \theta \text{ in radians}$$

Vector Relations

Any vector $\mathbf{A}$ with components A_x, A_y, A_z along the x, y, z directions may be written

$$\mathbf{A} = A_x \hat{\mathbf{x}} + A_y \hat{\mathbf{y}} + A_z \hat{\mathbf{z}}$$

Let θ be the smaller angle between the forward directions of $\mathbf{A}$ and $\mathbf{B}$. Then

Scalar (or Dot) Product:

$$\mathbf{A} \cdot \mathbf{B} = \mathbf{B} \cdot \mathbf{A} = |\mathbf{A}||\mathbf{B}|\cos\theta = AB\cos\theta$$
$$= A_x B_x + A_y B_y + A_z B_z$$

Vector (or Cross) Product:

$$\mathbf{A} \times \mathbf{B} = -\mathbf{B} \times \mathbf{A} = \begin{vmatrix} \hat{\mathbf{x}} & \hat{\mathbf{y}} & \hat{\mathbf{z}} \\ A_x & A_y & A_z \\ B_x & B_y & B_z \end{vmatrix}$$

$$= (A_y B_z - B_y A_z)\hat{\mathbf{x}} + (A_z B_x - B_z A_x)\hat{\mathbf{y}} + (A_x B_y - B_x A_y)\hat{\mathbf{z}}$$

$$|\mathbf{A} \times \mathbf{B}| = |\mathbf{A}||\mathbf{B}|\sin\theta = AB\sin\theta$$

Let $\hat{\mathbf{x}}, \hat{\mathbf{y}}, \hat{\mathbf{z}}$ be unit vectors in the x, y, z directions. Then

$$\hat{\mathbf{x}} \cdot \hat{\mathbf{x}} = \hat{\mathbf{y}} \cdot \hat{\mathbf{y}} = \hat{\mathbf{z}} \cdot \hat{\mathbf{z}} = 1$$
$$\hat{\mathbf{x}} \cdot \hat{\mathbf{y}} = \hat{\mathbf{y}} \cdot \hat{\mathbf{z}} = \hat{\mathbf{z}} \cdot \hat{\mathbf{x}} = 0$$
$$\hat{\mathbf{x}} \times \hat{\mathbf{y}} = \hat{\mathbf{z}} \qquad \hat{\mathbf{y}} \times \hat{\mathbf{z}} = \hat{\mathbf{x}} \qquad \hat{\mathbf{z}} \times \hat{\mathbf{x}} = \hat{\mathbf{y}}$$
$$\mathbf{A} \times (\mathbf{B} + \mathbf{C}) = (\mathbf{A} \times \mathbf{B}) + (\mathbf{A} \times \mathbf{C})$$
$$(s\mathbf{A}) \times \mathbf{B} = \mathbf{A} \times (s\mathbf{B}) = s(\mathbf{A} \times \mathbf{B}) \quad \left[\begin{array}{l} \text{where } s \text{ is} \\ \text{a scalar} \end{array} \right]$$

APPENDIX F
Fourier Analysis

The French mathematician François Fourier (1722–1836) showed that almost any periodic function,[1] such as that shown in Figure F-1, may be expressed as an infinite sum of sine and cosine functions and possibly a constant term. Such a sum is called a *Fourier series*, with the general form

$$f(t) = a_0 + \sum_{n=1}^{\infty} (a_n \cos n\omega t + b_n \sin n\omega t)$$

---

[1] Certain mathematical criteria must be met. The function representing the motion must be single-valued and continuous except for a finite number of finite discontinuities, and must not have an infinite number of maxima or minima in the neighborhood of any given point. However, all motions of real physical objects, electrical currents, and so forth meet these criteria, so we may always use this method for analyzing physical phenomena.

where $\omega = \dfrac{2\pi}{T}$

$$a_0 = \dfrac{1}{T}\int_0^T f(t)\,dt \qquad \begin{bmatrix}\text{the average}\\ \text{value of }f(t)\end{bmatrix}$$

$$a_n = \dfrac{2}{T}\int_0^T f(t)\cos n\omega t\,dt$$

$$b_n = \dfrac{2}{T}\int_0^T f(t)\sin n\omega t\,dt$$

Note that for even functions $[f(t) = f(-t)]$ all the b's are zero, and for odd functions $[f(-t) = -f(t)]$ all the a's are zero (except possibly a_0). Very often a function may be made even or odd by a shift of the origin, as shown in Figure F-2.

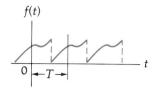

FIGURE F-1
A periodic function with a period T that may be expressed by an infinite sum of sine and cosine functions and a constant.

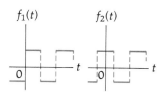

FIGURE F-2
A square wave may be expressed either as a sum of only sine terms or as a sum of only cosine terms. The Fourier series of $f_1(t)$ contains only sine terms (since it is an odd function), and $f_2(t)$ contains only cosine terms (since it is an even function).

Example F-1

The periodic function shown in Figure F-3 is written mathematically as

$$f(t) = \begin{cases} A, & 0 < t < \dfrac{T}{2} \\ -A, & \dfrac{T}{2} < t < T \end{cases}$$

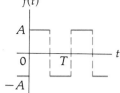

FIGURE F-3
Example F-1.

Its Fourier series is

$$f(t) = \dfrac{4A}{\pi}\left(\dfrac{\sin \omega t}{1} + \dfrac{\sin 3\omega t}{3} + \dfrac{\sin 5\omega t}{5} + \cdots\right)$$

Example F-2

The sawtooth waveshape of Figure F-4 is expressed mathematically as

$$f(t) = t, \qquad -\dfrac{T}{2} < t < \dfrac{T}{2}$$

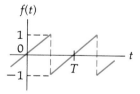

FIGURE F-4
Example F-2.

Its Fourier series is

$$f(t) = 2\left(\dfrac{\sin \omega t}{1} - \dfrac{\sin 2\omega t}{2} + \dfrac{\sin 3\omega t}{3} - \cdots\right)$$

The following illustrations show the Fourier series approximations made by combining, respectively, the first three, six, and nine terms of the series.

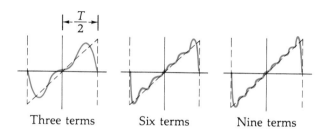

Three terms   Six terms   Nine terms

Example F-3

The waveshape of Figure F-5 is written mathematically as

$$f(t) = A\sin t, \qquad -\dfrac{T}{2} < t < \dfrac{T}{2}$$

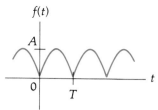

FIGURE F-5
Example F-3.

Its Fourier series is

$$f(t) = \dfrac{2A}{\pi} - \dfrac{4A}{\pi}\left(\dfrac{\cos 2\omega t}{1\cdot 3} + \dfrac{\cos 4\omega t}{3\cdot 5} + \dfrac{\cos 6\omega t}{5\cdot 7} + \cdots\right)$$

APPENDIX G
Calculus Formulas

In the following, a, b, c, and n are constants; u and v are functions of x; and x and y are functions of t. Logarithmic expressions are to the base $e = 2.71828\ldots$. All angles are measured in radians.

G-I Derivatives

1. $\dfrac{d}{dx}[cu] = c\dfrac{du}{dx}$

2. $\dfrac{d}{dx}[u + v] = \dfrac{du}{dx} + \dfrac{dv}{dx}$

3. $\dfrac{d}{dx}\left[\dfrac{u}{v}\right] = \dfrac{v\dfrac{du}{dx} - u\dfrac{dv}{dx}}{v^2}$

4. $\dfrac{d}{dx}[uv] = u\dfrac{dv}{dx} + \dfrac{du}{dx}v$

5. $\dfrac{d}{dx}[u]^n = nu^{n-1}\dfrac{du}{dx}$

6. $\dfrac{d}{dx}[e^{ax}] = ae^{ax}$

7. $\dfrac{d}{dx}[a^u] = (a^u \ln a)\dfrac{du}{dx}$

8. $\dfrac{d}{dx}[\sin ax] = a\cos ax$

9. $\dfrac{d}{dx}[\cos ax] = -a\sin ax$

10. $\dfrac{d}{dx}[\tan ax] = a\sec^2 ax$

11. $\dfrac{d}{dx}[\ln u] = \dfrac{1}{u}\dfrac{du}{dx}$

12. $\dfrac{du}{dt} = \dfrac{du}{dx}\dfrac{dx}{dt}$ (the "chain rule")

13. $\dfrac{du}{dv} = \dfrac{\left(\dfrac{du}{dx}\right)}{\left(\dfrac{dv}{dx}\right)}$

G-II Integrals

1. $\int a\,dx = ax + c$

2. $\int [u+v]\,dx = \int u\,dx + \int v\,dx + c$

3. $\int_a^b u\,dv = (uv)\Big|_a^b - \int_a^b v\,du$

4. $\int \dfrac{dx}{ax+b} = \dfrac{1}{a}\ln(ax+b) + c$

5. $\int x^n\,dx = \dfrac{x^{n+1}}{n+1} + c \qquad (n \neq -1)$

6. $\int e^{ax}\,dx = \dfrac{e^{ax}}{a} + c$

7. $\int xe^{ax}\,dx = \dfrac{e^{ax}}{a^2}(ax - 1) + c$

8. $\int x^n e^{ax}\,dx = \dfrac{x^n e^{ax}}{a} - \dfrac{n}{a}\int x^{n-1}e^{ax}\,dx + c$

9. $\int \sin ax\,dx = -\dfrac{1}{a}\cos ax + c$

10. $\int \cos ax\,dx = \dfrac{1}{a}\sin ax + c$

11. $\int \tan ax\,dx = -\dfrac{1}{a}\ln(\cos ax) + c$

12. $\int \sin^2 ax\,dx = \dfrac{x}{2} - \dfrac{\sin 2ax}{4a} + c$

13. $\int \cos^2 ax\,dx = \dfrac{x}{2} + \dfrac{\sin 2ax}{4a} + c$

14. $\int (\sin ax)(\cos ax)\,dx = \dfrac{\sin^2 ax}{2a} + c$

15. $\int \dfrac{dx}{\sqrt{a^2 + x^2}} = \ln[x + \sqrt{a^2+x^2} + c]$

16. $\int \dfrac{dx}{\sqrt{a^2 - x^2}} = \begin{cases} \sin^{-1}\left(\dfrac{x}{|a|}\right) + c \\ -\cos^{-1}\left(\dfrac{x}{|a|}\right) + c \end{cases}$

17. $\int \dfrac{dx}{a^2 + x^2} = \dfrac{1}{a}\tan^{-1}\left(\dfrac{x}{a}\right) + c$

18. $\int \dfrac{dx}{(a^2 + x^2)^{3/2}} = \dfrac{x}{a^2\sqrt{a^2+x^2}} + c$

19. $\int \dfrac{x\,dx}{(a^2+x^2)^{1/2}} = \sqrt{a^2+x^2} + c$

20. $\int \dfrac{x\,dx}{(a^2+x^2)^{3/2}} = \dfrac{-1}{\sqrt{a^2+x^2}} + c$

G-III Definite Integrals

1. $\int_0^\infty x^n e^{-ax}\,dx = \dfrac{n!}{a^{n+1}} \qquad \left(\begin{array}{l} n = \text{positive integer} \\ a > 0 \end{array}\right)$

2. $I_0 = \int_0^\infty e^{-ax^2}\,dx = \dfrac{1}{2}\sqrt{\dfrac{\pi}{a}} \qquad \left(\begin{array}{l} \text{Gauss's probability} \\ \text{integral} \end{array}\right)$

3. $I_1 = \int_0^\infty xe^{-ax^2}\,dx = \dfrac{1}{2a}$

4. $I_2 = \int_0^\infty x^2 e^{-ax^2}\,dx = -\dfrac{dI_0}{da} = \dfrac{1}{4}\sqrt{\dfrac{\pi}{a^3}}$

5. $I_3 = \int_0^\infty x^3 e^{-ax^2}\,dx = -\dfrac{dI_1}{da} = \dfrac{1}{2a^2}$

6. $I_4 = \int_0^\infty x^4 e^{-ax^2} dx = \dfrac{d^2 I_0}{da^2} = \dfrac{3}{8}\sqrt{\dfrac{\pi}{a^5}}$

7. $I_5 = \int_0^\infty x^5 e^{-ax^2} dx = \dfrac{d^2 I_1}{da^2} = \dfrac{1}{a^3}$

$\vdots$

8. $I_{2n} = (-1)^n \dfrac{d^n}{da^n} I_0$

9. $I_{2n+1} = (-1)^n \dfrac{d^n}{da^n} I_1$

The Definite Integral

Most of the use of integration in this text involves the *definite integral*, that is, the integration between two specific values of the variable. The procedure may be illustrated by the following example.

Consider a one-dimensional force $F(x)$, which varies as a function of distance x, as shown in Figure G-1. Let us find the work done by this force as it moves through a displacement from x_1 to x_2. We divide the total displacement into a large number N of small intervals, $\Delta x_1, \Delta x_2, \Delta x_3, \ldots, \Delta x_i, \ldots, \Delta x_N$. Since $F(x)$ is nearly constant during each small displacement, we may assume it has the average value $F(x_i)$ during the displacement Δx_i. Thus, the work ΔW_1 accomplished during the first interval Δx_1 is approximately

$$\Delta W_1 \approx F(x_1) \Delta x_1$$

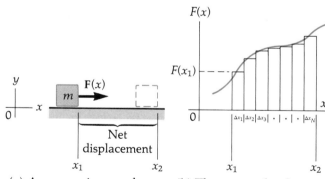

(a) A mass m is moved through a displacement $x_2 - x_1$ by the *variable* force $F(x)$.

(b) The area under the curve of $F(x)$ vs. x between the limits x_1 and x_2 equals the work done by the force $F(x)$.

FIGURE G-1
Illustration of the definite integral.

and so on for the rest of the intervals. The total work done in moving from x_1 to x_2 is therefore

$$W_{12} \approx \sum_{i=1}^{N} F(x_i)\Delta x_i$$

To make a better approximation, we divide the displacement into an even greater number of intervals, so that each Δx_i becomes smaller and the total number of intervals N becomes larger. Continuing to improve the approximation, we let the intervals become smaller and smaller as the total number of intervals becomes larger and larger.

The *exact* value for the work done is obtained as Δx shrinks to zero and N goes to infinity. This defines the *definite integral of $F(x)$ with respect to x from x_1 to x_2*. The notation is

THE DEFINITE INTEGRAL
$$\lim_{N\to\infty} \sum_{i=1}^{N} F(x_i)\Delta x_i = \int_{x_1}^{x_2} F(x)\,dx \qquad \text{(G-1)}$$

The definite integral is equal to the area under the curve of $F(x)$ vs. x between the limits x_1 and x_2.

There is a close connection between the *definite integral*, such as Equation (G-1), and the *indefinite integral*, such as $\int F(x)\,dx$ and those integrals listed in G-II. The connection is known as *the fundamental theorem of calculus*, which we state here without proof:

$$\int_{x_1}^{x_2} F(x)\,dx = \underbrace{\int F(x)\,dx}_{\substack{\text{Evaluated}\\ \text{at } x=x_2}} - \underbrace{\int F(x)\,dx}_{\substack{\text{Evaluated}\\ \text{at } x=x_1}} \qquad \text{(G-2)}$$

Thus, in calculating a definite integral between two limits, one merely evaluates the integral at the upper limit of x_2 and subtracts its value at the lower limit x_1. In the process, the constant of integration c is eliminated.

G-IV Differentiation and Integration of Vectors

To differentiate the dot and cross products of vectors, care must be taken (particularly with the cross product) to preserve the order of multiplication. The rule is similar to the derivative of an ordinary scalar product.

Thus:
$$\dfrac{d}{dt}(uv) = u\dfrac{dv}{dt} + \dfrac{du}{dt} v$$

$$\dfrac{d}{dt}(\mathbf{A}\cdot\mathbf{B}) = \mathbf{A}\cdot\dfrac{d\mathbf{B}}{dt} + \dfrac{d\mathbf{A}}{dt}\cdot\mathbf{B}$$

$$\dfrac{d}{dt}(\mathbf{A}\times\mathbf{B}) = \mathbf{A}\times\dfrac{d\mathbf{B}}{dt} + \dfrac{d\mathbf{A}}{dt}\times\mathbf{B}$$

To solve integrals involving the dot or cross products of vectors, the first step is to replace the dot or cross symbol by the appropriate sine or cosine function, thus changing the operation to a simple scalar integral.

$$\int \mathbf{F}\cdot d\mathbf{x} \quad \text{becomes} \quad \int F(\cos\theta)\,dx$$

$$\int (\mathbf{r}\times\mathbf{F})\,d\mathbf{r} \quad \text{becomes} \quad \int rF(\sin\theta)\,dr$$

where θ is the smaller angle between the forward directions of the vectors.

G-V Partial Derivatives

If a function depends upon more than one variable, we may take its derivative with respect to one of the variables, holding the other variables fixed. The notation $\partial f/\partial x$ means the derivative of f with respect to x, with other variables treated as constants. For example, if

$$f(x,y) = xy^2$$

then $\quad \dfrac{\partial f}{\partial x} = y^2 \quad$ and $\quad \dfrac{\partial f}{\partial y} = 2xy$

APPENDIX H
Finite Rotations

Finite rotations of an object cannot be represented by vectors. Consider a book that is to be turned through two rotations, one about a vertical axis and the other about a horizontal axis. Suppose a 90° rotation about the vertical axis is represented by the axial vector **V** and a 90° rotation about the horizontal axis by the axial vector **H**. The sum **V** + **H** is the vertical rotation followed by the horizontal rotation, as shown in Figure H-1(a).

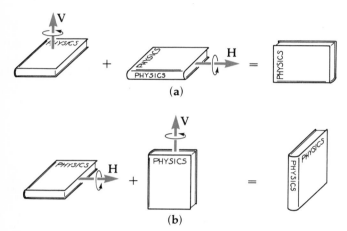

FIGURE H-1
Successive 90° rotations of a book.

The sum **H** + **V** is represented by the horizontal rotation followed by the vertical rotation, as shown in Figure H-1(b). The final orientation of the book depends on which rotation is first. Mathematically, **V** + **H** ≠ **H** + **V**. Therefore, finite rotations cannot be represented as axial vectors because vectors must have the property of commuting in addition.

However, as the angular displacements become smaller and smaller, the final orientation depends less and less upon the order of the rotations. In the limit of *infinitesimal* rotations, axial vectors may be used to describe such rotations because they commute in addition. Thus, while $\Delta\theta$ is not an axial vector, $d\theta$ is. For this reason, angular velocity $\omega = \lim_{\Delta t \to 0} (\Delta\theta/\Delta t) = d\theta/dt$ may be represented by the axial vector $\boldsymbol{\omega}$ according to the right-hand rule, as shown in Figure H-2. Angular acceleration $\boldsymbol{\alpha} = d\boldsymbol{\omega}/dt$ may similarly be defined.

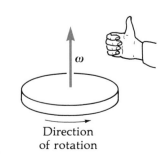

FIGURE H-2
Right-hand rule: If the fingers of the *right* hand are curled around in the rotational sense, the extended thumb points in the direction of $\boldsymbol{\omega}$. The axial vector $\boldsymbol{\omega}$ represents the angular velocity of rotation of the disk.

APPENDIX I
Derivation of the Lorentz Transformation

Suppose that at the instant two reference frames S and S' are coincident, a flashbulb is set off at the coincident origins O and O' (refer to Figure 41-3). At a later time, observers in each frame measure an expanding spherical wavefront that is centered at the origin of their respective frame of reference. The equation of a sphere of radius r in three dimensions is $x^2 + y^2 + z^2 = r^2$, so the equations for this light sphere are

In the S frame (at time t)	In the S' frame (at time t')	
$x^2 + y^2 + z^2 = c^2 t^2$	$x'^2 + y'^2 + z'^2 = c^2 t'^2$	(I-1)

We seek the relations between the primed and unprimed parameters. We require that these relations be linear in x and x'. If they were not, a single event in one frame would not necessarily be a single event in the other frame—clearly an unacceptable situation. The relations should also reduce to the familiar Galilean transformation $x' = (x - Vt)$ as $V \to 0$, which we know to be satisfactory in ordinary classical mechanics.

Because there is no relative motion in the y and z directions, we assume that

$$y = y' \quad \text{and} \quad z = z' \qquad \text{(I-2)}$$

As a simple possibility, we try the relation

$$x' = \gamma(x - Vt) \qquad \text{(I-3)}$$

where γ is a factor that can depend on V or c, but not on x or t. Now the equations of physics must have the same form in S and S'; so, changing the sign of V (to account for the difference in the direction of relative motion) and interchanging primes and unprimes, we obtain the inverse relation

$$x = \gamma(x' + Vt') \qquad \text{(I-4)}$$

We find the relation between t and t' by substituting the value of x' from Equation (I-3) into Equation (I-4):

$$x = \gamma^2(x - Vt) + \gamma V t'$$

Solving for t' gives

$$t' = \gamma t + \left(\frac{1 - \gamma^2}{\gamma V}\right) x \qquad \text{(I-5)}$$

We evaluate γ by considering a flash of light that starts at $t = t' = 0$ at the origins O and O'. Light travels at the same speed c in each frame, so at the later times t and t' the light will arrive along the x axis at

In the S frame (at time t)	In the S' frame (at time t')	
$x = ct$	$x' = ct'$	(I-6)

In the right-hand equation above, we substitute for x' and t' from Equations (I-3) and (I-5):

$$\gamma(x - Vt) = c\gamma t + c\left(\frac{1 - \gamma^2}{\gamma V}\right)x$$

Solving for x gives

$$x = \frac{c\gamma t + V\gamma t}{\gamma - c\left(\frac{1 - \gamma^2}{\gamma V}\right)} = ct\left[\frac{\left(1 + \frac{V}{c}\right)}{1 - \frac{c}{V}\left(\frac{1}{\gamma^2} - 1\right)}\right] \quad \text{(I-7)}$$

This will yield the expression $x = ct$ if the factor in brackets equals 1.

$$\left[\frac{\left(1 + \frac{V}{c}\right)}{1 - \frac{c}{V}\left(\frac{1}{\gamma^2} - 1\right)}\right] = 1 \quad \text{(I-8)}$$

Solving for γ yields

$$\gamma = \frac{1}{\sqrt{1 - \frac{V^2}{c^2}}} \quad \text{(I-9)}$$

Inserting this into the relevant equations above, we obtain the Lorentz transformation:

THE LORENTZ TRANSFORMATION (where $\beta \equiv V/c$)

$$\begin{cases} x = \dfrac{x' + Vt'}{\sqrt{1 - \beta^2}} & x' = \dfrac{x - Vt}{\sqrt{1 - \beta^2}} \\ y = y' & y' = y \\ z = z' & z' = z \\ t = \dfrac{t' + Vx'/c^2}{\sqrt{1 - \beta^2}} & t' = \dfrac{t - Vx/c^2}{\sqrt{1 - \beta^2}} \end{cases} \quad \text{(I-10)}$$

The notation $\gamma = 1/\sqrt{1 - V^2/c^2}$ has become standard, so the Lorentz transformation is frequently written as

THE LORENTZ TRANSFORMATION (relative velocity V)

$$\begin{cases} x = \gamma(x' + Vt') & x' = \gamma(x - Vt) \\ y = y' & y' = y \\ z = z' & z' = z \\ t = \gamma(t' + Vx'/c^2) & t' = \gamma(t - Vx/c^2) \end{cases} \quad \text{(I-11)}$$

APPENDIX J
Periodic Table of the Elements

IA																	VIIIA
1 H Hydrog. 1.007 94	IIA											IIIA	IVA	VA	VIA	VIIA	2 He Helium 4.002 602
3 Li Lithium 6.941	4 Be Beryll. 9.012 182											5 B Boron 10.811	6 C Carbon 12.011	7 N Nitrog. 14.006 74	8 O Oxygen 15.9994	9 F Fluorine 18.998 403 2	10 Ne Neon 20.1797
11 Na Sodium 22.989 768	12 Mg Magnes. 24.3050	IIIB	IVB	VB	VIB	VIIB	VIII			IB	IIB	13 Al Alumin. 26.981 539	14 Si Silicon 28.0855	15 P Phosph. 30.973 762	16 S Sulfur 32.066	17 Cl Chlorine 35.4527	18 Ar Argon 39.948
19 K Potass. 39.0983	20 Ca Calcium 40.078	21 Sc Scand. 44.955 910	22 Ti Titan. 47.88	23 V Vanad. 50.9415	24 Cr Chrom. 51.9961	25 Mn Mangan. 54.938 05	26 Fe Iron 55.847	27 Co Cobalt 58.933 20	28 Ni Nickel 58.69	29 Cu Copper 63.546	30 Zn Zinc 65.39	31 Ga Gallium 69.723	32 Ge German. 72.61	33 As Arsenic 74.921 59	34 Se Selen. 78.96	35 Br Bromine 79.904	36 Kr Krypton 83.80
37 Rb Rubid. 85.4678	38 Sr Stront. 87.62	39 Y Yttrium 88.905 85	40 Zr Zircon. 91.224	41 Nb Niobium 92.906 38	42 Mo Molybd. 95.94	43 Tc Technet. (98)	44 Ru Ruthen. 101.07	45 Rh Rhod. 102.905 50	46 Pd Pallad. 106.42	47 Ag Silver 107.8682	48 Cd Cadm. 112.411	49 In Indium 114.82	50 Sn Tin 118.710	51 Sb Antimon. 121.75	52 Te Tellur. 127.60	53 I Iodine 126.904 47	54 Xe Xenon 131.29
55 Cs Cesium 132.905 43	56 Ba Barium 137.327	57–71 Rare Earths	72 Hf Hafnium 178.49	73 Ta Tantal. 180.9479	74 W Tungst. 183.85	75 Re Rhen. 186.207	76 Os Osmium 190.2	77 Ir Iridium 192.22	78 Pt Platin. 195.08	79 Au Gold 196.966 54	80 Hg Mercury 200.59	81 Tl Thall. 204.3833	82 Pb Lead 207.2	83 Bi Bismuth 208.980 37	84 Po Polon. (209)	85 At Astatine (210)	86 Rn Radon (222)
87 Fr Franc. (223)	88 Ra Radium 226.0254	89–103 Acti- nides	104 (261)	105 (262)	106 (263)	107 (262)	108 (265)	109 (266)									

57 La Lanthan. 138.9055	58 Ce Cerium 140.115	59 Pr Praseody. 140.907 65	60 Nd Neodym. 144.24	61 Pm Prometh. (145)	62 Sm Samar. 150.36	63 Eu Europ. 151.965	64 Gd Gadolin. 157.25	65 Tb Terb. 158.925 34	66 Dy Dyspros. 162.50	67 Ho Holm. 164.930 32	68 Er Erbium 167.26	69 Tm Thulium 168.934 21	70 Yb Ytterb. 173.04	71 Lu Lutet. 174.967	Rare earths (Lanthanide series)
89 Ac Actin. 227.0278	90 Th Thorium 232.0381	91 Pa Protactin. 231.035 88	92 U Uranium 238.0289	93 Np Neptun. 237.0482	94 Pu Pluton. (244)	95 Am Americ. (243)	96 Cm Curium (247)	97 Bk Berkel. (247)	98 Cf Californ. (251)	99 Es Einstein. (252)	100 Fm Fermi. (257)	101 Md Mendel. (258)	102 No Nobel. (259)	103 Lr Lawrenc. (260)	Actinide series

[1] The atomic number (top left) is the number of protons in the nucleus. The atomic mass (bottom) is weighted by isotopic abundance in the earth's surface, relative to the mass of the carbon 12 isotope, which is assigned a mass of exactly 12 unified atomic mass units (u). Standard errors range from 1 to 9 in the last digit quoted. Relative isotopic abundances often vary considerably, both in naturally occurring specimens and in commercially available samples. Numbers in parentheses are mass numbers (the whole number nearest the atomic mass, in u) of the most stable isotope of that element. Some elements without stable nuclides nevertheless exhibit a range of characteristic terrestrial compositions of long-lived radionuclides such that a meaningful atomic weight can be given. Adapted from the Table of Standard Atomic Weights of the Elements, 1985 [*Pure and Applied Chemistry* **58**, 1677 (1986)].

APPENDIX K
Constants and Standards[1]

Quantity	Symbol	Value	Units	Uncertainty (parts per million)
Speed of light in vacuum	c	299 792 458	m/s	(exact)
Permeability of vacuum	μ_0	$4\pi \times 10^{-7}$	N/A^2	
		= 12.566 370 614 ...	$10^{-7}\ N/A^2$	(exact)
Permittivity of vacuum	ε_0	$1/\mu_0 c^2$		
		= 8.854 187 817 ...	10^{-12} F/m	(exact)
Gravitational constant	G	6.672 59(85)	$10^{-11}\ m^3/kg \cdot s^2$	128
Planck constant	h	6.626 075 5(40)	10^{-34} J·s	0.60
		= 4.135 669 2(12)	10^{-15} eV·s	0.30
$h/2\pi$	$\hbar$	1.054 572 66(63)	10^{-34} J·s	0.60
		= 6.582 122 0(20)	10^{-16} eV·s	0.30
Elementary charge	e	1.602 177 33(49)	10^{-19} C	0.30
Electron mass	m_e	9.109 389 7(54)	10^{-31} kg	0.59
		= 5.485 799 03(13)	10^{-4} u	0.023
		= 0.510 999 06(15)	MeV/c^2	0.30
Proton mass	m_p	1.672 623 1(10)	10^{-27} kg	0.59
		= 1.007 276 470(12)	u	0.012
		= 938.272 31(28)	MeV/c^2	0.30
Neutron mass	m_n	1.674 928 6(10)	10^{-27} kg	0.59
		= 1.008 664 904(14)	u	0.014
		= 939.565 63(28)	MeV/c^2	0.30
Proton–electron mass ratio	m_p/m_e	1836.152 701(37)		0.020
Rydberg constant, $\frac{1}{2}m_e c \alpha^2/h$	R_∞	10 973 731.534(13)	m^{-1}	0.0012
Avogadro constant[2]	N_A	6.022 136 7(36)	10^{23}/mol	0.59
Faraday constant,[2] $N_A e$	F	96 485.309(39)	C/mol	0.30
Molar gas constant[2]	R	8.314 510(70)	J/mol·K	8.4
Boltzmann constant, R/N_A	k	1.380 658(12)	10^{-23} J/K	8.5
		= 8.617 385(73)	10^{-5} eV/K	8.4
Stefan–Boltzmann constant, $(\pi^2/60)k^4/\hbar^3 c^2$	σ	5.670 51(19)	$10^{-8}\ W/m^2 \cdot K^4$	34

Non-SI units used with SI

Electron volt, $(e/C)\ J = \{e\}\ J$	eV	1.602 177 33(49)	10^{-19} J	0.30
Unified atomic mass unit, $1\ u = m_u = \frac{1}{12}m(^{12}C)$	u	1.660 540 2(10)	10^{-27} kg	0.59
		= 931.494 32(28)	MeV/c^2	0.30

[1] Compiled from the tables prepared by E. Richard Cohen and Barry N. Taylor under the auspices of the CODATA Task Group on Fundamental Constants, *CODATA Bulletin No. 63*, Nov. 1986. The digits in parentheses are the one-standard-deviation uncertainty in the last digits of the given value.

[2] In this text, "mol" (mole) means "gram-molecular weight" ($= 10^{-3}$ kg-molecular weight).

APPENDIX L
Terrestrial and Astronomical Data

Terrestrial

Equatorial radius	6.378×10^6 m
Polar radius	6.357×10^6 m
Radius of sphere having the earth's volume	6.371×10^6 m
Volume	1.083×10^{21} m^3
Mean orbital speed	2.977×10^4 m/s
Sidereal rotation period	86 164 s
Tangential speed of rotation at equator	465.1 m/s
Solar constant (average solar power incident perpendicularly on unit area)	
at top of atmosphere	1.37×10^3 W/m^2
at earth's surface (average)	0.84×10^3 W/m^2 = 1.13 hp/m^2

Astronomical

Solar power output (luminosity)	3.86×10^{26} W
Solar surface temperature	5780 K
Number of stars in the Milky Way	$\sim 1.6 \times 10^{11}$
Distance of sun from center of Milky Way	$\sim 2.2 \times 10^{20}$ m
Diameter of Milky Way	$\sim 7 \times 10^{20}$ m
Total mass of Milky Way	$\sim 1 \times 10^{41}$ kg
Number of galaxies in the observable universe	$\sim 10^{12}$
Distance to edge of the observable universe	$\sim 10^{26}$ m
Age of universe	$1.5 \pm 0.5 \times 10^{10}$ yr

Data of the Solar System*

Body	Mass (kg)	Equitorial radius of object (m)	Mean density (kg/m^3)
Sun	1.989×10^{30}	6.960×10^8	1.42×10^3
Mercury	3.303×10^{23}	2.439×10^6	5.42×10^3
Venus	4.870×10^{24}	6.050×10^6	5.25×10^3
Earth	5.976×10^{24}	6.378×10^6	5.52×10^3
Mars	6.418×10^{23}	3.397×10^6	3.94×10^3
Jupiter	1.899×10^{27}	7.140×10^7	1.31×10^3
Saturn	5.686×10^{26}	6.000×10^7	0.69×10^3
Uranus	8.66×10^{25}	2.615×10^7	1.19×10^3 (?)
Neptune	1.030×10^{26}	2.43×10^7	1.71×10^3
Pluto	$1. \times 10^{22}$ (?)	1.2×10^6 (?)	1.2×10^3 (?)
Moon	7.347×10^{22}	1.738×10^6	3.36×10^3

Body	Acceleration due to gravity at equator (m/s^2)	Period of revolution about the sun (days)	Mean distance from the sun (m)
Sun	274.4	—	—
Mercury	3.78	87.97	5.79×10^{10}
Venus	8.60	2.247×10^2	1.08×10^{11}
Earth	9.78	3.653×10^2	1.50×10^{11}
Mars	3.72	6.870×10^2	2.28×10^{11}
Jupiter	22.88	4.333×10^3	7.78×10^{11}
Saturn	9.05	1.076×10^4	1.43×10^{12}
Uranus	7.77	3.069×10^4	2.87×10^{12}
Neptune	11.00	6.019×10^4	4.50×10^{12}
Pluto	4.3 (?)	9.047×10^4	5.90×10^{12}
Moon	1.67	27.32†	3.84×10^8‡

* Planetary data from *Cambridge Atlas of Astronomy*, Cambridge University Press (1988). Values with (?) are uncertain by more than 10 percent.
† Revolution about the earth.
‡ Distance from the earth.

1 light year (ly)	= 9.461×10^{15} m
1 parsec (pc) = 3.262 1y	= 3.086×10^{16} m
1 astronomical unit (AU)	= 1.496×10^{11} m

APPENDIX M
SI Units

The General Conference on Weights and Measures has developed *Le Système International d'Unités*, abbreviated SI, a system of units that has been adopted by almost all industrial nations of the world. It is an outgrowth of the MKSA (*meter-kilogram-second-ampere*) system. SI units are divided into three classes: *base units*, *derived units*, and *supplementary units*. Although such a division is not logically essential, it does have certain practical advantages. The General Conference meets from time to time, occasionally revising or adding to the list of official standards. The following information is from the second revision (1973) of the publication NASA SP-7012, available from the Superintendent of Documents, U.S. Government Printing Office, Washington, D.C. 20402.

Names and Symbols of SI Units

Quantity	Name of Unit (Symbol)
SI Base Units	
length	meter (m)
mass	kilogram (kg)
time	second (s)
electric current	ampere (A)
thermodynamic temperature	kelvin (K)
luminous intensity	candela (cd)
amount of substance	mole (mol)
SI Derived Units	
area	square meter (m^2)
volume	cubic meter (m^3)
frequency	hertz (Hz, s^{-1})
mass density (density)	kilogram per cubic meter (kg/m^3)
speed, velocity	meter per second (m/s)
angular velocity	radian per second (rad/s)
acceleration	meter per second squared (m/s^2)
angular acceleration	radian per second squared (rad/s^2)
force	newton (N, $kg \cdot m/s^2$)
pressure (mechanical stress)	pascal (Pa, N/m^2)
kinematic viscosity	square meter per second (m^2/s)
dynamic viscosity	newton-second per square meter ($N \cdot s/m^2$)
work, energy, quantity of heat	joule (J, $N \cdot m$)
power	watt (W, J/s)
quantity of electricity	coulomb (C, $A \cdot s$)
potential difference, electromotive force	volt (V, W/A)
electric field strength	volt per meter (V/m)
electric resistance	ohm (Ω, V/A)
capacitance	farad (F, $A \cdot s/V$)
magnetic flux	weber (Wb, $V \cdot s$)
inductance	henry (H, $V \cdot s/A$)
SI Derived Units	
magnetic flux density	tesla (T, Wb/m^2)
magnetic field strength	ampere per meter (A/m)
magnetomotive force	ampere (A)
luminous flux	lumen (lm, $cd \cdot sr$)
luminance	candela per square meter (cd/m^2)
illuminance	lux (lx, lm/m^2)
wave number	1 per meter (m^{-1})
entropy	joule per kelvin (J/K)
specific heat capacity	joule per kilogram kelvin [$J/(kg \cdot K)$]
thermal conductivity	watt per meter kelvin [$W/(m \cdot K)$]
radiant intensity	watt per steradian (W/sr)
activity (of a radioactive source)	1 per second (s^{-1})
SI Supplementary Units	
plane angle	radian (rad)
solid angle	steradian (sr)

Definitions of SI Units

meter (m)
The *meter* is the length of the path traveled by light in vacuum during a time interval of 1/ 299 792 458 of a second.

kilogram (kg)
The *kilogram* is the unit of mass; it is equal to the mass of the international prototype of the kilogram. (The international prototype of the kilogram is a particular cylinder of platinum–iridium alloy which is preserved in a vault at Sèvres, France, by the International Bureau of Weights and Measures.)

second (s)
The *second* is the duration of 9 192 631 770 periods of the radiation corresponding to the transition between the two hyperfine levels of the ground state of the cesium-133 atom.

ampere (A)
The *ampere* is that constant current which, if maintained in two straight parallel conductors of infinite length, of negligible circular cross section, and placed 1 meter apart in vacuum, would produce between these conductors a force equal to 2×10^{-7} newton per meter of length.

kelvin (K)
The *kelvin*, unit of thermodynamic temperature, is the fraction 1/273.16 of the thermodynamic temperature of the triple point of water.

candela (cd)
The *candela* is the luminous intensity, in the perpendicular direction, of a surface of 1/600 000 square meter of a blackbody at the temperature of freezing platinum under a pressure of 101 325 newtons per square meter.

mole (mol)
The *mole* is the amount of substance of a system which contains as many elementary entities as there are carbon atoms in 0.012 kg of carbon-12. The elementary entities must be specified and may be atoms, molecules, ions, electrons, other particles, or specified groups of such particles.

newton (N)
The *newton* is that force which gives to a mass of 1 kilogram an acceleration of 1 meter per second per second.

joule (J)
The *joule* is the work done when the point of application of 1 newton is displaced a distance of 1 meter in the direction of the force.

watt (W)
The *watt* is the power which gives rise to the production of energy at the rate of 1 joule per second.

volt (V)
The *volt* is the difference of electric potential between two points of a conducting wire carrying a constant current of 1 ampere, when the power dissipated between these points is equal to 1 watt.

ohm (Ω)
The *ohm* is the electric resistance between two points of a conductor when a constant difference of potential of 1 volt, applied between these two points, produces in this conductor a current of 1 ampere, this conductor not being the source of any electromotive force.

coulomb (C)
The *coulomb* is the quantity of electricity transported in 1 second by a current of 1 ampere.

farad (F)
The *farad* is the capacitance of a capacitor between the plates of which there appears a difference of potential of 1 volt when it is charged by a quantity of electricity equal to 1 coulomb.

henry (H)
The *henry* is the inductance of a closed circuit in which an electromotive force of 1 volt is produced when the electric current in the circuit varies uniformly at a rate of 1 ampere per second.

weber (Wb)
The *weber* is the magnetic flux which, linking a circuit of one turn, produces in it an electromotive force of 1 volt as it is reduced to zero at a uniform rate in 1 second.

lumen (lm)
The *lumen* is the luminous flux emitted in a solid angle of 1 steradian by a uniform point source having an intensity of 1 candela.

radian (rad)
The *radian* is the plane angle between two radii of a circle which cut off on the circumference an arc equal in length to the radius.

steradian (sr)
The *steradian* is the solid angle which, having its vertex in the center of a sphere, cuts off an area of the surface of the sphere equal to that of a square with sides of length equal to the radius of the sphere.

Units Outside the International System

Though not official SI units, certain other units are in widespread use with the International System of Units.

Units in Use with the International System

Name	Symbol	Value in SI unit
minute	min	1 min = 60 s
hour	h	1 h = 60 min = 3600 s
day	d	1 d = 24 h = 86 400 s
degree	°	1° = $(\pi/180)$ rad
minute	′	1′ = $(1/60)° = (\pi/10\,800)$ rad
second	″	1″ = $(1/60)′ = (\pi/648\,000)$ rad
liter	L	1 L = 1 dm^3 = 10^{-3} m^3
tonne	t	1 t = 10^3 kg

Units Used with the International System Whose Values in SI Units Are Obtained Experimentally

One **electron volt** (eV) is the kinetic energy acquired by an electron in passing through a potential difference of 1 volt in a vacuum.

The **unified atomic mass unit** (u) is equal to the fraction $\frac{1}{12}$ of the mass of an atom of the nuclide ^{12}C (carbon-12).

The **astronomical unit** (AU in English) is the mean distance of the earth from the sun: 1 AU = $1.495\,978\,92 \times 10^{11}$ m (with an uncertainty of about 5 km).

The **parsec** (pc) is the distance at which 1 astronomical unit subtends an angle of 1 second of arc: 1 pc = 2.063×10^5 AU = 3.262 light-year.

Answers to Odd-Numbered Problems for Chapters 1–23

Chapter 2

- **2A-1** 200 km
- **2A-3** 1 light-year = 9.46×10^{15} m; 1 parsec = 3.09×10^{16} m
- **2A-5** 0.447%
- **2A-7** 49.4
- **2A-9** 3.15×10^7 s
- **2B-11** 7.40 min; the hour
- **2B-13** 7.37 m^2
- **2A-15** 6.67×10^{-22} s
- **2A-17** 1.63 cm/yr
- **2A-19** 55.4 s
- **2B-21** Impossible
- **2B-23** (a) 1.20 m/s (b) 7.00 s (c) -1.54 m/s (approx.)
- **2A-25** 62.5 m/s^2
- **2A-27** 2.65×10^4 m/s^2
- **2A-29** 2.59 m/s
- **2A-31** 0.639 s
- **2A-33** (a) 1.63 s (b) 9.96 m/s (c) 13.1 m
- **2A-35** (a) 5.00 s (b) 75.0 m
- **2B-37** 3.34 s
- **2B-39** 0.804 s; 0.0127 s
- **2B-41** (a) -1.5 m/s^2 (b) 4 s (c) 5.33 m
- **2B-43** (a) $6t^2$ (b) $3t$
- **2B-45** (a) 2 m; 3 m/s; 4 m/s^2 (b) $v = 3 - 8t$ (c) -8 m/s^2 (d) 0.375 s (e) 2.56 m
- **2B-47** (c) -4 m/s (d) 34.0 m
- **2B-49** $x(2) = 2$ m; $x(4) = 6$ m; $x(6) = 14$ m; $x(10) = 22$ m
- **2B-51** $a = \tfrac{1}{2}$; $b = -\tfrac{1}{2}$
- **2C-53** (a) 12.8 m/s (b) 5.90 m
- **2C-55** 12.2 m/s
- **2C-57** (a) 7 m/s (b) -5.35 m/s (c) -9.8 m/s^2
- **2C-59** (a) 46.2 s (b) 34.6 m/s
- **2C-61** 14.2 s
- **2C-63** 4.83×10^{-3} m/s^2
- **2C-65** (a) 26.4 m (b) 6.89%
- **2C-67** (a) 40.4 s (b) 69.3 ft/s (c) 34.6 ft/s
- **2C-69** (a) $v = 3At^2$ (b) $a = 6At$ (c) 0.0533 m/s^3

Chapter 3

- **3A-1** (a) 7 blocks (b) 5 blocks; 36.9° north of west
- **3A-3** (a) $\mathbf{C} = 6\hat{\mathbf{x}} + 5\hat{\mathbf{y}}$; $\mathbf{D} = -2\hat{\mathbf{x}} + 7\hat{\mathbf{y}}$
  (b) $\mathbf{C} = 7.81 \underline{/39.8°}$; $\mathbf{D} = 7.28 \underline{/106°}$
- **3B-5** $\mathbf{C} = 5.39$ at 21.8°; $\mathbf{D} = 6.08$ at 80.5°; $\mathbf{E} = 10.8$ at 248.2°
- **3A-7** (a) $\mathbf{C} = \hat{\mathbf{y}} - 2\hat{\mathbf{z}}$; 2.24 m (b) $\mathbf{D} = 4\hat{\mathbf{x}} + 5\hat{\mathbf{y}} - 6\hat{\mathbf{z}}$; 8.78 m
- **3B-9** 2.50 m/s
- **3B-11** (a) 4.87 km/s; 61.4° west of south (b) 23.3 m/s (c) 13.5 m/s; 61.4° west of south
- **3B-13** 16.1° below the horizontal
- **3A-15** 13.6 m
- **3B-17** 24.7 m/s
- **3B-19** 55.4 m/s
- **3B-21** (a) 11.1 m/s (b) 24.7 m/s; 26.5° from the vertical
- **3B-23** (a) 21.9 m (b) 2.74 s (c) 14.1 m (d) 21.4 m/s; 13.9° from the vertical
- **3C-25** (a) $6\hat{\mathbf{x}} - 2\hat{\mathbf{y}} + 2\hat{\mathbf{z}}$ (b) $2\hat{\mathbf{x}} + 4\hat{\mathbf{y}} - 6\hat{\mathbf{z}}$ (c) $6\hat{\mathbf{x}} + 5\hat{\mathbf{y}} - 8\hat{\mathbf{z}}$
- **3C-27** (2.44 m, 11.9 m)
- **3C-29** $R = \dfrac{v_0^2 \sin 2\theta}{g}$
- **3C-31** Answer given.
- **3C-33** $y_m = \dfrac{v_0^2 \sin^2 \theta}{2g}$
- **3C-35** Answer given.
- **3C-37** $y = (\tan \theta)x - \left[\dfrac{g}{2(v_0 \cos \theta)^2}\right] x^2$
- **3C-39** $\phi = \tan^{-1}\left(\dfrac{\tan \theta}{2}\right)$
- **3C-41** 23.7 ft

Chapter 4

- **4A-1** (a) 8.73×10^{-3} rad (b) 0.030 rad
- **4B-3** 91.7°
- **4A-5** 126 m/s
- **4A-7** 2.72×10^{-1} m/s^2
- **4A-9** 4.43 m/s
- **4A-11** (a) 87.0 m/s^2 (b) $8.88g$
- **4B-13** (a) 7.90×10^5 m/s^2 (b) 5.58×10^5 m/s^2
- **4B-15** (a) 18.3 m/s (b) $6.85 \times 10^4 g$
- **4B-17** 0.821 m/s^2; 62.4°
- **4B-19** (a) 1.25 m/s^2 toward the center of curvature of the road (b) -1.67 m/s^2 (c) 1.85 m/s^2; 64.4° back from the inward radial direction
- **4B-21** (a) 1.30 ft/s^2 (b) 16.4 ft/s^2
- **4C-23** Answer given.
- **4C-25** 0.851 m/s^2 (b) 5.34 m/s^2 (c) 5.41 m/s^2; 9.04° back from the radial inward direction
- **4C-27** 54.4 m/s^2

Chapter 5

- **5A-1** (a) 160 lb (b) 5.00 slug (c) 196 N (d) 20 kg (e) 160 lb (f) 196 N
- **5A-3** 282 kg
- **5A-5** (a) 4.00 m/s^2 (b) 8.00 m
- **5A-7** (a) 20.0 lb (b) 18.0 ft
- **5A-9** (a) 25.0 ft (b) 10.0 ft/s

5A-11   14.8°
5A-13   1.63 m/s^2
5B-15   (a) 6.00 lb   (b) 53.1° below the horizontal
        (c) a straight line
5B-17   (b) 359 N
5B-19   (a) 0.102 s   (b) 0.0255 m
5B-21   (a) 20.0 ft/s^2 (b) 1875 lb   (c) 1125 lb
5A-23   (a) 170 N   (b) 170 N
5A-25   300 lb
5A-27   6.39 N
5A-29   (a) 0.300 m/s^2 (b) 0.900 N
5B-31 $t = 2\pi \sqrt{\dfrac{\ell \cos \theta}{g}}$
5B-33   (a) 2.05 kg   (b) 16.0 N
5B-35   (a) 10.7 ft/s^2 (b) 5.33 lb   (c) 3.27 ft/s
5B-37   (a) 4.90 m/s^2 (b) 1.96 m/s^2
5B-39   4.70 kg
5A-41   (a) 8.40 N   (b) 15.7 N
5A-43   7.00 s
5A-45   0.364
5A-47   0.732
5B-49   28.7 m
5B-51   7.54 lb
5B-53   (a) 0.204   (b) 90.8 N
5B-55   4343 lb
5B-57   (b) gR/v^2
5B-59   Answer given.
5B-61   31.4 N
5B-63   (a) 600 N   (b) 1100 N
5C-65   (a) 4.92 N   (b) 16.7 N
5C-67   0.143 m
5C-69   Answer given.
5C-71   (a) 403 lb   (b) 11.4°   (c) 297 lb
5C-73   Answer given.
5C-75   0.209 rev/s

Chapter 6

6A-1 1.2×10^5 ft·lb
6A-3   180 ft·lb
6A-5   960 J
6A-7   (a) 417 N/m   (b) 3.00 J
6B-9   (b) $k_1/(k_1 + k_2)$
6A-11   121 ft
6B-13   (a) 60 J   (b) 10 J   (c) 7.75 m/s   (d) 3.16 m/s
6B-15   (a) 2.25×10^4 N   (b) 1.33×10^{-4} s
6A-17   (a) 9.75×10^4 N/m   (b) 3.12 J
6A-19   1390 J
6A-21   0.029 J
6B-23   (a) 6.86 m/s^2 (b) 6.41 m/s
6A-25   124 J
6A-27   115 J
6B-29   (a) 980 J   (b) 355 J
6B-31   1.68 m/s
6B-33   (a) 104 J   (b) 88.2 J   (c) 15.8 J   (d) 1.98 N
6A-35   1.27 hp
6B-37 $5.76
6A-39   14.4 hp
6B-41   141 kW
6A-43   39.2 kW
6A-45   57.5 hp
6B-47   48.6 hp
6A-49   4
6A-51   single pulley
6B-53 1.76×10^4 N
6B-55   280 N
6C-57   22.0 J
6C-59   (a) $mg \cos\left(\dfrac{s}{R}\right)$ (b) mgR
6C-61   Answer given.
6C-63 $\dfrac{k_1 \ell_1 + k_2(L - \ell_2)}{k_1 + k_2}$
6C-65 9.6×10^5 N/m^2
6C-67   0.303 m/s
6C-69   (c) $k_2/(k_1 + k_2)$
6C-71   Answer given.
6C-73   242 J
6C-75   Answer given.

Chapter 7

7A-1   (a) N·m^3 (b) $2C/r^3$
7A-3   (a) $-3ax^2 + 2bx$ (b) at $x = b/3a$
7A-5   8.26 m/s
7A-7 $2mg$
7A-9   (a) 5.42 m/s   (b) $3mg$
7A-11 $\sqrt{2g\ell(1 - \cos \theta)}$
7B-13   5.79 m/s
7B-15   (a) 3.61 m/s   (b) 1.74 N
7B-17   (b) $g\sqrt{3}/2$, g (c) $3mg/2$, radially inward
7B-19 $4.20mg$
7B-21 mg
7B-23   (a) 37.6°   (b) 36.3 N
7B-25   1.45 m
7B-27 $8d$
7B-29   (a) $-3ax^2 + b$ (b) $\sqrt{b/a}$ (c) $\tfrac{2}{3}\sqrt{b^3/3a}$
7A-31   17.0 m
7B-33   (a) 7.67 m/s   (b) 0.932
7B-35 1.12×10^5 J
7C-37   Answer given.
7C-39 -1.00 J
7C-41   (a) x_0 (b) $2bx_0$; $+x$ direction
7C-43   Answer given.
7C-45   Answer given.
7C-47   Answer given.
7C-49   0.344 m
7C-51   (a) $0.222g$ (b) 1.52×10^4 N
7C-53   (a) $0 \leq x \leq 2$ m   (b) 8 J

Chapter 8

8A-1   1.23 m/s
8A-3   34.3 J
8A-5   7.60 mi/h, approaching the train
8A-7   0.400 m/s

8B-9   7.22 m/s; $-48.4°$
8B-11 $\theta = \tan^{-1} k$
8B-13 2.93 ft/s; 47° north of east
8B-15 13.0 kg·m/s; 202.6° counterclockwise from the $+x$ direction
8B-17 Answer given.
8B-19 0.0466
8A-21 900 N, opposite to the particle's original velocity
8A-23 (a) 1.20×10^4 kg·m/s   (b) 2.40×10^4 N
8B-25 6.38 N·s upward
8B-27 7.80 m/s
8B-29 (a) 4.37×10^{-4} s   (b) 0.153 m   (c) 1.22×10^{-3} J   (d) 1.23×10^3 J
8B-31 (a) 7.80 kg·m/s; 22.6° above the horizontal   (b) 3900 N; 22.6° above the horizontal
8A-33 200 N
8A-35 (a) 1.88 N   (b) 3.75 N
8B-37 4.00×10^3 N
8A-39 535 m/s
8B-41 (a) 3.48×10^6 N   (b) 1659 kg/s
8C-43 (a) 338 m/s   (b) 56.3 N
8C-45 (a) $(M-m)/M$
8C-47 Answer given.
8C-49 $nmg\left(t + \sqrt{\dfrac{2h}{g}}\right)$
8C-51 $0.368M$

Chapter 9

9A-1   (a) -0.167 m/s   (b) 0.333 m/s
9A-3   (a) 42.9 m/s, 37° south of west   (b) 7720 J
9A-5   No; 2.80 J lost
9B-7   (a) $\sqrt{1.41}$ m/s   (b) 57.4 m/s   (c) 97.6%
9B-9   Answer given.
9B-11 1.81 m/s, 2.27 m/s
9A-13 0.200 m
9A-15 $(\frac{7}{13}\text{ m}, \frac{1}{13}\text{ m})$
9B-17 5.35 m
9A-19 (a) 30 m/s, horizontal   (b) 21.2 m/s, 45° below horizontal
9B-21 Answer given.
9B-23 Answer given.
9B-25 7.28 m/s
9B-27 (a) 3.00 m/s   (b) 3.00 m/s   (c) 608 J and 824 J   (d) 0 and 216 J
9B-29 216 J
9C-31 (a) $\left(\dfrac{M-m}{M+m}\right)$ (b) the same as (a)
9C-33 $4M\sqrt{g\ell/m}$
9C-35 (a) 65.2 m/s   (b) 0.458
9C-37 $\sqrt{1 - d/h}$
9C-39 (3.46 ft, 3.00 ft)
9C-41 Answer given.
9C-43 $v_A = -0.667$ m/s; $v_B = 0.800$ m/s
9C-45 2.21 m/s
9C-47 (a) $\frac{3}{8}v$ (b) $\frac{25}{32}\mu v$
9C-49 (a) $\frac{10}{30}v$ (b) $\frac{11}{30}v$ (This is a quantitative response to Chapter 8, Question 6.)

Chapter 10

10A-1 (a) 1.50 m   (b) 24.0 N·m
10A-3 (a) $2bF$ (b) $2bF$
10A-5 Answer given.
10A-7 $\ell/2$
10B-9 $R/12$
10B-11 $\frac{5}{16}mg\ell$
10A-13 6.19 ft
10A-15 (a) 200 N   (b) 173 N toward right, 100 N up
10A-17 3.17 ft
10A-19 Answer given.
10B-21 Answer given.
10B-23 $\theta = \tan^{-1}(f/\mu_s)$
10B-25 (a) 1011 N   (b) 854 N, 14.2° above horizontal
10B-27 (a) 214 N   (b) 369 N, 54.5° above horizontal
10B-29 $b(1 + \sqrt{3})$
10B-31 515 N
10B-33 (a) 277 lb   (b) 260° at 67.7° with respect to horizontal
10B-35 Answer given.
10B-37 $\theta = \tan^{-1}\mu_k$
10B-39 (a) $N_A = N_B = 60$ N   (b) 16.4 N
10C-41 15.9°
10C-43 446 lb
10C-45 Answer given.
10C-47 Answer given.
10C-49 (a) 17.6 N   (b) 42.9 N   (c) 13.3 N, 41.0° from vertical
10C-51 Answer given.
10C-53 Answer given.
10C-55 $(0, 3a/4)$
10C-57 Answer given.
10C-59 $1.04T$

Chapter 11

11A-1 (a) 3.14×10^{-4} m/s   (b) 1.75×10^{-3} rad/s
11A-3 hour hand: $\frac{1}{45} \times 10^{-4}$ rad/s; astronaut: 1.05×10^{-3} rad/s; minute hand: 1.75×10^{-3} rad/s; grindstone: 628 rad/s
11A-5 (a) 17.4 s   (b) 4.85 rev
11B-7 13.5 s
11B-9 Answer given.
11A-11 43.4 rad/s
11B-13 $\dfrac{v}{\pi D}\sqrt{\dfrac{2h}{g}}$
11C-15 $\delta = (R - r)\theta/r$

Chapter 12

12A-1 (a) $2m\ell^2$ (b) $m\ell^2$ (c) $2m\ell^2$ (d) $m\ell^2$
12B-3 Answer given.
12B-5 Answer given.
12A-7 3.16 cm
12A-9 8.50×10^{-32} kg·m²/s
12A-11 (a) 0.320 kg·m²   (b) 0.960 kg·m²/s   (c) 4.80 N
12B-13 32.9 N·m in the $-\hat{\mathbf{z}}$ direction
12A-15 0.480 N
12B-17 (a) 24.0 N·m   (b) 3.56×10^{-2} rad/s²   (c) 1.07 m/s²
12B-19 (a) $T_1 = 24.2$ N, $T_2 = 30.0$ N   (b) 4.30 m/s
12B-21 (a) $\sqrt{g/R}$ (b) $\dfrac{Mk^2}{4F}\sqrt{\dfrac{g}{R^3}}$

12B-23 (a) 6.25 rad/s² (b) 50 rad
12B-25 0.103 rad/s
12B-27 0.526 rad/s
12A-29 $L^2/2mR^2$
12B-31 2.87 rad/s²
12B-33 (b) 85.0 ft
12C-35 $\frac{3}{5}Ma^2$
12C-37 (c) $2M\ell^2/3$
12C-39 376 kg·m²
12C-41 (a) m/M (b) $\frac{1}{2}[mM/(m+M)]D^2\omega^2$
12C-43 Answer given.
12C-45 (a) 0.506 kg (b) 60.4 N
12C-47 1.97×10^{-3} lb
12C-49 Answer given.
12C-51 (a) $6t$ (in units of Newton·meters if t is in seconds)
(b) $0.060t^2$ (in units of radians if t is in seconds)
12C-53 Answer given.
12C-55 $\sqrt{3gD \sin \theta}$
12C-57 2.20 m/s
12C-59 $D\omega_0 \sqrt{\frac{3}{2}}$

Chapter 13

13A-1 (a) 4.00 J (b) 2.00 J (c) 6.00 J
13A-3 5.96 m/s
13B-5 $\theta = \tan^{-1}(7\mu_s/2)$
13B-7 Answer given.
13B-9 Answer given.
13B-11 (a) 66.7 rad/s (b) 8.84 rad/s
13A-13 $M\ell^2/3$
13B-15 $g \sin \theta/(1 + k_0^2/R^2)$
13B-17 35.3°
13B-19 (a) 7.48 m/s (b) 55.0°
13B-21 (a) 0.741 rad/s (b) counterclockwise
13C-23 $[11(R - r)]/4$
13C-25 $h = \frac{7}{5}R$
13C-27 $a = F/(M + m/3)$
13C-29 (b) $\dfrac{4T}{11}$
13C-31 $\sin^{-1}(r/R)$
13C-33 Answer given.
13C-35 $\frac{2}{3}\ell$
13C-37 (a) $5v_0/7$ (b) $12mv_0^2/49f$
13C-39 (a) $2Mg/3$, down incline (b) $R/2$ from CM
13C-41 Answer given.
13C-43 (a) $v_0/3$ (b) 0.500
13C-45 (a) $F_1 = 197$ N, $F_2 = 131$ N
13C-47 Right

Chapter 14

14A-1 2.2 m/s² upward
14A-3 16.0 ft/s²
14B-5 $g/2$
14B-7 11.7 m/s²
14B-9 (a) 20.6° (b) 3.20 lb
14B-11 5.5 N
14A-13 8.54 rpm
14A-15 50.4
14B-17 South, 60° above the horizon
14B-19 $\sqrt{g/R}$
14B-21 (a) a radially inward friction force: $f_r = 4.00 \times 10^{-3}$ N
(b) the above, plus an outward centrifugal force: $F_{cf} = 4.00 \times 10^{-3}$ N (c) TURNTABLE'S FRAME: the forces in (a) and (b) plus a Coriolis force $F_{Cor} = 8.00 \times 10^{-4}$ N toward the bug's right and an equal and opposite tangential friction force component $f_t = 8.00 \times 10^{-4}$ N toward the bug's left
(d) INERTIAL FRAME: only the two friction components: $f_r = 4.00 \times 10^{-3}$ N radially inward and $f_t = 8.00 \times 10^{-4}$ N tangentially toward the bug's left
14C-23 5 N/m
14C-25 (a) 10.6 ft
14C-27 $g \tan 2\theta$
14C-29 $F/(M + 2m/7)$
14C-31 Answer given.
14C-33 7.5 N, toward the left
14C-35 (a) 20 N, radially outward (b) 80 N, radially outward
(c) 180 N, radially inward
14C-37 (a) zero (b) $m\omega^2 R$, inward
14C-39 Answer given.
14C-41 (a) $4m\omega v$ (b) westward

Chapter 15

15A-1 (a) 0.020 m (b) 0.942 m/s, at midpoint
(c) 17.8 m/s², at extremities
15B-3 0.0356 m
15B-5 $4\pi^2 f^2 A/g$
15B-7 (a) 0.910 s⁻¹ (b) 0.588 N
15A-9 (a) 0.50 s (b) 79.0 lb/ft (c) 6.28 ft/s (d) 79.0 ft/s²
(e) 9.88 ft·lb (f) 5.45 ft/s (g) 39.5 ft/s²
15A-11 (a) 1.19 Hz (b) 0.210 s (c) 0.784 N downward
15B-13 (a) 8.17 cm (b) 1.42 s⁻¹
15B-15 (a) 0.0280 J (b) 1.03 m/s (c) 0.0158 J (d) 0.0123 J
15B-17 (a) 0.10 m (b) −0.0654 m (c) 0.262 s (d) 0.0160 J
(e) 0.0160 J
15A-19 (a) 0.136 Hz (b) 7.37 s
15B-21 Answer given.
15A-23 19.9 s
15B-25 1.58 s
15B-27 0.790 Hz
15B-29 (a) 3.559 Hz (b) 3.554 Hz; 1.38 s
15A-31 1.104 cm³
15A-33 952 N/m²
15B-35 AY/L_0
15C-37 (a) $3k$, $1.5k$ (b) $\sqrt{2}:1$
15C-39 $4mg/\ell$
15C-41 Answer given.
15C-43 Answer given.
15C-45 Answer given.
15C-47 $\pi b A^2 \omega$
15C-49 (a) 0.149 m (b) 132°
15C-51 (b) $(y/2)(\Delta L/L)^2$

Chapter 16

16A-1 (a) 3.32×10^{-5} N (b) 5.92×10^{-3} N
16A-3 $g/9$
16A-5 35.0 N

16A-7   18.8 mi/s
16A-9   2.41
16B-11 $4\pi^2/Gm$, 3.00×10^{-19} s^2/m^3
16B-13 (a) 84.4 min   (b) 7.90 km/s
16B-15 $\frac{128}{81}G\pi^2R^4\rho^2$
16B-17 8.74×10^7 m
16B-19 1.62×10^{27} kg
16B-21 $1.91 Gm/\ell^2$, toward diagonally opposite corner
16A-23 Answer given.
16B-25 (a) 1.32×10^{12} m/s^2 (b) 9.21×10^{13} N
         (c) 7.70×10^{-11} J
16B-27 $\sqrt{(GM/R)(2-\sqrt{2})}$
16A-29 2380 m/s
16A-31 $4R/3$
16B-33 $3Gm^2/\ell$
16B-35 $\sqrt{2Rg(1+R/r)}$
16B-37 Answer given.
16B-39 Answer given.
16C-41 $\sqrt{125\pi/3G\rho}$
16C-43 (b) $2\pi\sqrt{D^3/3GM}$
16C-45 (b) 6.54×10^{-3}
16C-47 $Gm^2/3L^2$
16C-49 $2\sqrt{R^3/GM}$
16C-51 $T = 2\pi\sqrt{R^3/GM} = 84.5$ min
16C-53 1.41 h
16C-55 $\dfrac{GMm}{2R}\left[3 - \left(\dfrac{r}{R}\right)^2\right]$
16C-57 $\dfrac{Gm}{R^2}\left(\dfrac{M}{4} - \dfrac{m}{3}\right)$

Chapter 17

17B-1   90.0%
17B-3   Answer given.
17A-5   50 lb
17A-7   20 cm
17B-9   (a) 5000 kg/m^3 (b) 667 kg/m^3
17B-11 (a) 2704 kg/m^3 (b) 59.8 N
17B-13 55.5 lb/ft^3
17B-15 4.00 mg
17B-17 $\Delta V/V = 0.0830$
17A-19 1.77×10^{-3} m^3/s
17A-21 40 cm/s
17B-23 7.71 lb/in.^{2}
17B-25 4.49 atm
17B-27 $\rho A v^2$
17B-29 (a) 7.67 m/s   (b) 2.80 mm
17B-31 Answer given.
17C-33 $a\ell/g$
17C-35 0.933
17C-37 $(1 - 1/\sqrt{2})$
17C-39 $(1 - 1/\sqrt{2})$
17C-41 Answer given.
17C-43 $T = 2\pi\sqrt{m/\rho Ag}$
17C-45 Answer given.
17C-47 27.3 cm^3/s
17C-49 $H/2$

Chapter 18

18A-1   (a) 2.27×10^{-3} s   (b) 0.782 m
18A-3   Answer given.
18A-5   8.33 cm
18B-7 $A = 7 \times 10^{-4}$ m, $k = 3.14$ m^{-1}, $\omega = 6.28 \times 10^{-3}$ s^{-1}
18B-9   (a) 1.27 Pa   (b) 170 Hz   (c) 2.00 m   (d) 340 m/s
18B-11 18.56 m
18B-13 860 m
18A-15 2.94×10^{-16} J/cm^3
18B-17 1.13 μW
18B-19 Answer given.
18B-21 (a) 565 Hz   (b) A sound of descending pitch.
18A-23 2.07 N
18A-25 (a) 515 Hz   (b) 4.13 cm
18A-27 (a) 0.773 m   (b) 1.55 m   (c) 330 Hz   (d) 220 Hz
18A-29 870 Hz, 2610 Hz
18B-31 (a) 34.8 m/s   (b) 0.977 m
18B-33 800 Hz
18A-35 19.9 m/s
18B-37 (a) 1091 Hz   (b) 1100 Hz   (c) 1000 Hz
18A-39 28.4°
18B-41 5.64 Hz
18C-43 Answer given.
18C-45 3.14 m/s, 9.87×10^3 m/s^2
18C-47 $B = 2.47 \times 10^{11}$ N/m^2
         $S = 1.25 \times 10^{11}$ N/m^2
18C-49 (b) $v = R\omega$
18C-51 (a) +6.99 dB   (b) 2.24
18C-53 $\mu = 4.00 \times 10^{-3}$ kg/m, 2.50 cm long
18C-55 12.6 m/s^2
18C-57 60.0 Hz
18C-59 0.335 cm

Chapter 19

19A-1   40.0°C
19B-3   Answer given.
19A-5   Add 7.20 mm
19A-7 3×10^{-5}/C°
19B-9 2.17×10^5 N
19B-11 0.0191 gal
19A-13 6.44 kJ
19A-15 0.463 kJ/kg·C°
19A-17 0.103 cal/g·C°
19A-19 0.122 kg
19B-21 0.126 kJ/kg·C°
19B-23 87.5
19B-25 Answer given.
19A-27 557 J/s
19A-29 1.38×10^8 J
19B-31 (a) 290 g   (b) 42.9 g
19B-33 Answer given.
19B-35 Answer given.
19B-37 (a) 8.44 kW;   (b) $162 (!)
19A-39 5.00 W/m^2·C°
19B-41 2.84 J/s
19A-43 (a) 61.1 kW·h   (b) $3.67
19A-45 (a) −28.3°C, 244.7 K   (b) 5.56°C, 267.4 K
         (c) 37.0°C, 310.0 K

19A-47 (a) $-32.8°F$ (b) $192.2°F$ (c) $-178.6°F$ (d) $167°F$
19C-49 Answer given.
19C-51 (a) 13.9 cm (b) 2.6×10^{-5} $(C°)^{-1}$
19C-53 8.0039 cm
19C-55 (a) $\dfrac{T_2 k_1 \Delta x_2 + T_1 k_2 \Delta x_1}{k_2 \Delta x_1 + k_1 \Delta x_2}$
19C-57 Answer given.
19C-59 3.52×10^4 s = 9.78 h
19C-61 Answer given.
19C-63 Answer given.

Chapter 20

20A-1 48.5 L
20A-3 (a) 4.48 m^3 (b) 5.60 kg
20A-5 1.63 ft^3
20A-7 12.0 L
20A-9 27.8 lb/in.^{2}
20A-11 38 100 lb, or about 19 tons!
20A-13 (a) 1.10×10^{30} electrons (b) 1.82×10^6 mol
20A-15 8.01 km
20B-27 (a) 0.489 atm (b) 0.888 kg/m^3
20B-19 244 ft^3
20B-21 on the average, 59.0 atoms
20B-23 on the average, 3.48 molecules
20B-25 (a) 2.56 atm (b) 16.1 m
20B-27 Answer given.
20B-29 (a) 4.14×10^{-16} J (b) 7.04×10^5 m/s
20A-31 5.80×10^9 K
20B-33 Answer given.
20B-35 (b) 10.8% of the escape speed
20B-37 $(8.28 \times 10^{-9}/\ell^3)$ N/m^2 (with ℓ in meters)
20C-39 8.22×10^{23} collisions/s
20C-41 $mv^2/3\ell^3$
20C-43 Answer given.
20C-45 $\omega = v\theta/x$
20C-47 385 m/s, 417 m/s
20C-49 (a) 1.77 cm (b) 12.6°C
20C-51 63.4°C

Chapter 21

21A-1 (a) 209 J (b) 209 J (c) 0 (d) 0.0896 L
21A-3 (a) 0.144 atm (b) 157 K
21A-5 (a) 0.160 atm (b) 131 K
21B-7 (a) 546 K (b) 4538 J (c) 1.13×10^4 J (d) 6806 J
21B-9 Answer given.
21B-11 2.09×10^4 J
21B-13 Answer given.
21A-15 2.93R
21B-17 (a) 216°C (b) 0.178 L

21B-19 4.14×10^{-21} J
21A-21 56.1
21C-23 Answer given.
21C-25 (a) 70.2 J (b) 36.0 J (c) 208.3 J (d) -53.6 J
(e) -36.0 J (f) 16.6 J
21C-27 (a) 47.3 J (b) 1.61×10^{-4} m^3 (c) 13.5 J (d) 33.8 J
21C-29 (b) $\frac{13}{11}$

Chapter 22

22A-1 150 J
22A-3 14.2%
22A-5 280 K
22A-7 5.76%
22B-9 (a) 44.6% (b) 25%
22B-11 $-5.40°C$
22B-13 Answer given.
22B-15 (a) 414 J (b) 4600 J
22B-17 (a) 0.99 J (b) 3.45 J
22B-19 1.97×10^5 J
22B-21 (a) 370 persons (b) $14 800.00 (c) $4.80
22B-23 (a) $\frac{4}{3}P_0 V_0$ (b) 22.2%
22B-25 $\frac{2}{13}$
22C-27 (a) 12.4 (b) 2.07×10^7 J (c) 6.00×10^7 J
(d) 2.32 L (e) 1.33 L
22C-29 173 W
22C-31 $\left(1 - \dfrac{V_1}{V_3}\right)^{(\gamma-1)}$
22C-33 (a) a: 4.92 L; b: 1.67 atm; c: 6.69 L, $T_e = 408$ K
(b) 52.7 J
22C-35 Answer given.
22C-37 300 N, 400 N

Chapter 23

23A-1 -24.2 J/K
23A-3 123 J/K
23A-5 5.27 J/K
23A-7 12.6 J/K
23B-9 Answer given.
23B-11 $\sim 5 \times 10^5$ J/K
23B-13 Answer given.
23C-15 3807 J
23C-17 Answer given.
23C-19 (b) $mc\left[(T_2 + T_1) - 2\sqrt{T_2 T_1}\right]$
23C-21 Answer given.
23C-23 Answer given.
23C-25 (a) 588 J (b) zero (c) 1.96 J/K (d) 1.96 J/K
23C-27 $8k \ln 2$
23C-29 2.40×10^{26} J/K·h

Answers to Odd-Numbered Problems for Chapters 24–45

Chapter 24

- **24A-1** 649 kg
- **24A-3** 2.27×10^{39}
- **24A-5** 110 N at 157° from the $+x$ axis
- **24B-7** 2.51×10^{-10} (or about 1 in 4 billion)
- **24B-9** at $x = 0.817$ m
- **24B-11** 9.55 electrons
- **24A-13** 4.90×10^{-3} C
- **24B-15** $y = (qE_0/2mv_0^2)x^2$
- **24A-17** 1.70×10^{-10} m
- **24B-19** $Q/4\pi\varepsilon_0 d(d + L)$
- **24B-21** Answer given.
- **24C-23** Answer given.
- **24C-25** Answer given.
- **24C-27** $W = Q^2/8\pi\varepsilon_0 R$
- **24C-29** 3.67 cm
- **24C-31** $2\pi\sqrt{m\ell/2qE}$
- **24C-33** $d = \dfrac{\pi}{2\omega}\left[v_1 + \dfrac{eE_0}{m\omega\sqrt{2}}\right]$ $\ell = \dfrac{3\pi}{2\omega}\left[v_1 + \dfrac{eE_0\sqrt{2}}{m\omega}\right]$
- **24C-35** (a) $E_y = \lambda L/4\pi\varepsilon_0 a\sqrt{L^2 + a^2}$
  (b) $E_x = (\lambda/4\pi\varepsilon_0)[(1/a) - (1/\sqrt{L^2 + a^2})]$
- **24C-37** Answer given.
- **24C-39** Answer given.
- **24C-41** $E = (Q/4\pi^2\varepsilon_0 R_2)\sin(\ell/2R)$, away from the remaining segment

Chapter 25

- **25A-1** (a) zero   (b) $3q/\varepsilon_0$ (c) $-2q/\varepsilon_0$
- **25A-3** 7.50×10^{-2} N·m²/C
- **25A-5** (a) zero   (b) σ/ε_0
- **25B-7** (a) $\rho x/\varepsilon_0$ (b) $\rho d/2\varepsilon_0$
- **25B-9** (a) σ/ε_0 (b) $\sigma/2\varepsilon_0$
- **25A-11** -1.15×10^{-9} C/m²
- **25B-13** (b) 1.38×10^7 electrons/m³, deficiency
- **25C-15** (a) C/m²   (b) $Q = \kappa 2\pi L(b - a)$ (c) $E = (\kappa/\varepsilon_0)(1 - a/r)$
- **25C-17** Answer given.
- **25C-19** $E(\pi R^2)$

Chapter 26

- **26A-1** (a) 2.05×10^6 m/s   (b) 12 eV, 1.92×10^{-18} J   (c) 3.89 ns
- **26A-3** (a) 2.19×10^6 m/s   (c) -13.6 eV
- **26A-5** $0.0415\ kq/a$; center
- **26B-7** Answer given.
- **26B-9** 1.01×10^{-19} N, 18.4° with respect to x axis
- **26B-11** $Q_0/2$, $3Q_0/2$
- **26B-13** Units apply after insertion of value for ε_0. Inside the spheres: $E = 0$, $V = -(140/3\pi\varepsilon_0)\ \mu V$
  between the spheres: $E = (80/4\pi\varepsilon_0 r^2)\ \mu N/C$, inward; $V = -(1/4\pi\varepsilon_0)(80/r - 40/0.5)\ \mu V$
  outside the spheres: $E = (40/4\pi\varepsilon_0 r^2)\ \mu N/C$, inward; $V = -(40/4\pi\varepsilon_0 r)\ \mu V$
- **26C-15** $\dfrac{Q}{4\pi\varepsilon_0 \ell}\ln\dfrac{\ell + \sqrt{\ell^2 + y^2}}{y}$
- **26C-17** (a) C/m⁴   (b) $Q = A\pi R^4$ (c) $E = Ar^2/4\varepsilon_0$
  (d) For $r \geq R$; $V = AR^4/4\varepsilon_0 r$
  For $r \leq R$; $V = (A/12\varepsilon_0)(4R^3 - r^3)$
- **26C-19** 2000 V
- **26C-21** Answer given.
- **26C-23** (a) 2.72×10^{-5} m
  (b) 2.65×10^7 m/s and 6.19×10^5 m/s
  (c) 2000 eV, 3.20×10^{-16} J
- **26C-25** (a) $(q/2\pi\varepsilon_0)\{[x^2 + (\ell/2)^2]^{-1/2} - 1/x\}$
  (b) $(q/4\pi\varepsilon_0)[(y + \ell/2)^{-1} + (y - \ell/2)^{-1} - 2/y]$

Chapter 27

- **27A-1** 0.885 pF
- **27A-3** (a) $\tfrac{3}{5}C$ (b) $3C$ (c) C
  (d) Capacitors are shorted by a conductor.
- **27B-5** C
- **27B-7** $C = \varepsilon_0 A(a + b)/[b(a - b)]$
- **27B-9** Answer given.
- **27B-11** Answer given.
- **27B-13** Answer given.
- **27B-15** (a) 400 pC   (b) 80 V
- **27B-17** 0.188 m²
- **27B-19** $C = \kappa C_0/[f = (1 - f)]\kappa$
- **27B-21** $2\pi\varepsilon_0 L\kappa_1\kappa_2\left[\kappa_2 \ln\left(\dfrac{b}{a}\right) + \kappa_1 \ln\left(\dfrac{c}{b}\right)\right]^{-1}$
- **27B-23** (a) 1.60 mJ   (b) 0.800 mJ
- **27B-25** Answer given.
- **27B-27** (a) 600 nC, decrease   (b) 30 μJ, decrease   (c) 30 μJ
- **27B-29** Answer given.
- **27B-31** Answer given.
- **27C-33** 267 V
- **27C-35** $C/L = \kappa 2\pi\varepsilon_0/[\ln(b/a)]$
- **27C-37** $1/(1 + \kappa)$
- **27C-39** $CV^2/2d$
- **27C-41** Answer given.
- **27C-43** 1.41×10^{-15} m

A-23

Chapter 28

- **28A-1** 3.12×10^{19} electrons/s
- **28B-3** (a) 5.86×10^{28} electrons/m^3 (b) 51.9 mA (c) 1.76×10^{-6} m/s
- **28A-5** 0.667 Ω
- **28A-7** 418°C
- **28A-9** 276°C
- **28B-11** 1.56R
- **28B-13** 1.66 V
- **28B-15** 5.25 W
- **28A-17** (a) 11.1 Ω (b) 1.08 A
- **28A-19** (a) 66.7% more power (b) No
- **28A-21** $\rho L/\pi(b^2 - a^2)$
- **28B-23** (a) 116.7 V (b) 12.8 kW (c) 366 W
- **28B-25** (a) 2.16 kW (b) 1.34 hp (c) 46.3%
- **28B-27** (a) 9.36×10^{11} particles/s (b) 6.00 W
- **28A-29** 6.00×10^{-15} s
- **28B-31** 4.17×10^6 A/m^2
- **28B-33** Answer given.
- **28C-35** Answer given.
- **28C-37** Answer given.
- **28C-39** (in SI units) (a) $4000V^{2/3}$; $(2.50 \times 10^{-4})V^{5/2}$
- **28C-41** Answer given.
- **28C-43** 8.32 h
- **28C-45** $(b - a)/4\pi ab\sigma$
- **28C-47** Answer given.
- **28C-49** Answer given.

Chapter 29

- **29A-1** 220 Ω
- **29B-3** (a) A (b) B (c) 4.50
- **29B-5** Answer given.
- **29B-7** $R_{AB} = \tfrac{7}{5}R$
- **29B-9** In watts: 10, 16, 24, 30, 40, $53\tfrac{1}{3}$, $66\tfrac{2}{3}$, 100, 160
- **29A-11** 9.20 V
- **29A-13** Answer given.
- **29A-15** Answer given.
- **29B-17** (a) 5.00 Ω (b) 6.00 A (c) 2.00 A
- **29B-19** 0.0860 Ω
- **29B-21** 2.67 mA in R_1; 2.50 mA in R_2; 0.167 mA in R_3
- **29B-23** Answer given.
- **29B-25** Answer given.
- **29B-27** (a) 2.41 kΩ (b) 2.46 kΩ
- **29B-29** (a) 0.517% (b) 0.103%
- **29B-31** $R_1 = 5.025 \times 10^{-3}$ Ω; $R_2 = 4.523 \times 10^{-2}$ Ω; $R_3 = 4.523 \times 10^{-1}$ Ω; $R_4 = 4.523$ Ω
- **29B-33** Answer given.
- **29A-35** Answer given.
- **29B-37** 0.587 MΩ
- **29B-39** Answer given.
- **29B-41** 1.44 μF
- **29C-43** $R_1 = (R_AR_B + R_BR_C + R_CR_A)/R_C$; $R_2 = (R_AR_B + R_BR_C + R_CR_A)/R_A$; $R_3 = (R_AR_B + R_BR_C + R_CR_A)/R_B$
- **29C-45** $R_A = R_1R_3/(R_1 + R_2 + R_3)$; $R_B = R_1R_2/(R_1 + R_2 + R_3)$; $R_C = R_2R_3/(R_1 + R_2 + R_3)$
- **29C-47** $R(1 + \sqrt{3})$
- **29C-49** Answer given.
- **29C-51** Answer given.
- **29C-53** Answer given.
- **29C-55** 201 Ω
- **29C-57** $R/2$
- **29C-59** 163 V; 1.43 MΩ
- **29C-61** 0.050 J in R_1; 0.0167 J in R_2
- **29C-63** 6.90 Hz

Chapter 30

- **30A-1** 1.86×10^{-6} m/s
- **30B-3** $\mathbf{F} = 1.44 \times 10^{-13}\hat{\mathbf{y}} - 3.36 \times 10^{-13}\hat{\mathbf{z}}$ (in newtons)
- **30A-5** 1.20 keV
- **30A-7** 0.357 T
- **30B-9** $R_\alpha = R_\rho = 42.8R$
- **30B-11** $R = \sqrt{2mV/qB^2}$
- **30A-13** 7.78×10^5 m/s
- **30B-15** 2.44×10^5 V/m
- **30B-17** (b) 0.708 T
- **30A-19** Answer given.
- **30B-21** $\tau = (-1.44 \times 10^{-3}\,\text{N·m})\hat{\mathbf{z}}$
- **30B-23** $\boldsymbol{\mu} = Iab\cos\theta\hat{\mathbf{x}} + Iab\sin\theta\hat{\mathbf{y}}$
- **30A-25** $\tfrac{1}{5}$
- **30A-27** (a) 37.7 mT (b) $(4.28 \times 10^{25})/$m^3
- **30A-29** 0.438 μW
- **30C-31** (a) 12.2 MHz (b) 35.4 MeV (c) 24.4 MHz, 70.8 MeV (d) 17.7 MeV (e) 1.60 T (f) 70.8 MeV (g) None
- **30C-33** Answer given.
- **30C-35** qBt/m
- **30C-37** IBR
- **30C-39** $mg/\pi rB_x$
- **30C-41** (a) 1.05×10^{-3} A (b) 9.27×10^{-24} A·m^2
- **30C-43** Answer given.
- **30C-45** Answer given.
- **30C-47** Answer given.
- **30C-49** Answer given.
- **30C-51** Answer given.

Chapter 31

- **31A-1** 1.43°
- **31A-3** $\mu_0NI/2\sqrt{2}R$
- **31A-5** Answer given.
- **31B-7** $\mu_0I(b - a)/4ab$ (out)
- **31B-9** (a) $\Phi_B = \mu_0I\ell(\ln 3)/2\pi$ (b) 1.32×10^{-7} A
- **31B-11** $\mu_0I2\sqrt{2}/\pi b$
- **31A-13** (a) 2.20×10^{-5} Wb (b) 5570 turns
- **31B-15** $B_{\text{inside}} = \mu_0Ir/2\pi a^2$
- **31C-17** Answer given.
- **31C-19** Answer given.
- **31C-21** Answer given.
- **31C-23** (a) $\mathbf{B} = -(\mu_0Ia/\pi[z^2 + a^2])$ (b) $\lim_{z \propto \ell}\mathbf{B} = -(\mu_0Ia/\pi z^2)\hat{\mathbf{z}}$
- **31C-25** $2RB_e\tan\theta/\mu_0N$
- **31C-27** (a) A/m^3 (b) 0 (c) $\mu_0k(r^3 - a^3)/3r$ (d) $\mu_0k(b^3 - a^3)/3r$
- **31C-29** $-(\mu_0I/6\pi a)\hat{\mathbf{x}}$, independent of y!
- **31C-31** $\mu_0\sigma\omega R$
- **31C-33** $(\mu_0I/2\pi w)\ln(1 + w/d)$

31C-35 Answer given.
31C-37 $\mu_0 NI/\ell$

Chapter 32

32B-1 30 V, clockwise
32B-3 $\mathcal{E} = \dfrac{2B\pi r^2}{t}$
32B-5 Answer given.
32B-7 3.38 A/s
32A-9 Answer given.
32A-11 $N\mu_0 \pi R/2$
32B-13 (a) 360 mV (b) 180 mV (c) 3.00 s
32A-15 (a) $\mu_0 N_1^2 A/\ell$, $\mu_0 N_2^2 A/\ell$ (b) $\mu_0 N_1 N_2 A/\ell$
32B-17 $M = \mu_0 A N_1 N_2/\ell$
32B-19 (a) V/L
32B-21 Answer given.
32A-23 145 J/m³
32B-25 (a) 20 W (b) 20 W (c) 0 (d) 20 J
32C-27 (a) 0.171 mV (b) East
32C-29 (b) 0.458 mV
32C-31 (a) b to a (b) $\Delta Q = N\Delta\Phi/R$ (c) $B = QR/NA$
32C-33 3.08 μC
32C-35 (a) $C\pi a^2 k$ (b) the top plate
32C-37 0.132 μA
32C-39 Answer given.
32C-41 Answer given.
32C-43 Answer given.
32C-45 $\mu_0 I^2/16\pi$
32C-47 Answer given.

Chapter 33

33A-1 88.6 mA
33A-3 318 A
33B-5 Answer given.
33B-7 (a) 0.0251 T (b) 10.0 A
33B-9 1.48 mC

Chapter 34

34A-1 Answer given.
34A-3 Answer given.
34B-5 (a) $v = 24.1 \sin 377t$
 (b) Plane of the loop is perpendicular to **B**.
34A-7 (b) 3.2×10^{-2} J
34B-9 (b) $v = 8.32 \sin(1000t + 33.7°)$ (in SI units)
34B-11 (a) 173 Ω (b) 8.66 V
34B-13 Answer given.
34B-15 $i = 2.11 \sin(10^5 t + 71.6°)$
34A-17 100
34A-19 46.5 pF to 419 pF
34B-21 Answer given.
34A-23 $v = 170 \sin(377t)$ V
34A-25 122 W
34B-27 Answer given.
34B-29 Answer given.
34B-31 (b) 141 V (c) 36.2 mA (d) 109 V (e) 90.5 V
34B-33 (a) 211 μF (b) 979 W
34B-35 (a) 5.00 A (b) 2.77 A (c) 2.77 A
34A-37 (a) 20.0 V (b) 0.660 A
34A-39 (a) 1.82×10^4 A (b) 909 A
34C-41 (b) 82.1 V (c) -70.8 V (d) 53.1 V (e) 64.4 V
34C-43 Answer given.
34C-45 Answer given.
34C-47 2000 A/s
34C-49 Answer given.
34C-51 Answer given.
34C-53 $i = 40.8 \sin(\omega t + 25.6°)$
34C-55 239 mH
34C-57 (a) 100 μF (b) 632 rad/s (c) 125 W (d) 39.5 V
 (e) 150 μF, in parallel
34C-59 Answer given.

Chapter 35

35A-1 30.0 cm (about one foot)
35B-3 Answer given.
35B-7 for $r < R$: $(2rC/R^2) \, dV/dt \times 10^{-7}$;
 for $r > R$: $(C/r) \, dV/dt \times 10^{-7}$
35A-9 Answer given.
35B-11 377 Ω
35A-13 (a) 1.67×10^{13} T (b) 3.32×10^{12} W/m²
35B-15 (a) $(2 \times 10^{-8}) \sin(kx - 10^{16}t)\hat{\mathbf{z}}$ (b) 1.88×10^{-7} m
 (c) 1.59×10^{-10} J/m³
35B-17 (a) 1.20 m (b) $u = 2.36 \times 10^5$ J/m³
 (c) $E_0 = 2.31 \times 10^8$ V/m
35B-19 Answer given.
35A-21 8.97×10^{-3} N
35A-23 5.60×10^{-6} N/m²
35B-25 (a) 1900 V/m (b) 5.00×10^{-11} J
 (c) 1.67×10^{-19} kg·m/s
35C-27 (a) $1.88 \times 10^{-10} \cos 377t$
 (b) $1.00 \times 10^{-4} \cos[(3.77 \times 10^8)t]$
35C-29 Answer given.
35C-31 Answer given.
35C-33 21.9 V/m
35C-35 (a) 292 nm
35C-37 Answer given.
35C-39 Answer given.
35C-41 (a) 22.6 h (b) 30.5 s

Chapter 36

36A-1 Answer given.
36B-3 Answer given.
36B-5 $(30, -40)$, $(-30, 40)$, $(-30, -40)$, all in cm
36A-7 (a) 1.09 cm inside (b) erect, virtual, $M = 0.273$
36A-9 (a) 7.50 cm (b) ∞
36B-11 9.23 cm
36B-13 8.00 cm
36C-15 Answer given.
36C-17 40.0 cm
36C-19 Answer given.
36C-21 for $p = 228$ cm, the image is inverted, real, and $M = -0.123$
 for $p = 21.9$ cm, the image is erect, virtual, and $M = 8.12$
36C-23 Answer given.
36C-25 (a) 30.0 cm (b) 1.67
36C-27 real, erect, unit magnification

Chapter 37

37A-1 $n = 1.52$
37B-3 Answer given.
37B-5 0.624 cm
37B-7 (a) 20.6° (b) 0.400 sr (c) 35.4°
37B-9 1.51
37B-11 2.14 sr
37B-13 17.0%
37A-15 R
37B-17 2.00
37B-19 3.57 mm outward
37A-21 26.7 cm
37B-23 $2f$
37A-25 (a) 0.436 mm (b) 0.0125
37B-27 (a) 17.2 cm (b) 51.7 cm (c) -51.7 cm (d) -17.24 cm
37B-29 (a) 42.0 cm (b) 14.0 cm
37A-31 (a) 24.0 (b) 48.1°
37B-33 (a) $+3.50$ diopters (b) 28.6 cm
37B-35 18.2 cm to 66.7 cm
37C-37 Answer given.
37C-39 Answer given.
37C-41 Answer given.
37C-43 (a) 20.8 km (b) 113 million; (c) 2.63 μs
37C-45 From front of sphere: (a) $2.67R$ (b) $1.80R$ (c) $0.960R$
37C-47 Answer given.
37C-49 $(L^2 - 4fL)^{1/2}$
37C-51 Answer given.
37C-53 Answer given.
37C-55 (a) 20 cm behind the lens, virtual, inverted, $M = -2$ (b) on the object side of the lens
37C-57 real, inverted image 0.174 m beyond the convergent lens; $M = -0.42$
37C-59 Answer given.

Chapter 38

38A-1 5.00 mm
38A-3 1.33 mm
38B-5 Answer given.
38B-7 (a) 1034.4827 wavelengths (b) 62.1°, lagging the uninterrupted beam
38B-9 six
38B-11 dark
38B-13 (a) $2.73E_0$, 30° (b) $2E_0$, 60° (c) 0, not defined
38A-15 (a) 105 nm (b) 1.30
38A-17 199 nm
38B-19 (a) green (b) red
38B-21 99.6 nm
38B-23 113
38B-25 1.31
38B-27 18.7 cm
38C-29 Answer given.
38C-31 Answer given.
38C-33 Answer given.
38C-35 Answer given.
38C-37 Answer given.
38C-39 (a) $0.155\lambda/d$ (b) $0.500\lambda/d$
38C-41 543 nm
38C-43 Answer given.
38C-45 1.000 30

Chapter 39

39A-1 0.396 mm
39B-3 18.0 mm
39B-5 (a) $\lambda_1/\lambda_2 = 2$
39B-7 0.684
39B-9 (a) 120 (b) 60
39A-11 11.5 km
39A-13 15.4
39B-15 420 m
39B-17 1.07×10^{-5} m (b) 1.97×10^{-5} m
39A-19 36.9°
39A-21 7.16×10^{-2} deg/nm (b) 25 000
39B-23 688 nm
39B-25 1.375×10^{-3} deg
39A-27 0.300 nm
39A-29 Answer given.
39B-31 17.0
39C-33 0.1233 rad
39C-35 See Footnote 2.
39C-37 Answer given.
39C-39 Answer given.
39C-41 Answer given.

Chapter 40

40A-1 $\frac{7}{8}$
40A-3 $\frac{1}{8}$
40A-5 32.0°
40A-7 49.2°
40B-9 $\tan \theta_p = \csc \theta_c$
40B-11 16.4 μm
40B-13 Answer given.
40B-15 68.4 mg/cm^3
40C-17 0° and 90°
40C-19 78.1%
40C-21 Answer given.
40C-23 Answer given.
40C-25 0.085 65 mm or 0.1199 mm
40C-27 Answer given.
40C-29 118°

Chapter 41

41B-1 1.5 cm/s
41A-3 (a) 2.31 min (b) 1.16 $c\cdot$min
41A-5 (a) $1 - \beta \approx 2.35 \times 10^{-7}$ (b) One $c\cdot$day
41B-7 22.5 m/c or $\frac{7}{5} \times 10^{-8}$ s
41B-9 6.17 ns
41B-11 (a) 60 m (b) 75 m/c (c) 45 m/c (d) 36 m (e) 45 m/c
41A-13 $0.946c$ and $-0.385c$
41A-15 $v_x = 0.994c$
41B-17 1.78
41A-19 $v = 0.866c$
41A-21 889 kg
41B-23 4.28×10^9 kg/s

41B-25 Answer given.
41B-27 Answer given.
41B-29 Answer given.
41B-31 Answer given.
41B-33 Clock in nose earlier by 270 m/c or 9.00×10^{-7} s
41B-35 (b) 80 m/c
41C-37 (a) $1.33c \cdot$ s (b) 3.00 s
41C-39 (a) 2.00 m/c (b) 2.50 m/c
41C-41 Answer given.
41C-43 Answer given.
41C-45 5.55×10^{-17} s
41C-47 $V = v\left(\dfrac{1 - \sqrt{1-\beta^2}}{\beta^2}\right)$ (where $\beta \equiv v/c$)
41C-49 (a) $K = 4E_0$ (b) $p = \sqrt{24}E_0/c$ (c) $\beta = \sqrt{\dfrac{24}{25}}$
41C-51 Answer given.
41C-53 Answer given.
41C-55 Answer given.

Chapter 42

42A-1 1.51 cm^2
42B-3 0.646%
42A-5 9660 nm
42A-7 5222 K
42A-9 2.43×10^{-12} m
42A-11 Answer given.
42B-13 3.54×10^6 m (about the distance between New York and London!)
42A-15 451 nm
42A-17 (a) 3.56×10^5 m/s (b) 432 nm
42A-19 4.85 pm
42A-21 128 MeV
42B-23 Answer given.
42B-25 Answer given.
42C-27 Answer given.
42C-29 Answer given.
42C-31 38.3 m
42C-33 Answer given.
42C-35 Answer given.
42C-37 (b) 2.27×10^{-13} J/m^3
42C-39 Answer given.
42C-41 288 keV
42C-43 Answer given.
42C-45 Answer given.
42C-47 Answer given.
42C-49 Answer given.

Chapter 43

43A-1 Answer given.
43B-3 Answer given.
43B-5 Answer given.
43B-7 Answer given.
43B-9 $1/9.12 \times 10^4$
43A-11 0.173 nm
43A-13 10.2 fm
43A-15 $\frac{1}{4}$
43B-17 (a) 0.158 nm (b) 47.2°
43B-19 5.71 V
43B-21 Answer given.
43B-23 1.03×10^{-3}
43B-25 956 pm
43B-27 (a) 5.86×10^{-30} m (b) 5.79 cm
43B-29 Answer given.
43C-31 (a) $r_n = (4\pi\varepsilon_0 \hbar^2/me^2)n^2$ (b) $E_n = -(6.80 \text{ eV})/n^2$ (c) 243 nm, 182 nm
43C-33 Answer given.
43C-35 Answer given.
43C-37 Answer given.

Chapter 44

44A-1 7
44B-3 54.7° and 125.3°
44B-5 32.3°, 59.5°, 80.3°, 99.7°, 120.5°, 147.7°
44B-7 32 states
44B-9 (b) 9.42×10^{-25} J (c) 3.34×10^{-49} J
44B-11 25.2
44B-13 Ge; [Zn]$3d^{10}4s^24p^2$
44B-15 Answer given.
44A-17 22.0 kV
44B-19 Answer given.
44B-21 3.04 photons
44B-23 (a) 4.20 mm (b) 4.53×10^{33} (c) 3.81×10^{31}
44C-25 $d = (\mu \cos\theta)(dB/dz)(x/v)^2(1/M)$
44C-27 (a) $E_1 = -15.5$ eV, $E_2 = -7.75$ eV, $E_3 = -5.16$ eV (c) 479 nm
44C-29 (a) Potassium (b) [ ]$3p^64p^1$, [ ]$3p^54p^2$
44C-31 (b) three; $4d_{5/2} - 3p_{1/2}$
44C-33 Answer given.
44C-35 Answer given.
44C-37 (a) $(1/96\pi a^5)r^2 e^{-r/a}$ (b) $(1/24a^5)r^4 e^{-r/a}$; $4a$
44C-39 (a) 1.18×10^{-33} (b) 32 800 K
44C-41 (b) $-325\,000$ K

Chapter 45

45A-1 13.7
45A-3 5.68
45B-5 825 GJ
45A-7 27.9 min
45B-9 3.73×10^{10} dis/s
45B-11 $1.78 t_0$
45B-13 0.565 cm
45B-15 1.71×10^{14} (!)
45B-17 Answer given.
45B-19 1.37 MeV
45B-21 Answer given.
45B-23 38.8 keV
45B-25 (a) 0.144 MeV (b) 0.288 MeV
45C-27 3785 yr
45C-29 (a) 86 s; 8.1×10^{-3} s^{-1}
45C-31 Answer given.
45C-33 1.61 W
45C-35 Answer given.
45C-37 Answer given.
45C-39 (a) 0.931 MeV, 5.49 MeV, and 12.86 MeV (b) 24.7 MeV (c) 27.7 MeV
45C-41 Answer given.
45C-43 0.247 GeV for both particles
45C-45 7.8 cm

Photograph and Illustration Credits

Chapter 2
Fig. 2-3: National Institute of Standards and Technology.

Chapter 3
Fig. 3-26: From *PSSC Physics*, 5th ed., D. C. Heath and Company.

Chapter 5
Fig. 5-5: Courtesy of the National Institute of Standards and Technology. **5-21:** General Motors Corporation. **5-28:** Six Flags Magic Mountain. **5-61:** NASA. **Unn., p. 75:** Bettmann Archive, Inc. **Unn., p. 76:** Bettmann Archive, Inc.

Chapter 6
Fig. 6-2: AIP Niels Bohr Library, E. Scott Bar Collection. **6-29:** Photo, A. Hudson. **6-27:** From *The Conversion of Energy*, by Claude M. Summers. Copyright ©September 1971 by *Scientific American*, Inc. All rights reserved. **6-28:** NASA.

Chapter 8
Fig. 8-5: ©Harold Edgerton, courtesy Palm Press, Inc.

Chapter 9
Fig. 9-6: Photo, A. Hudson. **9-4a:** From *PSSC Physics*, 5th ed., 1981, D. C. Heath and Company. **9-4b:** Lawrence Berkeley Laboratory, University of California.

Chapter 10
Fig. 10-24: Lacy Atkins/Los Angeles Times.

Chapter 13
Fig. 13-2a: From *PSSC Physics*, 2nd ed., 1965, D. C. Heath and Company and Education Development Center, Inc., Newton, MA. **13-2d:** AP/Wide World Photo. **13-15:** Photo, A. Hudson.

Chapter 14
Unn., p. 315: The Armand Hammer Collections, Los Angeles, CA. **14-15a:** United States Department of Commerce, National Oceanic and Atmospheric Division. **14-16a:** Annan Photo Features.

Chapter 15
Fig. 15-21a: United Press International. **15-21b:** Courtesy of Prof. S. Taneda. **15-22:** © G. Whitely/Photo Researchers, Inc.

Chapter 16
Fig. 16-13: From *PSSC Physics*, 2nd ed., 1965, D. C. Heath and Company and Education Development Center, Inc., Newton, MA.

Chapter 17
Fig. 17-1a: Photo by George Wuerthner. **17-1b:** Photo, A. Hudson. **17-5:** Deutsches Museum, Munchen. **17-10:** United Press International. **17-12a:** Courtesy of Education Development Center, Newton, MA. **17-12b:** From the film "Flow Visualization" by S. J. Kline, Stanford University, produced by the National Committee for Fluid Mechanics Films, Educational Services, Inc. **17-12c:** NASA. **17-20:** *American Scientist*, 63, May–June 1975, p. 286. **17-21:** The Boston Globe. **17-23:** From "Wind Effects on Buildings and Structures," Hans Thomann, *American Scientist*, Volume 63, May–June 1975, pp. 278–287. **17-36:** Photo, A. Hudson.

Chapter 18
18-18a: From the film "Vibrations of a Drum," Kalmia Company. **18-18b:** From Mary Walker, *Chladni Figures: A Study in Symmetry*, Bell and Hyman, London, 1961. **18-18c:** Photo, A. Hudson. **18-20:** U.S. Army Ballistic Research Laboratory, Aberdeen Proving Ground, MD. **18-22:** Photo, A. Hudson.

Chapter 19
Fig. 19-4a: Wide World Photos. **19-4b:** The Daily News Photo. **19-4c:** Photo, A Hudson Library. **19-13:** Photo by Tom Wolfe. **19-16b:** Photo, A. Hudson.

Chapter 22
Fig. 22-8: From Knut Schmidt-Nielson, *How Animals Work*, Cambridge University Press, 1972.

Chapter 23
Fig. 23-3: Photography by Lotte Jacobi, processed by Case Western Reserve University, Cleveland, OH. **23-5e:** M. C. Escher, Waterfall, 1961, © M. C. Escher, c/o Cordon Art, Baarn, Holland.

Chapter 24
Fig. 24-11a, b, c: From O. D. Jefimenko, *Electricity and Magnetism*, 2nd ed., Electret Scientific 1989.

Chapter 26
Fig. 26-16a: Professor O. Nishikawa, The Graduate School of Nogatsuta, Midon-Ku, Yokohama 227 Japan.

Chapter 27
Fig. 27-3: Photo, A. Hudson. **27-5:** Photo, A. Hudson.

Chapter 28
Fig. 28-8: Photo, A. Hudson. **28-10:** CRYENCO, Inc., 2929 S. Santa Fe Drive, Englewood, CO 80110; per Mr. Thomas Theil, Quality Control Director. **28-13a:** Marx Brook–New Mexico Tech. **28-13b:** Swiss High Voltage Research Committee, Zurich, Switzerland.

Chapter 30
Fig. 30-1: From *PSSC Physics*, 2nd ed., 1965, D. C. Heath and Company and Education Development Center, Newton, MA. **30-7c:** L. A. Frank, University of Iowa. **30-7d:** Photo by Lee Snyder, Geophysical Institute, University of Alaska, Fairbanks, ©1977. **30-9:** Fermilab Photograph.

Chapter 31
Fig. 31-3: From *PSSC Physics*, 5th ed., 1981, D. C. Heath and Company and Education Development Center, Newton, MA. **31-6a:** From *PSSC Physics*, 5th ed., 1981, D. C. Heath and Company and Education Development Center, Newton, MA. **31-8c:** From O. D. Jefimenko, *Electricity and Magnetism*, 2nd ed., Electret Scientific 1989. **31-9c:** ©Kodansha, Tokyo.

Chapter 32
Fig. 32-14: Photo, A. Hudson.

Chapter 33
Fig. 33-6b: Courtesy of AT&T Archives.

Chapter 34
Fig. 34-1b: Courtesy of Masto, Dagastine & Associates, Inc. **34-15:** *Bulletin of the Atomic Scientists*, Dec. 1974, p. 35.

Chapter 35
Fig. 35-12d: By permission of the *American Journal of Physics* [*American Journal of Physics* 40 (1972): 46]. **35-16:** *Los Angeles Times* Photo by Robert Gabriel. **35-17:** Photo, A. Hudson. **35-20:** Palomar Observatory Photograph.

Chapter 36
Fig. 36-1b: Helen Faye. **36-3:** Photo, A. Hudson. **36-4a:** From *PSSC Physics*, 2nd ed., 1965, D. C. Heath and Company and Education Development Center, Newton, MA. **36-12c:** NASA.

Chapter 37
Fig. 37-1: Hale Observatories. **37-2:** Adapted from Eugene Hecht/Alfred Zajac, *Optics*, ©1974, Addison-Wesley Publishing Co., Inc., Reading, MA. Reprinted by permission of the publisher. **37-5:** Photo, A. Hudson. **37-12a:** GTE. **37-11c, d:** Courtesy of American Optical Corporation. **37-26a:** Manfred Kage; copyright Peter Arnold, Inc. **37-26b:** Courtesy of E. R. Lewis, F. S. Werblin, and Y. Y. Zeeri.

Chapter 38
Fig. 38-1c: From *Fundamentals of Optics* by Jenkins and White; copyright 1976, McGraw-Hill. Reproduced with permission of the publisher. **38-4a:** KLINGER EDUC. PROD. CORP., College Pt., NY 11356. **38-13:** Francis W. Sears, *Optics*, 3rd ed., 1949, Addison-Wesley Publishing Co., Inc., Reading, MA. **38-15:** From *Atlas of Optical Phenomena* by M. Cagnet, M. Francon, and J. Thrien; copyright 1962, Springer-Verlag. **38-16a, b:** Bausch & Lomb. **38-17a:** Bausch & Lomb. **38-20:** Photo, A. Hudson. **38-21a, b:** From *Fundamentals of Optics* by Jenkins and White; copyright 1976, McGraw-Hill. Reproduced with permission of the publisher.

Chapter 39
Fig. 39-1: From *Atlas of Optical Phenomena* by M. Cagnet, M. Francon, and J. Thrien; copyright 1962, Springer-Verlag. **39-4:** From *Fundamentals of Optics* by Jenkins and White; copyright 1976, McGraw-Hill. **39-8:** From *Fundamentals of Optics* by Jenkins and White; copyright 1976, McGraw-Hill. Reproduced with permission of the publisher. **39-9:** From *Atlas of Optical Phenomena* by M. Cagnet, M. Francon, and J. Thrien; copyright 1962, Springer-Verlag. **39-12b:** From *Atlas of Optical Phenomena* by M. Cagnet, M. Francon, and J. Thrien; copyright 1962, Springer-Verlag. **39-13a, b:** From *Atlas of Optical Phenomena* by M. Cagnet, M. Francon, and J. Thrien; copyright 1962, Springer-Verlag. **39-14a:** The Arecibo Observatory is part of the National Astronomy & Ionosphere Center operated by Cornell University under contract with the National Science Foundation. **39-14b:** Courtesy of National Radio Astronomy Observatory/Associated Universities, Inc. **39-14c:** Courtesy of National Radio Astronomy Observatory/Associated Universities, Inc. **39-16a:** J. R. Eyerman, LIFE Magazine, ©Time, Inc. **39-16b, c:** Palomar Observatory Photograph. **39-19:** Francis W. Sears, *Optics*, 3rd ed.; copyright 1949, Addison-Wesley, Reading, MA. **39-25b:** From B. Rossi, *Optics*, Addison-Wesley, 1957. **39-25c:** Permission requested from M. H. F. Wilkins, Kings College, London. **39-25d:** Courtesy of Bell Laboratories, Mrs. M. H. Read. **39-26:** From *Atlas of Optical Phenomena* by M. Cagnet, M. Francon, J. Thrien; copyright 1962, Springer-Verlag, p. 21. **39-27:** From P. M. Rinard, *American Journal of Physics* **44**, 1 (1976): 70. **39-28a, b, c:** From *Atlas of Optical Phenomena* by M. Cagnet, M. Francon, J. Thrien; copyright 1962, Springer-Verlag. **39-28d:** Courtesy of M. E. Hufford. **39-28e, f, g:** Photo, Clifford Chen, Occidental College. **39-35:** Conductron Corporation. **39-37a:** Courtesy of Professor Stuart B. Elliott, Occidental College. **39-37b, c:** Courtesy of Dr. Ralph Wverker, TRW.

Chapter 40
Fig. 40-4: A. Hudson and R. Nelson. **40-8:** Photo, A. Hudson. **40-9a:** A. Hudson and R. Nelson. **40-10:** Photo, A. Hudson. **40-18c:** Photo, A. Hudson. **40-19a, b, c, e:** Courtesy of the Measurements Group Inc., Raleigh, NC, USA. **40-19d:** Frocht, *Photo Elasticity*, Vol. 1, John Wiley & Sons. 1948. **40-20:** Photo, A. Hudson.

Chapter 41
Fig. 41-5: AIP Niels Bohr Library. **41-10:** Adapted from W. Bertoozi, *American Journal of Physics* **32** (1964): 555, with permission of the *American Journal of Physics*. **41-11:** Courtesy of Stanford Linear Acceleration Center and the U.S. Department of Energy.

Chapter 42
Fig. 42-3: From *Exploring the Universe*, 4th ed., by W. M. Protheroe, E. R. Capriotti, and G. H. Newsom; copyright ©1989 Merrill Publishing Company, Columbus, OH. **42-18:** Miller, *College Physics*, 4th ed., Harcourt Brace Jovanovich, 1977; courtesy Dr. Albert Rose.

Chapter 43
Fig. 43-9: AIP, Niels Bohr Library, Margrethe Bohr Collection. **43-11:** AIP Niels Bohr Library. **43-16a, b:** R. E. Lapp. **43-16c, d:** Courtesy of Education Development Center, Newton, MA. **43-16e:** Courtesy of C. G. Shull. **43-17a:** From J. Valasek, *Introduction to Theoretical and Experimental Optics*, John Wiley and Sons, 1949. **43-17b:** H. Raether, "Elektron Interferenzen," *Handbuch Der Physik*, XXXII, Springer, Berlin, 1957. **43-17c:** From P. G. Merli et al., *American Journal of Physics* **44**, 3 (1976): 306. **43-20:** From the film by A. Goldberg, H. M. Schey, and J. L. Schwartz, "Scattering in One Dimension," described in *American Journal of Physics* **35** (1967): 177. **43-21:** Courtesy, International Business Machines Corporation, Almaden Research, San Jose, CA.

Chapter 44
Fig. 44-9: Drawings constructed from computer art generated by Matt Nelson. **44-10:** William P. Spencer, MIT. **44-11:** Graph adapted from Figure 12V in Harvey E. White, *Introduction to Atomic and Nuclear Physics*, Van Nostrand (1964), p. 156. **44-14:** AIP Niels Bohr Library, W. F. Maggers Collection. **44-17:** Photographs reprinted by permission from Nicholas George, Automatic Pattern Recognition (ARC, Inc., Pittsford, NY, 1986), p. 43.

Chapter 45
Fig. 45-1: Kevin C. Jones, RT; MRI Institute. **45-3:** Reproduced with permission from the *Annual Review of Nuclear and Particle Science*, Vol. 37, ©1987 by Annual Reviews, Inc. **45-10:** From "Plutonium Project Report on Nuclei Formed in Fission," J. M. Siegel et al., *Review of Modern Physics* **18** (1946): 539. **45-14:** Brookhaven National Laboratory. **45-16b:** Photo, A. Hudson. **45-17c:** Courtesy of the Princeton Plasma Physics Laboratory. **45-21:** Fermilab Photograph. **45-22:** Fermilab Visual Media Services.

Index

A

Aberrations of lenses, 870
Absolute pressure, 396
Absolute temperature scale, 465, 529
Absorption of heat, 451
AC circuits, 763
    amplitude variations, 764
    C only, 764
    impedance diagram, series, 772
    L only, 766
    Ohm's law, 771
    phase constant, 764, 770
    power, 781
    R only, 764
    RLC, parallel, 775
        resonance, 780
    RLC, series, 768
        impedance, 771, 772
        phase constant, 772
        resonance, 778, 780
    steady-state, 770
    transient term, 770
Acceleration, 18, 83
    angular, 252, 271
    average, 19
    center of mass, 205
    centripetal, 65, 66, 253
    Galilean relation, 208
    due to gravity, 22, 83
    instantaneous, 19, 252
    radial, 65, 66, 253
    tangential, 65, 253
    in three dimensions, 52
Accelerator, drift tube, 578
Accelerometer, 334
Action-at-a-distance, 77
Action, line of, 224
Activity, radioactive, 1066
Actual mechanical advantage, 147
Adiabatic process, 502
    lines, 503
Air components, table, 486
Allowed solution, wave equation, 1034
Alpha decay, 1070
    quantum mechanical tunneling, 1071
Alternating current. *See* AC circuits.
American customary units, 84
Ammeter, 665
Ampere (unit), 714
Ampère, André Marie, 717
Ampère's law, 716
    modified by Maxwell, 795
Amplitude
    AC variations, 764
    simple harmonic motion, 338
    in waves, 419

Analyzer, 929
Angle
    conical solid, 595
    plane, 64
    solid, 584
Angular
    acceleration, instantaneous, 252
    acceleration, vector, 271
    deviation (prism), 875
    frequency, 338, 420
    magnification, 862
        astronomical telescope, 867
    momentum, 269
        collisions, 278
        conservation of, 277
        particle, circular motion, 270
        vector form, 270
    position, 64, 251
    speed, instantaneous, 252
    speed, precessional, 306
    velocity, vector, 271
Antimatter, 1021
Antinodes, 430
Apothecary system, A-3
Approximations, mathematical, A-6
Archimedes' principle, 398
Area, vector element, 581
Arm, moment, 224
Astronomical data, A-14
Astronomical telescope
    angular magnification, 867
    Cassegrain reflector, 841
    exit pupil, 868
    eye relief, 868
Atmospheric pressure, standard, 396
    bar (unit), 396
Atom
    approximate mass, table, 484
    ionization energy, 1050
    models of, 1004
        Bohr, 1006
        Rutherford, 1005
        Thomson, 1005
        vector, 1037
    shell notation, 1042
Atomic mass unit, 81, 475, 1062
Atomic number, 1060
Atomic physics, 1033
Atwood's machine, 92
Aurora, 689
Avogadro's number, 475
Avoirdupois system, A-3
Axes
    Cartesian, 41, 48
    plane polar, 41, 64

    rotation about fixed, 264
    rotation about moving, 294
Axial vector, 228, A-10

B

B and **H**, 758
Babinet compensator, 942
Ballistic pendulum, 203
Balmer series (hydrogen), 1007
    Rydberg formula, 1030
Banked roads, 98
Bar (unit), 396
Barn (unit), 1079
Barometer, 396
Barrier tunneling, 1021
    applications, 1022
Baryons, 1093
Battery, 601
Bay of Fundy, 357
Beats, sound, 435
Bequerel (unit), 1068
Bernoulli's principle, 402, 403
    examples, 405
Beta decay, 1071
    β^-, 1072
    β^+, 1073
Big Bang, 1100
Binding energy, 383
    of nucleon, 1064
        graph, 1066
    of nucleus, 1062
Binomial theorem, A-6
Biot–Savart law, 711
Birefringence, 932
Black hole, 974
Blackbody, 462
    radiation, 982
Blind spot, human eye, 865
Bohr, Niels, 1009
Bohr
    complimentarity principle, 1027
    correspondence principle, 1010
    magneton, 709, 1037
    model of atom, 1006
        energy states, 1009
        postulates, 1008
        radii of orbits, 1009
    radius, hydrogen atom, 1043
Boiling, 451
Boltzman's constant, 483
Born's probability interpretation, 1018
Boson force-carriers, table, 1096
Bottom, b-quark, 1095
Boundary conditions, wave equation, 1034

Boyle's law, 474
Bragg, W. H. and W. L., 916
Bragg
    reflection of x-rays, 916
    scattering condition, 917
Brahe, Tycho, 369
Branch point, 658
Breakdown, electric field, 614
Breaking stress, 358
Breeder reactor, 1087
Bremsstrahlung (x-rays), 1050
Brewster's law, 931
Bright-line spectrum, 1004
Browning motion, 545
Btu (British thermal unit), 451
Bubble chamber, 203
Bulk modulus, 360
Buoyant force, 399

C

Cadmium-113 (^{113}Cd), neutron capture by, 1080
Calculus, formulas, A-8
Calorie, 451
    food calorie, 450
    thermochemical, 451
Calorimeter, 453
Camera, 869
    f-stop, 869
    iris diaphragm, 869
    pinhole, 872
Capacitance, 618
    combinations of, 623
        in parallel, 623
        in series, 624
    cylindrical capacitor, 620
    electrolytic capacitor, 626
    equivalent, 624
    parallel-plate capacitor, 619
    spherical capacitor, 621
    variable capacitors, 622
Capacitive reactance, X_C, 765
Capacitor, 618
    charged, 628
        energy stored in, 628
        charging of, 671
    combinations of
        in parallel, 623
        in series, 624
    cylindrical, 620
    discharge of, 672
    electrolytic, 626
    equivalent, 624
    parallel plate, 619
    spherical, 621
    variable, 622
Carbon cycle, 1088
Carnot
    cycle, 519
        steps, 520
    engine
        efficiency, 523
        table, 520
    refrigerator, 524
    theorem, 528

Cartesian coordinates, 41
    plane polar, 41, 64
Cassegrain reflector, 841
Cavendish experiment, 378
Cavity radiation, 982
    Planck's theory, 986
        radiation law, 987
    Rayleigh–Jeans
        radiation law, 986
        theory, 984
    spectral
        distribution curves, 983
        energy density, 983
    Wien's displacement law, 983
    Wien's radiation law, 984
Celsius temperature scale, 444
Center of gravity
    definition, 229
    "negative" mass method, 233
    x coordinate, 230
Center of mass, 213, 229
    acceleration, 205
    collisions, 205, 213
    frame of reference, 213
    kinetic energy, 211
    location, 205
    "negative" mass method, 233
    Newton's second law, 294
    velocity, 205
    zero-momentum frame, 212
Centimeters of mercury, 396
Central force, 371
Centrifugal force, 322
Centripetal acceleration, 65, 66, 253
Cerenkov radiation, 434
Cesium-137 decay scheme, 1074
Change of phase, 451, 454
    latent heat, 454
Characteristic line spectra, x-rays, 1051
Charles' and Gay-Lussac's law, 474
Charge
    electronic, 558
    by induction, 557
    negative, 556
    positive, 556
Charging, 671
Charm flavor, *c*-quarks, 1095
Chladni figures, 432
Circle of reference, simple harmonic motion, 344
Circuits
    AC (alternating current), 763
        amplitude variations, 764
        C only, 764
        impedance, 772, 775
        L only, 766
        Ohm's law, 771
        phase constant, 764, 770, 772
        power, 781
        R only, 764
        RLC parallel, 775
        RLC series, 768
        resonance, 778, 780
        steady-state, 770
        transient term, 770
    DC (direct current), 775

Kirchhoff's rules, 658
    junction rule, 659
    loop rule, 659
    multiloop, 658
    phase shifter, 790
    RC (with battery), 670
    RL (with battery), 741
    root-mean-square (rms) values, 783
Circular motion, 64
    kinematic equations, 254
Circular polarization, 934
Clocks
    nonsynchronism of moving, 966
    synchronization, 949
Coaxial cable, 636
Coefficient of performance, 524, 525
Coefficient of restitution, 219
Coherence, 879
Collider detector, Fermilab (CDF), 1102
Collision, 199
    angular momentum, 278
    center of mass, 205, 213
    elastic, 200
    inelastic, 200
Color
    confinement (particles), 1096
    by interference, 938
Combinations of capacitors
    in parallel, 623
    in series, 624
Combinations of resistors
    in parallel, 655
    in series, 655, 656
Comet, Mrkos, 817
Complex conjugate, 1043
Complimentarity principle, 1027
Compression and tension, 88
Compton
    effect, 994
    scattering, 995
    shift, 995
    wavelength, 995, 1056
Computerized tomography (CT), 1060
Concave mirror, 829
Condensation, 451
Conduction, heat, 456
Conductivity, 642, 648
    electron, 641
        drift speed, 641
        siemens, 642
    thermal, table, 457
Conductor, 557
    and Gauss's law, 591
    and superconductor, 647
Conical solid angle, 595
Conservation of
    angular momentum, 277
    energy, 156
        with friction, 170
    linear momentum, 180
    mechanical energy, 161
    momentum, system of particles, 211
Conservative force, 156
    definition, 157
    and potential energy, 159, 160
Conservative system, 159

Constant-volume gas thermometer, 463
Constants and standards, A-13
Constraint forces, 119
Containment, fusion reactor, 1090
    inertial confinement, 1090
    magnetic confinement, 1090
Continuity equation, 402
Continuous spectrum, x-rays, 1050
Convection
    coefficients, table, 458
    heat transfer, 458
Conventional current, 645
Convergent lens, 853
Conversion factors, A-2
Conversion of energy, 143
Conversion of units, 11
Convex mirror, 829
Conveyor belt, 189
Cooling, Newton's law, 472
Coordinate systems, 13
    Cartesian, 41, 48
    plane polar, 64
    right-handed, 48
    two-dimensional, 41
Coriolis force, 322, 325
Corner reflector, 828
    laser ranging retroreflector, 828
Correspondence principle, 1010
Coulomb (unit), 557
Coulomb's law, 557, 558
Couple, 302
Coupling
    L–S, 1039
        alternate quantum numbers, list, 1040
    spin–orbit, 1039
Covariance, principle of, 974
Critical mass, 1087
Critical temperature, 477, 647
Cross section, nuclear, 1079
    barn (unit), 1079
Curie (unit), 1068
Current, electric, 638
    conductivity, 648
    conventional, 645
    density J, 648
    direction, 639
    resistance, 641
Current loop
    in external magnetic field, 754
    magnetic field of, 715, 724
Curvature of spacetime, 974
Curvilinear motion, 69
Cutoff wavelength, x-rays, 1050
Cyclotron, 690
    frequency, 688
Cylindrical capacitor, 620

D

Damped oscillations, 352
    damping coefficient, 352
    equation of motion, 352
    resonant frequency, 355
Dark matter, 1101
Daughter nuclide, 1070
Davisson–Germer experiments, 1013

DC Circuits. *See* Circuits, DC.
de Broglie, Louis, 1011
de Broglie wave, 1011
    wavelength, 1012
        for electrons, 1015
        matter waves, 1012
        phase waves, 1012
Decay constant λ, 1066
Decay series, radioactive, 1078
Decibel, 424
Definite integrals, A-9
Deformation
    length, 359
    shear, 359
    volume, 359
Degree of freedom, 508
Del, 609
Delta–wye transformation, 680
Density, 394
    electric current, 648
    specific gravity, 394
    table, 394
    weight, 394
Derivative
    partial, 608
    of vectors, A-9
Detection of charged particles, 1084
Deviation, angle of (prism), 875
Dewar flask, 462
Diamagnetism, 754
Dielectric, 624
    constant, 626
        table, 626
    nonpolar, 625
    polar, 625
    strength, 626
        table, 626
Diesel engine, 526
Differentiation
    formulas, A-8
    of vectors, A-9
Diffraction
    Fraunhofer, 900
        circular aperture, 907
        half-wave zones, 901
        minimum angle of resolution, 907
        rectangular aperture, 905
        single-slit, 900, 902, 904, 905
    Fresnel, 900
        of circular aperture, 918
        of zone plate, 918, 920
    grating, 909
        dispersion, 914
        Fraunhofer lines, 910
        resolving power, 914
    hologram, 921
    Laue-spot pattern, 917
    pattern
        of circular aperture, 918
        of opaque disk, 918
        by particles, 1015
        of various objects, 919
        by x-rays, 1015
    single-slit
        formula, 902
        minima, 904

        pattern, 905
        phasors, 902
    x-ray
        Bragg reflection, 916
        Bragg scattering condition, 917
Dimensional analysis, 28
Diopter power, 856
Dipole antenna, 808
    pattern, 809
Dipole, electric, 565
    electric field of, 566, 568
    far-field approximation for, 567
    moment, 567
    in nonuniform field, 569
    potential, 609
    potential energy of, 568
    torque on, 568
Dipole, magnetic, 694, 798
    of Bohr magneton, 709
Direct current. *See* Circuits, DC.
Discharge of capacitor, 672
Dispersion, 844, 913
    of grating, 914
    of prism vs. grating, table, 914
    water waves, 426
Displacement, 14
    current, 795
    equation, 796
Distances, measurement, comparison
    table, 8
Divergent lens, 853
DNA molecule, 547
Domain
    magnetic, 756
    of physics, 1
Doppler shift, 432
    for light, 970
    sound, 433
Dosimeter, pocket, 559
Dot product, 118
Double refraction, 932
Double-slit interference, 878
    equation, 883
Downhill direction of heat flow, 543
Drift speed of electrons, 639, 641
Drift-tube accelerator, 578
Driving force, 354
Dulong and Petit law, 512
Dumbbell
    rigid, 507
    vibrating, 507
Dynamic imbalance, 285

E

Ear, human, 330
Eddy currents, 736
Effective resistance, transformer, 786
Effective values, root-mean-square, 783
Efficiency, 142
    Carnot, 523
    table of typical, 143
Einstein, Albert, 950
Einstein's
    general relativity, postulates, 974
    photoelectric equation, 992

Einstein's *(continued)*
 quantization of radiation, 991
 special relativity theory, 943
Elastic collision, 200
Elastic moduli
 bulk, 360
 shear, 360
 Young's, 360
Elastic properties of matter, 357
Electric current, 638
 conductivity, 648
 conventional, 645
 density, 648
 direction, 639
 resistance, 641
Electric dipole, 565
 comparison with magnetic dipole, 753
 electric field, 566, 568
 far-field approximation, 567
 moment, 567
 nonuniform field, 569
 potential, 609
 potential energy, 568
 torque on, 568
Electric equilibrium, 594
Electric field
 breakdown, 614
 conductor, 588
 continuous charge distributions, 569
 dipole, 566, 568
 and emf, 734
 energy density, 631
 energy, stored in, 630
 Gauss's law, 585, 798
 infinite line charge, 572
 lightning, 592, 650
 lines, 562
 of plane charge, 588
 of point charge, 563
 similarity to magnetic field, table, 721
 of surface charge, 588
 thunderstorm, 592, 650
Electric flux, 580
 definition, 581
 point charge, 583
Electric potential, 597
 differences, table, 598
 energy, 597
Electric quadrupole, 578
Electricity, laws, table, 795
Electrolytic capacitors, 626
Electromagnetic radiation, dual nature, 997
Electromagnetic spectrum, 806
Electromagnetic waves, 799
 and accelerated charge, 808
 energy, 809
  density, 809, 812
 equation for E and B, 802
 forces on electrons, 813
 intensity, 812
 momentum of, 812, 814
 plane, 803
 pressure, 815

production of, 807
 relation between E_y and B_z, 806
Electromagnetism and relativity, 971
Electromechanical analogues, table, 769
Electromotive force, 637
 seat of, 637
Electron
 capture, radioactivity, 1075
 charge, 558
 de Broglie wavelength, 1015
 drift speed, 639, 641
 radius, classical, 1056
 spin, 1038
  and fine structure, 1039
  quantum number, 1038
Electron-volt (unit), 603
Electroscope, 559
Electrostatic force, 159, 555
Electrostatics, 555
Electroweak theory, 1099
Elements
 ground-state configuration, table, 1048
 Paschen's triangle, table, 1049
 nuclear data, table, 1063
 periodic table, 1047
Elliptic polarization, 936
emf, 637
 back-emf, 737
 battery, 601
 and electric fields, 734
 motional, 730
Emissivity, 461
Emittance, radiation, 983
Energy
 availability, 357
 binding, 383
 conservation of, 156
  with friction, 170
  mechanical, 161
 conversion efficiency, table, 143
 in electric field, 630
 in electromagnetic waves, 809
 electron-volt, 603
 entropy and unavailable, 545, 548
 equipartition, 508
  variables, 506
 for the future, 144
 gravitational potential energy, 130, 382
 in inductors, 744
 internal, 132, 137, 450, 494
 ionization of atoms, 1050
 kilowatt-hour, 141
 kinetic, 124, 126, 211
  per mole, 483
  per molecule, 483
 mass-energies, particles, table, 962
 and momentum relation, relativistic, 964
 potential, 130
  and conservative forces, 159
 relativistic, 961
  kinetic, 961
  total, 963

rest-mass, 962
 rotational motion, 281
 satellite motion, 385
 simple harmonic motion, 345
 spring potential energy, 131
 stored
  in charged capacitor, 628
  in electric field, 630
 thermal, 132, 450, 494
 threshold, 1082
 wave motion, 426
 work–energy relation, 124. *See also names of kinds of energy.*
Energy density
 in E and B fields, 809
 electric field, 631
 in electromagnetic waves, 812
 in magnetic field, 745
 spectral, 983
Energy diagrams, 165
 hydrogen, 1042
Engine
 Carnot, 519
  efficiency, 523
  table, 521
 diesel, 526
 efficiency, 523
 internal combustion, 526
 jet, 192
 Stirling, 527
Enrichment, nuclear reactor, 1087
Entropy, 536
 and information, 549
  bits, 546
 macroscopic view, 536, 537, 540
 microscopic view, 540, 542
 negative, 549
 and probability, 540, 541
 and second law of thermodynamics, 543
  statement, 549
 state function, 537
 and unavailable energy, 545, 548
Entropy change
 free expansion, 538
 heat conduction, 540
 ice to water, 539
 mixture, 539
Equation
 of continuity, 402
 of motion, 352
 of state, 474, 537
Equilibrium, 235
 conditions, 236, 474
 dynamic, 235
 electrostatic, 594
 neutral, 235
 stable, 235
 static, 236
 thermodynamic, 493
 unstable, 235
Equipartition theorem, 508
 energy variables, 506
Equipotential surfaces, 610
Equivalence, principle of, 974

Equivalent
    capacitor, 624
        in parallel, 623
        in series, 624
    resistor
        in parallel, 657
        in series, 656
Escape velocity, 383
Ether, 825, 894
Event, point, 944
Expansion
    isothermal, ideal gas, 497
    thermal, 445
Extraneous roots, 30
Extraordinary ray, 932
Eye, human, 864
    accommodation of, 864
    blind spot, 865
    defects of, 866
    diagram, 863
    iris, 865
    near point, 864
    retina, 863
Eyeglasses, 866

F

f-stop, 869
Fahrenheit temperature scale, 444
Farad (unit), 619
Faraday, Michael, 619
Faraday cage, 615
Faraday's law of induction, 727, 728
    most general form, 733
Femtometer (unit), 1060
Fermat, Pierre de, 827
Fermat's
    last theorem, 827
    principle
        of reflection, 827
        of refraction, 847
Fermi, Enrico, 1061
Fermions, fundamental, table, 1096
Ferromagnetism, 755
Fictitious forces, 316
Field
    electric, 562
        in conductor, 588
        of continuous charge distributions, 569
        of dipole, 566, 568
        energy density, 631
        energy stored in, 630
        Gauss's law, 585
        of line charge, 572
        of lines, 562
        of plane charge, 588
        of point charge, 563
        of surface charge, 588
        in thunderstorm, 592
    gravitational, 379
        lines, 382
    magnetic, 685
Field ion microscope, 613

Figure of merit, 665
Filter, low-pass, 790
Fine structure
    constant, 1056
    spectral lines, 1038
        and electron spin, 1039
Finite rotations, A-9
First law
    Newton's, 74
        statement, 76
    of thermodynamics, 492, 494
        statement, 495
Fission, nuclear, 1086
    distribution of energy, table, 1077
    liquid-drop model, 1078
    power, 1086
    spontaneous, 1076
    yields, 1977
Fizeau experiment, 976
Flavors, quarks, 1095
Floating-coin illusion, 842
Flow
    laminar, 401
    streamline, 401
Fluid, 393
    laminar flow, 401
    in motion, 400
    pressure in, 396
    streamline flow, 401
Flux
    electric, 580
        definition, 581
        point charge, 583
    magnetic, 703, 728
Focal length
    of lenses, 855, 857
    of mirrors, 832
Focal point
    of lenses, 857
    of mirrors, 833
Force
    of absorbed radiation, 815
    buoyant, 399
    carriers, 1104
    central, 371
    centrifugal, 322
    conservative, 156, 157
        potential energy, 159
    contact, 82
    Coriolis, 322, 325
    on current-carrying conductor, 692
    electrostatic, 159, 555
    fictitious, 316
    gravitational, 373
    Hooke's law, 388
    inertial, 318
    line of action, 224
    Lorentz, 691
    moment of, 225
    nonconservative, 159
    noncontact, 82
    spring, 122
    work by constant, 116
    work by varying, 120
Forced harmonic motion, 354

Forced oscillation
    driving force, 354
    steady-state term, 354
    transient term, 354
Foucault pendulum, 328
Fourier analysis, A-6
Frames of reference, 7, 10, 13
    center of mass, 213
    inertial, 77
    linearly accelerated, 316
    Newton's second law, 325
    rotating, 321
    uniformly accelerated, 317
    zero-momentum, 213
Fraunhofer diffraction, 900
    of circular aperture, 907
        minimum angle of resolution, 907
    half-wave zones, 901
    lines, 910
    pattern
        of circular aperture, 907
        of rectangular aperture, 905
        of single-slit, 905
    single-slit, 900
        equation, 902
        minima, 904
        phasors, 902
Free-body diagram, 86
Free expansion, 538
Freezing, 451
Frequency
    angular, 338, 420
    cyclotron, 688
    damped, 338
    half-power, 790
    harmonic, 431
    natural, 352
    resonant, damped oscillations, 355
    simple harmonic motion, 419
    waves, 419
Fresnel
    biprism, 897
    diffraction, 900
        circular aperture, 918
    lens, 869
    zone plate, 918, 920
Friction, 93
    energy conservation, 170
    kinetic, 94
    static, 94
    thermal energy, 132
Fundamental forces in nature, 74
Fundamental interaction, particle physics, 1093
Fusion, nuclear, 1088
    carbon cycle, 1088

G

Galilean
    acceleration relation, 208
    relativity principle, 78, 944, 947
    transformation, 944
        equations, 945
    velocity addition, 208

Galvanometer, 664, 697
   tangent, 725
Gamma decay, 1074
Gas
   constant, universal, 475
   constant volume thermometer, 463
   ideal, 474, 479
      processes, 497
      specific heat, 499
   molar specific heats, table, 510
   standard conditions, 475
Gauge pressure, 396, 408
Gaussian surface, 583
Gauss's law, 583
   conductors, 591
   for electric fields, 585, 798
   example, 586
   for magnetic fields, 799
   and symmetry, 586
Gay-Lussac's law, 474
Geiger counter, 1084
General relativity, theory of, 973
   black hole, 974
   curvature of spacetime, 974
   postulates, 974
   principle of covariance, 974
   principle of equivalence, 974
Gluons, 1093, 1104
   properties, table, 1096
Grad, 609
Gradient of V, 608
Gram-molecular mass, 475
Grand unification, 1099
Grating spectroscope, 910
Gravimeter, superconducting, 647
Gravitation, 77
   acceleration due to, 22, 83
   action-at-a-distance, 77
   extended mass, 373
   field, 379
   field lines, 382
   inverse-square law, 371
   Newton's law of universal, 370
   potential energy, 130, 382
   shell theorems, 377
   universal constant, 82
Gravitational
   constant, universal, 82
   potential energy, 130, 382
Graviton, 964, 1099, 1104
Gravity, 77
   acceleration due to, 22, 83
   action-at-a-distance, 77
   center of, 229
      x coordinate, 230
   variations, 381
   work by, 119
Great American Revolution, 104
Greek alphabet, A-1
Ground-state configuration of elements, table, 1048
   Paschen's triangle, table, 1049
Gulliver's Travels, 138
Gyration, radius of, 268
Gyromagnetic ratio, 726
Gyroscope, 306

H

H and **B**, 758
h-bar, 1008
Hadrons, 1093
   quark structure of, table, 1097
Half-life, 1066
Half-power frequency, 790
Halfwave plate, 934
Halfwave zone, 901
Hall
   effect, 699
   potential, 700
Halley, Edmund, 371
Harmonic frequencies, 431
Harmonic motion
   circle of reference, 344
   damped, 352
   forced, 354
   simple, 166, 338
He–Ne gas laser, 1052
Heat, 132, 449
   absorption, 451
   conduction, 455
   downhill flow, 543
   of fusion, 454
   latent, table, 454
   phase changes, 451
   pump, 524
      coefficient of performance, 525
   reservoir, 493
   transfer by convection, 458
      coefficients, table, 458
   transfer by radiation, 459
   of vaporization, 454
Heavy water, 1087
Heisenberg's uncertainty principle, 1024
Helmholtz coil, 724
Henry (unit), 738
Hertz, Heinrich, 338
Hertz (unit), 338
Holography, 922
   applications, 922
   hologram, 921
Hooke, Robert, 123
Hooke's law, 123
   for oscillations, 338
   for vertical springs, 347
Horsepower, 141
Hubble constant, 35
Huygens' principle, 824
Hydrogen atom
   Balmer series, 1007
      Rydberg formula, 1030
   Bohr model, 1006
      energy states, 1009
      postulates, 1008
      radii of orbits, 1009
      radius, 1043
   energy-level diagram, 1042
   probability–density distributions, 1047
   quantum states, 1041
      based on quantum numbers n, ℓ, m_ℓ, m_s, table, 1041
      based on quantum numbers n, ℓ, j, and m_j, table, 1041
   wave functions, hydrogen atom, 1043
      normalization, 1043
      normalized, table, 1044
      probability density function, 1043, 1046
Hysteresis, 759

I

Ideal gas, 474, 476
   law, 475
   model, 475
   specific heat, 499
   thermodynamic relations, table, 506
Ideal liquid, 393
Ideal mechanical advantage, 147
Ignition, nuclear, 1091
Image
   characteristics, 837
   in plane mirror, 827
      size, 857
   in spherical mirror, 829
   virtual, 829
Impedance
   diagram, AC series, circuits, 772
   in parallel RLC, 775
   in series RLC, 771
      diagram, 772
Impedance matching, 793
Impulse, 185
Inch, definition, 10
Index of refraction, 844
   of materials, table, 844
   relative, 855
Inductance
   mutual, 739
   self, 737
      back emf, 737
   unit (henry), 738
Induction
   charging by, 557
   eddy currents, 736
   Faraday's law of, 728
      most general form, 733
   Lenz's law, 735
   mutual, 739
Inductive reactance, X_L, 768
Inductors, energy in, 744
Inelastic collision, 200
Inertia, 76
   moment of, 264
      calculation, 266
      table, various shapes, 266
Inertial confinement, fusion reactor, 1090
Inertial force, 318
Inertial frame of reference, 77
Inertial mass, 80
Information, 549
   bits, 546
   entropy and, 549
Infrared, 461
Initial phase angle, simple harmonic motion, 339
Instruments, optical, 862
   astronomical telescope, 867
      angular magnification, 867

Cassegrain reflector, 868
    exit pupil, 868
    eye relief, 868
  camera, 869
  eyeglasses, 866
  microscope, 868
    magnifying power, 869
  periscope, 841
  reversibility, principle of, 847
  simple magnifier, 862
Insulation R-value, 457
Insulator, 557
Intensity level, sound, 424
Interference
  colors by, 938
  constructive, 429
  criteria, 881
  destructive, 429
  double-slit, 878
    equation, 883
  multiple slit, 887, 911
    intensity equation, 912
    pattern, 888
  path difference, 882
  phase difference, 882
  superposition principle, 881
  by thin films, 888
  by thin wedges, 890
Interferometer
  Michelson, 892
    compensating plate, 892
  Pohl's, 898
Internal combustion engine, 526
Internal conversion, 1075
Internal energy, 132, 137, 450, 494
Internal kinetic energy, 213
Internal reflection, total, 848
    critical angle, 848
    light pipe, 849
Internal resistance, 668
Inverse-square law, 371
Ionization energy, atom, 1050
Ionosphere, 633
Iris diaphragm, 869
Iris, human eye, 865
Irreversible processes, 496, 535
Isobaric process, 500
Isochoric process, 500
Isolation diagram, 86
Isotherm, 477
Isothermal expansion, 497
Isothermal process, 498
Isovolumic process, 500

J

Jet engine, 192
Joule heating, 645
Joule, James, 117
Joule's law, 645
Junction, 658

K

Kelvin absolute temperature scale, 465, 529

Kepler's laws, 369
  planetary motion, 369
  second law, 290
Kilogram, international prototype, 80
Kilowatt-hour, 141
Kinematics, 24
  definition, 6
  equations
    comparison of linear and rotational, table, 284
    constant acceleration, 21
    derivation using calculus, 24
    linear motion, 20
    rotational motion, 253, 254
  graphical relations, 26
  rotational, 251, 254
Kinetic energy, 124, 126, 211
  center of mass, 211
  internal, 213
  per mole, 483
  per molecule, 483
  relativistic, 961
  rotational, 254, 265, 281
  system of particles, 211
  variable force, 127
Kinetic friction, 401
Kirchhoff's rules, 658
  junction rule, 659
  loop rule, 659

L

Laminar flow, 401
Land, Edwin H., 929
Large hadron collider, 1100
Laser, 1052
  He–Ne gas laser, 1052
  population inversion, 1052
  ranging retroreflector, 828
Latent heat, 454
  of fusion, 454
  phase change, 454
  table, 454
  of vaporization, 454
Lateral magnification, 835
  equation, 836
Laue diffraction pattern, 917
LCD (liquid crystal display), 938
Length
  contraction, 954
  interval of, 9
  proper length, 955
  standard, 9
Lens
  aberrations, 870
  combinations, 859
  convergent, 853
  diopter power, 856
  divergent, 853
  focal length, 855
  focal point, 857
  Fresnel, 869
  linear magnification, 858
  negative, 853
  positive, 853

  thin-lens, 852
    approximation, 853
    equation, 855
    image size, 857
    principal foci, 857
    ray tracing, 857
    sign convention, 856
  various types, 853
Lens-maker's formula, 855
Lenz's law, 735
Leptons, 1093
Lever arm, 224
Light
  coherence, 879
  Doppler shift, 970
  extinction length, 844
  polarized, 927
    circularly, 935
    elliptically, 936
    linearly, 927
  spectrum of visible, 845
  speed and $\mu_0 \epsilon_0$, 805
    defined exact, 806
  "tired," 844
  unpolarized, 928
  waves, superposition, 880
Light pipe, 849
Lightning, 592, 650
Lilliputians and Brobdingnagians, 138
Line of action, 224
Linear accelerator, Stanford (SLAC), 957
Linear expansion, thermal, 446
Linear magnification, 858
  of lens, 858
  of mirror, 835
Linear mass spectrometer, 701
Linear momentum, 180
  conservation, 180
Linear motion
  comparison with rotational motion, 284
  kinematic equations, 20
Linear polarization, 927
Liquid crystal display (LCD), 938
Liquid, ideal, 393
Liquid-drop model, 1078
Lloyd's mirror, 896
Longitudinal waves, 414
Loop, 658
Loop rule, Kirchhoff's, 659
Lorentz
  force, 691
  transformation, 949, A-10
    derivation, A-10
    equations, 949
Low-pass filter, 790

M

Mach, Ernst, 434
Mach
  cone, 434
  number, 434
Macroscopic view
  and entropy, 536, 537
  of matter, 473
Magdeburg sphere, 397

Magnetic bottle, 688
Magnetic confinement, fusion reactor, 1090
Magnetic dipole, 694, 798
    of Bohr magneton, 709
    comparison, electric dipole, 753
    moment, 695
        torque on, 695
    potential energy, 696
Magnetic field, 684
    Ampère's law, 716
    Biot–Savart law, 711
    cyclotron frequency, 688
    due to currents in
        Helmholtz coil, 724
        infinite sheet, 720
        long straight wire, 713, 717, 724
        loop, 715
        loop, along axis, 724
        parallel wires, 725
        solenoid, 718
        toroidal coil, 718
    energy density, 745
    flux, 703
    Gauss's law for, 799
    intensity, 758
    motion of charged particles in, 686
    right-hand rule for, 686, 713
    similarities to electric field, table, 721
    sources of, 711
    strength, 685
        search coil, 750
Magnetic flux, 703, 728
Magnetic force on a current-carrying wire, 692
Magnetic properties of materials, 752
    diamagnetism, 754
    ferromagnetism, 755
    hysteresis, 759
    paramagnetism, 753
    permeability, 758
Magnetic resonance imaging (MRI), 1060
Magnetic susceptibility, 757
    table, 758
Magnetism, laws, table, 795
Magneton, Bohr, 813
Magnetosheath, 689
Magnetostriction, 760
Magnification
    angular, 862
        of magnifier, 862
    lateral, 835
        equation, 836
    of lens, 858
    linear, 858
    of mirror, 835
Magnifier
    angular magnification, 862
    simple, 862
Malus, Etienne, 929, 933
Malus's law, 929
Mass, 80
    atomic, tables, 484, 1063, A-12
    center of, 213, 229
        acceleration, 405
        collisions, 205, 213

    kinetic energy, 211
    location, 205
    "negative" mass method, 233
    velocity, 205
    zero-momentum frame, 212
    comparisons, table, 81
    inertial, 80
    mass-energies, particles, table, 962
    molecular, 475
        table, 484
    number, 1060
    rest, 959
    standard, 80
    unified atomic mass unit, 81, 475
    units, 84
    on vertical spring, 347
    and weight, 83
Matching stub, 429
Mathematical
    approximations, expansions, and vector relations, A-6
    formulas, A-4
    symbols, A-1
Mathematics, role of, 4
Matter
    elastic properties, 357
    macroscopic view, 473
    microscopic view, 473
Matter waves, 1012
Maxwell distribution
    equation, 490
    graph, 485
Maxwell's equations
    and displacement current, 795
    in vacuum, table, 799
Measurements in relativity, 944
    length contraction, 954
    observer, 944
    proper length, 955
    proper time interval, 955
    rest mass, 962
    time dilation, 952
Mechanical advantage, 146
    actual, 147
    ideal, 147
Mechanical energy, conservation of, 161
Melting, 451
Mesons, 1093
"Message" of relativity, 968
Meter
    definition, 9, 894
    standard bar, 9
Method of mixtures, 453
Michelson, Albert, 892
Michelson interferometer, 892
    compensation plate, 892
Microscope, 868
    field ion, 613
    magnifying power of simple, 869
    scanning tunneling, 1022
Microscopic view
    and entropy, 540, 542
    of matter, 473
Microwave oven, 813
Milikan oil drop experiment, 577

Minimum angle of deviation, 875
Mirror
    concave and convex, 829
    equation, 832
        sign convention for, 832
        in terms of f, 833
        in terms of R, 832
    focal length, 832
    focal point, 833
    lateral magnification, 835
    Lloyd's, 896
    plane
        image location, 827
        ray-tracing, 827
        reflection by, 825
    spherical
        optic axis, 829
        ray-tracing, 829
        reflection, 828
Mixtures, method of, 453
Moderator, nuclear reactor, 1087
Moduli
    bulk, 360
    shear, 360
    Young's, 360
Molar specific heat, 500
    gases, table, 510
Mole, kinetic energy, 483
Molecular
    kinetic energy, 483
    mass, 475
        table, 484
    specific heat, 422
    weight, 475
Moment arm, 224
Moment of force, 225
Moment of inertia, 264
    calculation, 266
    table, various shapes, 266
Momentum, 81
    angular, 269
    continuous rate of change, 188
    conservation, system of particles, 210
    of electromagnetic waves, 812, 814
    linear, 180
    of photon, 995
    relativistic, 955, 957
    system of particles, 205
Mosley, Harry G., 1051
Mosley diagram, 1051
Motion
    Brownian, 545
    of charged particle in magnetic fields, 686
    circular, 64
    constant acceleration, 21
    curvilinear, 69
    equation of, 352
    extended object, 294
    in fluids, 400
    linear, kinematic equations, 20
    Newton's laws
        first, 76
        second, 81

third, 98
  summary, 103
one-dimensional, 6
perpetual, devices, 549
periodic, 167, 337
planetary
  Kepler's laws, 369
  Kepler's second law, 290
projectile, 54
rotational, kinematic equations, 254
satellite, 385
simple harmonic, 166, 338
steady-state, 355
in three dimensions, 50
Motional emf, 730
Mrkos comet, 817
Müller, Erwin, field ion microscope, 613
Multiloop circuits, 658
Multiple-slit interference, 887, 911
  intensity formula, 912
  pattern, 888
Mutual inductance, 739

N

Nanometer (unit), 822
Natural processes, 535
Negative lens, 853
Negative mass, method, 233
Neutrino, 1073
Neutron
  free, lifetime, 1086
  number, 1060
Neutron-activation analysis, 1083
Newton, Isaac, 75
Newton's first law, 74
  statement, 76
Newton's law of cooling, 472
Newton's law of universal gravitation, 370
Newton's laws of motion, summary, 103
Newton's rings, 891
  radii, 892
Newton's second law, 74, 81
  applications, 86
  rotating frames, 325
  rotational motion, 272
  statement, 81
  system of particles, 206
  translation of center of mass, 294
Newton's third law, 98
  statement, 99
Newton's third-law pairs, 99
Nodal lines, surfaces, 431
Nodes, 430
Nonconductor, 557
Nonconservative force, 159
Nonpolar dielectric, 625
Nonreflective coatings, 890
Nonsynchronism of moving clocks, 966
Normalization condition, 1018
  of ψ, 1018
Nuclear
  data, particles and elements, table, 1063
  fission, 1086

force
  strong, 1060, 1104
  weak, 1104
fusion, 1088
  carbon cycle, 1088
mass, 1062
  unified atomic mass unit, 1062
physics, 1059
potential, square-well, 1083
power, 1085
reactors, 1087
  breeder, 1087
  fission, 1087
  fusion, 1088, 1090
Nucleon, 1060
  binding energy, 1064
  graph, 1066
Nucleus, 1060
  atomic number, 1060
  binding energy, 1062
  cross section, 1079
    barn (unit), 1079
  data, particles, elements, table, 1063
  half-life, 1066
  mass, 1062
    number, 1060
  neutron number, 1060
  nucleon, 1060
    binding energy, 1064
  nuclide, 1060
  radioactive decay, 1066
  radius, 1061
  "size," 1061
  strong nuclear force, 1060, 1104
  unified atomic mass unit, 1062
  weak nuclear force, 1104
Nuclide, 1060
  daughter, 1070
  parent, 1070

O

Observer, in relativity, 944
Ohm (unit), 642
Ohm's law, 643
  alternative form, 648
Optic axis, 934
  of spherical mirror, 829
Optical activity, 932, 937
Optical fiber
  acceptance angle, 873
  cladding, 850
  communication, 850
Optical instruments, 862
  astronomical telescope, 867
    angular magnification, 867
    Cassegrain reflector, 841
    exit pupil, 868
    eye relief, 868
  camera, 869
  eyeglasses, 866
  microscope, 868
    magnifying power, 869
  periscope, 841

reversibility, principle of, 846
  simple magnifier, 862
Optical reversibility, 855
Ordinary ray, 932
Organ pipes, 431
Orthogonality, 611
Oscillations, 337
  amplitude, 338
  angular frequency, 338
  damped, 352
  forced, 354
  hertz, 338
  period, 338
  phase angle, 338
  steady-state term, 354
  transient term, 354
Oscillator, sawtooth, 683
Otto cycle, 526
  efficiency, 526

P

Pair production, 994, 996
Paradoxes, special relativity, 979
Parallel-axis theorem, 298, 299
Parallel combinations
  of capacitors, 623
  of resistors, 655
Parallel plate capacitor, 619
Parallel resonance, 780
Paramagnetism, 753
Paraxial ray, 830
Parent nuclide, 1070
Partial derivative, 608, A-9
Particle
  detection of charged
    Geiger counter, 1084
    scintillation counter, 1084
  diffraction by, 1015
  in a box
    normalized wave function, 1019
    energy states, 1020
  mass-energies, table, 962
  momentum of system, 205
  nuclear data, table, 1063
  wave nature of, 1004
  wave–particle duality, 1022, 1026
    complimentarity principle, 1027
Particle–antiparticle symmetry, 1094
Particle physics, 1092
  fundamental interactions, 1093
  particle–antiparticle symmetry, 1094
  particles
    elementary, table, 1094
    strange, 1095
  spin, 1093
Pascal (unit), 395
Pascal's principle, 398
Paschen's triangle, 1049
Path difference (optical), 882
Pauli exclusion principle, 1047
Pendulum
  ballistic, 203
  Foucault, 328
  physical, 350

Pendulum *(continued)*
   simple, 347
   torsional, 349
Period
   simple harmonic motion, 338
   in waves, 419
Periodic motion, 167, 337
Periodic table of the elements, 1047, A-12
Periodic wave train, 415
Periscope, 841
Permeability
   of free space, 712
      and speed of light, 805
   of magnetic materials, 758
Permittivity of free space, 558
   and speed of light, 805
Perpendicular-axis theorem, 313
Perpetual motion devices, 549
Perspective
   Chapters 1–5, 114
   Chapters 6–9, 222
   Chapters 10–18, 442
Phase, 418
   change, 451, 454, 889
      latent heat, 454
   velocity, 418
Phase angle, simple harmonic motion, 338
Phase constant
   in AC circuits, 764, 770
   in RLC circuits, 772
Phase difference, interference, 882
Phase of oscillation, 339
Phase-shifter, AC circuit, 790
Phase waves, 1012
Phasor
   AC circuits, 766
      diagrams, 766, 771
   optical, 902
Photoelasticity, 938
Photoelectric effect, 988
   Einstein's equation, 992
   threshold frequency, 990
   wave function, 991
Photomultiplier, 994
   secondary emission, 994
Photon
   Compton scattering, 995
   Compton shift, 995
   Compton wavelength, 995
   momentum of, 995
   pair production, 994, 996
Physical pendulum, 350
Pinhole camera, 872
Planck's
   constant, 987
   quantum hypothesis, 511, 986
   radiation law, 987
Plane mirror
   image location, 827
   ray tracing, 827
   reflection by, 825
Plane waves, 425
   for E and B, description, 803
Planetary motion
   Kepler's laws, 369
   Kepler's second law, 290

Plasma, 393
Pocket dosimeter, 559
Pohl's interferometer, 898
Point event, 944
Poisson's bright spot, 918
Polar coordinates, 64
Polar dielectric, 625
Polar vectors, 228
Polarimeter, 937
Polarization, 626
   Brewster's law, 931
   circular, 934
   direction, 927
   elliptic, 936
   linearly polarized wave, 927
   Malus's law, 929
   polarizer, 929
   polarizing angle (reflection), 931
   Polaroid, 929
   by reflection, 930
   by scattering, 931
Polaroid, 929
   analyzer, 929
   transmission axis, 929
Population inversion, laser, 1052
Position
   angular, 64, 251
   one-dimensional, 14
Position vector, 41, 42
Positive lens, 853
Positronium, 1032
Postulates
   of general relativity, 974
   of special relativity, 948
Potential barrier, classical, 167
Potential, electric, 597
   differences, 598
   energy, 597
Potential, V, dipole, 609
Potential energy, 130
   of charged capacitors, 628
   and conservative force, 160
   electric, 597
   gravitational, 130, 382
   of magnetic dipole, 696
   of spring, stressed, 131
Potential well, 167, 1083
Potentiometer, 667
Power, 140
   in AC circuits, 781
   average, 140
   definition, 140
   horsepower, 141
   nuclear, 1085
      fission, 1086
      fusion, 1088
   in resistors, 646
   transmitted by waves, 427
Power factor, 782
Poynting vector
   average value, 811
   instantaneous, 810
Precession, 306
   angular speed, 307
Prefixes, metric, table, 11
Prefixes, SI (Appendix A), 10

Pressure, 395
   absolute, 396
   of electromagnetic waves, 815
   fluid at rest, 396
   gauge, 396, 408
   radiation, from sun, 816
   standard atmospheric, 396
The *Principia*, 75
Principle of covariance, 974
Principle of equivalence, 974
Principle of relativity, Galilean, 78
Principle of reversibility, 846
Principle of superposition, 559, 665, 881
   DC circuits, 660
Prism
   dispersion of, vs. grating, table, 914
   minimum angle of deviation, 875
Probability, 1018
   Born's interpretation, 1018
   density distribution, hydrogen, 1047
   density function, 1018
      hydrogen, 1043, 1046
Problem solving, general procedures, 91
Processes
   irreversible, 535
   reversible, 535
   thermodynamic, 497
      summary, table, 506
Projectile motion, 54
Proper measurements, 955
   length, 955
   time interval, 955
Proton, free, lifetime, 1086
Pseudovectors, 228
P–T diagram, 477
P–V diagram, 477, 495
PVT surface, 476

Q

Q (resonance), 779
Quadrupole, electric, 578
Quantization of radiation, 991
Quantum
   chromodynamics (QCD), 1095
   electrodynamics (QED), 1016, 1095
   hypothesis
      Einstein, 991
      Planck, 511, 986
   mechanical tunneling, alpha decay, 1071
Quantum number
   alternate numbers for **L**–**S** coupling, list, 1040
   inner, 1040
   list, 1038
   magnetic, 1036
   orbital, 1036, 1040
   principal, 1036, 1040
   spin, 1038
Quantum physics, 1004
   chronology of theory, 1028
   probability interpretation of, 1018
Quantum radiation, 981
Quantum states
   energy-level diagram, hydrogen, 1042

ground-state configuration of elements, table, 1048
of hydrogen atom, 1014
based on quantum numbers n, ℓ, m_ℓ, and m_s, table, 1041
based on quantum numbers n, ℓ, j, and m_j, table, 1041
probability density distribution, 1047
selection rules, 1042
spectroscopic notation, 1041
Quarks, 1093
flavors
bottom, 1095
charm, 1095
down, 1095
strange, 1095
top, 1096
up, 1095
properties of, table, 1096
structure of hadrons, table, 1097
Quarterwave plate, 935
Quasi-static processes, 496

R

R-value, insulation, 457
Radial acceleration, 65, 66, 253
Radiation
black body, 982
cavity, 982
Planck's theory, 986, 987
Rayleigh–Jeans theory, 984, 986
spectral distribution curves and energy density, 983
Wien's displacement law, 983
Wien's radiation law, 984
Cerenkov, 434
Compton effect, 994
shift, 995
wavelength, 995
electromagnetic, dual nature of, 997
force of absorbed, 815
heat transfer, 458
photoelectric effect, 988
Einstein's equation, 992
threshold frequency, 990
work function, 991
Planck's
constant, 987
quantum hypothesis, 987
radiation law, 987
pressure, 815
from sun, 816
quantization, 991
quantum nature, 981
Stefan–Boltzmann law, 461, 983
emittance, 983
Radiometer, 815
Radio telescope, 908
Very Large Array (VLA), 908
Very Long Baseline Array (VLBA), 908
Radioactive dating, 1084
Radioactive decay, 1066
activity, 1066
alpha decay, 1070
quantum-mechanical tunneling, 1071
beta decay, 1071
β^-, 1072
β^+, 1073
cesium-137, 1074
daughter nuclide, 1070
decay constant λ, 1066
electron capture, 1075
gamma decay, 1074
half-life, 1066
internal conversion, 1075
modes, 1069
parent nuclide, 1070
processes, table, 1076
Q of reaction, 1070
series, 1078
spontaneous fusion, 1076
uranium-238 decay series, 1079
Radiocarbon dating, 1085
Radius, nucleus, 1061
Radius of gyration, 268
Rainbow, 846
Rankine temperature scale, 467
Ray
extraordinary, 932
ordinary, 932
paraxial, 830
and wavefronts, 832
Ray tracing, 829
and magnification, 835
plane mirror, 827
spherical mirror, 829
rays used for
mirrors, 836
thin lens, 857, 858
thin lens, 857
used in, 858
Rayleigh–Jeans
radiation law, 986
theory, 984
Rayleigh's criterion, 907, 915
RC circuits, 670
charging, 671
discharging, 672
RC time constant, 672
Reactance
capacitive, 765
inductive, 768
Reactor, nuclear, 1087
breeder, 1087
fission, 1086
fusion, 1088
inertial containment, 1090
magnetic containment, 1090
Reflection
by corner reflector, 828
laser ranging retroreflector, 828
diffuse, 826
Fermat's principle, 827
floating-coin illusion, 842
Huygens' principle, 824
laws of, 826
nonreflective coatings, 890
optical reversibility, 855
periscope, 841
phase change in, 889
by plane mirror, 825
image location, 827
ray tracing, 827
reversibility, principle of, 846
by spherical mirror, 828
image location, 829
ray tracing, 829
by thin films, 888
total internal, 848
critical angle, 848
light pipe, 849
of waves, 428
Refraction, 845
depth, apparent, 847
dispersion, 844
double, 932
index of materials, 844
optical reversibility, 855
by plane interface, 843
relative index, 855
reversibility, principle of, 846
Snell's law, 846
by spherical interface, 851
by thin lens, 852
Refractive index, 844
of materials, table, 844
relative, 855
Refrigeration, coefficient of performance, 524
Refrigerator, Carnot, 524
Relative velocity, geometrical method, 207
Relativistic
Doppler shift for light, 970
energy and momentum relations, 946
momentum, 955, 957
total energy, 963
velocity addition, 946
Relativity, general theory of, 973
black hole, 974
curvature of spacetime, 974
postulates, 974
principle of covariance, 974
principle of equivalence, 974
Relativity, special theory of, 78, 943
clocks
nonsynchronism, of moving, 966
synchronization of, 949
Doppler shift for light, 970
and electromagnetism, 971
energy, relativistic, 961
Fizeau experiment, 976
fundamental postulates, 948
Galilean
relativity principle, 944, 947
velocity addition, 946
kinetic energy, relativistic, 961
length contraction, 954
mass-energies, particles, table, 962
measurements, 944
length contraction, 954
observer, 944
proper length and time, 955
rest mass, 962
time dilation, 952

Relativity, special theory
of (continued)
"message" of, 968
momentum, 955, 957
paradoxes, 979
point event, 944
postulates, 948
principle, Galilean, 944, 947
rest energy, 962
rest mass, 962
Terrel effect, 971
time dilation, 952
transformation
Galilean, 944, 945
Lorentz, 949
twin paradox, 969
velocity addition
Galilean, 208, 946
relativistic, 959
Reservoir, heat, 493
Resistance, electrical, 641
equivalent
in parallel, 657
in series, 656
internal, 668
Resistivity, 642
ohm, 642
thermal coefficient of, 642
table, 643
Resistor
delta–wye transformations, 680
equivalent
in parallel, 657
in series, 656
in parallel, 655, 657
power in, 646
in series, 655, 656
wye–delta transformation, 680
Resolving power of grating, 914
Resonance, 778
frequency, 355
in parallel RLC, 780
in series RLC, 778
sharpness, Q, 779
Rest energy, 962
Rest mass, note about, 959
Retardation plates, 934
Retina, human eye, 863
Reversibility, principle of, 846
Reversible processes, 496
Right-hand rule
for magnetic fields, 686, 713
cross-product, 686
for torques, 226
for vector cross products, 226
Right-handed coordinate system, 48
Rigid body, rotational kinematics, 251
RL circuits, 941
RLC circuits, series, 768
phase constant, 772
Rocket, 190
Rolling
with slipping, 301
without slipping, 258
Root-mean-square (rms)
effective values, 783

speed, 483
values, AC circuits, 783
Rosa, E. B., and Dorsey, N. E., 806
Rotational dynamics
axes, fixed, 264
axes, moving, 294
kinematic equations, 254
kinetic energy, 281
radius of gyration, 268
of symmetrical objects, 271. See also
Rotational motion.
Rotational kinematics, 251, 254
angular acceleration, 252
angular position, 251
angular speed, 252
kinematic equations, 253, 254
rigid body, 251
Rotational kinetic energy, 254, 265, 281
kinematic equations, 254
derivation using calculus, 254
Rotational motion
comparison with linear motion, 284
energy, 281
equations, summary, 309
frames of reference, 321
kinematic equations, 254
kinetic energy, 281
linear analogies, table, 284
Newton's second law, 272
work, 281
work–energy relation, 282
Rotations, finite, A-10
Rowland ring, 761
Rutherford model of atom, 1005
Rydberg
constant, 1030
formula, 1030

S

Sagitta formula, A-5
Sailboat, 408
Satellite motion, energies, 385
Sawtooth oscillator, 863
Scalar, 118
Scalar product of vectors, 117
Scanning tunneling microscope, 1022
Schrödinger
time-independent wave equation, 1017
wave equation, 1035
quantum number, 1036
Scientific method, 3
Scintillation counter, 944, 1084
Search coil, 750
Seat of electromotive force, 637
Second law, Kepler's, 290
Second law, Newton's, 74, 81
applications, 86
rotating frames, 325
rotational motion, 272
system of particles, 206
translation of center of mass, 294
Second law of thermodynamics, 517
and entropy, 543
statement, 549

Clausius statements, 519
Kelvin–Planck statement, 519
Secondary emission, 994
Selection rules, 1042
Self-inductance, 737
back emf, 737
unit (henry), 738
Semiconductor, 557
Separation of variables, 742
Serendipity in science, 5
Series combinations
of capacitors, 624
of resistors, 655, 656
Series resonance, 778
Sharpness Q, 779
Shear modulus, 360
Shell notation, atom, 1042
Shell theorems, 377
SHM, 338. See also Simple harmonic
motion.
Shock waves, 434
mach cone, 434
mach number, 434
SI system, A-15–A-16
conversion factors, A-2
length, 9
prefixes, 10, A-1
time interval, 9
units, A-15
base and supplementary, 84
Siemen (unit), 642
Sign convention
for mirrors, 832
for thin lenses, 856
Significant figures, 12
Simple harmonic motion, 166, 338
amplitude, 338, 419
angular frequency, 338
circle of reference, 344
energy, 345
equations, 340
frequency, 419
hertz, 338
initial phase angle, 339
period, 338, 419
phase angle, 338
phase of oscillation, 339
steady-state motion, 355
Simple pendulum, 347
Single-slit diffraction, 900
Fraunhofer
formula, 902
minima, 904
phasors, 902
pattern, 905
Sinusoidal
wave train, 415, 418
waves, 415, 520
Size, comparison, table, 2
"Size" of the nucleus, 1061
Slide wire, 667
Snel van Royen, Willebrord, 846
Snell's law for refraction, 846
Solar wind, 689, 816
Solid angle, 376, 584, A-4
Solids, specific heat capacities, 512

Sonic boom, 434
Sound
    decibels, 424
    dispersion, 426
    Doppler shift, 433
    intensity
        average, 427
        level, 424
        table, 424
    organ pipes, 431
    speed in gases, 422
    timbre, 431
Sound waves
    antinodes, 430
    beats, 435
    nodes, 430
    shock waves, 434
    standing, 429
Space, free
    permeability of, 712
    permittivity of, 558
    speed of light in, 805
Space, homogeneous and isotropic, 227
Space travel, general limits of, 971
Spacetime curvature, 974
Special relativity, 943
    fundamental postulates, 948. *See also* Relativity, special theory of.
Specific gravity, 394
Specific heat
    capacity, 452
        solids, 512
        table, 452
    ideal gas, 499
    molar, 500
Spectral
    distribution curves, 983
    energy density, 983
    lines, fine structure, 1038
        and electron spin, 1039
    radiation curve, 461
Spectrometer, linear mass, 701
Spectroscope, grating, 910
Spectroscopic notation, quantum states, 1041
Spectrum
    bright line, 1004
    electromagnetic, 806
    Fraunhofer lines, 910
    visible light, 845
    x-rays, characteristic line spectra, 1051
Speed
    angular, instantaneous, 252
    average, 14
    instantaneous, 16, 252
    most probable, 484
    sound, in gases, 422
    transverse waves, 414, 420
    wave, 421
Speed of light, 805
    defined exact, 806
    and $\mu_0 \varepsilon_0$, 805
Spherical capacitor, 621
Spherical mirror
    optic axis, 829
    ray tracing, 829
    reflection, 828
Spin, electron, 1038
    and fine structure, 1039
    L–S coupling, 1039
    quantum number, 1038
    spin–orbit coupling, 1039
Spin, in particle physics, 1093
Spontaneous fission, 1076
Spring
    constant, 123
    forces, 122
    Hooke's law, 123, 347
    stressed, potential energy, 131
    vertical, 347
Square-well nuclear potential, 1083
Standard
    atmospheric pressure, 396
    cell, 667
    conditions, 475
    length, 9
    mass, 80
    time interval, 9
Standards and constants, A-13
Standing waves, 429
    antinodes, 429
    nodes, 429
Standing-wave solutions, wave equation, 1034
Stanford Linear Accelerator Center (SLAC), 957
State, variables, 537
State function
    entropy, 537
    variables, 495, 537
Static friction, 94
Statistical mechanics, 473
Steady-state conditions, AC circuits, 770
Steady-state motion, 355
    terms, 354
Steel yard, 249
Stefan–Boltzmann radiation law, 461, 983
Steiner's theorem, 298, 299
Step-up/step-down transformer, 786
Steradian (unit), 376, 584, A-4
Stereoisomers, 937
Stern–Gerlach experiment, 1039
Stirling heat engine, 527
Stirling's approximation, A-6
Strain, 357
Strange particles, 1095
Stream tube, 402
Streamlines, 401
Stress, 357
    breaking, 358
Strong nuclear force, 1060
Sublimation, 451
Superconducting gravimeter, 647
Superconducting Super Collider (SSC), 1100
Superconductivity, 647
Supernova 1987A, 964, 1071
Superposition principle, 429
    DC circuits, 660
    fields, 559
    light waves, 880

Synchronization of clocks, 949
Synchrotron, 690
System of particles
    conservation of momentum, 211
    kinetic energy, 211
    momentum, 205
    Newton's second law, 206

T

Tachyons, 957
Tacoma Narrows Bridge collapse, 356
Tangent galvanometer, 725
Tangential acceleration, 65, 253
Telescope
    astronomical, 867
        angular magnification, 867
        Cassegrain reflector, 841
        exit pupil, 868
        eye relief, 868
    radio, 908
        Very Large Array (VLA), 908
        Very Long Baseline Array (VLBA), 908
Temperature, 443
    absolute scale, 465, 529
    Celsius scale, 444
    conversion
        Celsius, 444
        Celsius–Kelvin, 467
        Fahrenheit scale, 444
        Rankine scale, 467
    critical, 477
        for superconductivity, 647
    Fahrenheit scale, 444
    gradient, 456
    Kelvin scale, 465
    Rankine scale, 467
Tension and compression, 88
Terminal voltage, 668
Terrell effect, 971
Terrestrial and astronomical data, A-14
Tesla, Nikola, 685
Tesla (unit), 685
Theory, 3
Therm, 451
Thermal
    coefficient of resistivity, 642
        table, 643
    conductivity, table, 457
    contact, 493
    energy, 132, 450, 494
    expansion, 445
        area, 445
        coefficients, table, 447
        linear, 445
        volume, 447
Thermochemical calorie, 450
Thermodynamic processes
    adiabatic, 502
    Carnot cycle, 519
    irreversible, 496
    isobaric, 500
    isochoric, 500
    isothermal, 497

Thermodynamic processes (continued)
  isovolumic, 500
  quasi-static, 496
  reversible, 496
  summary, table, 506
Thermodynamic system, 474
Thermodynamics, 492
  equation of state, 474
  equilibrium, 493
    conditions, 474
  first law, 492, 494
    statement, 495
  macroscopic view, 473
  microscopic view, 473
  relations for ideal gas, table, 506
  second law, 517
    and entropy, 543, 549
  state, 495
  system, 492
  third law, 530
  Zeroth law, 494
Thermography, 1001
Thermometer, 443
  constant-volume gas, 463
Thermos bottle, 462
Thin films, interference by, 888
Thin lens, 852
  approximation, 853
  equation, 855
  image size, 857
  principle foci, 857
  ray tracing, 857
    rays used, 858
  sign convention for, 856
Thin wedges, optical, 890
Third law
  Newton's, 98, 99
  thermodynamics, 530
Thomson, Benjamin (Count Rumford), 449
Thomson model of atom, 1005
Threshold
  energy, 1082
  frequency, 990
Thrust, effective, 191
Thunderstorm electricity, 592
Timbre, 431
Time
  constant, RC, 672
  dilation, 952
  proper time, 955
  standard, 9
Tippy tube, 248
"Tired" light, 844
  extinction length, 844
Tokamak, 1090
Tonne (unit), A-16
Torque, 224
  couple, 303
  on electric dipole, 568
  on magnetic dipole, 695
  right-hand rule, vector cross products, 226
  as a vector, 226
Torricelli's law, 404
Torsion balance, 556
Torsional pendulum, 349
Torsional waves, 425

Total internal reflection, 848
  critical angle, 848
  light pipe, 849
Transformation
  delta–wye, 680
  Galilean, 944
    equations, 945
  Lorentz, 949
    equations, 949
  wye–delta, 680
Transformer, 785
  effective resistance, 786
  step-up/step-down, 786
  turns ratio, 786
Transient term, 354
  AC circuits, 770
Transmission axis, 929
Transmission coefficient, quantum mechanics, 1031
Transverse waves, 414, 420
  speed, 420, 422
Trigonometric identities, A-4
Triple point, 477
  of water, 464
Triple product, 323
Troy system, A-3
Turns ratio, transformer, 786
Twin paradox, 969

U

Ultrasonic waves, 421
Uncertainty principle, 1022
  Heisenberg relation, 1024
Unification, grand, 1099
Unified atomic mass unit, 81, 475, 1062
Unit vectors, 68, 609
  in rectangular coordinates, 609
  in spherical coordinates, 609
Units
  American customary system, 84
  conversion of, 11
  mass, 84
  SI, 84, A-15–A-16
  weight, 84
Universal
  gas constant, 475
  gravitational constant, 82
Universal gravitation, Newton's law, 370
Uranium-238 decay series, 1079

V

Van Allen belts, 689
Vapor, 477
Vaporization, heat of, 454
Variable capacitor, 622
Variable force, 127
  work by, 120
Variables, separation of, 742
Variables of state, 537
Vector differentiation and integration, A-9
Vector relations, mathematical, A-6
Vector(s)
  addition of, 144
  angular acceleration, 271

  angular velocity, 271
  area element, 581
  axial, 228, A-10
  components of, 42
  displacement, 43
  model of atom, 1037
  multiplication, A-9
    scalar product, 117
    vector product, 226
  polar, 228
  position, 41, 42
  Poynting
    average value, 811
    instantaneous, 810
  product, A-9
  pseudovectors, 228
  rectangular, 609
  scalar product, 117, A-9
  spherical, 609
  subtraction of, 44
  three-dimensional, 48
  unit, 68, 609
Velikovsky problem, 292
Velocity
  angular, 271
  average, 14
  center of mass, 205
  escape, 383
  Galilean addition, 208, 946
  instantaneous, 16, 51
  relative, geometrical method, 207
  relativistic addition, 959
Velocity addition
  Galilean, 946
  relativistic, 959
Velocity filter (charged particles), 691
Vena contracta, 404
Venturi
  effect, 406
  meter, 406
Vertical spring, 347
Virtual image, 829
Viscosity, 393
Volta, Count Allesandro, 598, 638
Voltage
  phasor diagram, 771
  terminal, 668
Voltmeter, 664
  figure of merit, 665

W

Water
  triple point, 464
  waves, dispersion, 426
Watt (unit), 140
Watt, James, 140
Wave equation, 415, 417
  allowed solutions, 1034
  boundary conditions, 1034
  general solution, 417
  mechanical, 415
  particular solution, 418
  Schrödinger, time-independent, 1017
  sinusoidal wave train, 418
  standing-wave solutions, 1034

Wave function
  Born's probability interpretation, 1018
  of hydrogen atom, 1043
  normalization of, 1018, 1043
  normalized
    hydrogen atom, table, 1044
    particle in a box, 1019
  probability, 1018
    density function, hydrogen, 1046
Wave mechanics, 1016
  Born's probability interpretation, 1018
  Heisenberg's uncertainty principle, 1022
  relation, 1024
Wave nature of particles, 1004
Wave–particle duality, 1022, 1026
  complimentarity principle, 1027
Wave plates (optical), 934
Wavefront, 425
  and rays, 823
Wavelength, 419
  and color, table, 822
  Compton, 1056
  cutoff, x-rays, 1050
  de Broglie, 1012
    for electrons, 1015
    matter waves, 1012
    phase waves, 1012
Waves
  amplitude, 419
  beats, 435
  de Broglie, 1011
    wavelength, 1012, 1015
  dispersion, water, 426
  electromagnetic, 799
    and accelerated charge, 808
    energy density, 809
    equation for E and B, 802
    forces on electrons, 813
    intensity, 812
    momentum of, 812, 814
    plane, 803
    pressure, 815
    production of, 807
    relation between E_y and B_z, 806
  energy of, 426
  equation, 415, 417
  frequency, 419
  infrasonic, 421
  linearly polarized, 927
  longitudinal, 414
  number, 420
  period, 419
  periodic wave train, 415
  plane, 425
    for E and B, 803
  power transmitted, 427
  pulse, 415
  reflection, 428
  shock, 434
  sinusoidal, 415, 420
    wave train, 415, 418
  speed, 421
    compression, 422
    sound, 422
    transverse, 420, 422
  standing, 429
  superposition principle, 429
  torsional, 425
  transverse, 414
    speed, 420, 422
  traveling
    amplitude, 419
    frequency, 419
    period, 419
    wavelength, 419
  two and three dimensions, 423
  ultrasonic, 421
  water, 426
  wavelength, 419
Weak processes, nuclear, 1097
Weber (unit), 703
Weight
  density, 394
  and mass, 83
  units, 84
Wheatstone bridge, 666
Wien's law
  of displacement, 983
  of radiation, 984
Wind chill factor, 459
Wind tunnel, 408
Work, 116, 118, 126
  alternative form, 134
  area under F-vs.-x graph, 121
  constant force, 116
  by gravity, 119
  kinetic energy, 124
  in rotational motion, 281
  scalar product, 117
  in stretching spring, 123
  variable force, 120, 127
Work–energy relation, 124, 125, 134
  for rotation, 282
Work function, 991
Wye–delta transformation, 680

X

X_C, capacitive reactance, 765
X_L, inductive reactance, 768
X-ray, 1050
  bremsstrahlung, 1050
  characteristic line spectra, 1050
  continuous spectrum, 1050
  cutoff wavelength, 1050
  Mosley diagram, 1051
X-ray diffraction, 916
  Bragg reflection, 916
  Bragg scattering condition, 917
  pattern, 1015
    Laue spot, 917

Y

"y-delta" (wye–delta) transformation, 680
Young, Thomas, 878
Young's modulus, 360

Z

Z, AC impedance, 771
Zero-momentum frame, 213
Zeroth law of thermodynamics, 494

Some Solar System Data (See Appendix L for a more complete list.)

EARTH

Equatorial radius	6.378×10^6 m
Polar radius	6.357×10^6 m
Mass	5.976×10^{24} kg
Sidereal rotation period	86 164 s
Mean orbital speed	2.977×10^4 m/s
Solar constant (average solar power incident perpendicularly on a unit area)	
at top of atmosphere	1.37×10^3 W/m²
at earth's surface	0.84×10^3 W/m²

SUN

Equatorial radius	6.960×10^8 m
Mass	1.989×10^{30} kg
Power output (luminosity)	3.86×10^{26} W
Surface temperature	5780 K
Mean distance from earth	1.496×10^{11} m

MOON

Equatorial radius	1.738×10^6 m
Mass	7.347×10^{22} kg
Mean distance from earth	3.84×10^8 m
Period of revolution about the earth	27.32 d

Frequently Used Tables (See Index for additional tables.)

PAGE	NUMBER	TITLE
394	17-1	Densities of Selected Substances
447	19-1	Coefficients of Thermal Expansion
452	19-2	Specific Heat Capacities
454	19-3	Latent Heats
457	19-4	Thermal Conductivities
506	21-1	Thermodynamic Relations for an Ideal Gas
510	21-2	Molar Specific Heats
626	27-1	Dielectric Constants and Dielectric Strengths
643	28-1	Resistivities and Thermal Coefficients of Resistivity
758	33-1	Magnetic Susceptibilities
844	37-1	Refractive Indices
962	41-1	Mass-Energies of Fundamental Particles
1041	44-1	Wave Functions of the Hydrogen Atom
1063	45-1	Nuclear Data for Selected Particles and Elements